ArcGIS® 9

Using ArcMap™

DATA CREDITS

Quick-Start Tutorial Data: Wilson, North Carolina.
Population Density—Conterminous United States Map: U.S. Department of Census.
The African Landscape Map: Major Habitat Types—Conservation Science Program, WWF-US; Rainfall—ArcAtlas™, ESRI, Redlands, California; Population data from EROS Data Center USGS/UNEP.
Amazonia Map: Conservation International.
Forest Buffer Zone—100 Meters Map: U.S. Forest Service (Tongass Region).
Horn of Africa Map: Basemap data from ArcWorld™ (1:3M), ESRI, Redlands, California; DEM and Hillshade from EROS Data Center USGS/UNEP.
Mexico: ESRI Data & Maps CDs, ESRI, Redlands, California.
Mexico: 1990 Population: ESRI Data & Maps CDs, ESRI, Redlands, California.
Population Density in Florida (2001): ESRI Data & Maps CDs, ESRI, Redlands, California.
Rhode Island, the Smallest State in the United States: Elevation data from the USGS, EROS Data Center; other data from ESRI Data & Maps CDs, ESRI, Redlands, California.
Countries of the European Union: Member States and Candidate Country information from EUROPA (The European Union On-Line); ESRI Data & Maps CDs, ESRI, Redlands, California.
Mexico Population Density Map: ESRI Data & Maps CDs, ESRI, Redlands, California.
Health Care in the United States Map: Population data from U.S. Department of Census; Health Service Areas from the trustees of Dartmouth College; Service Providers data from Healthcare Financing Administration.
Clark County Land Use Map: Clark County Office, Washington State.
Southeast Asia Population Distribution Map: ArcWorld (1:3M), ESRI, Redlands, California.
Global 200—World's Biologically Outstanding Ecoregions Map: Ecoregions data from Conservation Science Program, WWF-US; Country boundaries from ArcWorld (1:3M), ESRI, Redlands, California.
Australia Map: Major Habitat Types data from Conservation Science Program, WWF-US; Basemap from ArcWorld (1:3M), ESRI, Redlands, California.
New Hampshire Telecom Map: Geographic Data Technology, Inc.
Redlands Image: Courtesy of Emerge, a division of TASC.

CONTRIBUTING WRITERS

Melanie Harlow, Rhonda Pfaff, Michael Minami, Alan Hatakeyama, Andy Mitchell, Bob Booth, Bruce Payne, Cory Eicher, Eleanor Blades, Ian Sims, Jonathan Bailey, Pat Brennan, Sandy Stephens, Simon Woo

Contents

Getting started

Displaying data

Querying data

14 Analyzing geometric networks 415

Map output

15 Laying out and printing maps 453

Customization

Getting started

Section 1

Welcome to ArcMap

1

IN THIS CHAPTER

- Visualizing information
- Working geographically
- Showing relationships
- Solving problems
- Creating and updating data
- Presenting results
- Developing mapping applications
- Tips on learning ArcMap

Welcome to ESRI® ArcGIS® ArcMap™, the premier software for desktop geographic information system (GIS) and mapping technology. ArcMap gives you the power to:

- **Visualize.** In no time you'll be working with your data geographically: seeing patterns you couldn't see before, revealing hidden trends and distributions, and gaining new insights.
- **Create.** It's easy to create maps to convey your message. ArcMap provides all the tools you need to put your data on a map and display it in an effective manner.
- **Solve.** Working geographically lets you answer questions such as "Where is...?", "How much...?", and "What if...?". Understanding these relationships will help you make better decisions.
- **Present.** Showing the results of your work is easy. You can make great-looking publication-quality maps and create interactive displays that link charts, tables, drawings, photographs, and other elements to your data. You'll find that communicating geographically is a powerful way to inform and motivate others.
- **Develop.** The ArcMap customization environment lets you tailor the interface to suit your needs or the needs of your organization, build new tools to automate your work, and develop standalone applications based on ArcMap mapping components.

The next few pages show you some of the things you can do with ArcMap. As you start making your own maps, you'll discover even more.

Visualizing information

Sometimes just looking at a map will tell you what you want to know. Maps not only tell you where things are, but also what's special about them. This population map shows you where people live in the United States. From it, you can easily see where the major metropolitan areas are located.

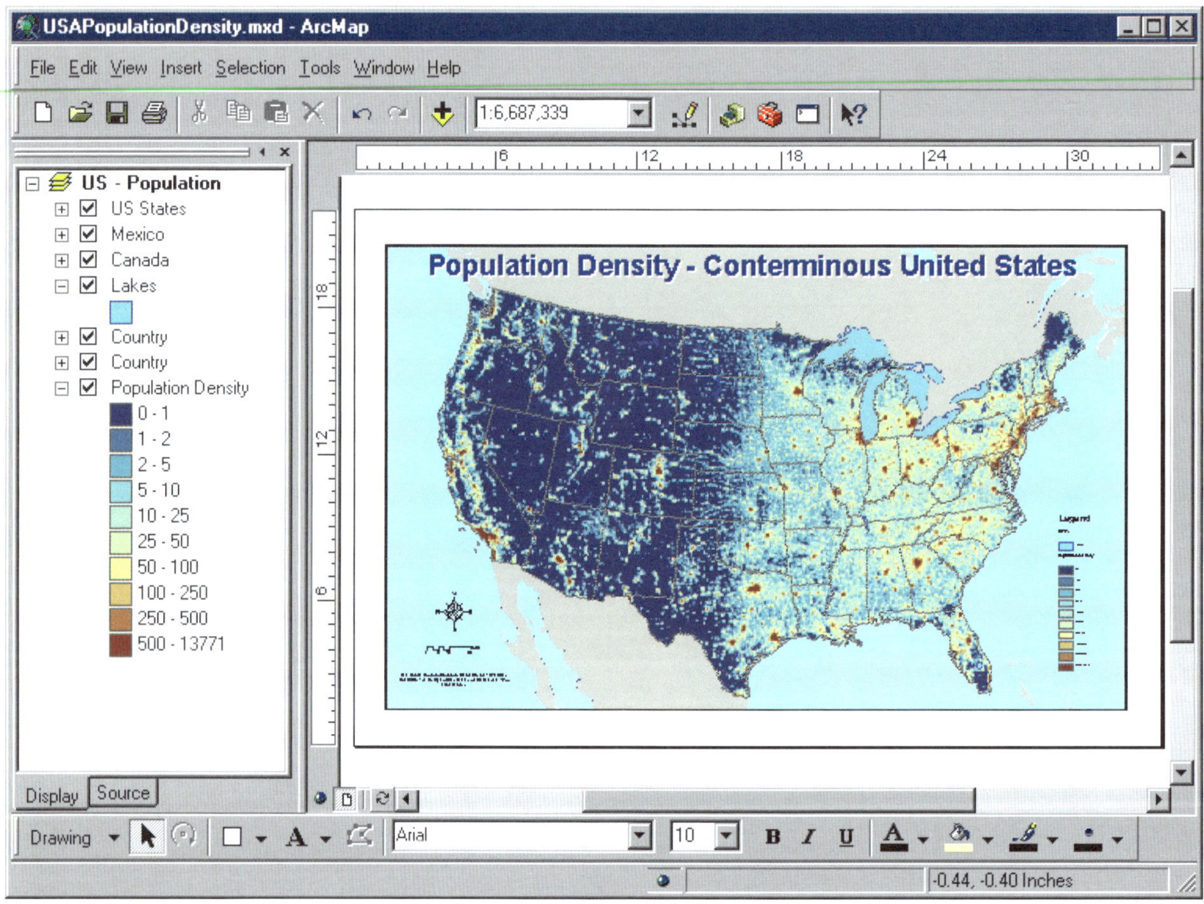

Do you live in a populated area? Areas drawn with dark blue have a lower population density than areas drawn with yellow and brown.

Working geographically

Maps are not static displays; they're interactive. You can browse a map—taking a closer look at a particular area—and point at features to find out more about them.

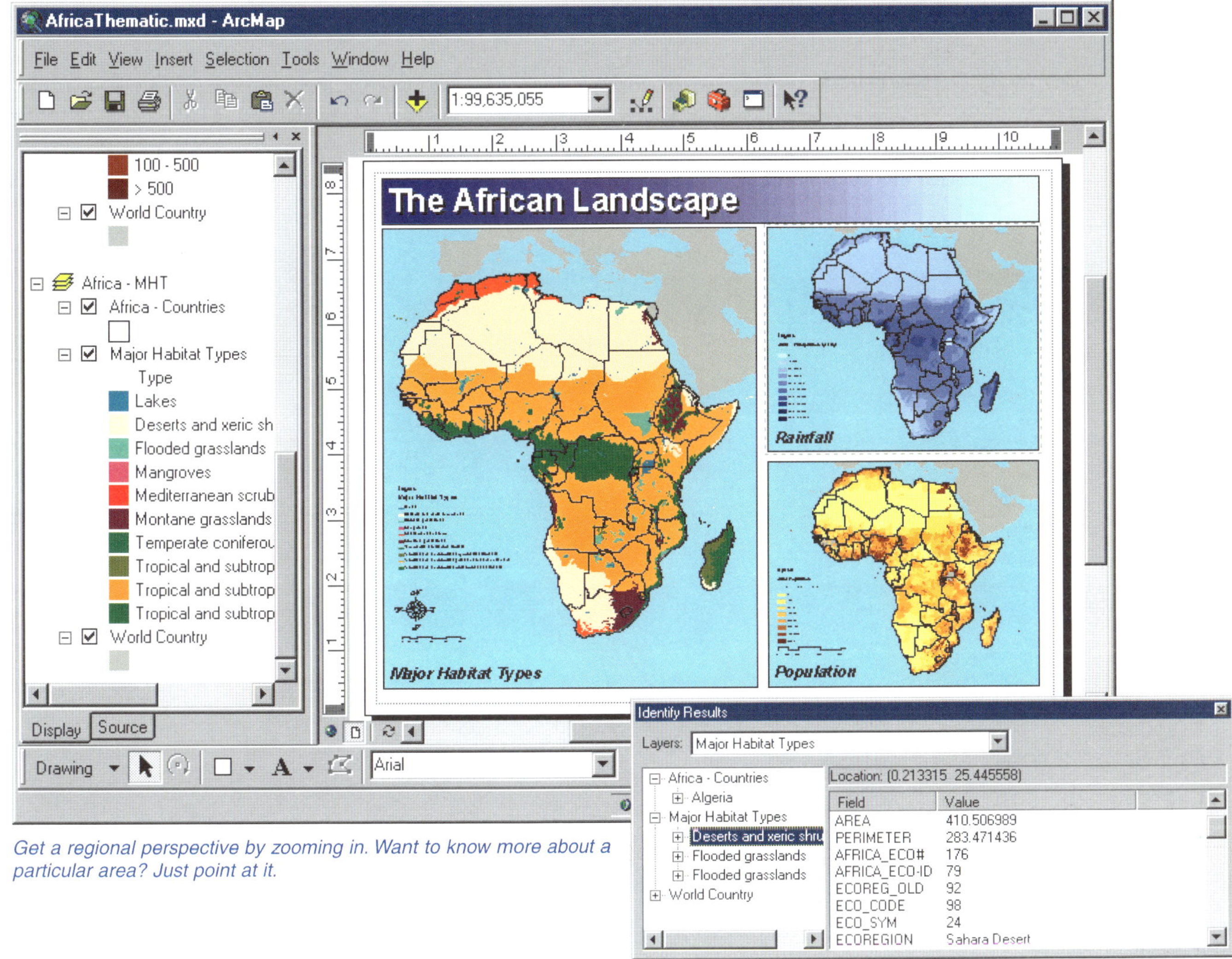

Get a regional perspective by zooming in. Want to know more about a particular area? Just point at it.

Showing relationships

You can show relationships between features by opening tables and creating charts, then adding these elements to the map.

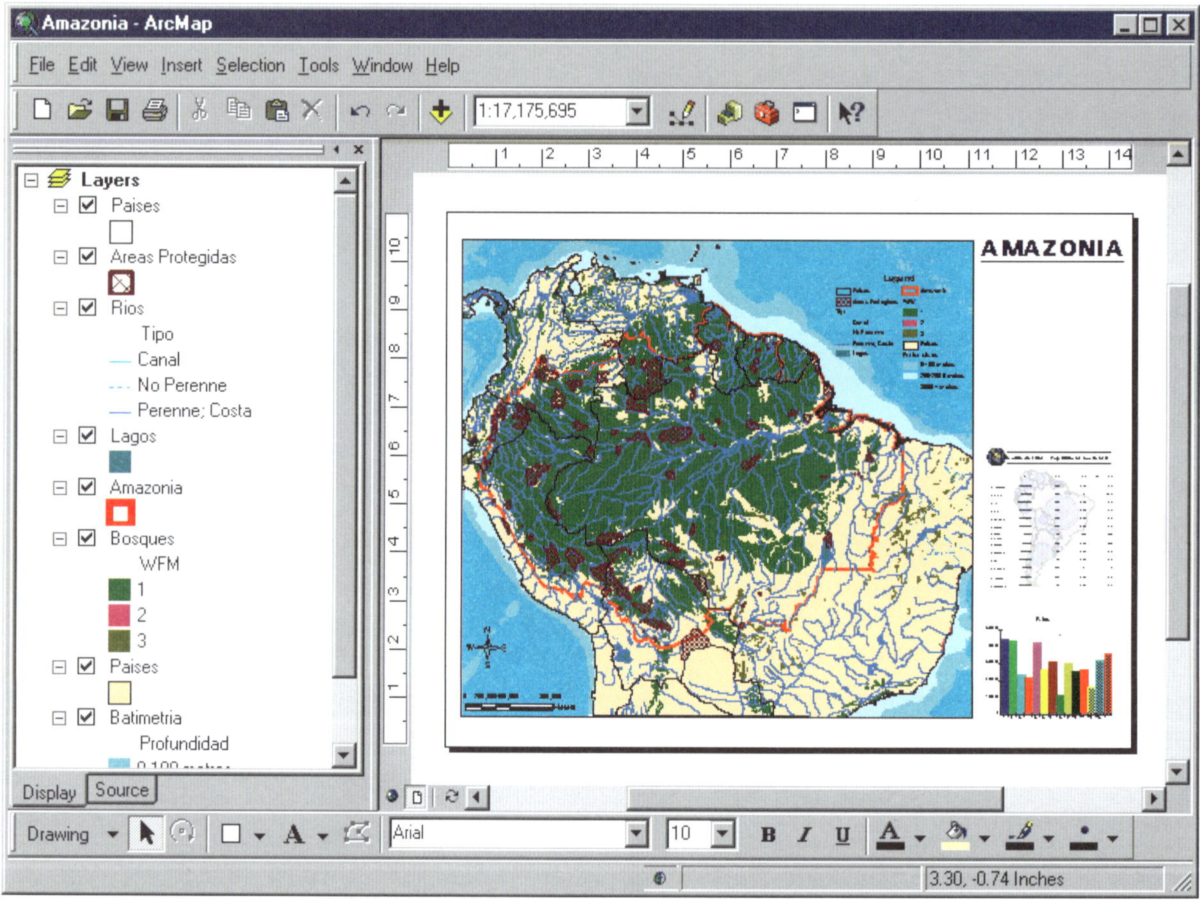

Charts and tables complement the map because they quickly summarize information that would otherwise take more time to understand.

Solving problems

You can search a map for features that meet particular criteria—for instance, find features by name, proximity, or characteristic.

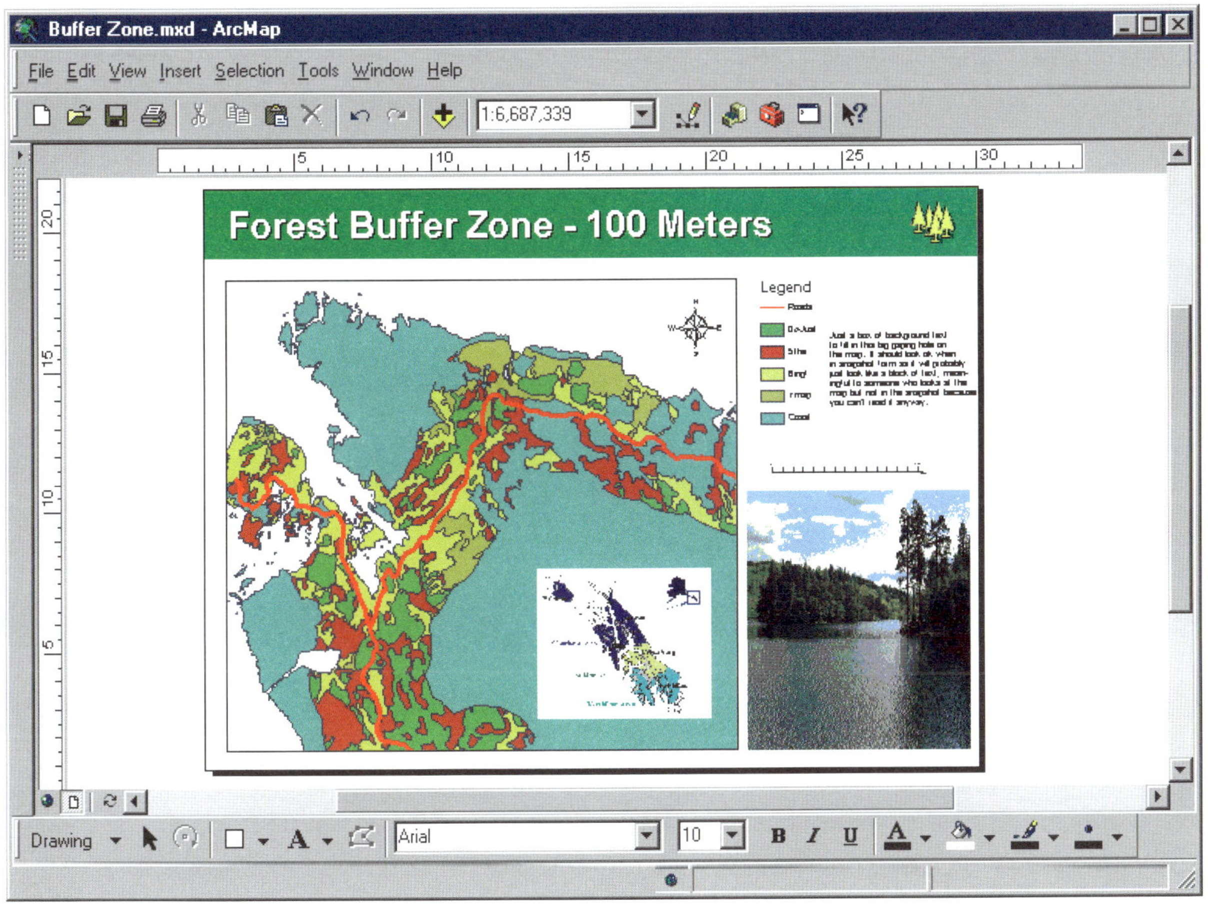

Finding forest habitats within 100 meters of roads aids in assessing environmental impact.

Creating and updating data

You can keep your data current with the latest information from the field. ArcMap has integrated editing tools to help you update data or create new data.

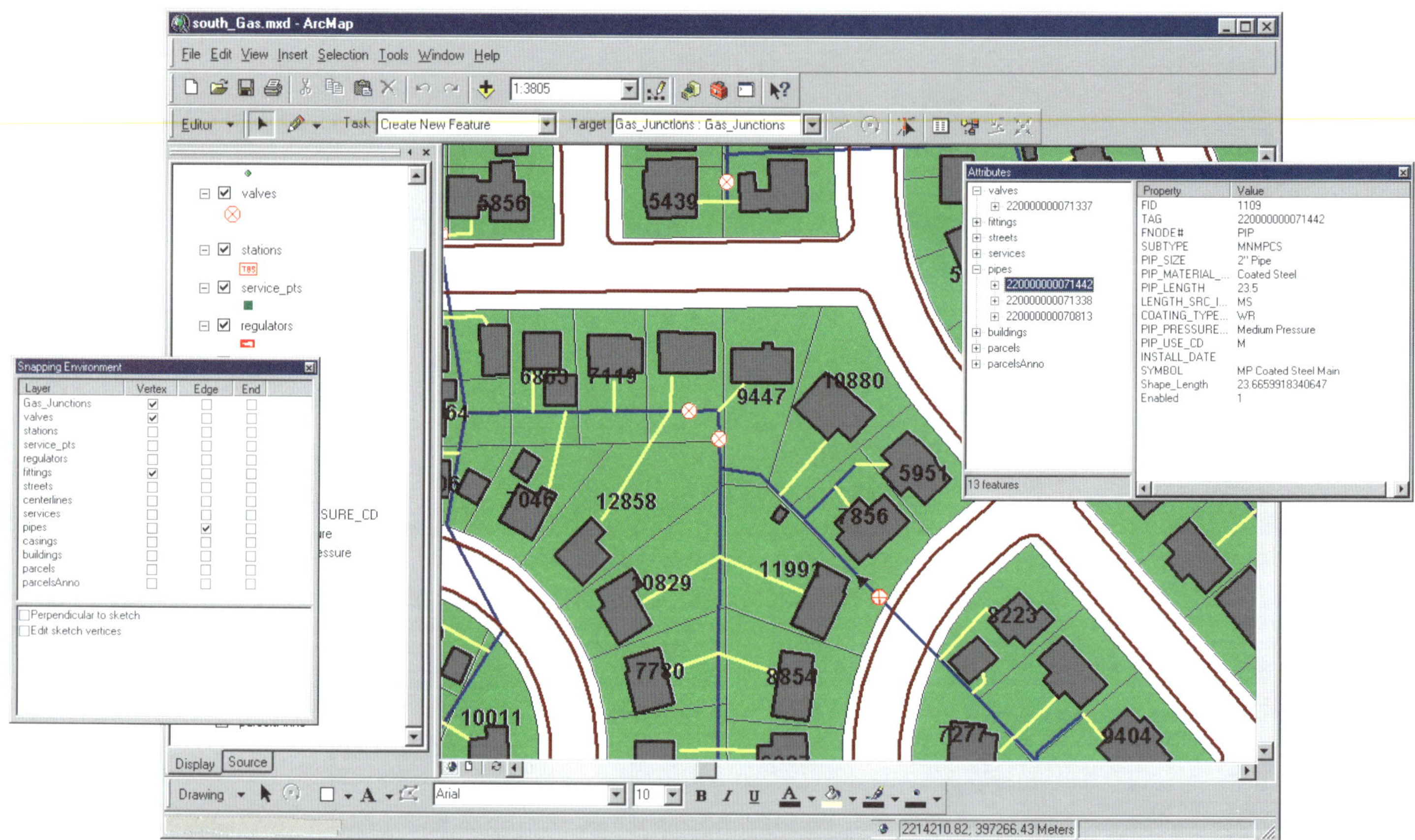

As a city grows, so too does its parcel database. ArcMap lets you edit both the geometry and attributes of features.

Presenting results

You can create high-quality maps and present them to others. Embed maps in reports, publish them on the Web, export them to standard formats, or print them out to hang on the wall.

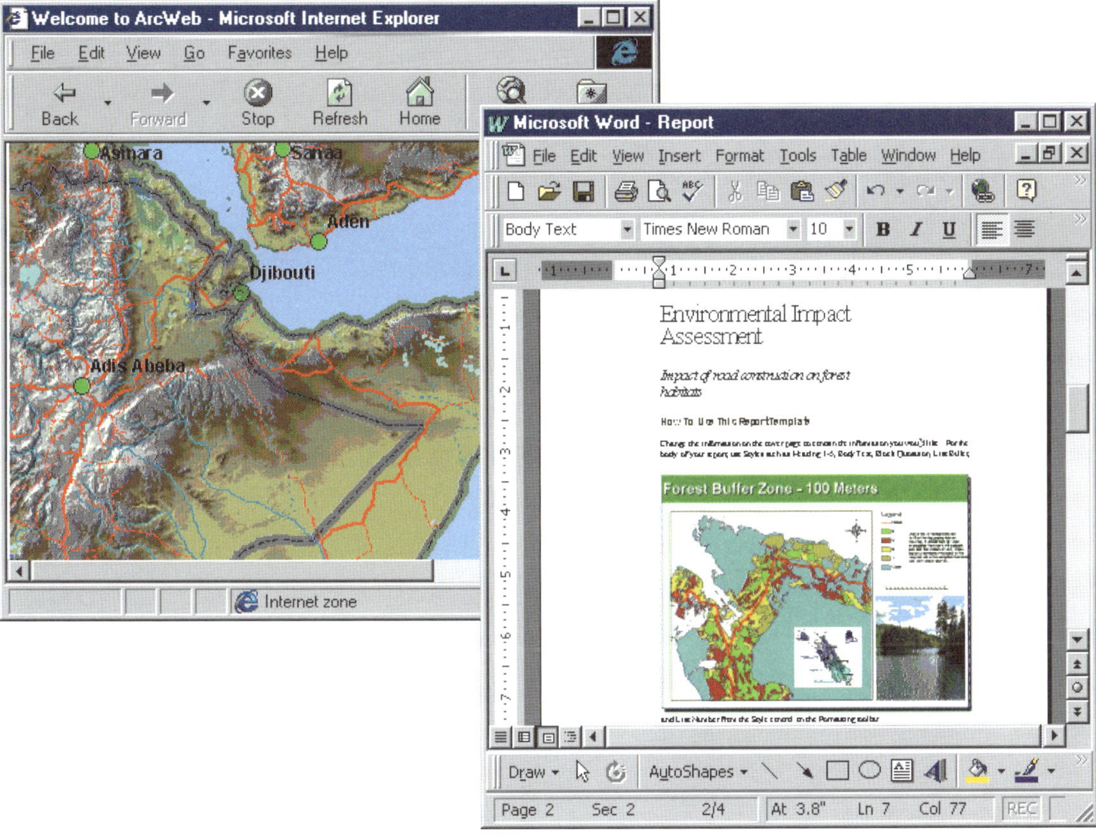

Developing mapping applications

You can develop custom mapping applications. Customize the out-of-the-box capabilities of ArcMap using the built-in Visual Basic® for Applications (VBA) programming environment or your favorite programming language. With ArcMap, you can customize the interface to suit your needs, write macros to automate work, or use ArcMap components to embed mapping capabilities into other software you create.

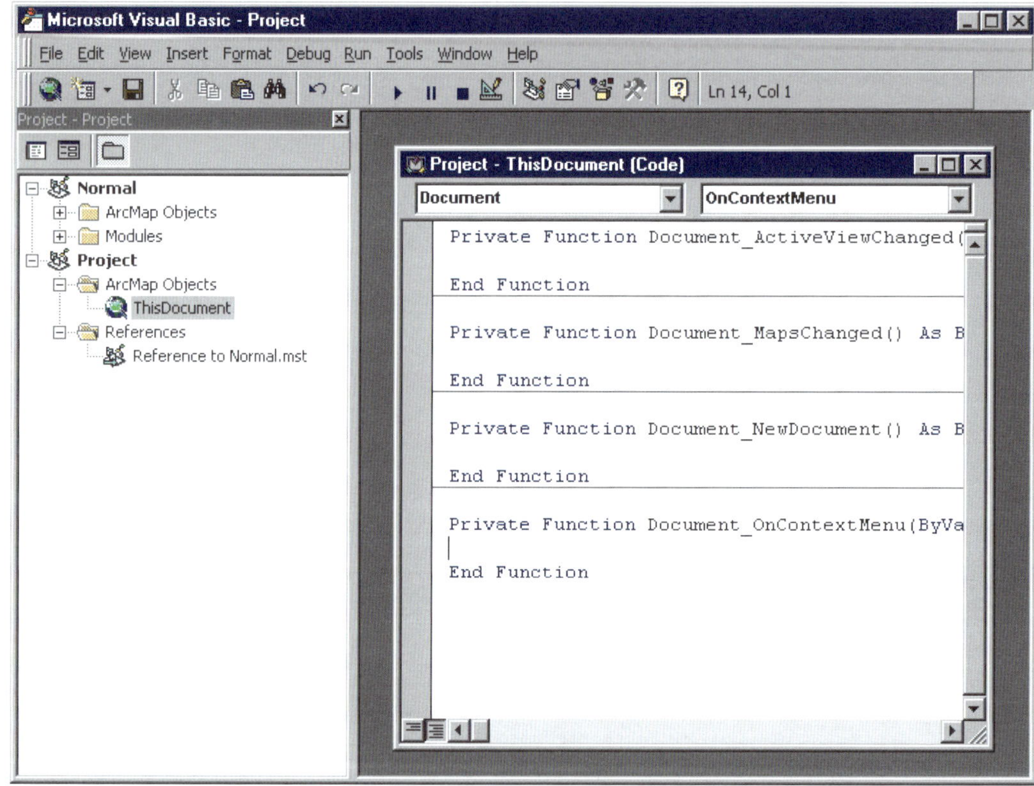

Automate your work with macros.

Tips on learning ArcMap

If you're new to GIS and mapping, remember that you don't have to learn everything about ArcMap to get immediate results. Begin learning ArcMap by reading the 'Quick-start tutorial'. This chapter shows you how quickly and easily you can make a map and gain insights into the steps you'll use to create your own. ArcMap comes with the data used in the tutorial, so you can follow along step by step at your computer. You can also read the tutorial without using your computer.

If you prefer to jump right in and experiment on your own, take a look at some of the maps distributed with ArcMap. Try browsing a map, changing symbols, and adding your own data.

When you're ready to build your own maps, you'll find that ArcMap comes with useful data you can use directly or as basemaps for your own data. If you don't find what you need, more data is available from ESRI, other organizations, and the Internet. ArcMap also comes with a lot of predefined symbols, North arrows, and scalebars to make building maps easier.

Finding answers to questions

Like most people, your goal is to complete your tasks while investing a minimum amount of time and effort on learning how to use software. You want intuitive, easy-to-use software that gives you immediate results without having to read pages of documentation. However, when you do have a question, you want the answer quickly so you can complete your task. That's what this book is all about—getting you the answers you need when you need them.

This book describes the mapping tasks—from basic to advanced—that you'll perform with ArcMap. Although you can read this book from start to finish, you'll likely use it more as a reference. When you want to know how to do a particular task, such as saving a map, just look it up in the table of contents or index. What you'll find is a concise, step-by-step description of how to complete the task. Some chapters also include detailed information that you can read if you want to learn more about the concepts behind the tasks. You may also refer to the glossary in this book if you come across any unfamiliar GIS terms or need to refresh your memory.

Getting help on your computer

In addition to this book, the ArcGIS Desktop Help system is a valuable resource for learning how to use the software. To learn how to use Help, see 'Getting help' in Chapter 3.

Learning about ArcMap extensions

ArcMap extensions are add-on programs that provide specialized GIS functionality. Extensions that come with ArcMap are covered in this book.

Contacting ESRI

If you need to contact ESRI for technical support, refer to 'Contacting Technical Support' in the 'Getting more help' section of the ArcGIS Desktop Help system. You can also visit ESRI on the Web at *www.esri.com* and *support.esri.com* for more information on ArcMap and ArcGIS.

ESRI education solutions

ESRI provides educational opportunities related to geographic information science, GIS applications, and technology. You can choose among instructor-led courses, Web-based courses, and self-study workbooks to find educational solutions that fit your learning style. For more information, go to *www.esri.com/ education*.

Quick-start tutorial

2

IN THIS CHAPTER

- **Exercise 1: Exploring your data**
- **Exercise 2: Working with geographic features**
- **Exercise 3: Working with tables**
- **Exercise 4: Editing features**
- **Exercise 5: Working with map elements**

The best way to learn ArcMap is to try it yourself. This tutorial guides you through some basic ArcMap skills as you create and print a set of maps for a county that is planning to expand its airport.

Residents of the county have identified several issues they are concerned about. These include noise affecting schools and houses near the airport and increased traffic along major roads. In this tutorial, you'll first create and print a map showing schools near the airport. Then you'll place this map—along with two other maps that show land use surrounding the airport and population density for the county—on a wall-sized poster for display.

In the tutorial, you'll learn how to:

- Display map features.
- Add data to your map.
- Edit geographic data.
- Work with data tables.
- Query and select geographic features.
- Create a summary chart.
- Lay out and print a map.

There are five exercises. Each exercise takes between 30 and 45 minutes to complete. You can work through the entire tutorial or complete each lesson one at a time.

Exercise 1: Exploring your data

In this exercise, you'll create a map showing locations of schools near the airport, along with a noise contour, to see which schools may be affected by noise from the airport. The noise contour is based on the 65 Community Noise Equivalency Level (CNEL), which indicates areas experiencing more than 65 decibels of noise, averaged over a 24-hour period. In many cases, buildings within the 65 CNEL will need soundproofing or other mitigation measures.

The exercises in this chapter use the tutorial data distributed with ArcMap. The default install location of the data is C:\ArcGIS\ArcTutor\Map. The exercises require that you have write access to this data. If you don't, you'll need to copy the data to a location that you do have write access to.

Starting ArcMap

ArcMap lets you explore your geographic data and create maps for display.

1. Click the Start button on the Windows taskbar.
2. Point to Programs.
3. Point to ArcGIS.
4. Click ArcMap.

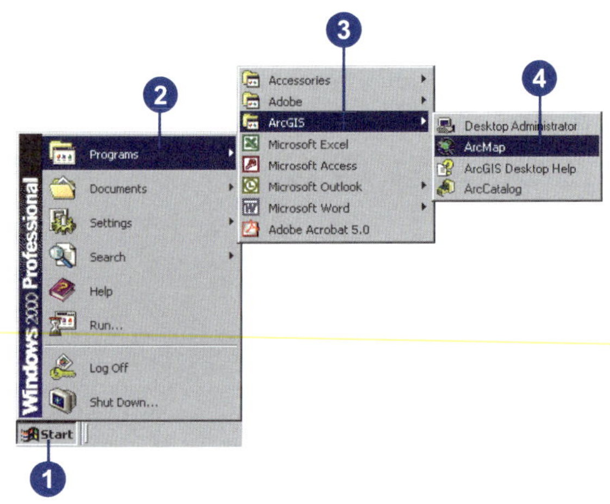

Opening an existing map document

The first time you start ArcMap, the Startup dialog box appears. The Startup dialog box offers you several options for starting your ArcMap session. For this exercise, you want to open an existing map document.

1. Double-click Browse for maps. If this is not the first time ArcMap has been started and the Startup dialog box does not appear, click File on the Main menu and click Open.

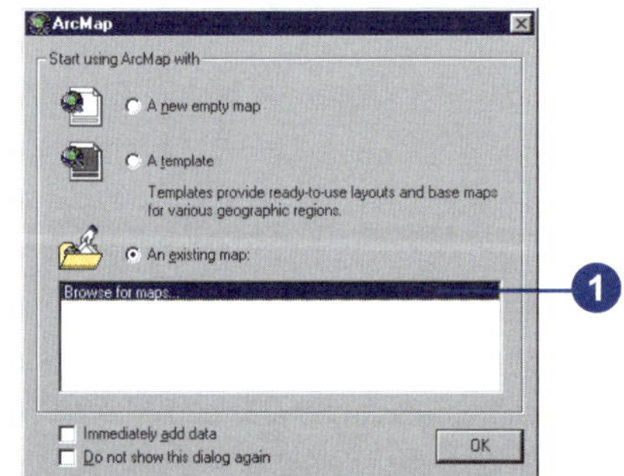

2. In the dialog box, click the Look in dropdown arrow, and navigate to the Map folder on the local drive where you installed the tutorial data (the default installation path is C:\ArcGIS\ArcTutor\Map).
3. Double-click airport.mxd. ArcMap opens the map.

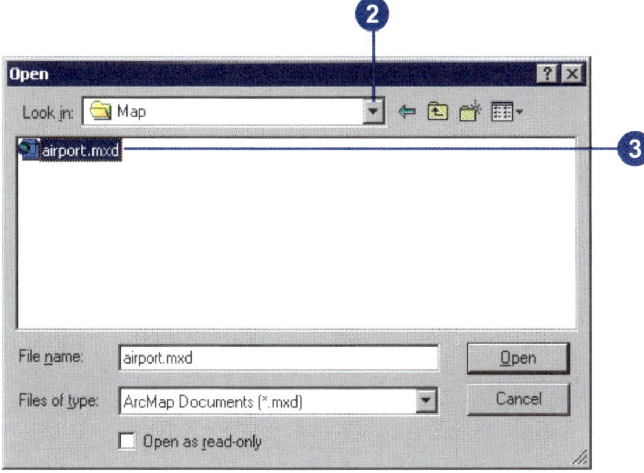

ArcMap stores a map as a map document (.mxd) so you can redisplay it, modify it, or share it with other ArcMap users. The map document doesn't store the actual data, but rather references the data stored on disk along with information about how it should be displayed. The map document also stores other information about the map, such as its size and the map elements it includes (title, scalebar, and so on).

To the left of the ArcMap display window is the table of contents, showing you which geographic layers are available to display. To the right is the map display area.

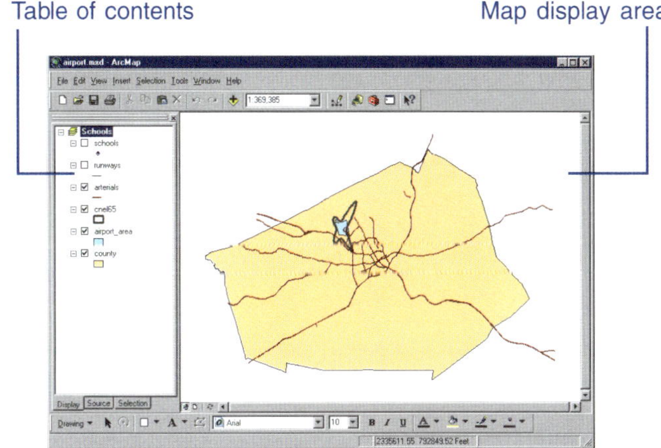

This particular map contains the following layers in a data frame called Schools:

schools	locations of elementary, middle, high, and private schools
runways	location of airport runways
arterials	major roads
cnel65	the noise contour
airport_area	the proposed airport expansion zone
county	the county boundary

The map currently displays the arterials, noise contour, airport area, and county boundary. Their boxes are checked in the table of contents.

Moving around the map

The Tools toolbar lets you move around the map and query the features on the map. Place your pointer over each icon (without clicking) to see a description of each tool.

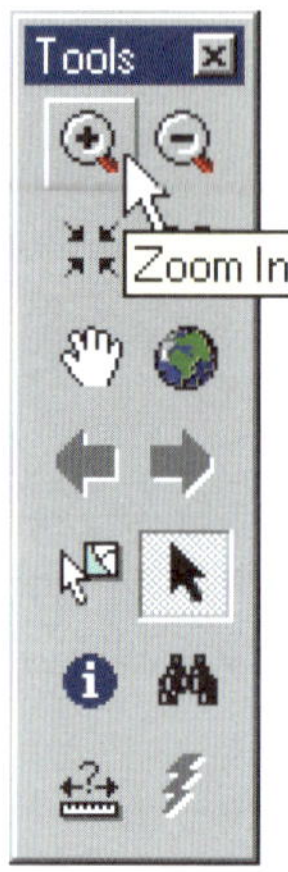

1. Using the Zoom In tool, draw a box around the noise contour to zoom in. Place the pointer on the upper-left part of the contour, press the mouse button, and hold it down while dragging to the lower right. You'll see the box drawn on the screen. When you release the mouse button, ArcMap zooms in to the area defined by the box.

2. If necessary, use the Pan tool (the hand) on the Tools toolbar to reposition the map so the noise contour is in the center of the display area (hold the mouse button down while dragging in the direction you want to move the features, then release the button).

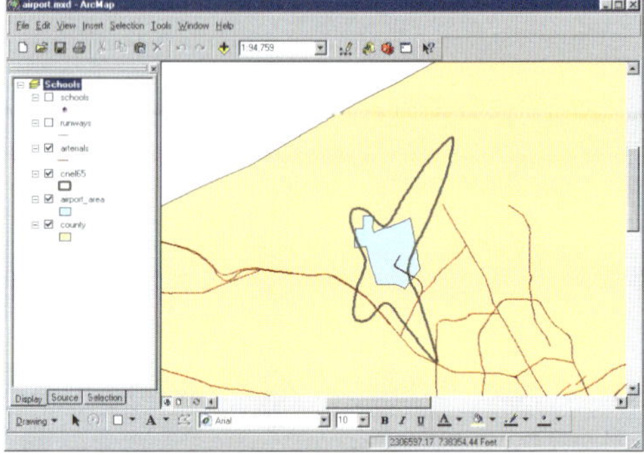

Displaying a layer

The table of contents lets you turn layers on and off in the display. To display a layer, check the box next to its name. To turn it off, uncheck it. Display the schools and runways by checking their boxes in the table of contents. For more information, see Chapter 5, 'Working with layers'.

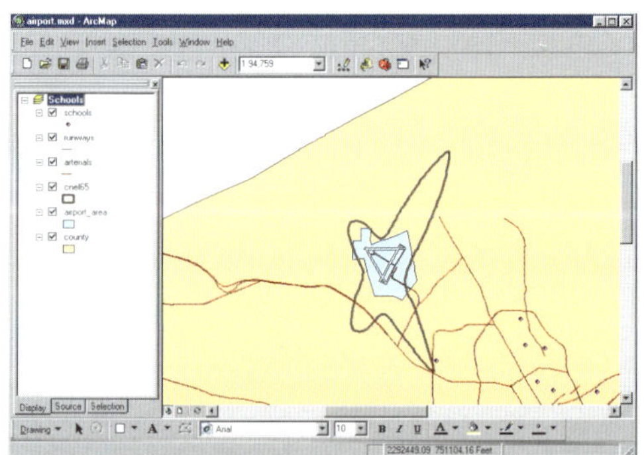

Changing the display symbol

ArcMap lets you change the colors and symbols you use to display features. You'll change the symbols for schools from a dot to a standard symbol used for schools on many maps.

1. Click the dot symbol in the table of contents to display the Symbol Selector window.

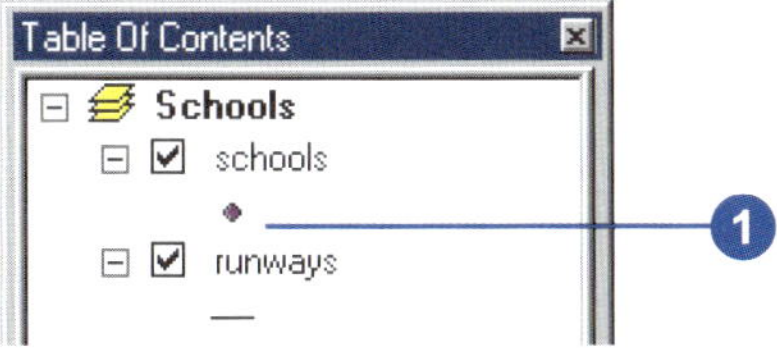

2. Scroll down until you find the School 1 symbol. Click it.

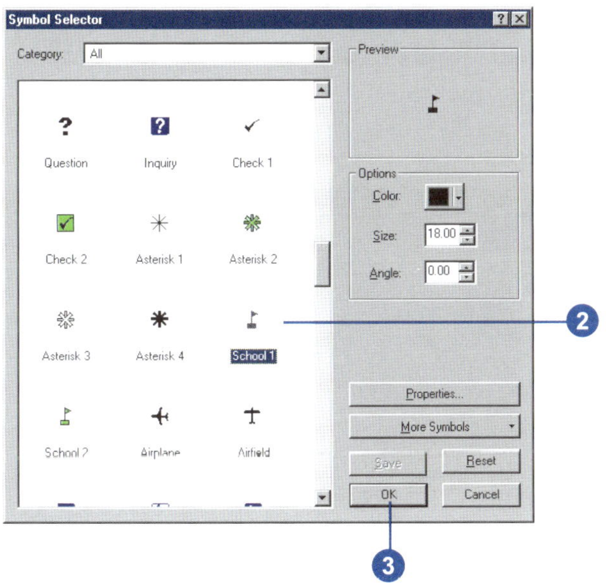

3. Click OK. The schools are drawn with the new symbol.

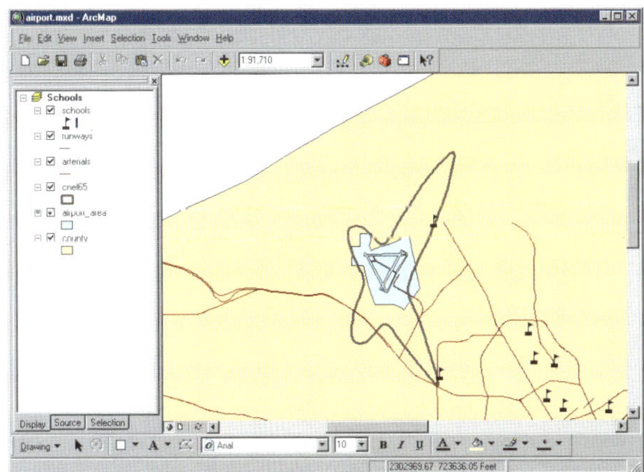

You can also open the symbol dialog box by right-clicking the layer name, choosing Properties from the menu that appears, and clicking the Symbology tab. To simply change the color of a symbol, right-click the symbol in the table of contents to display the color palette. For more information on changing display symbols, see Chapter 6, 'Symbolizing features'.

Identifying a feature

There is one school that may be within the noise contour around the airport.

1. Using the Zoom In tool, draw a box around the school to zoom in.

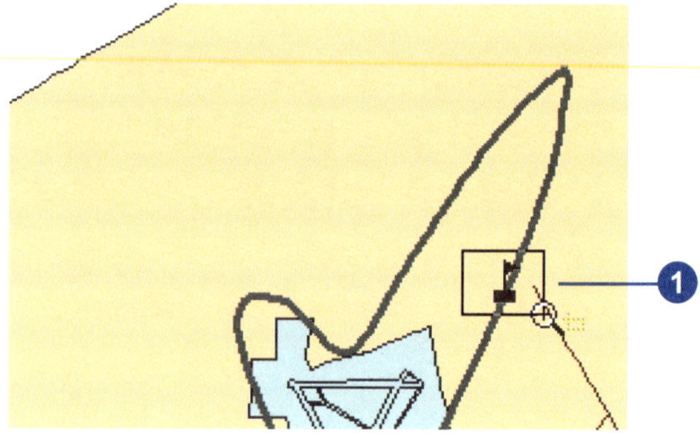

You can see that the school is indeed within the noise contour.

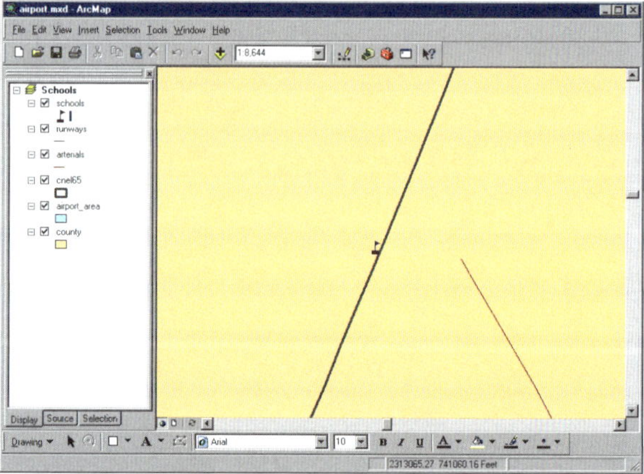

2. Click the Identify tool on the Tools toolbar. The Identify Results window appears.

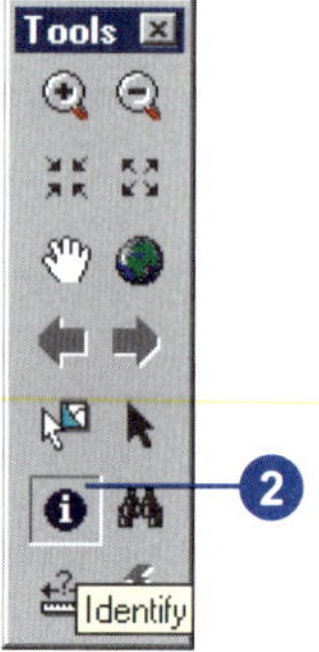

3. Move the mouse pointer over the school and click. The name of the school (Northwestern Prep) is listed in the Identify Results window. Notice that only the features in the topmost layer are identified. You can also identify features in other layers by choosing the specific layers you want to identify by clicking the Layers dropdown arrow in the dialog box.

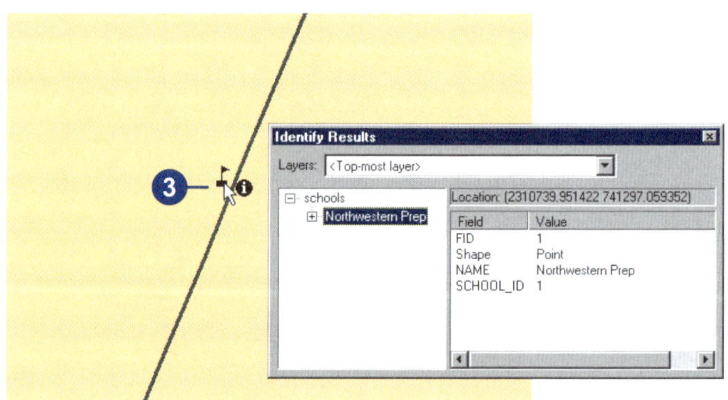

Close the Identify Results window.

4. Click the Back button on the Tools toolbar to return to your previous view.

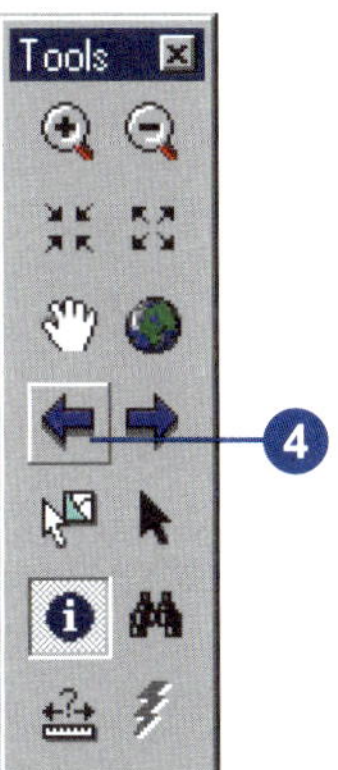

Adding graphics

You can add text and other graphics to your display using the Draw toolbar at the bottom of the ArcMap window.

1. Click the New Text button. The pointer changes to a crosshair with an **A**.

2. Move the mouse pointer near the school you identified and click.
3. In the text box that appears, type "Northwestern Prep" and press Enter.

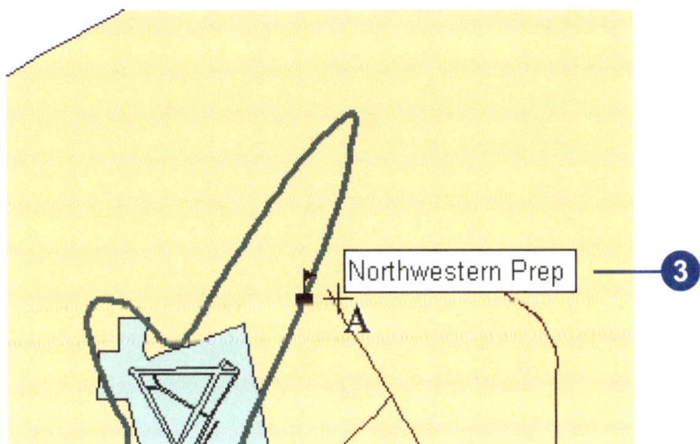

A blue dotted line surrounds the text, indicating it is currently selected. You can drag the text to a new position by clicking and holding down the mouse button while dragging the text, then releasing the button.

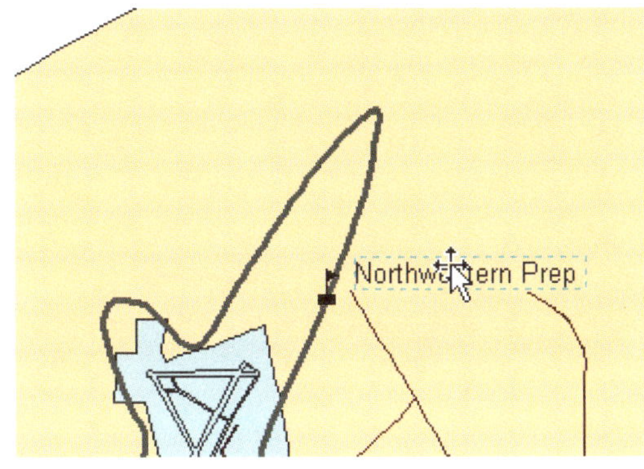

4. When you're finished positioning the text near the school, click outside the text box to deselect it.

For more information on working with text, see Chapter 7, 'Working with graphics and text'.

Laying out a map

ArcMap lets you work in data view or layout view. Data view focuses on a single data frame. Use data view when exploring or editing your data. Layout view shows you how the map page looks. Use layout view when composing and printing a map for display. You can also explore and edit your data in layout view if you want. All the tools and options available in data view are also available in layout view.

You can change the size and orientation of the page in layout view. In this case, you'll create a 16- by 12-inch map with a landscape orientation.

1. Click the Fixed Zoom Out button on the Tools toolbar several times to zoom to a smaller map scale.
2. Click the View menu and click Layout View. The Layout toolbar appears, and the display changes to show the page layout with rulers along the side.

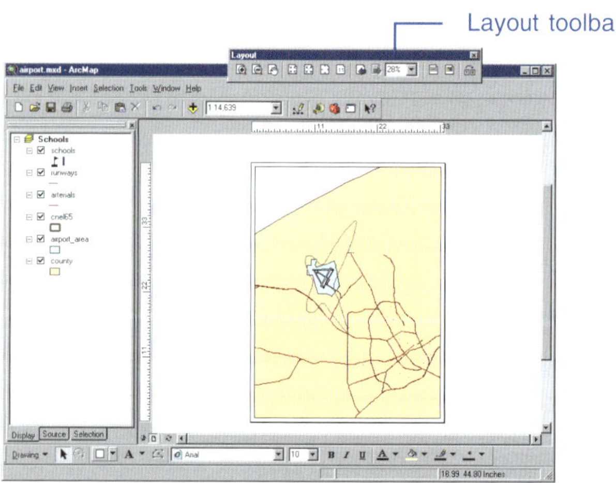

3. Right-click anywhere on the layout background and click Page and Print Setup. You can also access Page and Print Setup from the File menu.

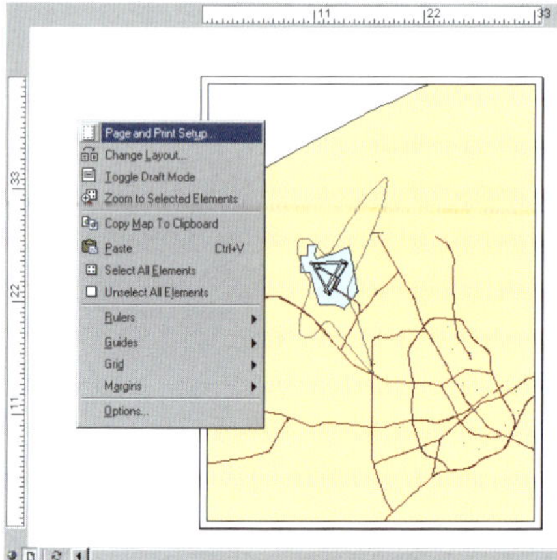

4. Make sure the Use Printer Paper Settings box is not checked—otherwise, the page size will default to be the same as your printer. If your printer does not print larger sizes, you can scale down the map when you print it, as you'll see later in this exercise.
5. Check Scale Map Elements proportionally to changes in Page Size. That way, the data will be rescaled to fit the page.
6. Set the Map Page Size Page Orientation to Landscape.
7. Set the page width to 16 and the height to 12 inches by clicking in each box and typing over the existing values.

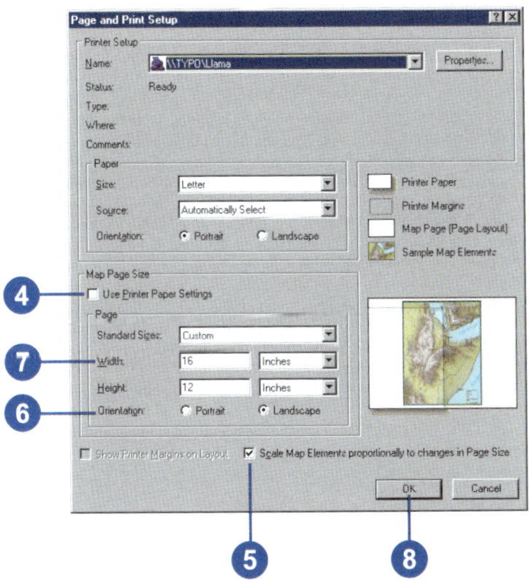

8. Click OK. The page display and rulers change to reflect the new size and orientation. You may need to resize your data frame manually to make it look like the map below. To do this, click the Select Elements tool on the Tools toolbar, click the data frame, and resize the data frame using the blue selection handles.

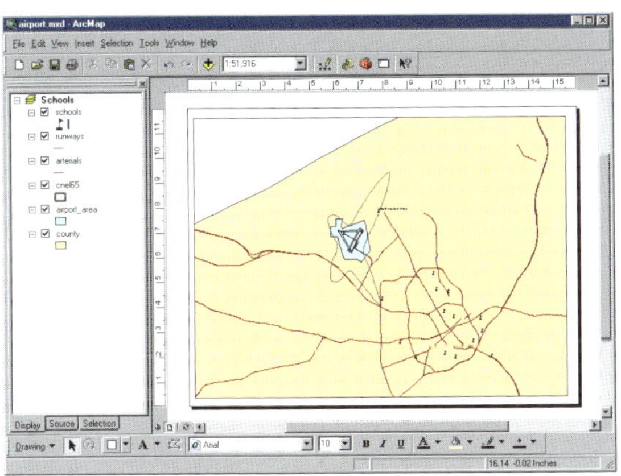

Zooming in on the page

The Layout toolbar controls your view of the scale and position of the whole map (as opposed to the data layers on the map). By default, the map size is set so you can see all of it. But at this scale it's hard to see the school name.

1. Click Zoom to 100% on the Layout toolbar. The page is displayed at the actual printed size so you can see the detail.

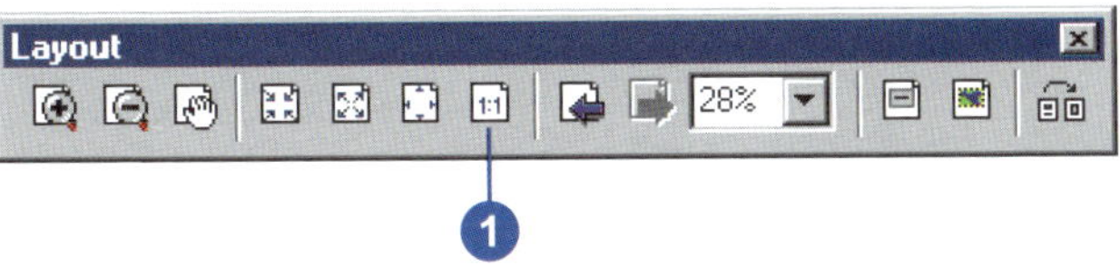

2. Click the Pan button on the Layout toolbar and drag the map to the lower left so you can see the name of the school.

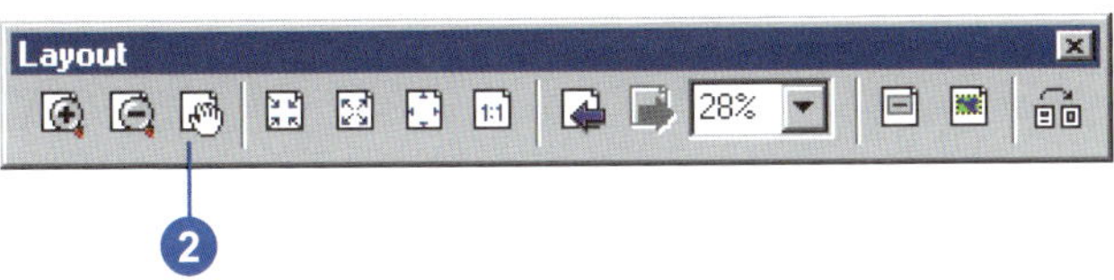

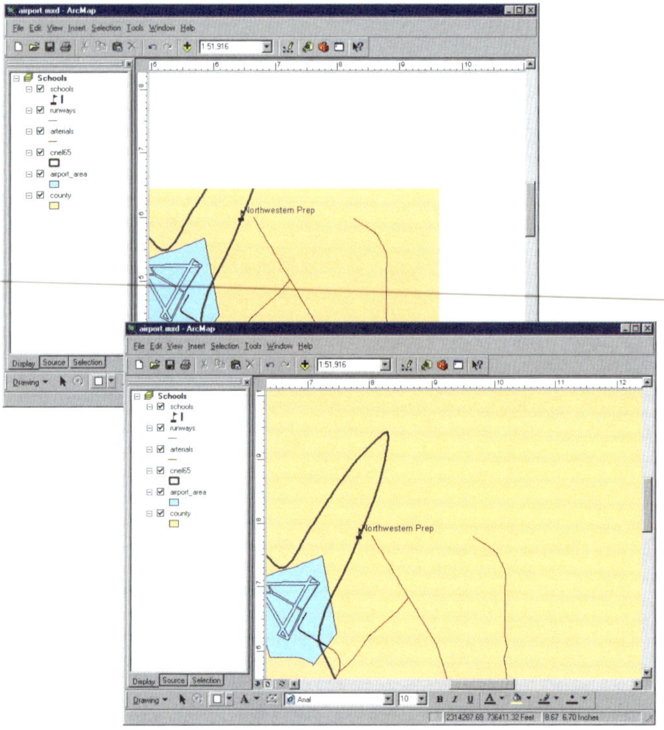

3. Click the Zoom Whole Page button on the Layout toolbar to see the entire page again.

Inserting map elements

ArcMap makes it easy to add titles, legends, North arrows, and scalebars to your map.

1. Click Insert on the Main menu and click Title. In the box that appears, type the title for your map, "Schools and Noise Contour", and press Enter.

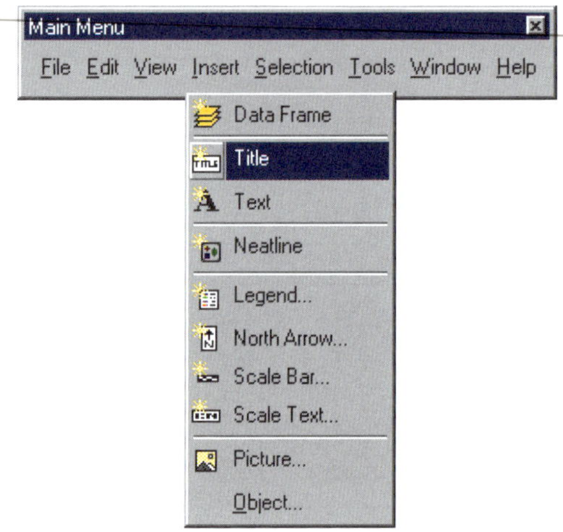

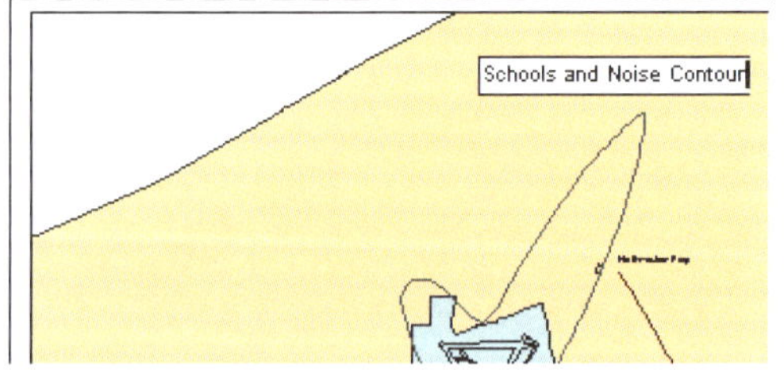

2. On the Draw toolbar at the bottom of the window, click the Text Size dropdown arrow and click 36 to change the title to 36 point.

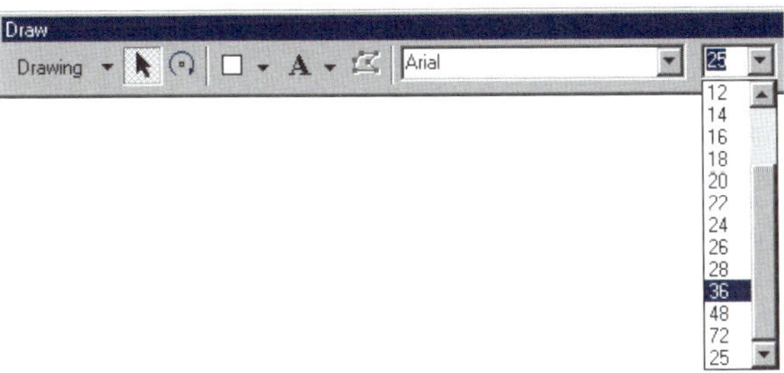

3. Click the title and drag it so it's centered at the top of the map.

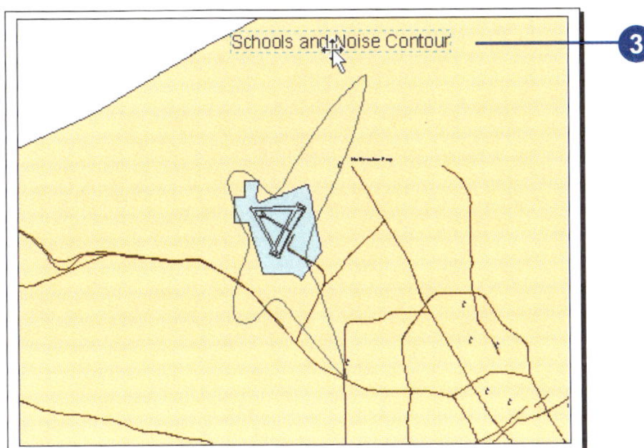

The Draw toolbar lets you add and change the format (font, size, color, and so on) of text and graphic elements—such as boxes, callout lines, or circles—on your map.

4. Click Insert and click Legend.

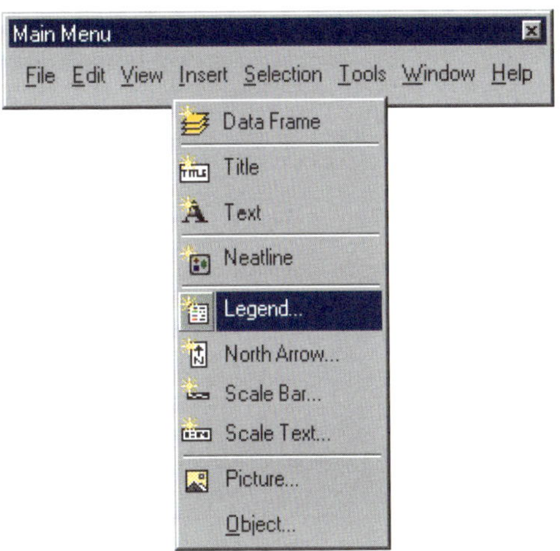

The Legend Wizard appears.

5. Click Next several times to step through the wizard accepting the default legend parameters. Click Finish when done.

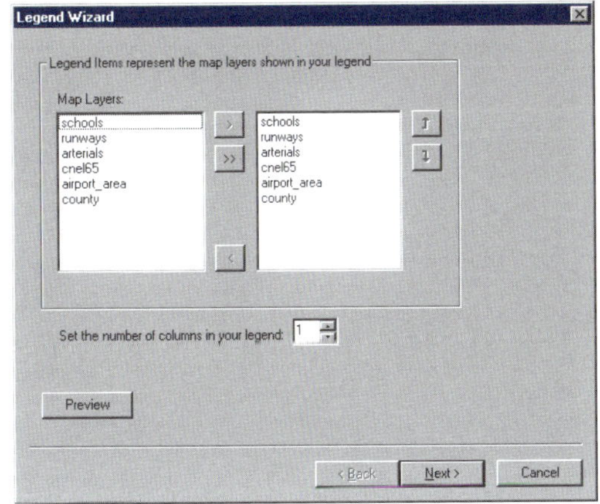

By default, ArcMap scales the legend to the page and includes all the layers that are currently displayed. You can modify the legend by right-clicking it and choosing Properties from the menu that appears. For now, just use the default legend.

6. Click and drag the legend to the lower-left corner of the map.

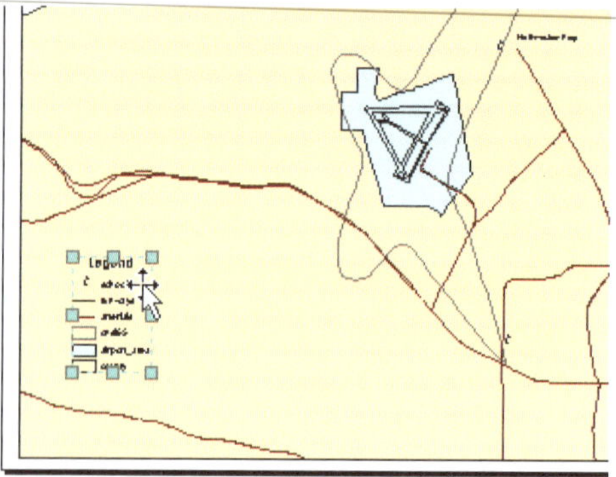

7. Click Insert and click North Arrow. The North Arrow Selector window appears.

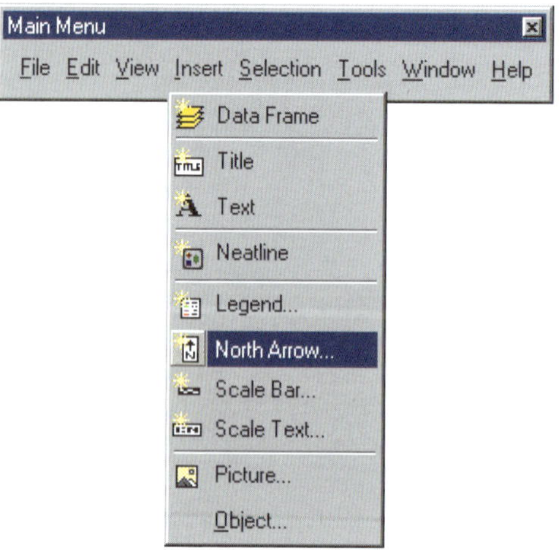

8. Click ESRI North 1 and click OK. Click and drag the North arrow so it is to the right of the legend.

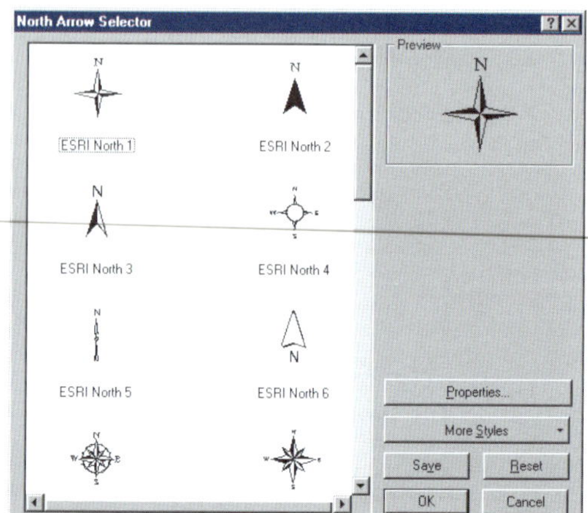

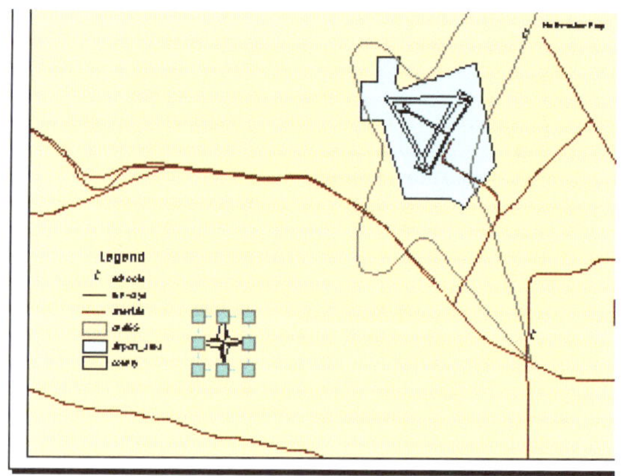

9. Now insert a scalebar from the Insert menu. Click Scale Line 1 in the Scale Bar Selector window and click OK.

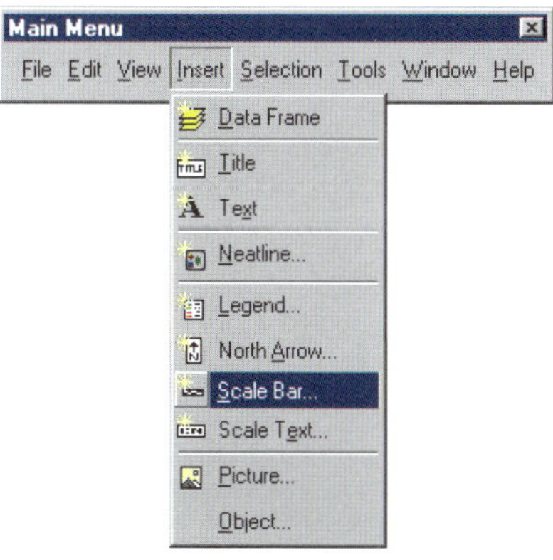

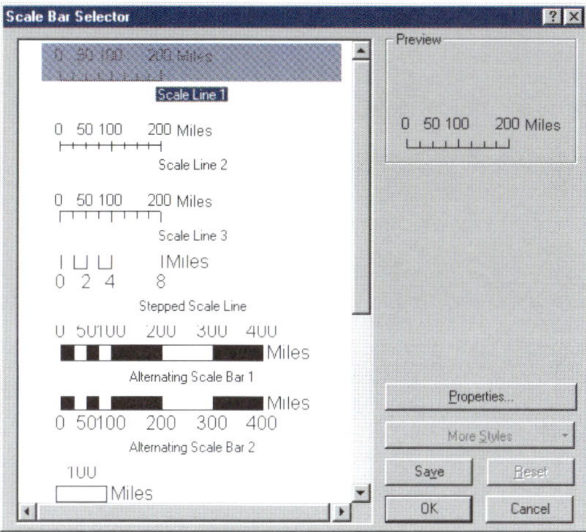

10. Click and drag the scalebar under the legend and North arrow.

11. Click the legend to select it; while holding down the Shift key, click the scalebar to select it as well.

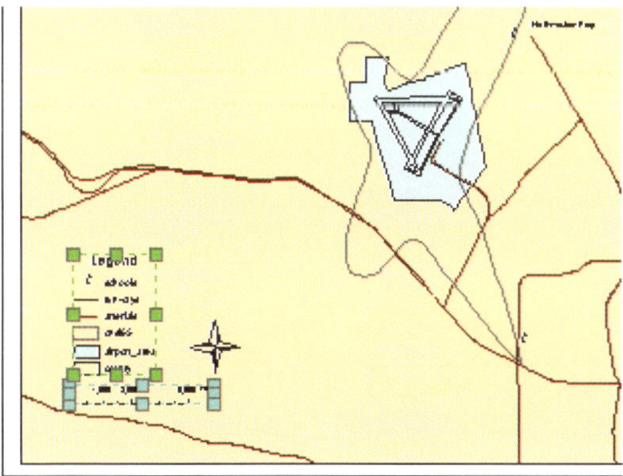

12. Click Drawing on the Draw toolbar, point to Align, and click Align Left from the menu that appears. The scalebar is now aligned with the left side of the legend.

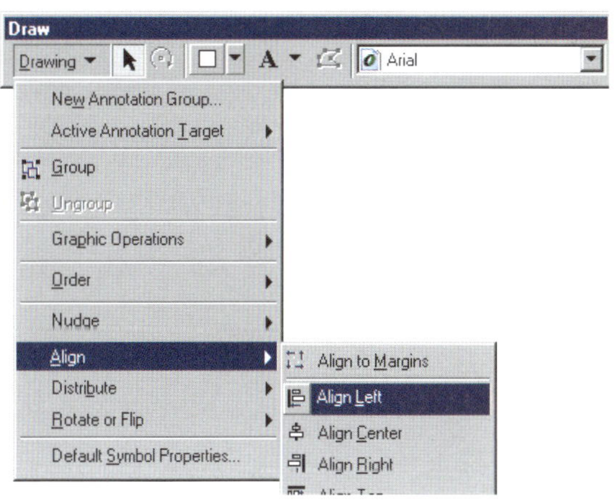

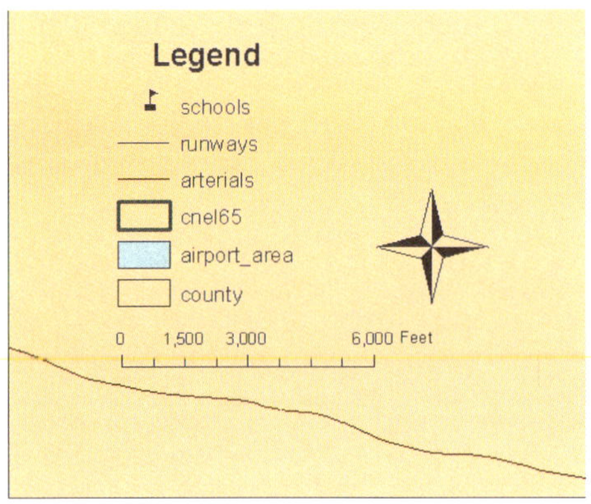

Printing a map

At this point, your first map is finished. If you have a printer connected to your computer, you can print the map.

1. Click File and click Print.

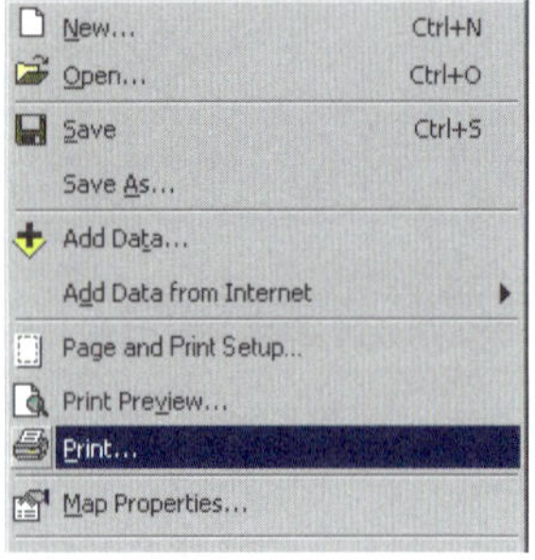

2. If the map (which is 16 by 12 inches) is larger than your printer paper, click Scale Map to fit Printer Paper.

3. Click Setup.

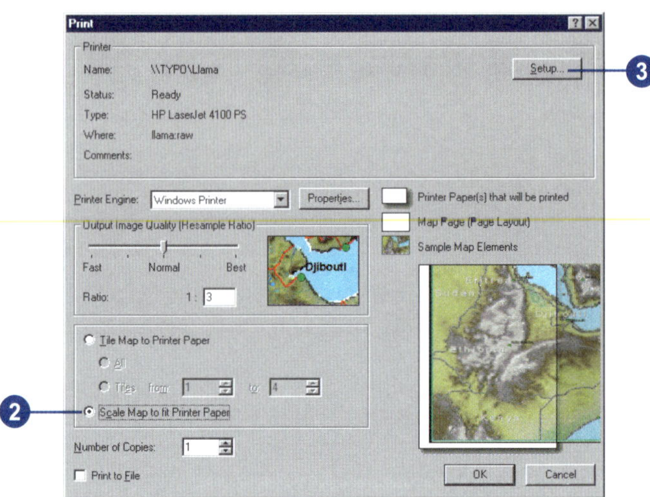

4. Click Landscape on the Printer Setup panel.
5. Click OK to close the Page and Print Setup dialog box.

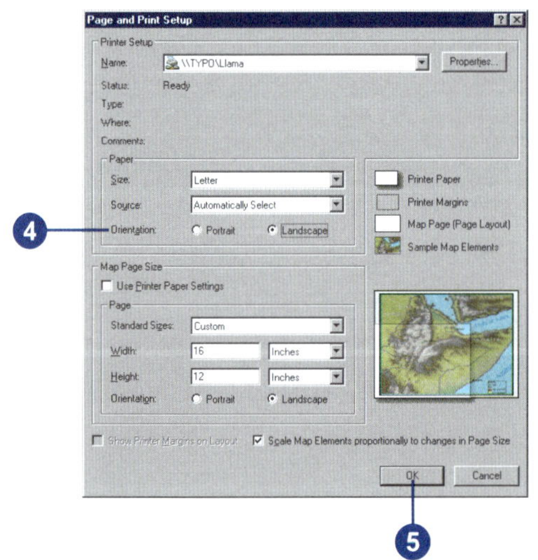

6. Click OK on the Print dialog box to print your map.

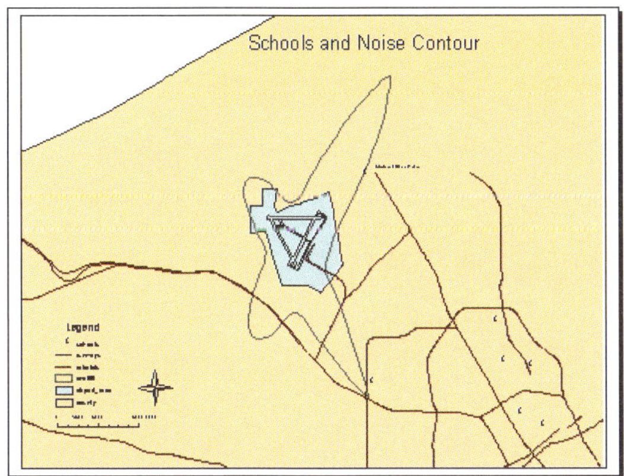

Saving a map

Save your map in the folder with the tutorial data. First, though, ensure that ArcMap uses the full pathname of the location of the data on your system. The airport map was created using relative pathnames, so ArcMap would find and display the data after the ArcTutor\Map folder is copied to your system.

1. Click File and click Map Properties.
2. Click Data Source Options on the Properties dialog box.

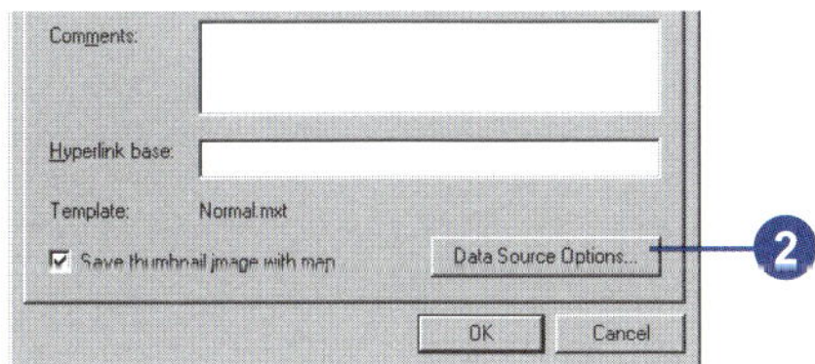

3. Click Store full path names and click OK.

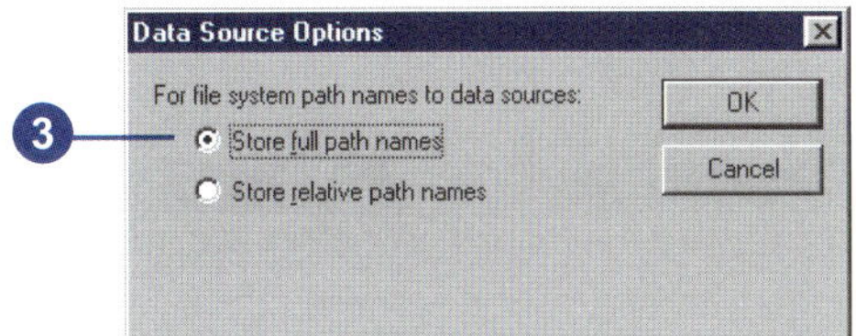

4. Click OK on the Map Properties dialog box.

Now save a copy of your map. You'll use this copy in the subsequent exercises.

1. Click File and click Save As.

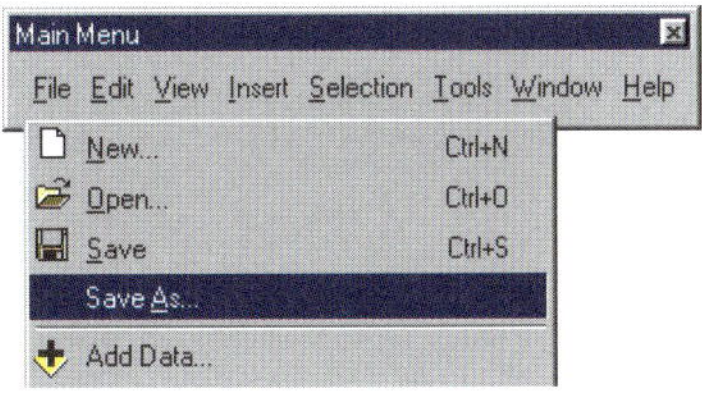

2. In the File name box, type "airport_ex".
3. Click Save.

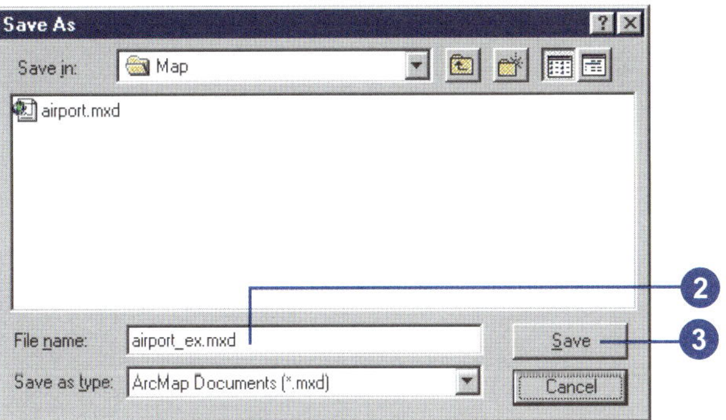

You can continue with the tutorial or stop and complete it at a later time.

Exercise 2: Working with geographic features

In this exercise, you'll map the amount of each land use type within the noise contour. You'll add data to your map, draw features based on an attribute, select specific features, and summarize them in a chart.

If necessary, start ArcMap, navigate to the folder where you saved the map from Exercise 1 (airport_ex), and open the map.

Changing the page layout

First, you'll create the map layout by changing the page size and orientation.

1. Make sure you're in layout view (click the View menu and click Layout View).

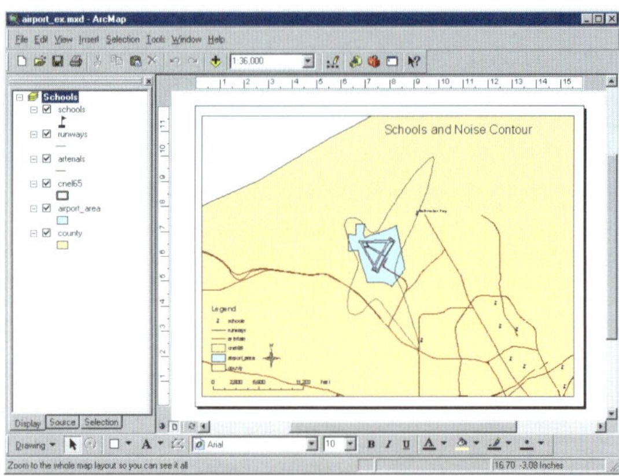

2. Click File and click Page and Print Setup.

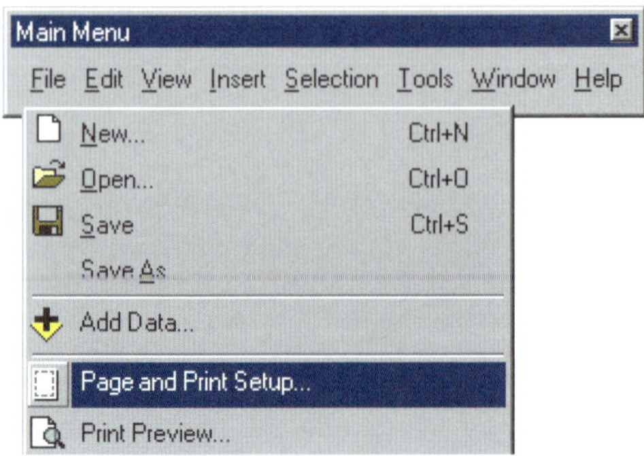

3. Click the Standard Sizes dropdown arrow and click ANSI E. That sets the width and height to a standard E-size page.
4. Click Portrait on the Map Page Size panel.
5. Uncheck Scale Map Elements proportionally to changes in Page Size (this way, the existing map of schools will remain the same size, rather than being scaled up to fit the page).

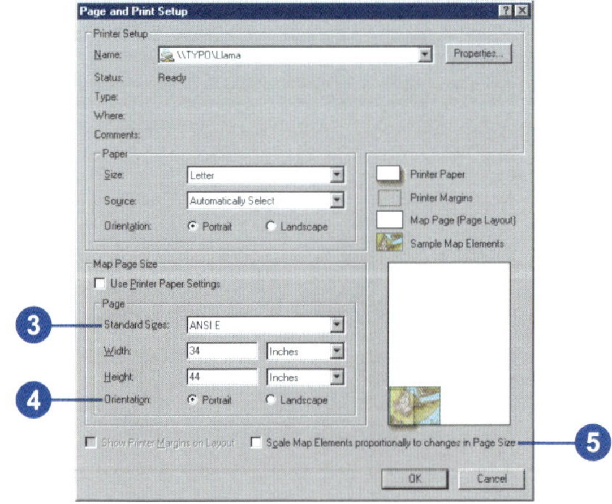

6. Click OK. The page size changes, and the existing map is displayed in the lower-left corner.

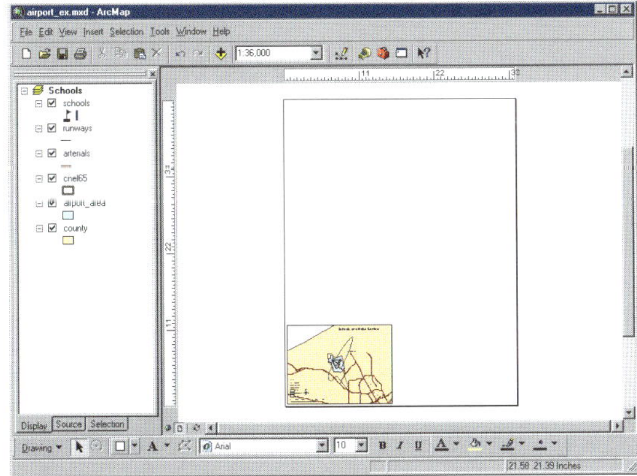

7. Click the Select Elements tool on the Tools toolbar.

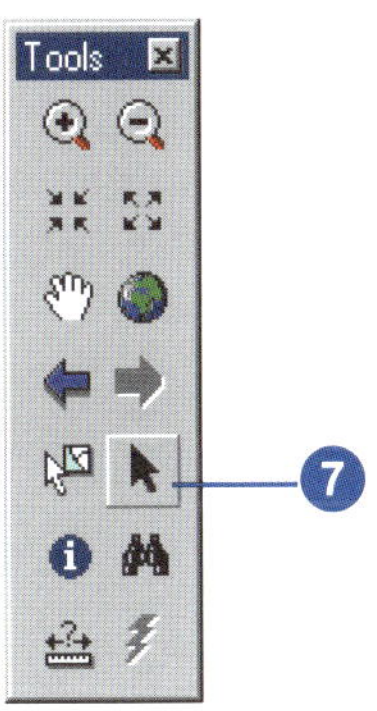

8. Click and drag a box around the elements to select them.

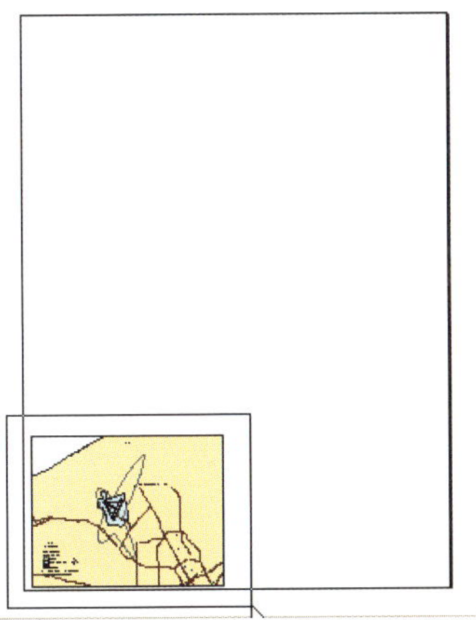

9. Click and drag the group of elements to the upper portion of the page.

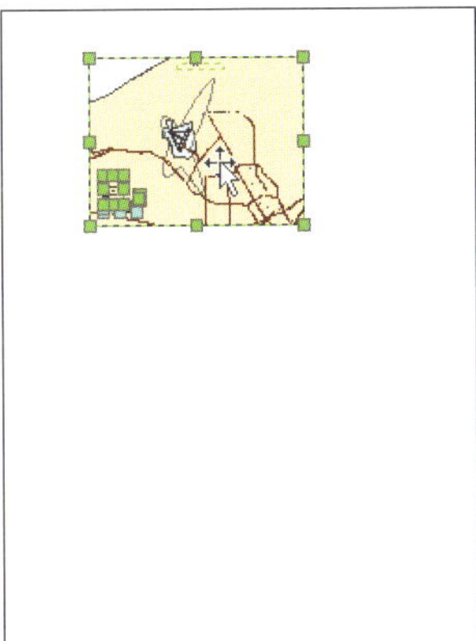

For more information on page layout, see Chapter 15, 'Laying out and printing maps'.

Creating a new data frame

A data frame is a way of grouping a set of layers you want to display together. Now you'll add a new data frame to show land use.

1. Click Insert and click Data Frame.

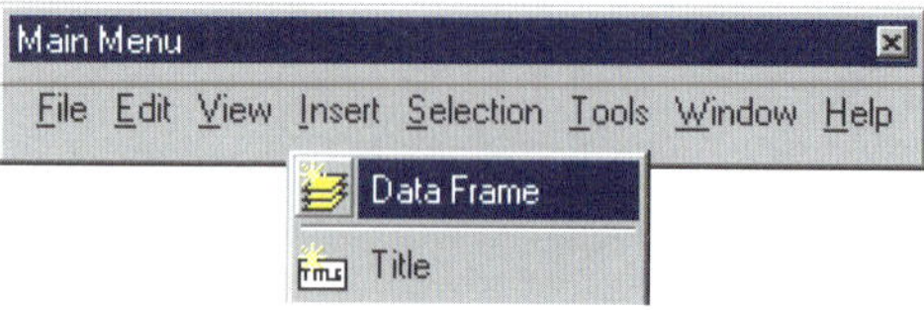

The frame appears on the layout and is listed in the table of contents.

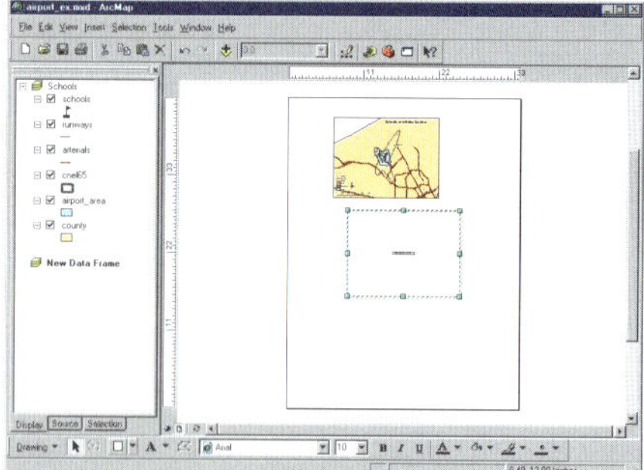

Adding a data layer

You'll map land use based on a code for each land parcel. First, add the parcels layer to the data frame.

1. Click the New Data Frame data frame on the page so only it is selected.

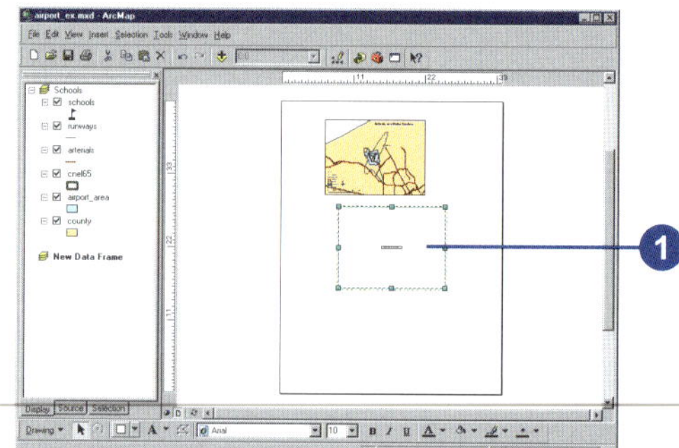

2. Click the Add Data button on the Standard toolbar.

3. Navigate to the Map folder on the local drive where you installed the tutorial data (the default installation path is C:\ArcGIS\ArcTutor\Map).
4. Double-click the airport geodatabase, airport.mdb.
5. Click the parcels layer and click Add.

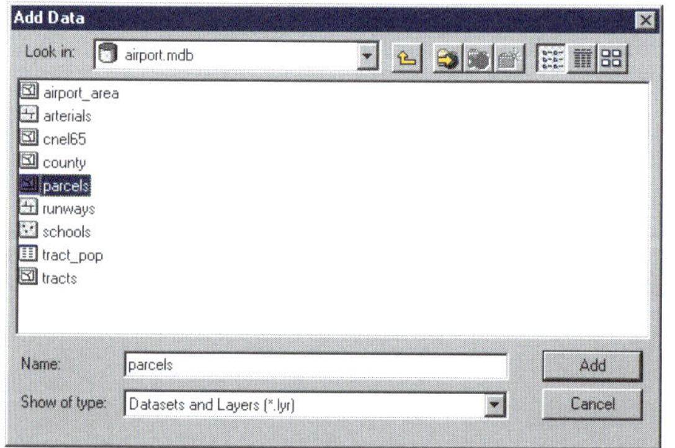

The data layer is added to the table of contents and displays in the layout (the parcels may be a different color on your map).

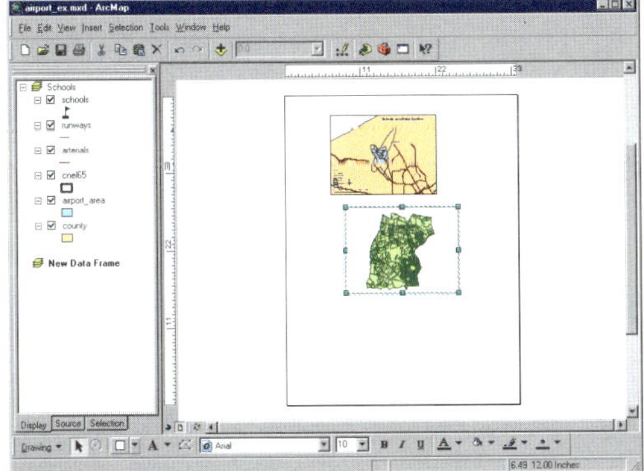

All the data used in this tutorial is stored in a geodatabase. ArcMap also lets you work with ArcInfo™ coverages, shapefiles, image files, and many other data formats. For more about geodatabases and other data formats, see *Using ArcCatalog*.

Setting properties of the data frame

1. Right-click New Data Frame in the table of contents and click Properties. You may need to hold your pointer over the arrow at the bottom of the menu to see Properties in the list.

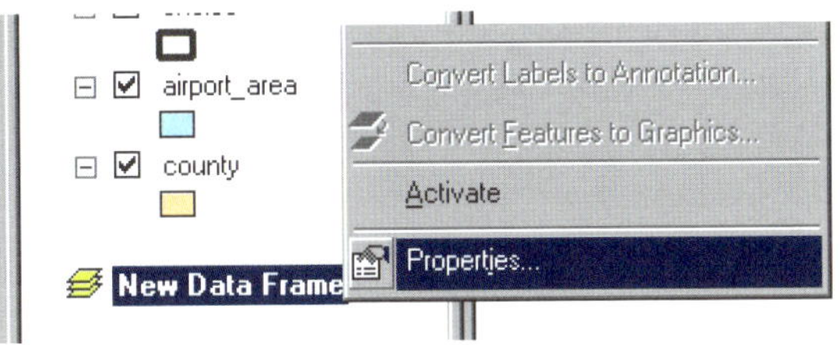

2. Click the General tab, highlight the existing text in the Name text box, and type "Land Use".
3. Click the Display dropdown arrow and set the display units to Feet. You can't change the map units because they are based on the data frame's coordinate system.
4. Click Apply.
5. Click the Size and Position tab.

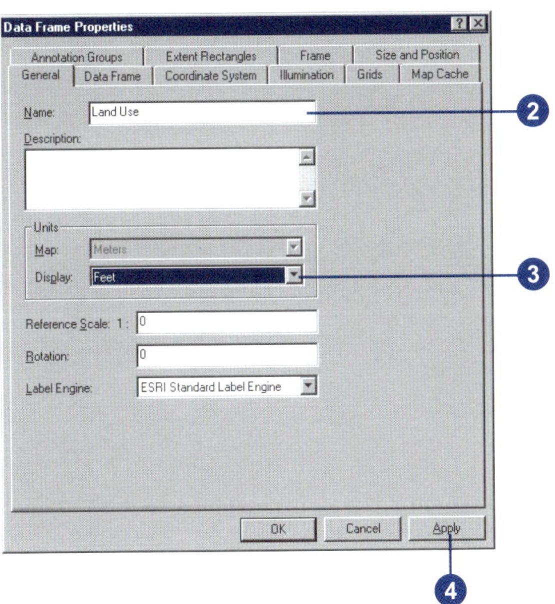

6. Set the X position to 15 and the Y position to 15 by typing in the text boxes. This sets how far the lower-left corner of the data frame is, in inches, from the lower-left corner of the page. (You can specify X,Y position for another location on the data frame by clicking the appropriate box on the diagram.)

   You can specify the position of any object on the page—the data frame itself, text, legends, and so on—either by selecting and dragging them or by setting the X and Y position explicitly.

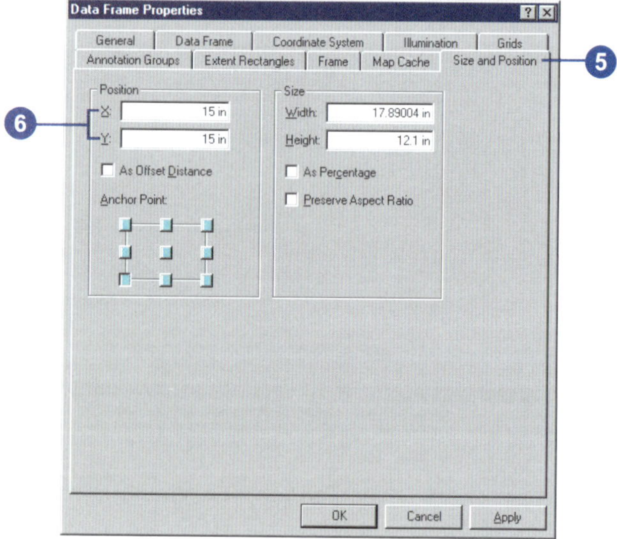

7. Click OK. The data frame is repositioned.

   The data frame is highlighted with a blue square, and its name is bold in the table of contents, indicating it is the frame you're currently working with.

8. Hold down the Shift key and click the top data frame on the page so both frames are selected.

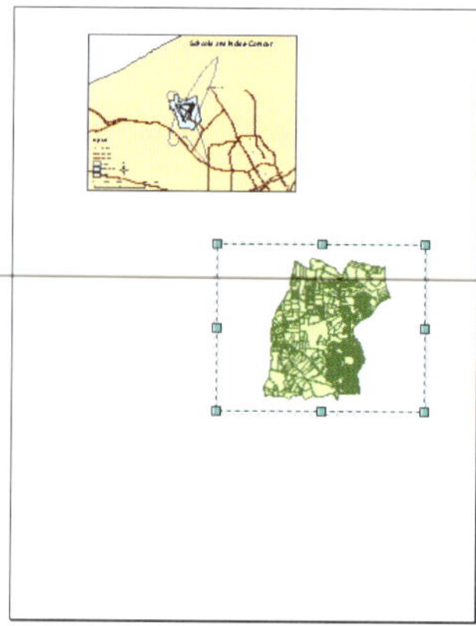

9. Click Drawing on the Draw toolbar, point to Distribute, and click Make Same Size.

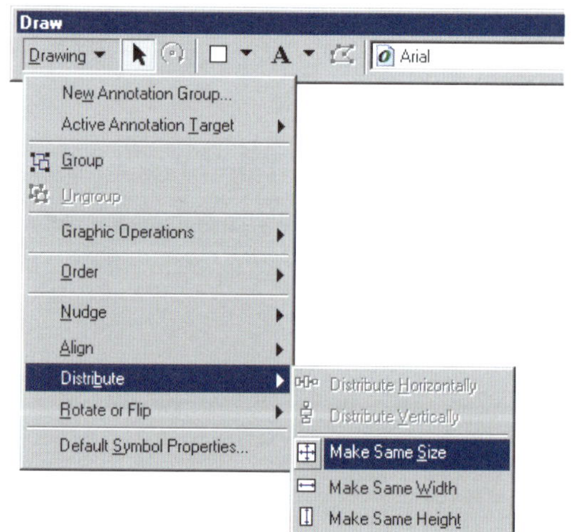

Both data frames are now the same size. Click the Land Use data frame on the page so it is the only data frame selected.

Copying a layer

You'll want to display the noise contour and airport area with the parcels. You can copy them from the Schools data frame. First, though, switch back to data view.

1. Click the View menu and click Data View. Now you're looking at only the area covered by the parcels, rather than the entire map.

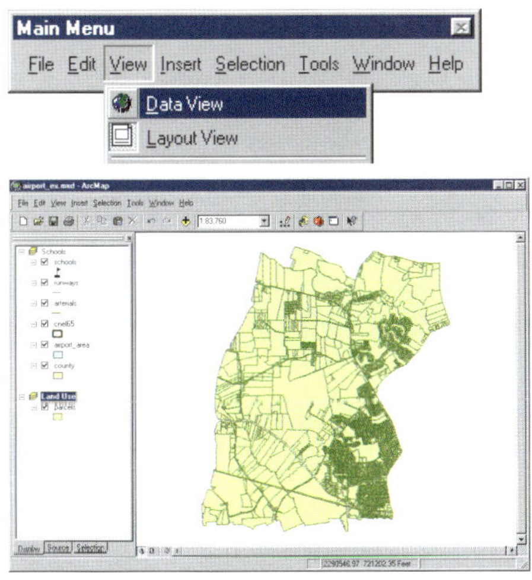

2. In the table of contents, right-click the airport_area layer under the Schools data frame and click Copy.

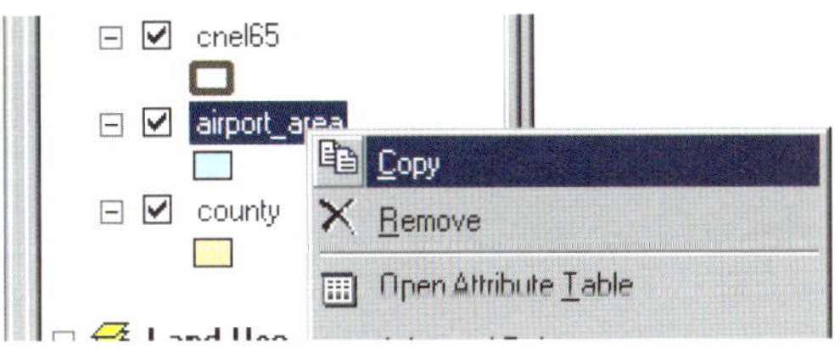

3. Right-click the Land Use data frame and click Paste Layer(s).

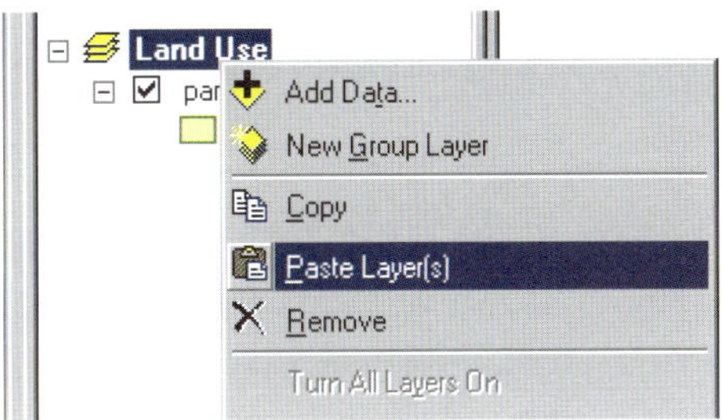

4. Copy the cnel65 layer the same way.

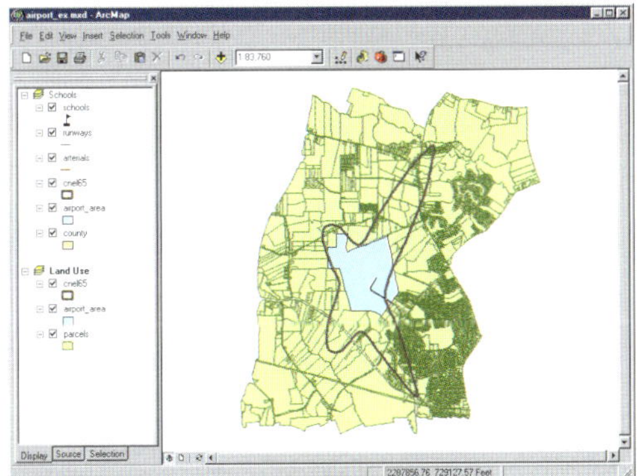

Displaying features by category

By default, all the parcels are drawn using the same symbol when you add them. You can also draw them based on an attribute (in this case, type of land use).

1. Right-click parcels in the table of contents and click Properties.

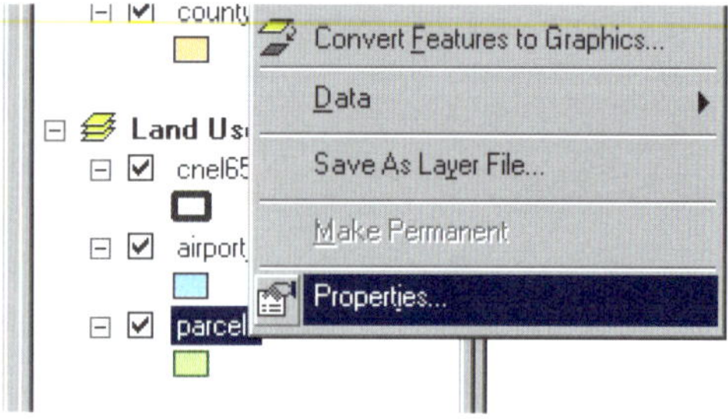

2. Click the Symbology tab. All parcels are currently drawn using the same symbol (the same solid fill color).
3. Click Categories in the Show box. Unique values is automatically highlighted.
4. Click the Value Field dropdown arrow and click LAND_USE as the field to use to shade the parcels.
5. Click Add All Values. A unique color is assigned to each land use type.

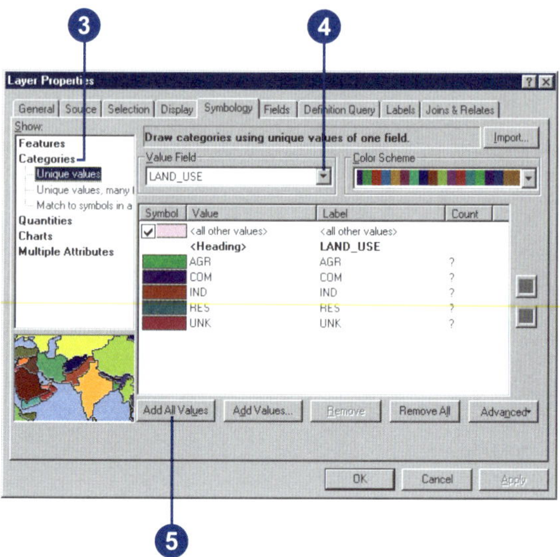

6. Click OK. The parcels are now drawn based on their land use type.

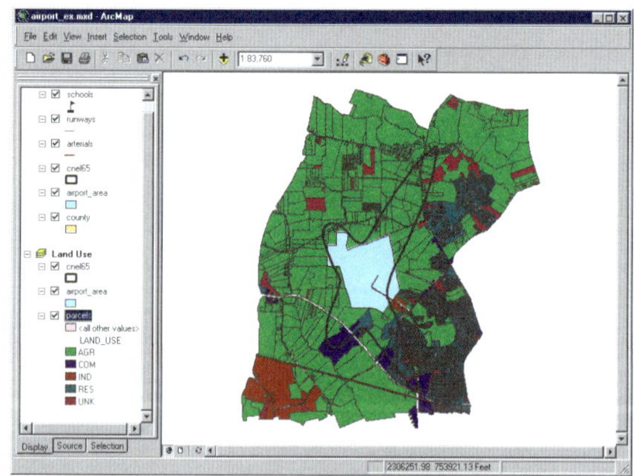

Using a style

ArcMap uses a random set of symbols to draw the land use types (although you can change the color scheme). You can change an individual color by double-clicking it and specifying a new color in the Symbol Selector, or you can specify a style to use predefined colors and symbols (a style is a set of elements, symbols, and properties of symbols stored in ArcMap, often specific to an application or industry). ArcMap provides some standard styles, and you can also create your own. You'll use a land use style created for this tutorial.

1. Right-click parcels in the table of contents and click Properties.
2. Click the Symbology tab.
3. Under Categories in the Show window, click Match to symbols in a style.
4. Click the Browse button and navigate to the Map folder on the local drive where you installed the tutorial data (the default installation path is C:\ArcGIS\ArcTutor\Map). Click the land_use style and click Open.
5. Click Match Symbols.
6. Click the check box to deselect and turn off the symbol displayed for <all other values>.

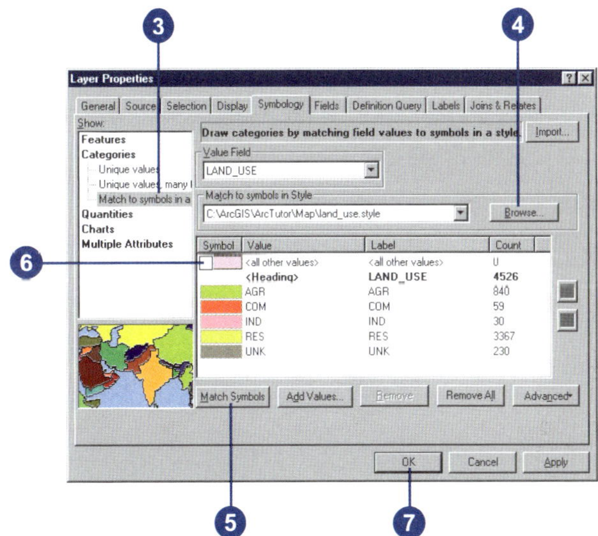

7. Click OK. The parcels will now be drawn using colors defined in the style.

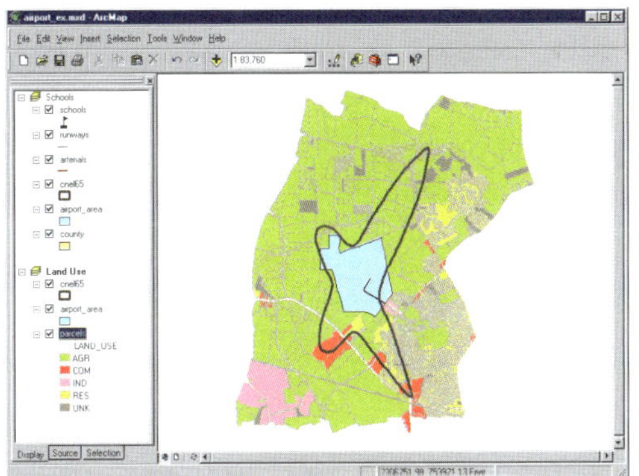

For more information on symbolizing and displaying features, see Chapter 6, 'Symbolizing features' and Chapter 8, 'Working with styles and symbols'.

Selecting features geographically

To find out how much of each land use is within the noise contour, select only those parcels within the contour.

1. Click Selection and click Select By Location.

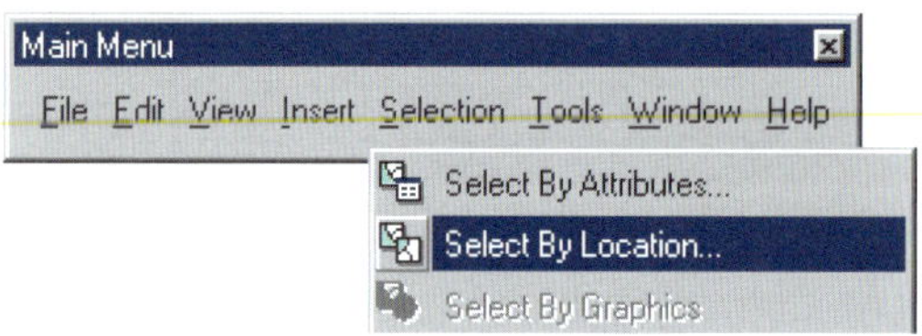

   The Select By Location dialog box guides you through creating a geographic query.

2. In the first box, click the dropdown arrow and click select features from.
3. In the second box, check parcels as the layer to select features from.
4. Click the dropdown arrow for the third box and click intersect. This will select those features in parcels that intersect the features of cnel65.
5. In the last box, click the dropdown arrow and click cnel65 as the layer to select by.
6. Click Apply. The selected parcels are outlined in a thick line.

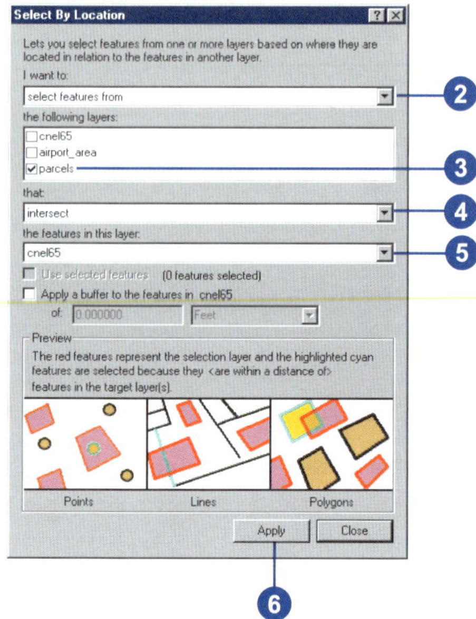

7. Close the Select By Location dialog box. Notice that any parcel even partially inside the contour is included.

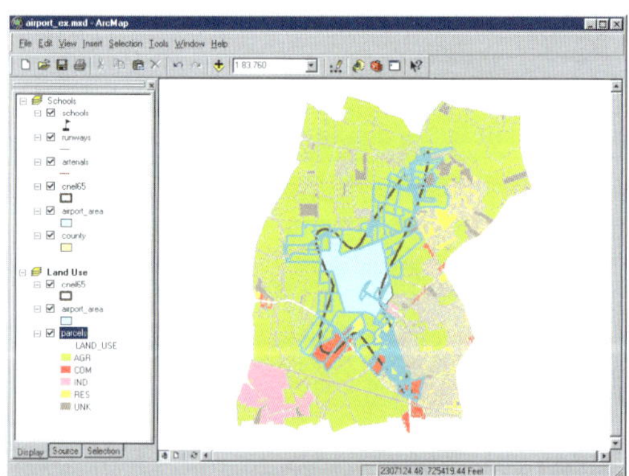

For more on selection, see Chapter 13, 'Querying maps'.

Exporting a layer

To find out how many parcels and how much land area of each land use type are within the noise contour, you'll create a new feature class and run statistics on its data table.

1. Right-click parcels in the table of contents, point to Data, then click Export Data.

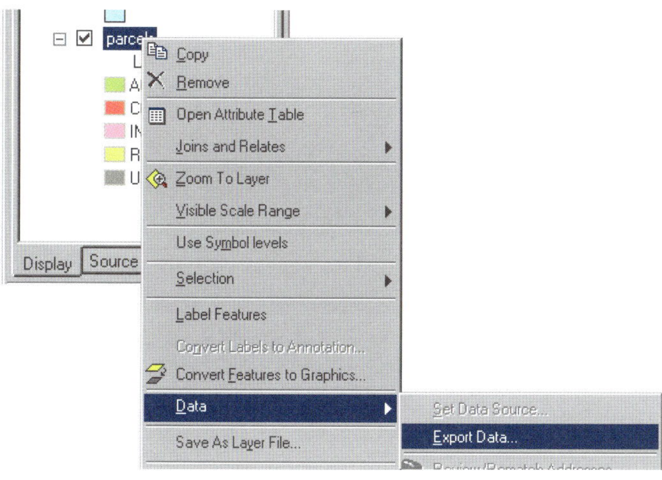

2. In the Export Data dialog box, click the Export dropdown arrow and click Selected features (to export only the selected parcels).
3. Save the selected features in the airport geodatabase as a feature class called parcels_sel. Type the path as shown below, substituting the install location of the tutorial data on your system. (The default installation path for the geodatabase is C:\ArcGIS\ArcTutor\Map\airport.mdb.)

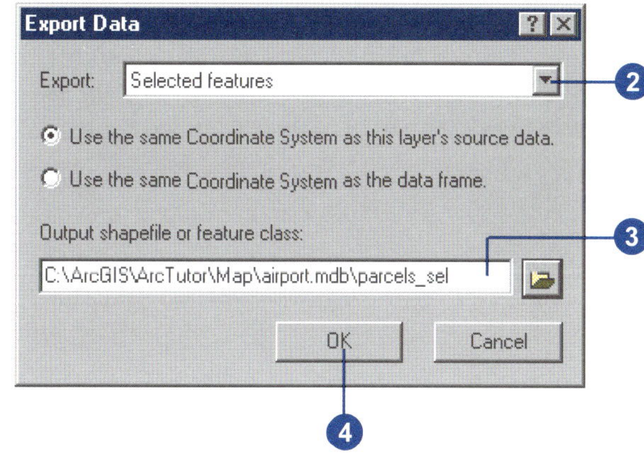

4. Click OK. ArcMap exports the parcels to a new feature class in the airport geodatabase.
5. Click Yes when prompted to add the exported data as a new layer on the map. The new layer contains only the selected parcels.
6. Right-click the original parcels layer, point to Selection, then click Clear Selected Features.

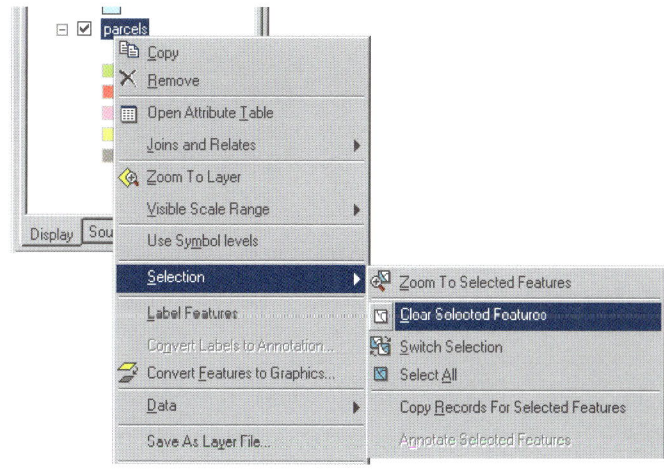

7. The new layer is displayed on top of the other layers. To see the noise contour and airport area, click parcels_sel in the table of contents and drag it down until the bar is above parcels. Then release the mouse button.

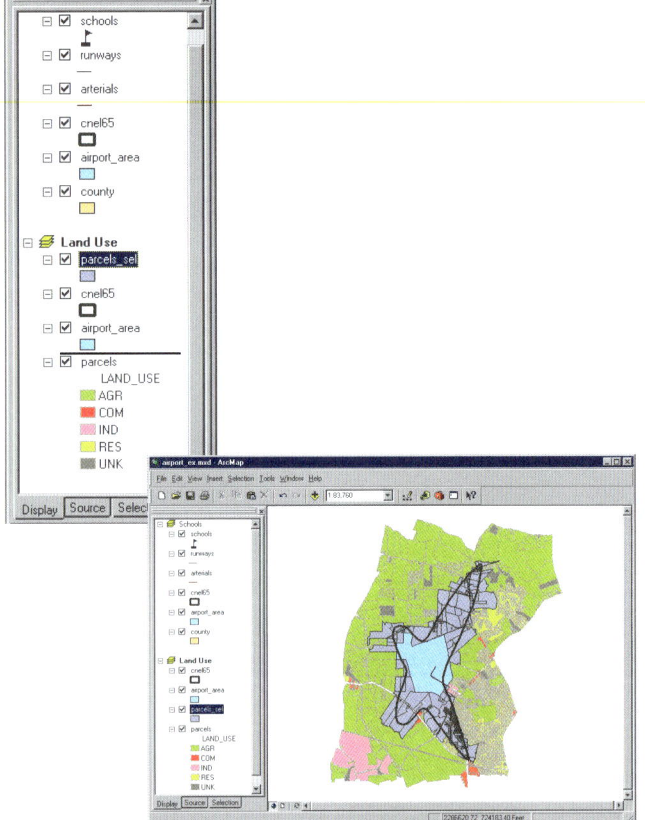

Creating summary statistics

ArcMap includes tools for statistical analysis. You'll create a table to summarize the number of parcels of each land use type within the noise contour and the total area of each type.

1. In the table of contents, right-click the parcels_sel layer and click Open Attribute Table.

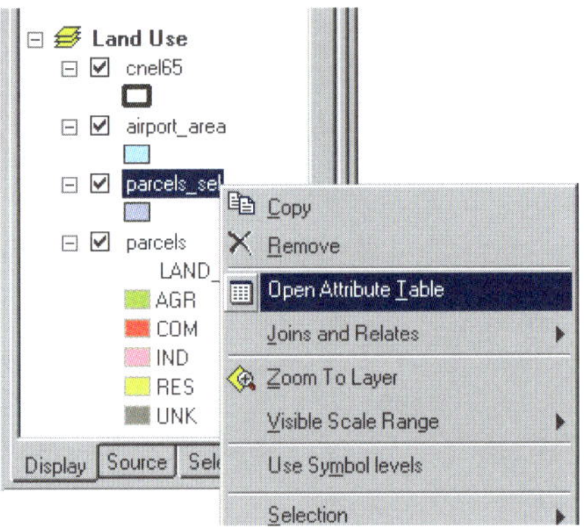

2. Right-click the LAND_USE field header and click Summarize.

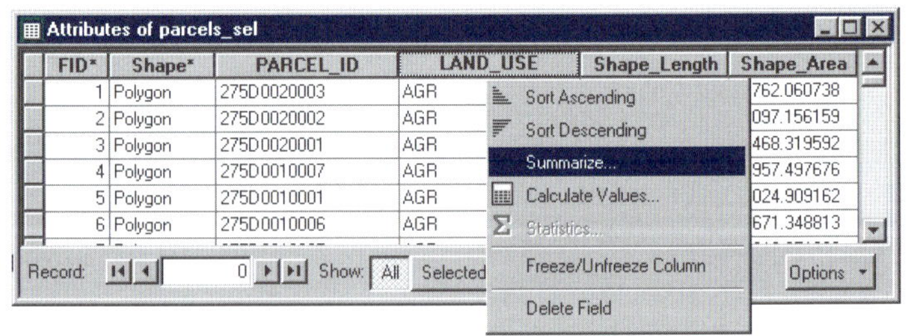

3. Make sure the field to summarize is LAND_USE.
4. Click the plus sign next to Shape_Area to expand it. Check Sum to summarize the area by land use type.
5. Create the output table in the airport geodatabase and name it lu_frequency.
6. Click OK. ArcMap creates a new table with a record for each land use type showing the number of parcels of that type and the total land area (in square feet).
7. Click Yes when prompted to add the resulting table to the map. Close the parcels_sel attribute table.

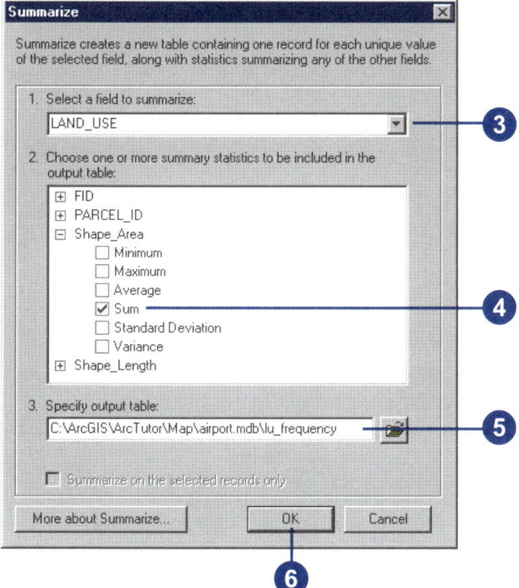

Opening a table

You may have noticed that when the table is added to the map, the table of contents switches from the Display tab to the Source tab (at the bottom of the table of contents). The Source tab shows the location of all data in the table of contents; this is useful when editing data in ArcMap because it shows you which layers are in the same workspace. (When you edit in ArcMap, you edit an entire workspace; that is, all the layers in the workspace are available for editing.) The Source tab also lists all tables. Tables don't show up when the Display tab is selected since a table is not a geographic feature that gets displayed on the map.

1. Right-click lu_frequency in the table of contents and click Open. You can see the number of parcels and the total area (in square feet) of each land use type.

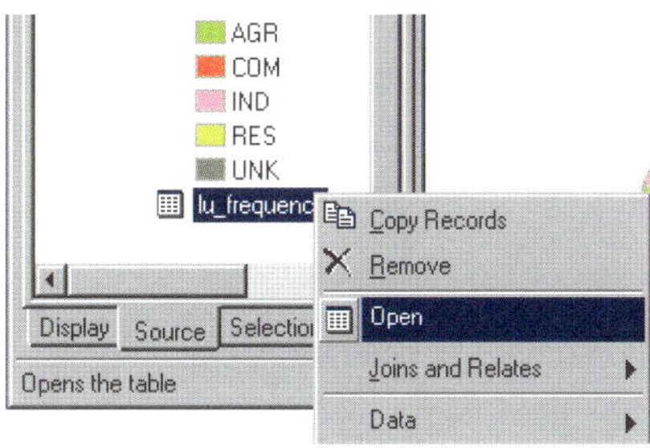

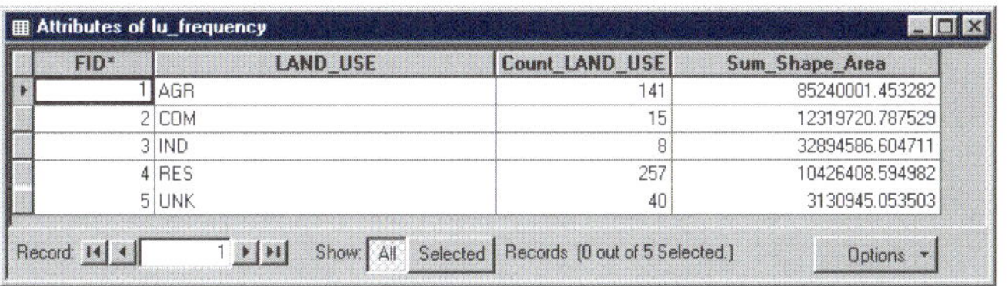
Attributes of lu_frequency

FID*	LAND_USE	Count_LAND_USE	Sum_Shape_Area
1	AGR	141	85240001.453282
2	COM	15	12319720.787529
3	IND	8	32894586.604711
4	RES	257	10426408.594982
5	UNK	40	3130945.053503

Record: 1 Show: All Selected Records (0 out of 5 Selected.) Options

2. Close the table window.

Making a graph

Next you'll create a column graph showing the number of parcels of each land use type.

1. Click the Tools menu, point to Graphs, and click Create. The Graph Wizard appears.

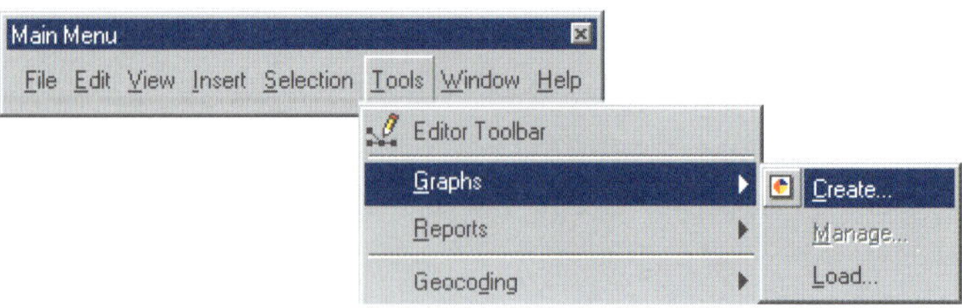

2. On the Graph Wizard dialog box, click the Column graph and click Next.

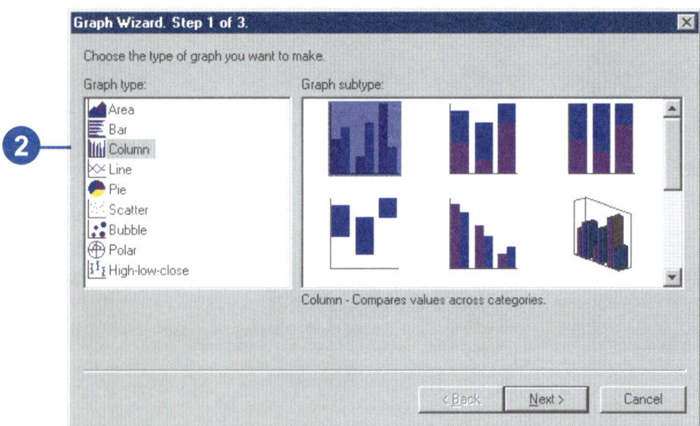

3. Click lu_frequency as the table containing the data to graph.
4. Make sure that Use selected set of features or records is not checked.
5. Check the field Count_LAND_USE as the field to graph.
6. Click Graph data series using Records and click Next.

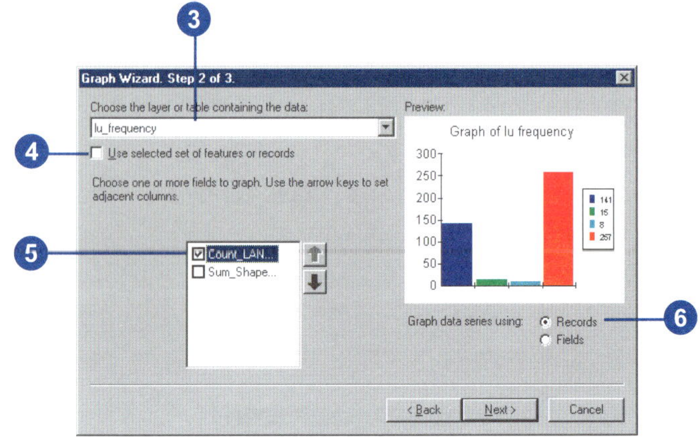

7. Type "Land Use in Noise Contour" as the title.
8. Check Label X Axis With and click LAND_USE as the labeling field.
9. Uncheck Show Legend.
10. Check Show Graph on Layout and click Finish.

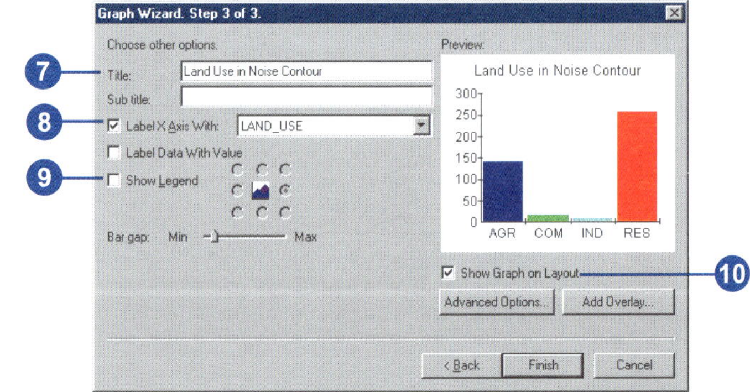

The graph appears on the layout. You can see that most of the parcels are residential.

11. Click the Select Elements tool on the Tools toolbar.

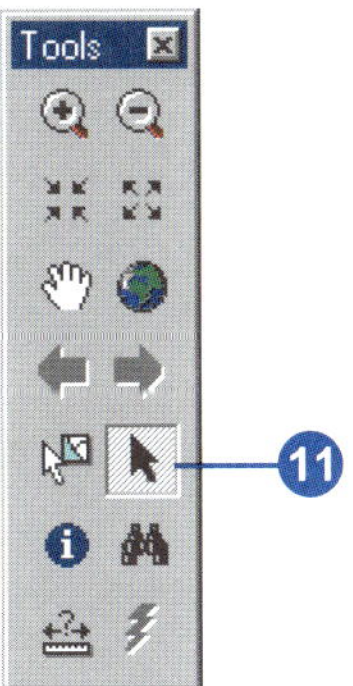

12. Click and drag the graph to the left of the parcel map.

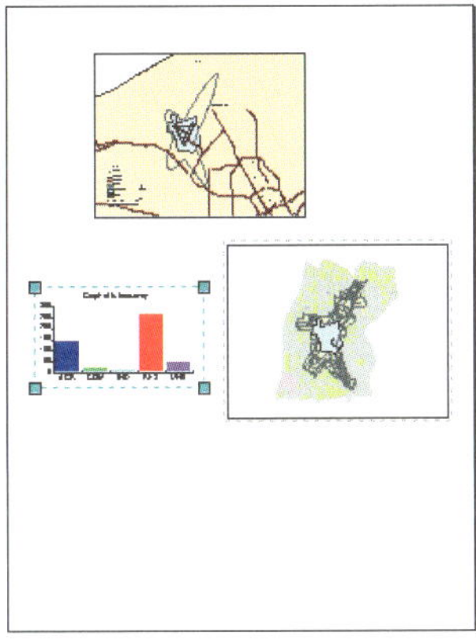

13. With the graph still selected, hold down the Shift key and click the land use map so both are selected.

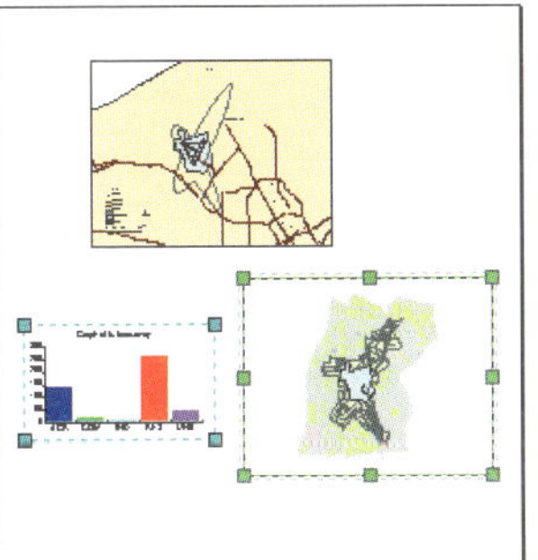

14. Click the Drawing dropdown arrow on the Draw toolbar, point to Align, and click Align Bottom to line up the graph and map.

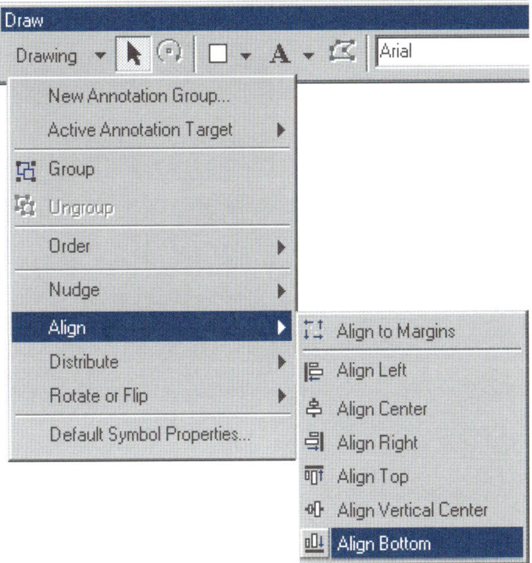

You can stop here or continue with the next exercise. Save your work by clicking Save on the File menu.

Exercise 3: Working with tables

In this exercise, you'll map population density for the county. A population density map shows where people are concentrated. First, you'll add population data for each census tract. Then you'll calculate population density for each tract and map it.

If necessary, start ArcMap, navigate to the folder where you saved the map from Exercise 2 (airport_ex), and open the map.

Creating a new data frame

As with the land use map, you'll start by creating a new data frame to display the data.

1. Switch to layout view, if necessary (click View and click Layout View).
2. Click Insert and click Data Frame.

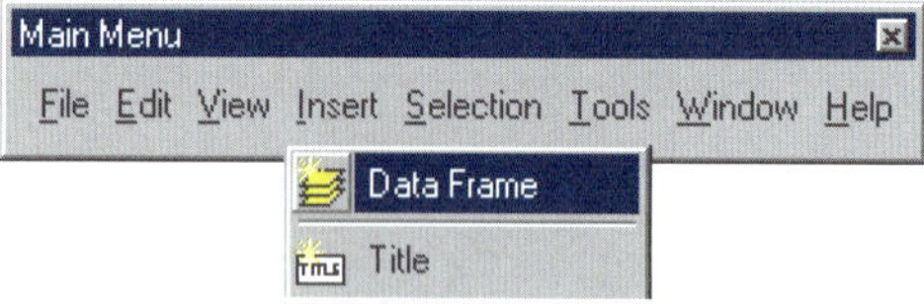

3. In the table of contents, right-click New Data Frame 2 and click Properties.

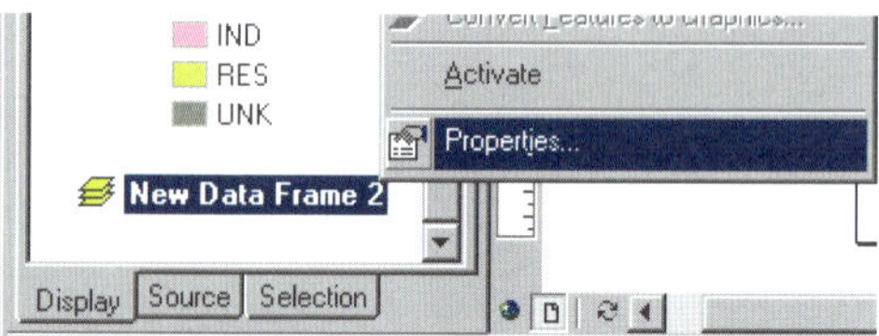

4. Click the General tab and type "Population Density" in the Name text box.
5. Click the Size and Position tab.

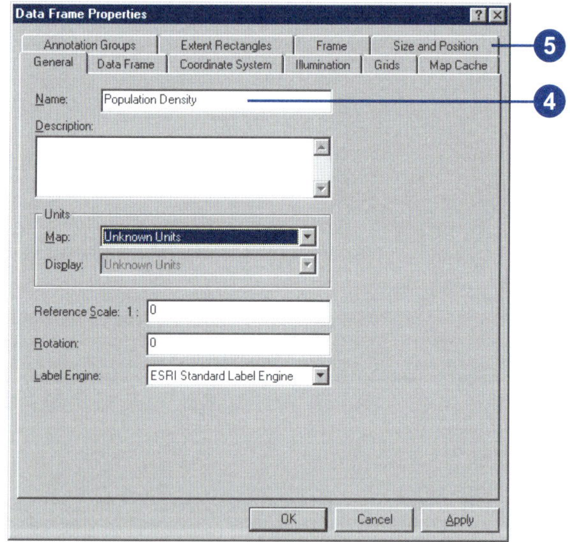

6. Set the X position to 9 and the Y position to 2.5.
7. Click OK.

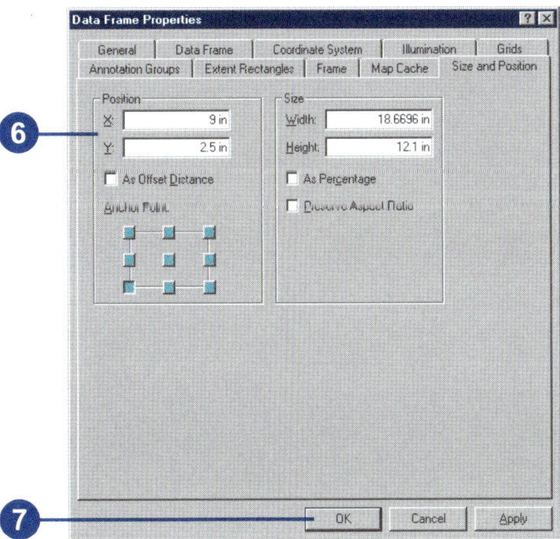

8. Hold down the Shift key and click the middle data frame (Land Use) on the page so both frames are selected.
9. Click Drawing on the Draw toolbar, point to Distribute, and click Make Same Size.

   The data frames are now the same size.

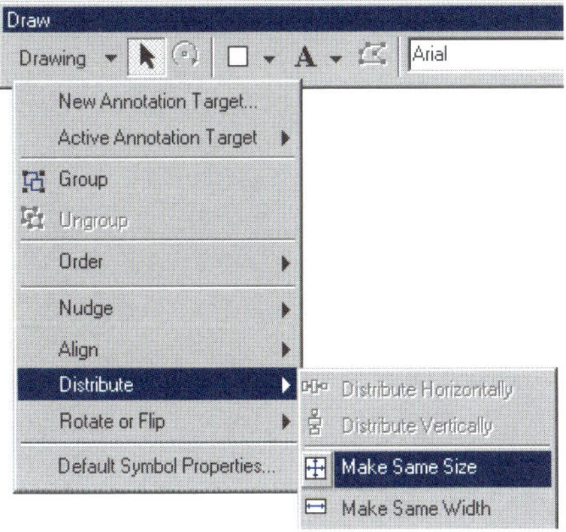

10. Click the Population Density data frame on the page so it is the only one selected.

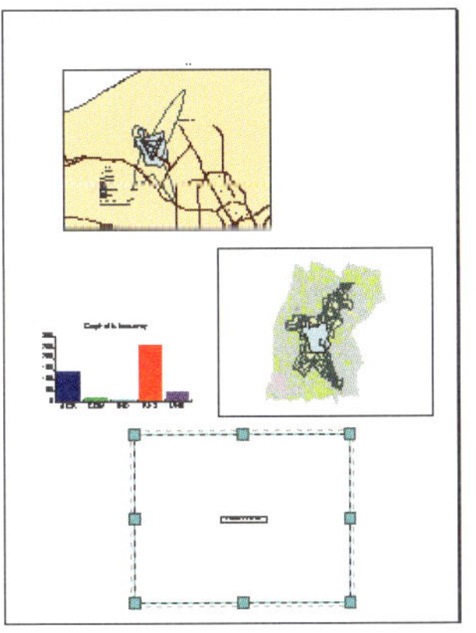

Adding data from ArcCatalog

You'll add the layers you need by dragging them from ArcCatalog™.

1. Start ArcCatalog by clicking the ArcCatalog button on the Standard toolbar in ArcMap. Position the ArcCatalog and ArcMap windows so ArcMap is visible behind the ArcCatalog window.

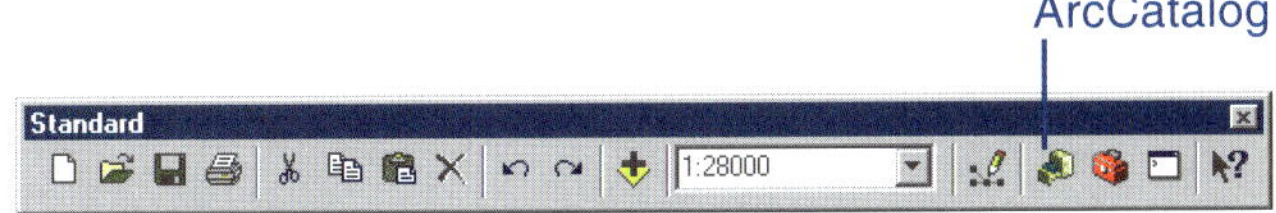

2. In ArcCatalog, navigate to the Map folder on the local drive where you installed the tutorial data (the default installation path is C:\ArcGIS\ArcTutor\Map).
3. Click the plus sign next to the Map folder to list the contents.
4. Click the airport geodatabase icon to display the contents in the right panel.
5. In the right panel, click arterials.

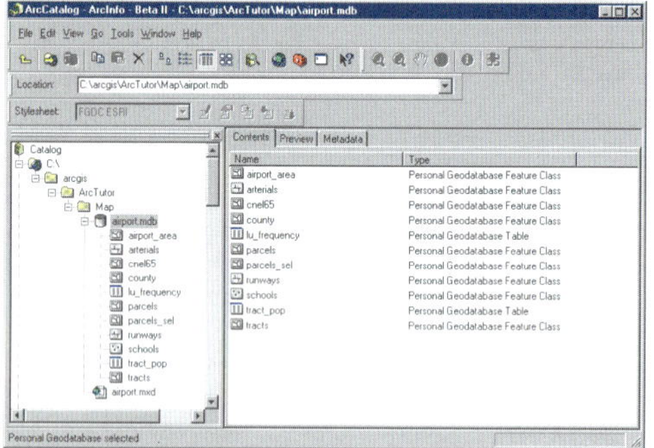

6. Hold down the Ctrl key and click tracts and airport_area to select them as well. The layers are highlighted as you select them.

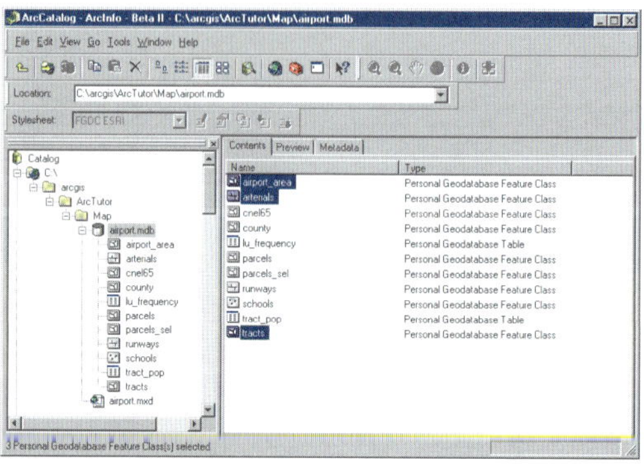

7. Point to arterials, hold down the left mouse button, and drag the pointer over the ArcMap layout view (anywhere is fine).

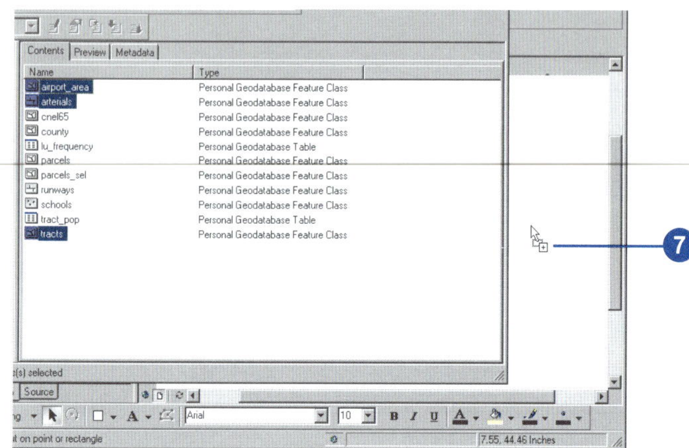

8. Release the mouse button. All three layers are added to the new data frame.
9. Close ArcCatalog.
10. Click tracts in the ArcMap table of contents so only it is selected. Right-click tracts and click Zoom To Layer. The map redraws to show all the tracts and centers them in the data frame.

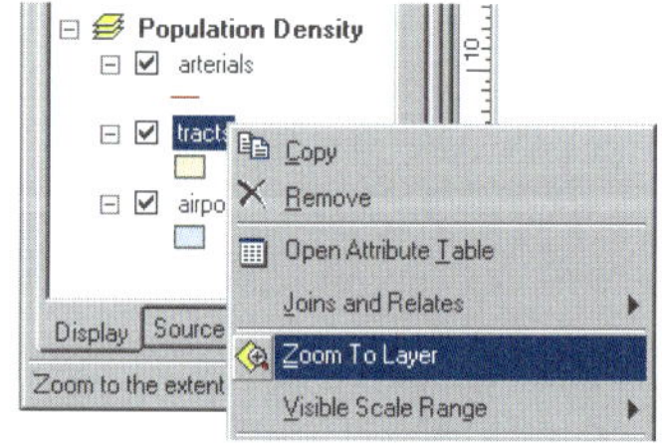

11. Right-click the Population Density data frame in the table of contents and click Properties.

12. Click the General tab, click the Display dropdown arrow, and set the display units to Feet. You can't change the map units because they are based on the data frame's coordinate system. Click OK.

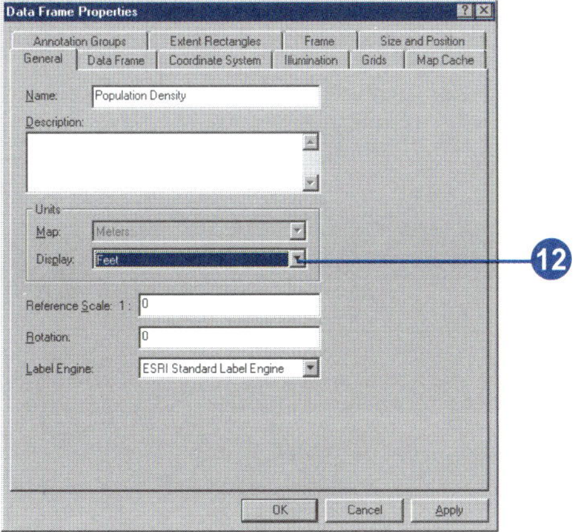

Adding tabular data

You also need to add the table containing the population data to your data frame.

1. In ArcMap, click the Add Data button.

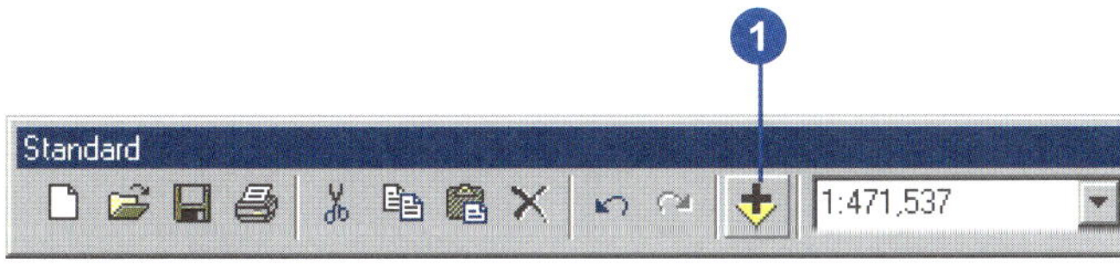

2. Navigate to the Map folder on the local drive where you installed the tutorial data (the default installation path is C:\ArcGIS\ArcTutor\Map) and double-click the airport geodatabase.

3. Click tract_pop (the icon looks like a table).

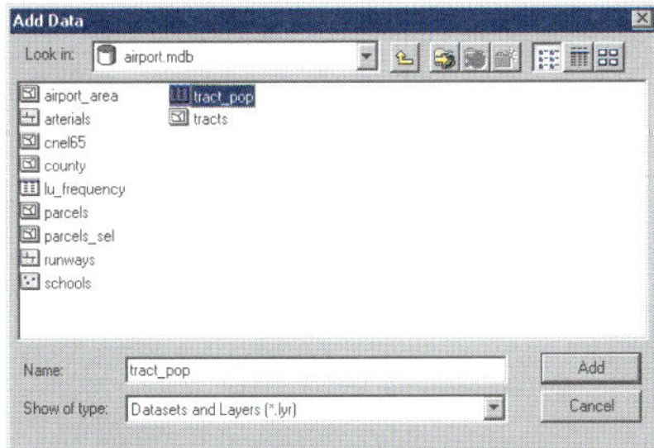

4. Click Add. The table is added to the Population Density data frame in the table of contents. ArcMap activates the Source tab so you can access the table.

Joining tables

The next step is to join the table containing the population data to the census tract data table. You'll do this using the census tract ID as the common field.

1. Right-click tracts in the table of contents and click Open Attribute Table to see the existing attributes including the census tract ID.

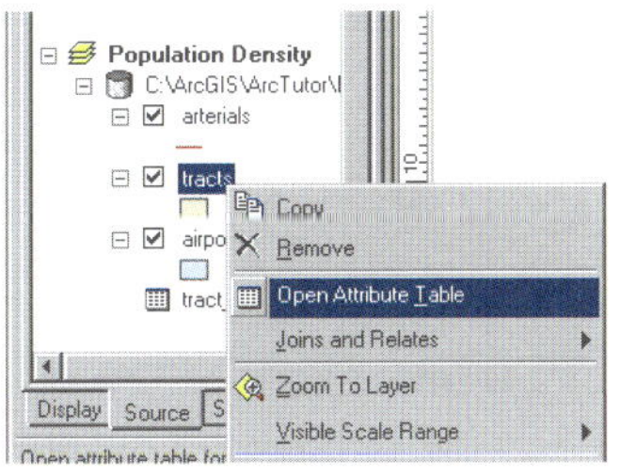

Attributes of tracts

FID	Shape	Shape_Length	Shape_Area	TRACT_ID
1	Polygon	23359.0646179392	29501864.0718	100
2	Polygon	20350.8213216268	17906796.4727	200
3	Polygon	19764.5068628924	17038547.9629	300
4	Polygon	71734.650763681	182638877.306	400
5	Polygon	41535.3888513427	101159098.343	500
6	Polygon	61452.6622484381	183391558.187	600
7	Polygon	91262.7436119726	292795476.836	700
8	Polygon	18980.4140028922	17437646.7340	801
9	Polygon	108657.097454915	452483831.055	802

Now right-click tract_pop in the table of contents and click Open. The table contains the TRACT_ID field and the population of each tract.

Close the tables before proceeding with the join.

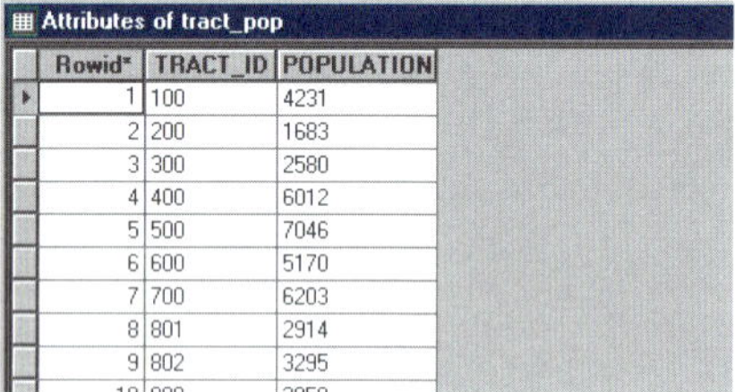

Attributes of tract_pop

Rowid*	TRACT_ID	POPULATION
1	100	4231
2	200	1683
3	300	2580
4	400	6012
5	500	7046
6	600	5170
7	700	6203
8	801	2914
9	802	3295

2. Right-click tracts in the table of contents again, point to Joins and Relates, and click Join.

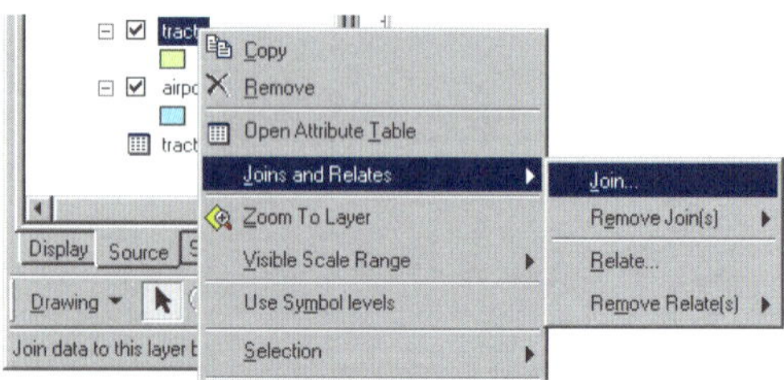

3. Click the dropdown arrow in the first text box and click Join attributes from a table.

4. Click the dropdown arrow in the next text box, scroll down, and click TRACT_ID as the field in the layer to base the join on.

5. Click the dropdown arrow in the next text box and click tract_pop as the table to join to the layer.

6. In the next text box, click TRACT_ID as the field in the table to base the join on.

7. Click OK to join the table to the layer. Click Yes if you are prompted to create an index.

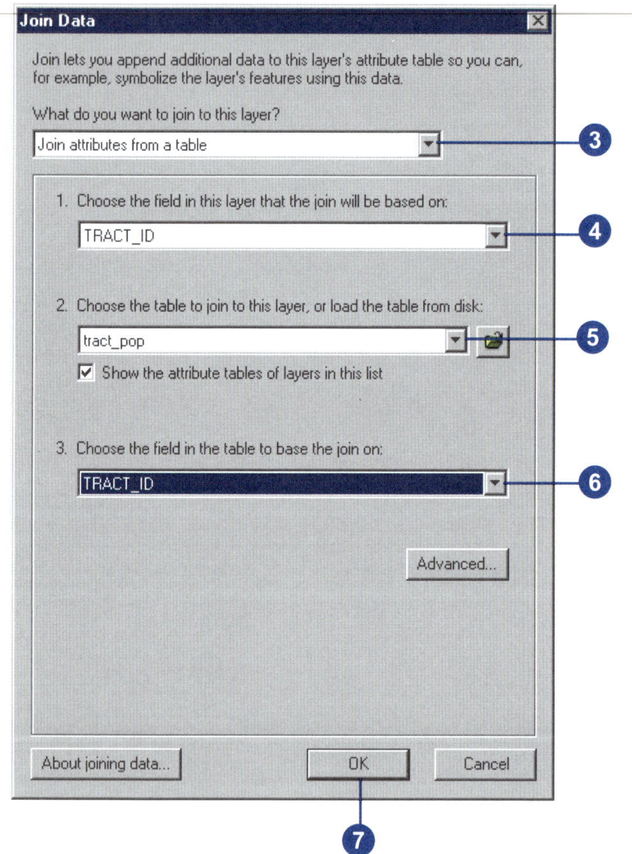

8. Right-click tracts and click Open Attribute Table. The population value has been added to each tract.

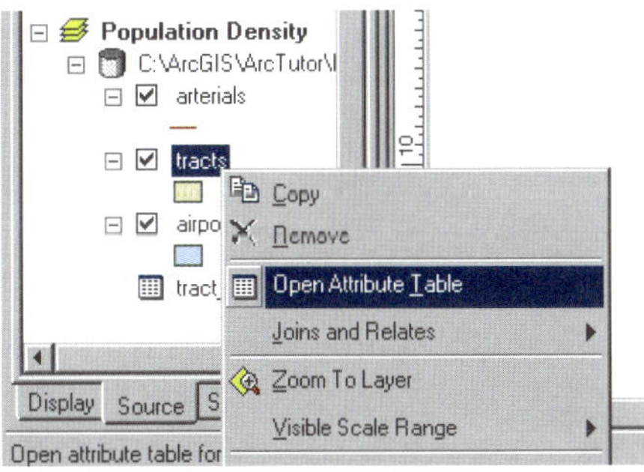

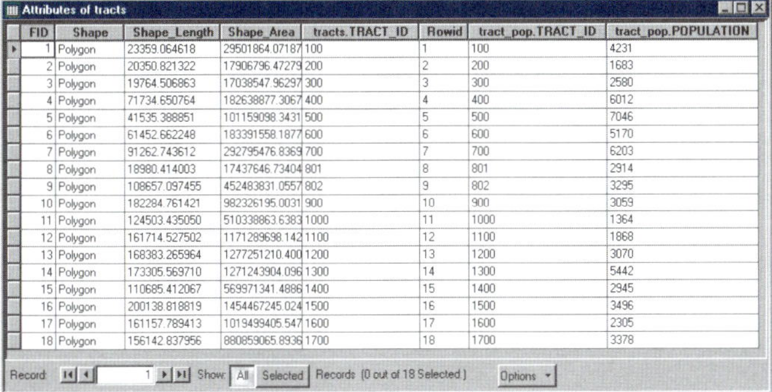

Attributes of tracts

FID	Shape	Shape_Length	Shape_Area	tracts.TRACT_ID	Rowid	tract_pop.TRACT_ID	tract_pop.POPULATION
1	Polygon	23359.064618	29501864.07187	100	1	100	4231
2	Polygon	20350.821322	17906796.47279	200	2	200	1683
3	Polygon	19764.506863	17038547.96297	300	3	300	2580
4	Polygon	71734.650764	182638877.3067	400	4	400	6012
5	Polygon	41535.388851	101159098.3431	500	5	500	7046
6	Polygon	61452.662248	183391558.1877	600	6	600	5170
7	Polygon	91262.743612	292795476.8369	700	7	700	6203
8	Polygon	18980.414003	17437646.73404	801	8	801	2914
9	Polygon	108657.097455	452483831.0557	802	9	802	3295
10	Polygon	182284.761421	982326195.0031	900	10	900	3059
11	Polygon	124503.435050	510338863.6383	1000	11	1000	1364
12	Polygon	161714.527502	1171289698.142	1100	12	1100	1868
13	Polygon	168383.265964	1277251210.400	1200	13	1200	3070
14	Polygon	173305.569710	1271243904.096	1300	14	1300	5442
15	Polygon	110685.412067	569971341.4886	1400	15	1400	2945
16	Polygon	200138.818819	1454467245.024	1500	16	1500	3496
17	Polygon	161157.789413	1019499405.547	1600	17	1600	2305
18	Polygon	156142.837956	880859065.8936	1700	18	1700	3378

Record 1 Show: All Selected Records (0 out of 18 Selected) Options

Adding a field to an attribute table

To map population density, you'll need to add a new field to the tracts layer. You'll use this field to store the population density of each tract.

1. Click the Options button at the bottom of the Attributes of tracts window and click Add Field.

   If a message appears indicating the table is in use by another user, make sure you closed ArcCatalog.

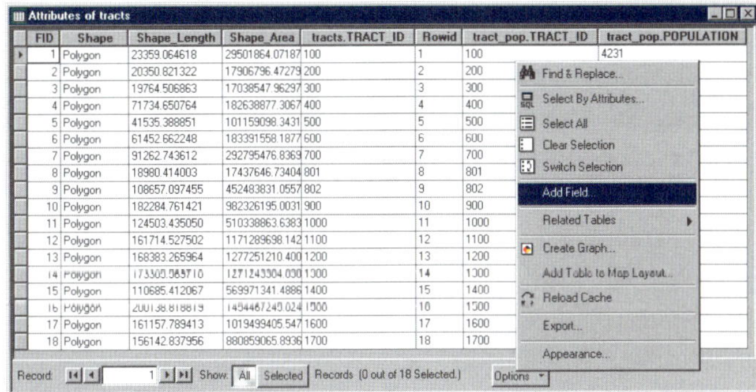

2. In the Add Field dialog box, type "POP_DEN" as the field name.
3. Click the Type dropdown arrow and click Long Integer.
4. Click OK.

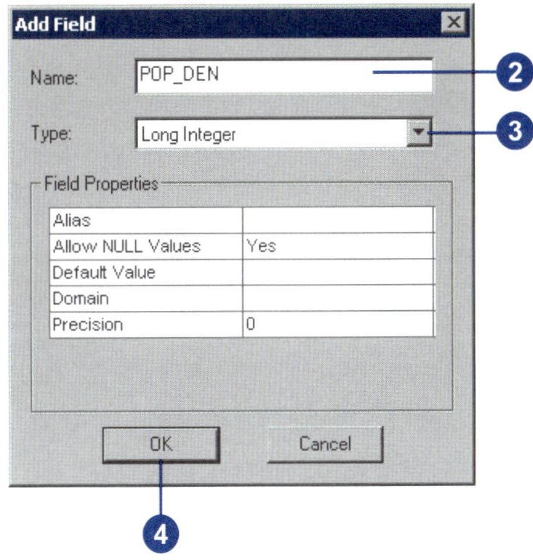

You should see the new field added to the attribute table. The field name you entered will be concatenated with tracts, to appear as tracts.POP_DEN.

Calculating attribute values

You'll calculate the population density for each tract by dividing the population by the area of each tract; this will give you the number of people per square mile. To do this, you'll use the editing functions of ArcMap to edit the census tract attributes. You can make calculations without being in an editing session; however, in that case, there is no way to undo the results. (In Exercise 4, you'll edit the geometry of a feature.)

1. Click the Editor Toolbar button on the Standard toolbar. The Editor toolbar appears.

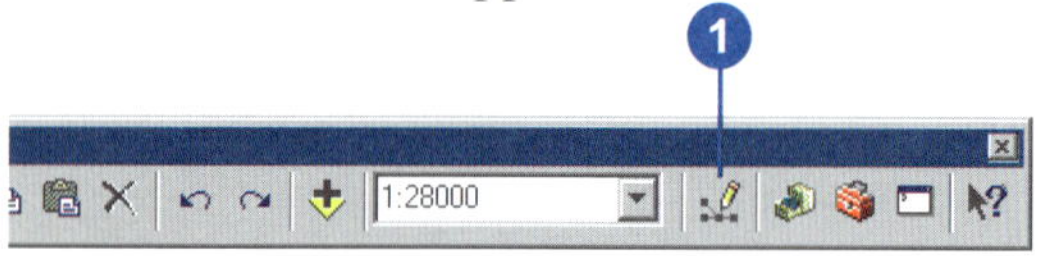

2. Click Editor and click Start Editing.

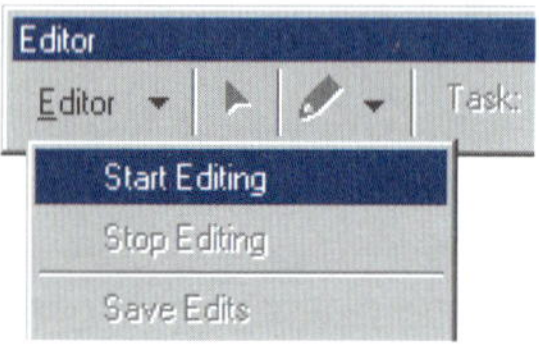

3. Right-click tracts.POP_DEN and click Calculate Values. The Field Calculator appears.

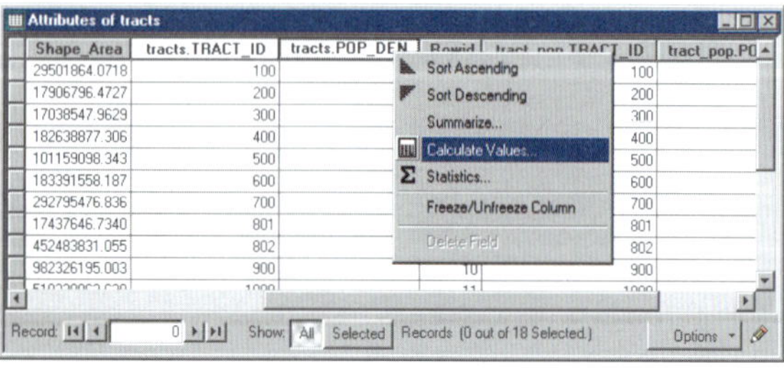

The first part of the formula is entered for you tracts.POP_DEN = . The full formula will look like this:

tracts.POP_DEN = [tracts_pop.POPULATION] / ([tracts.Shape_Area] / 27878400).

Dividing the area by 27,878,400 converts the area of each tract, stored in square feet, to square miles. You can type the formula right into the box or use the buttons on the dialog box. In this exercise, you'll use both.

4. Click tract_pop.POPULATION in the Fields list.
5. Click the division symbol.
6. Type a space and a left parenthesis from the keyboard.
7. Click tracts.Shape_Area from the field list.
8. Click the division symbol.
9. Type a space and type "27878400".
10. Type a space and a right parenthesis from the keyboard.
11. Click OK.

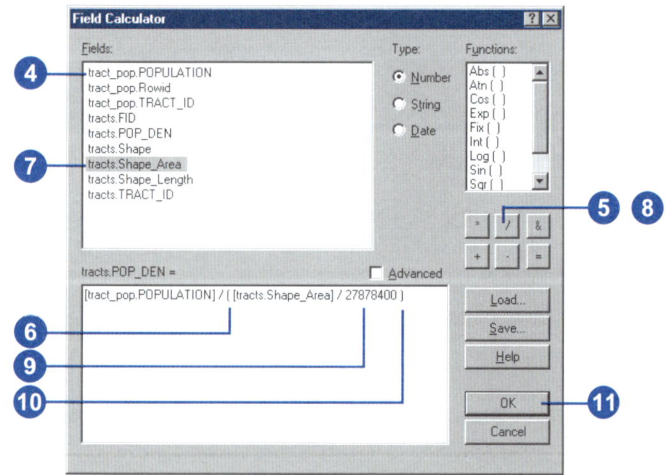

When the dialog box closes, you can see the population density values for each tract in people per square mile in the table.

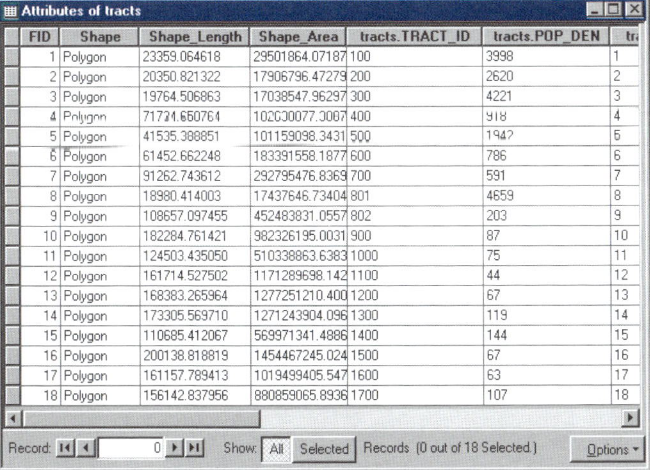
Attributes of tracts

FID	Shape	Shape_Length	Shape_Area	tracts.TRACT_ID	tracts.POP_DEN	tra
1	Polygon	23359.064618	29501864.07187	100	3998	1
2	Polygon	20350.821322	17906796.47279	200	2620	2
3	Polygon	19764.506863	17038547.96297	300	4221	3
4	Polygon	71734.660764	102030077.3087	400	918	4
5	Polygon	41535.388851	101159098.3431	500	1942	5
6	Polygon	61452.662248	183391558.1877	600	786	6
7	Polygon	91262.743612	292795476.8369	700	591	7
8	Polygon	18980.414003	17437646.73404	801	4659	8
9	Polygon	108657.097455	452483831.0557	802	203	9
10	Polygon	182284.761421	982326195.0031	900	87	10
11	Polygon	124503.435050	510338863.6383	1000	75	11
12	Polygon	161714.527502	1171289698.142	1100	44	12
13	Polygon	168383.265964	1277251210.400	1200	67	13
14	Polygon	173305.569710	1271243904.096	1300	119	14
15	Polygon	110685.412067	569971341.4886	1400	144	15
16	Polygon	200138.818819	1454467245.024	1500	67	16
17	Polygon	161157.789413	1019499405.547	1600	63	17
18	Polygon	156142.837956	880859065.8936	1700	107	18

Record: 0 Show: All Selected Records (0 out of 18 Selected.) Options

12. Click the Editor menu on the Editor toolbar and click Stop Editing.

13. Click Yes when prompted to save your edits.
14. Close the Editor toolbar and close the attribute table.

For more information on adding and calculating attributes, see Chapter 10, 'Working with tables'.

Classifying features by quantity

You can now map the tracts based on their population density values to see where people are concentrated in relation to the airport and to major roads.

1. Right-click tracts in the table of contents and click Properties.

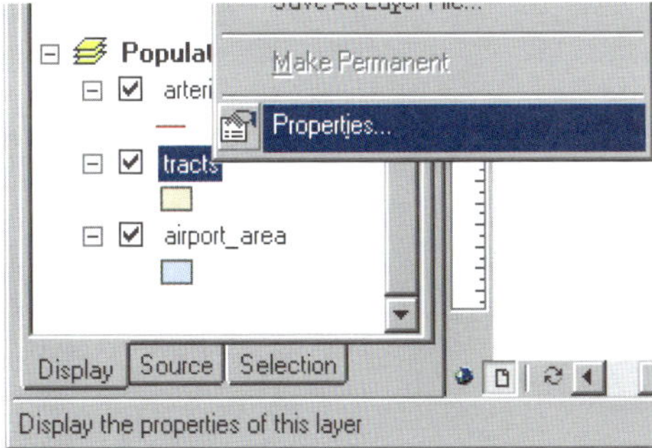

2. Click the Symbology tab. All tracts are currently drawn using the same symbol (the same solid fill color).

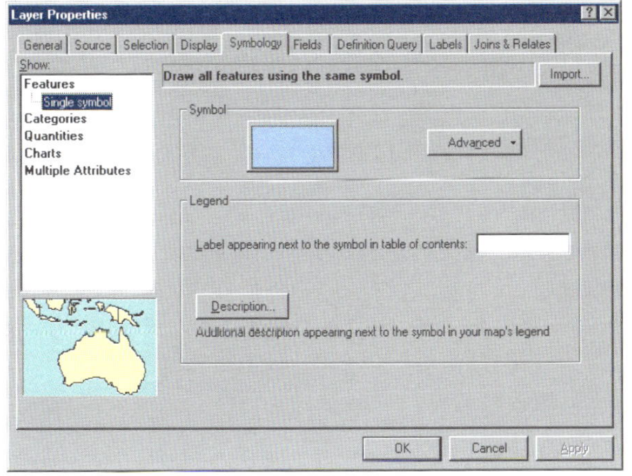

3. Click Quantities in the Show box. Graduated colors is automatically highlighted.
4. Click the Value dropdown arrow and click tracts.POP_DEN as the field to use to shade the tracts.
5. Click the Color Ramp dropdown arrow and click the blue color ramp.

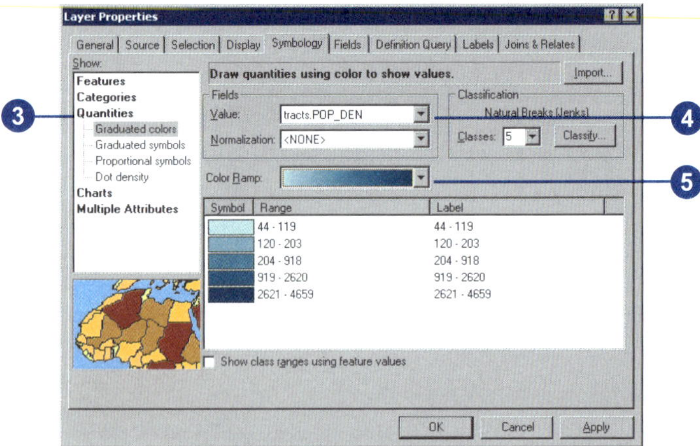

ArcMap chooses a classification scheme and the number of classes for you. You can modify these by clicking the Classify button in the Layer Properties dialog box. For now, just use the default classification.

6. Click OK.
7. Click the Display tab at the bottom of the table of contents.
8. Arterials should be at the top of the layers list. If not, click arterials in the table of contents and drag it to the top of the layers list in the Population Density data frame. Click airport_area and drag it so it is just below arterials. Now these layers draw on top of the tracts.

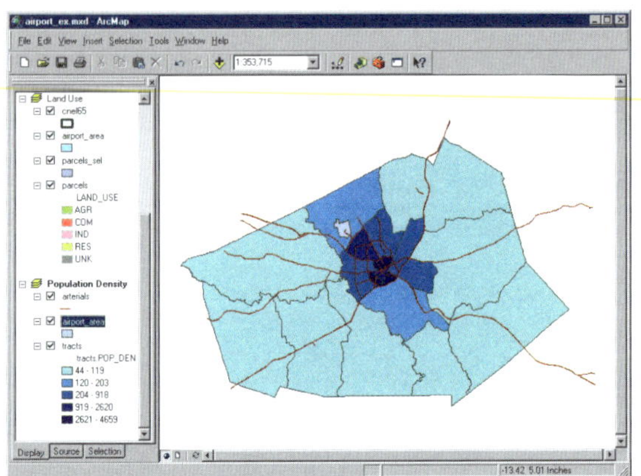

9. Switch to data view to get a closer look at the tracts. Click View and click Data View.

For more on classifying and displaying data, see Chapter 6, ‘Symbolizing features’.

You’ve now completed Exercise 3. You can continue with the next exercise or continue at a later time. Be sure to save your work by clicking Save on the File menu.

Exercise 4: Editing features

You can use ArcMap to edit your data as well as create maps. In this exercise you'll extend the airport road to create a new loop road joining an existing arterial road. This exercise is a brief introduction to editing, which is covered in more detail in *Editing in ArcMap*.

If necessary, start ArcMap, navigate to the folder where you saved the map from Exercise 3 (airport_ex), and open the map.

Exporting data

You'll be working with the Schools data frame. First, make a copy of the arterials data. That way, in case you need to, you can start over again with the original data.

1. Switch to data view by clicking the View menu and clicking Data View, if necessary.
2. Right-click the Schools data frame in the table of contents and click Activate.

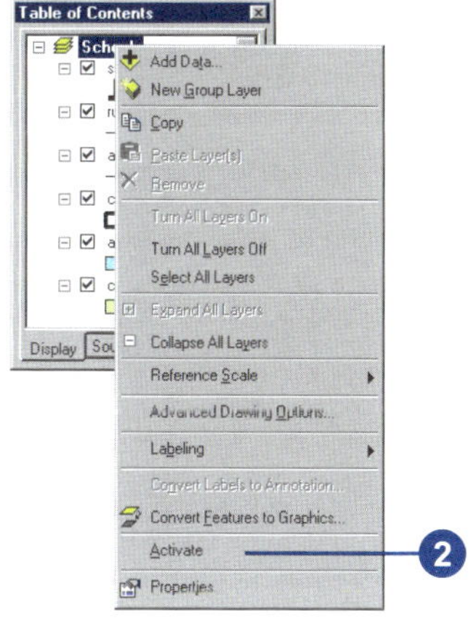

3. Right-click arterials, point to Data, and click Export Data.

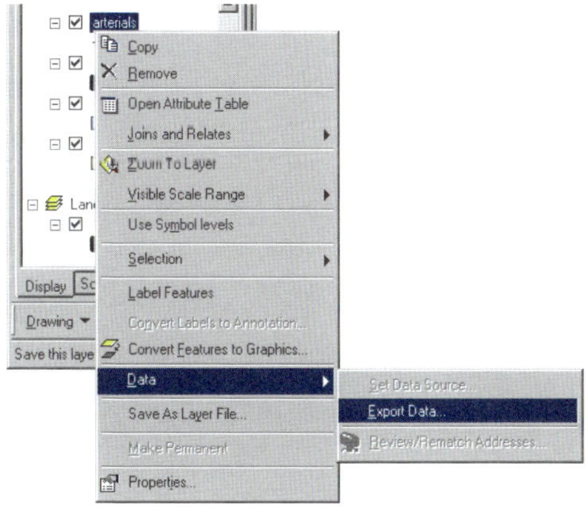

4. Click the Export dropdown arrow and click All features.
5. Click Use the same Coordinate System as this layer's source data.
6. Save the new feature class as arterials_new in the airport geodatabase (the default installation path is C:\ArcGIS\ArcTutor\Map\airport.mdb).

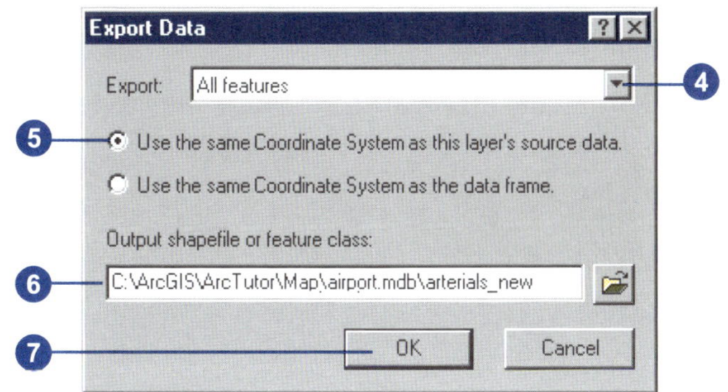

7. Click OK to export the data.
8. Click Yes when prompted to add the layer to the map.

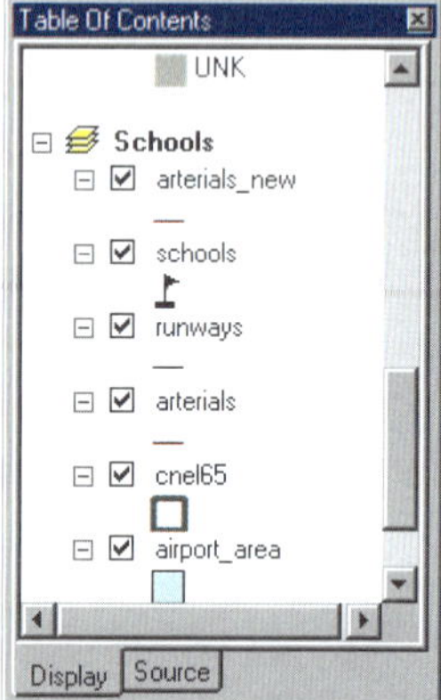

Using Export makes a copy of the data itself. If you'd chosen Copy from the menu, you'd be copying the layer, which is only a pointer to the underlying data and information about how the data is displayed.

Creating a new feature

You edit features in ArcMap using the Editor toolbar. All the layers in a workspace are available for editing within the same editing session. You specify which layer (the "target") new features will be added to.

1. Click the Zoom In button on the Tools toolbar and zoom in to the area around the existing road and the road you're adding.

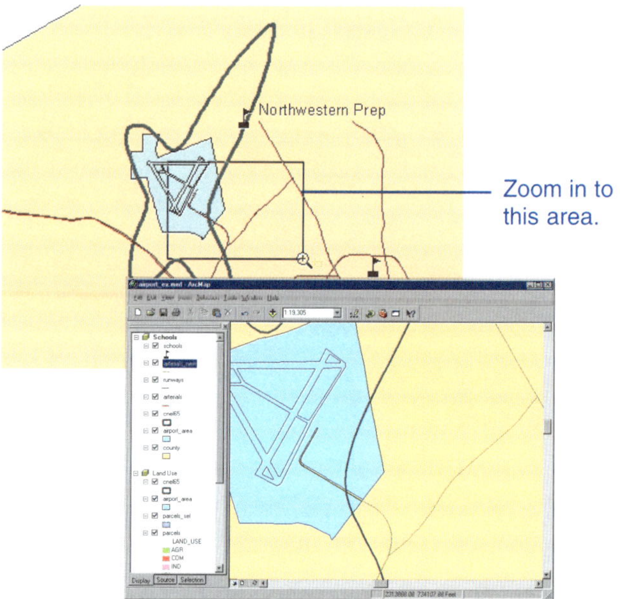

2. Turn off the cnel65 and airport_area layers by unchecking the boxes next to them in the table of contents so you can more easily see the existing roads.

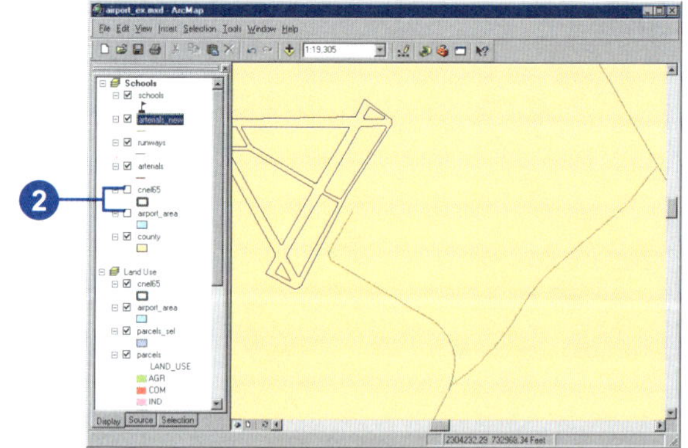

3. Click the Editor Toolbar button to display the Editor toolbar.

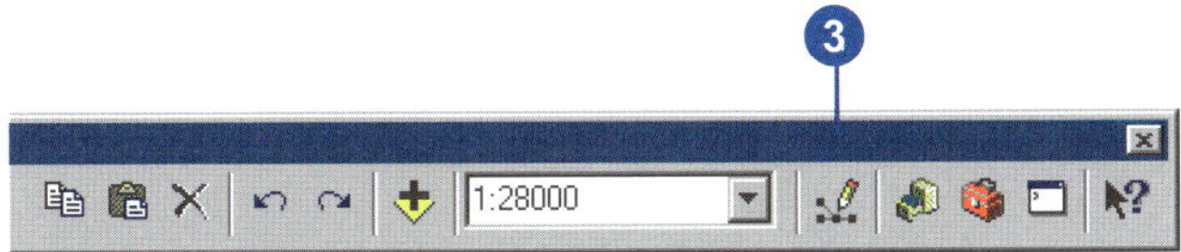

4. Click the Editor menu and click Start Editing.

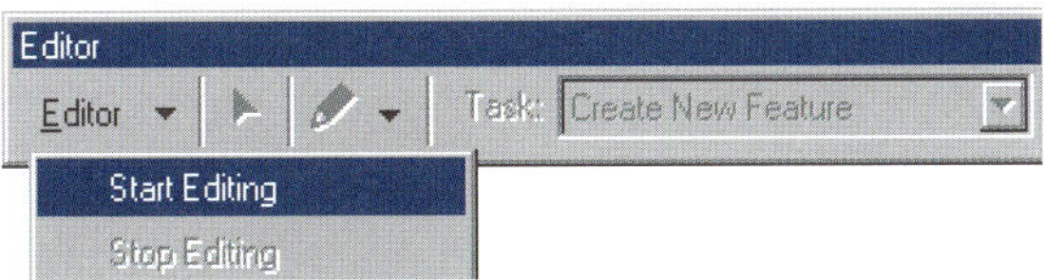

Setting snapping

Snapping lets you specify that new features connect to or align with existing features.

1. Click Editor and click Snapping.

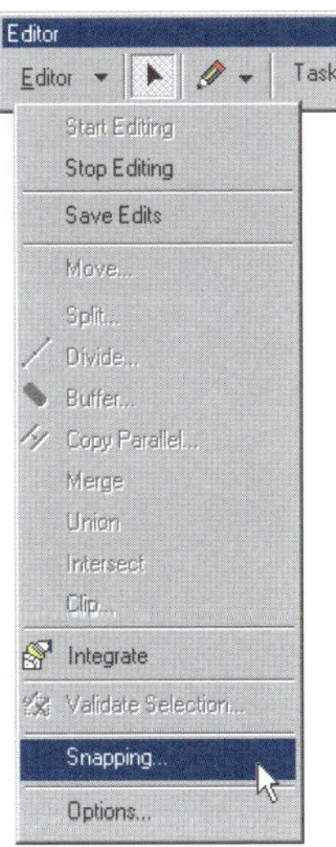

2. Check the boxes for Edge and End for arterials_new. This specifies that the new line you draw in the arterials_new dataset will snap to existing lines (edges) and endpoints of existing lines.

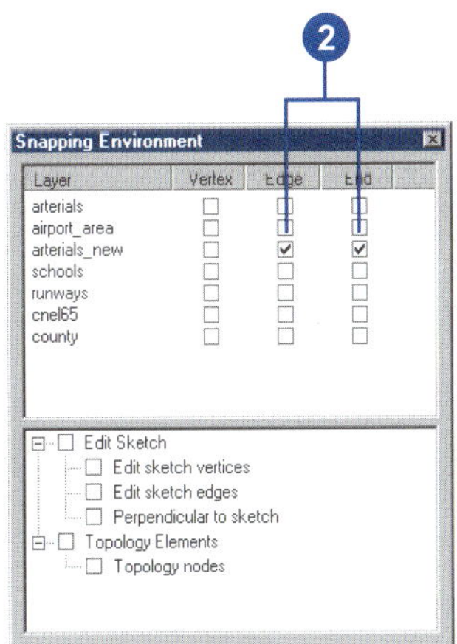

3. Close the Snapping Environment dialog box.

Digitizing a feature

1. Click the Target dropdown arrow and click arterials_new as the feature class you want to create new features in.
2. Click the Sketch tool on the Editor toolbar.

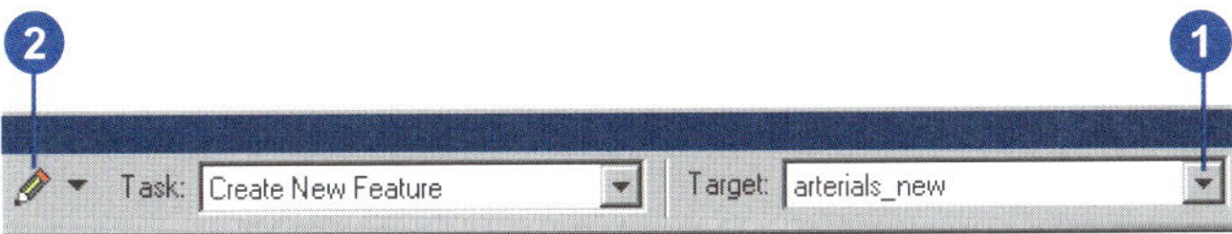

3. The pointer changes to a crosshair with a circle. Move the mouse pointer over the end of the existing road—the circle snaps to the end.

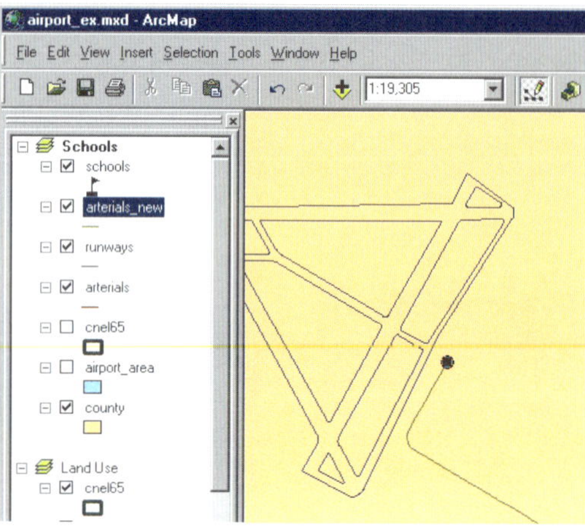

4. Click to start the new road.
5. Move the mouse pointer back over the existing road and right-click to display the context menu.

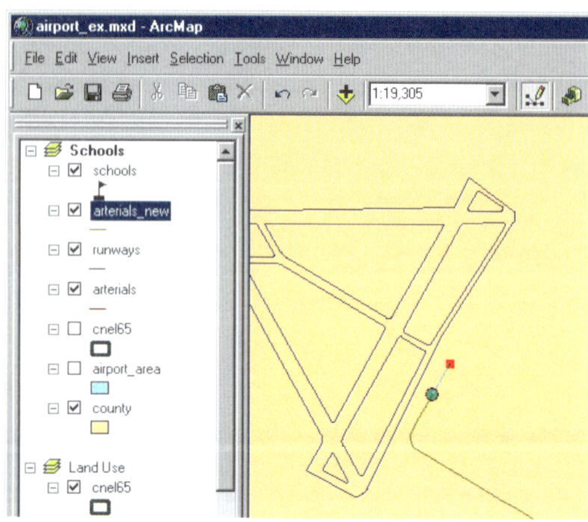

6. Click Parallel.

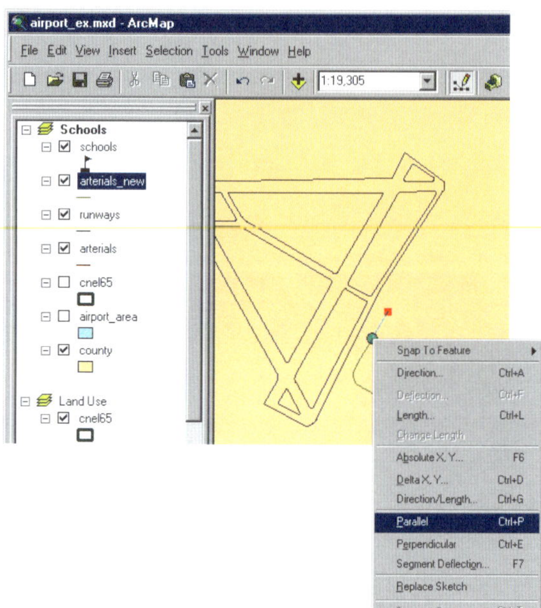

7. Move the mouse pointer in the direction you want the new road to go (up and to the right). Right-click and click Length.

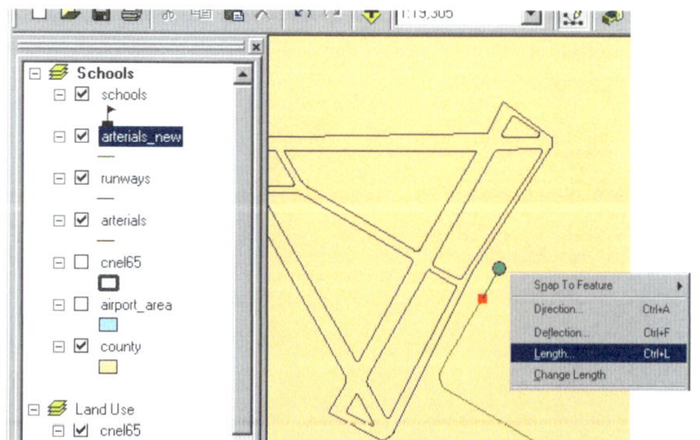

8. Type “900” (map units) and press Enter. ArcMap places a vertex at the correct location.

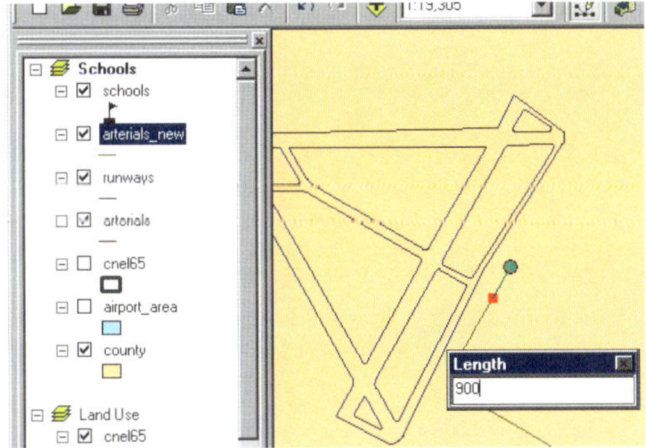

9. Right-click again and click Tangent Curve.

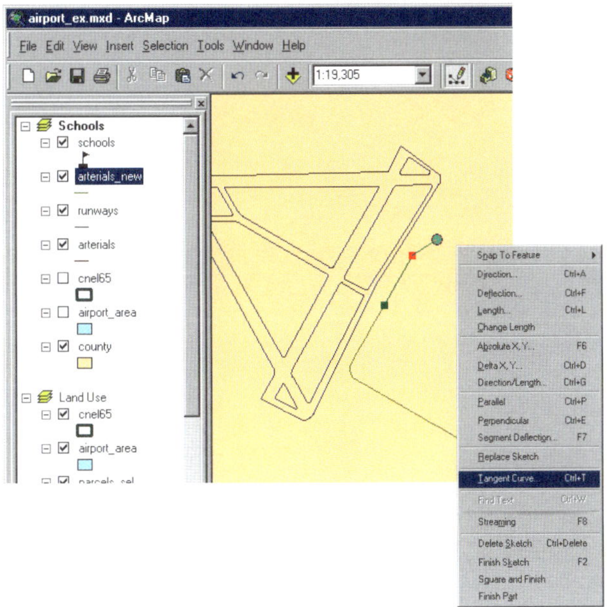

10. Click the dropdown arrow in the upper box and click Arc Length. Click the box to the right and type a length of “400”. In the lower box, click the dropdown arrow and click Delta Angle. Click the box to the right and type “90” (degrees). Click the button next to Right, if necessary. Then press Enter.

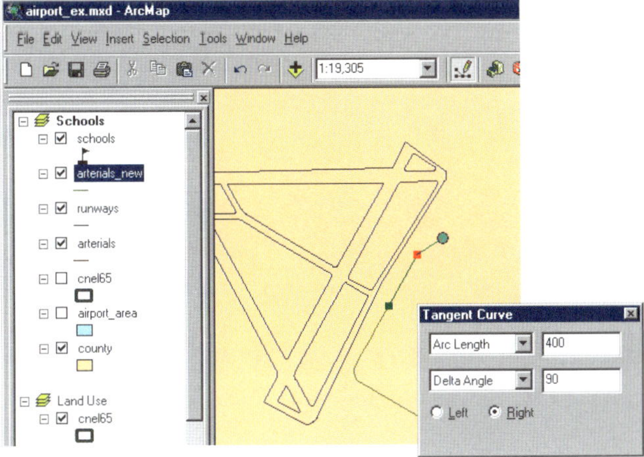

ArcMap draws the curve.

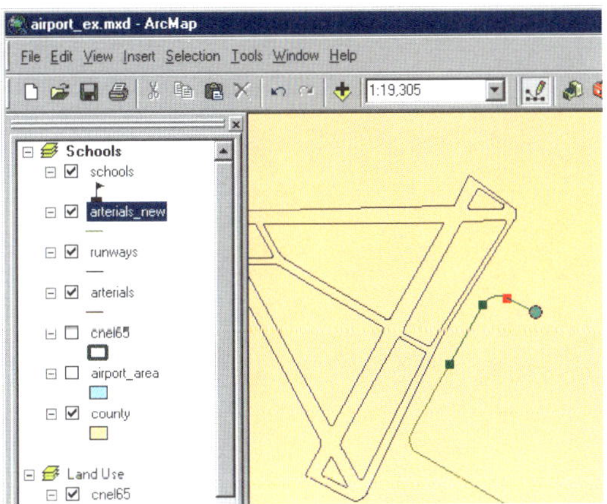

11. Move the mouse pointer so it snaps to the existing road, but don't click the mouse. You want the next segment of the new road to be parallel to the existing road.

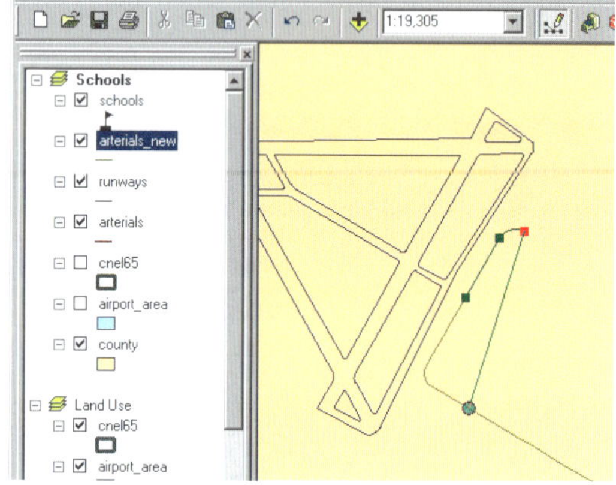

12. Right-click and click Parallel. The line is constrained to be parallel to the existing road.

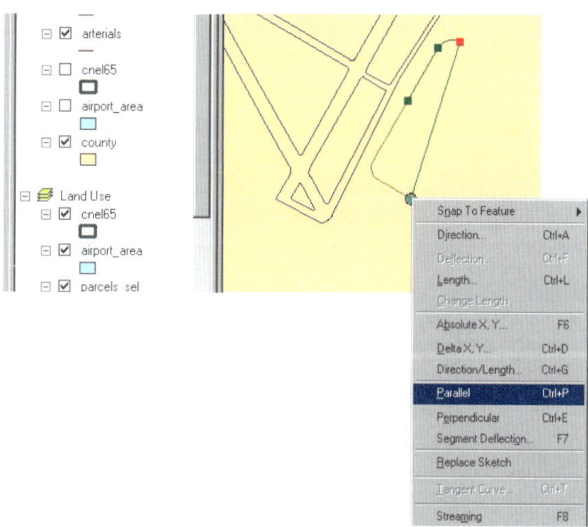

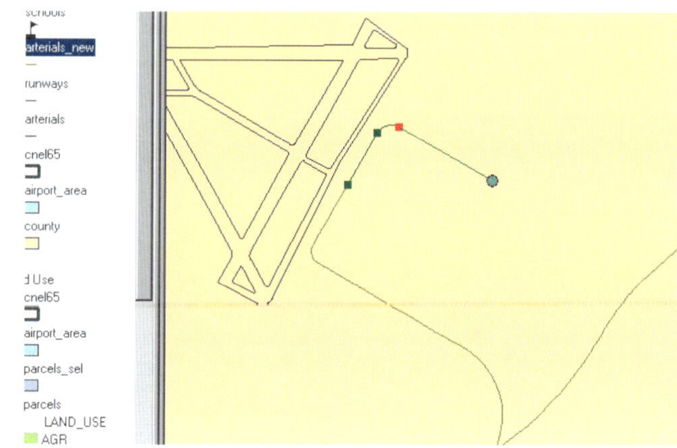

13. To finish the road, move the mouse pointer over the road that you want the new road to intersect, and make sure the circle snaps to it. Double-click to end the line.

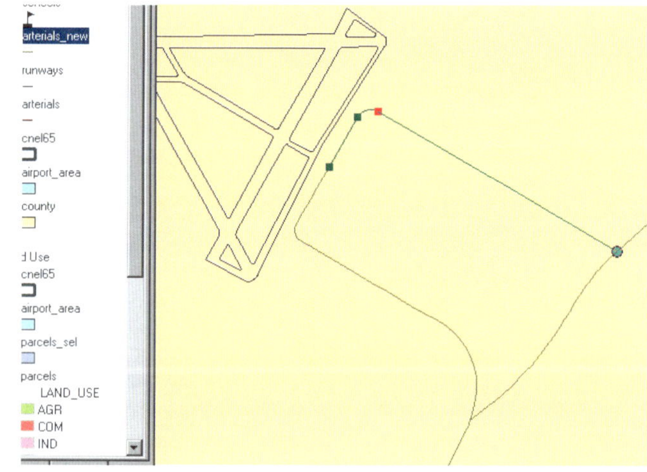

The new road is highlighted in a thick blue line.

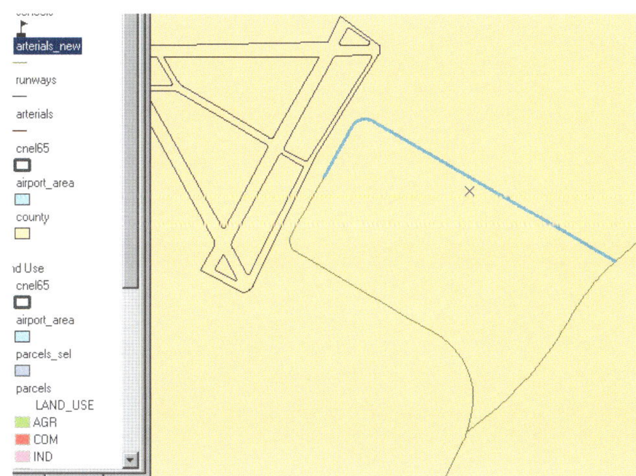

Adding attributes to new features

You can also add the name of the new road.

1. Click the Attributes button on the Editor toolbar.

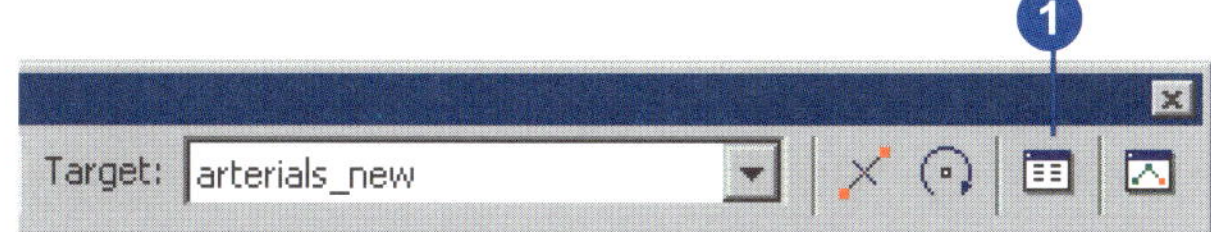

2. Click next to NAME on the list of attributes, type "AIRPORT DR", and press Enter.

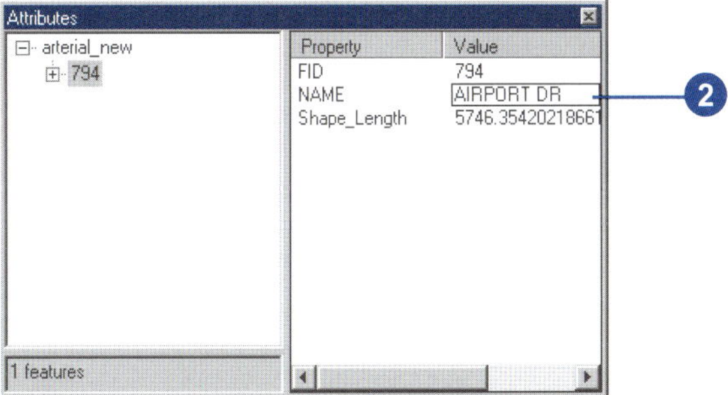

3. Close the Attributes window.
4. Click the Editor menu and click Stop Editing. Click Yes when prompted to save your edits.

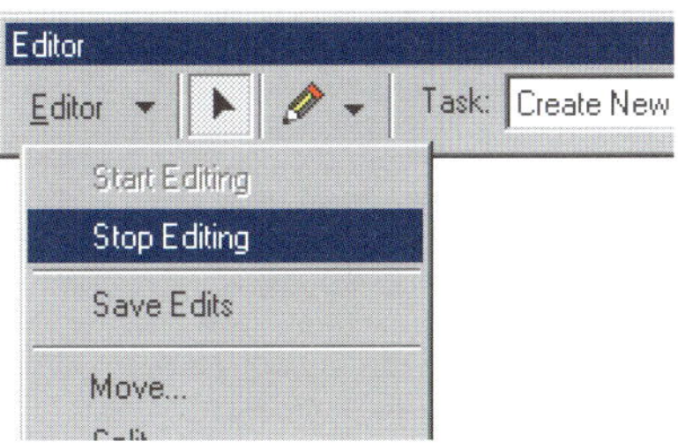

5. Close the Editor toolbar.

6. Right-click arterials_new in the table of contents and click Label Features. The road you added is labeled with its name.

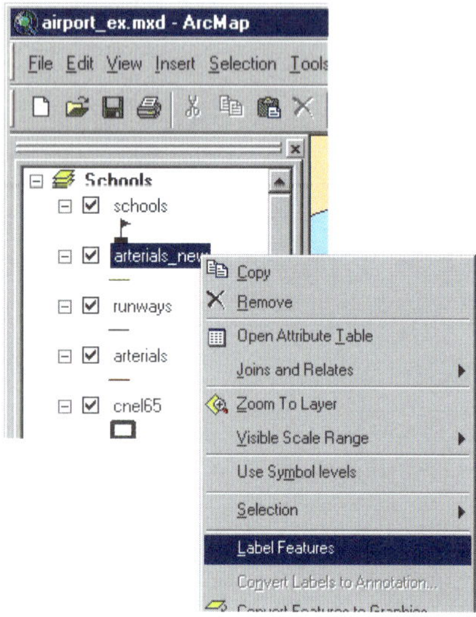

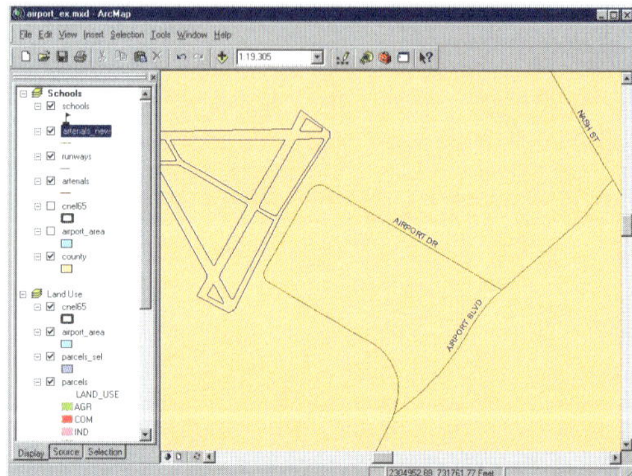

7. Turn the cnel65 and airport_area layers back on by checking their boxes in the table of contents.

8. Switch to layout view by clicking the View menu and clicking Layout View. You can see that the road has been added to your map.

9. You zoomed in for editing (when you switched to data view), so type "1:100,000" in the Map Scale text box on the Standard toolbar and press Enter to set the map scale.

Use the Pan tool on the Tools toolbar to place the noise contour in the center of the map.

You can continue with the final exercise or stop here. If you stop, be sure to save your work by clicking Save on the File menu.

Exercise 5: Working with map elements

In this exercise, you'll add additional map elements to complete your poster and print it.

If necessary, start ArcMap, navigate to the folder where you saved the map from Exercise 4 (airport_ex), and open the map.

Adding a background, titles, legends, and scalebars

1. Switch to layout view by clicking the View menu and clicking Layout View, if necessary.
2. Click the Land Use data frame on the page so it's highlighted. In the table of contents, uncheck the parcels_sel layer so it's not displayed (that way, the map will show the land use types within the noise contour).

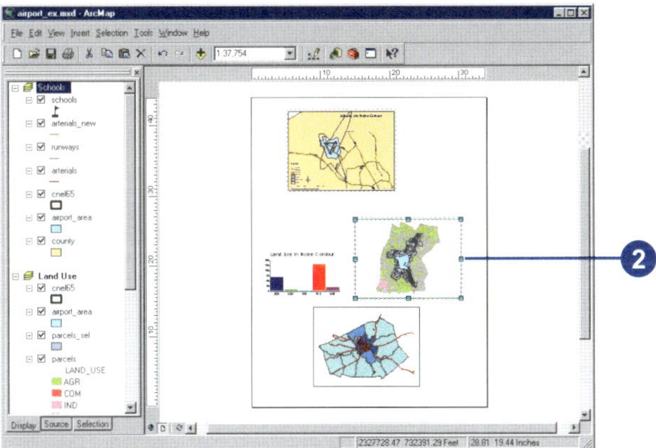

3. Right-click the data frame and click Properties.

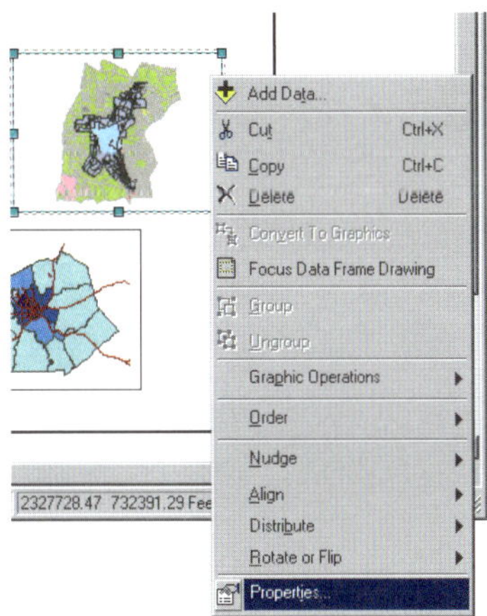

4. Click the Frame tab. Click the Background dropdown arrow and click Sand. Click OK.

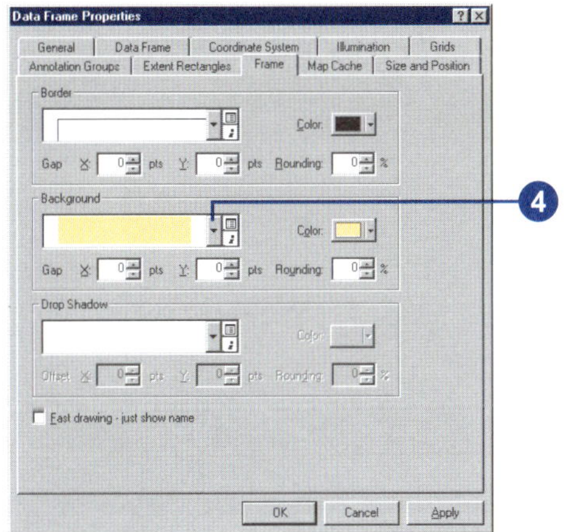

5. Click Insert and click Title.

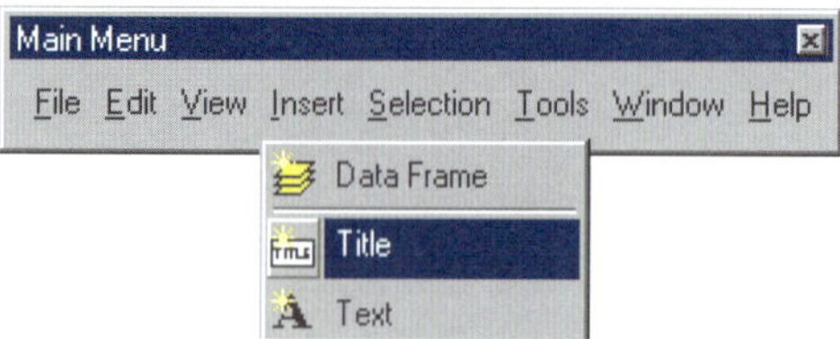

6. Type "Land Use within Noise Contour" in the text box and press Enter.

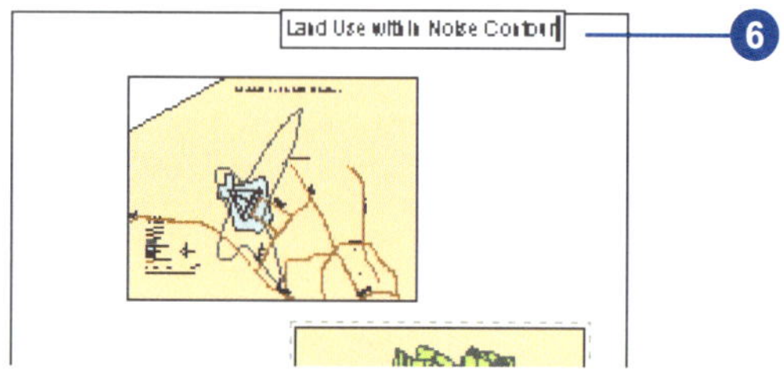

7. Click the Text Size dropdown arrow on the Draw toolbar. Click 36 to make the title 36 point.

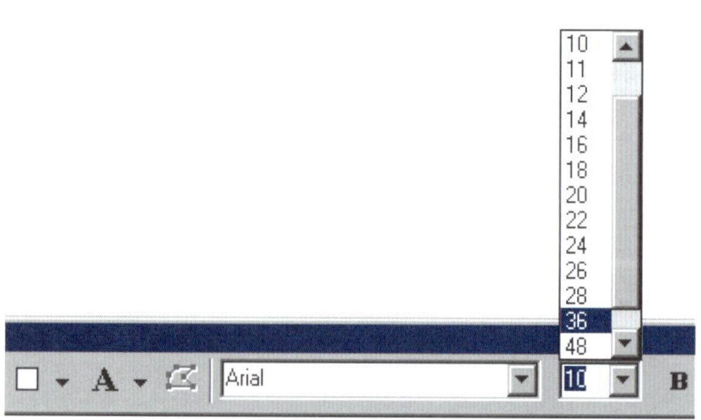

8. Drag the title onto the Land Use data frame, as shown below.

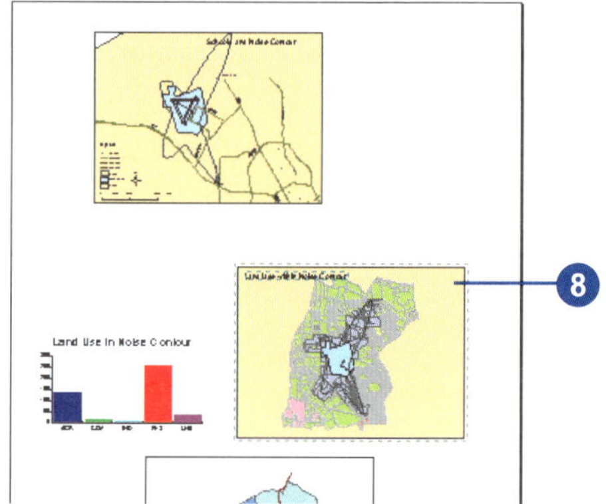

9. Click Insert and click Legend.

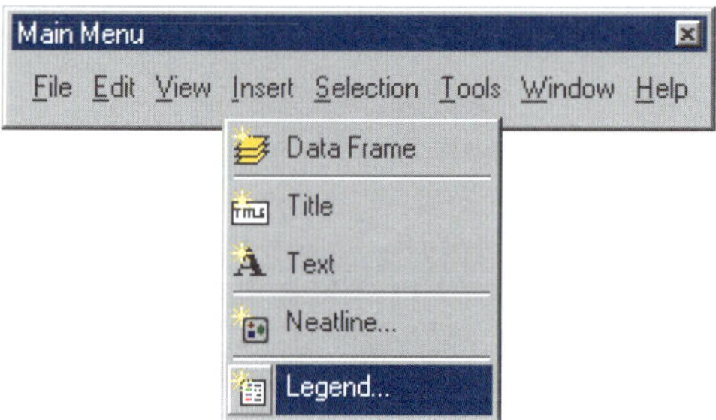

The Legend Wizard appears.

10. Click Next several times to step through the wizard, accepting the default legend parameters. Click Finish when you're done.

11. Drag the legend to the lower-left corner of the data frame, as shown below. Make it smaller by clicking the upper-right handle and dragging it down and to the left.

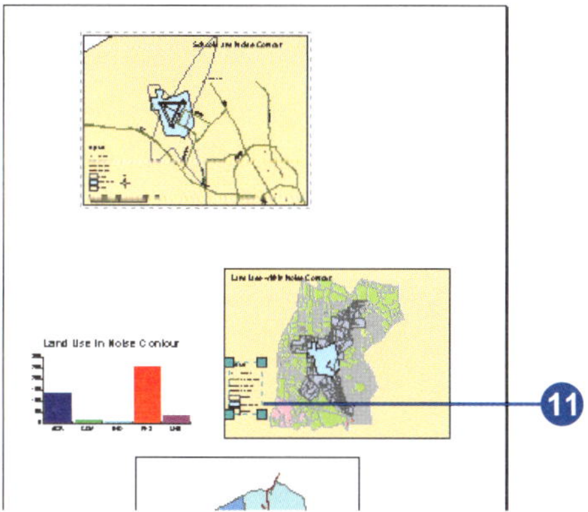

12. Click Insert and click Scale Bar.

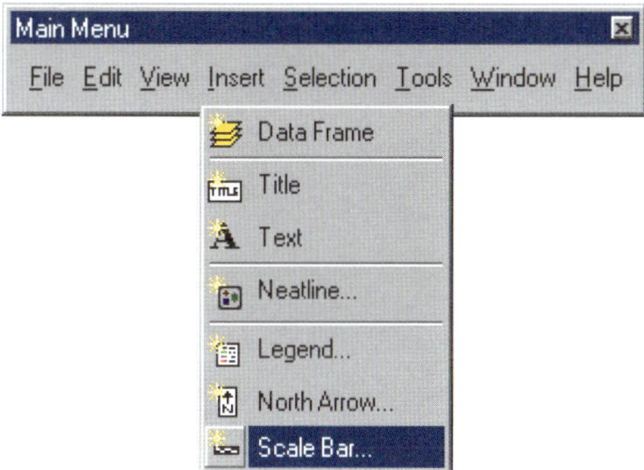

13. Click Scale Line 1 and click OK. Drag the scalebar under the legend and make it smaller.

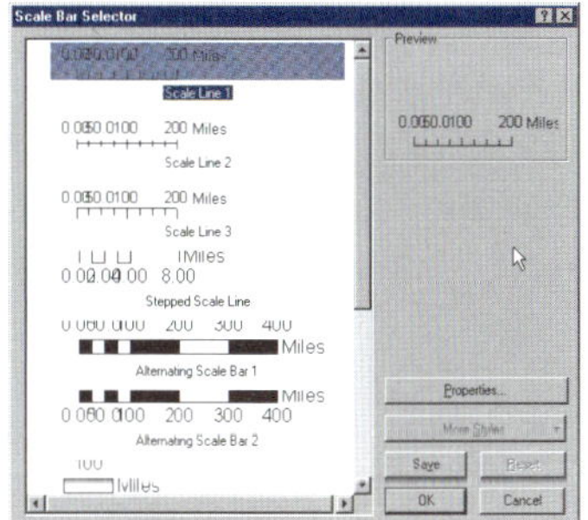

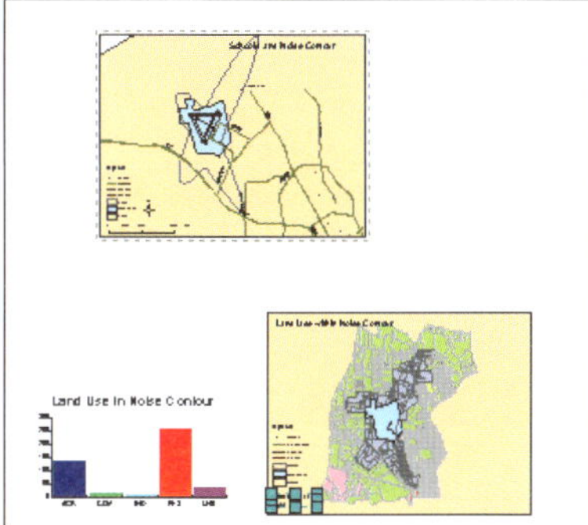

14. Now do the same for the Population Density data frame. First click to select it. Set the background to Sand, make the title read “Population Density”, and add a legend and scalebar. Place the legend in the upper-left corner of the data frame and place the scalebar in the lower-left corner.

15. Click the Schools data frame to select it and set the background to Olive.

16. You only need one North arrow since all maps are oriented in the same direction. Click the North arrow in the Schools data frame and enlarge it by dragging the upper-right handle. Then drag the North arrow to the lower-right corner of the page.

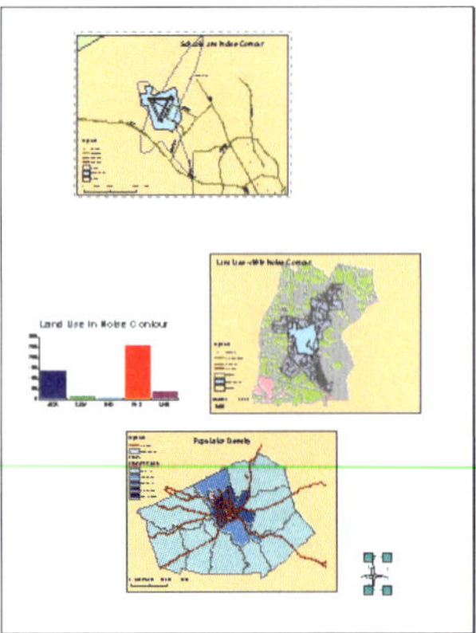

17. Click the New Text tool on the Draw toolbar and click the top of the page. Type "Proposed Airport Expansion" as the title and press Enter. Set the size to 72 point and make the title bold by clicking the Bold button. Position the title at the top and center of the page.

Adding drop shadows

You can add drop shadows to most of the graphic elements on the layout page. Add a drop shadow to each data frame.

1. Click the Population Density data frame to activate it.
2. Right-click the Population Density data frame and click Properties.

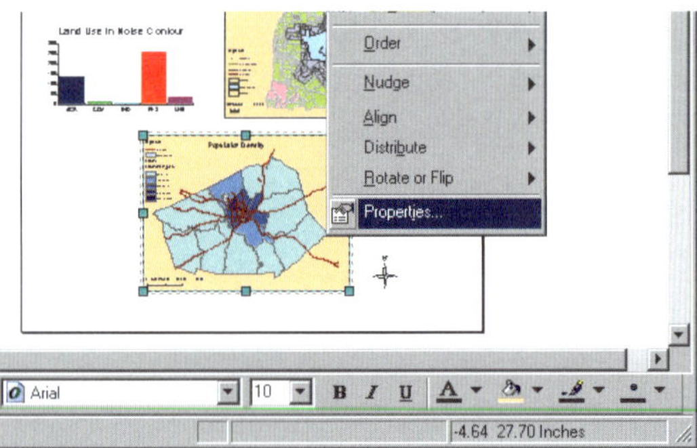

3. Click the Frame tab.
4. Click the Drop Shadow dropdown arrow and click Gray 30%.
5. Type "50" for the X Offset and "-50" for the Y Offset.
6. Click OK.

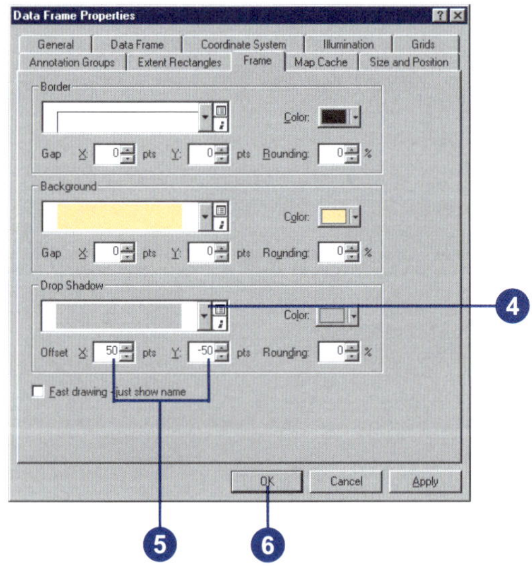

7. Repeat the steps above to add drop shadows to the Schools and Land Use data frames. When finished, your map should look like this:

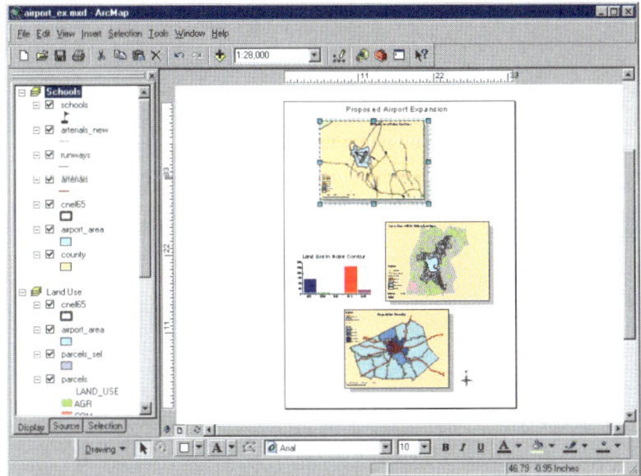

Adding a neatline

1. Click Insert and click Neatline.

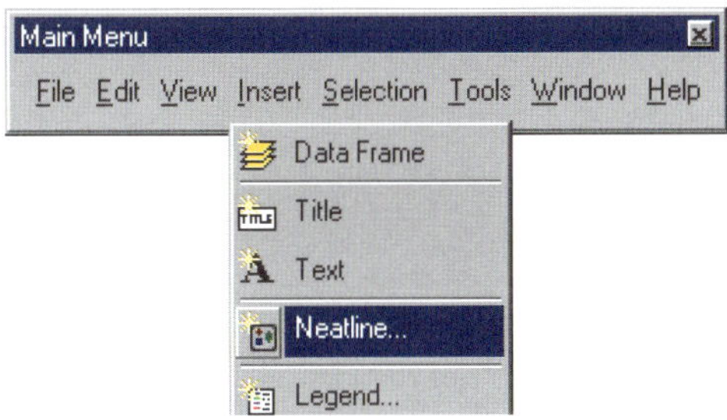

2. Click Place inside margins.
3. Type in a gap of 36 points. This places the neatline about one-half inch inside the margin of the page.

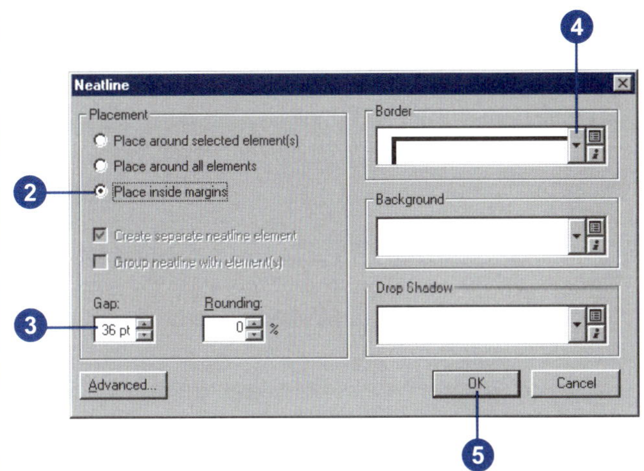

4. Click the Border dropdown arrow and click a border size of 3.0 points.
5. Click OK. Your map should look like this.

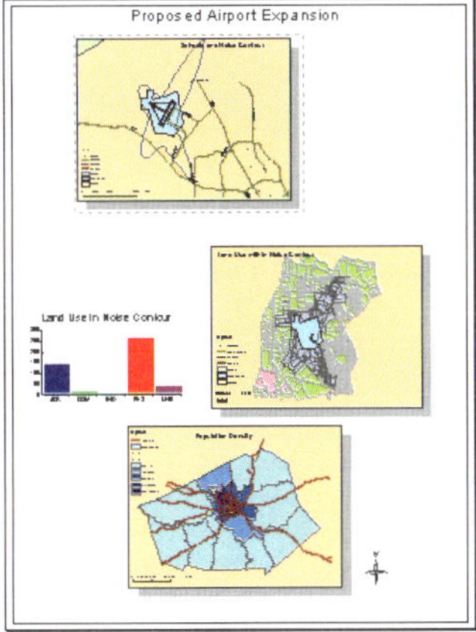

Printing a map

Your map is finished. You can print it if you have a printer connected to your computer. If your printer doesn't print the full size (34 by 44 inches), you can scale the map down to fit your printer.

1. Click the File menu and click Print.
2. If the map is larger than the printer paper, click Scale Map to fit Printer Paper. (Tile map to printer paper will print the map at full scale on separate sheets of paper so you can paste them together to display the full map.)
3. Click Setup.

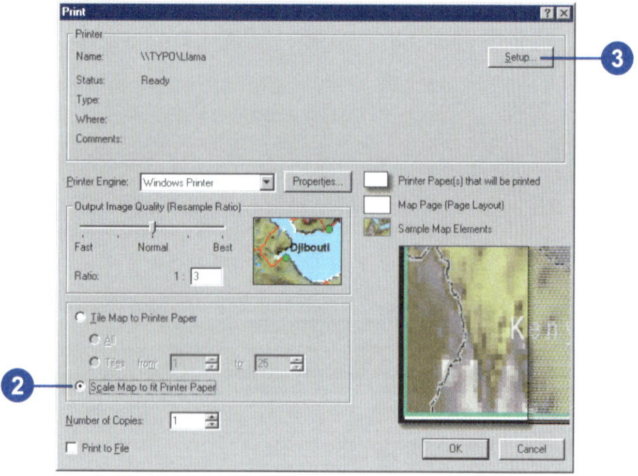

4. Click Portrait for the Paper Orientation.

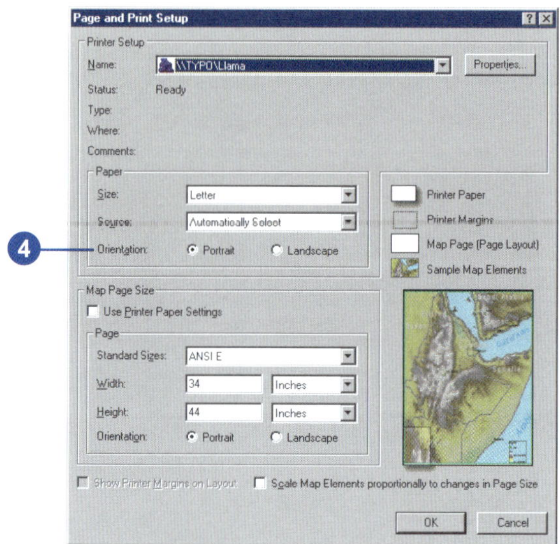

5. Click OK on the Page and Print Setup dialog box, then click OK on the Print dialog box to print the map.

For more information on adding graphics to your map, see Chapter 7, 'Working with graphics and text'. For more on map layout and composition, see Chapter 15, 'Laying out and printing maps'.

In this chapter, you've been introduced to many of the ArcMap tasks you'll often use. The rest of this book provides more detail on these tasks and shows you many more tasks you can perform using ArcMap.

ArcMap basics

3

IN THIS CHAPTER

- Basics of mapmaking, mapping, and GIS
- Layers, data frames, and the table of contents
- Starting ArcMap and opening a map
- Using the table of contents
- Data view and layout view
- Moving around the map
- Setting bookmarks
- Opening magnifier and overview windows
- Exploring data on a map
- Using the map cache
- Getting help
- Saving a map and exiting ArcMap
- Keyboard shortcuts in ArcMap

A traditional, printed map is considered to be a representation of a place or entity, such as a subway system, location of minerals, or a city. This is taken further by defining a *map* as the fundamental component you work with in ArcMap.

As it implies, ArcMap is used to make maps. It is a map-centric software application. In ArcMap, a *map document* (.mxd) is how a map is stored, shared, and managed on your computer. This map document contains not only the traditional cartographic elements of a map, but also the environment, or user interface, you use to work with that map. ArcMap is where you can perform your spatial analysis and querying along with editing, 3D analysis, data development, and display.

This chapter will discuss the basic traditional cartographic elements of a map—but most important, the concepts you need to understand an ArcMap map. These include the components of a map document and the basics you need to know to begin to operate this program.

Basics of mapmaking

Cartography can be described as the graphic principles supporting the art, science, or techniques used in making maps or charts. It was developed in a time before the computer and GIS. Thus we often have a paper-centered view of mapping. However, throughout its development (which continues) many critical principles have been established to advance cartography—such as Jacques Bertin's visual variables for symbology: size, value, texture, color, orientation, and form. He defined these variables to assist someone in representing one symbol differently from another. Also, other advancements have been developed through studies of human psychology and visual perception.

Traditionally, maps have been created to serve two main functions. The first function has been to store information. Creating a map has been a way to record information for future reference. The second function has been to provide a picture to relay spatial information to a user.

The purpose for designing a map is critical to its design. When designing a map, a map maker needs to know the answers to some fundamental questions, such as: What is being mapped? Who is the audience? How is this map being presented—on its own or as part of a report? What medium will be used to display this map?

Map formats

Generally maps are considered to be in two formats. One is a general *reference map* such as a U.S. Geological Survey (USGS) topographic map or the map of a city. In this form the map is providing information to convey where things are in relation to each other. The second is a *thematic map*, where the map is used to convey information about a particular theme or multiple themes, such as land use, population, or health statistics.

Reference Map

Thematic Map

Basic mapping principles

There are many kinds of maps, each with general and possibly specific requirements. While a skilled cartographer is usually required to make maps with specific or special requirements, anyone can make good general and informative maps by considering the following simple guidelines. These guidelines have been organized into seven areas that you can use as a checklist for creating or improving your maps.

Purpose—Typically, a map does not have more than one purpose. Trying to communicate too much in one map—having more than one purpose for the map—tends to blur the message and confuse the map reader. Using two or more maps, each focused on a single message, is always a better strategy.

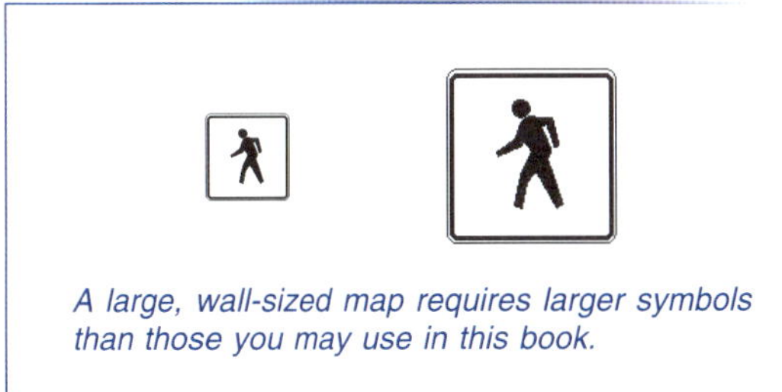

A large, wall-sized map requires larger symbols than those you may use in this book.

Audience—Who will be reading your map? Are you designing a map for a few readers or for a large audience of hundreds or millions of people? It's better to target your map to the person least prepared to understand your map's message.

Size, scale, and media—The physical size of a map relative to the geographic extent shown on the map will dictate the *scale* of the map and determine how you will represent the actual size and number of features shown on the map. Data is often collected at a particular scale. However, if you're not displaying the data at that scale, be sure your data "fits". For example, roads typically collected for 1:24,000 mapping will be far more detailed than needed for a smaller scale map (such as 1:2,000,000), so be sure to reduce the number of roads drawn on your map. Media also plays an import role, because a map printed on newspaper will not show fine details clearly, whereas one printed on high-quality paper will. In addition, the details on a digital map could vary depending on the viewing program. For example, a static map used in a Web page would be designed to encompass less information than one designed for browsing using a program such as ArcReader™.

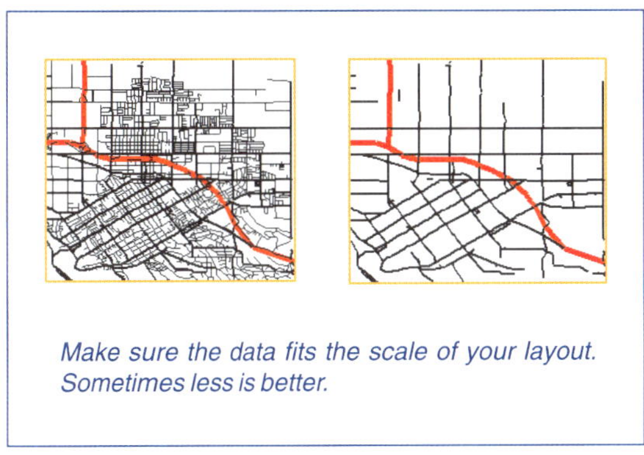

Make sure the data fits the scale of your layout. Sometimes less is better.

Focus—Refers to where the designer wishes the map reader to first focus. Typically cool colors (blues, greens, and light gray) are used for background information, and warm colors (red, yellow, black) are used to capture the reader's attention.

Integrity—You may want to cross-validate some of your information, such as the names or spelling of some features. And if the data was produced by another organization, it is often customary to give that organization credit on the map.

Balance—How does your map look on the page or screen? Are the parts of the map properly aligned? The body of the map should be the dominant element. Try to avoid large open spaces. Be flexible in where you place elements (for example, not all titles need to go at the top). Should some components on the map be contained within a border?

Completeness—A map generally should contain some basic elements, such as a title, legend, scalebar, and North arrow; however, there are exceptions. For example, if a graticule exists, it is not necessary to place a North arrow. Basically, place all the information you think your readers need to fully understand the map. An example of the basic elements to include on a map are shown in the diagram titled 'Map elements diagram'.

Before publishing your map, it is always advised to have someone else look it over—especially for spelling and overall appearance.

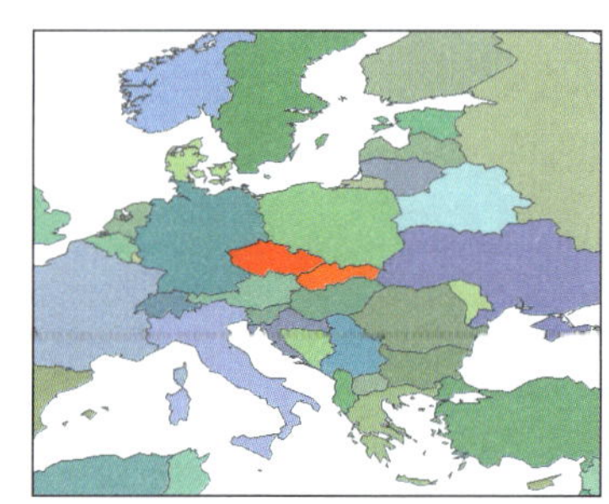

Color is used to help focus the reader on an area or symbol represented on the map.

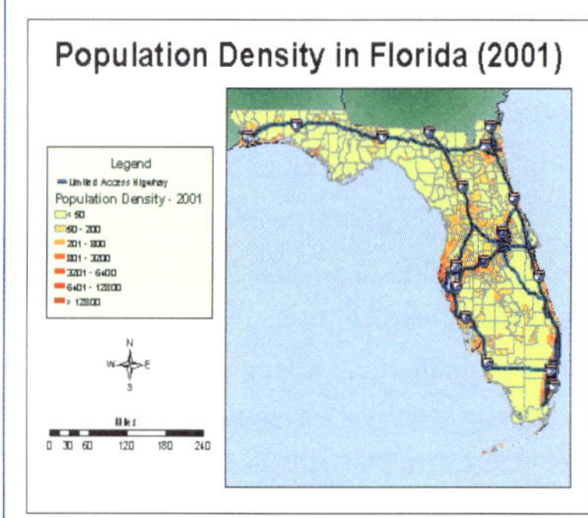

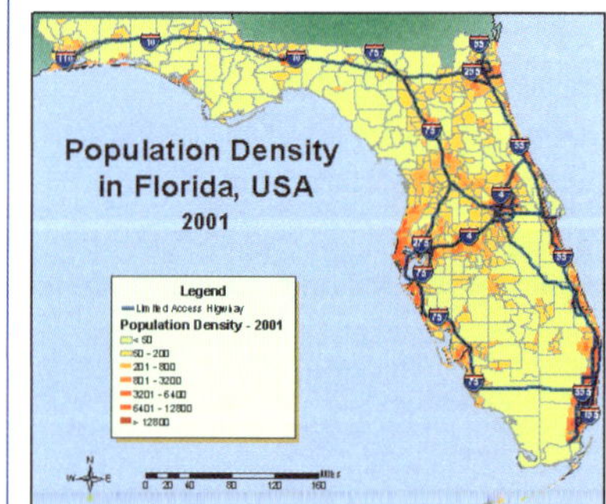

It is best to avoid a lot of unnecessary open spaces in your layout.

Map elements

Map body—The primary mapped area. You can display more than one image of your primary mapped area within your document. For example, you may want to portray change by showing several images with differing but related information, such as population maps of various years. Your map may also contain a locator map (a smaller-scale map used to help the reader in understanding where the main area of interest is located), an inset map (used to give more detailed information of an area within the main map that may not easily be understood), or an index map (often used to show where in a series of maps one map exists). All are used to assist in communicating your information to others. In ArcMap each of these mapped areas is referred to as a data frame.

Title—Is used to tell the reader what the map represents. This is often placed on a map *layout* as text.

Legend—Lists the symbology used within the map and what it represents. This can be created using the Legend Wizard in the layout—and edited further once created.

Scale—Provides readers with the information they need to determine distance. A map scale is a ratio, where one unit on the map represents some multiple of that value in the real world. It can be numeric (1:10,000), graphic (a scalebar), or verbal (one inch equals 10,000 inches). Maps are often referred to as large or small scale. This size reference refers to the ratio (or fraction). For example, a 1:100 scale map is larger than a 1:10,000 map, because 1/100 (0.01) is a larger value than 1/10,000 (0.0001). A smaller-scale map displays a larger area, but with less detail. The scale is inserted in the map layout view.

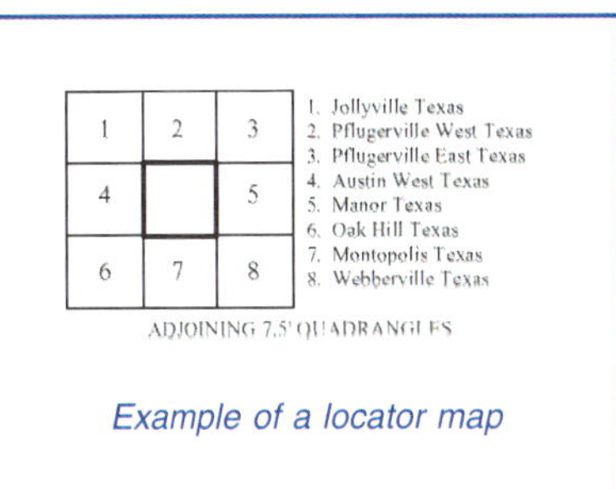

Example of a locator map

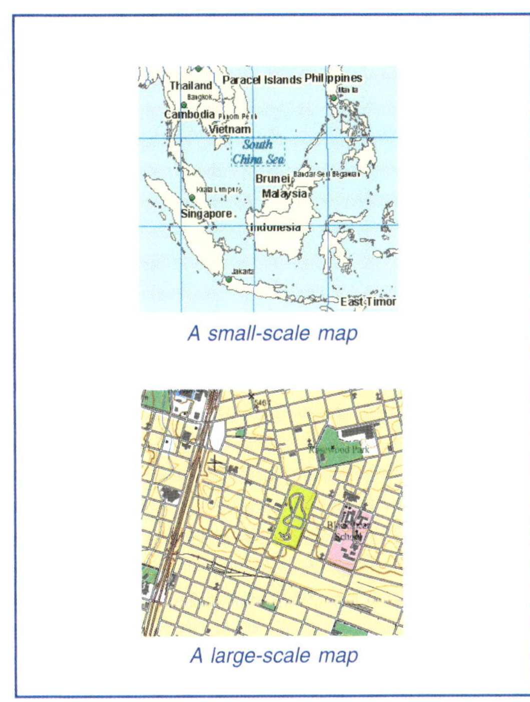

A small-scale map

A large-scale map

Projection—Is a mathematical formula that transforms feature locations from the earth's curved surface to a map's flat surface. Projections can cause distortions in distance, area, shape, and direction; no projection can avoid some distortion. Therefore, the projection type is often placed on the map to help readers determine the accuracy of the measurement information they infer from the map.

For more information on projections, see 'About coordinate systems' in Chapter 4 or *Understanding Map Projections*.

Direction—This is shown using a *North arrow*. A map may show true north and magnetic north. This element is inserted in the map layout view.

Data source—The bibliographic information for the data used to develop the map.

Other map components include (but are not limited to) dates, pictures, graticules or grids, reports, tables, additional text, neatlines, and authorship.

Grids and graticules

ArcMap contains several types of grids and graticules that can be added to a map layout using the Grids and Graticules Wizard.

You can place a grid that expresses location in geographic coordinates (degrees of latitude and longitude) by selecting the Graticule type.

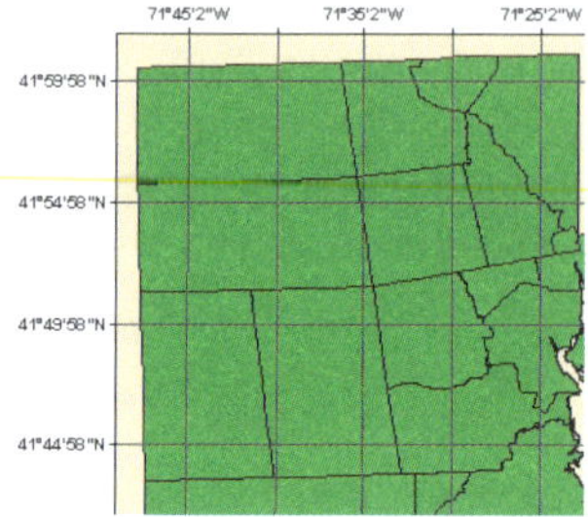

By selecting the Measured Grid type, you can place a grid that expresses location using projected coordinates, such as Universal Transverse Mercator or State Plane grids. ArcMap allows you to select a coordinate system for the grid that is different from the underlying coordinate system for the data frame.

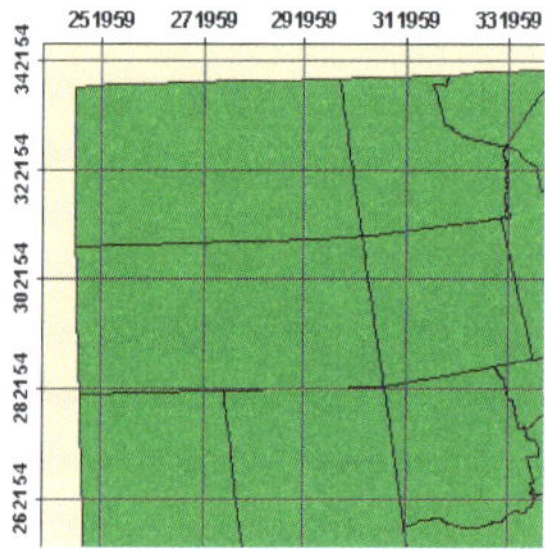

You can also place a grid that divides a map into a specified number of rows and columns by selecting the Reference Grid type. Often, the row and column labels of a reference grid identify locations listed in a map index.

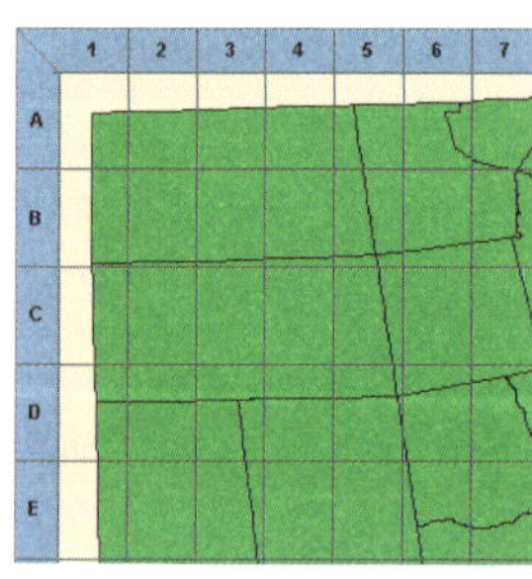

Other custom grids or military grid reference systems can be added using the Style Manager in ArcMap.

Map elements diagram

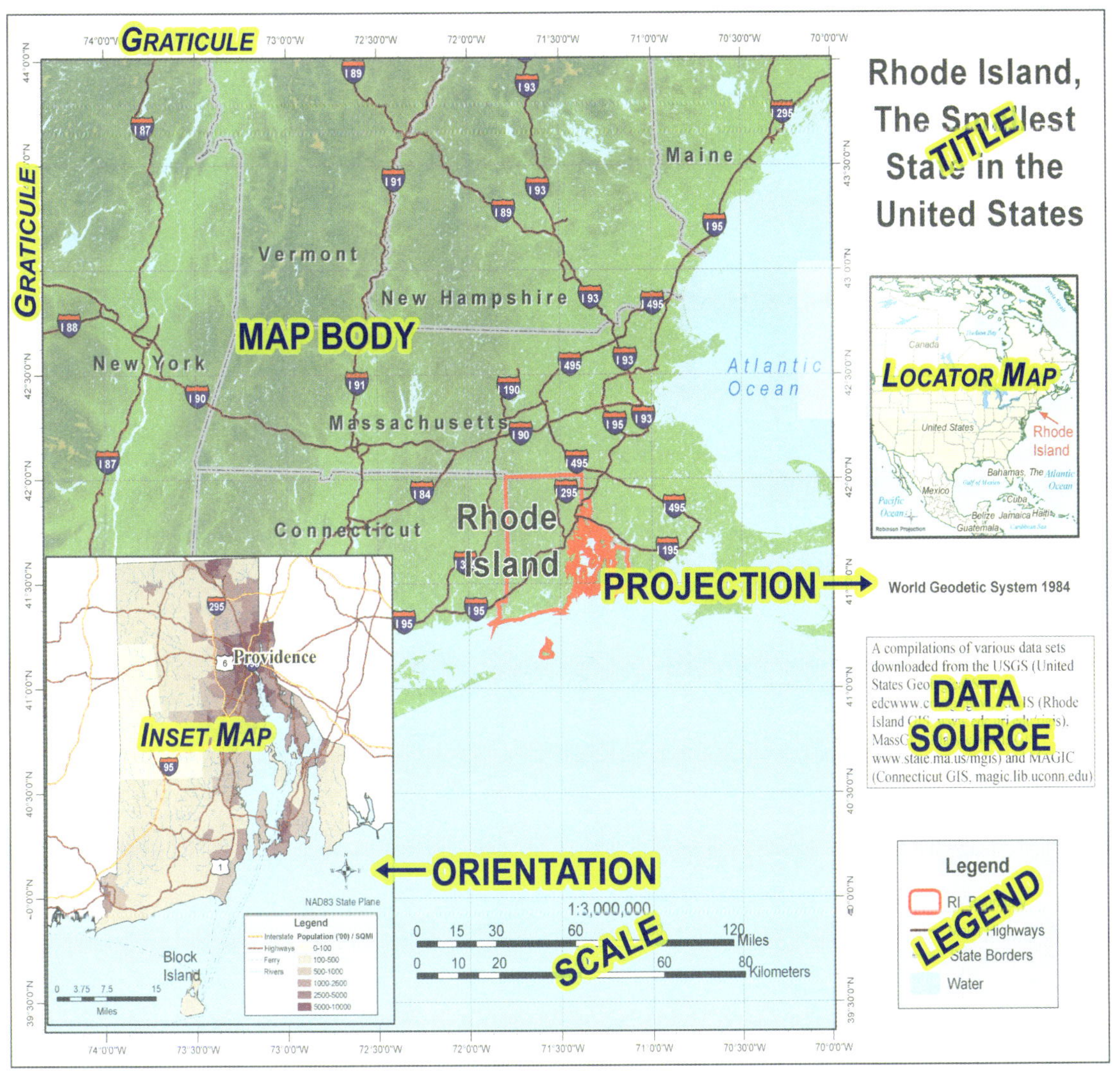

Mapping and GIS

So how are mapping and GIS related? For starters, GIS has its roots in mapping—both involve maps and attributes, and both use geographic data involving map scales, projections, and coordinate systems.

There are three basic users for both mapping and GIS—the viewer, the maker, and the designer. Before outlining their roles, it must be stated that these users can overlap—a viewer may be the maker, or a maker the designer—or they can all be the same person.

Viewer—This person, who can be described as the end user, is generally the reason for the existence of the map or data. The viewer is the person using a Web mapping program to determine the route from his or her home to the museum, the oil company executive who needs to locate potential drilling sites, a person planning a mountain hike, or a newspaper reader who is making the association between a new industrial site and his or her home.

Designer—The designer can be involved at the beginning, end, and throughout the GIS or mapping product creation process. This person determines what data to use, what tools to use, how to acquire the data, and so on. The designer can be the person who creates the output by defining what questions will be answered and how data is going to be displayed on the map. The map designer could also be the person specifying which symbols to use to represent roads or the project developer who is writing the contract proposal for an environmental assessment using GIS.

Maker—This is the person working with the data by editing, creating, acquiring, querying, or analyzing it. The maker is the person who is merging the road dataset, digitizing the rivers, editing the parcels, attaching the address locations, buffering the protected lands, analyzing the population projections, importing the elevation models, or identifying the new school locations.

How else are mapping and GIS related? GIS is used for display, analysis, storage, and retrieval. The mapping output is used to display and store information. From this a person can retrieve information and use that information in analysis. Mapping and GIS are becoming closer to one another through technological advancements. For example, a map is no longer a static product and visualization is not dependent on a printable medium. Viewers have been given the ability to interact with the display.

Basically, the cartographic principles of mapping can be thought of as the rule for output, and GIS can be considered the tool to bring information together. The rest of this chapter will discuss the map document design in ArcMap and the basics you need to know to use the program.

Layers, data frames, and the table of contents

You display geographic information on a map as *layers*, where each layer represents a particular type of feature, such as streams, lakes, highways, political boundaries, or wildlife habitats. A layer doesn't store the actual geographic data; instead, it references the data contained in coverages, shapefiles, geodatabases, images, grids, and so on. Referencing data in this way allows the layers on a map to automatically reflect the most up-to-date information in your GIS database.

The *table of contents* lists all the layers on the map and shows what the features in each layer represent. The check box next to each layer indicates whether it is currently turned on or off, that is, whether it is currently drawn on the map or not. The order of layers within the table of contents is also important; the layers at the top draw on top of those below them. Thus, you'll put the layers that form the background of your map, such as the ocean, at the bottom of the table of contents.

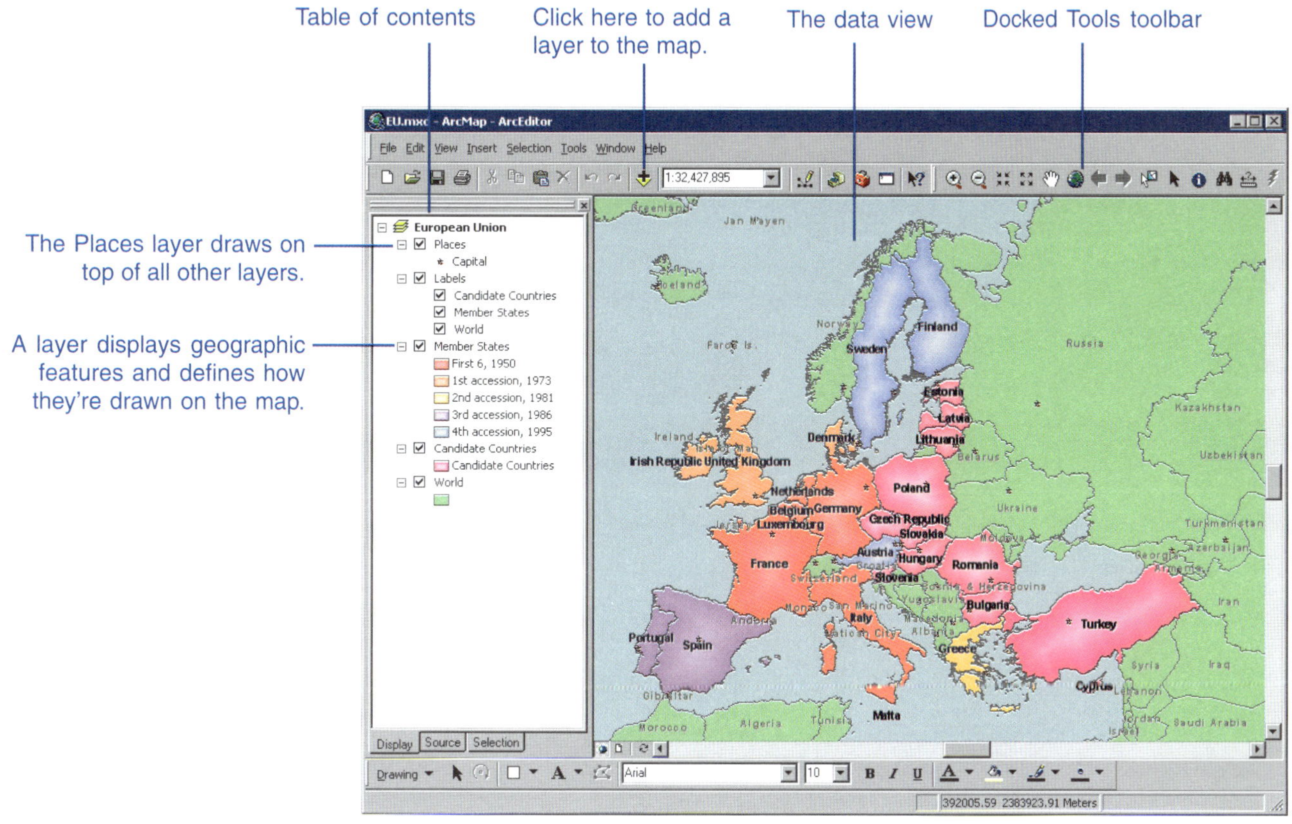

Layers in the table of contents can be further organized into *data frames*. A data frame simply groups, in a separate frame, the layers that you want to display together. You always get a data frame when you create a map; it's listed at the top of the table of contents as "Layers", but you can change the name to something more meaningful if you like. For many of the maps you make, you won't need to think much more about data frames; you'll just add layers to your map and, depending on how you order them in the table of contents, some layers will draw on top of others. You will want to think more about data frames—and adding additional ones—when you want to compare layers side by side or create insets and overviews that highlight a particular location or attribute, as shown in the map below.

When a map has more than one data frame, one of them is the *active data frame*. The active data frame is the one you're currently working with. For example, when you add a new layer to a map, it is added to the active data frame. You can always tell which data frame is active because it's highlighted on the map and its name is shown in bold text in the table of contents. Of course, if a map has only one data frame, it's always the active one.

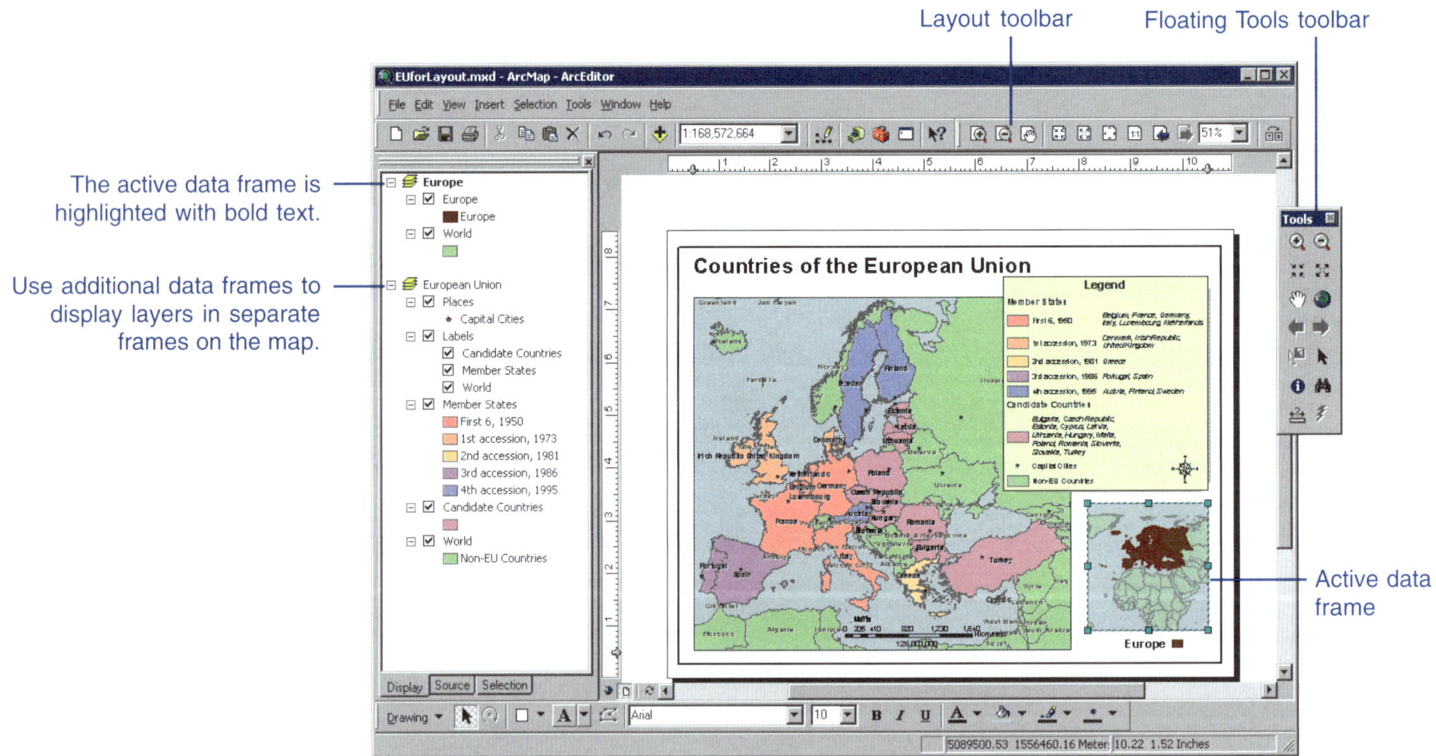

The layout view

Starting ArcMap

Starting ArcMap is the first step to exploring your data. You can access ArcMap from the Start button on the Windows taskbar. Each ArcMap session can display one map at a time. You can work with several maps by starting additional ArcMap sessions.

After you first start ArcMap, you can decide whether or not you want to see the splash screen and Startup dialog box. If you don't want to see these, you can easily turn them off by choosing Tools > Options > General from the program's Main menu bar.

Tip

Starting ArcMap by opening an existing map

Double-clicking a map in ArcCatalog or the Windows Explorer will launch ArcMap and display the map.

Tip

Working on one map at a time

You can only work on one map at a time in an ArcMap session. ArcMap will close any open map before opening another one.

Starting ArcMap from the Start menu

1. Click the Start button on the Windows taskbar.
2. Point to Programs.
3. Point to ArcGIS.
4. Click ArcMap.

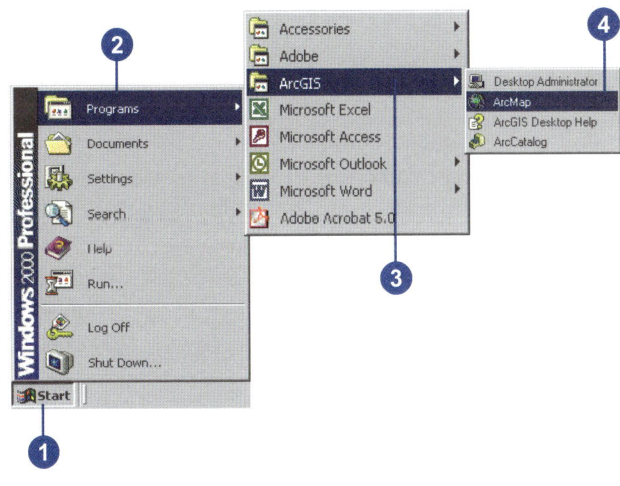

Starting ArcMap from ArcCatalog

1. Click the Launch ArcMap button on the Standard toolbar.

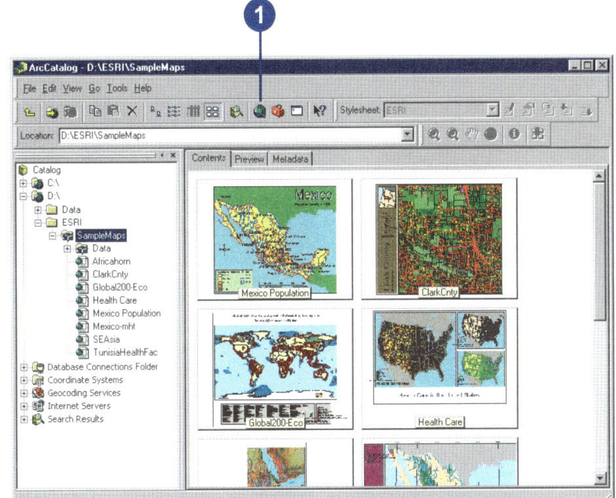

The ArcMap window

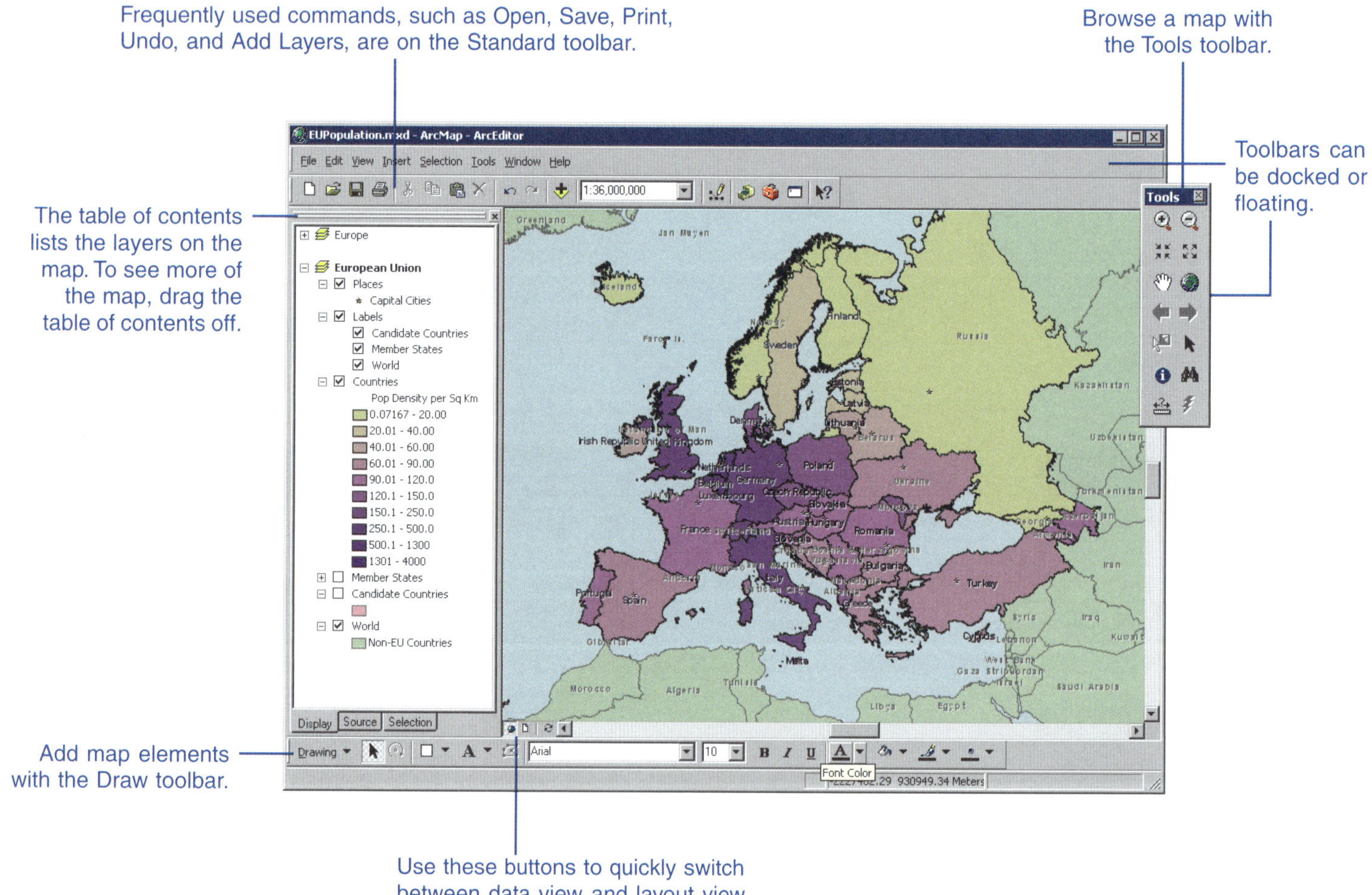

Opening a map

To work on a map, you open it in ArcMap. If you know its location on disk, you can navigate to it with ArcCatalog and open it in ArcMap. If you already have ArcMap running, you can open it directly within that session.

If you're not sure where your map is located, use ArcCatalog to find it by browsing for it in the folders in your database. Because ArcCatalog lets you preview a map before you open it, you'll always open the right one.

A map doesn't store the spatial data displayed within it. Instead, it stores references to the location of these data sources—such as geodatabases, coverages, shapefiles, and rasters—on disk. Thus, when you open a map, ArcMap checks the links to the data. If it can't find some data—for instance, if the source data for a layer has been deleted or renamed or a network drive is not accessible—ArcMap lets you locate it.

If the data is currently unavailable, you can ignore the broken link and display the map without the layer. The layer will still be part of the map and listed in the table of contents; it simply won't display.

Opening a map from ArcMap

1. Click the Open button on the Standard toolbar.
2. Click the Look in dropdown arrow and navigate to the folder that contains the map.
3. Click the map you want to open.
4. Click Open.

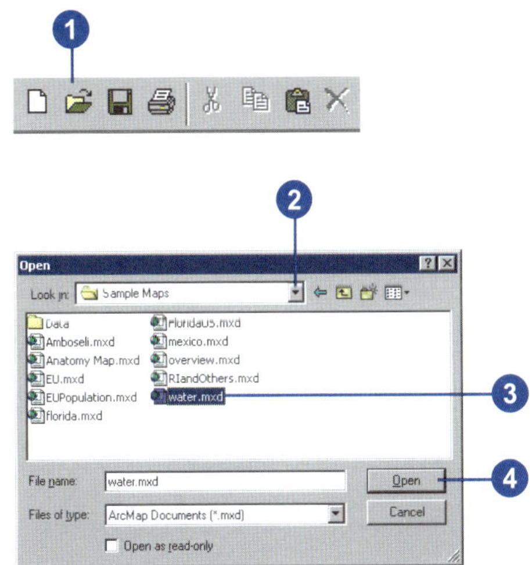

Opening a map from ArcCatalog

1. Start ArcCatalog if it isn't already running.
2. In ArcCatalog, navigate to the folder that contains your map.
3. Click the Thumbnails button to look at the maps the folder contains.
4. Double-click the map to open it in ArcMap.

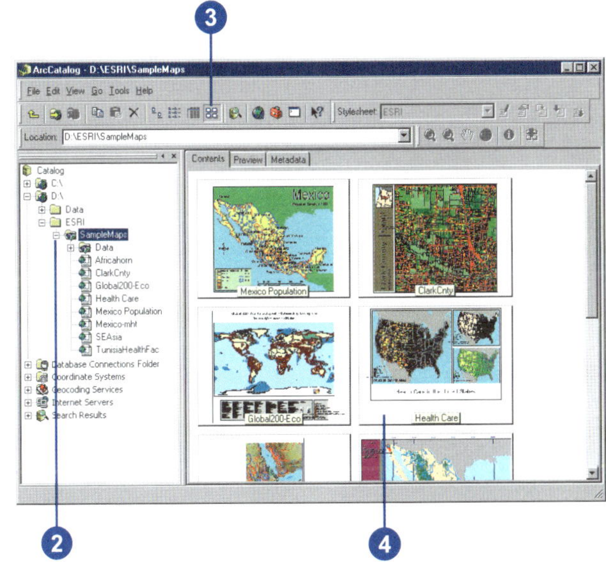

Using the table of contents

Every map has a table of contents. The table of contents shows you what layers the map contains and also how the map presents the geographic features in those layers.

Some maps display all the layers in one data frame. Others, such as those with insets and overviews, will have more than one data frame. The table of contents shows how the layers are organized into data frames.

When viewing a map, you'll use the table of contents primarily to turn layers on and off. As you begin building your own maps, you'll find that the table of contents is the focal point for many tasks, such as adding and deleting layers and determining how to draw layers.

You can choose to display the table of contents with either the Display, Source, or Selection tabs.

Tip

Need to see more of your map?

You can drag the table of contents off the ArcMap window.

Showing the table of contents

1. Click Window on the Main menu.
2. Click Table Of Contents.

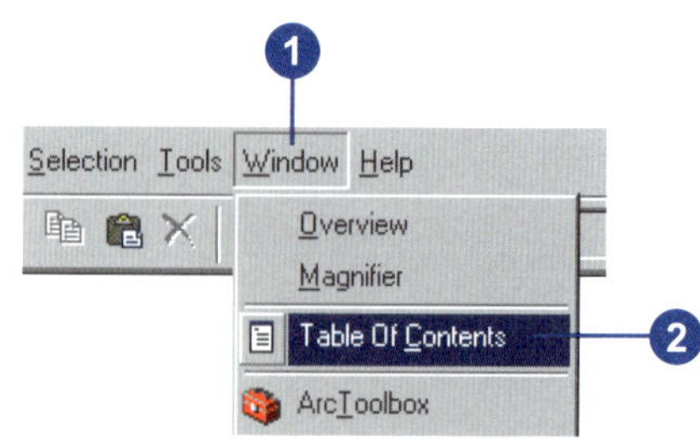

Turning a layer on or off

1. In the table of contents, check the box next to the layer's name.

   The layer should appear on your map. If you can't see the layer, it may be hidden by another layer or display only at a particular scale.

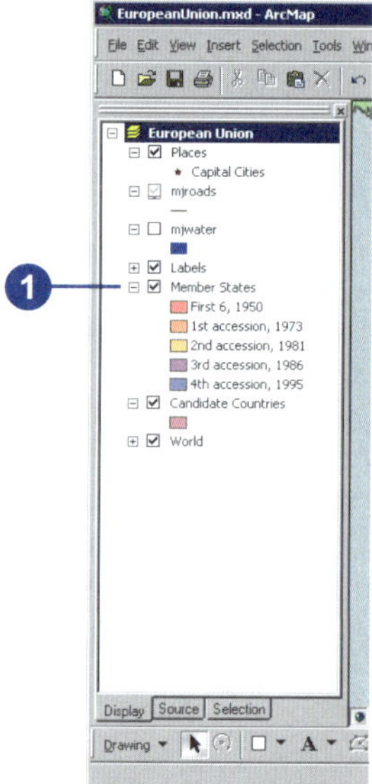

Tip

Drawing layers

Double-click a layer in the table of contents to see its properties. From there you can change how you draw the layer.

Tip

Changing colors

You can quickly change the color of a particular feature by right-clicking on the color in the table of contents.

Tip

Why isn't my layer drawing?

The layer may have a visible scale range set. If you see a gray scalebar underneath the layer's check box in the table of contents, it's not drawing because it's outside of a visible scale range. You'll need to zoom in or out to see it.

If you see a red exclamation point, the link to the layer's data source is broken. Right-click the layer, point to Data, and click Set Data Source to fix the link.

Showing a layer's legend

1. Click the plus or minus sign to the left of the layer name in the table of contents to show or hide its legend.

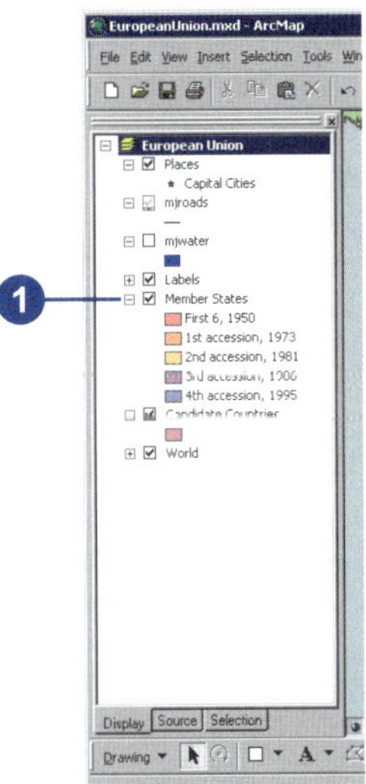

Showing the contents of a data frame

1. Click the plus or minus sign to the left of the data frame in the table of contents to show or hide the list of layers it contains.

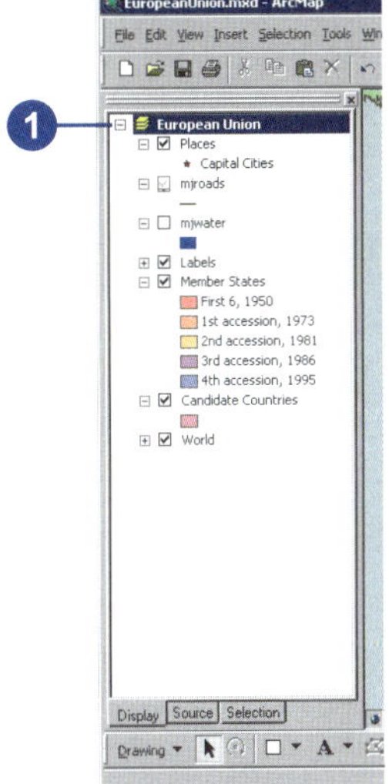

Data view and layout view

ArcMap provides two different ways to view a map: data view and layout view. Each view lets you look at and interact with the map in a specific way.

When you want to browse the geographic data on your map, choose data view. *Data view* is an all-purpose view for exploring, displaying, and querying the data on your map. This view hides all the map elements on the layout—such as titles, North arrows, and scalebars—and lets you focus on the data in a single data frame, for instance, to do editing or analysis.

When you're preparing your map to hang on the wall, put in a report, or publish on the Web, you'll want to work with it in layout view. *Layout view* is for laying out your map. In layout view, you'll see a virtual page upon which you can place and arrange map elements. In layout view, you can do almost everything you can in data view, plus design your map.

Switching to data view

1. Click the View menu on the Standard toolbar.
2. Click Data View.

   The ArcMap window displays the active data frame.

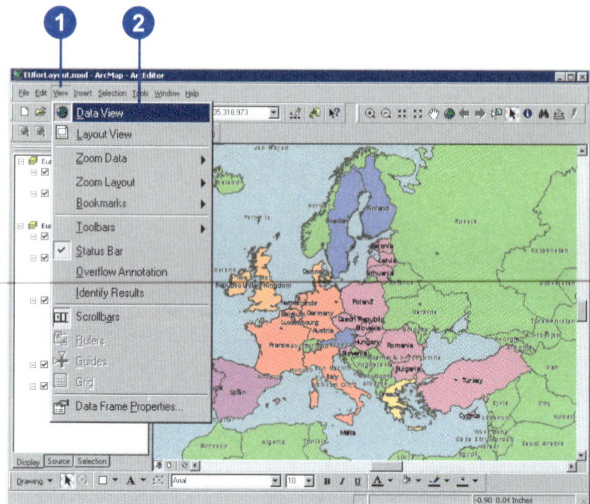

Switching to layout view

1. Click the View menu on the Standard toolbar.
2. Click Layout View.

   The ArcMap window displays the entire map.

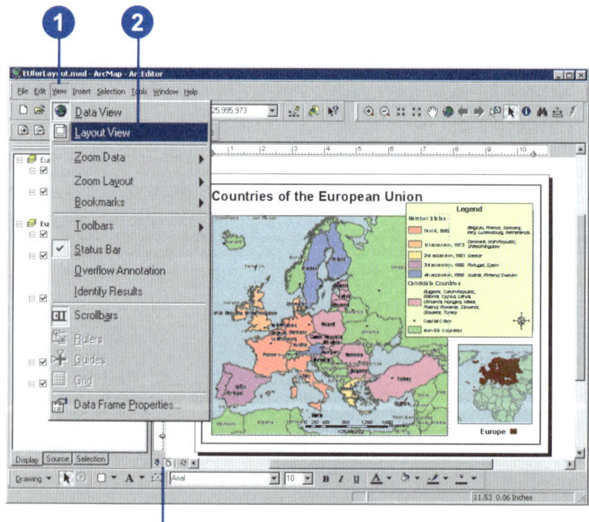

You can also use these buttons to quickly switch between the data and layout view.

Moving around the map

As you work with a map, you can easily change how you view the data it contains. When you're just browsing a map, you might want to pan and zoom around the data to investigate different areas and features. When you're creating a map to hang on the wall, displaying data at a specific scale may be important.

Most of the tools for navigating your data are found on the Tools toolbar.

Tip

Panning using the scrollbars

In data view, you can also pan the map by dragging the scrollbars.

Tip

Panning and zooming in maps with more than one data frame

If your map has more than one data frame, panning and zooming will occur in the active data frame. In layout view, clicking a data frame will activate it.

Zooming in or out

1. Click the Zoom In or Zoom Out button on the Tools toolbar.
2. Move the mouse pointer over the map display and click once to zoom around a point. Alternatively, click and drag a rectangle defining the area you want to zoom in or out on.

Zooming to the full extent of the data

3. Click the Full Extent button on the Tools toolbar.

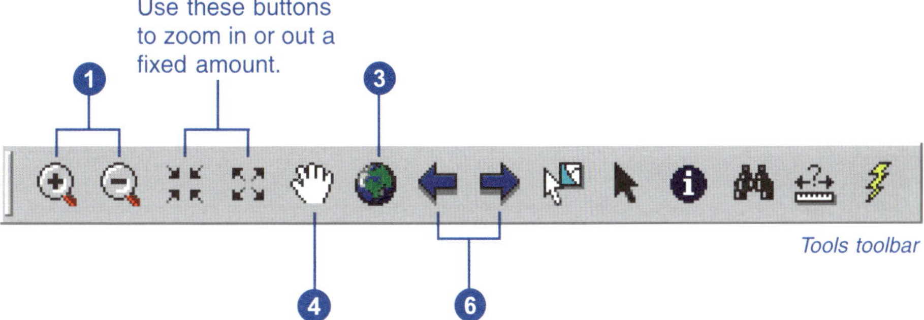

Tools toolbar

Panning

4. Click the Pan button on the Tools toolbar.
5. Move the mouse pointer over the map display and click and drag the pointer.

Moving back or forward one display

6. Click the Back or Forward Extent buttons on the Tools toolbar.

Tip

Selecting layers in the table of contents

Click a layer to select it. Hold down the Shift or Ctrl key to select multiple layers.

Tip

Why doesn't a layer draw when I zoom in or out?

The layer probably has a visible scale range set that prevents the layer from displaying on the map at certain scales. You can clear the scale range by right-clicking the layer in the table of contents, pointing to Visible Scale Range, and clicking Clear Scale Range.

Tip

Pausing drawing in ArcMap

The Pause Drawing command allows you to temporarily suspend all drawing in ArcMap so you can make changes to your map, such as changing the symbology of several layers, without having the map redraw after each change. The Pause Drawing command works in data view or layout view and can be added to any ArcMap menu or toolbar from the Customize dialog box. See Chapter 16, 'Customizing ArcMap', to learn more about the Customize dialog box.

Zooming to the extent of a layer

1. Right-click the layers you want to zoom to.
2. Click Zoom To Layer.

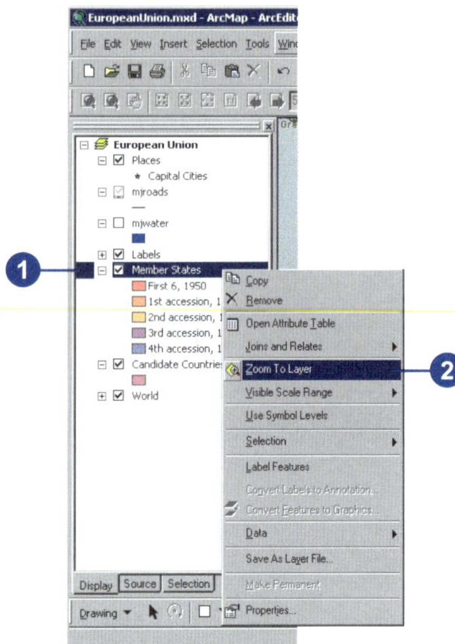

Zooming to a specific scale

1. Type the desired scale on the Standard toolbar.

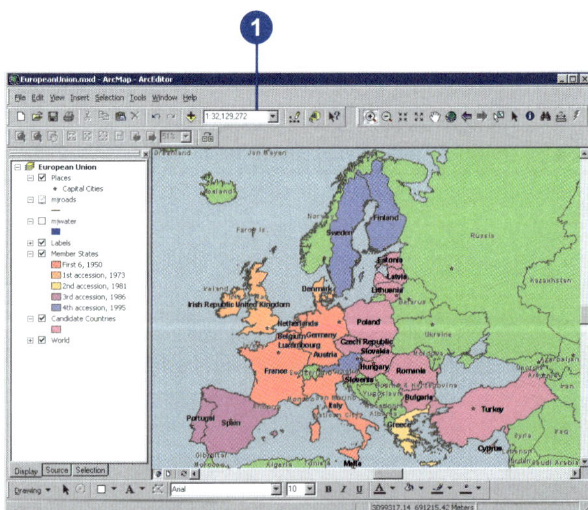

Setting bookmarks

A *spatial bookmark* identifies a particular geographic location that you want to save and refer to later. For example, you might create a spatial bookmark that identifies a study area. That way, as you pan and zoom around your map, you can easily return to the study area by accessing the bookmark. You can also use spatial bookmarks to highlight areas on your map you want others to see.

You can create a spatial bookmark at any time. As a shortcut, you can also create bookmarks when you find and identify map features. Spatial bookmarks, however, can only be defined on spatial data; they can't be defined on an area of the page in layout view.

Each data frame on your map maintains its own list of bookmarks. In layout view, the list reflects the bookmarks of the active data frame.

Creating a spatial bookmark

1. Pan and zoom the map to the area for which you want to create a bookmark.
2. Click the View menu, point to Bookmarks, and click Create.
3. Type a name for the bookmark.
4. Click OK.

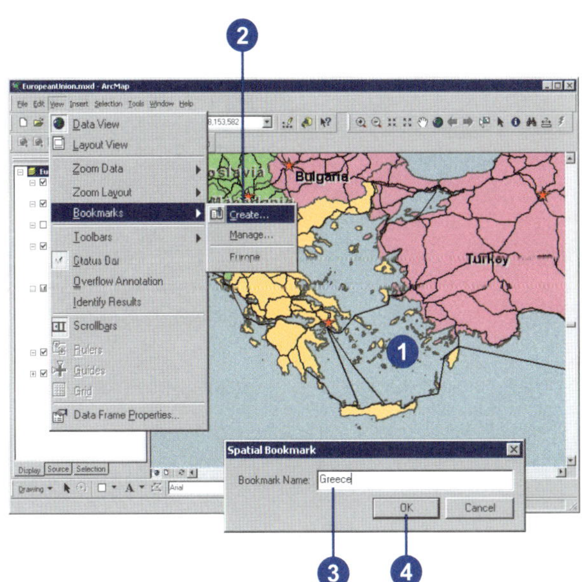

Using a spatial bookmark

1. Click the View menu, point to Bookmarks, and click the name of the bookmark you want to use.

   The bookmarked display appears.

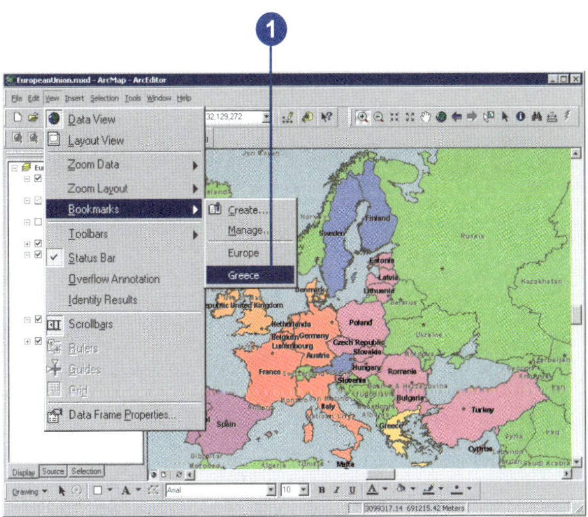

Creating a spatial bookmark from the Identify Results dialog box

1. Click the Identify button on the Tools toolbar.
2. Click the mouse pointer over the map feature to identify.
3. Right-click the identified feature in the Identify Results dialog box.
4. Click Set Bookmark.

   The bookmark is named after the feature.

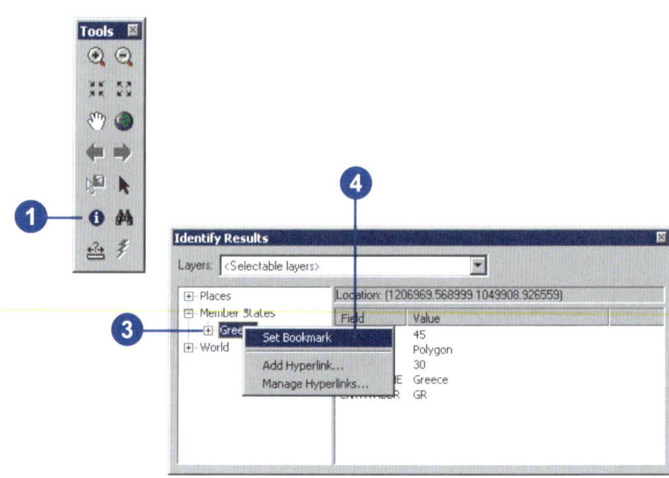

Creating a spatial bookmark from the Find dialog box

1. Click the Find button on the Tools toolbar.
2. Fill in the dialog box to find the features you want.
3. Right-click the Value in the Find Results list.
4. Click Set Bookmark.

   The bookmark is named after the feature.

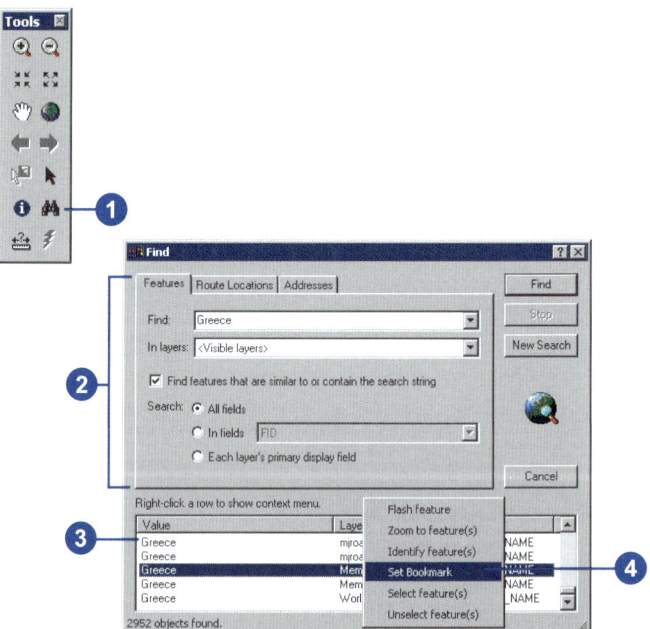

Tip

Removing more than one bookmark at a time

Hold down the Shift key to select more than one bookmark and click Remove.

Removing a spatial bookmark

1. Click the View menu, point to Bookmarks, and click Manage.
2. Click a bookmark.
3. Click Remove.

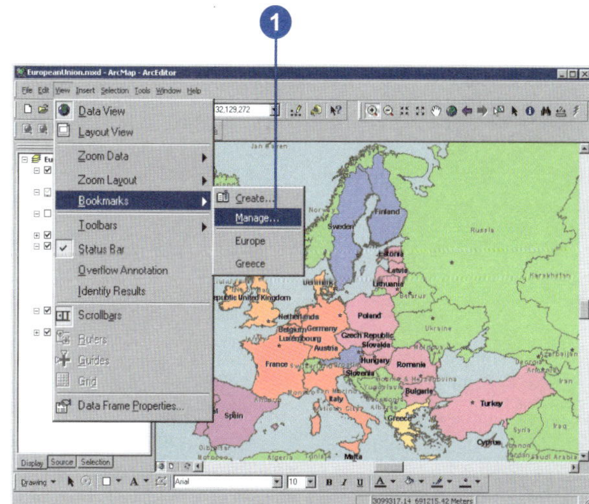

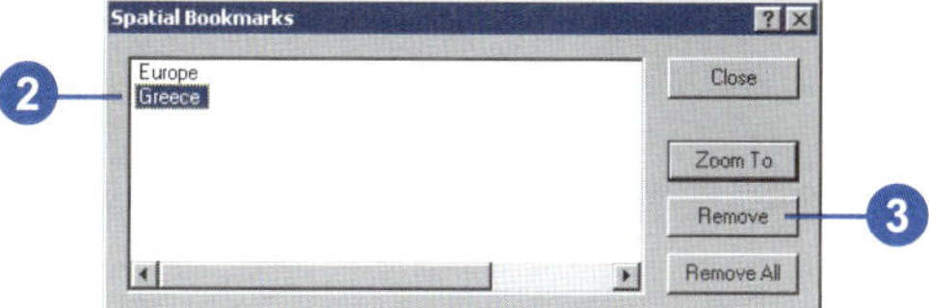

Opening magnifier and overview windows

When you don't want to adjust your map display, yet you want to see things a bit differently—see more detail or get an overview of an area—open another window. ArcMap provides two additional ways to explore the spatial data on your map—overview and magnifier windows.

The *magnifier window* works like a magnifying glass: as you pass the window over the data, you see a magnified view of the location under the window. Moving the window around does not affect the current map display.

The *overview window* shows you the full extent of the data. A small box in the overview window represents the currently displayed area on the map. You can move this box around to pan the map and also shrink or enlarge it to zoom in or out.

Both windows operate only in data view.

Opening a magnifier window and setting the view

1. Click the Window menu and click Magnifier.

   You must be viewing the map in data view to display a magnifier window.

2. When the magnifier window appears, drag it over the data to see a magnified view.
3. Right-click the title bar and click Snapshot to lock the view.

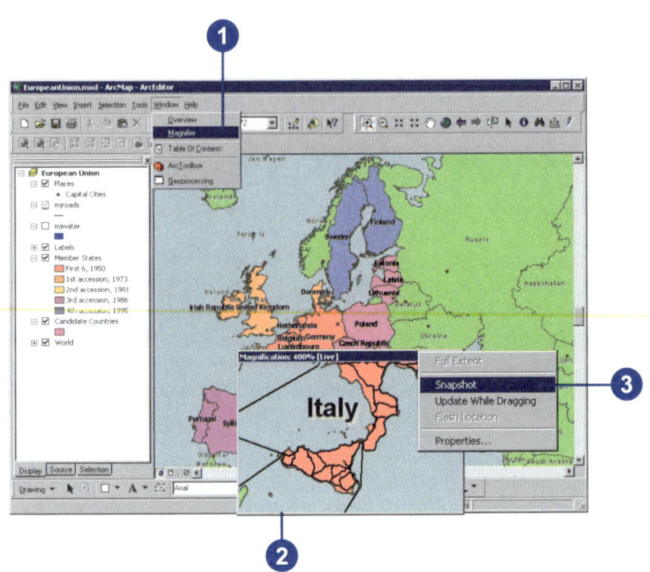

Opening an overview window to pan and zoom the map

1. Click the Window menu and click Overview.

   You must be viewing the map in data view to display an overview window.

2. Drag, shrink, or expand the box in the overview window to change the map display in the active data frame.

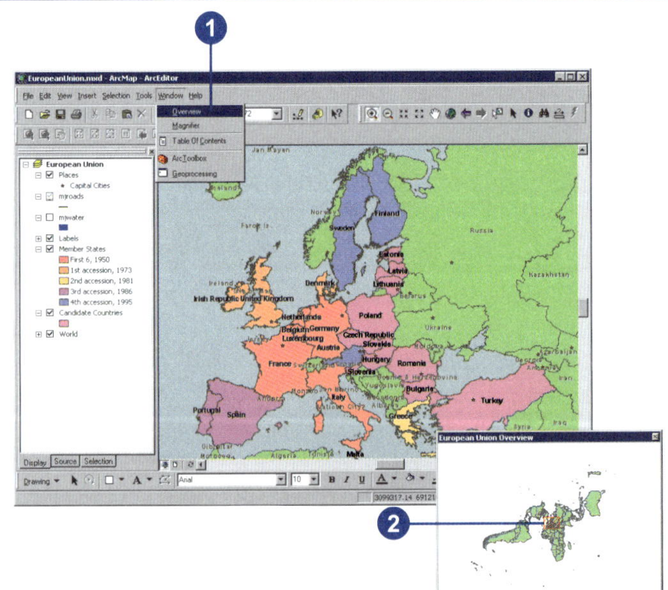

Exploring data on a map

Sometimes just looking at a map isn't enough. You need to query data to solve problems. ArcMap lets you explore the data on the map and get the information you need.

You can point at features to find out what they are, find features that have a particular characteristic or *attribute*, examine all the attributes of a particular layer, and measure distances on the map. *MapTips* also provide a quick way to browse map features. Like ToolTips for toolbar buttons, MapTips pop up as you pause the mouse pointer over a feature.

See Also

For powerful ways to explore your data, see Chapter 13, 'Querying maps'.

Tip

I can't see the MapTips

If you can't see MapTips even after you've enabled them, make sure that the layer is turned on and the features in the layer are not being hidden by features in overlapping layers.

Identifying features by pointing at them

1. Click the Identify button on the Tools toolbar.
2. Click the mouse pointer over the map feature you want to identify.

   The features in all visible layers under the pointer will be identified.

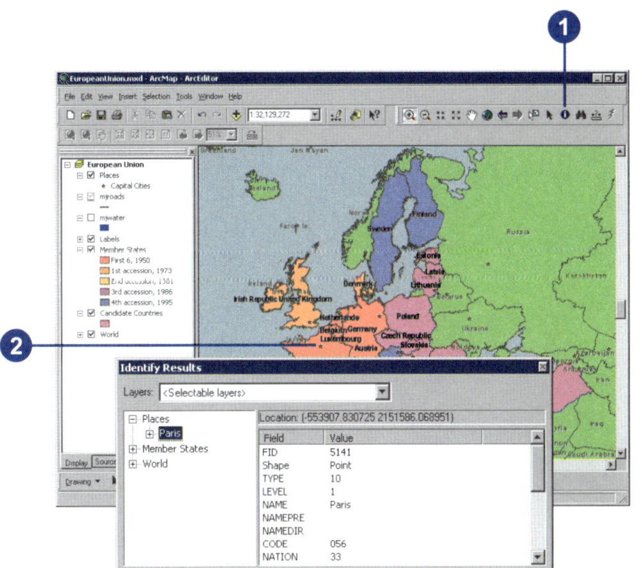

Displaying MapTips

1. In the table of contents, right-click the layer for which you want to display MapTips and click Properties.
2. Click the Display tab and check Show MapTips.
3. Click the Fields tab.
4. Click the Primary Display Field dropdown arrow and click the attribute field you want to display as the MapTip.
5. Click OK.
6. Move the mouse pointer over a feature to see the MapTip.

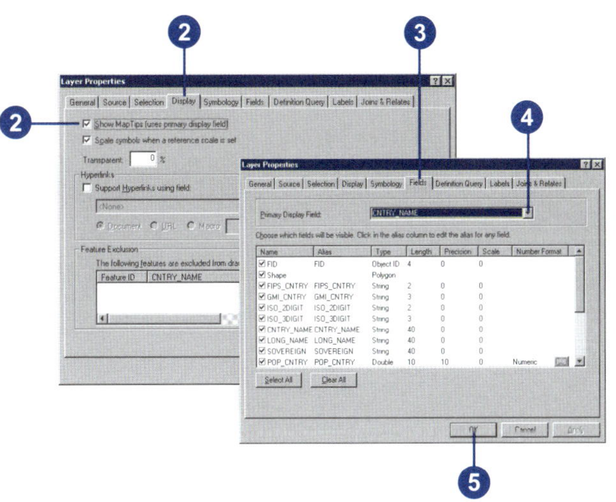

See Also

For more information on working with attribute tables, see Chapter 10, 'Working with tables'.

Viewing a layer's attribute table

1. In the table of contents, right-click the layer for which you want to display the attribute table.
2. Click Open Attribute Table.

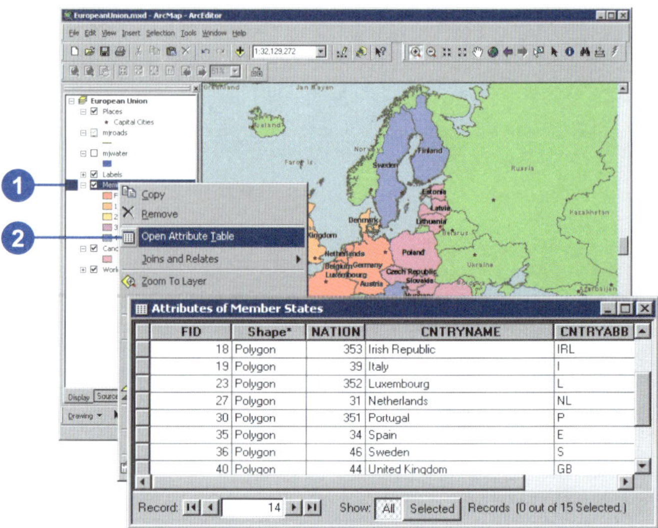

Tip

The primary display field

The primary display field is the field that contains the name or identifying characteristic of the feature. For example, on a map of the world, you might set the primary display field to the field that contains the country names. The primary display field is set through layer properties.

Finding features with particular attributes

1. Click the Find button on the Tools toolbar.
2. Type the string you want to find in the Find text box.
3. Click the In layers dropdown arrow and click the layer you want to search.
4. Uncheck Find features that are similar to or contain the search string if the string must match exactly.
5. Search for the string in all fields, in a specific field, or in the primary display field.
6. Click Find.

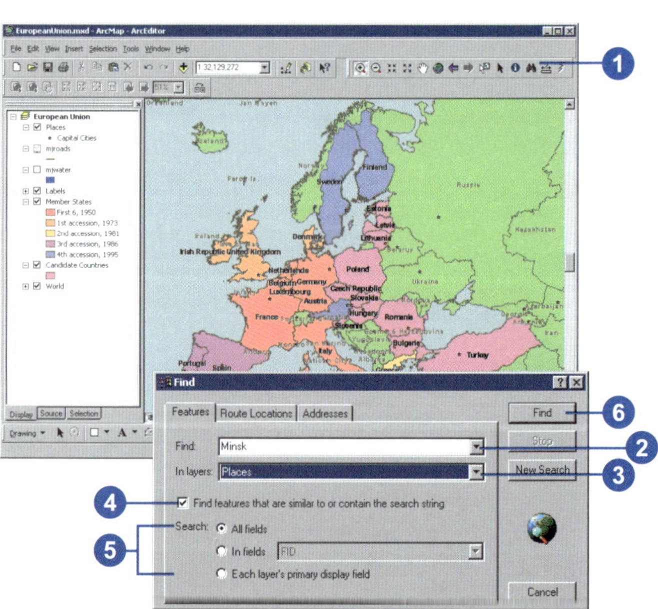

Tip

Do you want to measure distance in kilometers, miles, meters, or feet?

Each data frame can display distance measurements using whichever units you need. Set the distance units on the General tab of the Data Frame Properties dialog box.

Measuring distance

1. Click the Measure button on the Tools toolbar.
2. Use the mouse pointer to draw a line representing the distance you want to measure. The line can have more than one line segment.
3. Double-click to end the line.

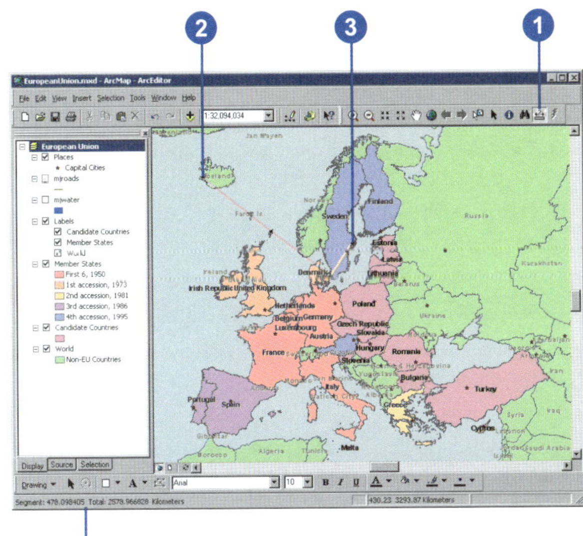

The measurement displays here on the status bar.

Using the map cache to improve geodatabase performance in ArcMap

What is the map cache?

The *map cache* allows you to temporarily store the features in the current map display extent in ArcMap in your local machine's memory. Because retrieving the features from local memory is a fast operation, using the map cache will often result in performance improvements in ArcMap for common tasks completed on features in databases.

For example, if you are working with data in a multiuser geodatabase that serves features over a network, features in the current extent must be retrieved from the source database each time your display is updated. Building a map cache, however, can reduce the load on your network and the geodatabase since ArcMap instead accesses this information from your computer's RAM. Since features are cached on the client, it cuts down the number of queries that the client needs to execute on the server.

ArcMap has tools to build and help you work with the map cache. These tools are found on the Map Cache toolbar.

Using the map cache

Drawing large or complex datasets, labeling, editing, selecting features, retrieving the same features for multiple layers on a map, and drawing features using a definition query are some of the common activities that can often benefit from building a map cache. However, the map cache only stores features in geodatabases, so no data from rasters, coverages, or shapefiles is cached.

Labeling, for example, can be a slow and costly process for the geodatabase, requiring multiple roundtrips to the geodatabase as the label engine attempts to place the maximum number of labels on the map. To learn more about labels, see Chapter 7, 'Working with graphics and text'.

Maximizing the map cache's performance advantages

In an enterprise environment, consistent use of the map cache in ArcMap can significantly improve the overall performance of the system by reducing the number of queries to the geodatabase, the number of features retrieved from the geodatabase, and the overall network traffic.

While the advantages of the map cache are most pronounced when the data source is a multiuser geodatabase, you may also benefit from creating a map cache if you're working with large amounts of data over a network or in a geodatabase.

There may be some performance improvement for personal geodatabase feature classes, although it is generally a small gain. One case where the map cache may be useful with personal geodatabase features is when you are editing features with large numbers of vertices in the current display extent with snapping enabled. Personal geodatabases accessed over a network may also show a performance gain.

A map cache is also most useful when you will be working within a specific area of a map. Work that requires frequent panning and zooming across a large area will not usually benefit from a map cache.

Working with the map cache

If you're working with data stored in a geodatabase, building a map cache can often speed up common ArcMap tasks, such as drawing, selecting, labeling, and editing features.

The map cache holds the features in the current map extent in memory on your local machine. A map cache results in faster processing because ArcMap doesn't have to retrieve data from the server.

The Map Cache toolbar contains the tools you'll need to create and work with the map cache. You can create a map cache by clicking the Build Map Cache button. You can also use the automatic cache (auto-cache) function to automatically update the map cache whenever you move outside of the currently cached extent.

The auto-cache can be useful if you are going to be working in a series of different geographic areas and you don't want to rebuild the cache for each area. It is also convenient when you don't know the exact bounds of the area you want to cache. Since auto-caching may cost performance too much, you should set an auto-cache minimum scale.

Adding the Map Cache toolbar

1. Click View and point to Toolbars.
2. Click Map Cache.

   The Map Cache toolbar appears.

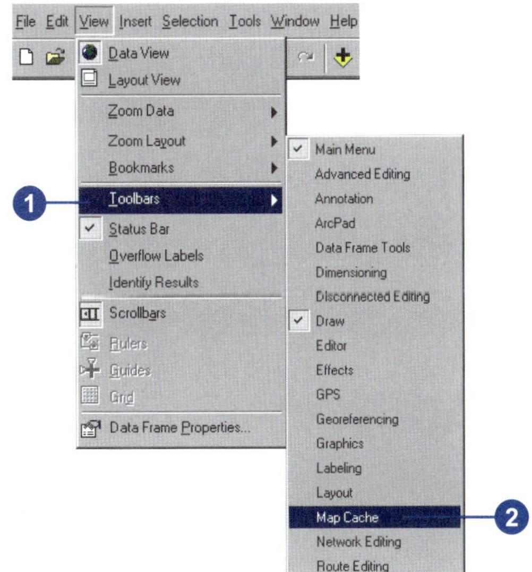

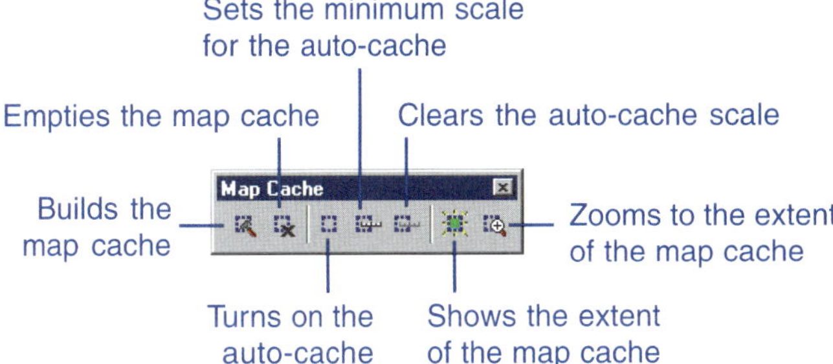

Tip

Canceling the build of a map cache

Because the map cache is stored in your desktop computer's RAM, building a cache for a large area with many features may consume a large amount of memory and can take a while. You can cancel building the map cache by pressing the Esc key.

Tip

Zooming to your map cache extent

You can quickly return to your map cache extent at any time in your ArcMap session. Click the Zoom to Map Cache button on the Map Cache toolbar.

Tip

Using the Show Map Cache button

You can click the Show Map Cache button to find out if you are in the extent of the current cache. If the button turns red, it means you are partially outside the map cache's extent and are no longer using the map cache.

Building a map cache

1. Add data stored in a geodatabase to your map.
2. Pan or zoom to the area on the map that you want to work with.
3. Click the Build Map Cache button on the Map Cache toolbar.

   The features visible in the current extent are held in memory locally.

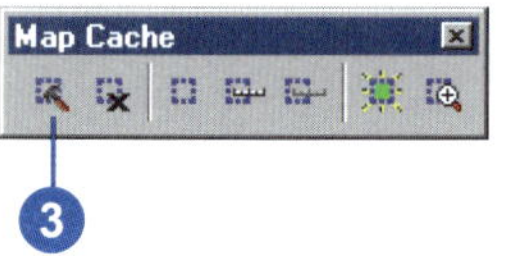

Setting the auto-cache minimum scale

1. Zoom out just beyond the scale at which you'll be working.
2. Click Set Auto-Cache Scale on the Map Cache toolbar.

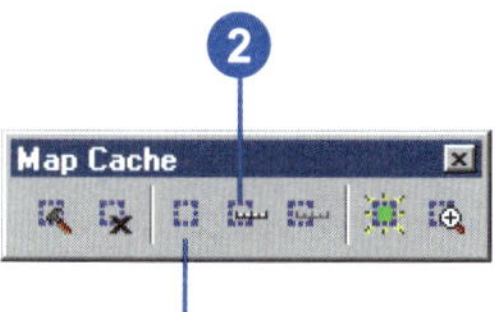

You can click the Auto-Cache button to turn auto-caching on or off.

Tip

Turning off the auto-cache

You might want to turn off the auto-cache when you begin working at a fixed display extent and are no longer moving around the map.

Tip

When is the auto-cache rebuilt?

When you move around the map and are still within the cached extent, the cache is not rebuilt. When the display extent is not completely within the cache, the cache will be rebuilt—provided you don't zoom out beyond the minimum auto-cache scale.

Tip

Setting a minimum auto-cache scale

To avoid inadvertently building an auto-cache of your whole geodatabase, set a minimum scale for the auto-cache. The auto-cache won't be rebuilt if you zoom out beyond the minimum scale.

Setting auto-cache options

1. Click View and click Data Frame Properties.
2. Click the Map Cache tab.
3. To use the auto-cache, check Automatically create map cache.
4. Check the Set minimum scale for auto-cache box.
5. Type a scale in the box or click the Use Current Scale button to set the current scale as the minimum scale.
6. Click OK.

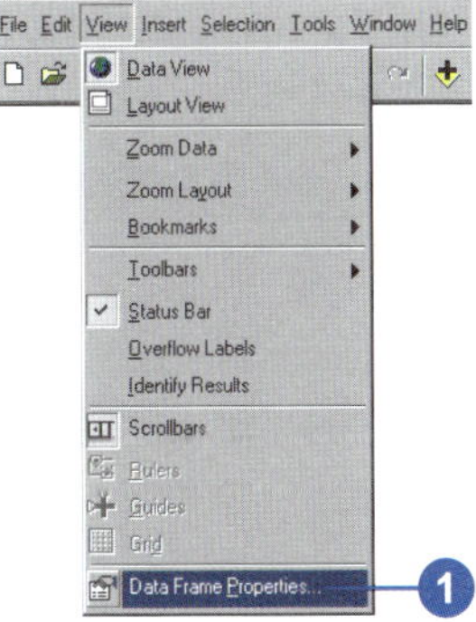

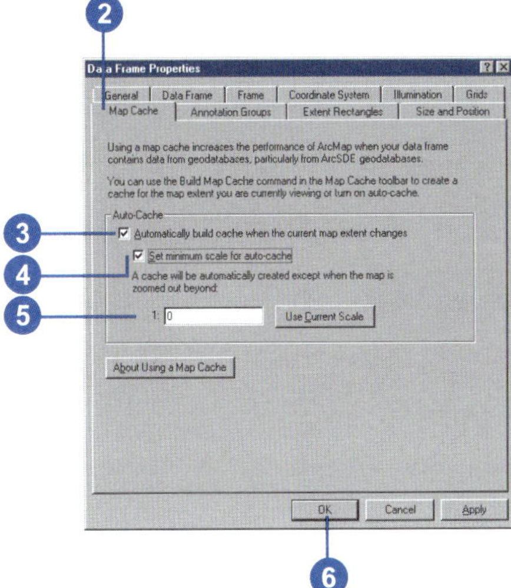

Clearing the auto-cache scale

1. Click the Clear Auto-Cache Scale button on the Map Cache toolbar.

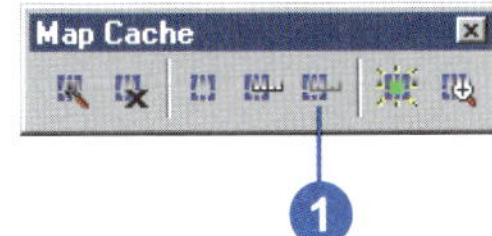

Tip

Why isn't the Show Map Cache button available?

The Show Map Cache button is enabled only when you have built a map cache and your map's current extent intersects the extent of the map cache.

Tip

Working outside the cached extent

If any part of your current display extent is outside the cached area, you are no longer using the data cached on your computer. To use the map cache again, you'll need to build a new cache, use auto-cache, or return to the cached extent.

Tip

Zooming to your map cache extent

You can quickly return to your map cache extent at any time by clicking the Zoom to Map Cache button on the Map Cache toolbar.

Seeing the extent of the cached area

1. Click the Show Map Cache button on the Map Cache toolbar.

   The currently cached area will flash on the map.

   If the Show Map Cache button is red, part of your current display extent is outside the cached area.

2. Optionally, if you plan to work at this extent, click the Build Map Cache button to build a new cache or click the Auto-Cache button to use the auto-cache.

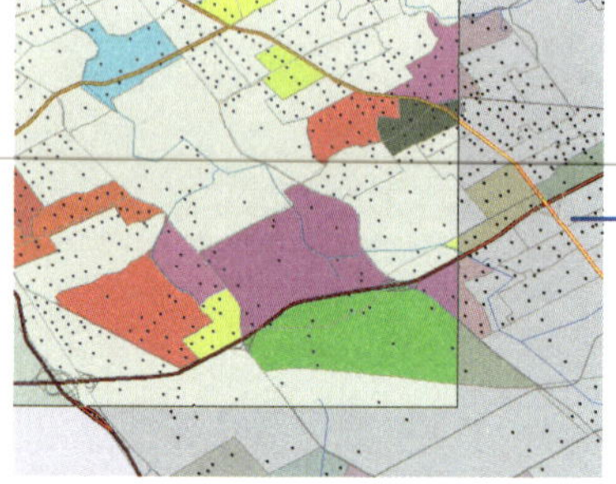

The area outside the rectangle is not within the current cache extent, so the Show Map Cache button is red.

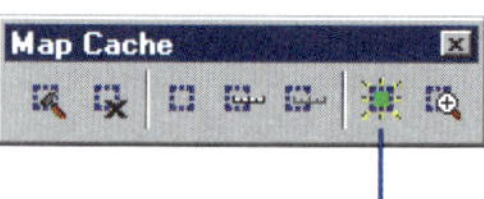

A green Show Map Cache button indicates that you are completely within the cached extent and are using cached data.

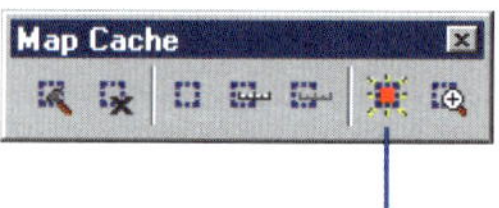

A red Show Map Cache button indicates that you are partially outside the cached extent. You are no longer using cached data.

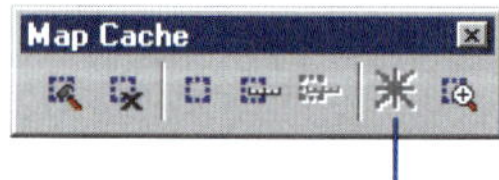

When the Show Map Cache button is unavailable, you are completely outside the cached extent. You are no longer using cached data.

Getting help

A quick way to learn what ArcMap can do is to get help about the buttons and menu commands you see on the interface. After clicking the What's This? button, you can click an item in the window to display a popup description of it.

You can also get help in some dialog boxes. When you click the What's This? button in the upper-right corner and click an item in the dialog box, a description of the item pops up. Sometimes a dialog box will also have a Help button on the bottom; clicking it opens a Help topic with detailed information about the task you're trying to accomplish.

Much of the information in this book is available in the ArcGIS Desktop Help system. The Help topics are organized around the main tasks you want to complete as well as the concepts behind the tasks.

You can look up general Help topics in the Help Contents. You can search the Index for specific tasks and issues. You can also use the Find tab to look up Help topics that have specific words or phrases.

Getting help in the ArcMap window

1. Click the What's This? button.
2. With the Help pointer, click the item in the ArcMap window about which you want more information.
3. Click anywhere on the screen to close the Help description box.

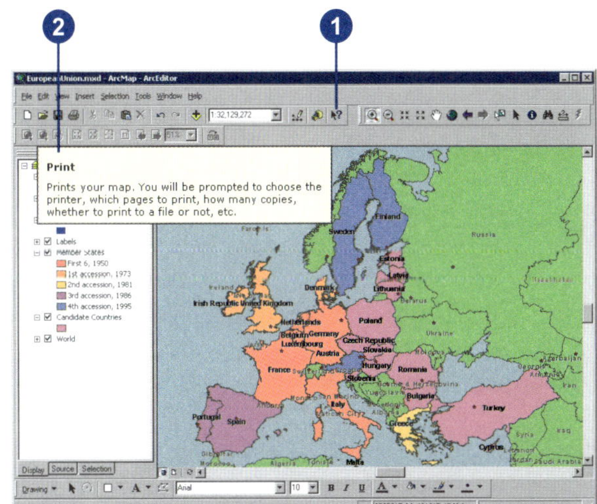

Getting help in a dialog box

1. Click the What's This? button.
2. With the Help pointer, click the item in the dialog box about which you want more information.
3. Click anywhere on the screen to close the Help description box.

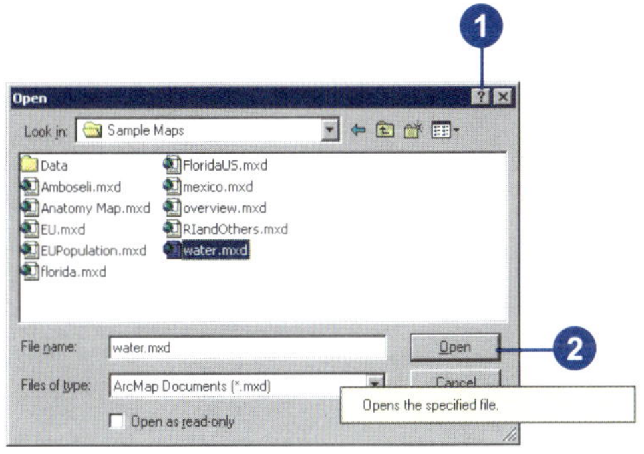

Tip

Tips for buttons and menus

When you pause the mouse pointer over a button, the button's name appears in a small box called a ToolTip. When you position the mouse pointer over a button or menu command, a description of what it does appears in the status bar.

Tip

Another way to get help in a dialog box

Sometimes a dialog box will also have a Help button on the bottom; clicking it opens a Help topic with detailed information about the task you're trying to accomplish.

Using the Help Contents to get help

1. Click the Help menu and click ArcGIS Desktop Help.
2. Click the Contents tab.
3. Double-click a book to see a list of the topics in that category.

   Double-clicking an open book closes its list.
4. Click the topic you want to read.

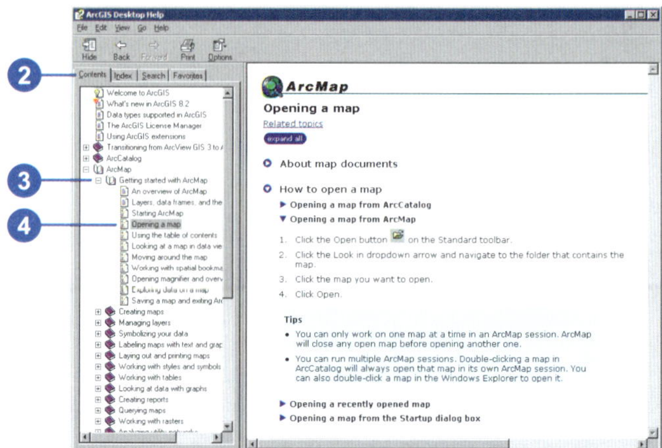

Searching the Index for help

1. Click the Help menu and click ArcGIS Desktop Help.
2. Click the Index tab.
3. Type the subject about which you want information.
4. Double-click the topic you want to read.

   If several topics are related to your selection, the Topics Found dialog box appears. Simply double-click the topic you want to read.

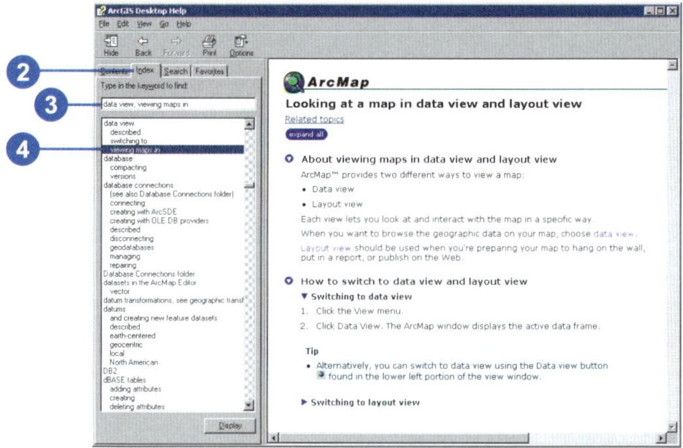

Finding Help topics containing specific words

1. Click the Help menu and click ArcGIS Desktop Help.
2. Click the Search tab.
3. Type the word that should be contained in the topics you want to find.
4. Click List Topics.
5. Double-click the topic you want to read.

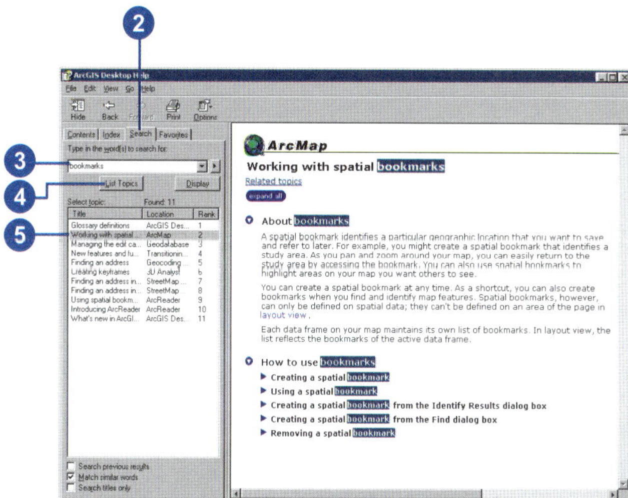

Saving a map and exiting ArcMap

After you finish working on a map, you can save it and exit ArcMap. You save a map as a document and store it on your hard disk. If you haven't saved the map before, you'll need to name it, preferably with a name that adequately describes its contents. ArcMap automatically appends a file extension (.mxd) to your map document name.

The data displayed on a map is not saved with it. Map layers reference the data sources in your GIS database. This helps to keep map documents relatively small in size. So if you plan to distribute your map to others, they'll need access to both the map document and the data your map references.

In general, it's a good idea to save your map periodically while editing it just in case something unexpected happens.

Tip

Opening a map will close the current one

In ArcMap, you work with one map at a time. If you need to work with more than one, start another ArcMap session.

Saving a map

1. Click the Save button on the Standard toolbar.

   If you haven't saved the map before, you'll need to provide a name for it.

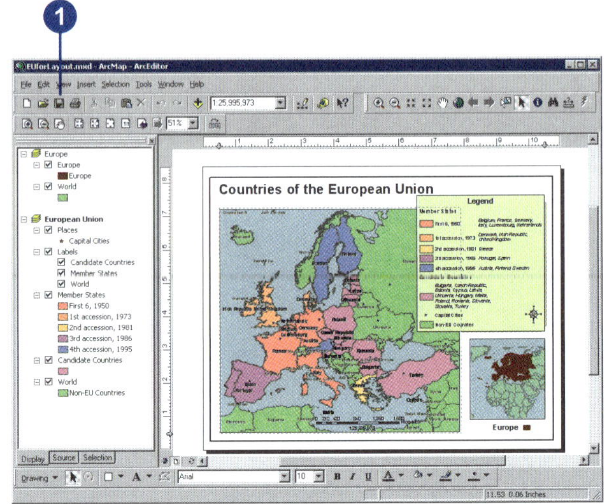

Saving a map as a new map

1. Click the File menu and click Save As.
2. Navigate to the location to save the map document.
3. Type a filename.
4. Click the Save as type dropdown arrow and click ArcMap Documents.
5. Click Save.

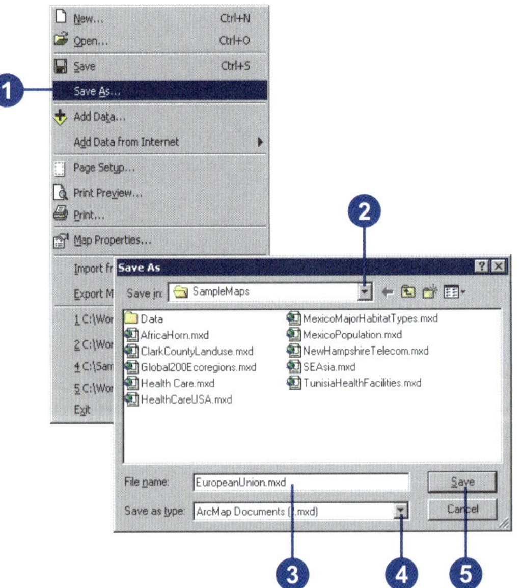

Tip

Differentiating a map template from a map document

Map templates have a .mxt file extension. Map documents have a .mxd file extension.

Tip

Saving a map with thumbnail images

Thumbnails are small images that provide an overview of the geographic data in your map. Thumbnails appear in ArcCatalog and when you add data in Thumbnails view in ArcMap. To save a thumbnail for your map, in ArcMap click File and click Map Properties. Check the Save thumbnail image with map box, then save the map.

See Also

See Using ArcCatalog *for more information on thumbnails.*

See Also

For more information on creating maps, see Chapter 4, 'Displaying data in maps'.

Saving a map as a map template

1. Click the File menu and click Save As.
2. Navigate to the location to save the map template.
3. Type a filename.
4. Click the Save as type dropdown arrow and click ArcMap Template.
5. Click Save.

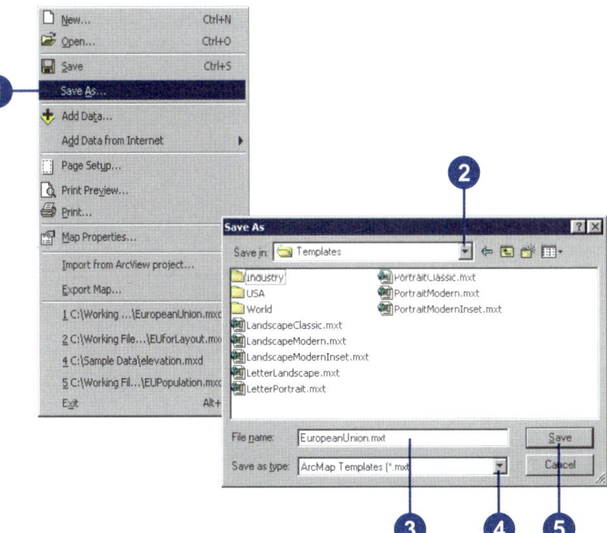

Exiting ArcMap

1. Click the File menu and click Exit.
2. Click Yes to save any changes, No to discard any changes, or Cancel to continue working on your map.

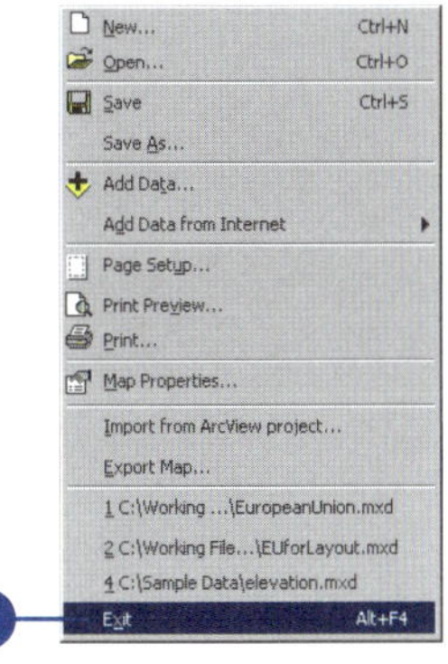

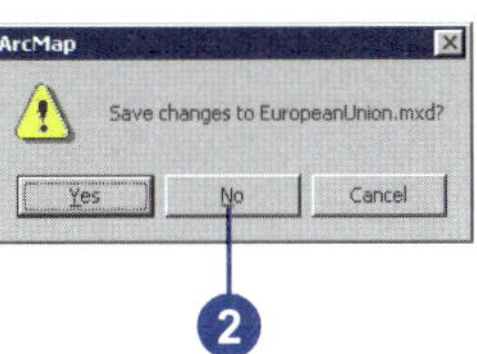

Keyboard shortcuts in ArcMap

Accessing ArcMap menu commands

Menu	Command	Shortcut
File	New	Ctrl + N
File	Open	Ctrl + O
File	Save	Ctrl + S
File	Exit	Alt + F4
Edit	Undo	Ctrl + Z
Edit	Redo	Ctrl + Y
Edit	Cut	Ctrl + X
Edit	Copy	Ctrl + C
Edit	Paste	Ctrl + V
Edit	Delete	Delete
Help	ArcGIS Desktop Help	F1
Help	What's This?	Shift + F1

To access the Main menu, press **Alt** and use the arrow keys to move through the menus; press **Enter** to make a selection.

Use **Esc** to close a menu or dialog box.

Docking and undocking

Hold down **Ctrl** while dragging a toolbar or dockable window to prevent it from docking.

Dragging and dropping

You can drag and drop or copy and paste multiple layers in the table of contents and between ArcMap sessions. You can also drag and drop or copy and paste data frames between ArcMap sessions.

Layers that are dragged and dropped between data frames and ArcMap sessions are copied; hold down **Ctrl** while dragging and dropping to move layers between data frames and ArcMap sessions.

Data frames that are dragged and dropped are moved; hold down **Ctrl** while dragging and dropping to copy them.

Layers that are dragged and dropped inside a data frame are moved; hold down **Ctrl** while dragging and dropping to copy them.

Use drag and drop to move layers in and out of a group layer within a data frame.

Navigating the table of contents with the keyboard

F3 or clicking inside the table of contents puts the keyboard focus on the table of contents so you can navigate and interact with it.

Esc or clicking the map puts the keyboard focus on the map.

Home selects the first item in the table of contents.

End selects the last item in the table of contents.

Up/Down arrows move through the items in the table of contents.

Left/Right arrows or the **+** and **-** keys expand or contract selected items. They also switch between the tabs at the bottom of the table of contents when they have keyboard focus.

Spacebar turns drawing of the selected layer(s) on or off.

F2 renames the selected item.

F12 or **Enter** opens the selected item's property dialog box.

Shift + F10 (or the Application key) opens the context menu for the selected item.

Use **Shift + F1** or **F1** to obtain context help when an item has keyboard focus or when the properties dialog box tab or a table of contents tab is selected.

F11 activates a selected data frame, or hold down **Alt** and click a data frame to activate it.

Selecting items in the table of contents

Ctrl + click selects or deselects multiple layers or data frames.

Shift + click selects all layers or data frames between two layers or data frames within the same table of contents level.

Using mouse shortcuts in the table of contents

Ctrl + click on an expansion control (+/-) to expand or contract all the items at that level. If any items are currently selected, only the selected items are expanded or collapsed.

Ctrl + click on a check box to turn all the layers on or off at that level. If any items are currently selected, only the selected items are turned on or off.

When dragging layers, hovering over an expansion control with the drop cursor expands or collapses any item.

Right-clicking features, layers, and data frames always opens a context menu.

Displaying data

Section 2

Displaying data in maps

4

IN THIS CHAPTER

- **Creating a new map**
- **Adding layers**
- **Adding coverages, shapefiles, and geodatabases**
- **Adding data from the Internet**
- **Adding data from a GIS server**
- **Adding TINs as surfaces**
- **Adding CAD drawings**
- **Adding x,y coordinate data**
- **Adding route events**
- **Creating and adding a new feature class**
- **About coordinate systems**
- **Specifying a coordinate system**
- **Referencing data on a map**

Before you sit down to create a map, you need to think about its purpose. What do you want your map to show? Will the map be displayed by itself, or will it be part of a larger presentation? Who is the audience for the map? Answering these and other similar questions will help you determine how to organize and present the information on your map—for instance, what level of detail you need to show; what colors and symbols you should use to draw features; and whether you need to create an interactive map people use at the computer, one you simply print out and display on a wall, or both.

The first step to creating a map is to locate the data you want to put on it. Finding data may be as simple as using ArcCatalog to browse your organization's GIS database or the spatial data distributed with ArcMap. The Internet is also an excellent resource for finding data—you can add data directly from the Internet using the Geography Network℠ Web site at *www.geographynetwork.com*. Many government agencies distribute data to the public at minimal or no cost. Commercial data vendors also package data for a wide variety of applications from business to natural resources. If you have special data requirements, you might create your own data (see *Editing in ArcMap*) or contact one of the many service bureaus or GIS consulting companies that can produce data for you. Even if you don't think you have any spatial data, you just might.

Creating a new map

No matter what kind of map you want to make, you begin the same way—by creating a new map document. You can either create an empty map with nothing on it or use a *map template* as a starting point. Map templates typically contain a predefined page layout that arranges map elements such as North arrows, scalebars, and logos on the virtual page. This means you can just add your data and immediately print the map. Templates can also contain data (as layers), special symbols and styles, custom toolbars, and macros such as VBA forms and modules.

ArcMap comes with many predefined templates to choose from when making your maps. Also, any map you make can be saved as a template. Templates provide an ideal method for defining the standard maps your organization needs.

Creating a new map from the Startup dialog box

1. Start ArcMap.
2. Click to create a new empty map, create a map from a template, or browse for an existing map.
3. Click OK.

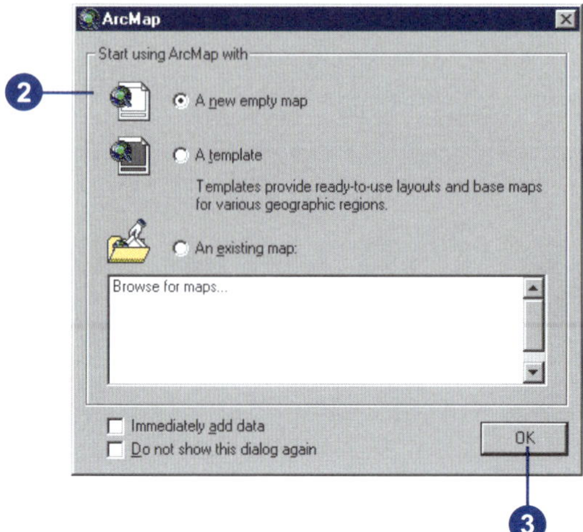

Creating a new empty map

1. Click the New button on the Standard toolbar to create a new empty map.

   If you have a map open already, you'll be prompted to save your changes.

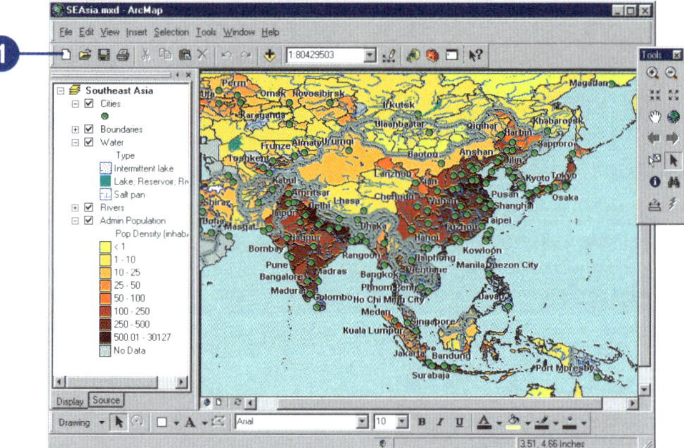

Tip

Organizing templates

You can create your own templates and organize them into folders on your computer. These folders appear as tabs on the New dialog box (lower right). Create folders in the \Bin\Templates folder where you've installed ArcGIS.

Using a map template

1. Click the File menu and click New.
2. Click the tab that corresponds to the type of map you want to make.

   The tabs you see will depend on how you've organized custom templates.
3. Click the template you want.

   Some of the templates included with ArcMap contain data. You can add your data right on top.
4. Click Document to create a new map document.
5. Click OK.

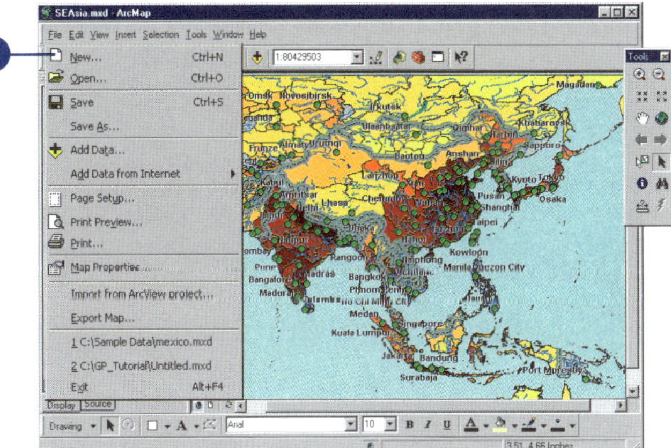

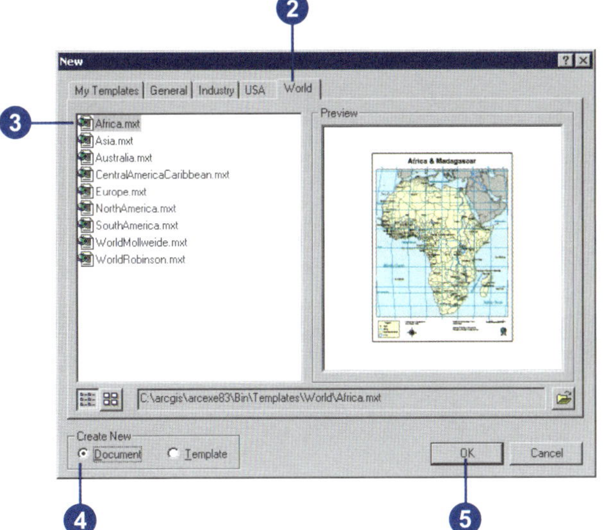

Adding layers

Geographic data is represented on a map as a layer. A layer might represent a particular type of feature, such as highways, lakes, or wildlife habitats, or it might represent a particular type of data, such as a satellite image, a *computer-aided design (CAD) drawing*, or a terrain elevation surface in a *Triangulated irregular network (TIN)*.

You don't need to know much about data to add a layer to a map. Simply drag one from ArcCatalog—or copy and paste one from another map—onto the map you're working on. The layer will draw as it was previously symbolized.

A layer doesn't store geographic data itself; instead, it references the data stored in coverages, shapefiles, rasters, and so on. Thus a layer always reflects the most up-to-date information in your database. If you don't have a layer, you can easily create one as described on the following pages. For example, you might create several layers that highlight different aspects of your data and distribute them to others in your organization.

Adding a layer from ArcCatalog

1. Start ArcCatalog from the Start menu.
2. Arrange the ArcCatalog and ArcMap windows so you can see both on the screen.
3. Navigate to the layer you want to add to the map.
4. Click and drag the layer from ArcCatalog.
5. Drop the layer over the map display in ArcMap.

   The layer is copied to the map. Any subsequent edits made to the layer on disk will not be reflected on this map.

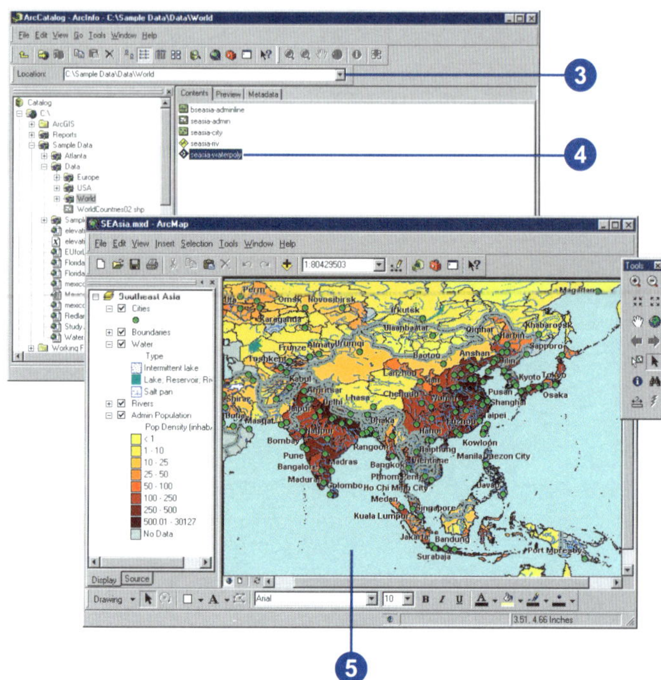

Adding a layer from the Add Data button

1. Click the Add Data button on the Standard toolbar.
2. Click the Look in dropdown arrow and navigate to the folder that contains the layer.
3. Click the layer.
4. Click Add.

   The new layer appears on your map.

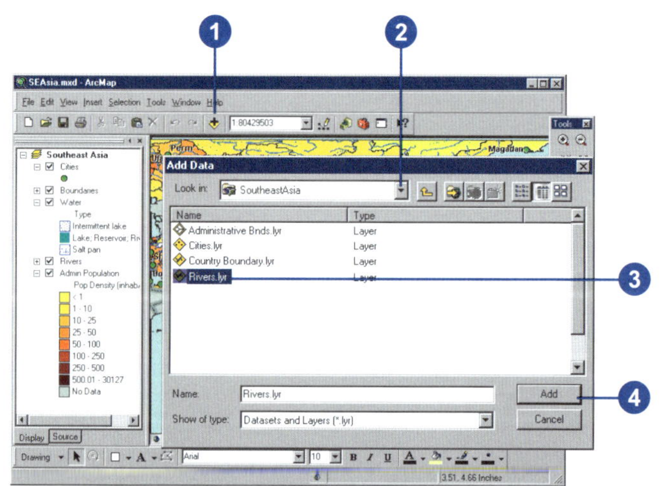

Tip

To see a layer on a map, you must have access to its data source

Even though you have access to a layer on disk, it won't draw on your map unless you also have access to the data source the layer is based on.

Tip

Adding a layer from the Catalog

When you add a predefined layer to your map from the Catalog, a copy is placed on the current map. Your current map will not change if the original layer is modified.

Adding a layer from another map

1. Open the map that contains the layer you want to copy.
2. In the table of contents, right-click the layer and click Save As Layer File.
3. Click the Look in dropdown arrow and navigate to the folder where you want to save the layer.
4. Type a name for the layer.
5. Click Save.
6. Click the Open button on the Standard toolbar to open the map you want to add the layer to.
7. Click the Add Data button.
8. Click the Look in dropdown arrow and navigate to the folder that contains the layer.
9. Click the layer.
10. Click Add.

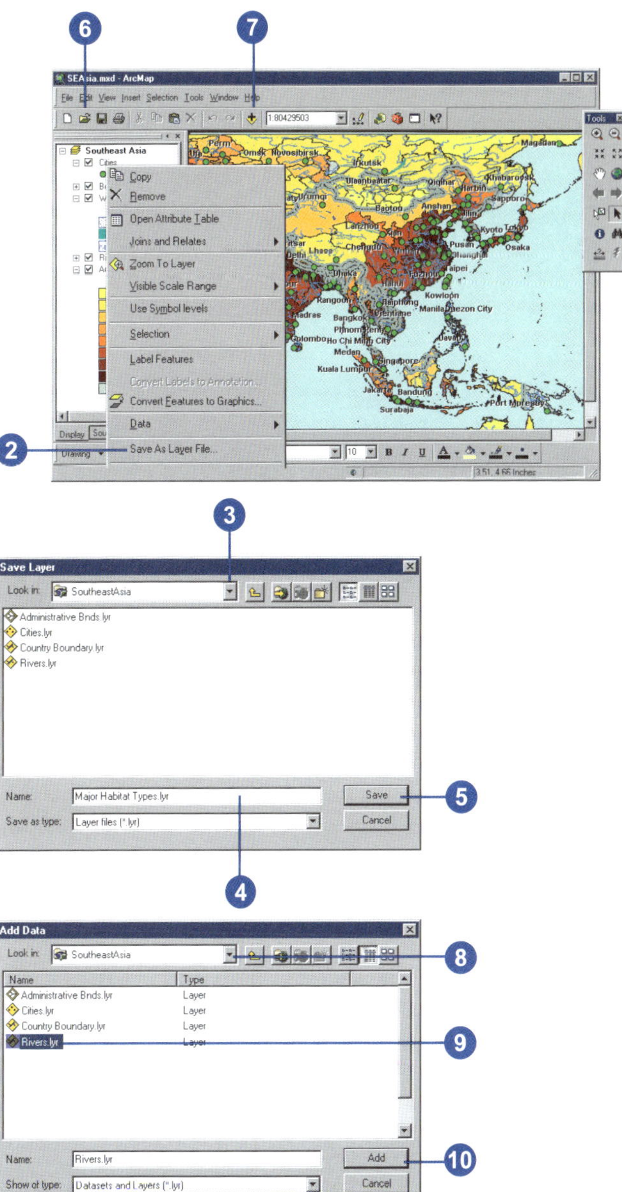

Adding coverages, shapefiles, and geodatabases

When you don't have a predefined layer at your disposal, you can create one directly from a data source such as a *shapefile*. To create a layer, add the data source to your map; ArcMap creates a new layer that references the data source.

Once a layer is part of a map, you can decide whether or not to display it, the scale at which it should be visible, what features or subset of features to display, and how to draw those features. You can also join other tabular information you have about features to the layer and group layers so they appear as one layer on the map.

The data you display on a map comes in a variety of types—such as raster, vector, and tabular—and can be stored in different formats. If your data is stored in a format supported by ArcMap, you can add it directly to your map as a layer. If your data isn't in a supported format, you can use the data conversion utilities in the ArcToolbox™ or other third party data conversion products to convert practically any data you have and display it on a map.

Adding data from ArcCatalog

1. Start ArcCatalog from the Start menu.
2. Arrange the ArcCatalog and ArcMap windows so that you can see both on the screen.
3. Navigate to the data source you want to add to the map.
4. Click and drag the data source from ArcCatalog.
5. Drop the data source over the map display in ArcMap.

   ArcMap creates a new layer on the map that references the data source.

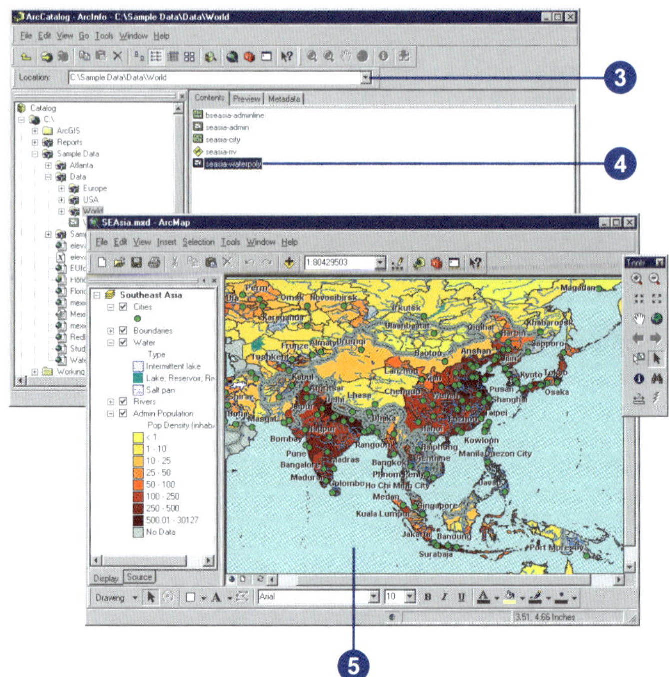

Adding data in ArcMap

1. Click the Add Data button on the Standard toolbar.
2. Click the Look in dropdown arrow and navigate to the folder that contains the data source.
3. Click the data source.
4. Click Add.

   ArcMap creates a new layer on the map that references the data source.

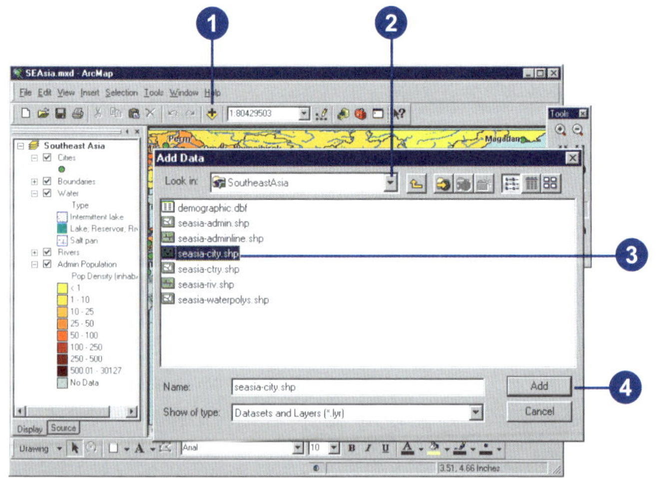

Tip

Creating a layer in ArcCatalog

In addition to creating a layer on the fly in ArcMap, you can create one in ArcCatalog.

Tip

More than one layer can reference the same data source

When you create a layer, you specify a data source that the layer references.

See Also

For more information on how to draw a layer once you've added it to a map, see Chapter 6, 'Symbolizing features'.

See Also

For more information on the syntax for building a Definition Query expression, see Chapter 13, 'Querying maps'.

See Also

For more information on adding raster data to your map, see Chapter 9, 'Working with rasters'.

Displaying a subset of the features in a layer that meet some criteria

1. In the table of contents, right-click the layer and click Properties.
2. Click the Definition Query tab.
3. Type an expression or click Query Builder.

   The Query Builder lets you create an expression to identify the particular features in the layer you want to display. For example, you might choose to display only those cities with a population greater than 1,000,000.
4. Click OK.

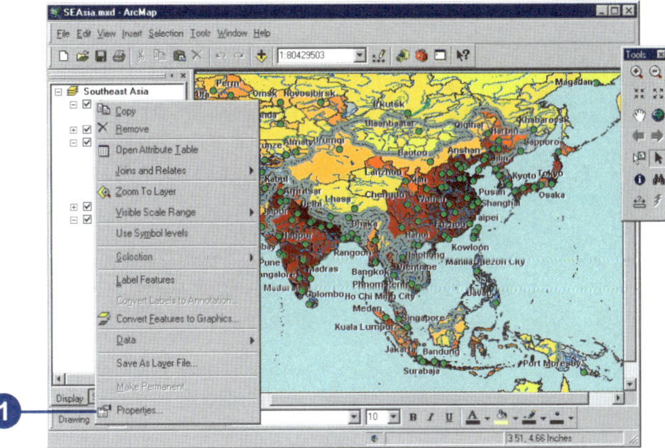

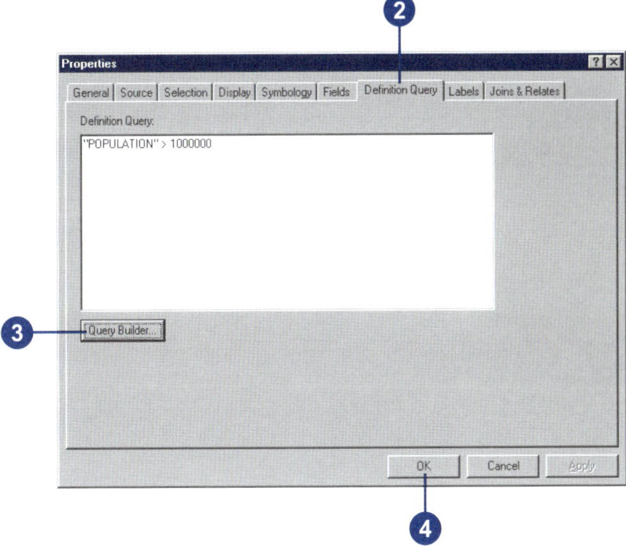

Adding data from the Internet

The Internet is a vast resource for geographic data. The Add Data from Internet option in ArcMap allows you to go directly to any Web site you choose, using a browser to explore Metadata Explorer sites. Generally, the data you add to your maps is accessed directly from the organization providing the data.

An example of a widely used source is the Geography Network (*www.geographynetwork.com*). The Geography Network is a global community of data providers committed to making geographic content available to the public. Published from sites around the world, it gives you immediate access to the latest maps, data, and related services over the Internet. Use the Geography Network to search for and explore maps and other geographic content. When you find what you want, add it directly to your maps in ArcMap.

In some cases there will be a lock symbol on the Add to ArcMap button. This indicates it is a service with which you need to be registered in order to use the data.

Adding data from an Internet site

1. Click the File menu, point to Add Data from Internet, and click Add Website.
2. Type the URL address of the Web site you wish to access.
3. Click OK.

   An Internet browser opens giving you access to the location you entered. From here you can search the site for the data you want to add to your map.

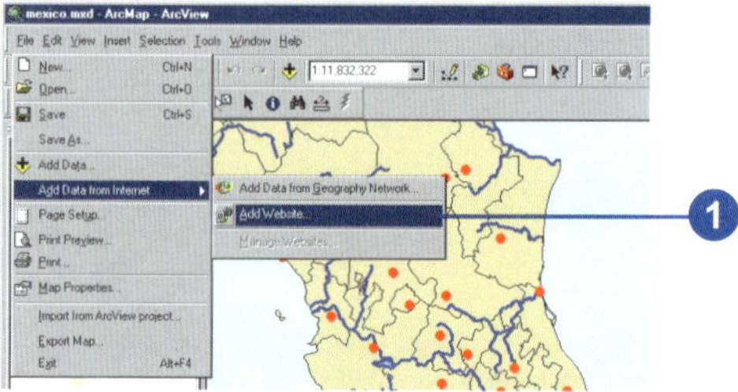

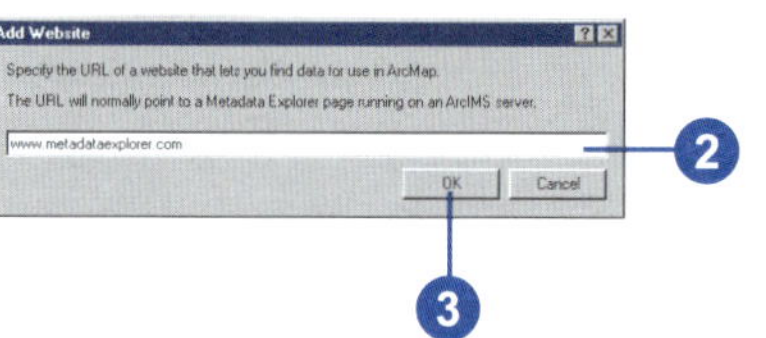

Adding data through the Geography Network

1. Click the File menu, point to Add Data from Internet, and click Add Data from Geography Network.
2. Browse the Geography Network to find the data you want.

   The Geography Network allows you to search, for example, by data provider, data type, and geographic extent. Once you find data, you can read a description of it, view it over the Internet, and add it to your map in ArcMap.

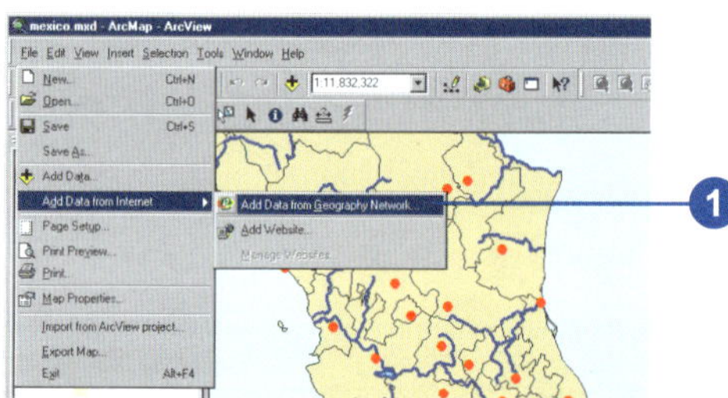

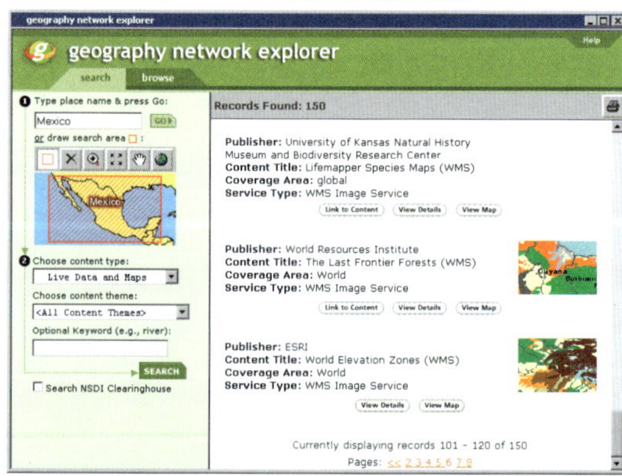

Adding data from a GIS server

You can use data available on the Internet or over a network directly in your maps. Any data being served over a network or the Internet, using ArcIMS® or ArcGIS Server, can be added to ArcMap as a layer. The ability to work with Internet data is built into ArcMap and doesn't require any additional software. To learn more about connecting to a GIS server, see *Using ArcCatalog*. Once you are connected to a GIS server, you can browse through its available data. You can add the data to ArcMap as a layer—by either dragging and dropping from ArcCatalog or by using Add Data in ArcMap.

When you work with data being served from a GIS server, you don't download the data to your computer; you work with a live service over the Internet or network. When you draw a layer based on a GIS service, ArcMap automatically retrieves the data for the service over the Internet or network. This saves you from having to store and manage the data yourself, but it also means that the layer will become unavailable if you go offline (unless you export the data locally).

The ArcIMS server layers

ArcIMS provides three types of map services that are opened as layers in ArcMap: an Image Service, an ArcMap Image Service, and a Feature Map Service. The layers provided from an Image Service and an ArcMap Image Service are essentially snapshots of a map on a server, which are delivered to you as an image. These snapshots are sent as a compressed JPEG, PNG, or GIF files. The Image Service layer is created within ArcIMS. In the case of the ArcMap Image Service layer, there is an ArcMap component embedded with ArcIMS that allows ArcIMS to serve up maps created from the ArcMap documents (.mxd or .pmf), allowing you to retrieve maps that use the advanced cartographic and open data access capabilities of ArcMap. A Feature Map Service layer receives actual vector features, providing additional functionality because the geographic features are directly accessible by ArcMap.

The ArcGIS Server layer

An ArcGIS Server provides a Map Service layer. This layer is associated with a single data frame in an actual map document (.mxd or .pmf). Like the ArcIMS Image and ArcMap Image Service layers, the Map Service layer is received as an image representation of the existing ArcMap map document.

Using data from a GIS server in ArcMap

GIS data served using a GIS server looks like any other layer on your map. There are a few differences depending on how the host decided to serve the data. When you add either Image Service layer to ArcMap, you'll see a new group layer containing feature class layers in your table of contents. This layer has been set up and saved by its creator using the symbology and organization he or she determined to be most useful; however, you can customize its appearance by turning on or off the individual feature class layers or by renaming them. You can also use the Identify tool with these layers.

A Feature Map Service layer is more like the regular layers you may have worked with in ArcMap. It allows you to do a little more customization than the Image Service layers, including changing the drawing order, spatial selection, and symbology.

Any of the services can exist as a background to your own data.

To learn more about ArcGIS server connections, see the ArcGIS Desktop Help.

To learn more about ArcIMS or ArcGIS Server, check the products page on *www.esri.com* or *support.esri.com*.

Tip

Connecting to a GIS server

If the GIS server you need is not listed, you can use either Add ArcGIS Server or Add ArcIMS Server to connect to a new server.

Tip

Why is there a red "x" on my GIS server?

The red "x" implies that the connection needs to be refreshed. You can do this by double-clicking the connection.

See Also

For more information on connecting to a GIS server, see Using ArcCatalog.

Adding data from a GIS server

1. Click the Add Data button on the Standard toolbar.
2. Click the Look in dropdown arrow and navigate to the GIS Servers folder.
3. Double-click the server with data you want to access.

   If you don't see the server you want, double-click Add ArcGIS Server or Add ArcIMS Server. For more information on connecting to a GIS server, see *Using ArcCatalog*.
4. Click the Details button so you can see the details of the types of map servers or server objects to which you can connect.
5. Click the name to choose the one you want to use.
6. Click Add.

   ArcMap creates a new layer on the map that references the data source.

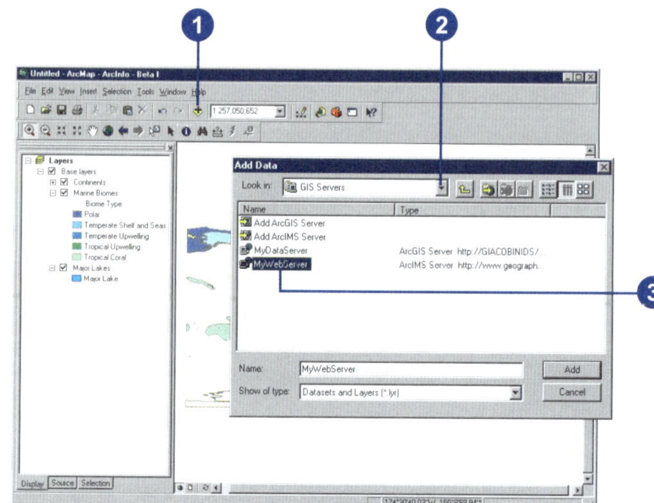

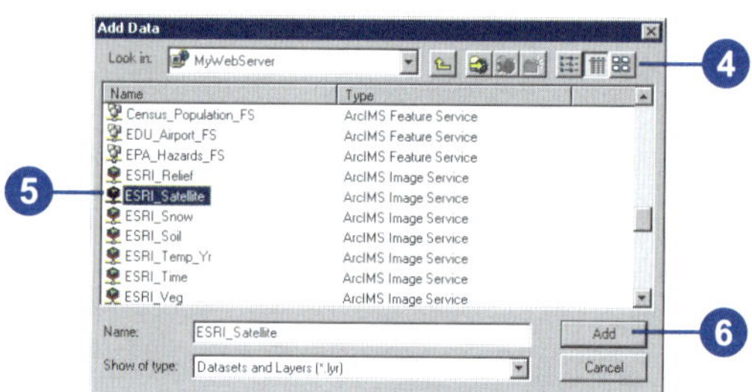

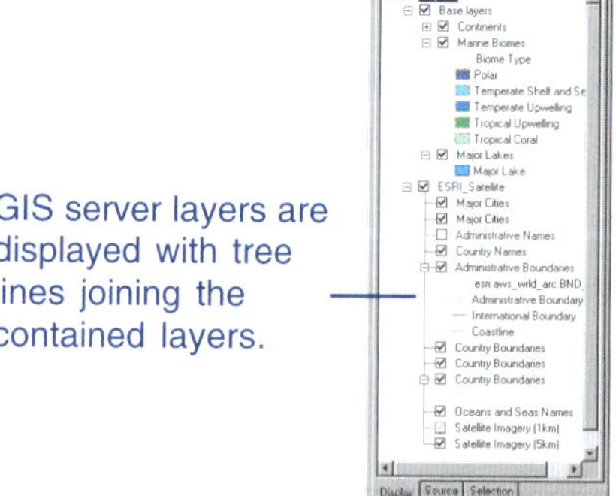

GIS server layers are displayed with tree lines joining the contained layers.

Adding TINs as surfaces

Data that varies continuously across an area—elevation, rainfall, and temperature—is often represented on a map as a *surface*. Surface data comes from a variety of sources and in many formats. Aerial photographs, radar, sonar, and similar sources generate information used to build surfaces. This data is processed into formats, such as SDTS raster profiles, digital elevation models (DEMs), DTED grids, vector coverages, and raw text files, all of which you can convert into TINs that display as surfaces on your map.

A TIN is built from a series of irregularly spaced points with values that describe the surface at that point (for example, an elevation). From these points, a network of linked triangles is constructed. Adjacent triangles, sharing two nodes and an edge, connect to form the surface.

A height can be calculated for any point on the surface by interpolating a value from the nodes of nearby triangles. In addition, each triangle face has a specific *slope* and *aspect*. You can display any one of these surface characteristics—slope, aspect, and elevation—or the internal structure of the TIN.

Adding TIN data from ArcCatalog

1. Start ArcCatalog from the Start menu.
2. Arrange the ArcCatalog and ArcMap windows so you can see both on the screen.
3. Navigate to the TIN data source you want to add to the map.
4. Click and drag the TIN data from ArcCatalog.
5. Drop the TIN data over the map display in ArcMap.

   ArcMap creates a new layer on the map that references the TIN data source.

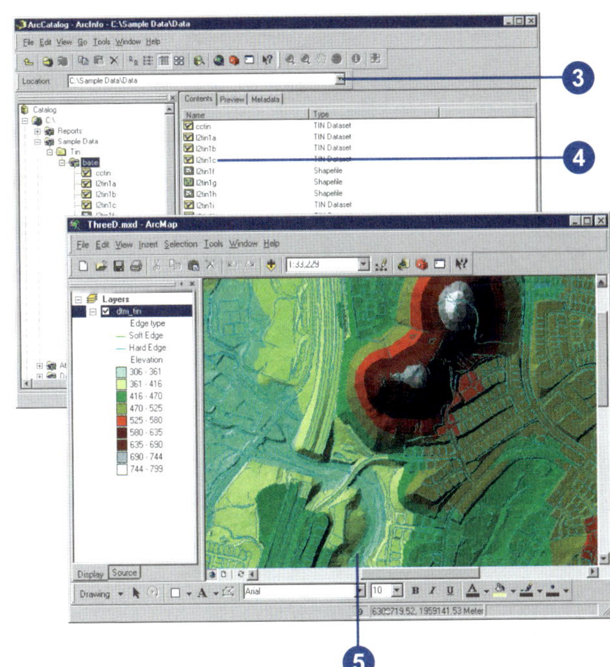

Adding TIN data in ArcMap

1. Click the Add Data button on the Standard toolbar.
2. Click the Look in dropdown arrow and navigate to the folder that contains the TIN data source.
3. Click the desired TIN.
4. Click Add.

   ArcMap creates a new layer on the map that references the TIN data source.

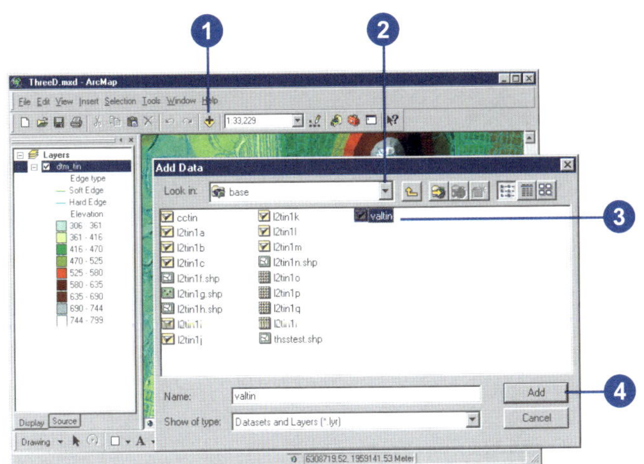

Adding CAD drawings

If your organization has existing CAD drawing files, you can use these immediately on your maps. You don't need to convert the data, but you do need to decide how you plan to use the data.

If you simply want to see the CAD drawing with your other data, you can add the CAD drawing as a layer for display only. The entities will draw as defined in the CAD drawing file. Alternatively, if you want to control how entities draw on the map or perform geographic analysis, you need to add the CAD data as features ArcMap can work with—specifically, point, line, or polygon features. When you browse for a CAD drawing to add to your map, you'll see two representations of the data: a CAD drawing file and a CAD dataset. Use the drawing file for display only and the dataset for display and geographic analysis.

CAD drawing files typically store different types of entities on different layers in a drawing file. One layer might contain building footprints, another streets, a third well locations, and a fourth textual annotation. CAD drawing files, however, ▸

Adding a CAD drawing from ArcCatalog

1. Start ArcCatalog from the Start menu.
2. Arrange the ArcCatalog and ArcMap windows so you can see both on the screen.
3. Navigate to the CAD drawing you want to add to the map.
4. Click and drag the CAD drawing from ArcCatalog.
5. Drop the CAD drawing over the map display in ArcMap.

   ArcMap creates a new layer on the map that references the CAD drawing.

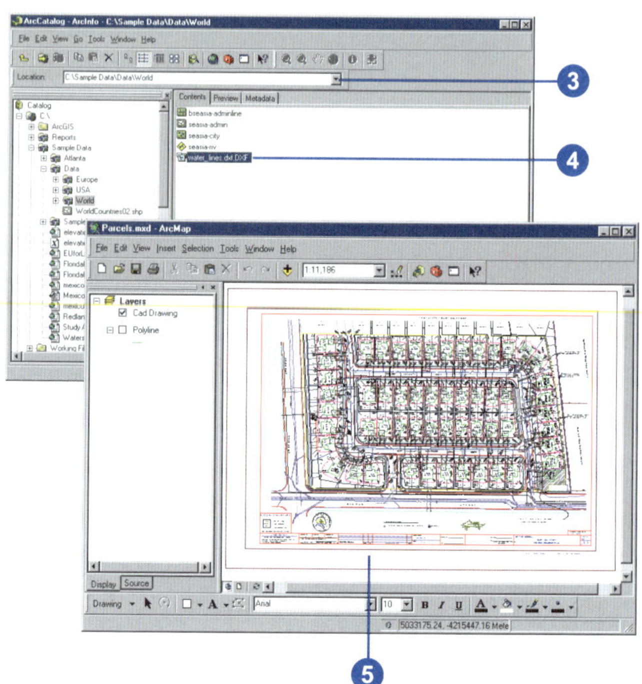

Adding a CAD drawing in ArcMap

1. Click the Add Data button on the Standard toolbar.
2. Click the Look in dropdown arrow and navigate to the folder that contains the CAD drawing.
3. Click the CAD drawing.
4. Click Add.

   ArcMap creates a new layer on the map that references the CAD drawing.

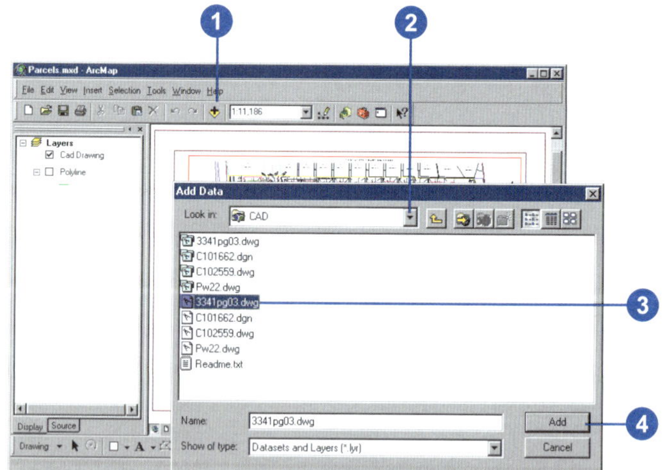

do not restrict the type of entities you can have on a drawing layer. Thus building footprints might be on the same drawing layer as streets. When working with a CAD drawing as features, you'll likely add several ArcMap layers from the same CAD drawing file and adjust what features display in those layers.

Adding a CAD dataset for display and analysis

1. Click the Add Data button on the Standard toolbar.
2. Click the Look in dropdown arrow and navigate to the folder that contains the CAD dataset.
3. Double-click the CAD dataset and click the CAD feature you want to add.
4. Click Add.

   Only the subset of features in the layer will display.

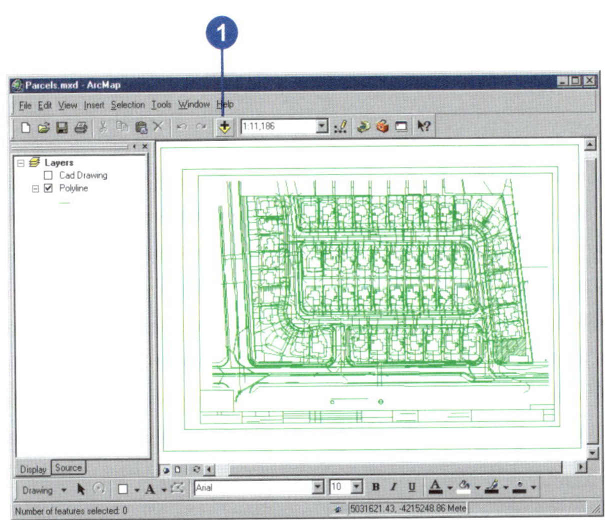

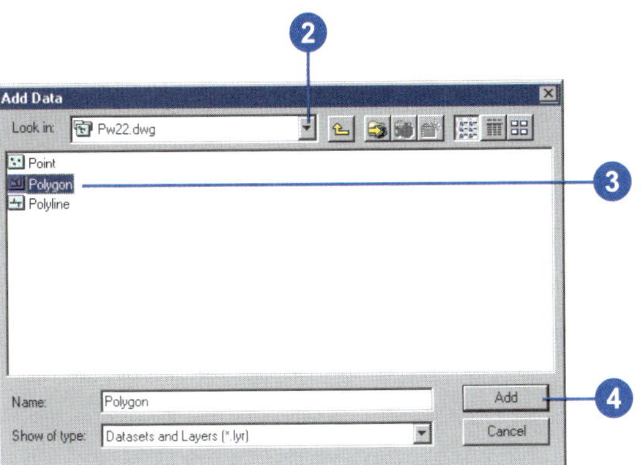

Adding x,y coordinate data

You don't always have to have a data source, such as a shapefile, to add data to your map. If you have some tabular data that contains geographic locations in the form of x,y coordinates, you can add this to a map as well.

Sets of x,y coordinates describe discrete locations on the earth's surface, such as the location of fire hydrants in a city or the points where soil samples were collected. You can easily collect x,y coordinate data using a *global positioning system (GPS)* device.

In order to add a table of x,y coordinates to your map, the table must contain two fields, one for the x coordinate and one for the y coordinate. The values in the fields may represent any coordinate system and units such as latitude and longitude or meters.

Once you have added the data to your map, the layer behaves just as any other feature layer. For instance, you can decide whether or not you want to display it, symbolize it, set the visible scale, or display a subset of features that meet some criteria.

Adding a table with x,y coordinates

1. Click Tools on the Main menu and click Add XY Data.
2. Click the table dropdown arrow and click a table that contains x,y coordinate data. If the table is not on the map, click the Browse button to access it from disk.
3. Click the X Field dropdown arrow and click the field containing x coordinate values.
4. Click the Y Field dropdown arrow and click the field containing y coordinate values.
5. Click Edit to define the coordinate system and units represented in the x and y fields.

   The x,y coordinates will be automatically transformed to match the coordinate system of the data frame.
6. Click OK.

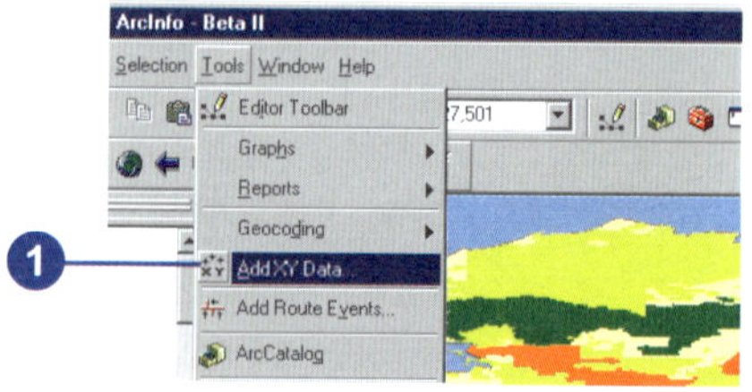

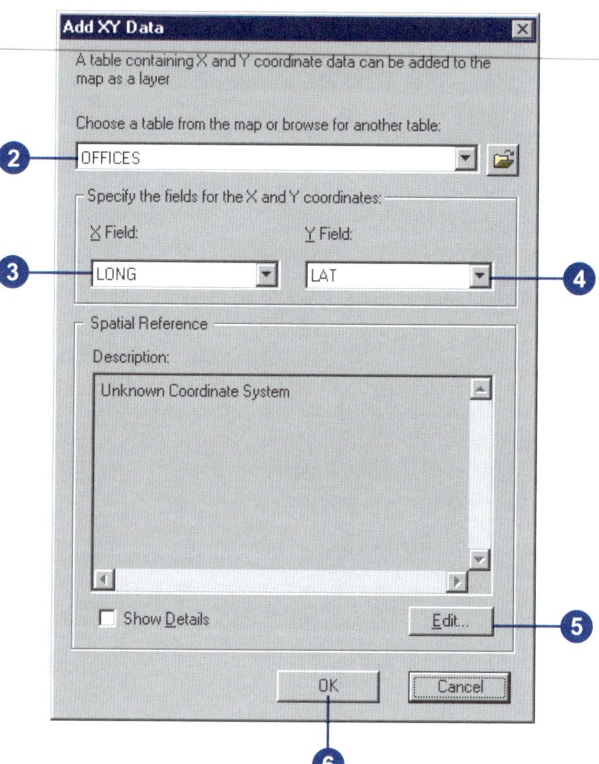

Adding route events

A *route event* is an attribute that describes a portion of a route or a single location on a route. Route events are organized into tables based on a common theme. For example, route event tables for highways might include speed limits, year of resurfacing, present condition, signs, and accidents. Route events use route and measure information to reference the attributes of particular locations in a route feature class.

There are two types of route events: point and line. Point events occur at precise locations along on a route. They are referenced to a location along a route using a single measure value. Line events describe portions of routes. They differ from point events in that they use two measure values to describe the measure location of the event.

A route event table has at least two fields: an event key and one or more measure locations. The event key field identifies the route to which an event belongs. A measure location is either one or two values describing the positions on the route where the event occurs.

1. Click Add Route Events from the ArcMap Tools menu.
2. Click the Route Reference dropdown arrow and click the route reference layer.

   Alternately, click the Browse button and navigate to the route reference feature class.
3. Click the Route Identifier dropdown arrow and click the route identifier field if necessary.
4. Click the Event Table dropdown arrow and click the event table.

   Alternately, click the Browse button and navigate to the event table.
5. Click the Route Identifier dropdown arrow and click the route identifier.
6. Click the type of events the route event table contains.
7. For point events, click the Measure dropdown arrow and click the Measure field.

   For line events, click the From-Measure dropdown arrow and click the from-measure field. Click the To-Measure dropdown arrow and click the to-measure field.
8. Optionally, click the Offset dropdown arrow and click the offset field.
9. Click Advanced Options. ▸

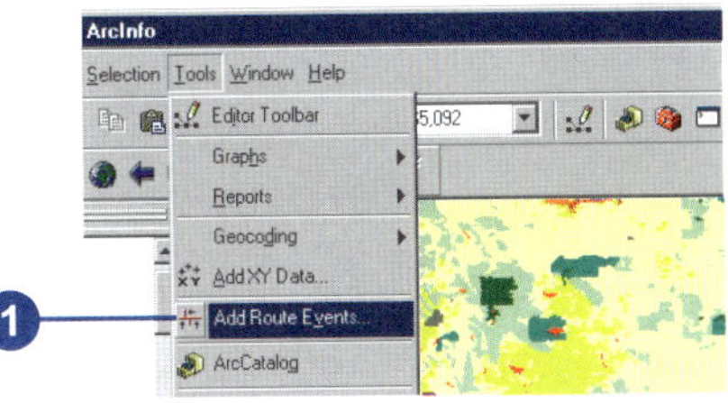

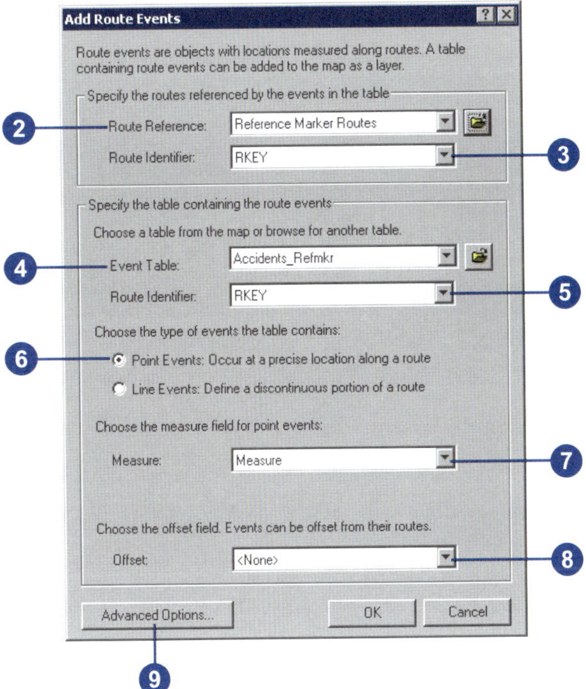

Tip

Route identifier field

The Add Route Events dialog box will pay attention to a route layer's route identifier field.

Tip

Displaying route events

Right-click an event table in the table of contents and click Display Route Events.

Tip

Layer files

You can save event layers as a layer file (.lyr). The next time you bring this file into ArcMap, the dynamic segmentation process will automatically be performed for you.

10. Click all of the advanced dynamic segmentations you want applied to the route event source.
11. Click OK.
12. Click OK.

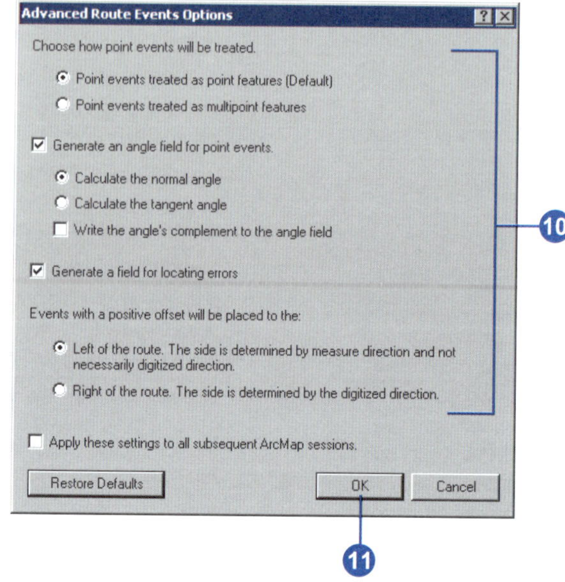

Creating and adding a new feature class

It is a three-step process to create and add a new *feature class* to your ArcMap document. The first step is to create a new geodatabase or select an existing one. You then create your new feature class, which is contained in the geodatabase. Both of these steps are performed in ArcCatalog.

The final step is to add this feature class to your ArcMap project. You can either add this feature by dragging and dropping it from the ArcCatalog window onto the ArcMap window or by selecting the Add Data button in the ArcMap window and choosing the new feature class file.

Creating a new geodatabase

1. Click the ArcCatalog button in the ArcMap Standard toolbar to open ArcCatalog.
2. Click a folder or create a new folder to create the geodatabase in.
3. Right-click the folder, point to New, then click Personal Geodatabase.
4. The folder name is already highlighted so you can change the name to something more appropriate.

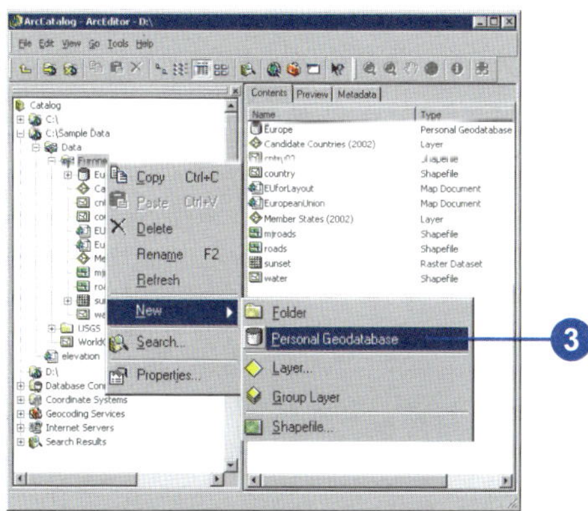

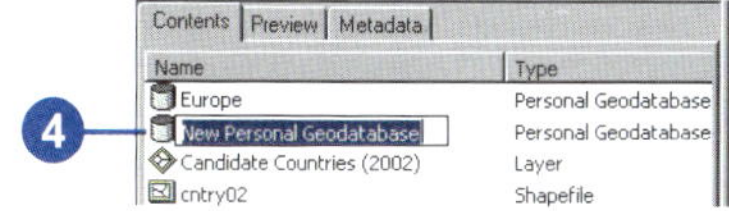

See Also

For information on how to create a new shapefile or table, see Using ArcCatalog.

Creating a new feature class

1. Right-click the geodatabase where you want to create the feature class. Point to New, then click Feature Class.
2. Enter a name for the feature.
3. Choose thc type of feature class to create.
4. Click Next.
5. Specify the database storage configuration. Choosing Default is often sufficient.
6. Click Next.
7. Click Shape under Field Name.
8. Under Field Properties, choose the Geometry Type by clicking in the space beside it and clicking the option you want from the dropdown menu.

   If you do not choose a Geometry Type, the default is Polygon.
9. Click Finish.

   To add this layer to ArcMap, you can click and drag it as previously described in 'Adding a layer from ArcCatalog'.

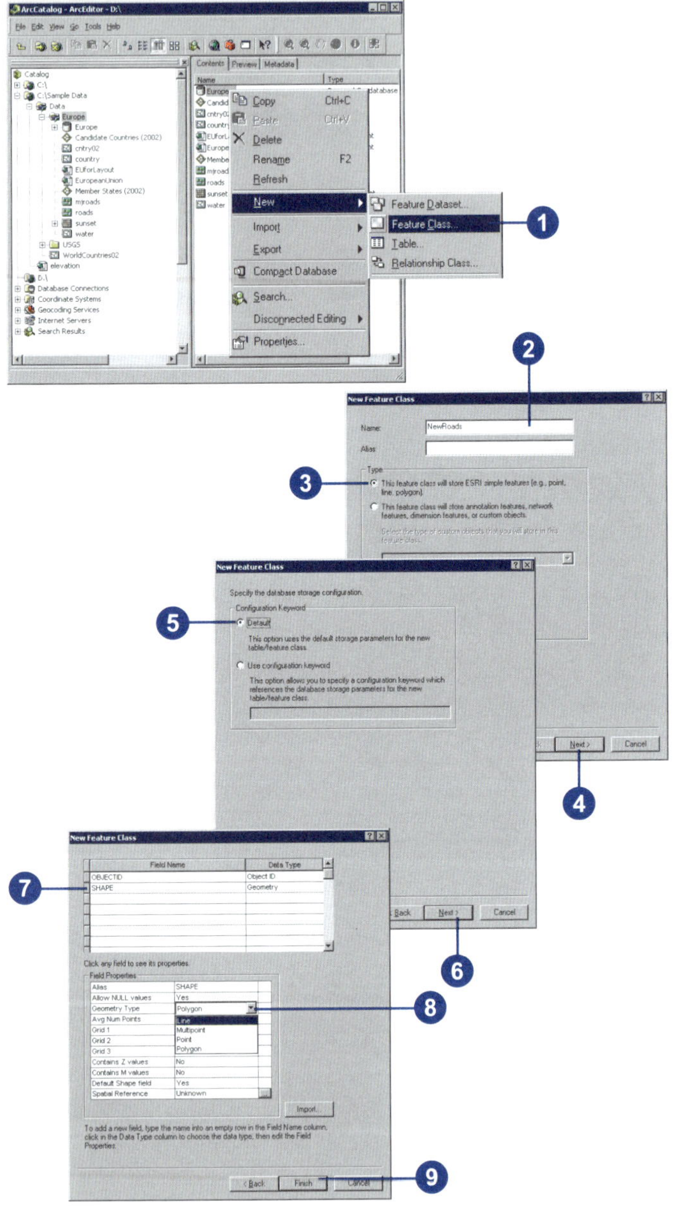

About coordinate systems

The features on a map reference the actual locations of the objects they represent in the real world. The positions of objects on the earth's spherical surface are measured in degrees of latitude and longitude, also known as *geographic coordinates*. While latitude and longitude can locate exact positions on the surface of the earth, they are not uniform units of measure; only along the equator does the distance represented by one degree of longitude approximate the distance represented by one degree of latitude. To overcome measurement difficulties, data is often transformed from the three-dimensional geographic coordinate system to the two-dimensional planar surface in a *projected coordinate system*. Projected coordinate systems describe the distance from an origin (0,0) along two separate axes—a horizontal x-axis representing east–west and a vertical y-axis representing north–south.

Because the earth is round and maps are flat, getting information from the curved surface to a flat one involves a mathematical formula called a *map projection*. A map projection transforms latitude and longitude to x,y coordinates in a projected coordinate system.

This process of flattening the earth will cause distortions in one or more of the following spatial properties: distance, area, shape, and direction. No projection can preserve all these properties and, as a result, all flat maps are distorted to some degree. Fortunately, you can choose from many different map projections. Each is distinguished by its suitability for representing a particular portion and amount of the earth's surface and by its ability to preserve distance, area, shape, or direction. Some map projections minimize distortion in one property at the expense of another, while others strive to balance the overall distortion. As a mapmaker, you can decide which properties are most important and choose a projection that suits your needs.

Do you need to display your data with a projected coordinate system?

If your spatial data references locations with latitude and longitude—for example, decimal degrees—you can still display it on your map. ArcMap draws the data by simply treating the latitude/longitude coordinates as planar x,y coordinates. If your map doesn't require a high level of locational accuracy—if you won't be performing queries based on location and distance or if you just want to make a quick map—you might decide not to transform your data to a projected coordinate system.

If, however, you need to make precise measurements on your map, you should choose a projected coordinate system. When displaying and performing analysis with datasets, they should be in the same coordinate space and in the same projection. If two datasets are in different coordinate systems, the values of the coordinates are on different scales. Errors will occur when comparing such datasets because they will represent different locations.

Reasons for using a projected coordinate system

- You want to make accurate measurements from your map and be sure that spatial analysis options you use in ArcMap calculate distance correctly. Latitude/Longitude is a good system for storing spatial data but not very good for viewing, querying, or analyzing maps. Degrees of latitude and longitude are not consistent units of measure for area, shape, distance, and direction.
- You are making a map in which you want to preserve one or more of these properties: area, shape, distance, and direction.
- You are making a small-scale map such as a national or world map. With a small-scale map, your choice of map projection

determines the overall appearance of the map. For example, with some projections, lines of latitude and longitude will appear curved; with others they will appear straight.

- Your organization mandates using a particular projected coordinate system for all maps.

What is an on-the-fly projection?

ArcMap can perform what is commonly known as an on-the-fly projection. This means ArcMap can display data stored in one projection as if it were in another projection. The new pseudoprojection is for display and query purposes only. The actual data is not altered. Data is projected on the fly anytime a data frame contains a layer whose coordinate system is defined as something different from the coordinate system definition of the data frame. A data frame's coordinate system can be defined by adding data with a defined coordinate system or by manually setting the coordinate system (by accessing the data frame's properties).

ArcMap will not project data on the fly if the coordinate system for the dataset has not been defined. A dataset with an undefined coordinate system will simply be displayed in its native coordinate system. The coordinate system for any dataset can be defined using ArcCatalog.

The first layer added to the data frame defines its coordinate system. This is true whether the data is projected or geographic. For example, if the first layer added contains a Lambert Conformal Conic projected coordinate system, all other layers will project on the fly to match this. Similarly, if the first layer added to the data frame contains data that uses a WGS84 geographic coordinate system, all other layers will adjust to match this. Even data that uses a projected coordinate system will unproject on the fly.

For information on projecting rasters on the fly, see Chapter 9, 'Working with rasters'.

What type of map projection should you choose?

Here are a few things to consider when choosing a projection:

- Which spatial properties do you want to preserve?
- Where is the area you're mapping? Is your data in a polar region? An equatorial region?
- What shape is the area you're mapping? Is it square? Is it wider in the east–west direction?
- How big is the area you're mapping? On large-scale maps, such as street maps, distortion may be negligible because your map covers only a small part of the earth's surface. On small-scale maps, where a small distance on the map represents a considerable distance on the earth, distortion may have a bigger impact, especially if you use your map to compare or measure shape, area, or distance.

Answering these questions will determine which map projection and thus which projected coordinate system you'll want to use to display your data.

Map projections can generally be classified according to which spatial attribute they preserve (distance, area, shape, or direction).

Equal area *projections preserve area and are also called equivalent projections. Most thematic maps should use an equal area projection. The Albers Equal Area Conic projection is commonly used for the United States; common projections for the world are Equal Area Cylindrical and Sinusoidal.*

Conformal *projections preserve angles and are useful for navigational charts and weather maps. Shape is preserved for small areas, but the shape of a large area such as a continent will be significantly distorted. Common conformal projections are the Lambert Conformal Conic and the Mercator.*

Azimuthal *projections preserve direction from one point to all other points. This property can be combined with preserving either area, angles, or distance. Thus, it is possible to have an Equal Area Azimuthal projection, such as Lambert, or an Equidistant Azimuthal projection.*

Equidistant *projections preserve distances, but no projection can preserve distances from all points to all other points. Instead, distance can be held true from one point (or a few points) to all other points or along all meridians or parallels. If you will be using your map to find features that are within a certain distance of other features, you should use an equidistant map projection.*

Compromise *projections minimize overall distortion but preserve none of the four properties. The Robinson projection, for example, is neither equal area nor conformal but is aesthetically pleasing and useful for general mapping.*

For more information on coordinate systems, see *Understanding Map Projections*.

Specifying a coordinate system

If all the data you want to display on your map is stored in the same coordinate system—for example, you're using your organization's database—you can just add it to a map and not consider whether the layers will overlay properly; they will. If, however, you've collected data from a variety of sources, you'll need to know what coordinate system each dataset uses to ensure ArcMap can display them together.

When you add a layer to an empty data frame, that layer sets the coordinate system for the data frame; you can change it later if necessary. As you add subsequent layers, they are automatically transformed to the data frame's coordinate system as long as there's enough information associated with the layer's data source to determine its current coordinate system. If there isn't enough information, ArcMap will be unable to align the data and display it correctly. In this case, you'll have to supply the necessary coordinate system information yourself.

ArcMap expects coordinate system information to be stored with the data source. For each ►

Finding out which coordinate system your data is currently displayed with

1. Right-click the data frame that you want to determine the coordinate system of and click Properties.
2. Click the Coordinate System tab.

   The details of the current data frame coordinate system display in the dialog box.

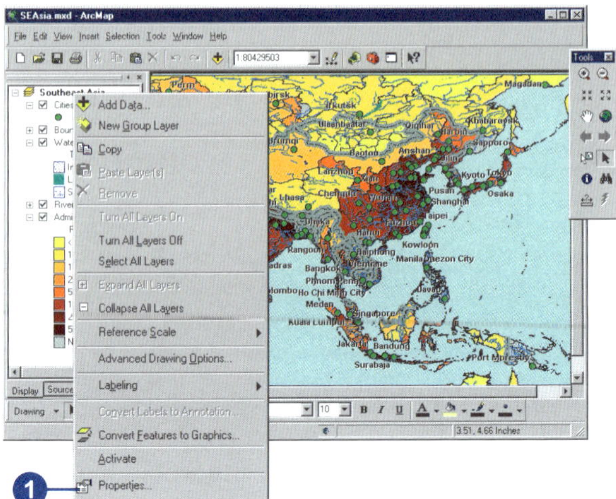

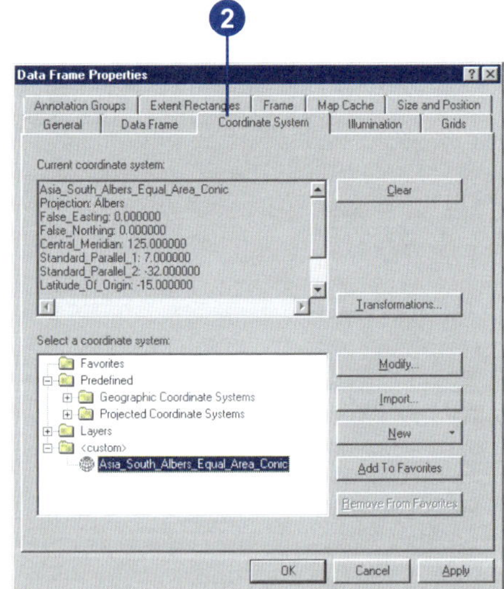

layer in a geodatabase, this information is part of the layer's metadata. For coverages, shapefiles, and rasters, it's stored on disk in a separate file named after the data source but with a .prj file extension (for example, streets.prj). These files are optional files; thus you may still need to define the coordinate system for one of these data sources. You can create a .prj file with ArcCatalog.

If no coordinate system information is associated with a data source, ArcMap will examine the coordinate values to see if they fall within the range: -180 to 180 for x-values and -90 to 90 for y-values. If they do, ArcMap assumes that these are geographic coordinates of latitude and longitude. If the values are not in this range, ArcMap simply treats the values as planar x,y coordinates.

Tip

Changing the coordinate system of a data frame

Changing the coordinate system of a data frame does not alter the coordinate system of the source data contained in it.

Displaying data with a predefined coordinate system

1. Right-click the data frame that you want to set the coordinate system of and click Properties.
2. Click the Coordinate System tab.
3. Double-click Predefined.
4. Navigate through the folders until you find the coordinate system you want and click it.
5. Click OK.

   All layers in the data frame will now be displayed with that coordinate system.

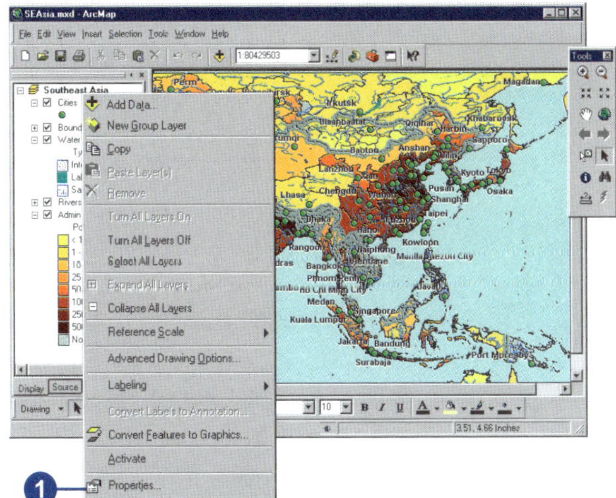

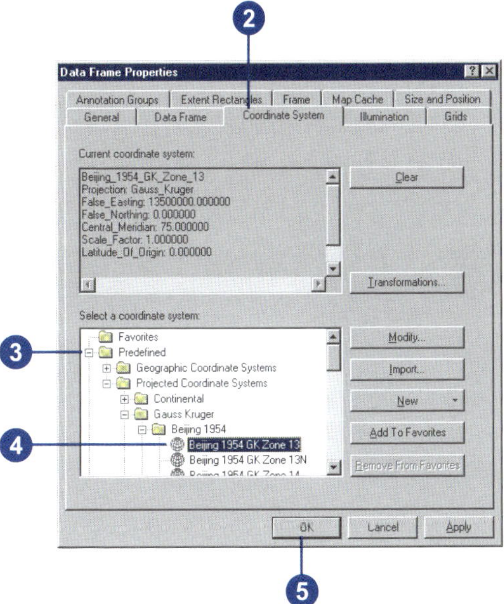

See Also

For more information on coordinate systems, see Understanding Map Projections.

Modifying the parameters of a coordinate system

1. Right-click the data frame whose coordinate system you want to modify and click Properties.
2. Click the Coordinate System tab.
3. Click Modify.
4. Adjust the coordinate system properties as appropriate.
5. Click OK.
6. Click OK on the Data Frame Properties dialog box.

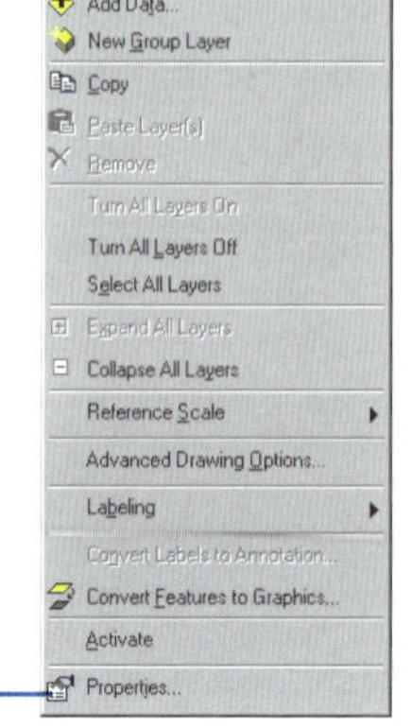

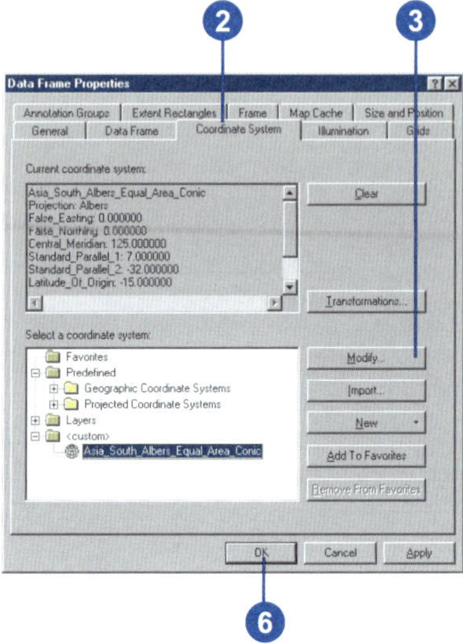

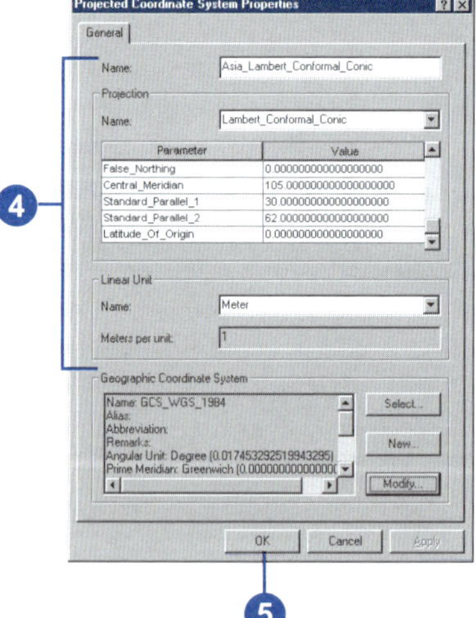

Tip

Do you want to see meters, miles, or feet?

When you measure lengths or find places by their coordinates, you can choose which units you want to use. Set the Display Units property as needed.

Tip

Why can't I set the map units?

Map units are a property of the coordinate system defined with your data. You can change the map units by modifying the coordinate system. Right-click the data frame containing your data and click the Coordinate System tab. Here you can modify the parameters of the coordinate system.

Setting the units for reporting lengths and displaying coordinates

1. Right-click the data frame and click Properties.
2. Click the General tab.
3. Click the Map dropdown arrow and click the appropriate units.

   The map units option is only available when your data has no coordinate system information associated with it.
4. Click the Display dropdown arrow and click the appropriate units.
5. Click OK.

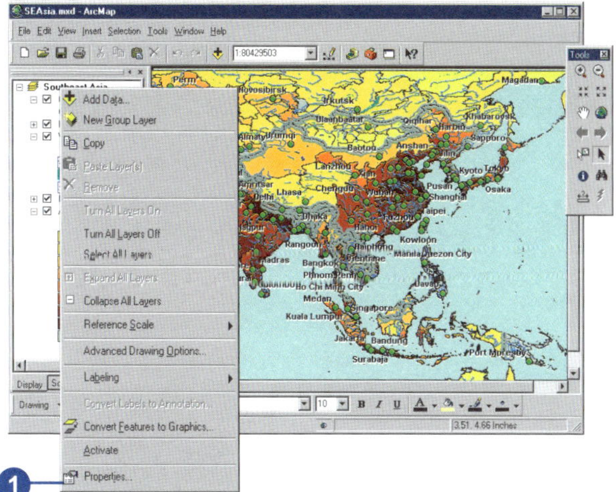

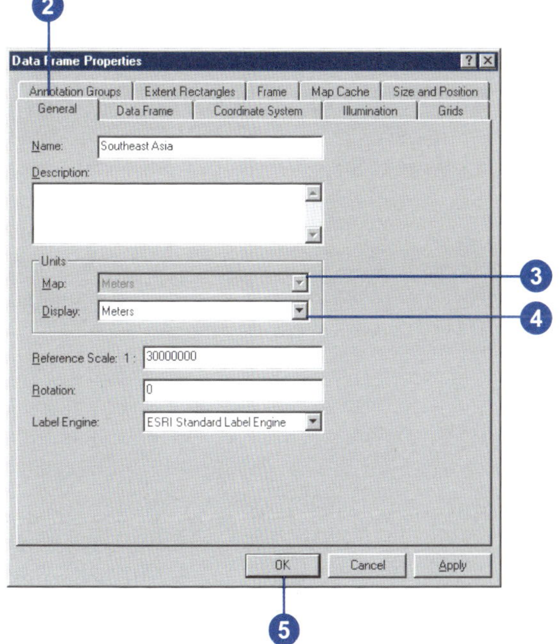

How to reference data on a map

ArcMap can reference data that is stored in databases (personal or ArcSDE®) or as files on a disk. If you plan to distribute your maps to others or if the location of your data has changed, you may need to change how your map references data so others will not need to repair layers. ArcMap has several options for referencing file-based data: full paths, relative paths, or Universal Naming Convention (UNC) paths.

Full paths

An example of a full (absolute) path is: C:\GIS\Project1\Boundary.shp. To share maps saved with paths to data with the full path option, everyone who uses the map must either do so on the same computer or have the data on their computer in exactly the same folder structure.

Relative paths

An example of a relative path is: \Project1\Boundary.shp. Relative paths in a map specify the location of the data contained in the map, relative to the current location on disk of the map document (.mxd file) itself. Since relative paths don't contain drive names, they enable the map and its associated data to be moved to any disk drive without the map having to be repaired. As long as the same directory structure is used at the new location, the map will still be able to find its data by traversing the relative paths.

Relative paths allow you to share maps that you made with data on your local "F:\" drive, for example, with people who only have a "C:\" drive. This also allows you to easily move the map and its data to a different hard drive on your computer or give the map and its data to others to copy to their computer.

Data referenced by a relative path can be in the same folder as the map or in a folder above or below the folder containing the map. To reference data in a folder that's above the folder containing the map, a relative path will contain \..\ for each level up in the folder structure that must be traversed. If layers in a map do not meet that criterion, they will not be saved with relative pathnames, but will instead retain their full pathname. You can reference relative paths by clicking the File menu, Map Properties, then Data Source Options.

UNC paths

An example of a UNC path is: \\GISServer\GIS\Project1\Boundary.shp. Using UNC paths allows you to make a map referencing data on a computer on your organization's network so the map can be shared with others without requiring that they map the network computer as a disk drive on their local machine. The network computer is referenced directly by name in the path.

One thing to keep in mind is that a personal geodatabase may have data stored within a feature dataset, which means its path will contain that additional information. For example, "C:\GIS\Project1\FieldData.mdb\stations\Feb10" indicates that there is a stations feature dataset with a Feb10 feature class. Paths to data that resides on a database server contain the connection properties (server name, database, username, and possibly the password), the feature dataset name, and the feature class name.

Referencing data on a map

When you add a layer to your map, ArcMap references the data source the layer is based on. When you save the map, the data references are stored with it. The next time you open your map, ArcMap locates the data based on the references. If ArcMap can't find a data source, you'll need to either locate the data source yourself or ignore the reference, in which case the layer won't be drawn.

If you plan on distributing your maps to others, they'll need access to the data referenced on it. If they have access to the data—for example, data stored on a server—they can simply update the references to the data if necessary. If they don't have access to the data, you'll probably have to distribute the data with your map.

To help make it easier to distribute data with your map, ArcMap allows you to store relative pathnames to data sources referenced on a map. This lets you, for example, distribute your map and data in the same directory. The references stored in the map are correct regardless of where they are placed on disk.

Storing relative pathnames to data

1. Click the File menu and click Map Properties.
2. Click Data Source Options.
3. Click Store relative path names.
4. Click OK.
5. Click OK on the Map Properties dialog box.

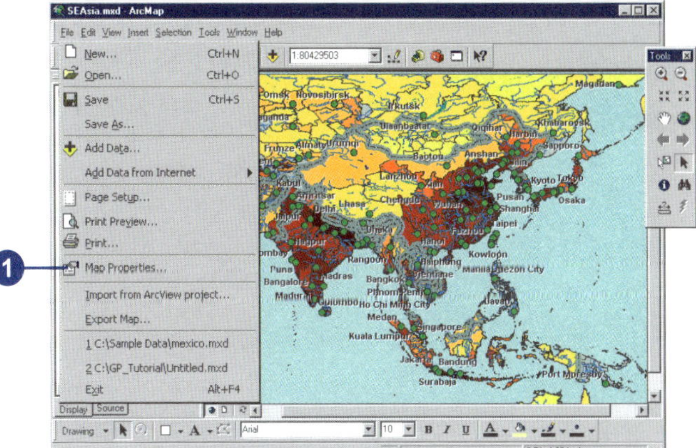

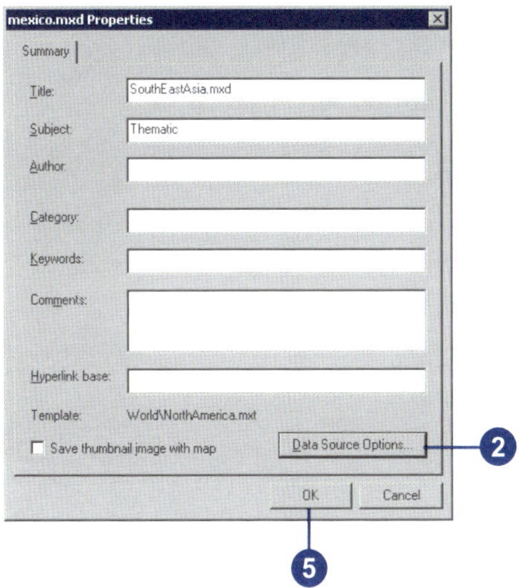

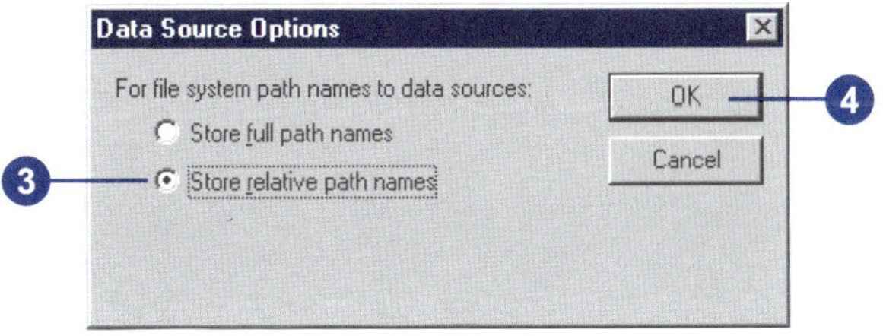

Repairing and updating data links

When you first open a map, ArcMap searches for the data referenced by the layers on the map. If it can't find the data—for example, because the data has been moved—the layer won't display.

You can immediately tell which layers on your map have broken links because you'll see a red exclamation mark next to their names in the table of contents. If you know the new location of the data, you can repair the link.

You can also update the link to a layer's data source. You might want to do this when, for example, you want to retain the layer's display properties but use an updated attribute table. For more information, see 'Changing a layer's source data' in Chapter 5.

Tip

Storing relative pathnames to data

If you plan on distributing your maps to others, you might choose to reference data using relative pathnames. See the topics 'How to reference data on a map' and 'Referencing data on the map' in this chapter for more information.

Repairing broken data source links

1. Locate the layer with the broken link in the table of contents. It will have a red exclamation mark next to it.
2. Right-click the layer, point to Data, and click Set Data Source.
3. Click the Look in dropdown arrow and navigate to the location of the data source.
4. Click the data source.
5. Click the Add button.

   The link to the data source is now updated.

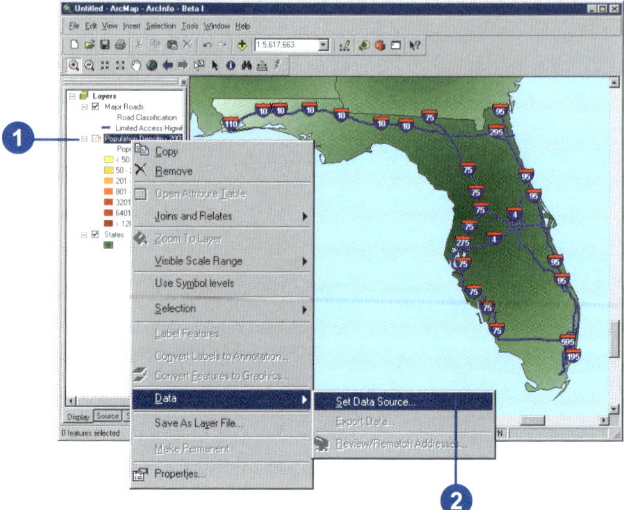

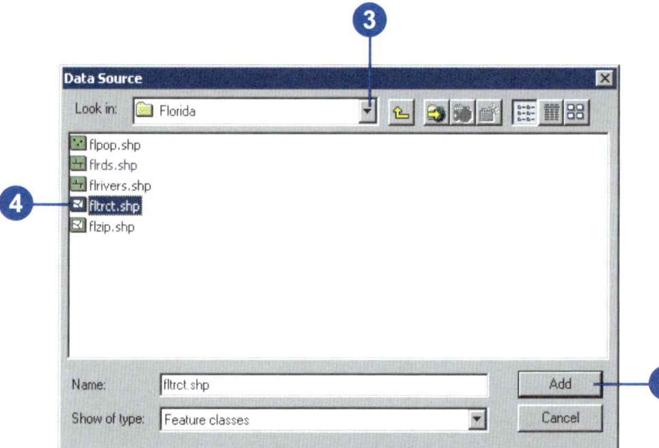

Updating a link to a data source

1. In the table of contents, right-click the layer and click Properties.
2. Click the Source tab.
3. Click Set Data Source.
4. Click the Look in dropdown arrow and navigate to the data source.
5. Click the data source.
6. Click Add.

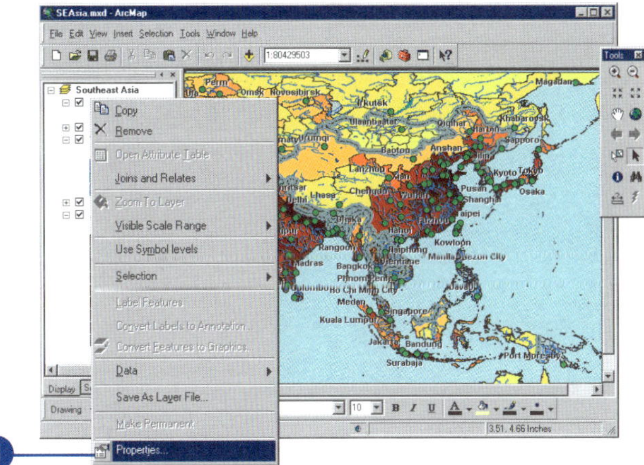

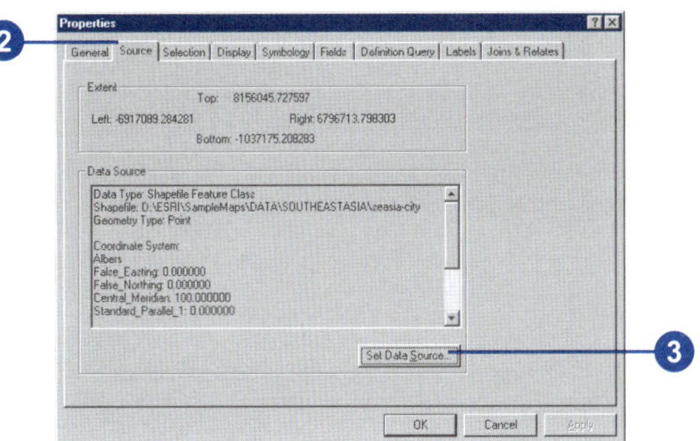

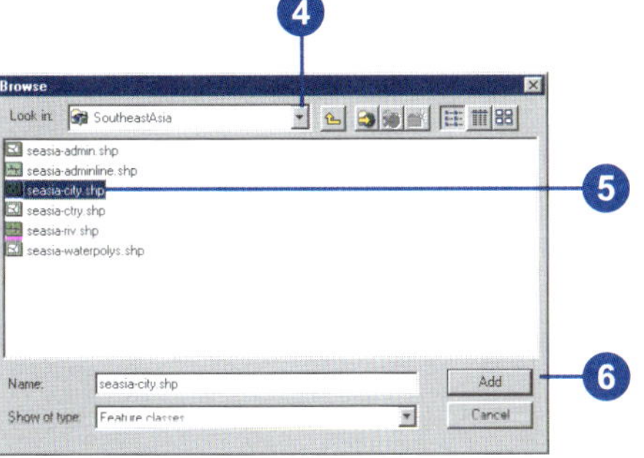

Working with layers

5

IN THIS CHAPTER

- **Description of a layer file**
- **Layer property functionality**
- **Adding layers**
- **Changing a layer's text**
- **About the drawing order**
- **Copying layers**
- **Removing layers from the map**
- **Grouping layers**
- **Saving a layer to disk**
- **Accessing layer properties**
- **Displaying a layer at specific scales**
- **Creating a transparent layer**
- **Changing a layer's source data**
- **Changing the appearance of the table of contents**
- **Using data frames to organize layers**

You can think of a layer as a specific way to display and work with geographic data. Layers exist within maps and can be saved independently of maps in a database or as a layer file (.lyr). In fact, many organizations save layers for their staff to access rather than providing direct access to the organization's geographic data. This helps to ensure that those accessing the data are viewing the same information.

As you saw in the previous chapter, it is easy to create layers just by adding data to your map. As you add different kinds of data to your map, ArcMap knows how to create specific kinds of layers by knowing what kinds of functionality that data can support. Thus, there are different types of layers—for example, shapefiles are used to create feature layers; CAD data can be used to create feature layers and a special kind of layer called a CAD layer. Adding *raster* data (or bitmaps) to a map will result in a raster layer. The diagram on the next page shows types of layers and what ArcMap can do with them.

Description of a layer file

This diagram shows the variety of data that can be used to create different types of layers.

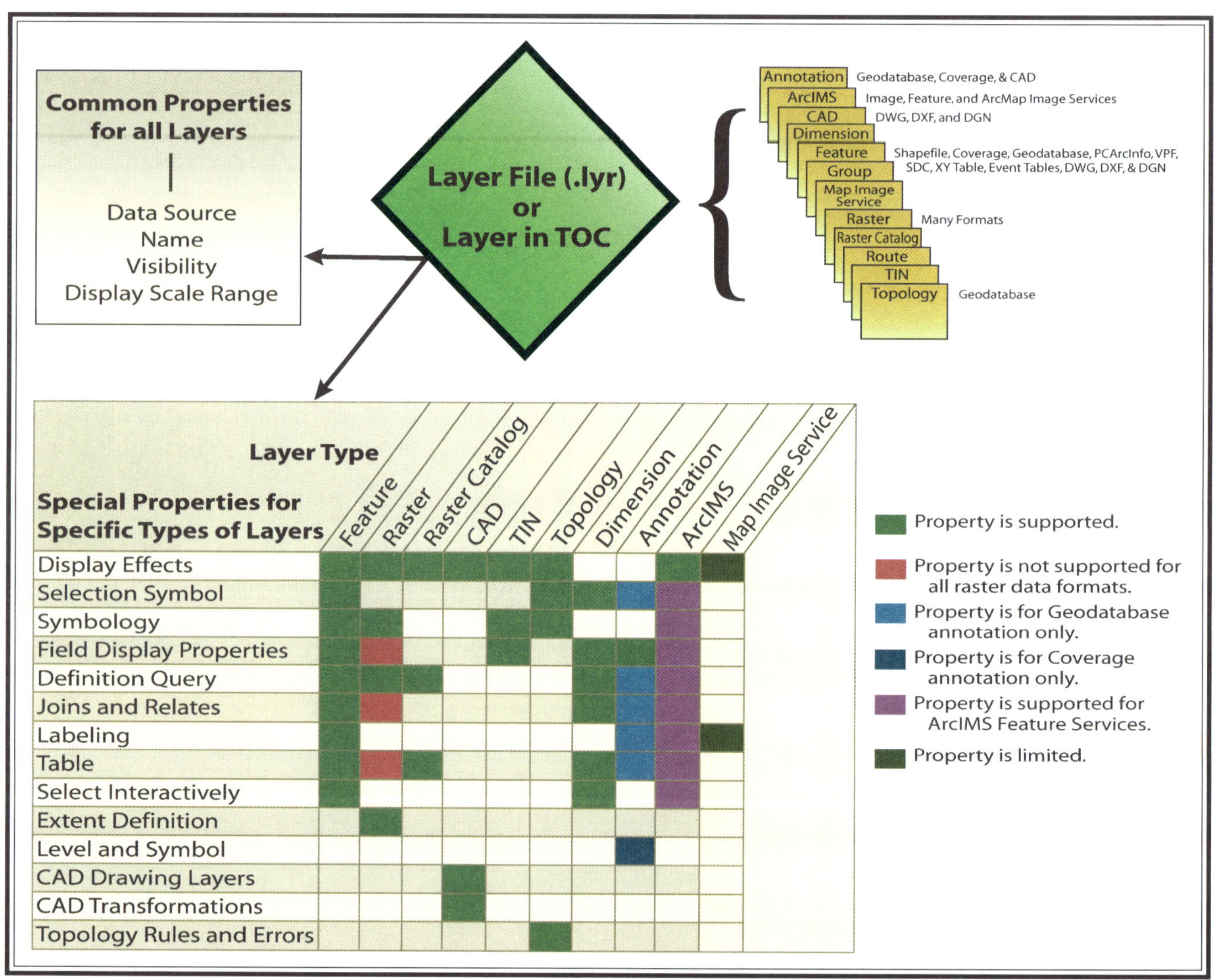

Layer property functionality

Display effects

The Display tab controls how your data is displayed as you move in the view. Options include making a layer transparent, adding MapTips and hyperlinks, and restoring excluded features.

Selection symbol

The Selection tab allows you to set how features in a specific layer will look when they are selected. Selection property changes in a specific layer override the default Selection Options settings.

Symbology

This tab offers methods for representing your data. Options include drawing features in one symbol, proportional symbols, categories, quantities, color ramps, or charts.

Field display properties

The Fields tab provides characteristics about attribute fields. You can also create aliases, format numbers, or make fields invisible.

Definition Query

This tab allows you to display a subset of your data that meets some criteria without altering the data. With the Query Builder, you can create an expression to display particular features of a layer.

Joins and Relates

This tab allows you to join (include within ArcMap) or relate (associate) data to the layer's attribute table. You can also remove joins or relates.

Labeling

The Labels tab allows you to turn on a layer's labels; build label expressions; manage label classes; and set up the labeling options, which include label placement and symbology.

Table

A table includes descriptive information that is stored in rows and columns in a database and can be linked to map features. Each row represents an individual entity, record, or feature, and each column represents a single field or attribute value.

Select interactively

This function allows you to make selections using the Select Features tool.

Extent definition

This appears on the Source tab and shows the bounding extent of your data along with its source and coordinate system.

Level and Symbol

These tabs, specific to coverage, PC ARC/INFO®, Spatial Database Engine™ (SDE®) 3.x, and vector product format (VPF) annotation, allow you to adjust annotation display properties. Annotation for these formats, however, is read-only.

CAD Drawing Layers

This tab, which is specific to CAD data, allows you to specify which CAD drawing layers are visible.

CAD Transformations

This tab is specific to CAD layers. It allows you to transform a CAD layer so it matches your coordinate system.

Topology rules and errors

These functions are specific to topology layers. The Feature Classes tab lists the feature classes in the topology and their ranks. The Rules tab shows and describes which topology rules are imposed on your data. The Errors tab allows you to generate a summary of the errors found in the participating feature classes.

Adding layers

Adding a layer is as straightforward as adding other data. A layer file contains the file extension .lyr and is identified by a diamond-shaped icon.

Tip

Dragging and dropping

If you drag and drop the layer file onto the table of contents, it will be drawn according to the place in the table of contents where you dropped it. However, if you drag and drop the layer into the window, it will be placed in the table of contents according to the default layer placement.

The default layer placement (top to bottom) typically is:

1. Annotation

2. Features

i. Point

ii. Line

iii. Polygon

iv. TIN

v. Raster

When adding a new layer, it will automatically be placed above a similar feature type. For example, a new line feature will be placed above the other line features.

Adding a layer file in ArcMap

1. Click the File menu and click Add Data, or click the Add Data button on the Standard toolbar.
2. Click the Look in dropdown arrow and navigate to the folder containing your layer file.
3. Click the layer file.
4. Click Add.

   ArcMap adds the layer to your table of contents using the special properties it was saved with.

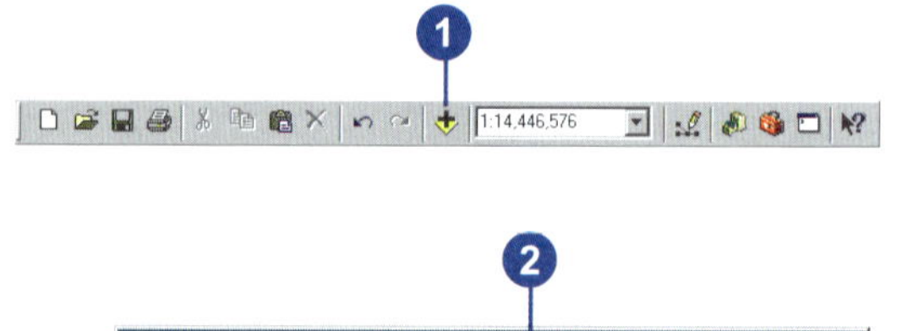

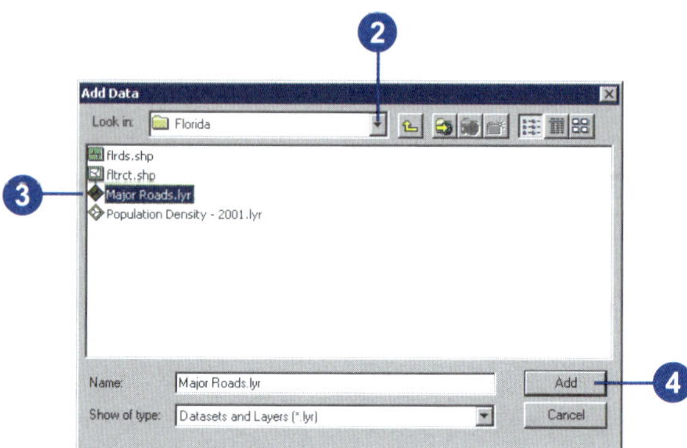

Adding a layer file from ArcCatalog

1. Start both ArcCatalog and ArcMap.
2. Arrange the ArcCatalog and ArcMap windows so you can see both on your screen.
3. In ArcCatalog, navigate to the layer file you are interested in.
4. Click and drag the file from ArcCatalog.
5. Drop the file on the ArcMap window.

   ArcMap adds the layer to your table of contents using the special properties it was saved with.

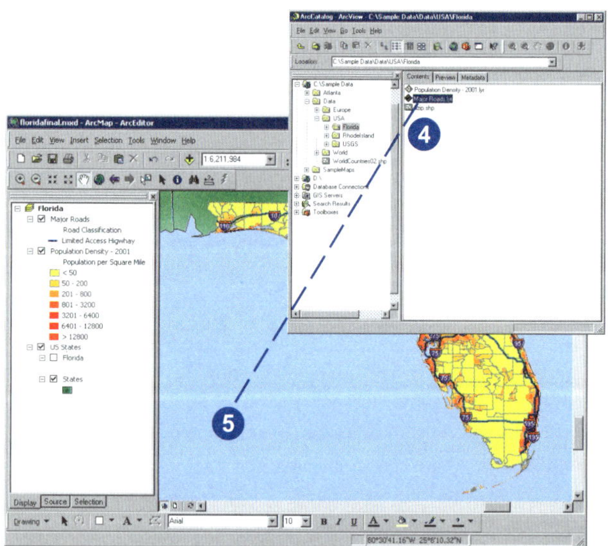

Changing a layer's text

A few pieces of descriptive text display beside each layer in the table of contents. One text string is the layer's name. The others describe what the features in the layer represent—the meanings of the symbols in the *legend*.

By default, when you add data to a map, the resulting layer is named after its data source. Often, the name of the data source is abbreviated. You can give a layer a more meaningful name without changing the name in the source data. This will make it easier to understand what layers are on the map.

When you draw the features of a layer, you use the attribute values in a particular field to symbolize them. These attribute values appear by default next to the symbol in the table of contents. As they don't usually provide a good label description for the features in your layer either, you'll likely want to change them as well.

Tip

Using a shortcut to rename layers

You can press the F2 key to rename a layer that is selected in the table of contents.

Changing the name of a layer

1. In the table of contents, click the layer to select it.
2. Click again over the name.

   This will highlight the name and allow you to change it.
3. Type the new name and press Enter.

   NOTE: This does not change the actual filename.

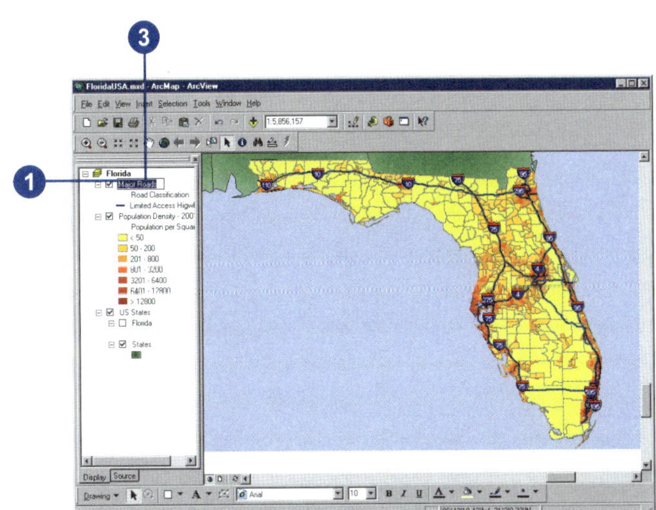

Changing the map feature label

1. In the table of contents, click the text you want to change.
2. Click again over the text string.

   This will highlight the string and allow you to change it.
3. Type the new description and press Enter.

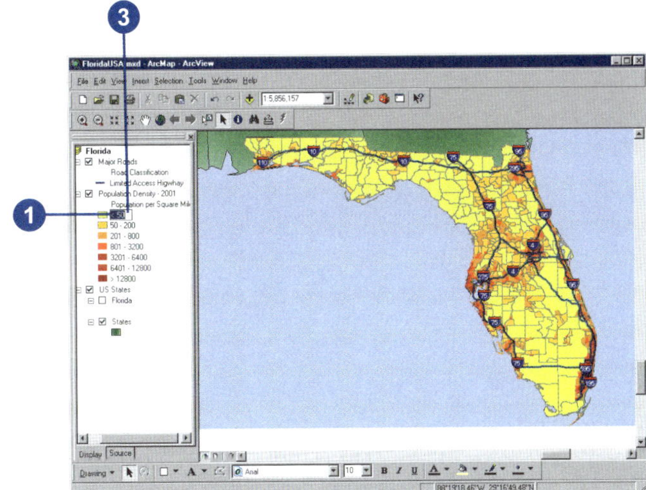

About the drawing order

The order of layers in the table of contents determines how layers are drawn on a map. Within a data frame, the layers listed at the top will draw over those listed below them, and so on down the list. You can easily move layers around to adjust their drawing order or organize them in separate data frames. However, no matter where the annotation layers or selections are in the table of contents, they will always draw after features.

Tip

Ordering your data before it's in the table of contents

For data of the same type, the order you clicked the layers in the Add Data dialog box or in ArcCatalog is the order they appear in the table of contents, with the first layer clicked being first in the table of contents.

Moving a layer to change its drawing order

1. In the table of contents, click and drag the layer up or down.

   A black bar indicates where the layer will be placed. This bar indents to reflect the position in the layer hierarchy where the drop will occur.

2. Release the mouse pointer to drop the layer in its new position.

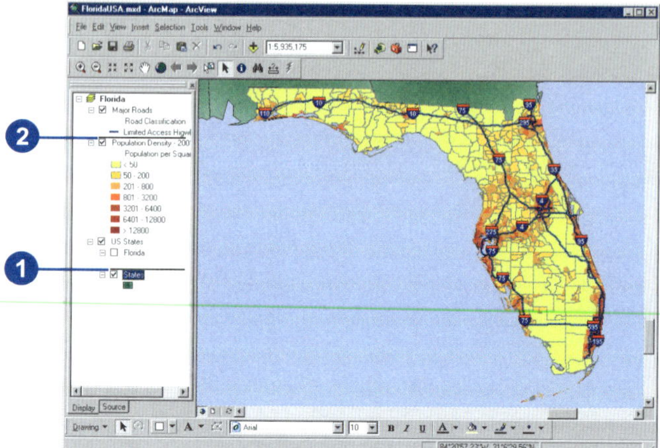

Copying layers

A quick way to build maps that reference the same data source is to copy and paste layers within a map or between maps. For example, suppose you want to show the change in population for an area over time. You can add a layer to a map and display it using one population attribute, then copy the layer to another map (or another data frame in the same map) and display it using the second population attribute.

Copying layers from a map to disk is a convenient way to let others access the layers you've created. Once you've defined how to draw a layer, that information is saved with the layer. Thus, anyone who adds the layer to a map will see it exactly as you created it. For information on copying a layer to disk, see 'Saving a layer to disk' in this chapter.

Tip

Opening one map document at a time

You can work on only one map per ArcMap session, although you can run multiple ArcMap sessions concurrently. Double-clicking a map in ArcCatalog or Windows Explorer always opens that map in its own ArcMap session.

Copying a layer between data frames

1. Right-click the layer or layers you want to copy to another data frame and click Copy.
2. Right-click the data frame you want to copy the layer or layers into and click Paste Layer(s).

   You can also drag and drop a layer from one data frame to another.

   If you want to move your layer instead of copying it while you drag and drop, hold down the Ctrl key when you drop the layer.

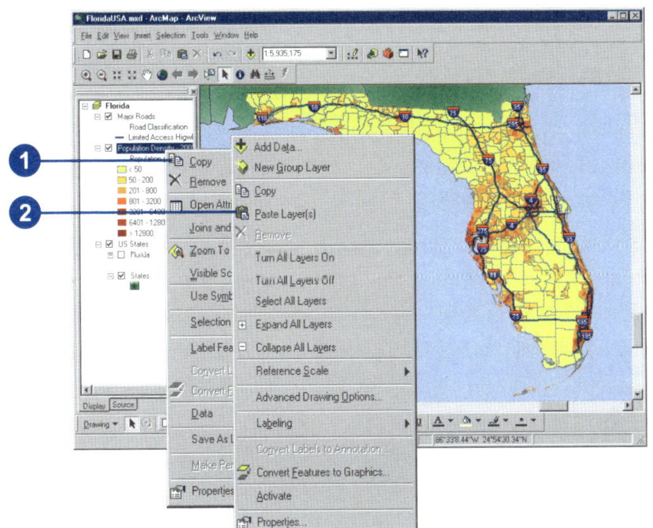

Copying a layer to another map

1. Right-click the layer or layers you want to copy to another data frame and click Copy.
2. Click the Open button on the Standard toolbar and open the map you want to copy the layer or layers into.
3. Right-click the data frame you want to copy the layer into and click Paste Layer(s).

   If you don't want to close your original map, open another ArcMap session.

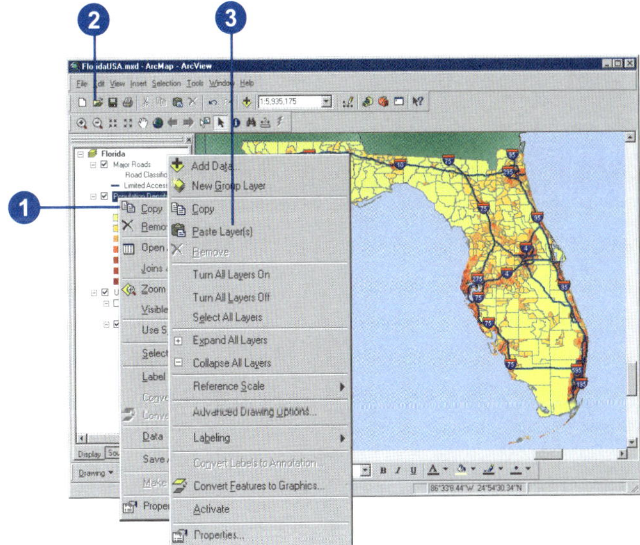

Removing layers from the map

When you no longer need a layer on your map, you can delete it. Deleting a layer from a map doesn't delete the data source on which the layer is based.

Tip

Deleting a data source

You can delete a data source, such as a coverage, in ArcCatalog.

Tip

Selecting more than one layer at a time

While you're clicking, you can hold down the Shift or Ctrl key to select multiple layers.

Removing a layer

1. In the table of contents, right-click the layer or layers you want to remove.
2. Click Remove.

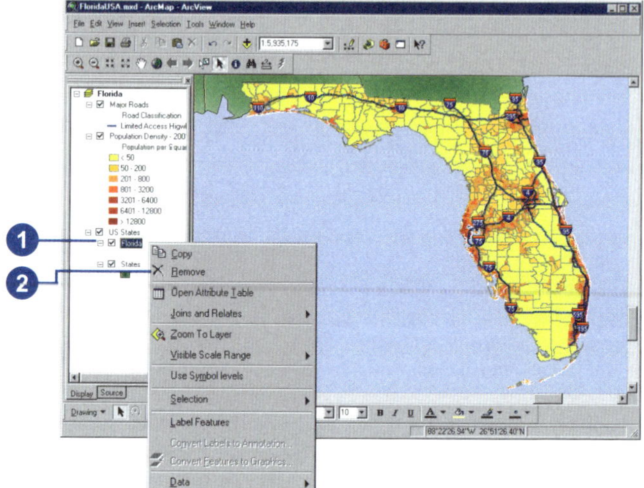

Removing several layers

1. In the table of contents, click the first layer you want to remove.
2. Hold down the Shift or Ctrl key and click to select additional layers.
3. Right-click the selection and click Remove.

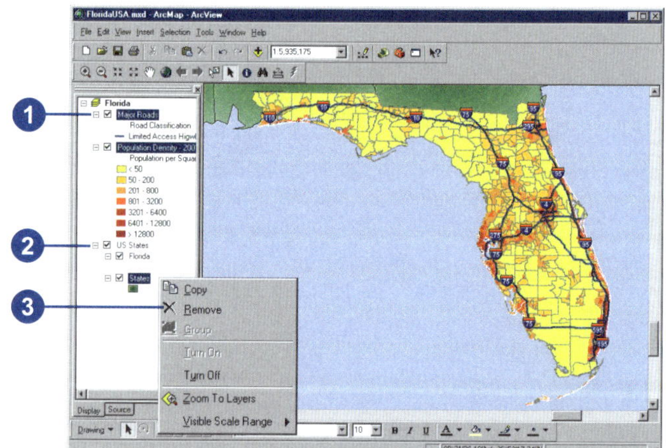

Grouping layers

When you want to work with several layers as one layer, you gather them together into a *group layer*. For example, suppose you have two layers on a map representing railroads and highways. You might choose to group these layers together and name the resulting layer "transportation networks".

A group layer appears and acts like an individual layer in the table of contents. Turning off a group layer turns off all its component layers. The properties of the group layer override any conflicting properties of its constituent layers. For example, a visible scale range set on a layer will be overridden by a visible scale range set on the group layer. If you need to, you can even create groups of group layers.

You can still work with the individual layers in the group. For instance, you can change how an individual layer is drawn, adjust the scale at which it is displayed, and control whether it is drawn as part of the group. You can change the drawing order of the group and add and remove layers as needed.

Creating a group layer

1. Right-click the data frame in which you want to create a group layer.
2. Click New Group Layer.

   A new group layer appears in the table of contents.

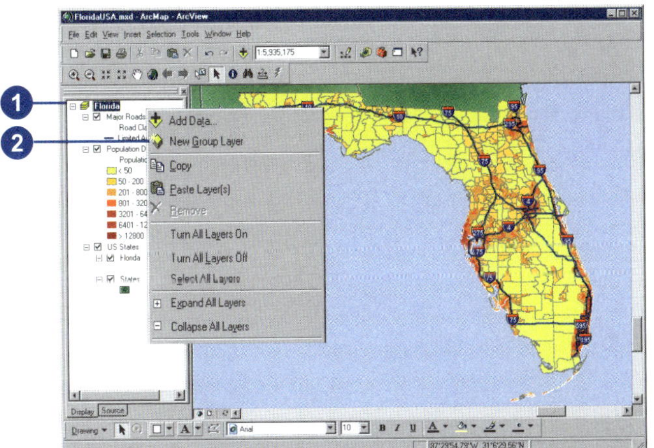

Grouping layers in the table of contents

1. Hold the Ctrl key and highlight multiple layers in the table of contents.
2. Right-click one of the chosen layers.
3. Click Group.

   A new group layer appears in the table of contents containing the selected layers.

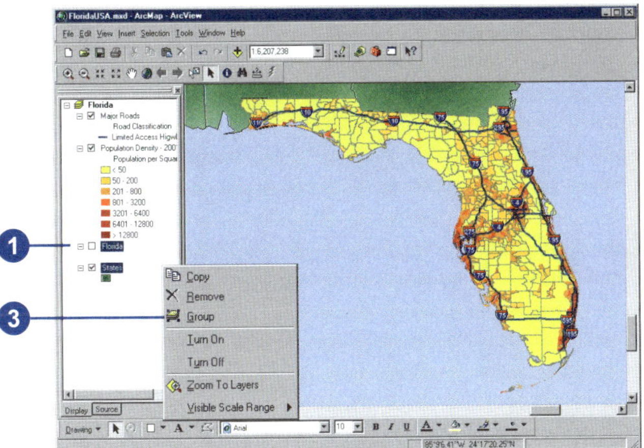

Tip

Changing the drawing order of layers in a group

The layers listed at the top of a group layer are drawn over those beneath it. You can also drag and drop the layer to a new position to change the drawing order of the group.

Tip

Using ArcCatalog to create group layers

You can also create your group layer in ArcCatalog.

See Also

For information on how to draw the individual layers in a group, see, Chapter 6, 'Symbolizing features'.

Adding layers to a group layer

1. Double-click the group layer in the table of contents to display its properties.
2. Click the Group tab.
3. Click Add.
4. Click the Look in dropdown arrow and navigate to the data source you want to add to the group.
5. Click the data source.
6. Click Add.

   Tip: If the layer you want to add to a group is already on the map, you can drag and drop it in the group.

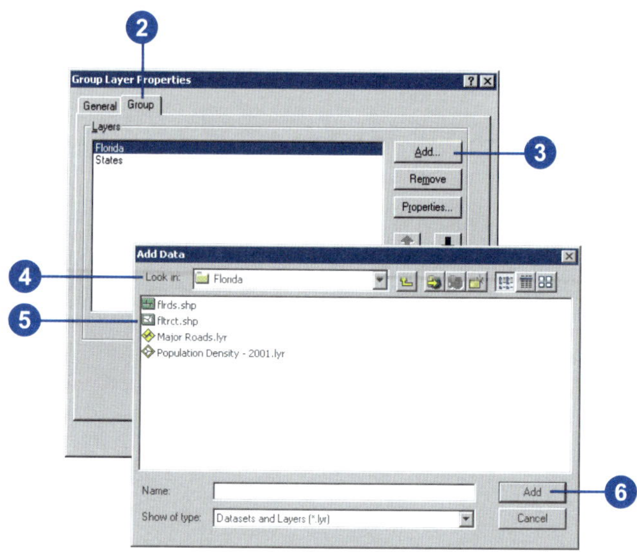

Changing the layer order in a group layer

1. Double-click the group layer in the table of contents to display its properties.
2. Click the Group tab.
3. Click the layer you want to move.
4. Click the appropriate arrow button to move the layer up or down.
5. Click OK.

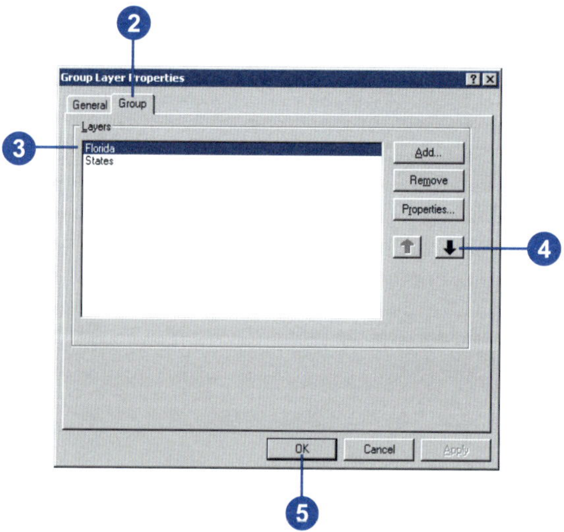

Displaying the properties of a layer in a group layer

1. Double-click the group layer in the table of contents to display its properties.
2. Click the Group tab.
3. Click the layer for which you want to display properties.
4. Click Properties.

   You can now modify the layer's properties—for example, you can change the drawing properties.
5. Click OK.

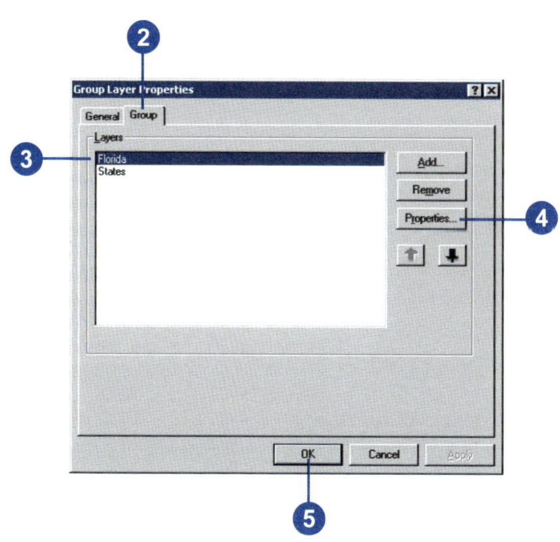

Removing a layer from a group layer

1. Double-click the group layer in the table of contents to display its properties.
2. Click the Group tab.
3. Click the layer that you want to remove.
4. Click Remove.
5. Click OK.

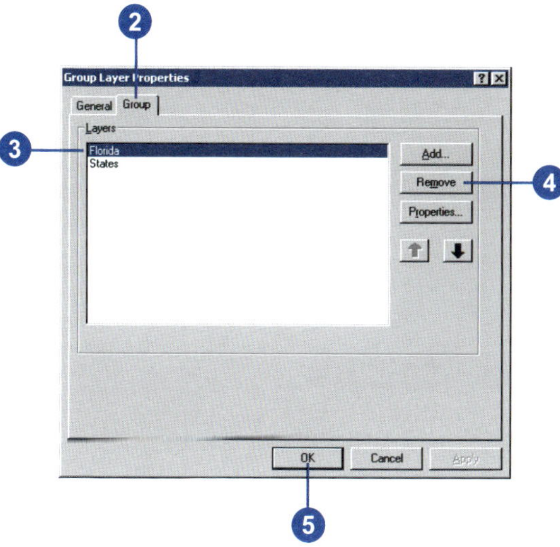

Tip

Removing several layers from a group layer

Hold down the Shift or Ctrl key to select more than one layer in the group.

Saving a layer to disk

One of the main features of a layer is that it can exist as a file in your GIS database. This makes it easy for others to access the layers you've built.

When you save a layer to disk, you save everything about the layer. When you add the layer to another map, it will draw exactly as it was saved. This is very convenient when others at your organization need to make maps but don't know how to represent or access the data in your database. All they need to do is add the layer.

Tip

About layer filenames

The filename you provide when you save a layer to disk does not have to be the same as its name on the current map. The layer name on the map, not the layer filename, will be displayed whenever the layer is added to another map.

Tip

Passing layers to others

Don't forget, the layer file is just a link to the actual data source. So if you are passing the layer to others, be sure they also have the data referred to in the layer.

1. In the table of contents, right-click the layer and click Save As Layer File.
2. Click the Look in dropdown arrow and navigate to the location where you want to save the layer.
3. Optionally, change the layer name.
4. Click Save.

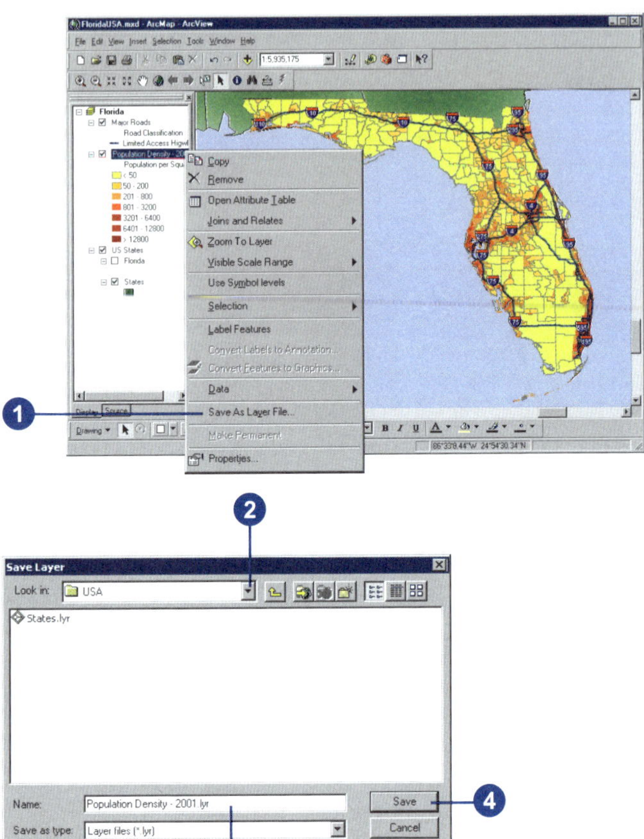

Accessing layer properties

You control all aspects of a layer with its *properties*. From here, you can define how to draw the layer, what data source the layer is based on, whether to label the layer, and what attribute fields the layer contains.

The Layer Properties dialog box is different for different layer types. See the layer file diagram at the beginning of this chapter to determine the properties of your layers.

Tip

Displaying layer properties

You can also double-click a layer in the table of contents to display its properties.

Displaying layer properties

1. In the table of contents, right-click the layer and click Properties.
2. Click the tab containing the properties you want to adjust.
3. When finished, click OK.

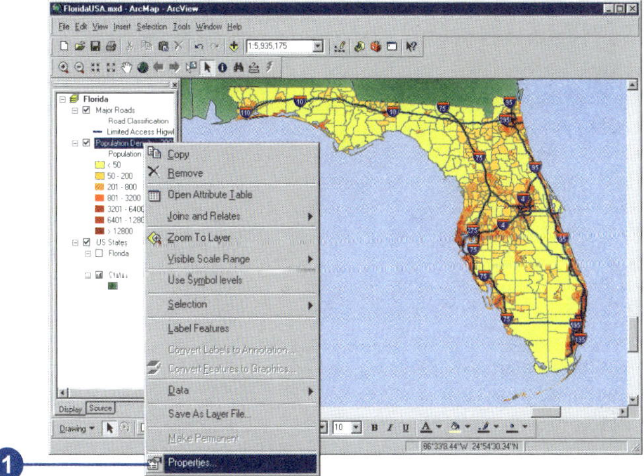

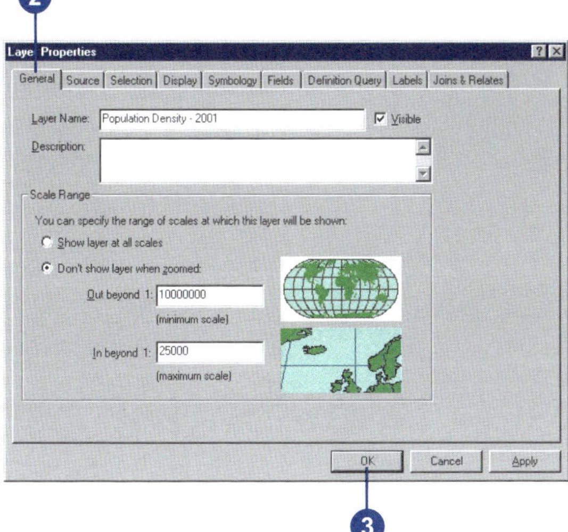

Displaying a layer at specific scales

As long as a layer is turned on in the table of contents, ArcMap draws it, regardless of the map *scale*. As you zoom out, it may become harder to distinguish features in layers that contain more detailed information. While you can turn off a layer, this may be inconvenient, especially if your map contains several layers and you change scale frequently as you work.

To help you automatically display layers at the appropriate scale, you can set a layer's *visible scale range* to define the range of scales at which ArcMap draws the layer. Whenever the scale of the data frame is outside the layer's visible scale range, the layer won't draw. In this way, you can control how the map looks at various scales.

For example, you can hide a detailed layer that might otherwise clutter your map when you zoom out. You can also progressively display more detailed layers as you zoom in on an area, that is, as the scale of the data frame gets larger. Setting a visible scale range is ▸

Setting the minimum visible scale for a layer

1. Right-click the layer in the table of contents and click Properties.
2. Click the General tab.
3. Click Don't show layer when zoomed.
4. Type a minimum scale for the layer.

   If you zoom out beyond this scale, the layer will not be visible.
5. Click OK.

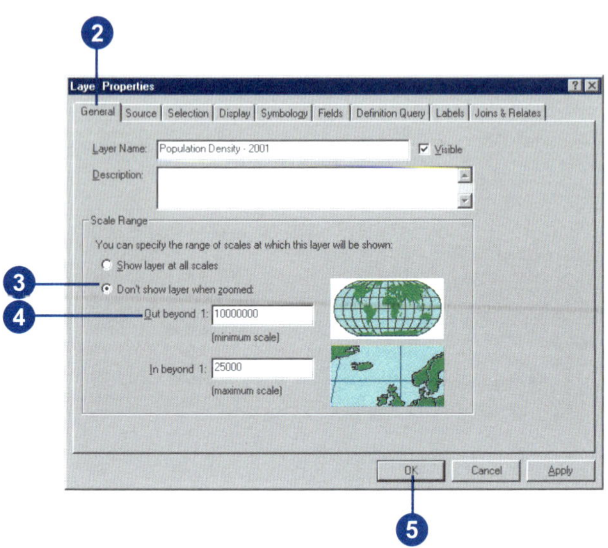

Setting the maximum visible scale for a layer

1. Right-click the layer in the table of contents and click Properties.
2. Click the General tab.
3. Click Don't show layer when zoomed.
4. Type a maximum scale for the layer.

   If you zoom in beyond this scale, the layer will not be visible.
5. Click OK.

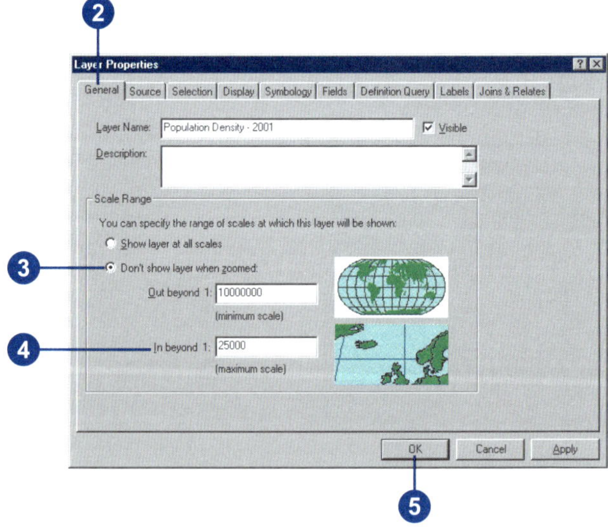

especially useful if you are creating a map for others to use because it makes browsing the map easier.

Tip

How can I tell if a layer is not drawing because of a visible scale range?

If a layer isn't drawing because it has a visible scale range set, you'll see a gray scalebar under the layer's check box in the table of contents.

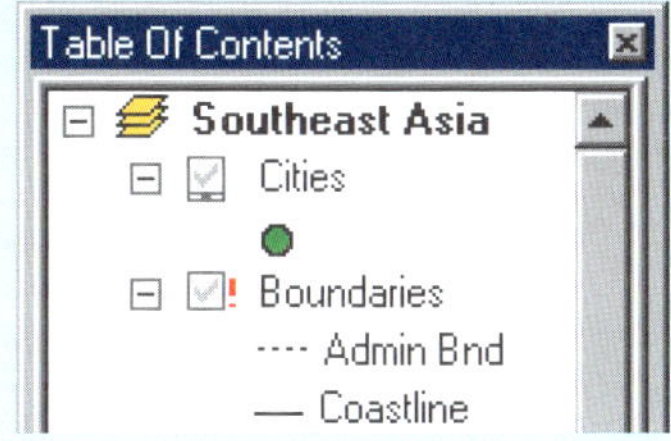

Setting a visible scale based on the current scale

1. Adjust the data frame display to the appropriate scale.
2. Right-click the layer for which you want to set a visible scale.
3. Point to Visible Scale Range and click Set Maximum Scale or Set Minimum Scale.

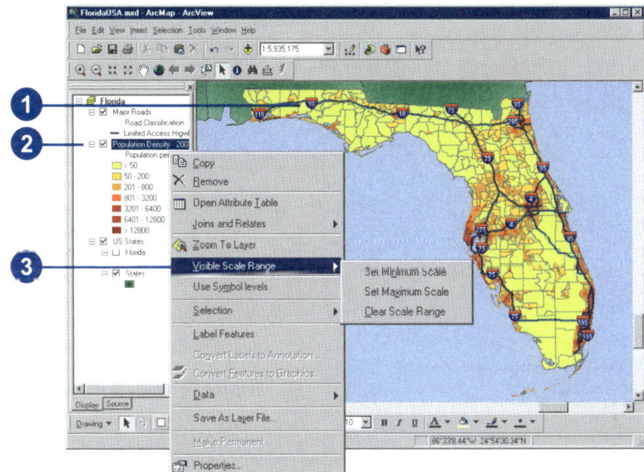

Clearing a layer's visible scale

1. Right-click the layer for which you want to clear a visible scale range.
2. Point to Visible Scale Range and click Clear Scale Range.

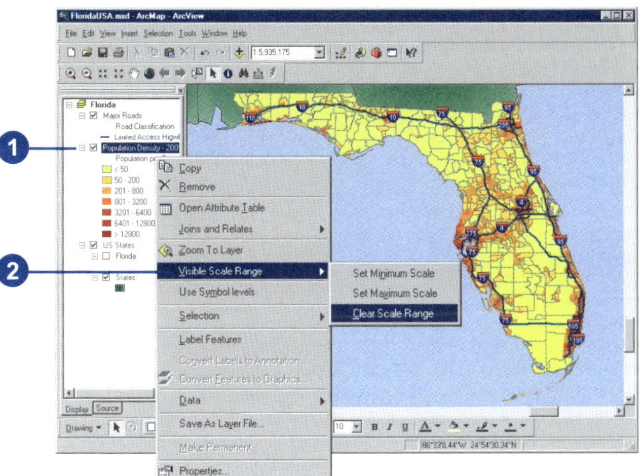

Creating a transparent layer

Giving a layer transparency is an easy way to show varying and overlapping information. One situation in which transparency might be useful is if you needed to see the information displayed in two overlapping polygon layers. Similarly, another use is to allow you to see an image under a polygon layer.

To learn about raster transparency see Chapter 9, 'Working with rasters'.

Setting the layer's transparency

1. Right-click the layer and click Properties.
2. Click the Display tab.
3. Enter the transparency value, in percent, you wish this layer to be displayed with.
4. Click OK.

   If you are uncertain how transparent you want the layer, click Apply before OK, and keep changing the transparency value before you close the Properties dialog box.

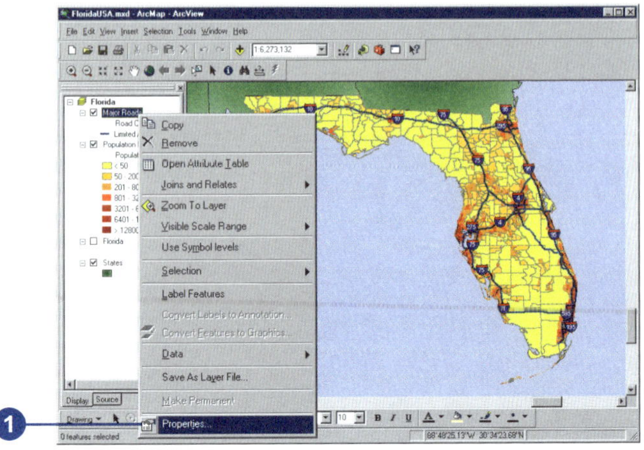

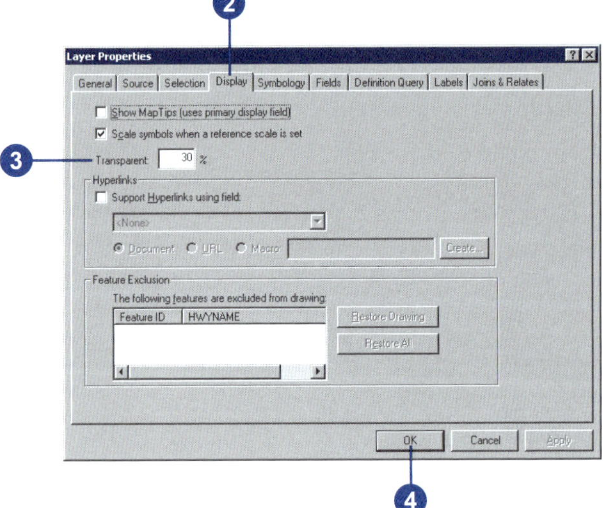

Changing a layer's source data

The properties saved within a layer file can be applied to other data sources at times. This data needs to be in the same format with the same attribute information.

You might do this when using updated census data, for example. This data will be in the same format; therefore, the data should display using the same layer properties as the previous data. You might also change source data to allow for a consistent presentation between counties using data with the same table and feature classes.

See Also

For more information on referencing and changing source data links, see Chapter 4, 'Displaying data in maps'.

Updating source data in a layer

1. Right-click the layer and point to Properties.
2. Click the Source tab.
3. Click Set Data Source.
4. Identify the new data source, and click Add.
5. Click OK.

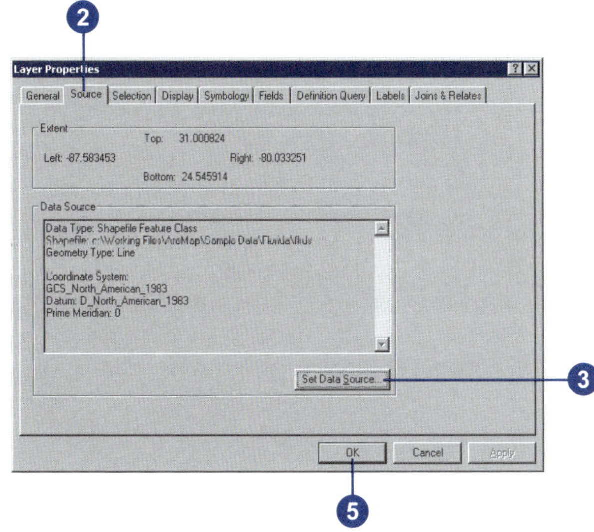

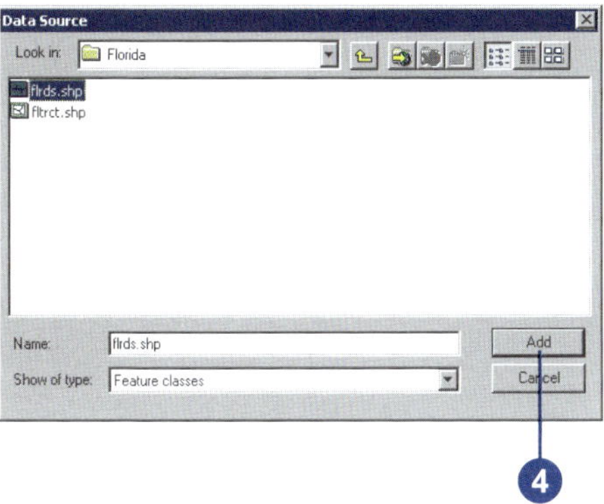

Tip

Using the table of contents Source tab

This tab displays the layers organized by data source. This is especially helpful if you have many data layers. You can also look here to see what data you need to gather when passing a layer or map document along.

Tip

Using the table of contents Selection tab

This allows you to override the global selection properties. The Selection choices can also be located at Selection > Set Selectable Layers.

Tip

Where are the table of contents tab choices saved?

Your choice of tabs is saved with the application, not the map document. However, when you save a map document it will remember the table of contents tab you were last working with.

Showing the Display, Source, and Selection tabs

1. Click the Tools menu on the Standard toolbar and click Options.
2. Click the Table Of Contents tab.
3. Check the boxes to show the Display, Source, and Selection tabs.
4. Click OK.

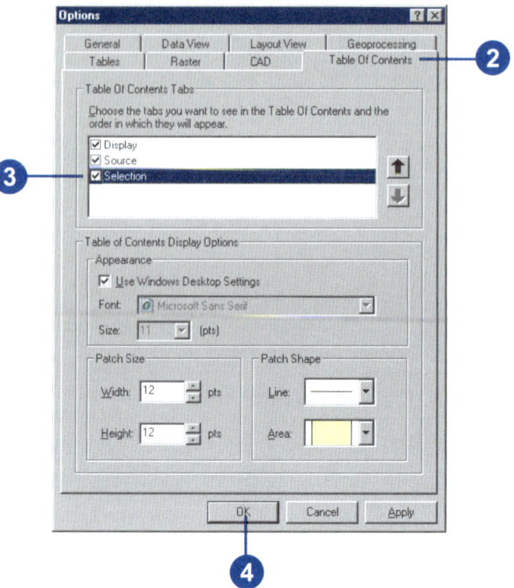

Viewing the data source in the table of contents

1. Click the Source tab in the table of contents.

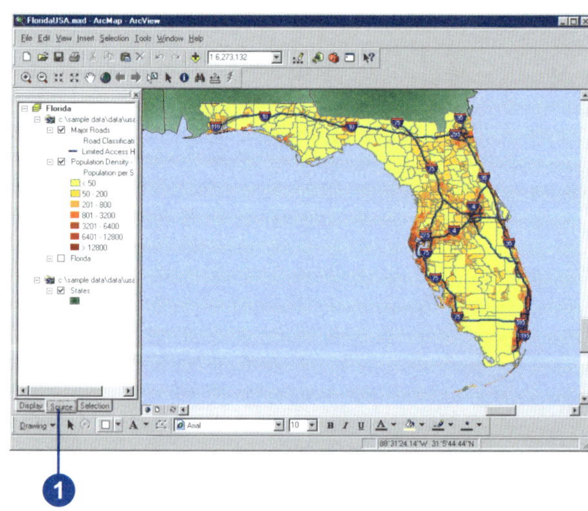

Changing the appearance of the table of contents

You can adjust the look of the table of contents to suit your needs. For example, you might change the text size and font so it makes a greater visual impact or is easier to read. You also might want to change the shape of the lines and patches that represent the features on a map.

The table of contents has three tabs at the bottom, a Display tab, Source tab, and a Selection tab. The Display tab shows the drawing order of the layers and allows you to change the order. The Source tab sorts layers by where they're stored on disk. The Selection tab shows a list of the layers in the active data frame and lets you check the ones you want to make selectable.

Seeing the Source tab is useful during editing, when you edit all layers in a given folder or database. If you're not planning on using your map for editing, you can hide the Source tab. You can't change the drawing order of layers from the Source tab.

Setting the text font for layers

1. Click the Tools menu on the Standard toolbar and click Options.
2. Click the Table Of Contents tab.
3. Uncheck Use Windows Desktop Settings.
4. Click the Font dropdown arrow and click the font you want to use.
5. Click the Size dropdown arrow and click the font size.
6. Click OK.

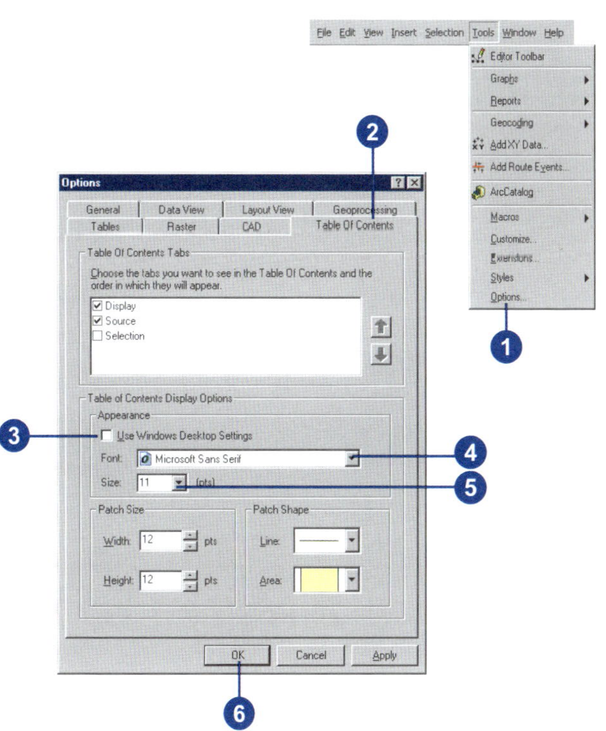

Setting the line and patch for layer symbology

1. Click the Tools menu on the Standard toolbar and click Options.
2. Click the Table Of Contents tab.
3. Click the Line or Area dropdown arrow and click the appropriate shape.
4. Click OK.

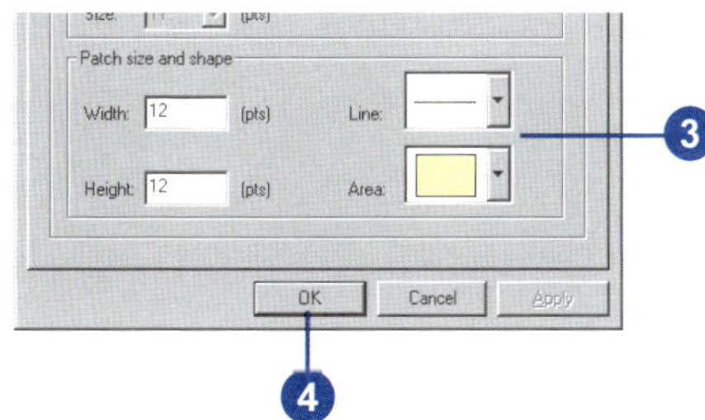

Using data frames to organize layers

A data frame is simply a frame on your map that displays layers. When you create a map, it contains a default data frame listed in the table of contents as Layers. You can immediately add layers to this data frame or give it a more meaningful name.

The layers in a data frame display in the same coordinate system and may overlap. When you want to display layers separately and not have them overlap—for example, to compare layers side by side or create insets and overviews that highlight an area—add additional data frames to your map.

When a map has more than one data frame, one of them is the active data frame. The active data frame is the one you're currently working with—for instance, adding layers to or panning and zooming. The active data frame is highlighted on the map in layout view or is the displayed data frame in data view. The name of the active data frame is also shown in bold text in the table of contents.

Once on a map, a data frame acts like any other map element. You can change its size, move it around, or delete it.

Adding a data frame

1. Click the Insert menu.
2. Click Data Frame.

   The new data frame will appear in the table of contents and in the center of the layout when in layout view.

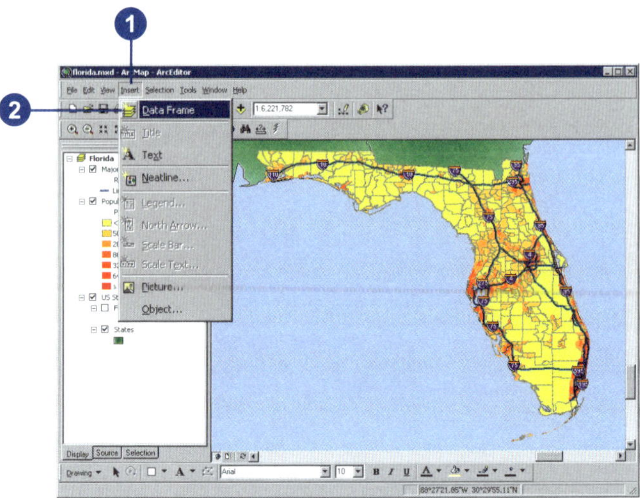

Making a data frame active

1. Right-click the data frame in the table of contents.
2. Click Activate.

   You can also click the frame in layout view to activate it.

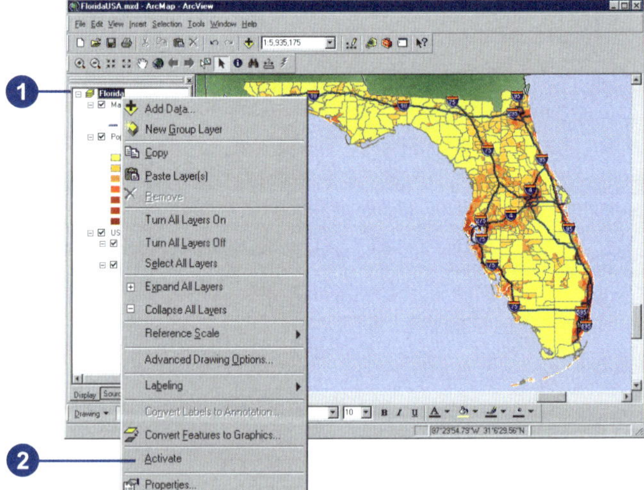

Tip

A map always has one data frame

A map must have at least one data frame on it. You can't delete the last data frame on a map.

Removing a data frame

1. Right-click the data frame in the table of contents that you want to remove.
2. Click Remove.

   In layout view, you can select the data frame and press the Delete key on the keyboard.

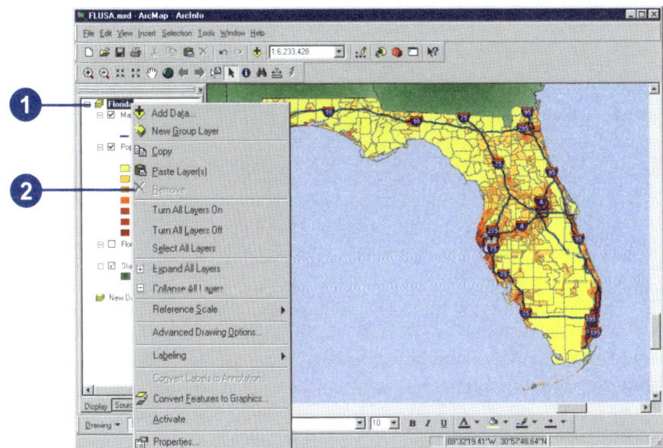

Tip

Rotating using a specific angle

If you want to rotate your data frame by a specific angle, you can type in the angle on the Data Frame Tools toolbar.

Rotating the data in a data frame

1. Click View on the Standard toolbar, point to Toolbars, and click Data Frame Tools.
2. Click the Rotate Data Frame tool.
3. Click and drag the mouse over the data frame to rotate its contents.

   Rotating the data in this manner does not alter the original source data, just its display in the data frame.

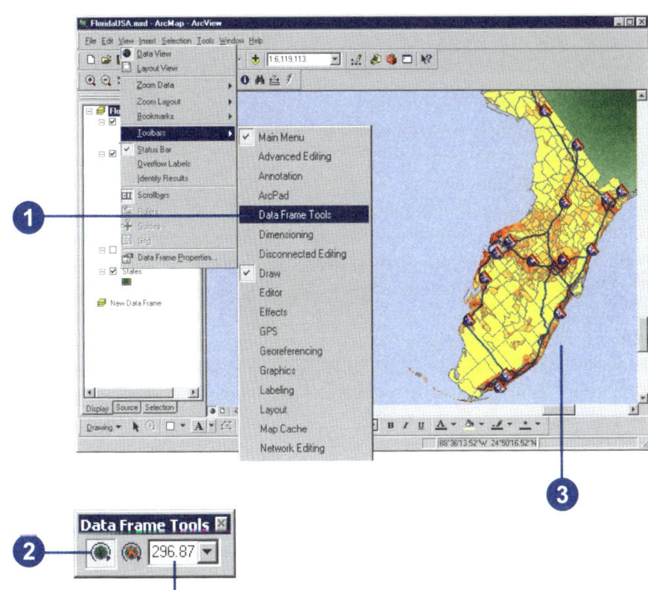

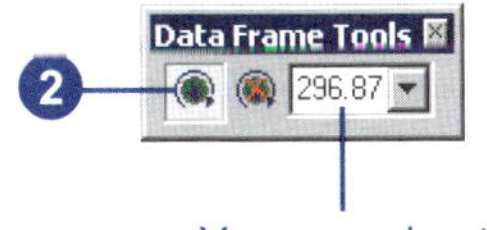

You can also type an angle here.

Symbolizing features 6

IN THIS CHAPTER

- A map gallery
- Drawing all features with one symbol
- Drawing features to show categories such as names or types
- Managing categories
- Ways to map quantitative data
- Standard classification schemes
- Setting a classification
- Representing quantity with color, graduated or proportional symbols, and charts
- Drawing features to show multiple attributes
- Drawing TINs as surfaces
- Drawing CAD layers
- Working with advanced symbolization

Choosing how to represent your data on a map may be the most important mapmaking decision. How you represent your data determines what your map communicates.

On some maps, you might simply want to show where things are. The easiest way to do this is to draw all the features in a layer with the same symbol. On other maps, you might draw features based on an attribute value or characteristic that identifies them. For example, you could map roads by type to get a better sense of traffic patterns or map the wildlife habitat suitability of a particular bird species, ranking suitability from least to most.

In general, you can draw map features:

- With a single symbol
- To show a category such as a type (unique values maps)
- To represent a quantity such as population (graduated color, graduated symbol, and dot density maps)
- To show multiple attributes that are related (multivariate and chart maps)

You can also draw images and rasters, TINs representing a three-dimensional surface, or CAD drawing files.

Browse the map gallery on the next few pages to see the various ways you can symbolize your data.

A map gallery

Single symbol map

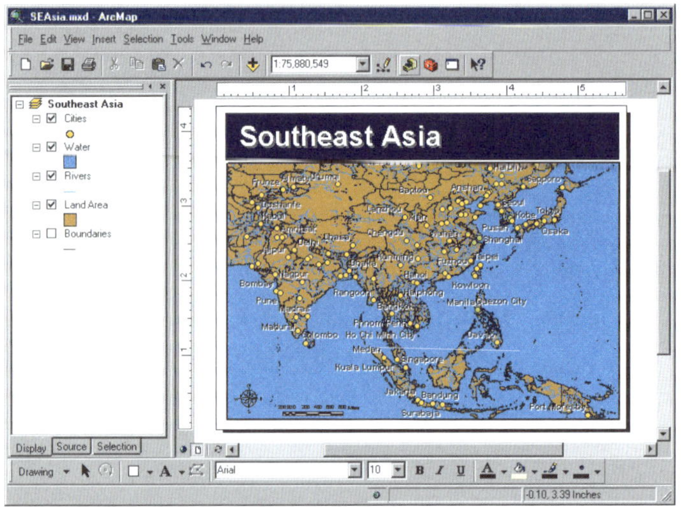

Drawing your data with just a single *symbol* gives you a sense of how features are distributed—whether they are clustered or dispersed—and may reveal hidden patterns.

In the map above, you can easily see where people live and conclude that some areas are more densely populated based on the number of cities clustered together.

Unique values map

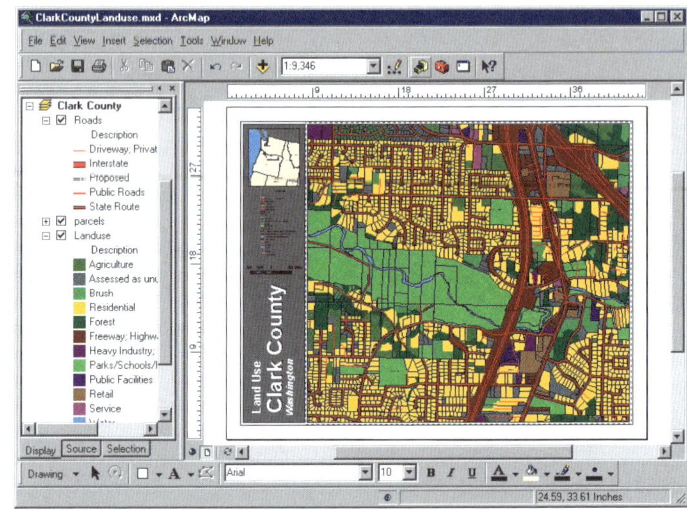

On a unique values map, you draw features based on an attribute value, or characteristic, that identifies them. In the map above, each land use type is drawn with a specific color. Typically, each unique value is symbolized with a different color. Drawing features based on unique attribute values shows the following:

- How similar features are distributed—whether they are grouped or dispersed
- How different feature types are located in relation to each other
- How much of one category there is compared to other categories

Graduated color map

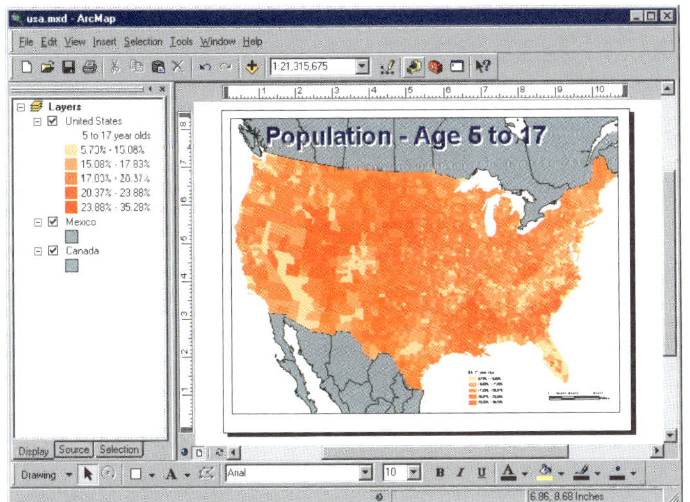

When you need to map quantities or amounts of things, you might choose to use a graduated color map. On a graduated color map, colors correspond to the values of the particular attribute. Graduated color maps are most useful for showing data that is ranked (for example, 1 to 10, low to high) or has some kind of numerical progression (for example, measurements, rates, percentages).

The map above uses different shades of color—in a graduated *color ramp*—to represent different amounts of people. Here, darker shades indicate a greater number of people.

Graduated symbol map

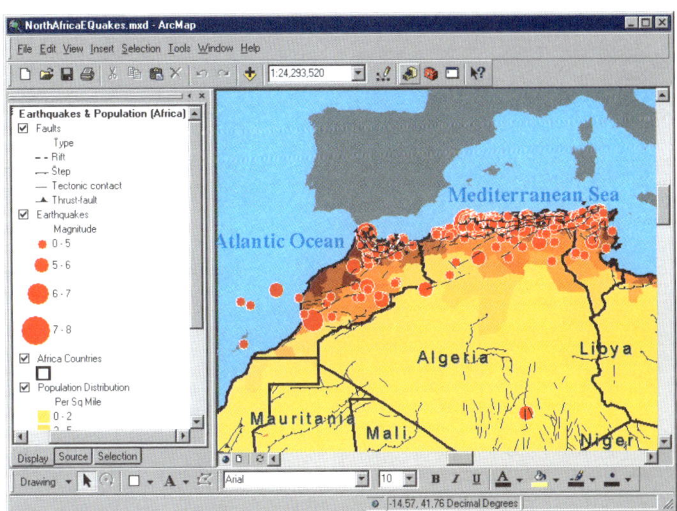

Another way to represent quantities is to vary the size of the symbol a feature is drawn with. The graduated symbol map above uses a larger symbol to show earthquakes with a larger magnitude. Similar to graduated color maps, graduated symbol maps are most useful for showing ranks or a progression of values. However, instead of using color to represent the differences in values, the size of the symbol varies.

When making a graduated symbol map, it is important to choose the range of symbol sizes carefully. The largest symbols need to be small enough that neighboring symbols don't completely cover one another. At the same time, the range in size from the smallest to the largest needs to be great enough that the symbol for each class is distinct.

Multivariate map

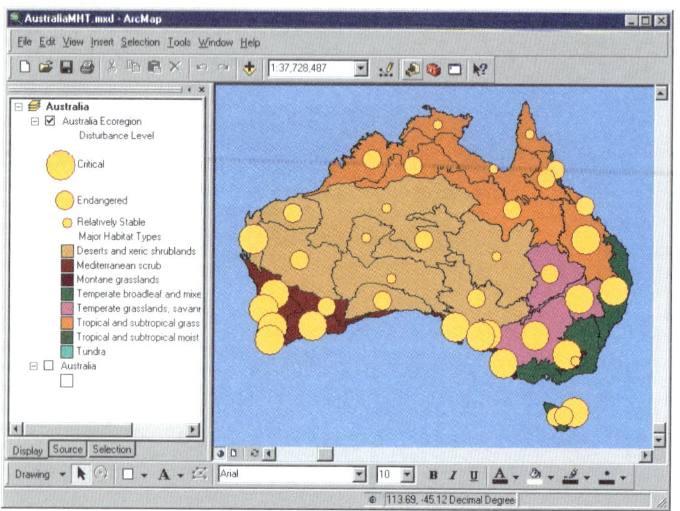

The maps on the previous pages display one attribute, or characteristic, of the data—for example, a type or an amount. Multivariate maps display two or more attributes at the same time. The map above illustrates the level of human impact on the natural landscape of Australia. Major habitat types are shown with unique colors, and the level of disturbance for each habitat is shown with a graduated symbol. The larger the symbol, the higher the human impact is on the particular habitat.

Chart map

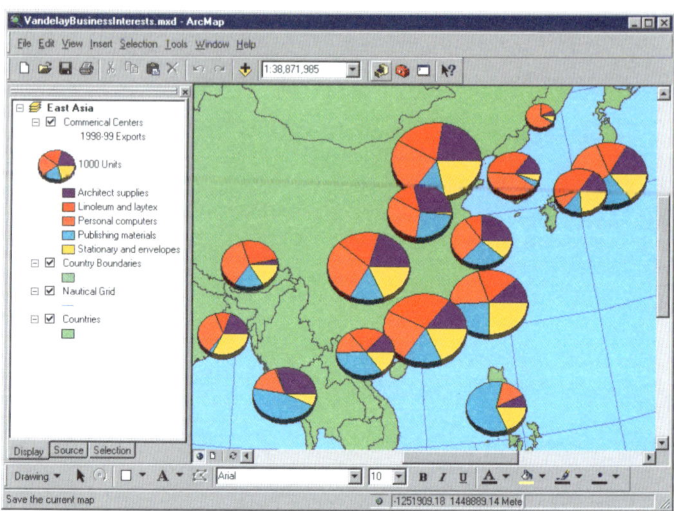

Chart maps allow you to symbolize multiple attributes on one map to communicate the relationship among different attributes. Chart maps display bar and pie charts over features. The map above illustrates the volume and type of goods distributed by an exporter throughout Asia.

Pie charts show relationships between parts and the whole and are particularly useful for showing proportions and ratios. Bar charts compare amounts of related values and are well suited to showing trends over time. Stacked bar charts can show the relative relationship between data as well as allow for absolute comparisons.

Density map

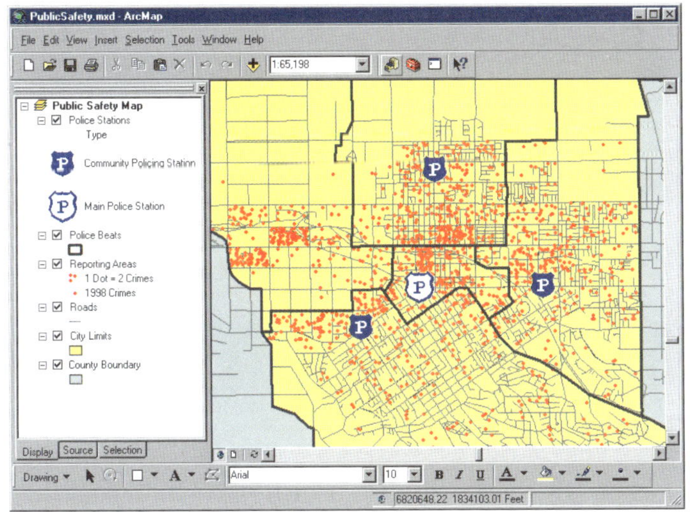

Mapping the density of features lets you see where things are concentrated. This helps you find areas that require action or meet some criteria. For example, the map above shows where the highest concentrations of crimes occur in a city. Using this map, the city may choose to increase the number of police patrols in the areas of high density.

One way to map density is with a dot density map. This type of map symbolizes features using dots drawn inside polygons to represent a quantity. Each dot represents a specific value. For example, on the crime map, each dot might represent two incidents of crime. When creating a dot density map, you specify how many features each dot represents and how big the dots are. You may need to try several combinations of amount and size to see which one best shows the pattern.

Raster map

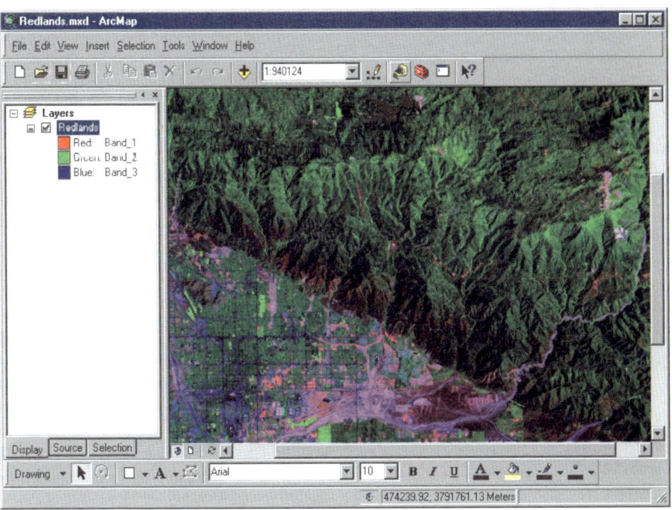

Much of the most readily available geographic data is in the form of rasters. A raster can represent almost any geographic feature, though most rasters you'll work with in ArcMap will probably be scanned maps, photographs of the earth's surface, or DEMs. You might add an aerial photograph to your map to provide a realistic background to your other data, or you might use satellite imagery to add up-to-the-minute information about weather conditions or flood levels. You can even use a raster as a guide for editing—for example, you might scan in a map and digitize features from it.

For more information on displaying raster data, see Chapter 9, 'Working with rasters'.

TIN surface map

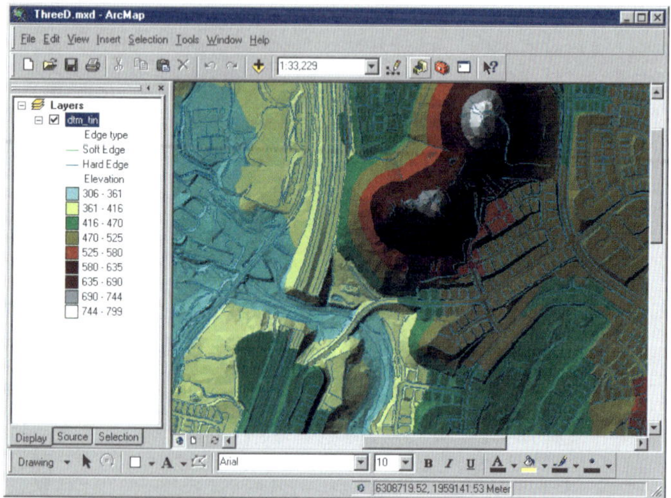

One of the ways you can represent a continuous surface, such as terrain elevation or temperature gradient, is to display the surface as a TIN. TINs can be shown with color-shaded relief. This type of map displays elevation ranges in graduated colors, and it shades ridges, valleys, and hillsides using a simulated light source. The shading adds a realistic effect that makes the surface look as though you are viewing it from high above. The combined use of color for elevation and shading for surface morphology results in a highly informative, yet easy to interpret, view of the surface.

Computer-aided design map

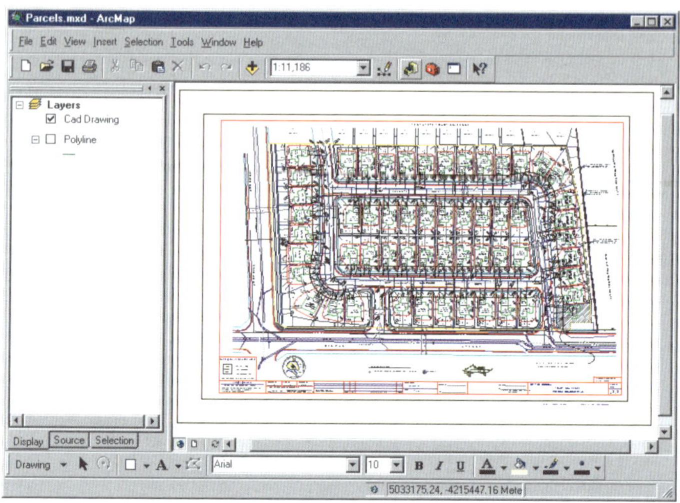

You can integrate CAD drawings into your maps seamlessly, without having to convert these files into other GIS formats. This is particularly useful if your organization has existing CAD data. For example, some departments in your organization may be using a CAD package to help manage facilities and other infrastructure. You can let ArcMap draw these layers as they appear in the CAD package, or you can override their symbology and draw them as you wish.

Drawing all features with one symbol

Often, seeing where something is—and where it isn't—can tell you exactly what you need to know. Mapping the location of features reveals patterns and trends that can help you make better decisions. For example, a business owner might map where his or her customers live to help decide where to target his or her advertising.

The easiest way to see where features are is to draw them using a single symbol. You can draw any type of data this way. When you create a new layer, ArcMap by default draws it with a single symbol.

Drawing a layer using a single symbol

1. In the table of contents, right-click the layer you want to draw with a single symbol and click Properties.
2. Click the Symbology tab.
3. Click Features.

   Because Single symbol is the only option, ArcMap automatically selects it.
4. Click the Symbol button to change the symbol. ►

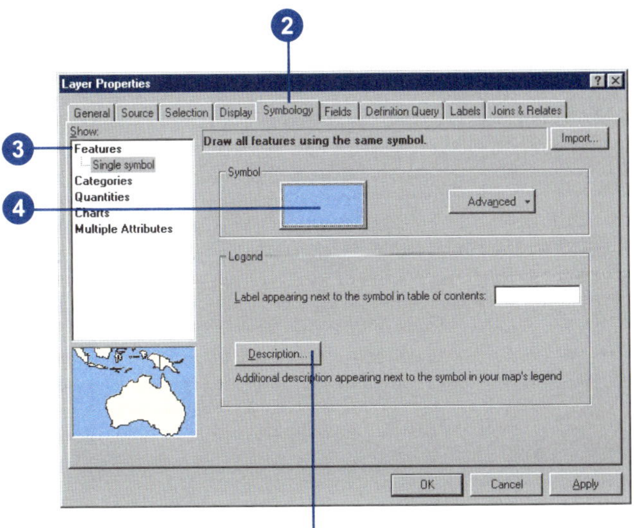

Click Description if you want an additional description of your layer to appear in your legend. You can press the Ctrl and Enter keys together in the Description for Legend dialog box to insert a line break in your description. For more information on working with legends, see Chapter 15, 'Laying out and printing maps'.

Tip

Changing the symbol

To change the symbol that features are drawn with quickly, click the symbol in the table of contents to display the Symbol Selector.

Tip

Changing the color

To change the color of a symbol quickly, right-click the symbol in the table of contents to display the Color Selector.

5. In the Symbol Selector dialog box, click a new symbol or change specific properties of the symbol.
6. Click OK on the Symbol Selector dialog box.
7. Type a label for the feature.

   The label appears next to the symbol in the table of contents.
8. Click OK.

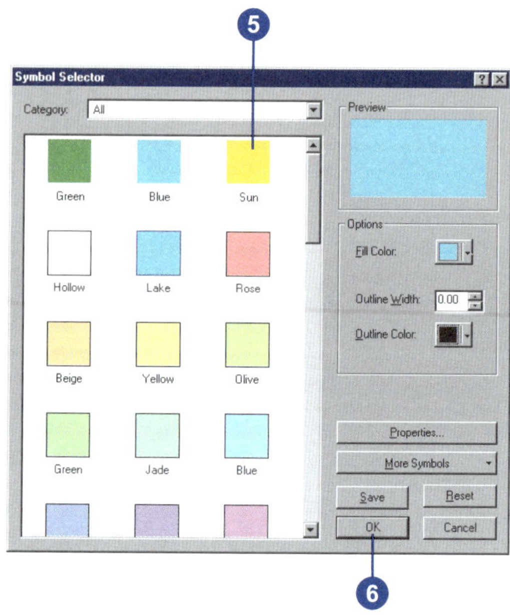

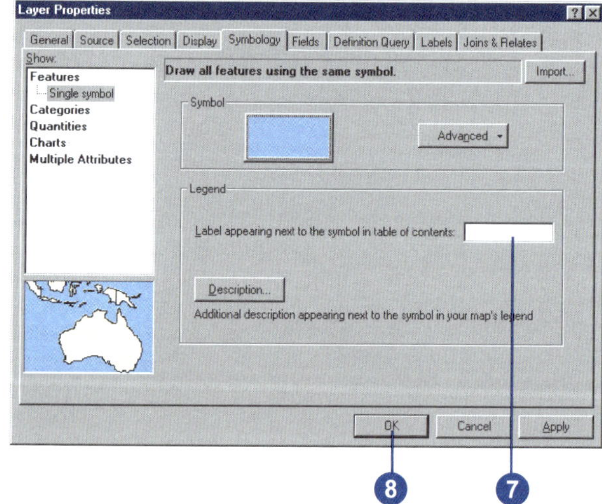

Drawing features to show categories such as names or types

A category describes a set of features with the same attribute value. For example, given parcel data with an attribute describing land use, such as residential, commercial, and public areas, you can use a different symbol to represent each land use. Drawing features this way allows you to map features and what category they belong to. This can be useful if you're targeting a specific type of feature for an action or policy. For instance, a city planner might use the land use map to target areas for redevelopment.

In general, look for these kinds of attributes when mapping by category or unique value:

- Attributes describing the name, type, or condition of a feature.
- Attributes that uniquely identify features; for example, a county name attribute could be used to draw each county with a unique color.

You can let ArcMap assign a symbol to each unique value based on a color scheme you ▸

Drawing a layer showing unique values

1. In the table of contents, right-click the layer you want to draw showing unique values and click Properties.
2. Click the Symbology tab.
3. Click Categories.

   ArcMap automatically selects the Unique values option.
4. Click the Value Field dropdown arrow and click the field that contains the values you want to map.
5. Click the Color Scheme dropdown arrow and click a color scheme.
6. Click Add All Values.

   This adds all unique values to the list. Alternatively, click the Add Values button to choose which unique values to display.
7. If you want to edit the default label so more descriptive labels appear in your legend and the table of contents, click a label in the Label column and type the label you want.
8. Click OK.

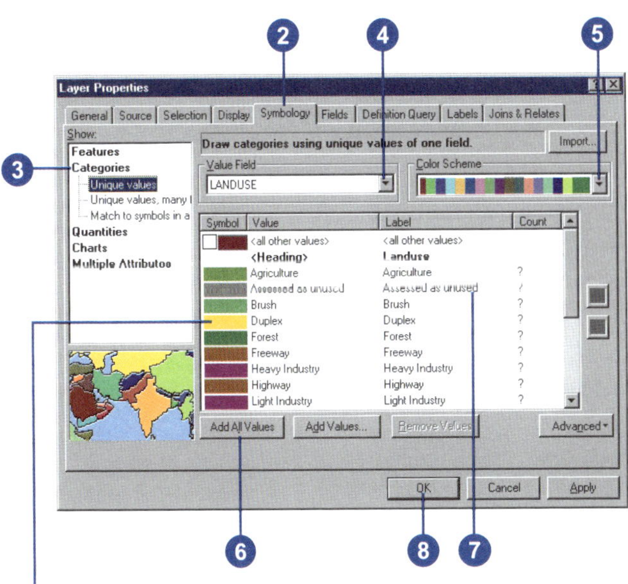

Double-click a symbol to change it.

choose, or you can explicitly assign a specific symbol to a specific attribute value.

To draw features with specific symbols based on attribute values, you need to create a style beforehand that contains symbols named after the attribute value they represent. For example, if you have a dataset that categorizes roads as either major or minor, then you would need to have line symbols within that style named “major” and “minor”. ArcMap will match the attribute value to the line symbol name to draw the feature. Features that don’t have a matching line symbol won’t be drawn. This way of drawing features is especially useful if you want to draw your data the same way on different maps.

See Also

For more information on creating styles, see Chapter 8, ‘Working with styles and symbols’.

Drawing features by referencing specific symbols in a style

1. In the table of contents, right-click the layer you want to draw showing unique values and click Properties.
2. Click the Symbology tab.
3. Click Categories.
4. Click Match to symbols in a style.
5. Click the Value Field dropdown arrow and click the field that contains the values you want to map.
6. Click the Match to symbols in Style dropdown arrow and click the style that contains symbol names that match attribute values. If the style you want is not displayed in the list, click Browse to search for it on disk.
7. Click Match Symbols.

   This adds all unique values that have a matching symbol in the style. Alternatively, click the Add Values button to choose which unique values to display.
8. If you want more descriptive labels to appear in the legend and the table of contents, click a label in the Label column and type the label you want.
9. Click OK.

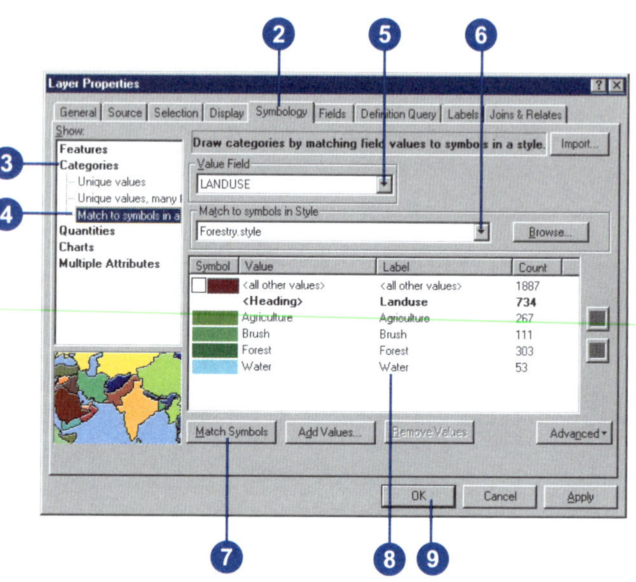

Reversing the sort of unique values

1. In the table of contents, right-click the layer whose unique values you want to sort and click Properties.
2. Click the Symbology tab.
3. Click the Value column to show a context menu.
4. Click Reverse Sorting to reverse the alphanumeric sorting of the entire list of classes.
5. Click OK.

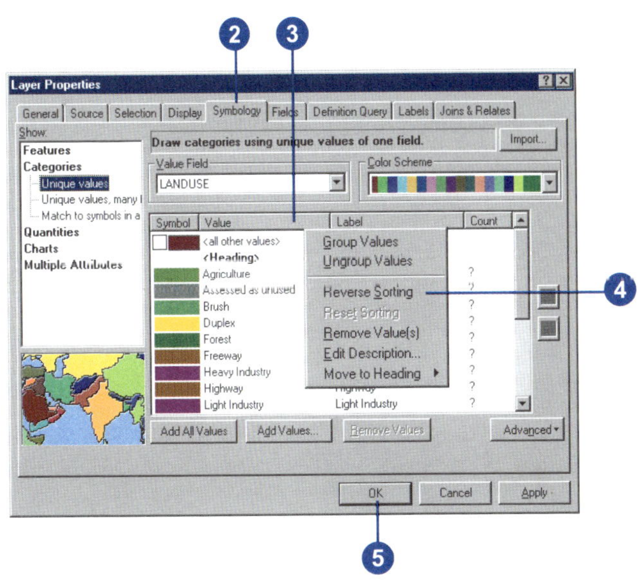

Tip

Moving values within a list

Clicking the arrow buttons within the Values list only moves the values within a heading. To learn more about using headings, see 'Adding headings' in this chapter.

Ordering unique values

1. In the table of contents, right-click the layer whose unique values you want to reorder and click Properties.
2. Click the Symbology tab.
3. Click the value you want to move up or down in the list.
4. Use the up and down arrows to either promote or demote the value in the list.
5. Click OK.

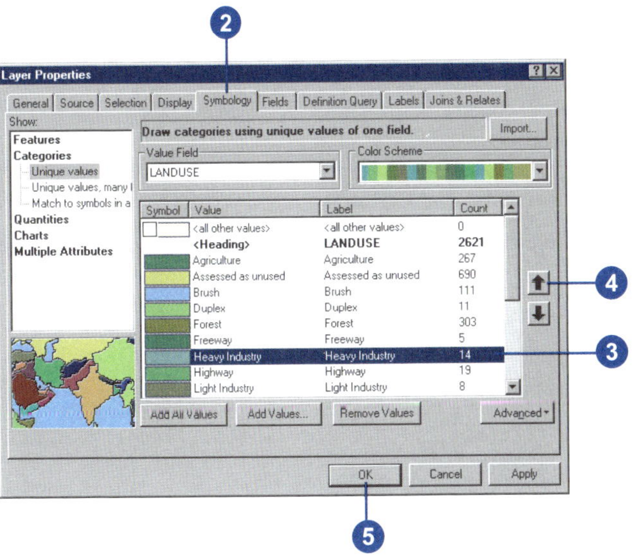

Managing categories

If you're drawing features by category, the number of categories you display will affect what patterns are revealed on the map. Most people can discern between five and seven categories for a given layer. The more advanced the audience, the more categories it will be able to identify and the more easily it can interpret complex patterns. Conversely, a less advanced audience may benefit more from a map with fewer categories.

When displaying your data, you can control how you organize and display categories for a layer. If you want to display fewer categories, you can combine similar categories into one category to help make the patterns more apparent—for example, combine two detailed land use categories into a more general one. However, the trade-off is that some information will be lost.

Alternatively, instead of reducing the number of categories, you might want to organize individual categories into groups that you define. This allows you to work with and view them as a group. ▸

Combining two or more categories into one

1. In the table of contents, right-click the layer drawn with unique values you want to combine categories for and click Properties.
2. Click the Symbology tab.

   If you don't already see categories in the scrolling list, follow the steps for 'Drawing a layer showing unique values' earlier in this chapter.
3. Click the first of the values you want to combine. Hold down the Shift or Ctrl key and click the additional values that you want to combine.
4. Right-click over the values and click Group Values. The selected values will now be combined into one category.
5. If you want more descriptive labels to appear in your legend and the table of contents, click a label in the Label column and type the label you want.
6. Click OK.

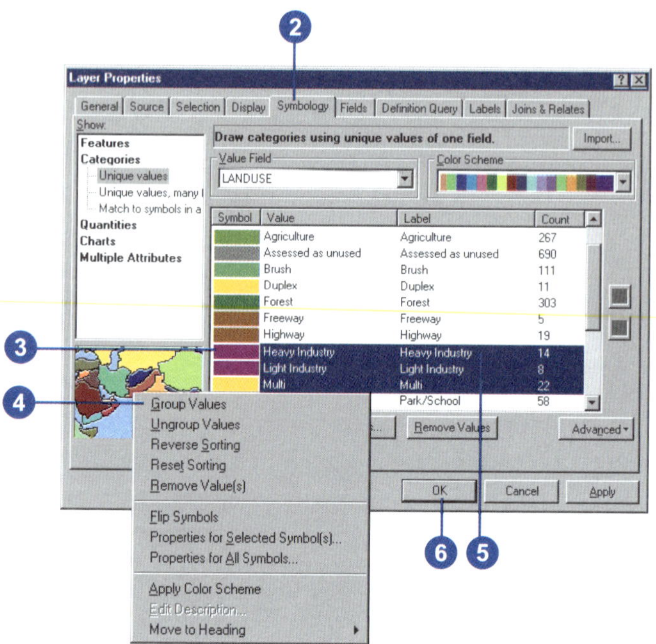

In addition, a map reader will see the groups listed in the table of contents.

You can also organize your unique values by adding headings. For example, if you were working with a land use dataset, you could create a set of broad land use category headings and organize similar categories of values into them. You might have a heading of Commercial and include land uses such as light industry, heavy industry, and retail within that heading. Headings appear in the legend and in the table of contents.

Tip

Deleting groups

ArcMap will automatically delete groups that contain no attribute values in them.

Tip

Renaming groups

Click the group heading in the table of contents and type a new name.

Ungrouping combined categories

1. Right-click a combined category in the scrolling list.
2. Click Ungroup Values.

   The symbols for each of the ungrouped values are the same as when they were grouped.

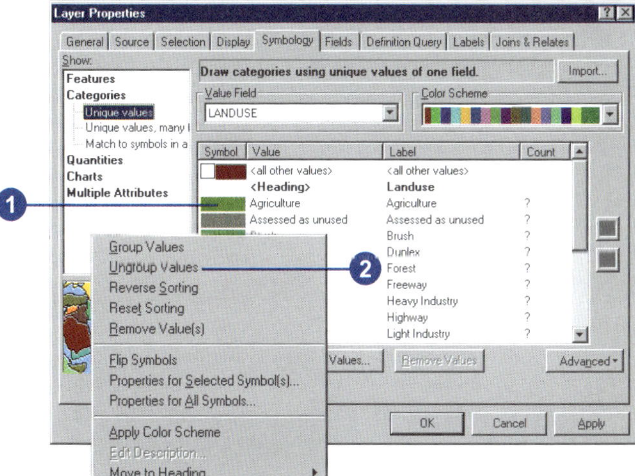

Tip

Ordering unique value headings

You can arrange the headings for unique values. Just select a heading and use the arrow keys to move it.

Adding headings

1. In the table of contents, right-click the layer drawn with unique values for which you want to organize categories in groups and click Properties.
2. Click the Symbology tab.

   If you don't already see categories in the scrolling list, follow the steps for 'Drawing a layer showing unique values' earlier in this chapter.
3. Click the first value you want to move to a new heading. Hold down the Shift or Ctrl key and click the additional values that you want to move to that heading.
4. Right-click a selected value, point to Move to Heading, and click New Heading.
5. Type a name for the new heading.
6. Click OK.
7. Click OK on the Layer Properties dialog box.

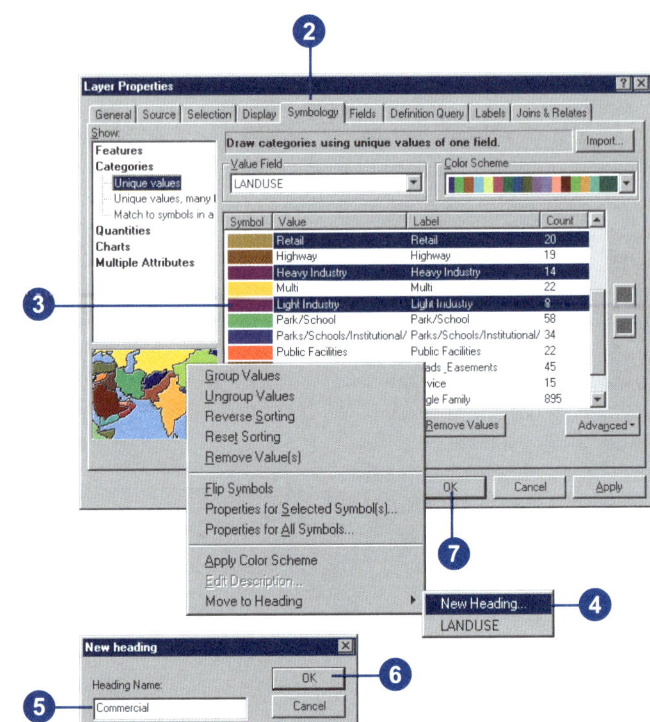

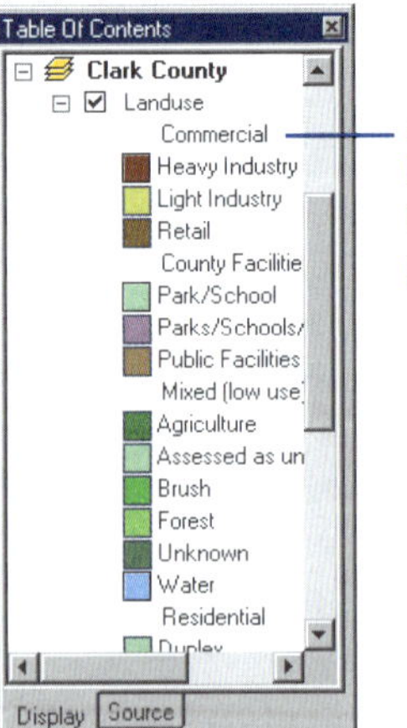

A new heading now appears in the table of contents with the values you selected grouped in it.

Ways to map quantitative data

Quantitative data is data that describes features in terms of a quantitative value measuring some magnitude of the feature. Unlike categorical data, whose features are described by a unique attribute value such as a name, quantitative data generally describes counts or amounts, ratios, or ranked values. For example, data representing precipitation, population, and habitat suitability can all be mapped quantitatively.

Which quantitative value should you map?

Knowing what type of data you have and what you want to show will help you determine what quantitative value to map. In general, you can follow these guidelines:

- Map counts or amounts if you want to see actual measured values as well as relative magnitude. Use care when mapping counts as the values may be influenced by other factors and could yield a misleading map. For example, when making a map showing the total sales figures of a product by state, the total sales figure is likely to reflect the differences in population among the states.
- Map ratios if you want to minimize differences based on the size of areas or number of features in each area. Ratios are created by dividing two data values. Using ratios is also referred to as *normalizing* the data. For example, dividing the 18- to 30-year-old population by the total population yields the percentage of people aged 18–30. Similarly, dividing a value by the area of the feature yields a value per unit area, or density.
- Map ranks if you're interested in relative measures and actual values are not important. For example, you may know a feature with a rank of 3 is higher than one ranked 2 and lower than one ranked 4, but you can't tell how much higher or lower.

Should you map individual values or group them in classes?

When you map quantitative data, you can either assign each value its own symbol or you can group values into classes using a different symbol for each class.

If you're only mapping a few values (less than 10), you can assign a unique symbol to each value. This may present a more accurate picture of the data, since you're not predetermining which features are grouped together. More likely, your data values will be too numerous to map individually and you'll want to group them in classes, or *classify* the data. A good example of classified data is a temperature map you might find in a newspaper. Instead of displaying individual temperatures, these maps show temperature bands, where each band represents a given range in temperature.

Ways to classify your data

How you define the class ranges and breaks—the high and low values that bracket each class—will determine which features fall into each class and thus what the map will look like. By changing the classes you can create various maps. Generally, the goal is to make sure features with similar values are in the same class.

Two key factors for classifying your data are the classification scheme you use and the number of classes you create. If you know your data well, you can manually define your own classes. Alternatively, you can let ArcMap classify your data using standard classification schemes. The standard classification schemes are natural breaks, quantile, equal interval, defined interval, and standard deviation. These are described on the following pages.

Standard classification schemes

Natural breaks (Jenks)

Classes are based on natural groupings inherent in the data. ArcMap identifies break points by picking the class breaks that best group similar values and maximize the differences between classes. The features are divided into classes whose boundaries are set where there are relatively big jumps in the data values.

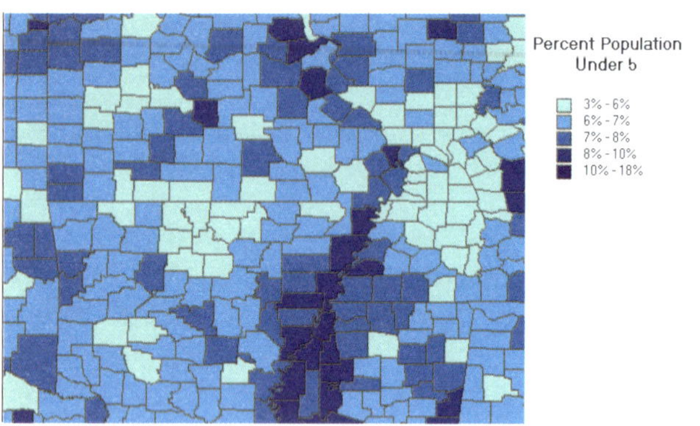

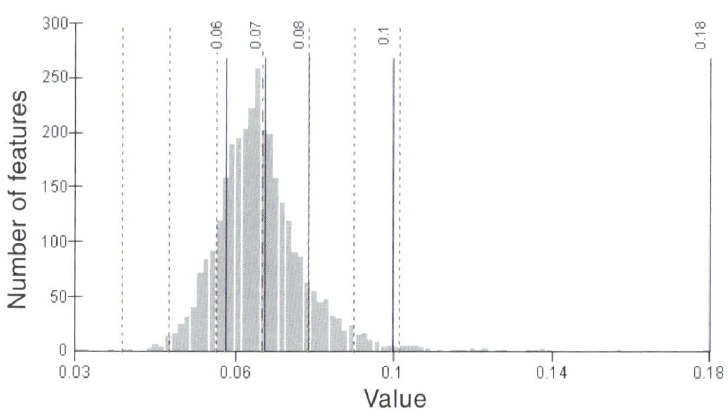

Quantile

Each class contains an equal number of features. A quantile classification is well suited to linearly distributed data. Because features are grouped by the number in each class, the resulting map can be misleading. Similar features can be placed in adjacent classes, or features with widely different values can be put in the same class. You can minimize this distortion by increasing the number of classes.

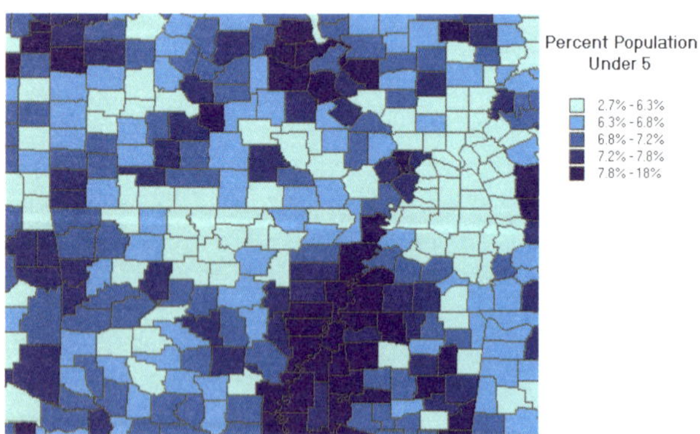

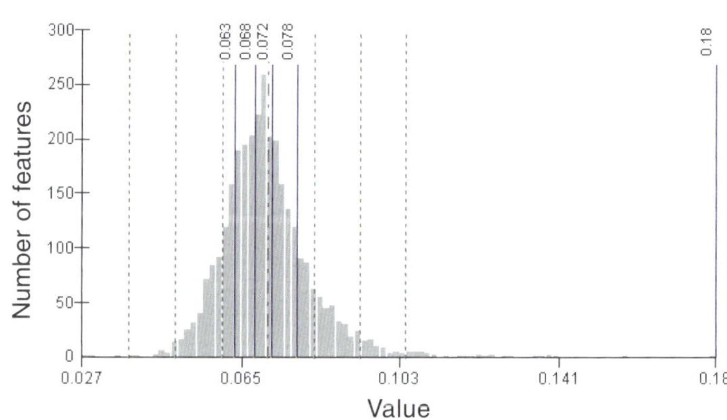

Equal interval

This classification scheme divides the range of attribute values into equal-sized subranges, allowing you to specify the number of intervals while ArcMap determines where the breaks should be. For example, if features have attribute values ranging from 0 to 300 and you have three classes, each class represents a range of 100 with class ranges of 0–100, 101–200, and 201–300. This method emphasizes the amount of an attribute value relative to other values, for example, to show that a store is part of the group of stores that made up the top one-third of all sales. It's best applied to familiar data ranges such as percentages and temperature.

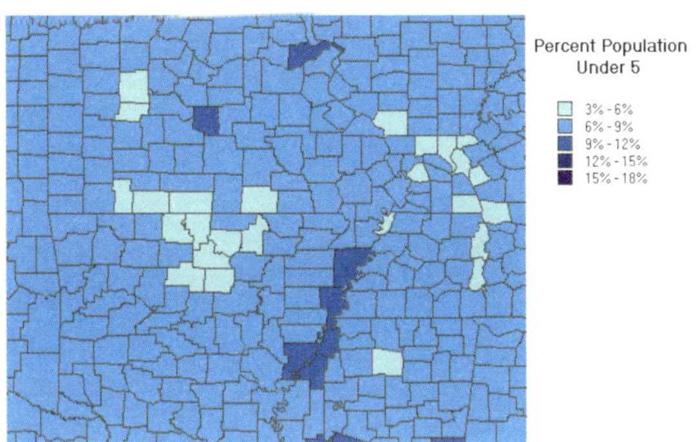

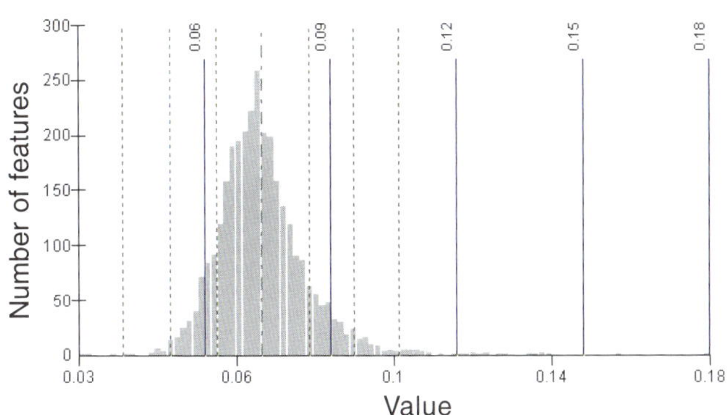

Defined interval

This classification scheme allows you to specify an interval by which to equally divide a range of attribute values. Rather than specifying the number of intervals as in the equal interval classification scheme, with this scheme, you specify the interval value. ArcMap automatically determines the number of classes based on the interval. The interval specified in the example below is 0.04 (or 4 percent).

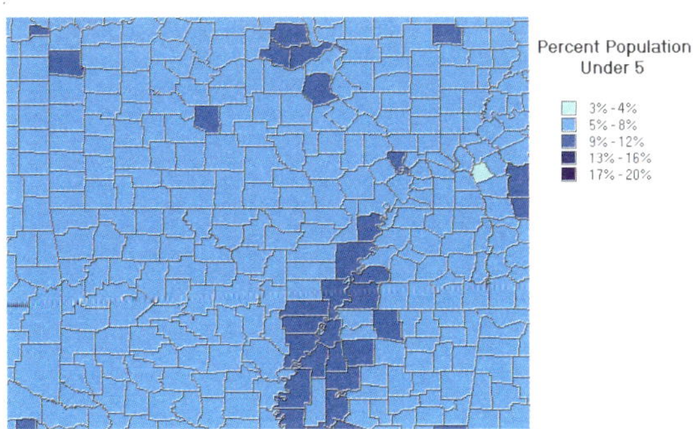

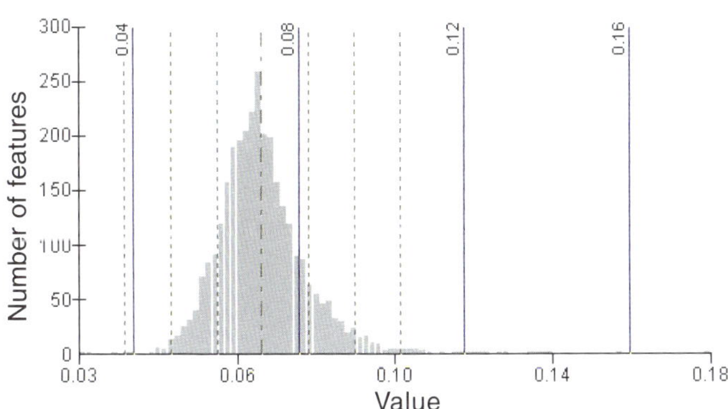

Standard deviation

This classification scheme shows you how much a feature's attribute value varies from the mean. ArcMap calculates the mean value and the standard deviations from the mean. Class breaks are then created using these values. A two-color ramp helps emphasize values above (shown in blue) and below (shown in red) the mean.

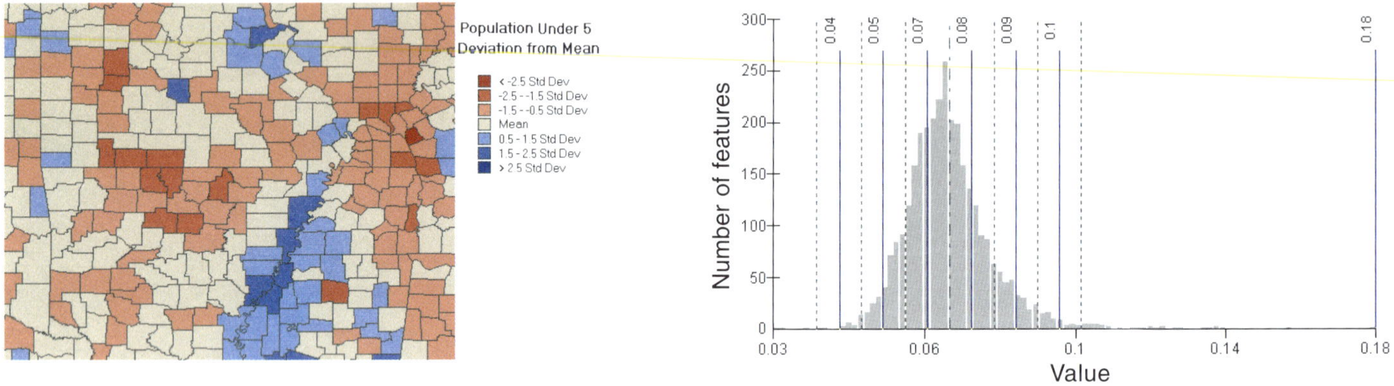

Setting a classification

When you classify your data, you can either use one of the standard classification schemes ArcMap provides or create custom classes based on class ranges you specify. If you want to let ArcMap classify the data, simply choose the classification scheme and set the number of classes. If you want to define your own classes, you can manually add class breaks and set class ranges that are appropriate for your data. Alternatively, you can start with one of the standard classifications and make adjustments as needed.

Why set class ranges manually? There may already be certain standards or guidelines for mapping your data. For example, temperature maps are often displayed with 10-degree temperature bands. Or you might want to emphasize features with particular values, for example, those above or below a threshold value. Whatever your reason, make sure you clearly specify what the classes mean on the map.

Setting a standard classification method

1. In the table of contents, right-click the layer that shows a quantitative value for which you want to change the classification.
2. Click the Symbology tab.
3. Click Quantities.
4. Click the Value dropdown arrow and click the field that contains the quantitative value you want to map.
5. To normalize the data, click the Normalization dropdown arrow and click a field.

   ArcMap divides this field into the Value to create a ratio.
6. Click Classify.
7. Click the Method dropdown arrow and click the classification method you want.
8. Click the Classes dropdown arrow and click the number of classes you want to display.
9. Click OK on the Classification dialog box.
10. Click OK on the Layer Properties dialog box.

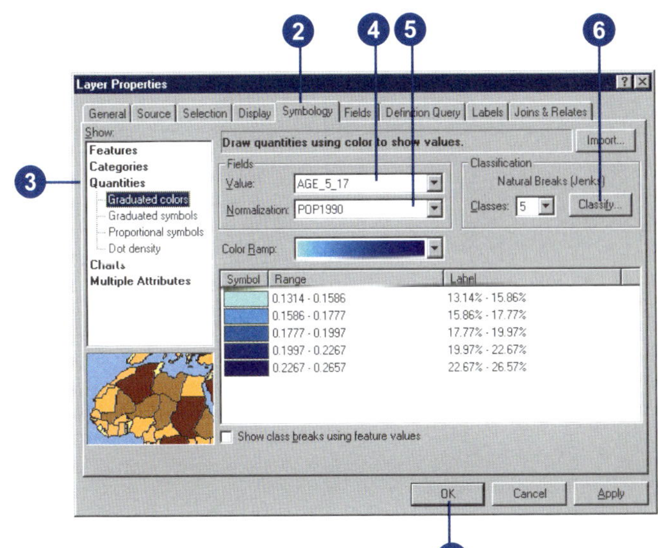

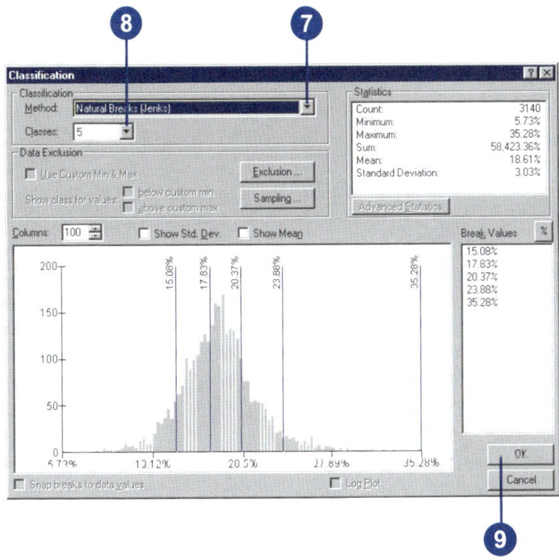

Tip

Adjusting the range using the Classification dialog box

Moving the blue vertical bars on the histogram changes the range. You can also type new breaks in the Break Values list on the right side of the Classification dialog box.

Tip

Switching to a manual classification scheme

Anytime you insert, delete, or move class breaks, the classification scheme automatically switches to manual, no matter what scheme you started with.

Tip

Seeing summary statistics

Statistics such as Minimum, Maximum, Sum, and Standard Deviation appear on the Classification dialog box. If you check Show Mean, the mean is plotted on the histogram; checking Show Std. Dev. superimposes standard deviation lines on the histogram.

Tip

Snapping to data values

Checking Snap breaks to data values uses actual data values as class breaks when you insert or move a class break. This option is only available when using a manual classification method.

Inserting your own class breaks and setting a range

1. In the table of contents, right-click the layer for which you want to set class breaks and click Properties.
2. Click the Symbology tab.

   If you have not already set a classification method, follow the steps in 'Setting a standard classification method' earlier in this chapter.
3. Click the Range you want to edit.

   Make sure to click the Range, not the Label.
4. Type a new value. This sets the upper value of the range.
5. Click OK.

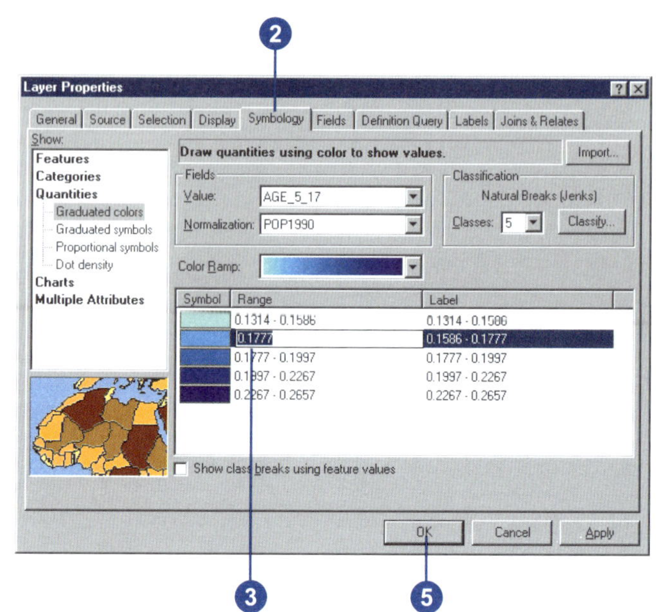

Deleting a class break

1. Click Classify from the Symbology tab of the Layer Properties dialog box.
2. Click the class break you want to delete.

   The selected break is highlighted.
3. Right-click over the histogram and click Delete Break.

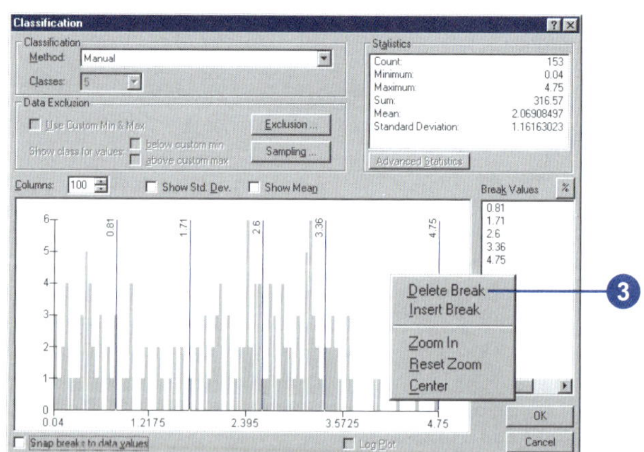

Tip

Seeing more data values plotted on the histogram

Increase the number of columns shown to see more data values in the histogram.

Tip

Loading, verifying, and saving SQL expressions

You can click Load to add an existing expression into the Data Exclusion Properties dialog box. You can also check for errors in your expression by clicking the Verify button. If you want to reuse your expression in another query, click Save.

See Also

For more information on building query expressions, see Chapter 13, 'Querying maps'.

Excluding features from the classification

1. Click Classify from the Symbology tab of the Layer Properties dialog box.
2. Click Exclusion.
3. Double-click the field you're using to draw the layer.
4. Click an operator.
5. Click Get Unique Values and double-click the value you want to exclude.
6. Click OK to execute the expression and exclude values.

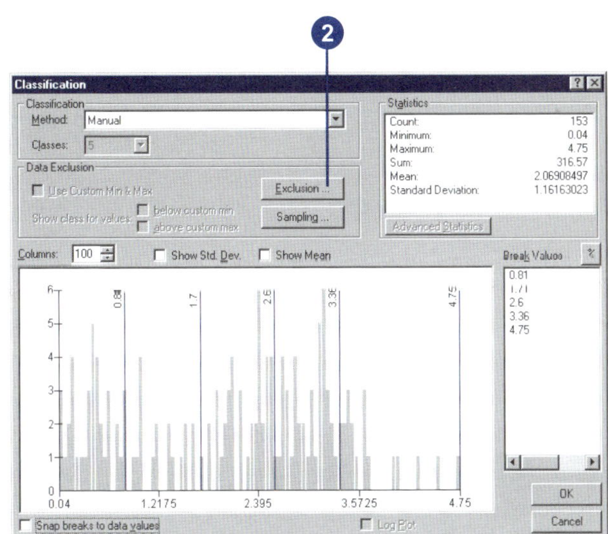

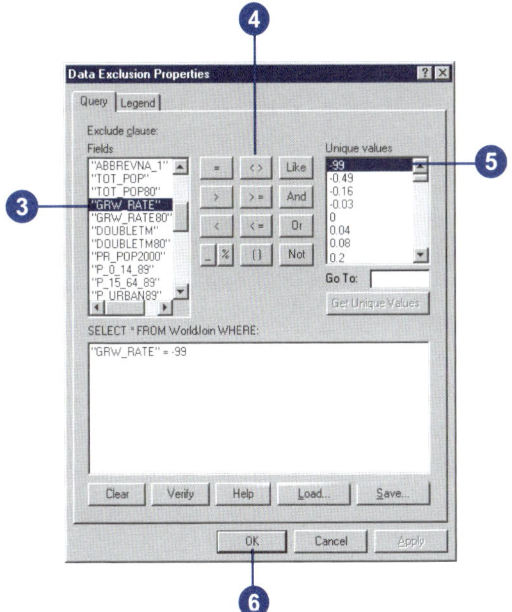

Representing quantity with color

When you want your map to communicate how much of something there is in a feature, you need to draw those features using a quantitative measure. This measure might be a count; a ratio such as a percentage; or a rank such as high, medium, and low.

You can represent quantities on a map by varying colors. For example, you might use darker shades of blue to represent higher rainfall amounts.

Before you display your data, you may need to classify it in order to group features with similar values into discrete classes that are displayed with the same symbol. After choosing a classification scheme and specifying the number of classes you want to show, you can add more classes, delete classes, or redefine class ranges.

It's always a good idea to examine your data before you map it. For instance, you may find that you have a few extremely high or low values or null values where no data is available. These values can skew a classification and thus the patterns on the map. ▸

Symbolizing data with graduated colors

1. In the table of contents, right-click the layer you want to draw showing a quantitative value and click Properties.
2. Click the Symbology tab.
3. Click Quantities.

   ArcMap automatically selects Graduated colors.
4. Click the Value dropdown arrow and click the field that contains the quantitative value you want to map.
5. To normalize the data, click the Normalization dropdown arrow and click a field.

   ArcMap divides this field into the Value to create a ratio.
6. Click Classify.
7. Click the Method dropdown arrow and click the classification method you want.
8. Click the Classes dropdown arrow and click the number of classes you want to display.
9. Click OK on the Classification dialog box.
10. Click the Color Ramp dropdown arrow and click a ramp to display the data with.
11. Click OK on the Layer Properties dialog box.

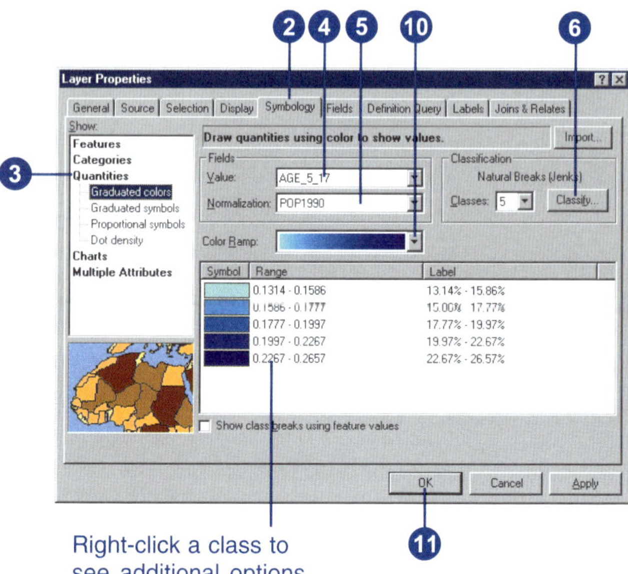

Right-click a class to see additional options, such as sorting and number formatting.

You may choose to exclude these values before you classify your data.

You may also want to normalize your data before you map it. When you normalize data, you divide it by another attribute to create a ratio. Often, ratios are easier to understand than the raw data values. For example, dividing total population by area yields the number of people per unit area, or density. Dividing a store's sales figure by the total sales for all stores yields a percentage of sales at that store.

In addition, normalizing data minimizes differences in values based on the size of areas or numbers of features in each area. When mapping total values with graduated colors, varying sizes of areas may alter the map's appearance or obscure uniform distributions. Therefore, normalizing data is a common practice when mapping quantities for areas.

See Also

For more information on creating and managing styles, see Chapter 8, 'Working with styles and symbols'.

Creating your own color ramp for a layer

1. In the table of contents, right-click the layer that shows a quantitative value and click Properties.
2. Click the Symbology tab.
3. Click Quantities.
4. Double-click the top symbol in the list and set the start color for the ramp.
5. Double-click the bottom symbol and set the end color.
6. Optionally, double-click any middle symbol to set its color.

   This lets you create a multipart color ramp.
7. Click all the middle symbols you've set the color of.

   By selecting one or more middle symbols, the color of those symbols is included in the new ramp. Otherwise, ArcMap only uses the top and bottom symbols.
8. Right-click a symbol and click Ramp Colors.
9. Optionally, if you want to use the new color ramp on another layer, right-click the Color Ramp dropdown and click Save to style to save your new ramp to your default style.
10. Click OK.

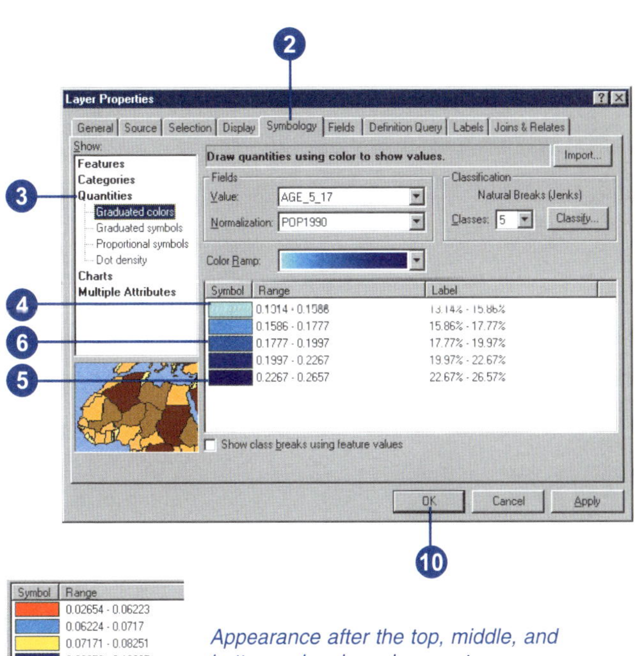

Appearance after the top, middle, and bottom colors have been set.

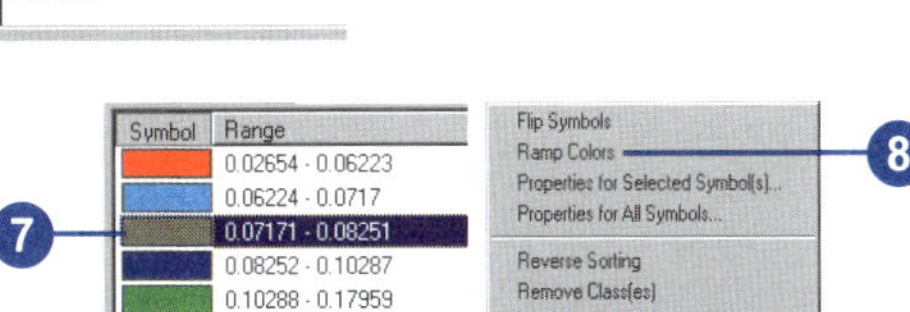

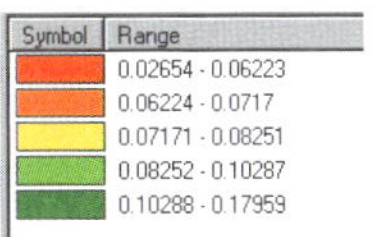

Resulting ramp goes from red to yellow to green.

Representing quantity with graduated or proportional symbols

You can represent quantities on a map by varying the symbol size you use to draw features. For example, you might use larger circles to represent cities with larger populations.

When you draw features with graduated symbols, the quantitative values are grouped into classes. Within a class, all features are drawn with the same symbol. You can't discern the value of individual features; you can only tell that its value is within a certain range.

Proportional symbols represent data values more precisely. The size of a proportional symbol reflects the actual data value. For example, you might map earthquakes using proportional circles, where the radius of the circle is proportional to the magnitude of the quake.

The difficulty with proportional symbols arises when you have too many values; the differences between symbols may become indistinguishable. In addition, the symbols for high values can become so large as to obscure other symbols.

Representing quantity with graduated symbols

1. In the table of contents, right-click the layer you want to draw showing a quantitative value and click Properties.
2. Click the Symbology tab.
3. Click Quantities and click Graduated symbols.
4. Click the Value dropdown arrow and click the field that contains the quantitative value you want to map.
5. To normalize the data, click the Normalization dropdown arrow and click a field.

   ArcMap divides this field into the Value to create a ratio.
6. Type the minimum and maximum symbol sizes.
7. Click Classify.
8. Click the Method dropdown arrow and click the classification method you want.
9. Click the Classes dropdown arrow and click the number of classes you want.
10. Optionally, click Exclusion to remove unwanted values from the classification (for example, null values or extreme outliers).
11. Click OK on the Classification dialog box.
12. Click OK on the Layer Properties dialog box.

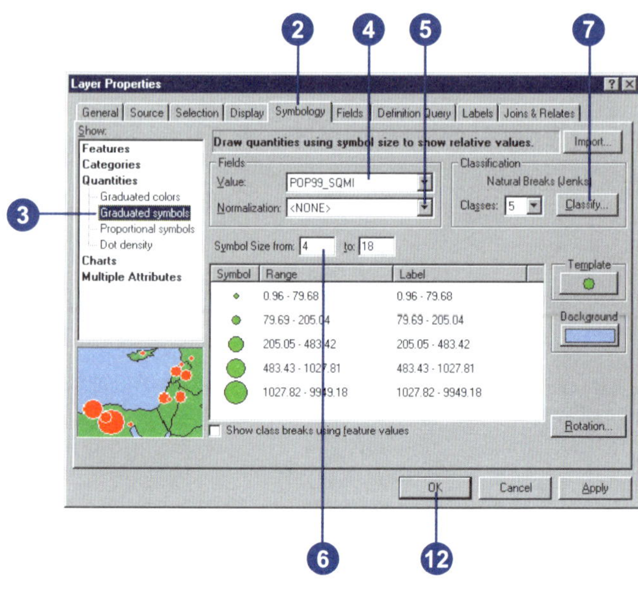

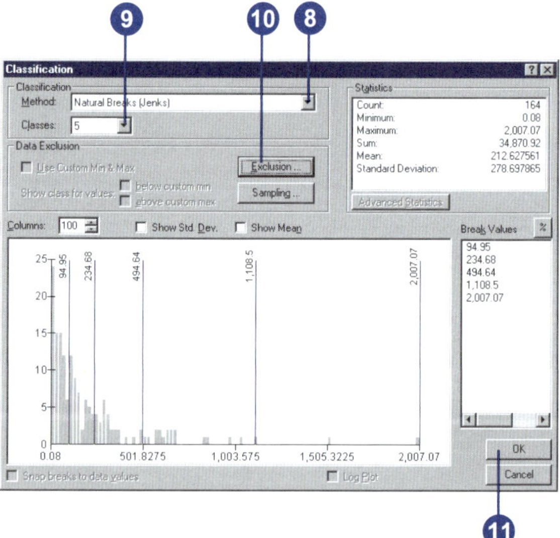

Tip

Changing the symbology and background of graduated symbols

Click the Template button to choose another marker. If you click Background, you can choose the fill symbol that will be used behind your graduated symbols.

Tip

Why don't the symbols get bigger when I zoom in?

If you want your graduated symbols to get bigger when you zoom in, set a reference scale. Right-click the data frame and click Set Reference Scale. Your symbols will appear at the same size on screen as they are when your map is printed.

Tip

What color should my symbols be?

All the symbols should be the same color. Make sure they contrast enough to be seen easily.

Tip

Formatting numbers in your class range labels

You can set the number format properties for the class range labels—which appear in the legend and table of contents—by clicking the Label column heading and clicking Format Labels.

Representing quantity with proportional symbols

1. In the table of contents, right-click the layer you want to draw showing a quantitative value and click Properties.
2. Click the Symbology tab.
3. Click Quantities and click Proportional symbols.
4. Click the Value dropdown arrow and click the field that contains the quantitative value you want to map.
5. To normalize the data, click the Normalization dropdown arrow and click a field.

   ArcMap divides this field into the Value to create a ratio.
6. Optionally, click Background and Min Value to change the symbol properties and background of the proportional symbols.
7. Click OK.

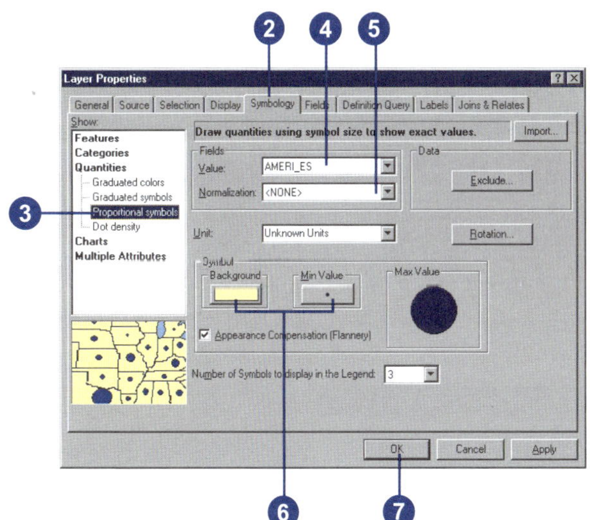

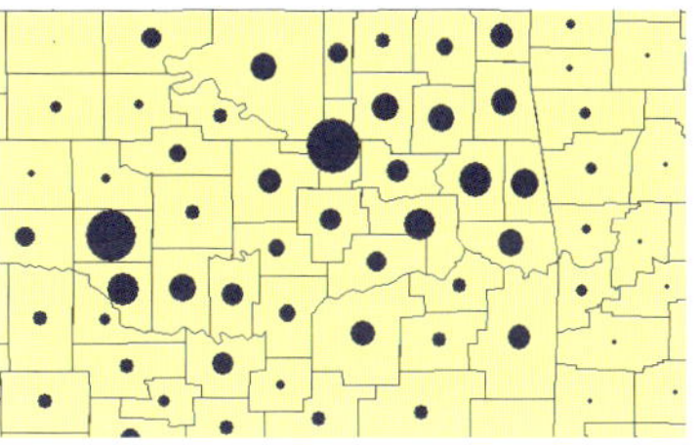

How the proportional symbols appear on your map when the Background is yellow with a black outline and the Min Value symbol is blue with a black outline

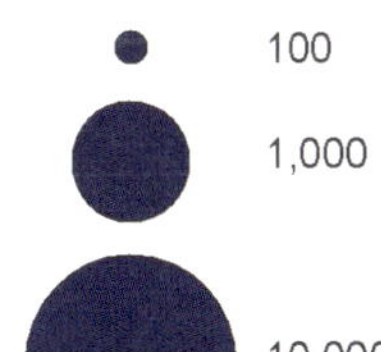

Proportional symbols appear in the legend and table of contents as a stack of progressively larger blue circles.

Tip

Using proportional symbol appearance compensation

Checking Appearance Compensation in the Symbol box turns on Flannery compensation, a technique that adjusts larger symbol sizes upward to account for the fact that map readers tend to underestimate the larger sizes of circular symbols.

Tip

Changing the symbology and background of proportional symbols

Click the Min Value button to choose another marker for your proportional symbols. If you click Background, you can choose the fill symbol that will be used behind your proportional symbols.

Tip

What if the maximum value symbol is too large?

If the symbol for the maximum value fills the space on the dialog box, it will probably be too big on the map. Try reducing the symbol size for the minimum value, normalizing the data, or excluding some values. If it's still too large, use graduated symbols instead.

Representing quantity with proportional symbols when values are measurements on a map

1. In the table of contents, right-click the layer you want to draw showing a quantitative value and click Properties.
2. Click the Symbology tab.
3. Click Quantities and click Proportional symbols.
4. Click the Value dropdown arrow and click the field that contains the quantitative value you want to map.
5. To normalize the data, click the Normalization dropdown arrow and click a field.

   ArcMap divides this field into the Value to create a ratio.
6. Click the Unit dropdown arrow and click a unit.
7. Click Square or Circle as the symbol. Optionally, change the color of the symbol and click Background to change the background of the proportional symbols.
8. Click Radius or Area.

   For example, click Radius if your data represents the distance an earthquake was felt from its epicenter. Click Area if the value represents an area.
9. Click OK.

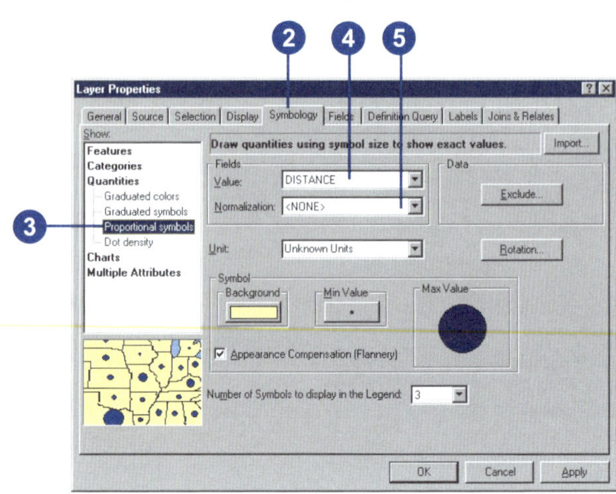

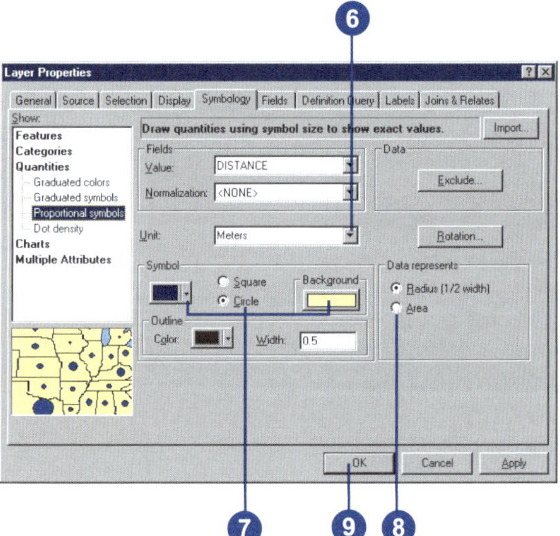

Representing quantity with dot densities

Another method of representing quantities is with a dot density map. You can use a dot density map to show the amount of an attribute there is within an area. Each dot represents a specified number of features, for example 1,000 people or 10 burglaries within an area.

Dot density maps show density graphically, rather than showing density value. The dots are distributed randomly within each area; they don't represent actual feature locations. The closer together the dots are, the higher the density of features in that area. Using a dot density map is similar to symbolizing with graduated colors, but instead of the quantity being shown with color, it is shown by the density of dots within an area.

When creating a dot density map, you specify how many features each dot represents and how big the dots are. You may need to try several combinations of dot value and dot size to see which one best shows the pattern. In general, you should choose value and ►

Drawing a dot density map

1. In the table of contents, right-click the layer you want to draw showing a quantitative value using dot densities and click Properties.
2. Click the Symbology tab.
3. Click Quantities and click Dot density.
4. Click the field or fields under Field Selection that contain the quantitative values that you want to map.
5. Click the arrow button to add fields to the field list.
6. Double-click a dot symbol in the field list to change its properties.
7. Type the dot size or click the slider to adjust the size.
8. Type the dot value or click the slider to adjust the value.
9. Check Maintain Density to preserve the dot density.

   When checked, as you zoom in, the dot size will increase so a given area will visually appear as dense. Otherwise, the dot size will remain constant. ►

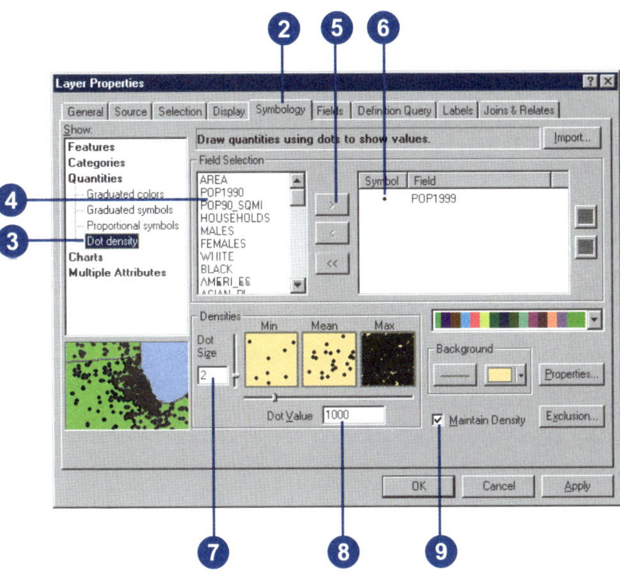

size combinations that ensure the dots are not so close as to form solid areas that obscure the patterns or so far apart as to make the variations in density hard to see.

In most cases, you'll only map one field using dot density maps. In special cases, you may want to compare distributions of different types and may choose to map two or three fields. When you do this, you should use different colors to distinguish between the attributes.

You have two options for placing dots within an area. Non-fixed Placement—the default option—indicates that the dots will be placed randomly each time the map is refreshed, while Fixed Placement freezes the placement of dots, even if the map is refreshed.

Tip

Excluding values

You can click the Exclusion button to type or load a SQL expression to exclude any values you don't want to be mapped.

10. Optionally, click Properties to set the dot placement options and use Masking.
11. Click OK.

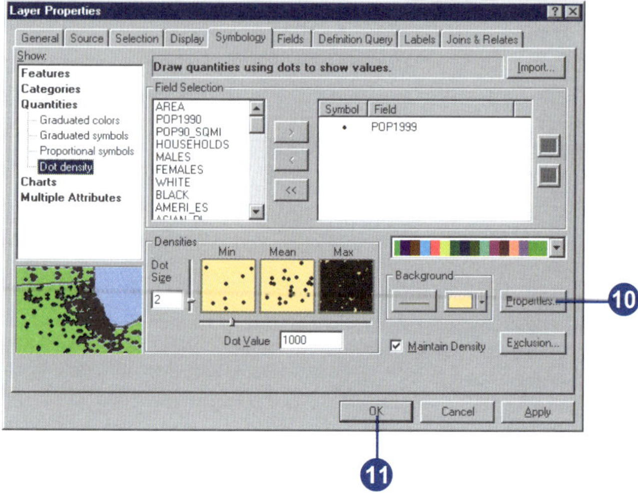

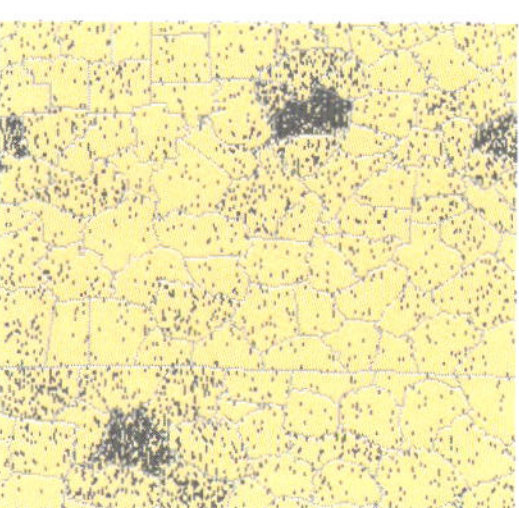

How the dot density map appears when the dots are black and the Background has a gray county boundary outline and a sand background

Dot density appears in the legend and table of contents as a square patch with randomly placed dots.

Representing quantity with charts

Pie charts, bar charts, and stacked bar charts can present large amounts of quantitative data in an eye-catching fashion. For example, if you're mapping population by county, you can use a pie chart to show the percentage of the population by age for each county.

Generally, you'll draw a layer with charts when your layer has a number of related numeric attributes that you wish to compare. Use pie charts when you want to show the relationship of individual parts to the whole. Use bar charts to show relative amounts, rather than a proportion of a total. Use stacked charts to show relative amounts as well as the relationship of parts to the whole.

Tip

Using symbol appearance compensation in pie charts

Checking Appearance Compensation on the Pie Chart Size dialog box turns on Flannery compensation, a technique that adjusts larger symbol sizes upward to account for the fact that map readers tend to underestimate the larger sizes of circular symbols.

Drawing pie charts

1. In the table of contents, right-click the layer you want to draw showing quantitative values with charts and click Properties.
2. Click the Symbology tab.
3. Click Charts and click Pie.
4. Click the fields under Field Selection that contain the quantitative values that you want to map.

   Choosing more than one field shows the relationship to the whole.
5. Click the arrow button to add fields to the field list.
6. Click the Color Scheme dropdown arrow and click the colors you want to use or double-click a symbol in the list to change its properties.
7. Check the box to prevent the charts from overlapping.
8. Click Size.
9. Click the Variation Type you want. You can either draw all pies the same size or vary the size based on the sum of the attributes or a particular attribute value.
10. Type a size or click the arrows to set the size.
11. Click OK.
12. Click OK.

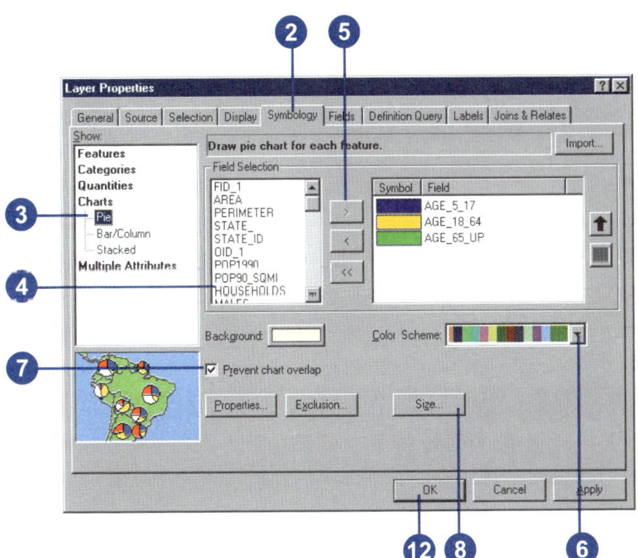

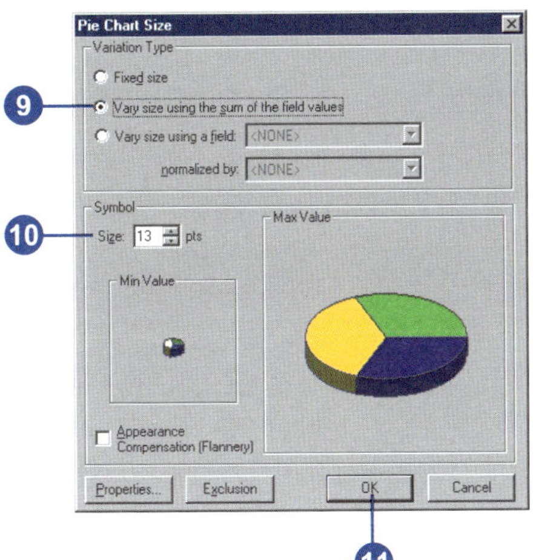

Tip

Using charts effectively

Charts are most effective when mapping no more than 30 features. Otherwise, the patterns on the map will be difficult to see. It's best to use five categories or less on a chart; if you want to show more categories, a series of maps showing each category will work better.

Tip

Choosing colors for your charts

Since the wedges or bars in a chart represent categories, not relative amounts, draw the bars or wedges using different colors rather than shades of one color.

Tip

Excluding values

You can click the Exclusion button to type or load a SQL expression to exclude any values you don't want to be charted.

Drawing bar or column charts

1. In the table of contents, right-click the layer you want to draw showing quantitative values with bar or column charts and click Properties.
2. Click the Symbology tab.
3. Click Charts and click Bar/Column.
4. Click one or more fields under Field Selection that contain the quantitative values that you want to map.
5. Click the arrow button to add fields to the field list.
6. Click the Color Scheme dropdown arrow and click the colors you want to use.

   You can double-click an individual symbol in the list to change its properties.
7. Check the box to prevent the charts from overlapping.
8. Click Size.
9. Type a maximum length or click the arrows to set the length.
10. Click OK.
11. Click OK.

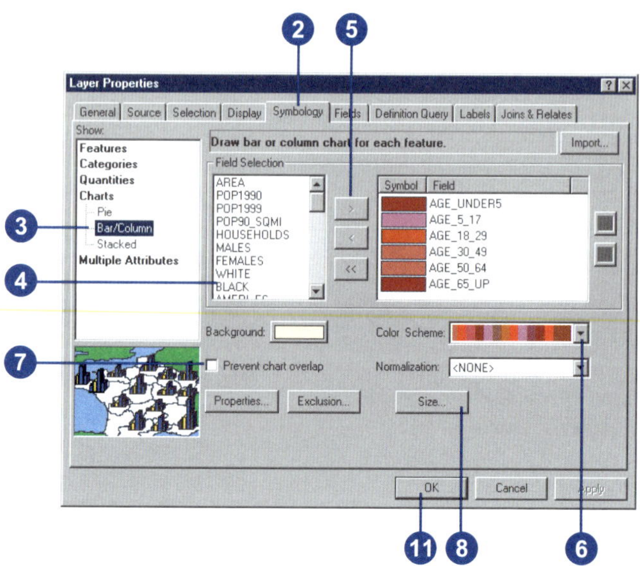

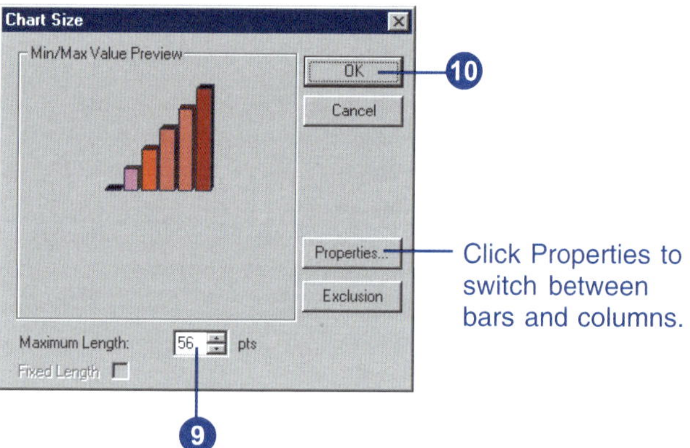

Click Properties to switch between bars and columns.

Tip

Charting negative values

Avoid using pie or stacked bar charts with data containing negative values because those values won't show up.

Tip

Specifying a length for stacked charts

You can specify a maximum length for your stacked charts in the Chart Size dialog box. If you check the Fixed Length box, all your charts will be the maximum length.

Drawing stacked charts

1. In the table of contents, right-click the layer you want to draw showing quantitative values with stacked charts and click Properties.
2. Click the Symbology tab.
3. Click Charts and click Stacked.
4. Click the fields under Field Selection that contain the quantitative values that you want to map.

   Choosing more than one field shows the relationship to the whole.
5. Click the arrow button to add fields to the field list.
6. Click the Color Scheme dropdown arrow and click the colors you want to use.

   You can double-click an individual symbol in the list to change its properties.
7. Check the box to prevent the charts from overlapping.
8. Click Size.
9. Type a maximum length or click the arrows to set the length.
10. Click OK.
11. Click OK.

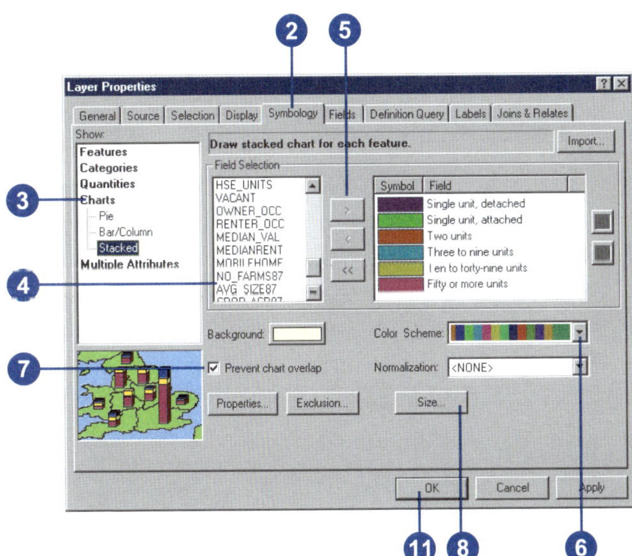

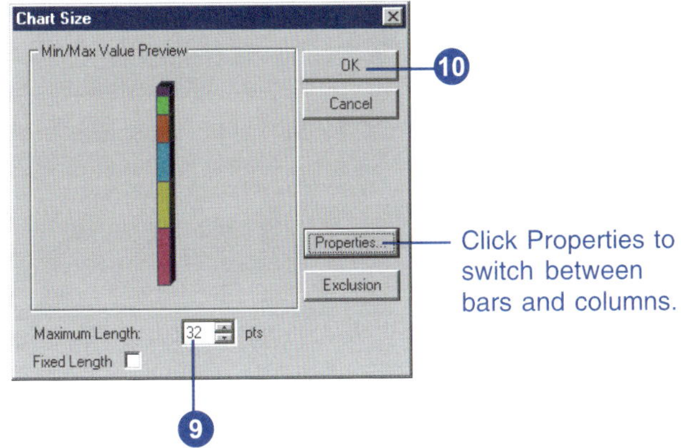

Click Properties to switch between bars and columns.

Drawing features to show multiple attributes

Geographic data usually has a number of different attributes that describe the features it represents. While you'll commonly use one of the attributes to symbolize the data—for example, showing categories or quantities—you may sometimes want to use more than one.

For example, you might display a road network using two attributes: one representing the type of road and the other representing the traffic volume along it. In this case, you could use line colors to represent the different types of roads and also use line width to indicate traffic volume along each road.

When you symbolize your data using more than one attribute, you create a multivariate display. Symbolizing your data this way can allow you to display more information; however, it can also make your map more difficult to interpret. Sometimes it might be better to create two separate displays than to try to display the information together.

Drawing a layer to show both categories and quantities

1. In the table of contents, right-click the layer you want to draw showing multiple attributes and click Properties.
2. Click the Symbology tab.
3. Click Multiple Attributes.

   ArcMap automatically selects Quantity by category.
4. Click the first Value Fields dropdown arrow and click the field that contains the values you want to map.
5. Click the Color Scheme dropdown arrow and click a color scheme.
6. Click Add All Values.
7. Click Symbol Size or Color Ramp, depending on how you want to symbolize the quantitative value. This example shows Symbol Size.
8. Click the Value dropdown arrow and click the quantitative value you want to map.

   Set other options as described in the 'Representing quantity' sections of this chapter.
9. Click OK.
10. Click OK.

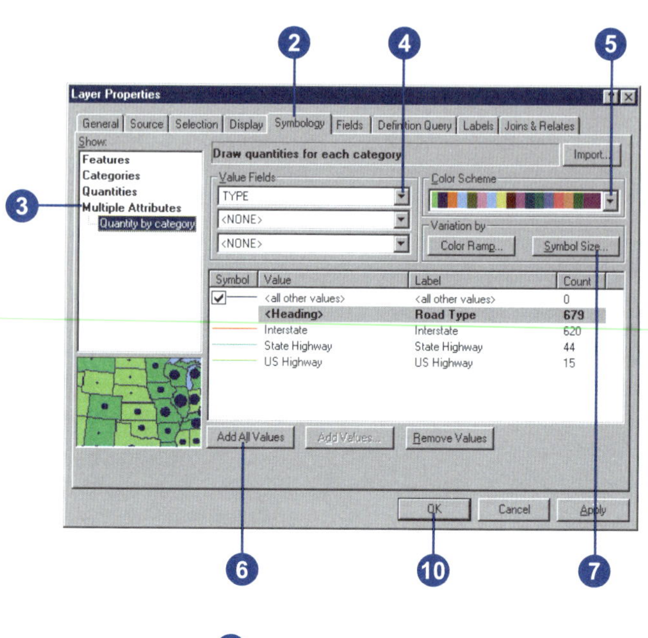

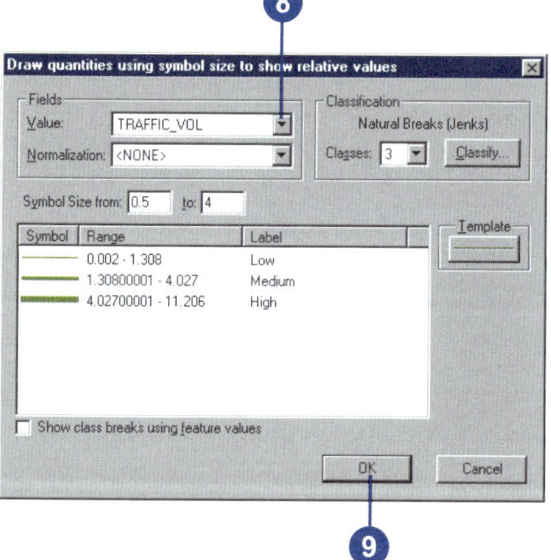

Drawing TINs as surfaces

TINs represent continuous surfaces such as terrain elevation or temperature gradient. The surface is represented as a set of facets formed by connecting data points at nodes to create adjacent triangles. Typically, you display a TIN using color-shaded relief to depict elevation. Shaded relief simulates the sun's illumination of the earth's surface. Adding color to this lets you easily see the ridges, valleys, and hillsides and their respective heights. Seeing the data this way can help explain why other map features are where they are.

You can display any one of three surface characteristics—elevation, slope, and aspect—on your map, and you can also simulate shaded relief.

Geographic features that cross the surface—such as a river, road, or shoreline—can be explicitly represented in a TIN with a breakline. These features form the edges of triangles and, therefore, influence the surface at their location. Since the underlying triangulation defines the surface, you might want to take a closer look at it. You can also display the internal ►

Drawing a color-shaded relief surface

1. In the table of contents, right-click the TIN layer that you want to draw and click Properties.
2. Click the Symbology tab.

   By default, ArcMap displays the face elevation and breakline edges of the TIN.
3. Click an entry in the Show list to see its symbolization properties.
4. Modify the symbolization properties as necessary. For example, set a new color ramp or change the number of classes.
5. Click the Add button to draw additional elements of the TIN—for example, nodes.
6. Click the renderer that represents the TIN feature you want to draw.
7. Click Add.
8. Click Dismiss when you are finished adding renderers.

   The list will update to show what you want to draw. ►

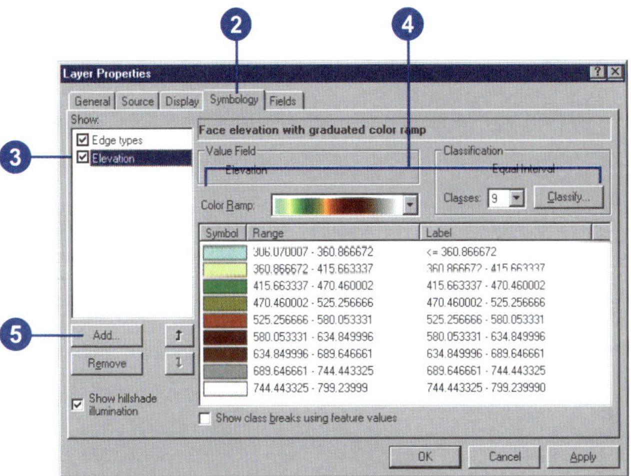

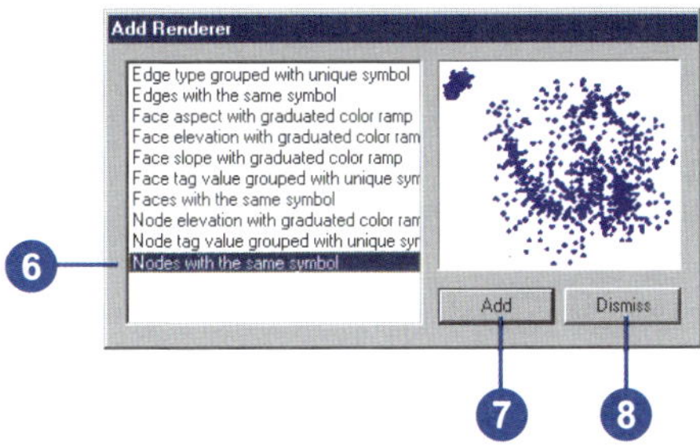

structure of a TIN—for example, nodes and breaklines—independently or on top of the shaded relief display.

Tip

How are slope and aspect measured?

Slope values range between 0 and 90 degrees, where 0 indicates no slope. Aspect is also measured in degrees. North is 0 degrees, east is 90 degrees, south is 180 degrees, and west is 270 degrees.

Tip

Making TINs look more 3D

Checking the Show hillshade illumination effect in 2D display box makes a TIN dataset look more 3D when it's displayed in 2D because it simulates the sun's illumination of the earth's surface.

See Also

You can click the Classify button to change the method of classification of the elevation, slope, or aspect. For more information on classifying data, see 'Setting a classification' in this chapter.

See Also

To learn how to symbolize raster datasets, see Chapter 9, 'Working with rasters'.

9. Click an element in the Show list.
10. Click the Up or Down arrow to change its draw order.

    The TIN features at the top of the list will draw on top of those below them.
11. Click OK.

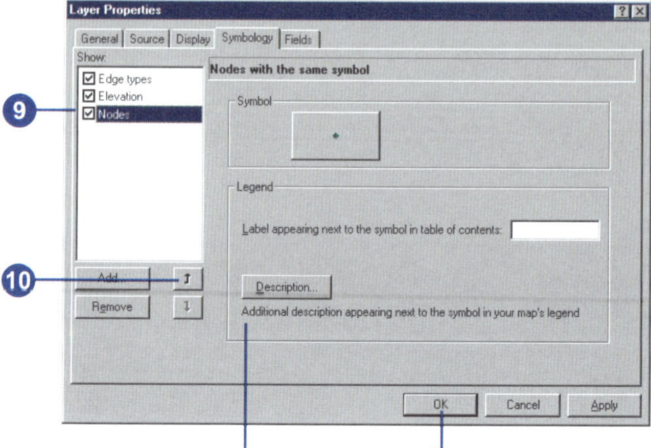

You can type descriptive text to display next to an edge or node symbol in the legend or the table of contents. Otherwise, no text will appear next to the symbol.

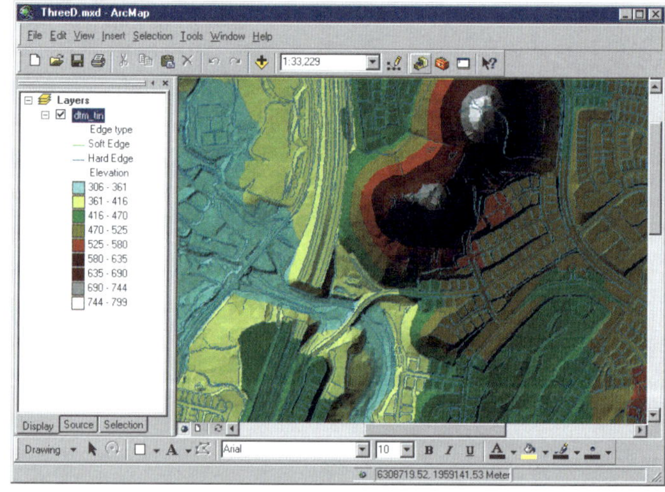

Drawing CAD layers

You can display CAD drawings on your map just like other data types. You can decide which CAD layers to draw and how to draw the entities on the layer.

Depending on how you added the CAD data to your map, you have two display options:

- If you added the CAD drawing file for display only, you can only choose which CAD layers to show or hide. ArcMap draws all entities according to the color specified in the drawing file. You can't override this drawing behavior.
- If you added the CAD drawing as features—point, line, or polygon—you can use the Symbology tab of the Layer Properties dialog box to access all the symbolization options available to other feature layers.

See Also

For more information on adding CAD layers, see 'Adding CAD drawings' in Chapter 4.

Displaying a CAD drawing file

1. In the table of contents, right-click the CAD drawing layer and click Properties.
2. Click the Display tab.
3. Click and drag the sliders to adjust the CAD display
4. Click the Drawing Layers tab.
5. Check the CAD layers that you want to display.
6. Click OK.

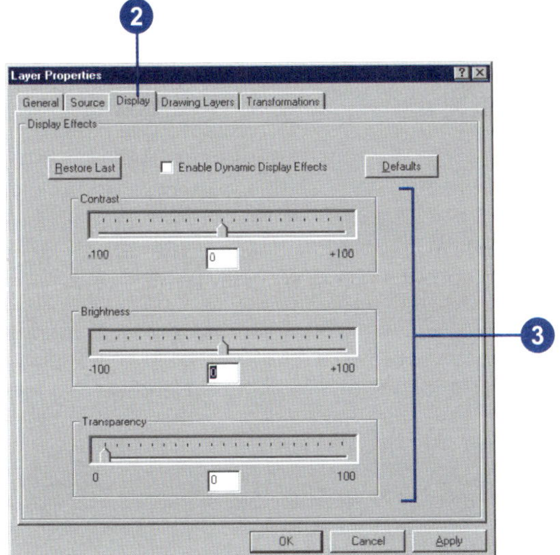

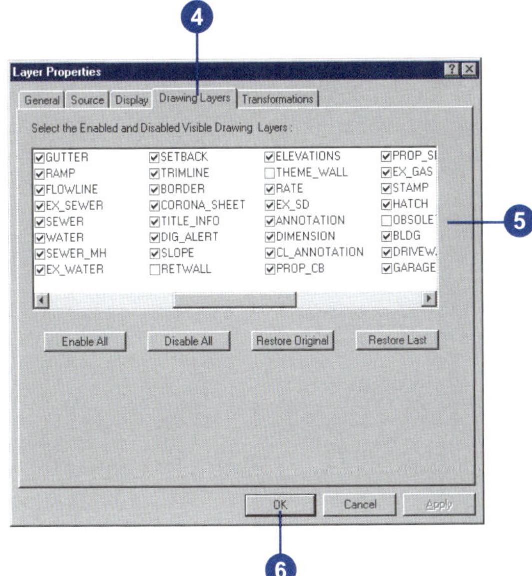

Tip

Adjusting transparency

You can use the Effects toolbar to adjust the transparency of CAD layers.

See Also

For more information on symbolizing the features in a CAD dataset, see 'Drawing features to show categories such as names or types' in this chapter.

Drawing CAD features as points, lines, or polygons

1. In the table of contents, right-click the CAD dataset and click Properties.
2. Click the Symbology tab.

   The drawing options available to you are the same as for other feature layers.
3. Modify the drawing properties as necessary.

   See the previous topics in this chapter for more detailed instructions.
4. Click the Drawing Layers tab.
5. Check the CAD layers that you want to display.
6. Click OK.

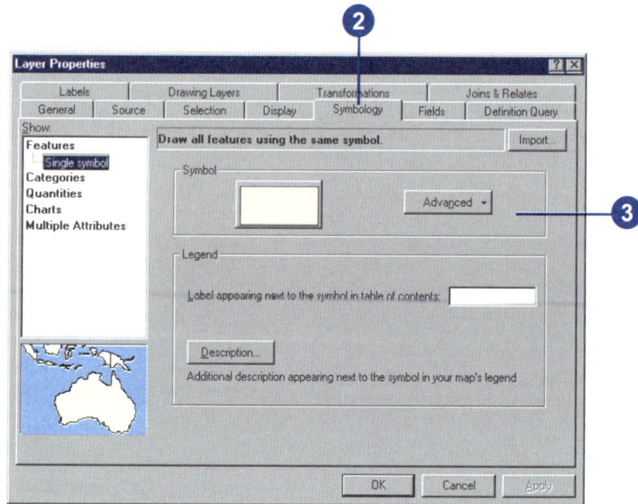

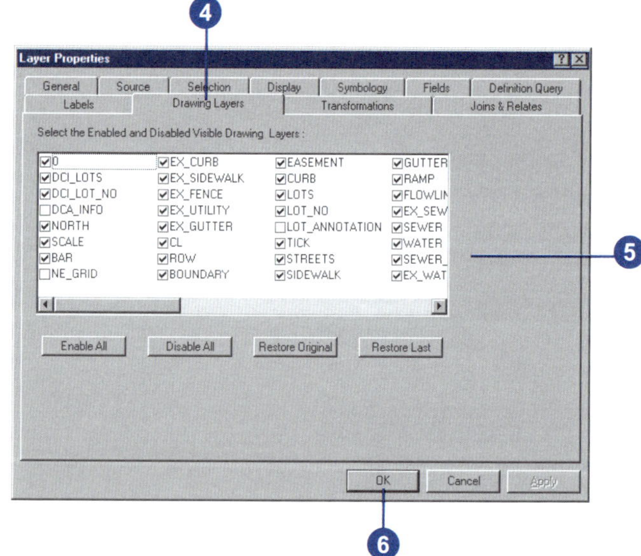

Working with advanced symbolization

ArcMap provides a few other tools that let you control how layers draw. You can:

- Draw layers transparently.
- Set a data frame reference scale for symbols to see how they will look at their true size on screen or paper.
- Use symbol level drawing to control the order in which feature symbology is drawn.
- Use variable-depth masking to hide parts of layers.

Adding transparency

Transparency can be used for any symbolization type, but it is especially useful for drawing raster layers with other layers on your map. This allows you to see the raster layer while still viewing underlying layers. For more information on using transparency with rasters, see Chapter 9, 'Working with rasters'.

Setting a data frame reference scale

With a *reference scale*, you define the scale at which text and symbols will appear at their true size. If you zoom in or out, the text and symbols will change scale along with the display. Symbols and text will appear larger as you zoom in on your data frame and smaller as you zoom out on your data frame. Setting a reference scale is like freezing the symbol and text sizes used in your data frame; the way they look at the reference scale is maintained at all scales. Unless you explicitly set a reference scale, the current scale is your reference scale.

One reason to set a reference scale is if you want the detail in your data frame to look the same onscreen in data view as it will when you print it out. If you're going to print a map, you should set a reference scale to maintain a quality appearance. Let's say you are creating a map for publication that will be printed at a scale of 1:25,000. If you set your data frame's scale to be 1:25,000 and choose Set Reference Scale, the symbols and text sizes in your data frame will appear onscreen at the same size in relation to each other as in your printed map.

When a reference scale is set, all feature symbology, labels, and graphics in the current data frame will be scaled relative to the reference scale. However, you can disable scaling for individual layers: double-click the layer, click the Display tab, and uncheck Scale symbols when a reference scale is set. Since geodatabase annotation and dimension features have their own reference scales, they are not affected by a data frame reference scale.

Using symbol level drawing

Symbol level drawing allows you to achieve special cartographic effects by giving you control over the drawing order of feature symbology. You specify the order that symbols and symbol layers for multilayer symbols are drawn on your map—overriding the default ArcMap drawing sequence. With symbol level drawing, the drawing order is based on each feature's symbol and the position of that symbol in the symbol level drawing order. Symbol level drawing parameters may be set individually for each feature layer or group layer.

Using symbol level drawing is useful for achieving some graphical effects that can give your maps a polished cartographic look. You can use symbol level drawing to symbolize overlapping and intersecting line features with cased line symbols. For example, on a large-scale reference map with intersecting streets, you can create high-quality representations of streets. Where streets intersect, you can blend the symbology for connectivity; otherwise you will represent an overpass. You can work with Join and Merge to achieve these cartographic connectivity effects.

You can work in default view or switch to advanced view to gain additional control over symbol level drawing. Advanced view breaks each symbol into its component layers and allows you to enter numeric values to specify where each symbol layer will be positioned in the draw order. More combinations are possible in advanced view than default view.

Working with variable-depth masking

Variable-depth masking is a drawing technique for hiding parts of one or more layers.

You can use any polygon feature class as a mask. Using the ArcMap Advanced Drawing Options dialog box, you can set which layers are affected by the mask and which are not. The masking settings are created at the data frame level and are saved as part of the map document (.mxd).

One common use for masking is to clarify the legibility of a map that is densely packed with text and feature symbology. You can create a polygon mask layer based on an annotation layer, then mask out some feature symbology to make the map more readable. With variable-depth masking, only some layers are hidden by the masks. For example, on a contour map, you might create text masks that mask out sections of contour lines but do not mask out the elevation shading appearing behind those layers.

Masking will be maintained on any of the ESRI-supported map export formats.

Creating a masking layer

You can use any polygon feature class as a masking layer; however, you may want to create specific masks from the symbology or annotation of a layer. The ArcGIS toolbox provides the Feature Outline Masks tool to perform this task.

This tool works on the associated symbology of a layer. Therefore, this tool uses a layer open in ArcMap or a saved layer file (.lyr). The output is a polygon feature class saved within a geodatabase.

The mask is created by identifying a margin (the area between the feature and the edge of the mask) and using an outline method. The outline methods include:

- Exact—The mask is created to represent the shape, including internal holes, and will follow characters exactly.
- Convex Hull—The mask is created to represent the shape, not including internal holes. For example, it will represent words, not each letter within a word.
- Box—The mask represents the shape as a rectangular bounding box.

Using a masking layer

You can use any polygon feature class to create masks. This feature class must be added to your map as a layer in the table of contents.

From a data frame's Advanced Drawing Options dialog box, you can turn the masking on or off by checking or unchecking the box to Draw using masking options specified below.

The masking layers can also be set up to mask one or more layers. Each layer to be masked can be checked in the Masked Layers list when the mask layer is highlighted in the Masking Layers list.

Drawing a layer transparently

1. Click the View menu, point to Toolbars, and click Effects.

   The Effects toolbar appears.

2. Click the layer dropdown arrow and click the layer you want to appear transparent.
3. Click the Adjust Transparency button.
4. Drag the slider bar to adjust the transparency.

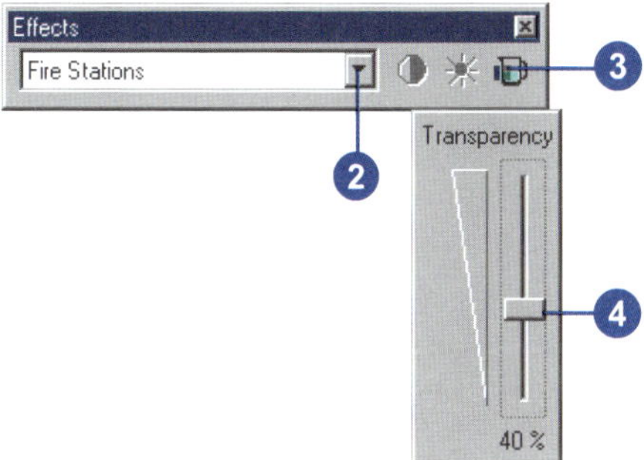

Fire station layer before (left) and after adjusting transparency

Setting a reference scale for symbols

1. Set the scale of the data frame to the scale you want to use as the reference scale.
2. Right-click the data frame in the table of contents, point to Reference Scale, and click Set Reference Scale.

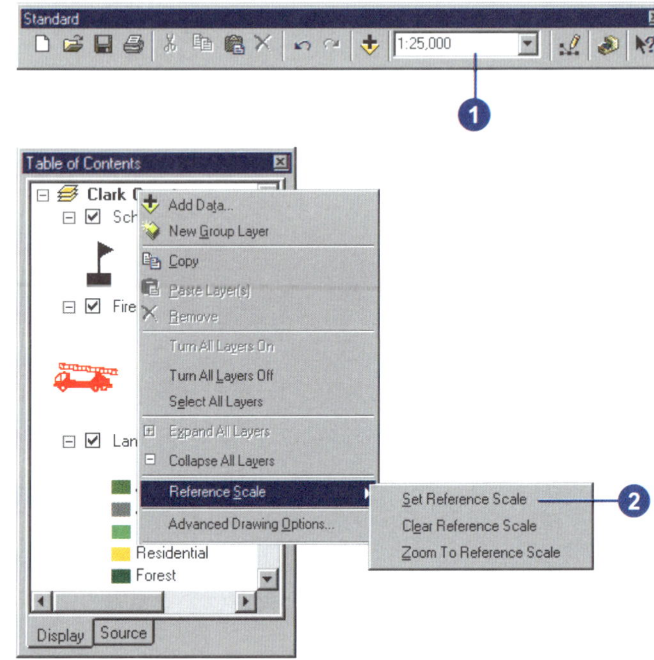

Clearing a reference scale

1. Right-click the data frame in the table of contents, point to Reference Scale, and click Clear Reference Scale.

With (left) and without (right) a reference scale set

Tip

Using a shortcut to turn on symbol level drawing

In the table of contents, right-click the feature layer or group layer and click Use Symbol Levels.

Tip

Upgrading a map created in ArcGIS 8 that uses advanced drawing options

You need to upgrade maps created in ArcGIS 8 that use symbol level drawing to be able to use ArcGIS 9 advanced drawing options. Right-click the data frame, click Advanced Drawing Options, and click the button to upgrade the map.

Tip

What symbology types can use symbol level drawing?

Symbol level drawing is available for features symbolized with Single Symbol, Unique Values, Graduated Colors, and Graduated Symbols. Join and Merge are only available for multilevel symbols. It helps to use unique values when drawing intersecting features.

Tip

Using symbol level drawing

Symbol level drawing may result in slower drawing. You can turn it on or off by right-clicking a layer in the table of contents and clicking Use Symbol Levels.

Turning on symbol level drawing

1. In the table of contents, right-click the layer or group layer you want to draw using Symbol Levels and click Properties.
2. Click the Symbology tab.

   If you're working with a group layer, click the Group tab.
3. Click the Advanced button and click Symbol Levels from the dropdown list.

   If you're working with a group layer, click the Symbol Levels button.
4. Check Draw this layer using the symbol levels specified below.
5. Click OK.

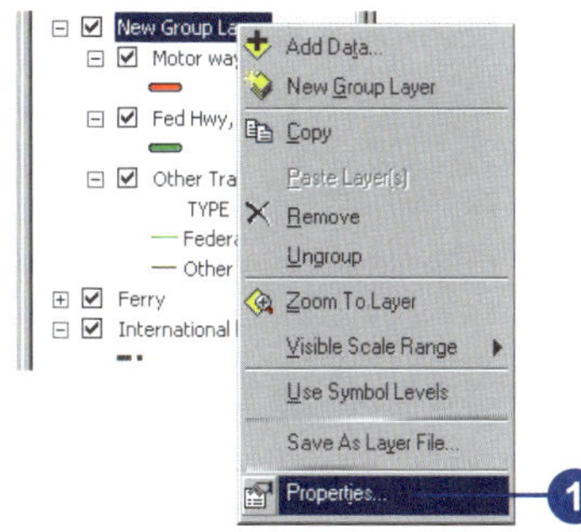

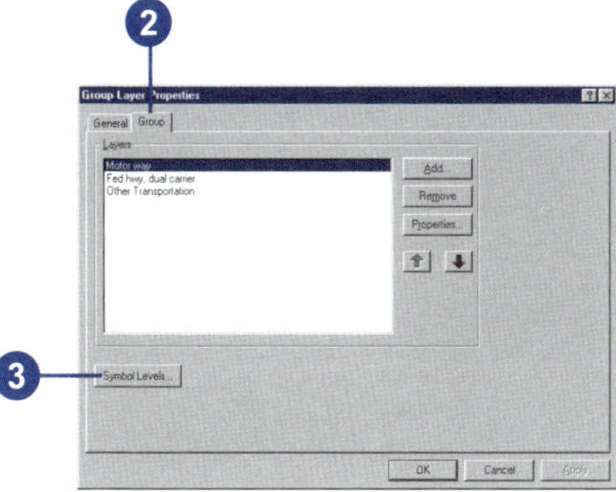

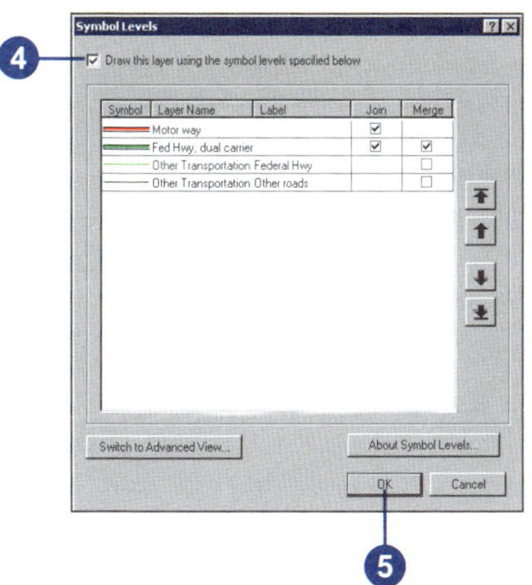

Tip

Quickly opening the Symbol Levels dialog box

You can add the Set Symbol Levels command to the layer or group layer context menu to open the Symbol Levels dialog box directly. This way, you can also apply your changes without having to close dialog boxes. You can find the command in the Layer category of the Customize dialog box.

Tip

Using keyboard shortcuts with Join or Merge

In default view, hold down Ctrl and click the check boxes to turn Join or Merge on or off for all symbols in the list. If symbols are selected, Join or Merge is toggled only for the selected symbols.

Tip

Getting help about symbol level drawing

You can click the About Symbol Levels button on the Symbol Levels dialog box to learn more about symbol level drawing. You can also see the ArcGIS Desktop Help.

Setting symbol levels in default view

1. Open the Symbol Levels dialog box for your layer or group layer and turn on symbol level drawing.

   If you're working in advanced view, click Switch to Default View.

2. Double-click a symbol to change its properties.
3. Change the symbol drawing order by moving symbols up and down in the symbol list with the arrows or drag and drop.

   Symbols at the bottom of the list are drawn before symbols at the top.

4. Check the Join box next to a symbol to achieve a blending effect for all features drawn with the symbol.
5. Check the Merge box next to a symbol to blend that symbol with the symbol directly above it in the symbol list. Use Merge to achieve a blending effect for features drawn with different symbols.
6. Click OK to close the Symbol Levels dialog box.
7. Click OK to apply these changes to your layer.

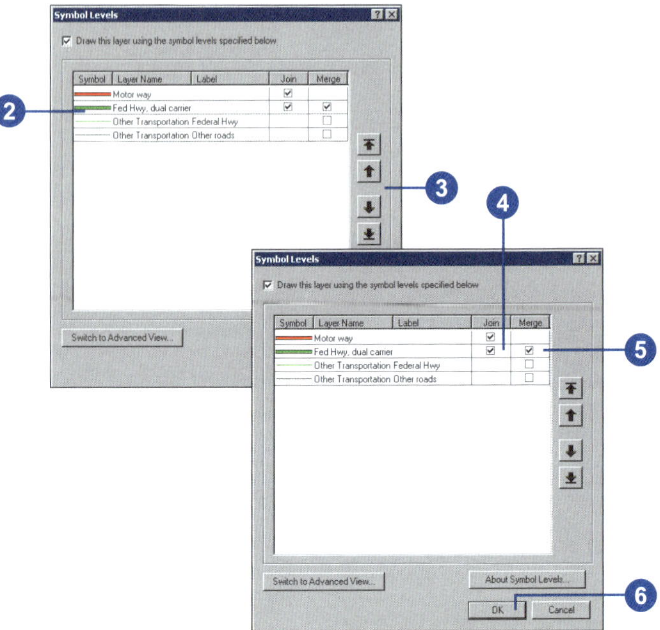

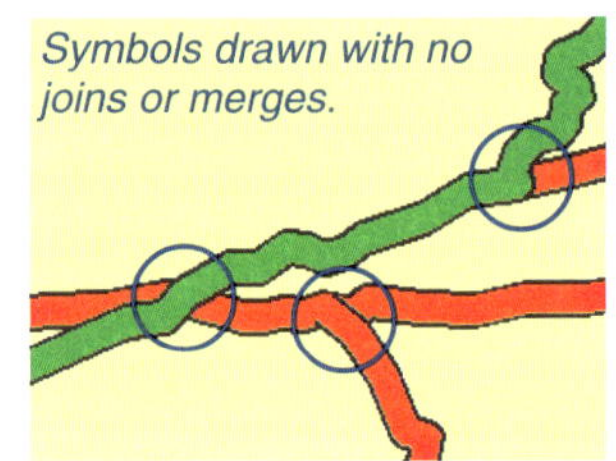

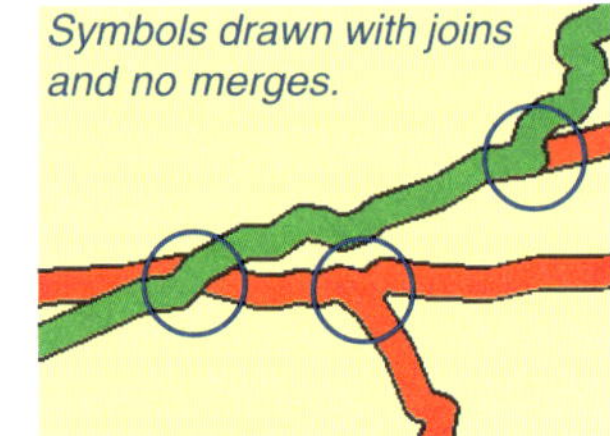

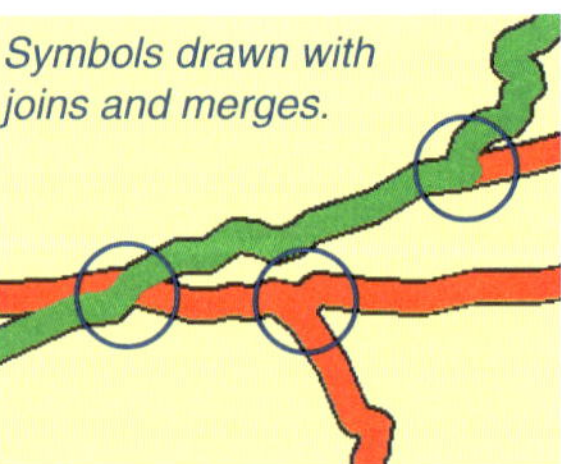

The blue circles highlight road intersections that are symbolized with various join and merge options.

Tip

Working in advanced view

Use advanced view if you need more control over symbol level drawing. All settings are preserved when you switch from default to advanced view, but some may be lost going from advanced to default view.

Tip

Setting options for advanced view

In advanced view, you can click a column heading to sort the column or change the row height.

Setting symbol levels in advanced view

1. Open the Symbol Levels dialog box for your layer or group layer and turn on symbol level drawing.

   If you're working in default view, click Switch to Advanced View. Advanced view contains a matrix of symbols and their layers.

2. Double-click a symbol to change its properties.
3. Type numbers into the **Layer columns** to set the drawing order. **Layers with lower level numbers are drawn before layers with higher level numbers**. A layer with a level number of 0 will be drawn first.

   In advanced view, drawing order is determined by the values you type in the columns, not the order of the symbol list.

4. If you have multilayer symbols, you can click the arrow in the Symbol column and view the layers. When you click a layer of a multi-layer symbol, its cell in the Layer column is selected so you can enter a value to set its drawing order.
5. Click OK to close the Symbol Levels dialog box.
6. Click OK to apply these changes to your layer.

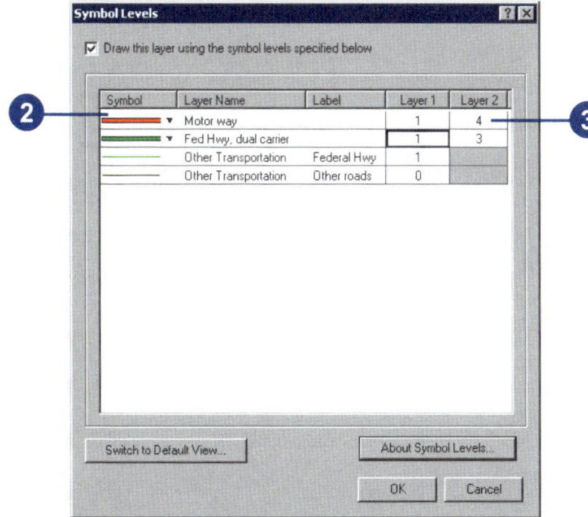

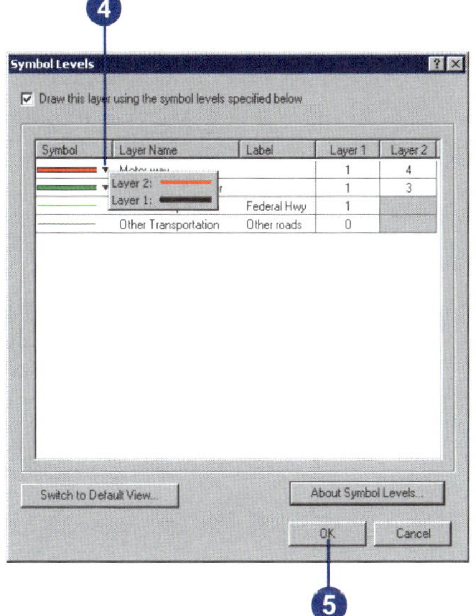

Creating a masking layer

1. Click the Show/Hide ArcToolbox button on the Standard toolbar to show ArcToolbox.
2. Expand the Cartography Tools.
3. Expand the Masking Tools.
4. Double-click the Feature Outline Masks tool.
5. Click the Input Layer dropdown arrow and click a layer, or click the Browse button to choose a layer file.
6. Type the name and location for the Output Feature Class, or click the Browse button to create it.
7. Type a value in the Reference Scale parameter.

   If you're using an annotation layer, the reference scale will automatically be set.
8. Click the Spatial Reference Query button to set or change the spatial reference.
9. Type a value in the Margin parameter and click the Margin dropdown arrow and select a unit of measure.
10. Click the Mask Kind dropdown arrow and click a mask kind.

    If using a point or annotation layer, you have three choices; otherwise the default is Exact.
11. Optionally, check Create masks for unplaced annotation.
12. Click OK.

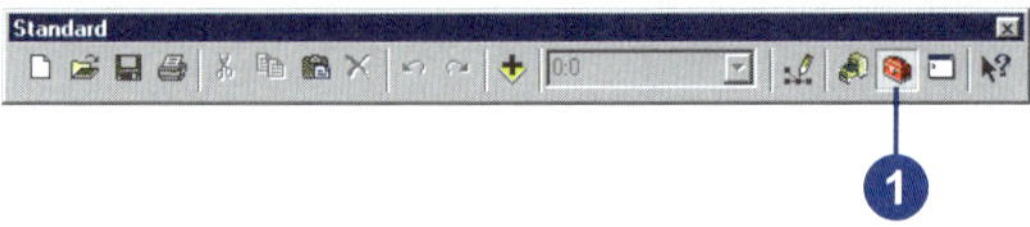

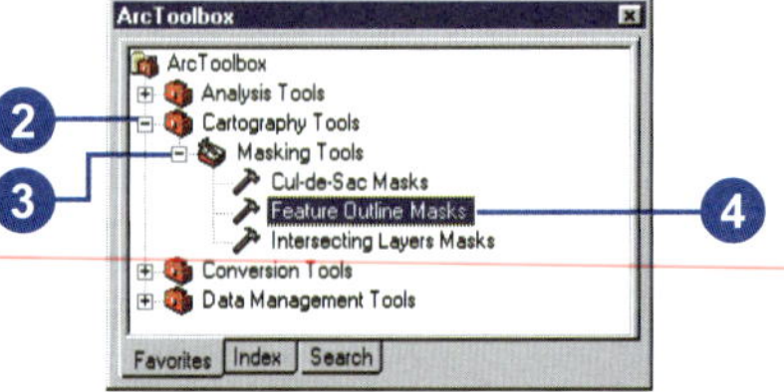

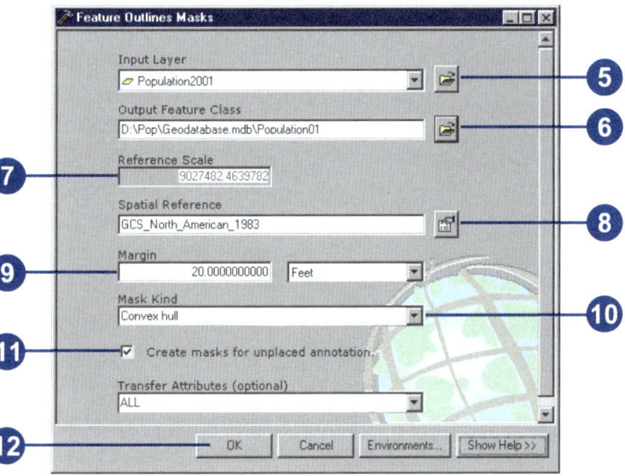

Using a masking layer

1. Click Add Data to add the masking layer to the data frame.
2. Right-click the data frame in the table of contents and click Advanced Drawing Options.
3. Check Draw using masking options specified below.

   Uncheck if you want to turn off masking.
4. Click the layer you want to use as a mask in the Masking list.
5. Check the layers you want to have affected by the mask in the Data Frame Layers list.
6. Click OK.

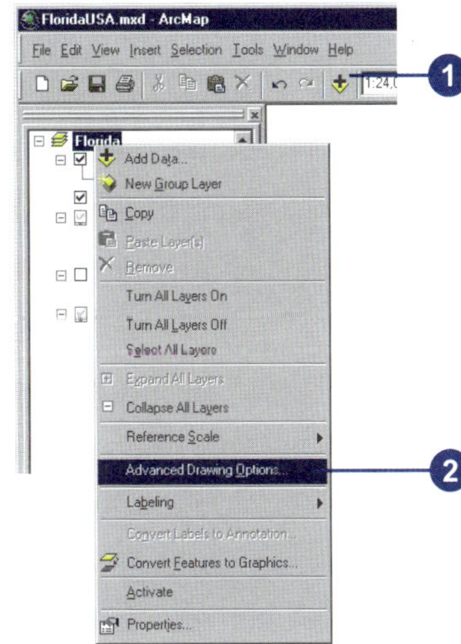

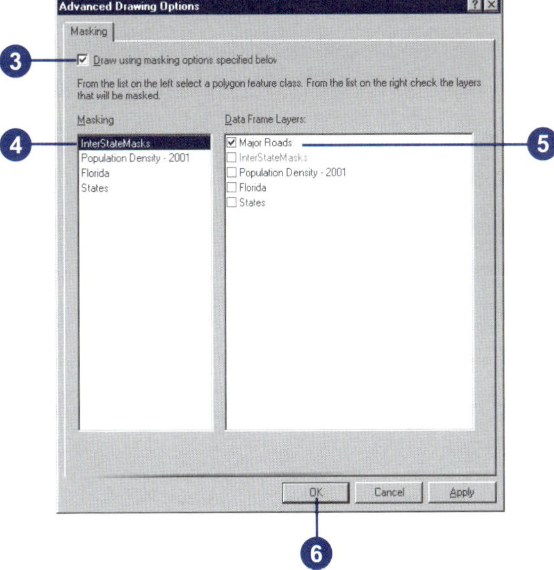

Working with graphics and text 7

IN THIS CHAPTER

- Working with graphics
- Drawing points, lines, and circles
- Moving, rotating, and ordering graphics
- Aligning, distributing, and grouping graphics
- Storing graphics as annotation
- Working with text in ArcGIS
- Working with labels
- Specifying the text of labels
- Building label expressions
- Prioritizing and positioning labels
- Converting labels to annotation
- Working with annotation
- Using text formatting tags

Maps convey information about geographic features, and adding graphics or text can often enhance your map's overall presentation. For example, you might add a few graphics or pieces of text to draw attention to particular features. In this manner, you could outline a study area with a polygon, point out potential locations for new stores, or label city streets with their names.

Because text serves many different mapping purposes, ArcMap offers several different kinds of text—graphic text, labels, and annotation. For example, on a map of Africa, you might add some text that indicates the general location of the Sahara Desert. You could also use text to describe individual features on your map by showing the name of each major city.

In this chapter, you'll learn how to create and work with graphics and text. You'll also learn about the different types of text available in ArcMap and in which situations you should use each kind of text.

Working with graphics

Adding graphics to your map

Adding graphics to your map can clarify the information that it conveys about the geographic features that are present and the geographic phenomena that are at work. For example, you may add circles on top of the data on your map to draw attention to particular features, outline a study area with a polygon, or add lines that point to potential locations for new stores.

Working with graphics

ArcMap contains a set of tools that works on graphics and graphic text. These tools are found on the Draw toolbar. You can use the tools on the Draw toolbar to create and edit graphics. You can quickly draw new squares, lines, circles, and polygons, for example, and change how they appear just by clicking tools. In fact, you can convert map elements, such as a legend, to graphics so you can edit their properties and position them more easily. You can also align and distribute multiple graphics or move a graphic above or below another graphic on the page. One advantage to graphics is that you can easily manage different geometry types together.

Graphics can be placed either in layout view, along with cartographic elements such as scalebars and North arrows, or in data view so the graphics resize with your data as you change the extent of your map. For more information on working in layout view with other types of map elements such as neatlines and pictures, see Chapter 15, 'Laying out and printing maps'.

Graphics are generally map-specific, which means that you will probably want to store your graphics in individual map documents (.mxd files). However, the geodatabase does support the storage of graphics in annotation feature classes. Use this option if you want to use the same set of graphics in many maps, if you want controlled multiuser editing, or if you need to store your graphics in a geodatabase.

Remember that graphics aren't features. Graphics have no associated attribute table and, therefore, cannot be queried as easily as geographic features.

Working with graphic text

Adding text to your map is another way to improve how your map communicates its message. You can use many of the graphic tools to display and edit text that is added to the layout view. For more information on text, see 'Working with text in ArcGIS' in this chapter.

Drawing points, lines, and circles

Points, lines, circles, polygons, and rectangles are among the graphic shapes you'll use to highlight features in your data and draw cartographic elements on your layout. Once you've added a graphic to your map, you can move it, resize it, change its color, align it with other graphics, and so on.

If you want to add a graphic as part of the map layout, add it in layout view. If you want the graphic to display with your data, add it in data view. For example, suppose you want to draw a circle representing a buffer around a feature. Instead of drawing the circle over the data frame in layout view, draw it directly over your data in data view. Then, as you pan and zoom your data, the circle pans and zooms with it.

Tip

Keeping tools active

You can make the graphics construct tools stay active after you complete a graphic by changing the options found on the Symbols/ Graphics tab of the Advanced ArcMap Settings utility. This application is installed in the Utilities folder of the \ArcGIS directory.

Adding a graphic

1. On the Draw toolbar, click the type of graphic you want to add. (See the tools in the table to the right.)
2. Move the mouse pointer over the display and click to add the graphic.

   Some graphics require more than a click. For example, you'll need to click and drag the mouse to add a rectangle.

Drawing tools

	New Circle		Select Elements
	New Curve		Edit Vertices
	New Ellipse		Rotate
	New FreeHand		Fill Color
	New Line		Line Color
	New Marker		Marker Color
	New Polygon		Zoom to Selected Elements
	New Rectangle		

Changing the size of a graphic

1. Click the Select Elements button on the Draw toolbar and click the graphic you want to resize.
2. Move the mouse pointer over one of the blue selection handles and click and drag the handle.

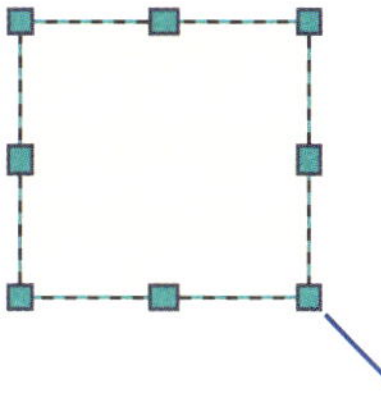

Click and drag a selection handle to resize the graphic. Use the Shift key to resize as a square or the Ctrl key to maintain the aspect ratio.

Deleting a graphic

1. Click the Select Elements button on the Draw toolbar and click the graphic you want to delete.
2. Press the Delete key on the keyboard.

Tip

Adding graphics to a data frame while in layout view

When you add a graphic to a map while in layout view, ArcMap will, by default, add it to the layout. To add the graphic to a data frame, click the Select Elements button on the Draw toolbar and double-click the data frame. Then click a drawing tool to add a graphic to the data frame.

Tip

Working with color

If you don't find the exact color you want in the array of colors on the Color Palette, you can mix your own. Click More Colors to open the Color Selector and mix with the sliders.

Editing vertices of a graphic

1. Click the Select Elements button on the Draw toolbar and click the graphic you want to edit the vertices of.
2. Click the Edit Vertices button on the Draw toolbar.

   If this button is unavailable, you can't edit the vertices of the selected graphic.
3. Right-click over the line and click Add Vertex to add a vertex, or right-click over a vertex and click Delete Vertex to delete it. Click and drag a vertex to move it.

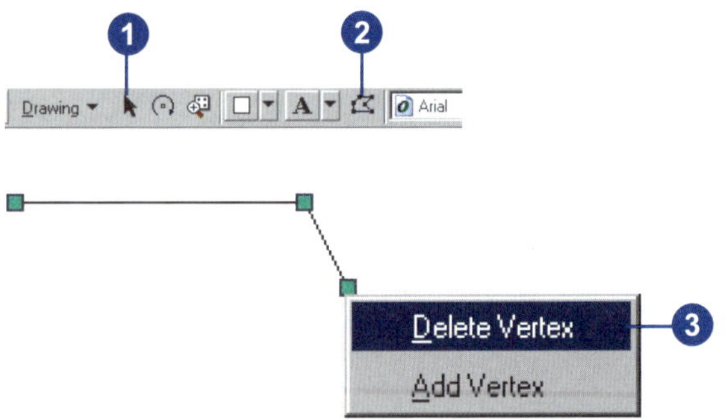

Using the Draw toolbar to make quick symbol changes

1. In layout view, click the Select Elements button on the Draw toolbar and click the graphic you want to modify.
2. Click the appropriate shortcut button on the Draw toolbar.
3. Click the new property.

   Your changes are immediately applied.

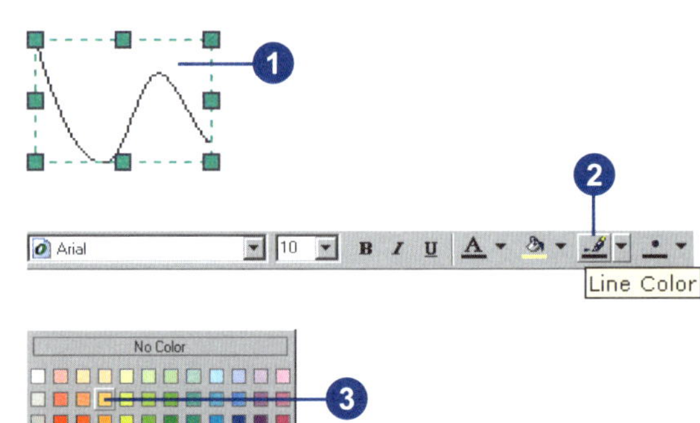

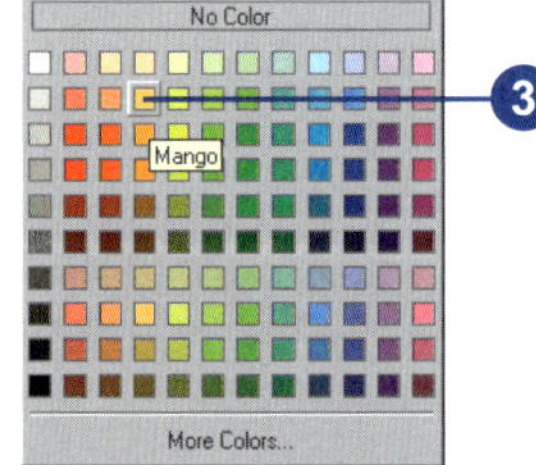

Tip

Displaying the Graphics toolbar

The Graphics toolbar provides quick access to frequently used tools for manipulating graphic elements. To display it, click the View menu, point to Toolbars, and click Graphics.

See Also

For more information on adding other elements, such as scalebars and North arrows, see Chapter 15, 'Laying out and printing maps'.

Changing the color or symbol of a graphic

1. Click the Select Elements button on the Draw toolbar and double-click the graphic to display its properties.

   The properties vary depending on the type of graphic you've selected.

2. To change the fill color, click the Fill Color dropdown arrow and click a new color.
3. Click OK.

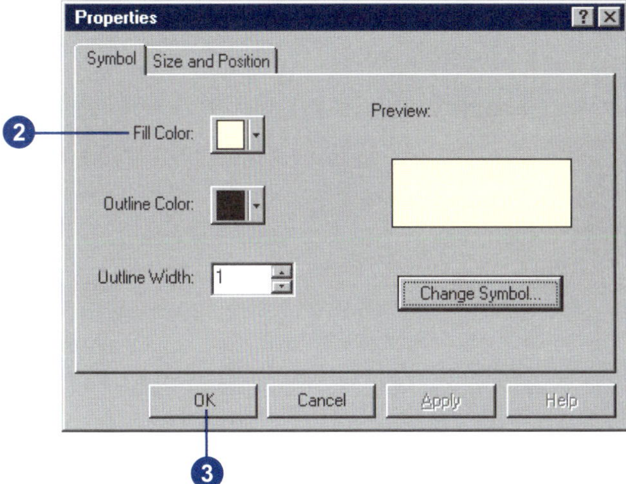

Setting the default symbol properties for new graphics created with the Draw toolbar

1. On the Draw toolbar, click Default Symbol Properties.
2. Click the appropriate button to set the symbol properties for that type of graphic element.
3. Click OK.

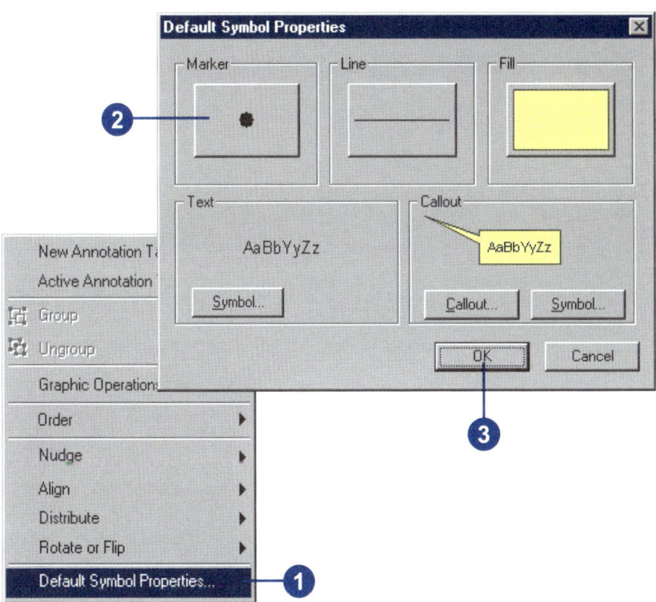

Tip

Why convert features to graphics?

You can convert the features in a layer to graphics that can be moved, resized, and edited on the map. This option is useful if you want to change the location of features relative to each other for cartographic purposes, such as generalization, but you don't want to edit the source data that your layer represents.

Converting features to graphics

1. Right-click the layer in the table of contents that you want to convert to graphics and click Convert Features to Graphics.
2. Click all to convert all features or selected to convert the selected features.
3. Click the Target dropdown arrow and click the target you want to add the graphics to—the location where the graphics will be stored. The default target saves the graphics into your map document.
4. Click OK.

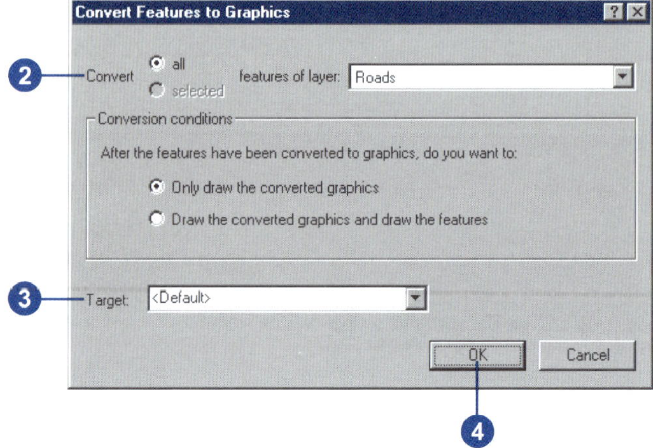

Selecting graphics

You can work with a graphic when it is selected. You can, for example, change its size, color, or shape. By selecting more than one graphic, you define a selected set that you can work with as a group. For example, you might align, move, or delete them.

You select graphics with the Select Elements tool. Select an individual graphic by clicking it, or select a group by dragging a rectangle around the graphics. Hold down the Shift key while selecting to add graphics to or remove graphics from the current selection.

You can tell when a graphic is selected because ArcMap draws selection handles around it. When more than one graphic is selected, one graphic has blue handles and the others have green handles. The blue handles indicate the dominant graphic, or the one that ArcMap will use, for example, to align other graphics with.

To change the dominant graphic, hold down the Ctrl key and click the selected graphic that you want as the dominant one.

Selecting graphics one at a time

1. Click the Select Elements button on the Draw toolbar.
2. Move the mouse pointer over the graphic you want to select and click the graphic.

   ArcMap draws selection handles around the selected graphic.

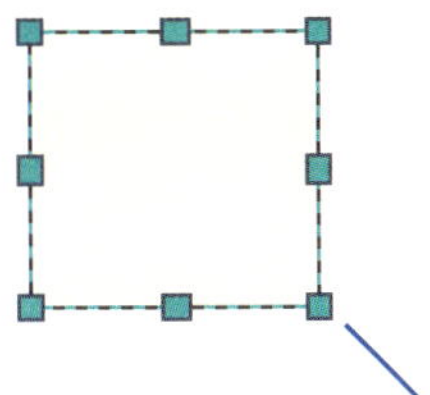

Move the mouse pointer over the graphic and click it. Hold down the Shift key and click to add to the current selected graphics.

Selecting all graphics

1. Click the Edit menu and click Select All Elements.

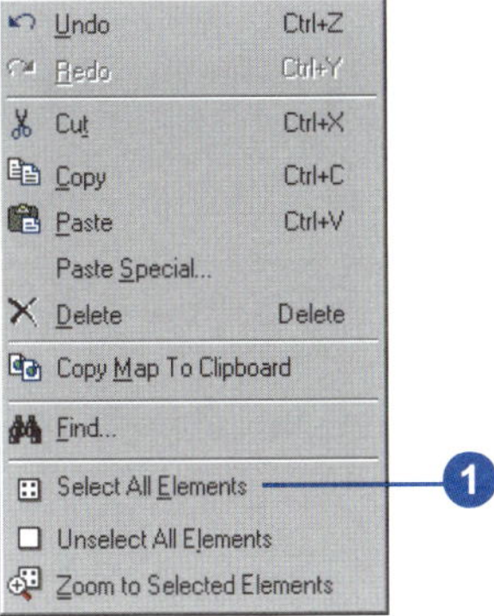

Zooming to selected graphics

1. Select the graphics you want to zoom to.
2. Click the Zoom to Selected Elements button on the Draw toolbar.

   The map display extent zooms to the selected graphics.

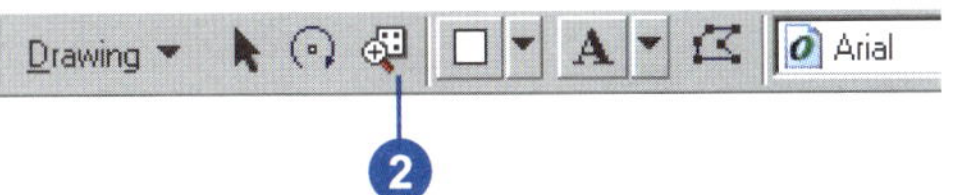

Moving, rotating, and ordering graphics

Much of the work you do while building your map involves arranging graphics and other elements on it. For instance, you might want to orient graphics around the associated features in a data frame or position map elements, such as titles, neatlines, and North arrows, on the layout.

ArcMap provides a number of tools that let you position and orient graphics. You can move graphics by dragging them with the mouse or, when you need more precise control, you can nudge them up, down, left, or right. You can also position graphics to a coordinate location you specify. You can move one graphic on top of another one, rotate it, and flip it horizontally or vertically.

Tip

Specifying coordinates

You can position a graphic to a location that you specify. In layout view, you are specifying x,y coordinates relative to the lower-left corner of the layout. In data view, you are specifying x,y coordinates in the units your data is stored in.

Moving a graphic

1. Click the Select Elements button on the Draw toolbar and click the graphic you want to move.
2. Click and drag the selected graphic to its new position.

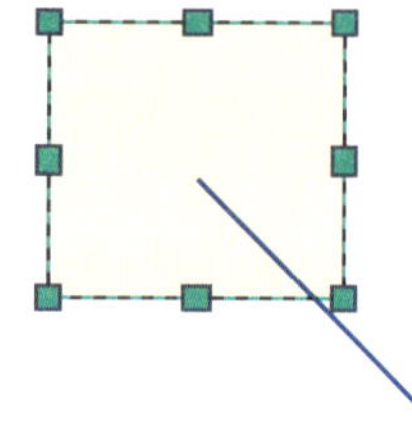

Move the mouse pointer over the graphic and click and drag it.

Nudging a graphic

1. Click the Select Elements button on the Draw toolbar and click the graphic you want to move a small amount.
2. On the Draw toolbar, click Drawing, point to Nudge, and click the direction you want to nudge the graphic.

   The graphic moves one pixel in the nudge direction.

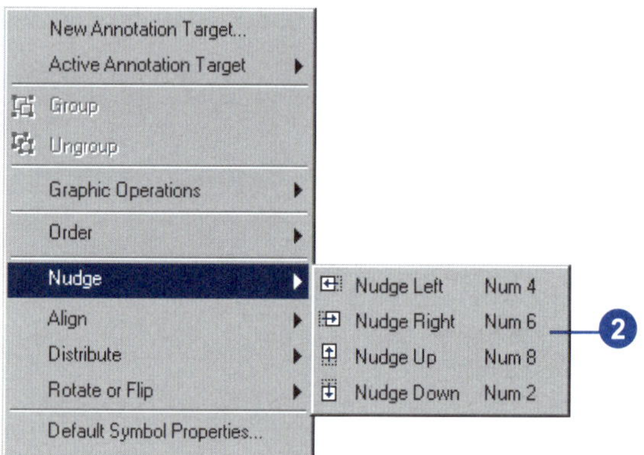

Positioning a graphic at a specific location

1. Click the Select Elements button on the Draw toolbar and double-click the graphic you want to position.
2. Click the Size and Position tab.
3. Type an X and Y position.
4. Click OK.

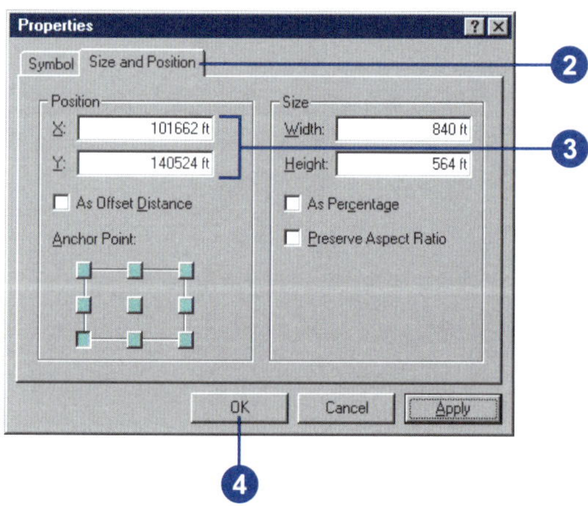

Tip

Working with a graphic in a data frame while in layout view

If you want to work with a graphic in a data frame while in layout view, click the Select Elements button on the Draw toolbar and double-click the data frame. Then click the appropriate tool to modify the graphic in the data frame.

Tip

Rotating by 90 degrees

To rotate a graphic by 90 degrees left or right, click Drawing on the Draw toolbar, point to Rotate or Flip, and click Rotate Left or Rotate Right.

Tip

Right-clicking to reveal the graphics context menu

You can right-click a graphic or a group of selected graphics to open the graphics context menu, which includes shortcuts to many graphics operations.

Ordering a graphic

1. Click the Select Elements button on the Draw toolbar and click the graphic you want to place in front of or behind other graphics.
2. On the Draw toolbar, click Drawing, point to Order, and click the ordering option.

Rotating a graphic

1. Click the Select Elements button on the Draw toolbar and click the graphic you want to rotate.
2. Click the Rotate button on the Draw toolbar.
3. Position the mouse pointer over the "x", which indicates the rotation point, and move it as necessary.
4. Click and drag the mouse to rotate the graphic.

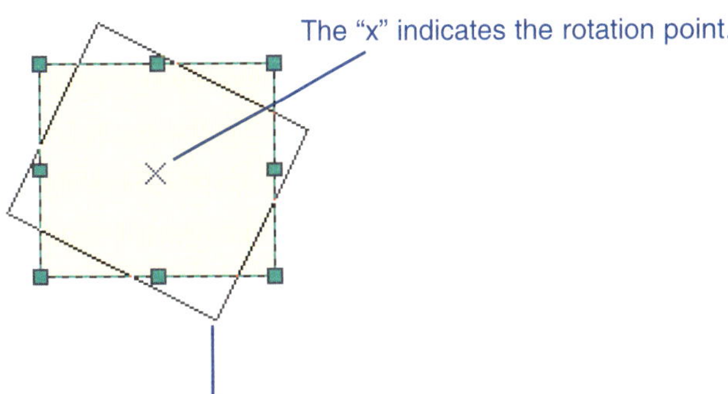

As you drag the mouse to rotate, ArcMap draws an outline of the graphic.

Tip

Displaying the Graphics toolbar

The Graphics toolbar provides quick access to frequently used tools for manipulating graphic elements. To display it, click the View menu, point to Toolbars, and click Graphics.

Flipping a graphic horizontally or vertically

1. Click the Select Elements button on the Draw toolbar and click the graphic you want to flip.
2. On the Draw toolbar, click Drawing, point to Rotate or Flip, and click Flip Horizontally or Flip Vertically.

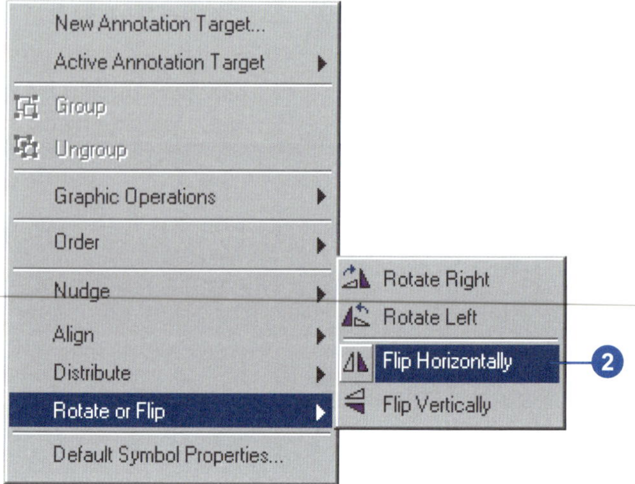

Making graphics the same size

1. Click the Select Elements button on the Draw toolbar and select the graphics you want to make the same size.
2. Click Drawing on the Draw toolbar, point to Distribute, and click Make Same Size.

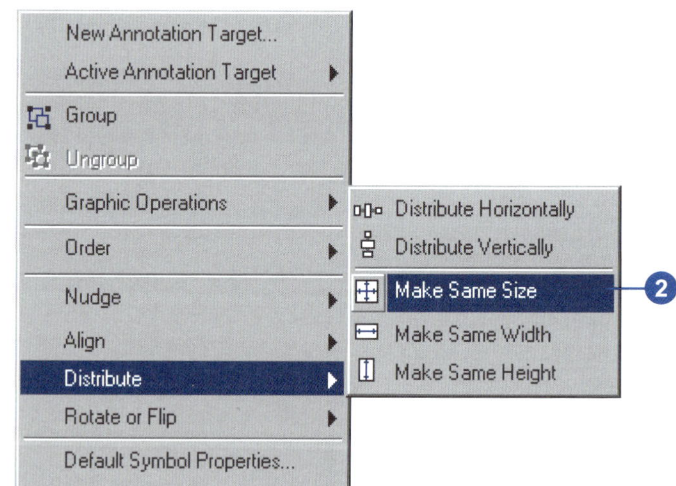

Aligning, distributing, and grouping graphics

Most of the time you'll probably just drag a graphic where you want it to be. However, you can arrange them more precisely when you need to.

You can align graphics with other graphics—using the sides, middles, or top or bottom edges. You can arrange graphics so they are equidistant from each other—distributing them either vertically or horizontally.

Once you've arranged the graphics, you may want to group them together so you can move them as a group and maintain their alignment.

Tip

What do the blue selection handles indicate?

When you have more than one graphic selected, the blue handles indicate the dominant graphic, or the one that ArcMap will use, for example, to align other graphics with. To change the dominant graphic, hold down the Ctrl key and click the selected graphic that you want to be dominant.

Aligning graphics

1. Click the Select Elements button on the Draw toolbar and select the graphics you want to align.
2. The dominant graphic has blue selection handles around it. To change the dominant graphic, press and hold the Ctrl key and click the graphic you want as the dominant one.
3. Click Drawing on the Draw toolbar, point to Align, and click the alignment you want.

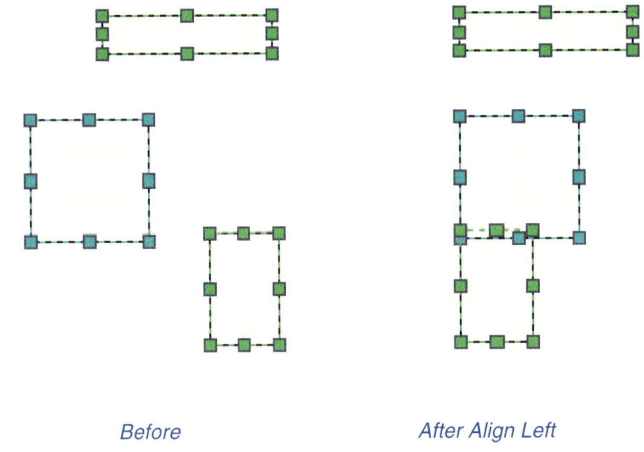

Distributing graphics

1. Click the Select Elements button on the Draw toolbar and select the graphics you want to distribute.
2. Click Drawing on the Draw toolbar, point to Distribute, and click the distribution method you want.

   The graphics are distributed with equal spacing between the centroid of each graphic.

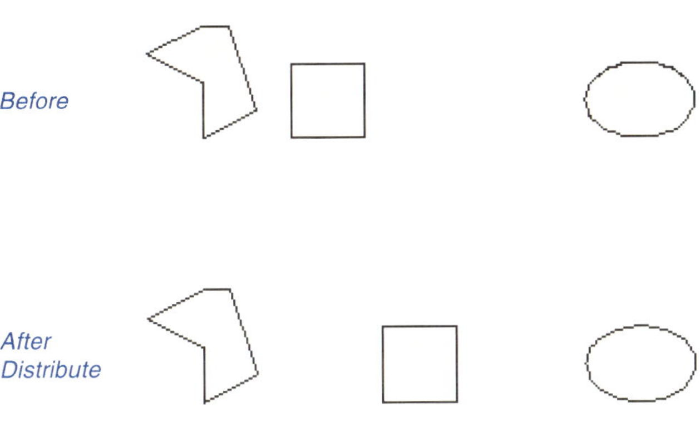

Tip

Displaying the Graphics toolbar

The Graphics toolbar provides quick access to frequently used tools for manipulating graphic elements. To display it, click the View menu, point to Toolbars, and click Graphics.

Tip

Right-clicking to reveal the graphics context menu

You can right-click a graphic or a group of selected graphics to open the graphics context menu, which includes shortcuts to many graphics operations.

Tip

Grouping grouped graphics

You can also group graphics that are already grouped.

Grouping graphics

1. Click the Select Elements button on the Draw toolbar and select the graphics you want to group.
2. Click Drawing on the Draw toolbar and click Group.

   The individual graphics now form a group.

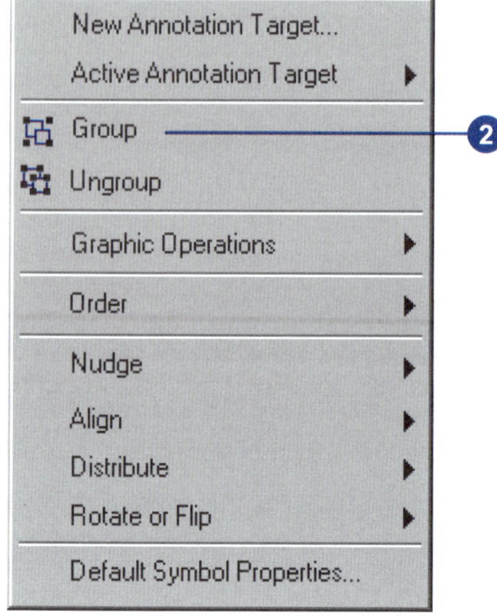

Ungrouping graphics

1. Click the Select Elements button on the Draw toolbar and click the graphics you want to ungroup.
2. Click Drawing on the Draw toolbar and click Ungroup.

   Each graphic formerly in the group is now independent.

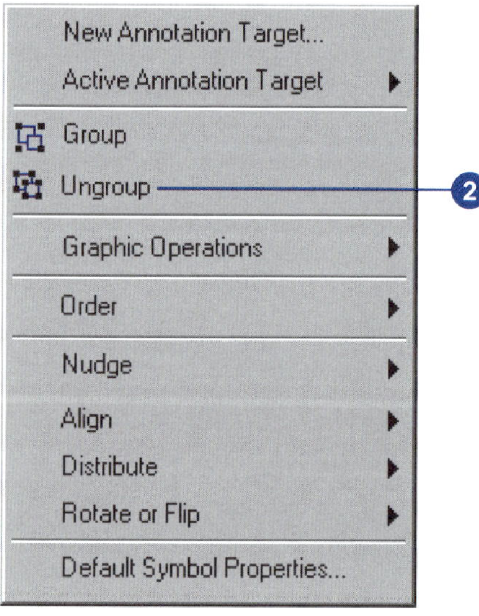

Joining graphics

You can join two or more polygon graphics you've drawn on your map to form a new graphic that is a combination of the input graphics. The graphic operations you can perform are:

- Union—joins all graphics into one large graphic. Where the graphics overlap, the boundaries are removed.
- Intersect—creates a new graphic from the shared area of the input graphics.
- Remove Overlap—creates a new graphic from the nonoverlapping areas of two input graphics.
- Subtract—creates a new graphic by subtracting the overlapping area of one graphic from another.

1. Select the polygon graphics on the map you want to join.
2. Click Drawing on the Draw toolbar, point to Graphic Operations, and click the method you want to use.

   The graphics will be joined.

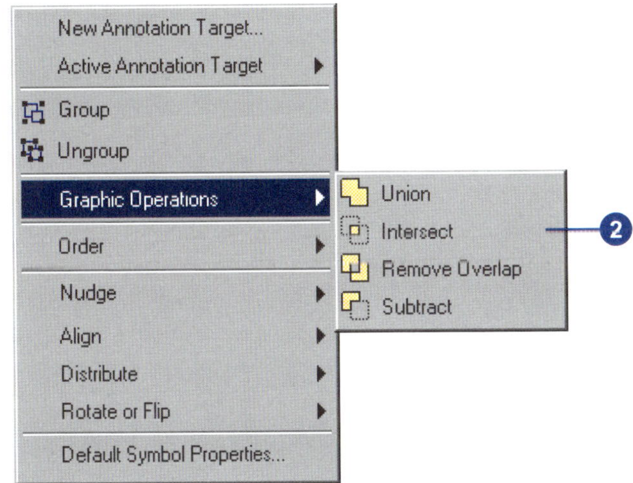

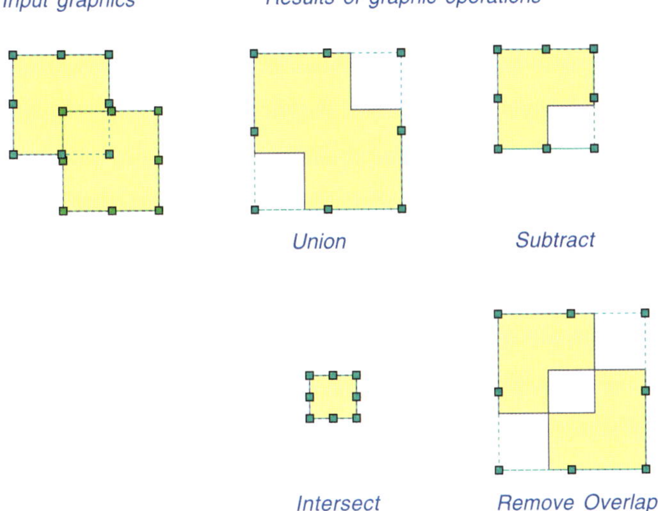

Storing graphics as annotation

Storing your graphics as annotation in groups or feature classes helps you organize them. For example, you could store all the graphics you use to annotate a particular layer in the same annotation group or feature class.

Annotation provides more control over how graphics draw in relation to each other. Annotation can be used to draw graphics only when a particular layer is visible, for example. Only graphics in geographic space can be stored as annotation.

You choose the *annotation target* where your annotation is saved. You can create *annotation groups*, which are useful for organizing a large number of graphics because you can turn them on and off individually. The Draw toolbar lets you create new annotation targets for map document annotation. Your map will always have a <Default> target—an annotation group that cannot be removed or deleted.

If you want to use your annotation in other maps or have multiple users editing in an enterprise system, store it in a geodatabase in an ▸

Creating an annotation group

1. On the Draw toolbar, click Drawing and click New Annotation Group.
2. Type a name for the annotation group.
3. Optionally, choose the layer the annotation group is associated with.
4. Optionally, set a reference scale.
5. Optionally, set the range of scales in which the annotation group is visible.
6. Click OK.

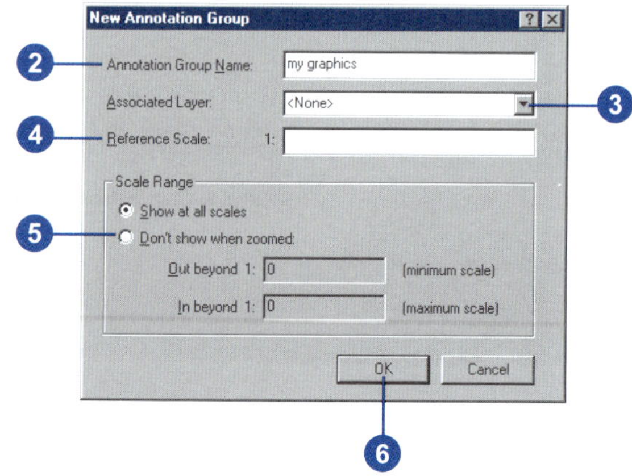

Setting the active annotation target

1. On the Draw toolbar, click Drawing and point to Active Annotation Target.
2. Click the name of the annotation group to which you want new annotation to be added. The <Default> target saves the graphics into your map document.

   The target can be an annotation group or an annotation feature class. If the annotation target is stored in a geodatabase, you need to start editing to add graphics to it.

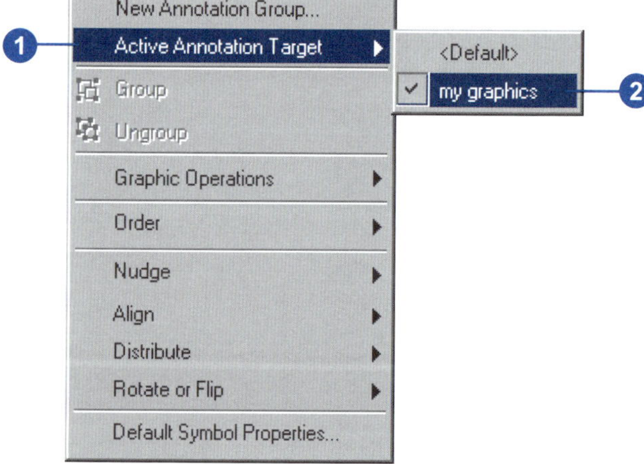

annotation feature class. You can create new annotation feature classes in ArcCatalog.

For more information on annotation and geodatabases, see 'Working with text in ArcGIS' and 'Working with annotation' in this chapter.

Tip

Which graphics can be stored as annotation?

Only text or graphics that are in geographic space can be stored as annotation. Layout graphics and map elements, such as scalebars and North arrows, cannot be stored as annotation.

Tip

Managing annotation groups

You can turn annotation groups on or off, create new groups, delete groups, and edit annotation group properties on the Annotation Groups tab of the Data Frame Properties dialog box.

Tip

Creating new geodatabase annotation targets

The Draw toolbar is used to create new map document annotation targets. For new geodatabase annotation targets, use ArcCatalog to create new annotation feature classes. See Building a Geodatabase *for more information.*

Adding new graphics to an annotation group

1. Set the active annotation target.
2. Add graphics to your map.

Moving graphics between annotation groups

1. Click the Select Elements button on the Draw toolbar and click the graphic or graphics you want to move between annotation groups.
2. Click the Edit menu and click Cut. You can also right-click the graphic and click Cut.
3. Set the active annotation target to the annotation group where you want to move the graphic.
4. Click the Edit menu and click Paste.

   The graphic is pasted slightly to the right and below the graphic's original position.

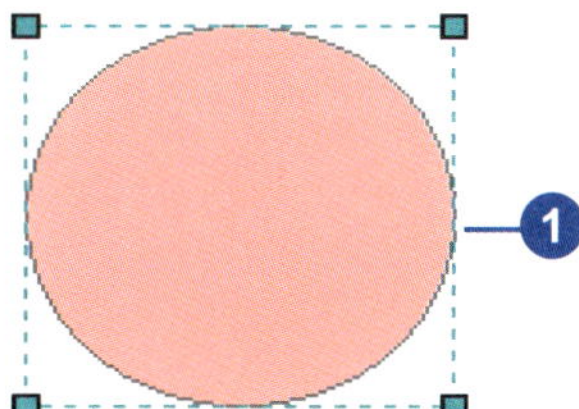

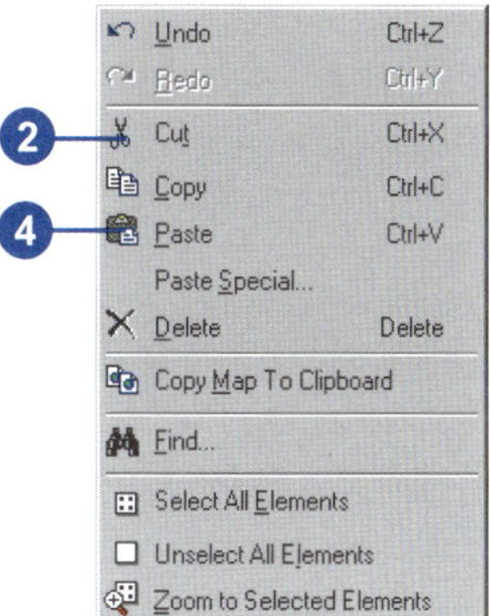

Working with text in ArcGIS

Adding text to your map

Maps convey information about geographic features, yet displaying only features on a map—even with symbols that convey their meaning—isn't always enough to make your point. In fact, most maps will not be useful without at least some textual information.

In general, there are different kinds of text you can add to your map. First, descriptive text can be placed near individual map features. For example, your map shows the name of each major city in Africa. You can also add just a few pieces of text to draw attention to a particular area of the map, such as adding text to indicate the general location of the Sahara Desert. Finally, you can add text that improves the presentation of your map. For example, a map title provides context; you might consider adding other information such as map author, data source, and date.

Using different kinds of text

Because text serves so many different mapping purposes, ArcGIS offers several different types. The main types are labels, annotation, and graphic text. A *label* is a piece of text in ArcMap that is automatically positioned and whose text string is based on feature attributes. Labels offer the fastest and easiest way to add descriptive text to your map for individual features. For example, you can turn on dynamic labeling for a layer of major cities to quickly add city names to your map of Africa. Because labels are always based on attribute fields, they can only be used to add feature descriptive text.

The second main option when working with text is to use *annotation*. Annotation can be used for describing particular features as well as to add general information to the map. You can use annotation, much like labels, to add descriptive text for many map features or just to manually add a few pieces of text to describe an area of your map. Unlike labels, each piece of annotation stores its own position, text string, and display properties. Compared to labels, annotation provides more flexibility over the appearance and placement of your text because you can select individual pieces of text and edit them. You can use ArcMap to convert labels to annotation.

Annotation can be further divided based on where you store it—in a map document or in a geodatabase. Map document annotation is in data space and organized into groups, stored in map documents, and edited with the graphic tools on the Draw toolbar. Geodatabase annotation is also in data space but is stored in a geodatabase as text or graphics and edited using the ArcMap editing tools.

In ArcGIS, some types of annotation can be displayed but not edited. These types include ArcInfo Workstation coverage, SDE 3.x, CAD, and VPF annotation. Annotation in these formats is read-only, but ArcGIS provides tools to convert to geodatabase annotation and map document annotation, both of which are editable formats.

Graphic text is useful for adding information on and around your map that exists in page space—as opposed to annotation, which is stored in geographic space. If you want to place text information on your map page that does not move as you zoom and pan on your map, you should use graphic text. Graphic text can only be added to ArcMap in layout view.

Options for storing your text

Before you begin working with text, you should take a moment to understand the text storage options in ArcGIS.

First, labels are not stored, at least not in the sense that annotation and graphic text are stored. Labels are generated dynamically and only labeling properties are stored—the settings used to create labels on the fly. If you are working in a map, your labeling properties will be saved when you save your map document (.mxd). Labeling properties can also be stored in layer files (.lyr). Use layer files, for example, to transfer labels between two maps without having to set up labeling again in the new map.

ArcGIS provides two main options for storing annotation. Geodatabase annotation is stored in a geodatabase in annotation feature classes. You can think of geodatabase annotation as a special type of geographic feature, stored together with other geographic data in a geodatabase. Like point, line, and polygon feature classes, annotation feature classes can be used in many different maps.

Map document annotation is stored in map documents in annotation groups within each data frame. Choose map document annotation if you only want to use your text in one particular map. You can use annotation groups to organize map document annotation, or you can put all of your annotation into a single <Default> annotation group that automatically exists in every map document data frame.

Graphic text is always stored in a map document. Like map document annotation, graphic text is added to a particular map. Graphic text is stored on the map layout's page and cannot be organized into groups.

Keep in mind that both annotation and graphic text are forms of graphics, and you can use the tools on the Draw toolbar to create and edit these types of text. In addition, specific tools are available in ArcMap for working with geodatabase annotation. For more information on creating and editing geodatabase annotation, see *Building a Geodatabase* or *Editing in ArcMap*.

What kind of text should I use?

The type of text that you should use is based on where you are starting from with your text and how you want to use text on your map. If you only want to add a few pieces of text and what you want to identify might not be based on attributes, then you can just use graphic text or map document annotation.

If, however, you want to have much feature-descriptive text, then you may want to use a different method. If you already have the text, such as if you have some existing coverage annotation you want to use in a new map, then you can simply add the text layer to ArcMap. You can use labels if you want to add text based on your feature attributes.

If you are starting from scratch with your features and text, then you can create a new feature class and a feature-linked annotation class. You'll be able to build your annotation automatically as you create your data.

Adding text

Text serves a variety of purposes on a map, and ArcMap supports three types of text that you can use: labels, annotation, and graphic text. To add labels to your map based on an attribute value, see 'Displaying labels' in this chapter.

ArcMap has several tools for creating new annotation and graphic text on your map. You can enter horizontal text, text that curves, and text that has a callout or leader line. To speed the task of adding descriptive text for features, you can use the Label tool to click a feature and automatically add text to annotate it. Once you have text on your map, you can use the tools on the Draw toolbar to change its position, appearance, and text string.

This section focuses on creating and editing map document annotation and graphic text. While you can follow these steps to create and edit geodatabase annotation, there are powerful, easy-to-use editing tools in ArcMap designed specifically for working with geodatabase annotation. If you are working with geodatabase annotation, see *Editing in ArcMap* for instructions on creating and ►

Adding text at a point

1. Click the New Text button on the Draw toolbar.
2. Click the mouse pointer over the map display and type the text string.

   The text will be horizontal.

Text tools on the Draw toolbar

New Text		Select Elements
New Splined Text		Edit Vertices
Label		Rotate
Callout		
New Polygon Text		
New Rectangle Text		
 New Circle Text		

Adding text along a curved line

1. Click the New Splined Text button on the Draw toolbar.
2. Click the mouse pointer over the map to add vertices along which the text should be splined.
3. Double-click to end the line.
4. Type the text string.

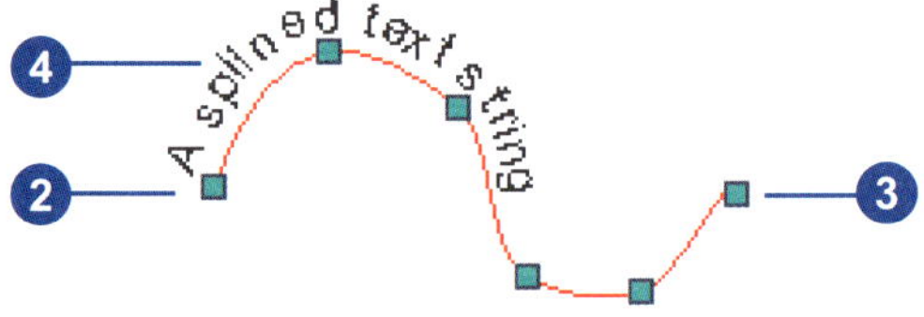

Click the Edit Vertices button on the Draw toolbar to edit the vertices of the splined text.

editing text stored in this format.

When using the tools on the Draw toolbar to add text, unless you specify otherwise, new text will be added to the <Default> annotation group of your data frame. You can change this by setting the Active Annotation Target. To learn more about this, see 'Storing graphics as annotation' earlier in this chapter.

Tip

Specifying default text symbol properties

You can set the default symbol properties for new text by clicking Drawing on the Draw toolbar and clicking Default Symbol Properties.

Adding text with a callout box and leader line

1. Click the Callout button on the Draw toolbar.
2. Click a start point for the leader line. Drag the mouse pointer and release the mouse button where you want the callout and text to be placed.
3. Type the text string.

You can click and drag the endpoint of the callout to position it correctly.

Adding text that flows within a graphic

1. Click one of the Paragraph Text tools on the Draw toolbar. Choose between New Polygon, New Rectangle, and New Circle.
2. Using your mouse pointer, click and drag to create the graphic shape. Then, double-click to complete the shape.
3. Double-click the graphic to change the text and modify its display properties.

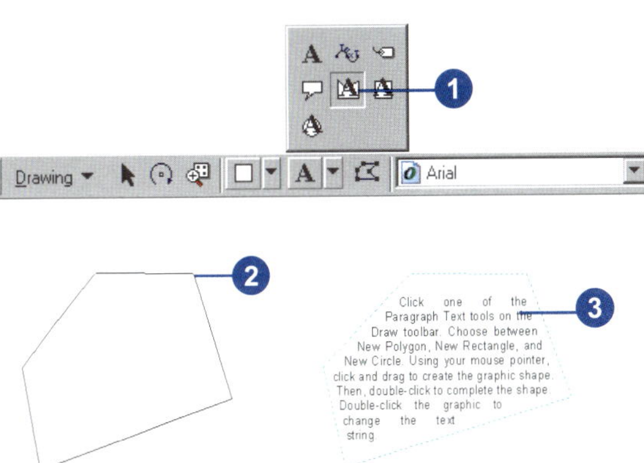

Tip

Adding text to a data frame while in layout view

When you add graphic text to a map while in layout view, ArcMap will, by default, add it to the layout. To add text to a data frame, click the Select Elements button on the Draw toolbar and double-click the data frame. Click the New Text tool to add graphic text to the data frame.

Tip

Setting the active annotation target

You can change the target location where your text is stored. For more information, see 'Storing graphics as annotation' earlier in this chapter.

See Also

You can add text to your map by annotating selected features. For more information, see Editing in ArcMap *and* Building a Geodatabase.

Adding text by clicking a feature

1. In the table of contents, right-click the layer you want to label and click Properties.
2. Click the Labels tab.
3. Click the Label Field dropdown arrow and click the field you want to use as a label.
4. Click OK.
5. On the Draw toolbar, click the Label button.

   You may have to click the dropdown arrow to choose the Label button.
6. Click Place label at position clicked.

   If you click Automatically find best placement, ArcMap finds the best location for the label.
7. Click Choose a style and click the label style you want.
8. Click the mouse pointer over the feature you want to label.

   ArcMap labels the feature.

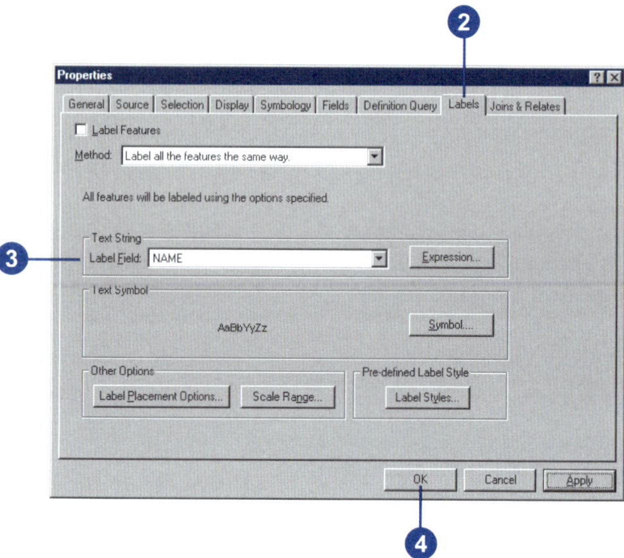

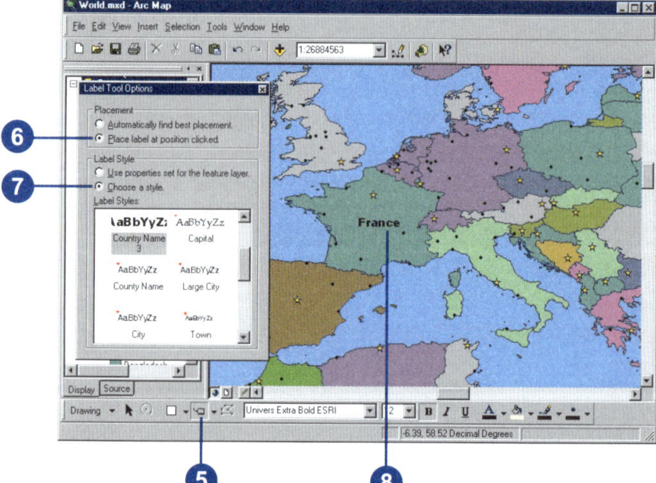

Tip

Using the Draw toolbar with geodatabase annotation

To add text to a geodatabase annotation feature class with the tools on the Draw toolbar, you must first start an edit session and set your active annotation target to the geodatabase annotation feature class.

Tip

Accessing the text Properties dialog box

You can also right-click a text element and click Properties to open the Properties dialog box.

See Also

For more information on creating text symbols, see Chapter 8, 'Working with styles and symbols'.

Changing the font, color, and size of text with the Draw toolbar

1. Click the Select Elements button on the Draw toolbar and click the text elements you want to edit.
2. Click the appropriate button on the Draw toolbar to modify a particular characteristic of the text.

Changing text with the Draw toolbar

Button	Action
B	Bold
I	Italic
U	Underline
A	Change font color
10	Change font size
Arial	Change font

Changing text properties

1. Click the Select Elements tool on the Draw toolbar and double-click the text element you want to edit.
2. Type a new text string.
3. Click Change Symbol to modify additional properties.
4. Click OK on all dialog boxes.

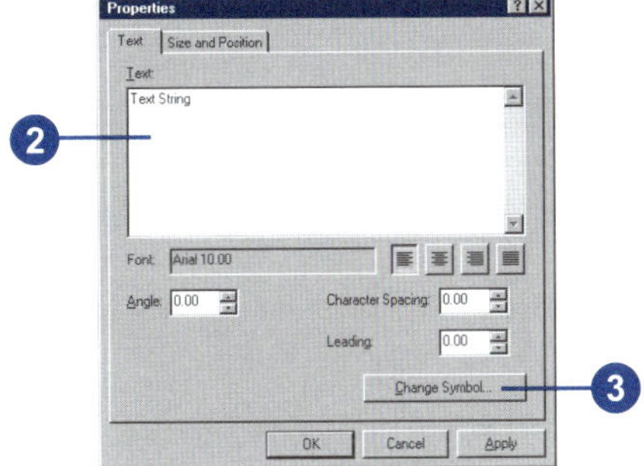

Editing a text string

1. Click the New Text button on the Draw toolbar.
2. Click the existing piece of text you want to edit.
3. Type a new text string and press the Enter key.

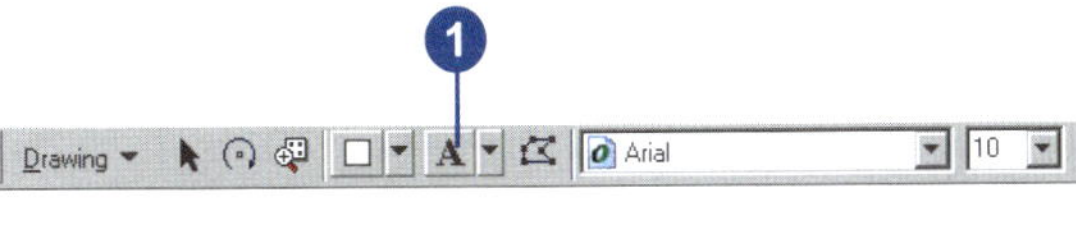

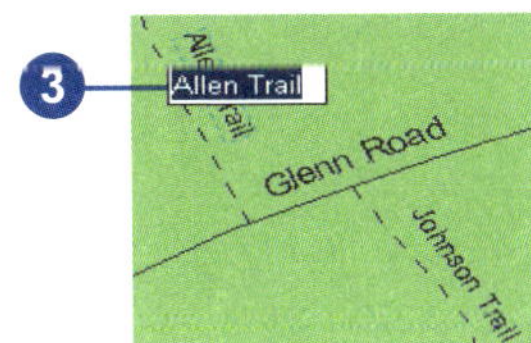

Working with labels

What is labeling?

Generally, labeling is the process of placing descriptive text onto or next to features on your map. In ArcGIS, labeling refers specifically to the process of automatically generating and placing descriptive text for map features. A label is a piece of text on the map that is derived from one or more feature attributes. Labels are not selectable, and you cannot edit the display properties of individual labels.

Labeling is useful to add descriptive text to your map for many features. Labeling can be a fast way to add text to your map, and it avoids you having to add text for each feature manually. In addition, ArcMap labeling dynamically generates and places text for you. This can be useful if your data is expected to change or you are creating maps at different scales.

Annotation is an alternative to labeling. If you only want to add descriptive text for a few features, if you want to reuse your text and make it appear in the same place all the time, or if you do not have attributed features, it will be better to add your text as annotation. Labeling your features is a good way to create annotation. To learn more about adding text as annotation, see 'Adding text' earlier in this chapter or *Editing in ArcMap*.

Displaying dynamic labels

To display labels for a layer, you simply specify the attribute or attributes of the feature you want to base your labels on—for example, a street name or soil type—then turn on labeling. ArcMap automatically places labels on or near the features they describe. You can also control the font, size, and color of the text to help differentiate labels for different types of features. A map of Europe, for example, could have both country and major city labels, each displayed with a different text symbol.

When you turn on dynamic labeling, ArcMap places as many labels on the map as possible without any overlap. In areas where features are tightly clustered together, some features may not be labeled. As you zoom in on your map, more labels will dynamically appear.

Controlling which features are labeled

Dynamic labeling is a fast and easy way to add text to your map, and this holds true even for complicated maps. Simply turn on labeling for your layer or layers, and as you pan and zoom around your map, ArcMap dynamically adjusts the labels to fit the available space. To gain more precise control over which features are labeled and where labels are placed, you need to work with more advanced labeling properties. Specifically, you can adjust which features are labeled and where labels are placed with respect to features.

There are three ways to control which features are labeled:

- Set the *label priority*, which controls the order that labels get placed on the map.
- Set *label weights* and *feature weights* to establish a ranking system for labels when there is a conflict—that is, overlap—on the map with other labels or features.
- Use *label classes* to be able to specify different labeling properties, including priority, weights, and placement properties, for features in the same layer.

Label priority, label weights, and feature weights work together to control which features are labeled. These settings also affect where labels are placed.

Consider a map of Europe on which you are labeling both country names and cities. Depending on the scale of your map, there may not be room for all features to be labeled. You decide that the country labels are more important than the city labels. To make your map labels reflect this, you first change the label priority to make sure that country labels are placed before city labels.

Label priority can work on a layer by layer basis, but you can also specify label priority within layers by further dividing a layer's labels into label classes. For example, you could divide your city labels into two label classes: major cities and secondary cities. Then, because major city labels are more important, you could give the major city labels a higher priority and a higher label weight than the secondary labels.

You can further refine your map by adjusting the feature weights of your city label classes. The general rule with weights is that a feature cannot be overlapped by a label with an equal or lower weight. Continuing with the example, you could increase the feature weights of your major cities class from None to High, which is the highest weight. Doing this will result in a map where labels can overlap secondary city symbols but not major city symbols.

To learn more about label priority and label and feature weights, see 'Prioritizing and positioning labels' in this chapter. To learn more about label classes, see 'Displaying labels' in this chapter.

Controlling where labels are placed

To control where labels are placed, you should use label placement properties. Like label priority and weights, these settings work on a layer basis, or you can use label classes to subdivide features in the same layer and assign them different placement properties. Label placement properties let you specify where each label is placed on the map with respect to the feature being labeled. ArcMap has different label placement options for point, line, and polygon labels. In addition, installing and enabling the Maplex for ArcGIS extension will give you a different, enhanced set of label placement properties.

To learn more about label placement properties, see 'Prioritizing and positioning labels' in this chapter.

Converting labels to annotation

If you need exact control over where a given label is placed on your map, you should convert your labels to annotation. Text stored as annotation is editable, which means that you can select and move individual pieces of text, as well as change their display properties (font, size, color, and so on). For example, you might want to convert labels to annotation so you can manually move a few pieces of text to make room for one piece that ArcMap was unable to place due to space constraints. When you convert labels to annotation, ArcMap provides you with a list of all the labels that weren't placed and lets you interactively place them on your map as needed. For more information, see 'Converting labels to annotation' in this chapter.

The Labeling toolbar and the Label Manager

The Labeling toolbar is where you start labeling in ArcMap. From here you can control the labeling process and open the *Label Manager*, which lets you view and change labeling properties for all the labels in your map. Additional controls are added to the Labeling menu when you install the Maplex for ArcGIS extension. See *Using Maplex for ArcGIS* for more information.

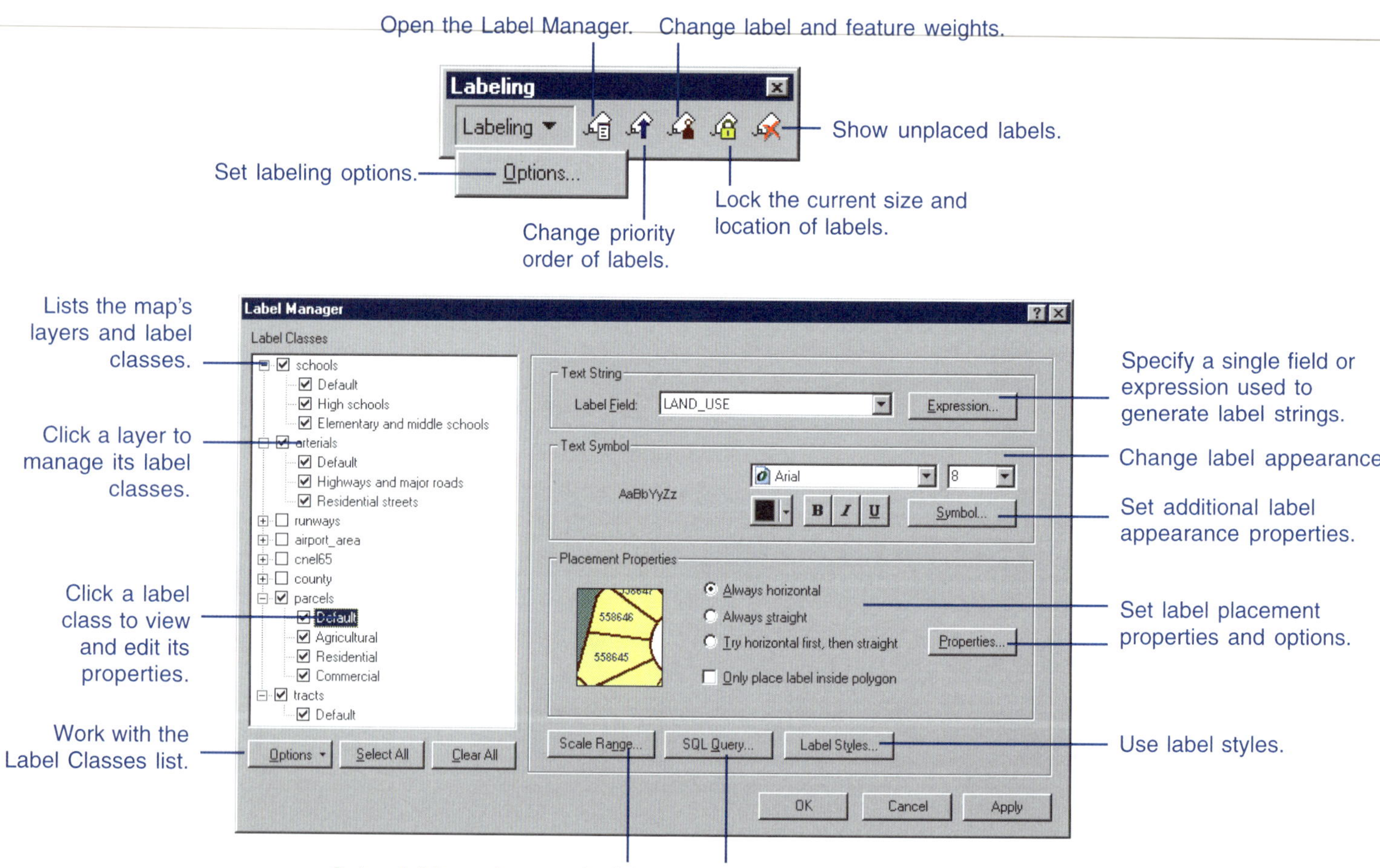

Making a map with labels

Follow these steps to create a high-quality map with labels:

1. Start ArcMap and create a new map or open an existing one. If necessary, add the data you want to label to your map.
2. Add the Labeling toolbar to ArcMap.

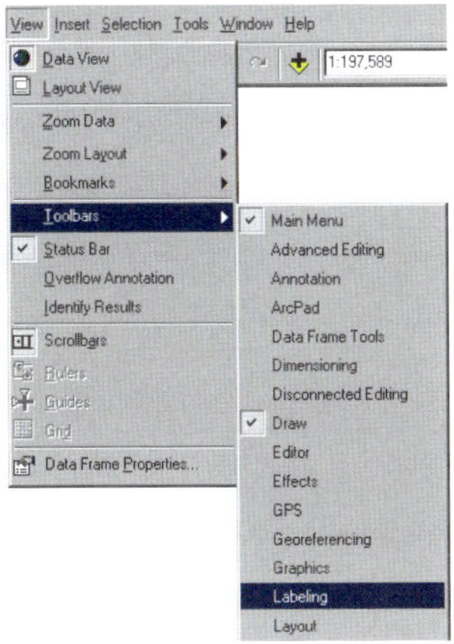

3. Open the Label Manager.

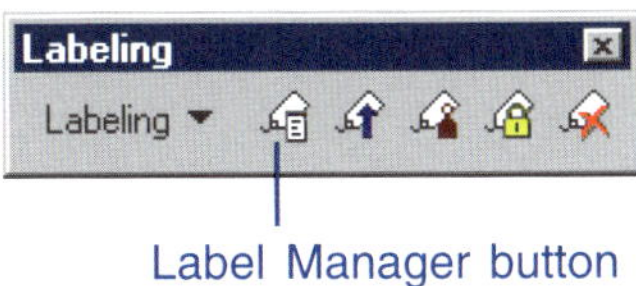

Label Manager button

4. Turn on labeling for your layer's default label class.

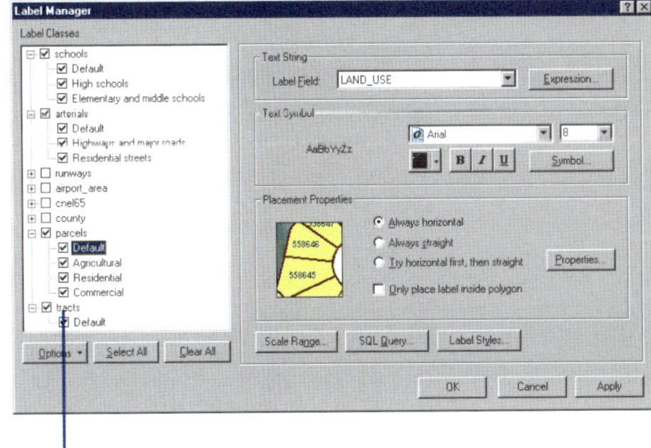

Check the box next to the label class that you want to label.

5. Create additional label classes in the Label Manager if you want to specify different labeling properties for the features in your layer.

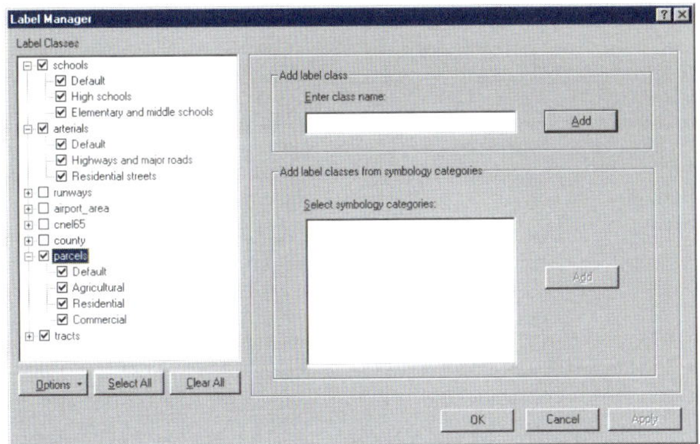

Click a layer on the left panel and add a new label class to it on the right panel.

6. Use the Label Manager to polish your map by changing the label expression, label text symbol, and label placement options.

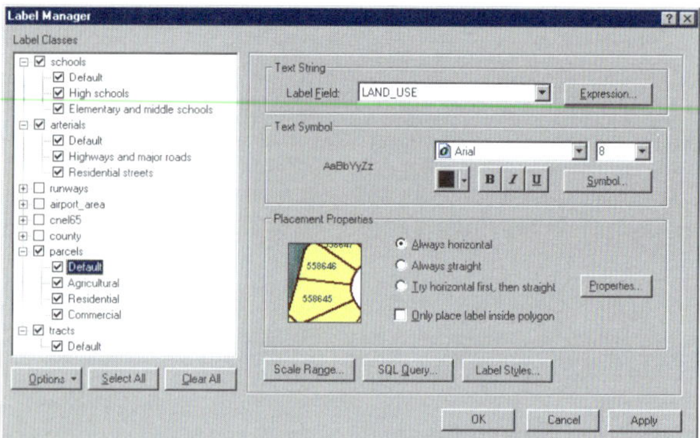

Work with the right panel of the Label Manager when formatting your labels.

7. Use the Label Priority and Label Weight Ranking dialog boxes (both accessed from the Labeling toolbar) to ensure that the most important features on your map are labeled and that their labels are in the best positions.

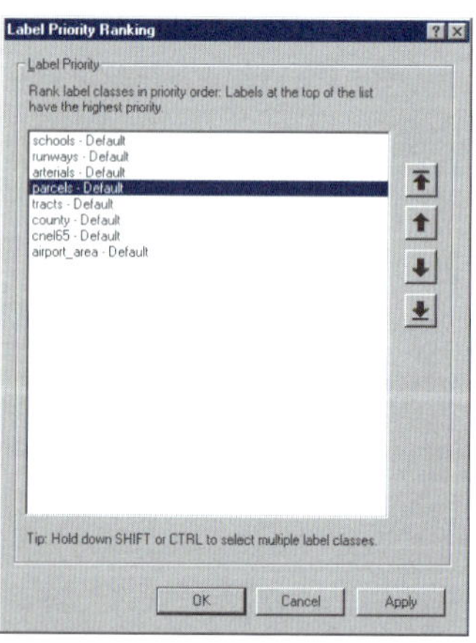

Label Priorities dialog box

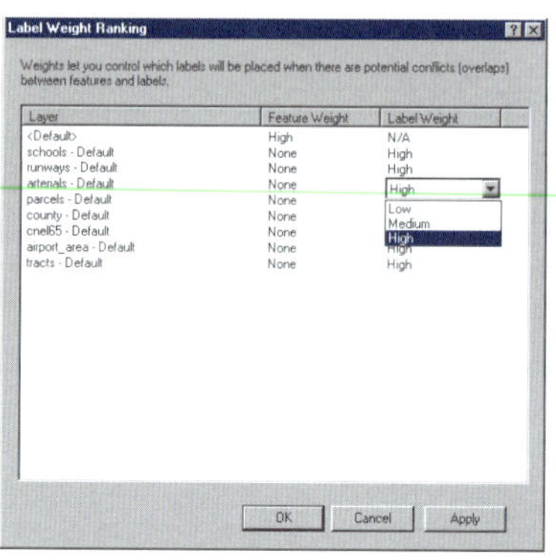

Label and Feature Weights dialog box

8. Convert your labels to annotation if you want to be able to position some pieces of text manually or if you want the text to always appear in the same position. For more information, see 'Converting labels to annotation' in this chapter.

Displaying labels

Labeling is an easy way to add descriptive text to features on your map. Labels are dynamically placed, and label text strings are based on feature attributes. You can easily turn labels on or off and can even lock them so their locations stay fixed as you zoom or pan on your map.

You can use dynamic labeling for all features in a layer, or alternatively, you can use label classes to specify different labeling properties for features within the same layer. For example, in a layer of cities, you might label those with a population greater than 100,000 with a larger font size and those cities with a population less than 100,000 with a smaller font size. In addition, if the features in your layer are symbolized with different symbols, you can create label classes from your symbology classes. Building your label classes from your symbol classes is a fast way to create maps with a consistent look.

Labels are not editable. If you want to be able to position individual pieces of text manually, you should convert labels to annotation. To learn more, see 'Converting labels to annotation' in this chapter.

Adding the Labeling toolbar and opening the Label Manager

1. Click the View menu, point to Toolbars, and click Labeling.
2. Click the Label Manager button.

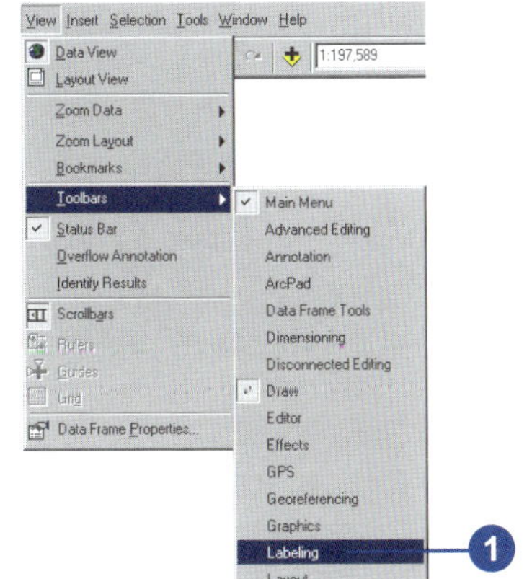

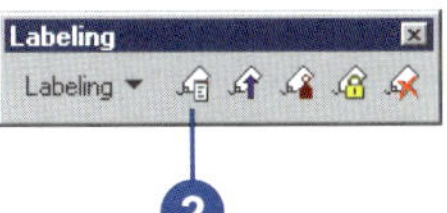

Using the Label Manager

1. Open the Label Manager by clicking the Label Manager button on the Labeling toolbar.
2. Check the box next to the layer you want to label.
3. Choose a label class under the layer.
4. Click the Label Field dropdown arrow and click the attribute field you want to use as a label.
5. Click OK.

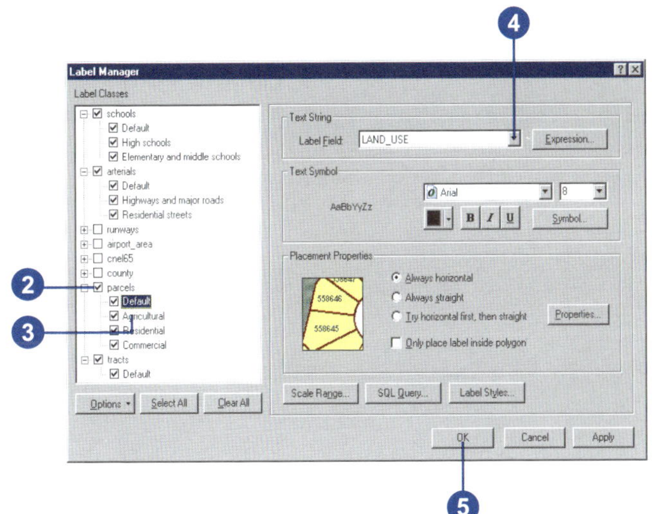

Tip

Automatically removing duplicate labels

ArcMap automatically removes duplicate labels. You might want to disable this behavior when labeling features, such as soil types or land use categories, where several features can have the same attribute value. Click Properties on the Label Manager's Placement Properties frame, and work with the options for label placement and conflicts (overlaps).

Tip

Using symbology classes

You can only use symbology classes to build label classes if your data is symbolized using unique values; unique values, many fields; match to symbols in a style; graduated colors; or graduated symbols.

See Also

To get more precise control over which features are labeled and where labels are positioned, see 'Prioritizing and positioning labels' in this chapter. For full control, convert your labels to annotation. See 'Converting labels to annotation' in this chapter.

Changing label symbols

1. Open the Label Manager.
2. Click a label class in the Label Classes list.
3. Click the buttons and dropdown menus in the Text Symbol box to set the font, size, color, or other symbol properties of your labels.
4. Optionally, click the Symbol button to change other properties or to choose an existing text symbol for your labels.
5. Click OK.

Building label classes from symbology classes

1. Open the Label Manager.
2. Click a layer in the Label Classes list.
3. In the Select symbology categories box, check the box next to the symbology class you want to use to make a new label class.
4. Click the Add button.
5. Click Yes or No on the Overwrite label classes dialog box, depending on what you want to do with your existing label classes.
6. Click OK.

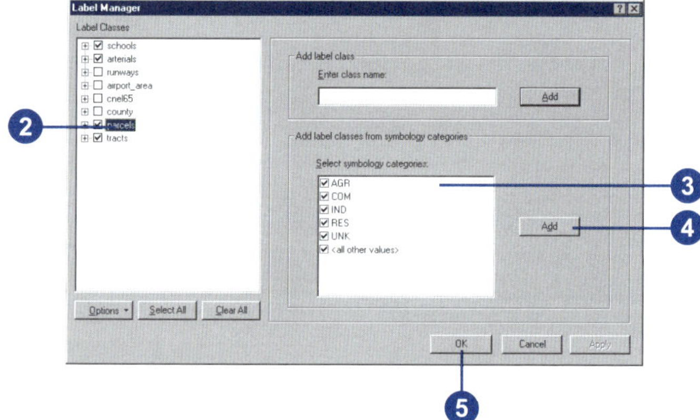

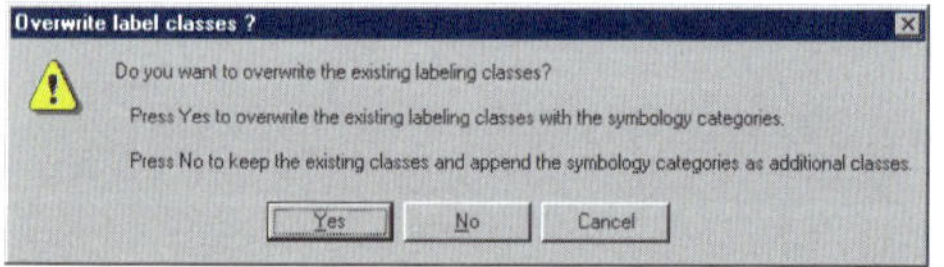

Tip

Locking labels

To lock the current size and location of labels, click the Lock Labels button on the Labeling toolbar. This turns off the labeling process, and as you pan and zoom, labels will stay in place. Click the Lock Labels button again to return to dynamic placement. When you lock labels, text will scale the same as when you set a reference scale.

Tip

Working with label styles

A label style consists of both a text symbol and a set of label placement properties. When you choose a label style, its text symbol replaces the current label symbol.

Tip

Creating text within a highway shield marker

You can label road line features with text inside a highway shield by using a label style and choosing one of the standard highway shield styles.

See Also

For more information on using styles and saving to your personal style folder, see Chapter 8, 'Working with styles and symbols'.

Using label styles

1. Open the Label Manager.
2. Click a label class in the Label Classes list.
3. Click the Symbol button.
4. Click a standard label style from the left pane of the Symbol Selector dialog box.
5. Optionally, modify the properties of a label style and click Save to save your new symbol in your personal style folder.
6. Click OK on the Symbol Selector dialog box.
7. Click OK on the Label Manager dialog box.

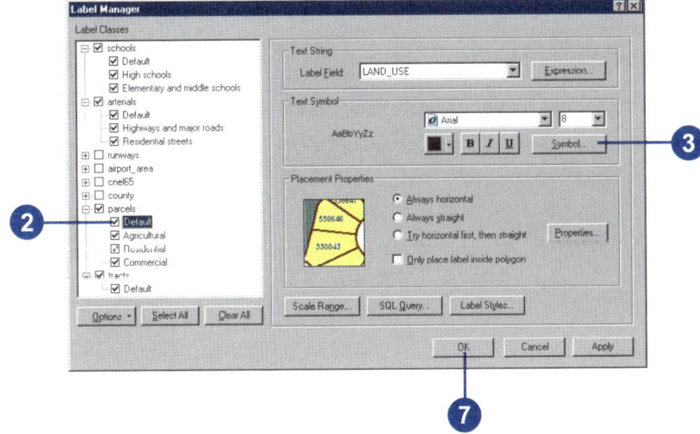

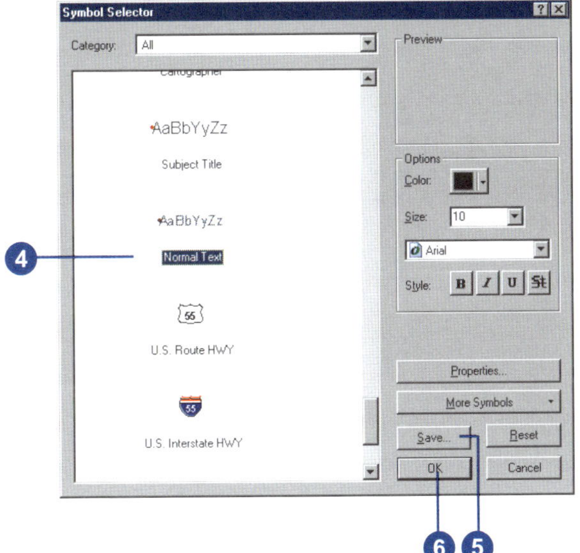

Tip

Using different text symbols to label features in a single layer

Label classes allow you to use a different text symbol to label different types of features in the same layer. For example, you could label cities with a large population in a larger font than those with a smaller population.

Tip

Displaying coverage annotation

If you have a coverage with annotation, you can display the annotation as a layer in the table of contents. Add the layer as you would any feature layer.

Tip

Using the Layer Properties Labels tab

You can still use the Labels tab on the Layer Properties dialog box to set the symbol and placement properties for your labels.

Tip

Making the labels get bigger when you zoom in

As you zoom in and out on your map, the size of the labels does not change. If you want the text to scale with the map, set a reference scale. Right-click the data frame and click Reference Scale.

Using label classes to label features from the same layer differently

1. Open the Label Manager.
2. Click the layer in the Label Classes box for which you want to create label classes.
3. Type a name for your new label class in the Enter label name box.
4. Click Add.
5. Uncheck the Default label class to avoid labeling some features twice.
6. Right-click the new label class in the Label Classes list and click SQL Query.
7. Click the operators to build an expression that identifies the subset of features you want to label.
8. Click OK.
9. Click the Label Field dropdown arrow and click the attribute field you want to use as a label.
10. Click the buttons and dropdown menus to define the symbol and placement properties of your labels.
11. Repeat steps 2 through 10 if you want to create additional label classes.
12. Click OK.

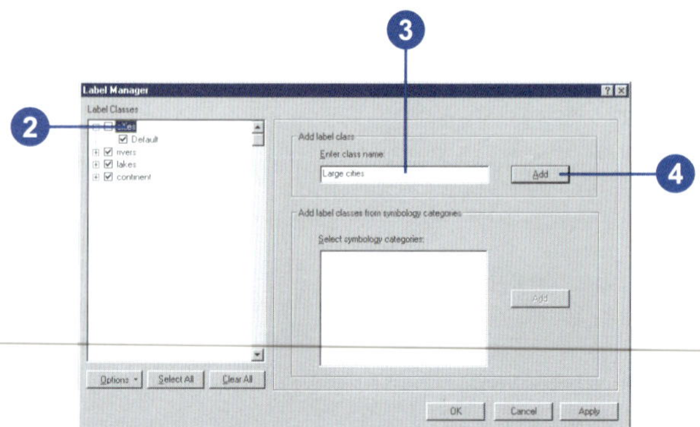

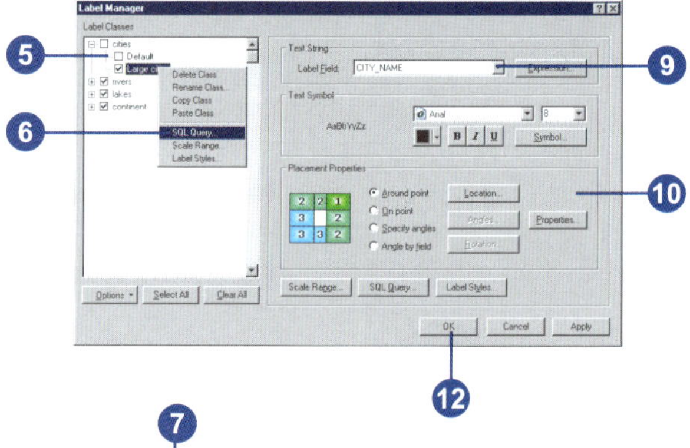

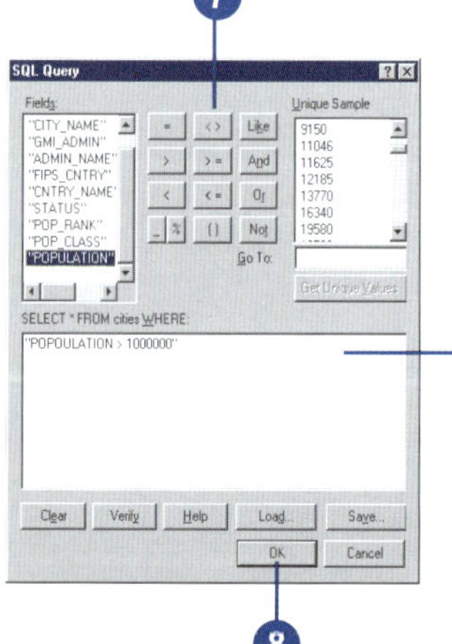

In this example, cities with a population greater than 1,000,000 will be labeled.

Specifying the text of labels

Label text strings are derived from one or more feature attributes. Labels are dynamic, so if feature attribute values change, then the labels will also change. When you turn on labeling, features are initially labeled based on one field—such as labeling city features based on a field that stores the city name.

You can also label based on an expression, which can have multiple fields, contain extra characters, and include VBScript or JScript functions that format your labels. So, you might label each city with both its name and its population, and use a special VBScript character to stack the city name on top of the population.

Using advanced label expressions is an even more powerful option. Using these you can add conditional logic, looping, and other scripting syntax to your label expressions. For example, you could produce labels that have only the first letter of each word capitalized from city names that are stored in all capital letters.

Label expressions are a useful place to use ArcGIS text formatting tags. To learn more ►

Labeling based on a single attribute field

1. Open the Label Manager.
2. Click a label class in the Label Classes list.
3. Click the Label Field dropdown arrow and click the field you want to use as a label.
4. Click OK.

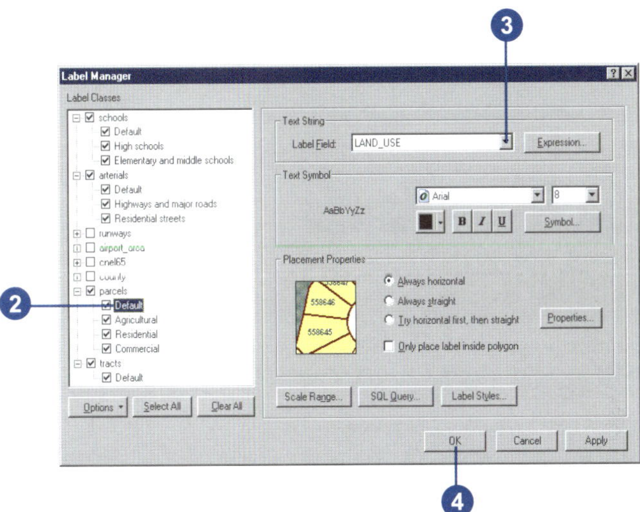

about label expressions and text formatting tags, see 'Building label expressions' and 'Using text formatting tags' in this chapter.

Tip

Adding characters to expressions

Characters must be enclosed in double quotes and connected to the rest of the label expression using the & character.

Tip

Writing expressions

Your expressions can include any valid statement supported by the scripting language chosen in the Parser dropdown list.

Tip

Writing an expression with multiple lines of code

The expression is limited to a single line of code unless you check the Advanced box on the Label Expression dialog box. Checking the Advanced box allows you to enter a function containing programming logic and spanning multiple lines of code.

Labeling based on multiple attribute fields

1. Open the Label Manager.
2. Click a label class in the Label Classes list.
3. Click Expression.
4. Click a label field and click Append to use the text of that field in your labels.
5. Optionally, use the Expression box to add additional characters you want to appear in your labels or add VBScript or JScript functions to format your labels.
6. Click Verify to make sure that there are no syntax errors and to preview your label string. Close the Label Expression Verification dialog box.
7. Click OK.
8. Click OK.

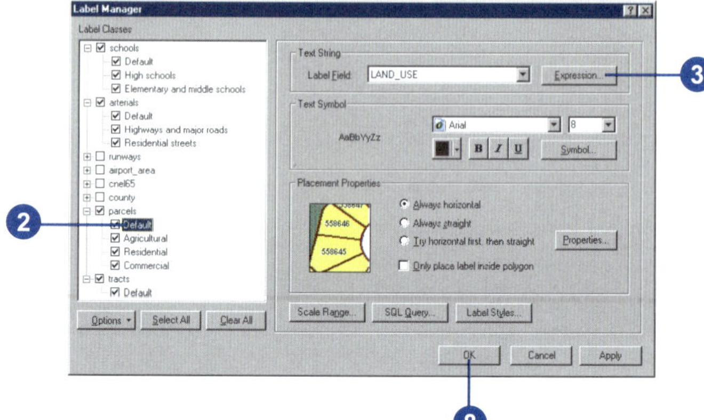

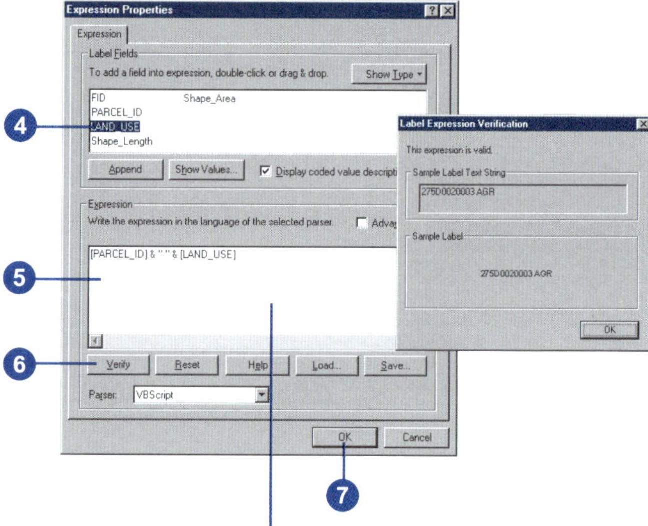

To create stacked text, use the VBScript constant, vbNewLine, between the field names—[PARCEL_ID] & vbNewLine & [LAND_USE]. You can also click the Help button for more information on syntax and building label expressions.

Building label expressions

About label expressions

You can use *label expressions* to adjust the formatting of your labels. In addition to inserting characters and scripting functions, you can also use ArcGIS formatting tags in label expressions. These are special characters that you can use to change the appearance of all or part of your labels. For example, you might use the bold formatting tag to make the first line bold in a stacked, multiline label.

A label expression is limited to a single line of code, unless you check the Advanced box on the Label Expression dialog box. Checking the Advanced box allows you to enter a function containing programming logic and spanning multiple lines of code. See 'Specifying the text of labels' earlier in this chapter for more information on applying label expressions.

Examples are shown below for common uses of VBScript functions as well as ArcGIS formatting tags in label expressions. In addition, a complete reference of the ArcGIS formatting tags is provided.

Examples of label expressions

The following are examples of label expressions:

Use the VBScript & operator to concatenate strings. For example, this expression creates a label where the value of the PARCELNO field is preceded by the text `"Parcel no: "`:

```
"Parcel no: " & [PARCELNO]
```

To control how decimal numbers are displayed, use the VBScript `Round` function. For example, this expression displays a field called Area rounded to one decimal place:

```
Round ([AREA], 1)
```

To convert your text labels to all uppercase or lowercase, use the VBScript `UCase` and `LCase` functions. For example, this expression makes a Name field all lowercase:

```
LCase ([NAME])
```

To create stacked text, use the VBScript `vbNewLine` or `vbCrLf` constants between the field names:

```
"Name: " & [NAME] & vbNewLine & [ADDRESS_1] &
vbNewLine & [ADDRESS_2]
```

Use the VBScript format functions to format your labels. For example, this expression displays the label as currency:

```
"Occupancy Revenue: " & FormatCurrency
([MAXIMUM_OC] * [RATE])
```

This VBScript function labels cities with their name only if their population exceeds 250,000:

```
Function FindLabel ([NAME], [POPULATION])
   if ([POPULATION] > 250000) then
   FindLabel = [NAME]
   end if
End Function
```

For more information, see the Microsoft® VBScript Language Reference or the Microsoft JScript Language Reference.

ArcGIS text formatting tags

Labels will be drawn using the symbol specified in the Label Manager or on the Labels tab of the Layer Properties dialog box. You can modify or override the appearance of this symbol for particular portions of the expression by inserting ArcMap *text formatting tags* into the expression as text strings. This lets you create mixed-format labels where, for example, one field in a label is underlined. You can even use tags with curved text.

The tags that you can use are listed in the table below. Acceptable values for Color (RGB) are red, green, blue = 0–255, and acceptable values for Color (CMYK) are cyan, magenta, yellow, black = 0–100; missing color attributes are assumed to be 0. The defaults are 0 percent for Character spacing (no adjustment), 100 percent for Character width (regular width) and Word spacing (regular spacing), and 0 points for Line leading (no adjustment).

Formatting action	Tag syntax
Font	"<FNT name='Arial' size='18'>" & [LABELFIELD] & "</FNT>" "<FNT name='Arial' scale='200'>" & [LABELFIELD] & "</FNT>"
Color (RGB)	"<CLR red='255' green='255' blue='255'>" & [LABELFIELD] & "</CLR>"
Color (CMYK)	"<CLR cyan='100' magenta='100' yellow='100' black='100'>" & [LABELFIELD] & "</CLR>"
Bold	"<BOL>" & [LABELFIELD] & "</BOL>"
Italic	"<ITA>" & [LABELFIELD] & "</ITA>"
Underline	"<UND>" & [LABELFIELD] & "</UND>"
All capitals	"<ACP>" & [LABELFIELD] & "</ACP>"
Small capitals	"<SCP>" & [LABELFIELD] & "</SCP>"
Superscript	"^{" & [LABELFIELD] & "}"
Subscript	"_{" & [LABELFIELD] & "}"
Character spacing	"<CHR spacing='25'>" & [LABELFIELD] & "</CHR>"
Character width	"<CHR width='150'>" & [LABELFIELD] & "</CHR>"
Word spacing	"<WRD spacing='150'>" & [LABELFIELD] & "</WRD>"
Line leading	"<LIN leading='12'>" & [LABELFIELD] & "</LIN>"
Un-Bold	"<_BOL>" & [LABELFIELD] & "</_BOL>"
Un-Italic	"<_ITA>" & [LABELFIELD] & "</_ITA>"
Un-Underline	"<_UND>" & [LABELFIELD] & "</_UND>"
Un-Superscript	"<_SUP>" & [LABELFIELD] & "</_SUP>"
Un-Subscript	"<_SUB>" & [LABELFIELD] & "</_SUB>"

Tag syntax

The following syntax rules apply to tags in label expressions.

Just like other static text in label expressions, formatting tags must be surrounded by double quotes and concatenated to other parts of the expression using the & operator:

```
"<BOL>" & [LABELFIELD] & "</BOL>"
```

Tags are not interpreted by VBScript/JScript. Instead, they are passed on to the ArcMap framework as plain text, to be dynamically formatted as they are drawn. You don't need to quote tags included inside quoted strings:

```
"Current <BOL>status</BOL> of parcel: " &
[LABELFIELD]
```

The ArcMap text formatting tags follow Extensible markup language (XML) syntax rules. Each start tag must be accompanied by an end tag. Tags can be nested, but you must close the inner tag before closing an outer tag:

```
"<BOL><UND>" & [LABELFIELD] & "</UND></BOL>"
```

The case of tag pairs must match exactly. So <BOL>...</BOL> is valid, as is <bol>...</bol>, but <Bol>...</bol> is invalid.

In label expressions, tag attributes must be surrounded either by single quotes (as shown in the table above) or by two sets of double quotes. The following expression is equivalent to the FNT entry in the table:

```
"<FNT name=""Arial"" size=""18"">" & [LABELFIELD]
& "</FNT>"
```

& and < are special characters and are not valid in your text if formatting tags are used. Use the equivalent character codes & and < instead. For example, this expression displays the values of the label field inside < > characters:

```
"<ITA>&lt;" & [LABELFIELD] & "></ITA>"
```

If you have special characters embedded in the values of the label field, you can replace them dynamically using a simple label script:

```
Function FindLabel ([LABELFIELD])
   NewString = Replace([LABELFIELD],"&","&amp;")
   FindLabel = "<ITA>" & NewString & "</ITA>"
End Function
```

Formatting tags can be embedded in the values of the field you use to label a layer's features, whether or not you use a label expression. In this way, you can change the format of any portion of a particular value in a label field. In order to embed formatting tags, the label field must be of string type. Tags and tag attributes used in field values do not need to be surrounded by quotes, so the following are valid values for a label field:

```
<ITA>Rochester</ITA>
<FNT size='18'>C</FNT>olorado
```

Tags aren't recognized by the table of contents, Attribute table window, or Identify Results window, so tags added to field values appear unformatted as raw text in these windows.

For more information on working with formatting tags, see 'Using text formatting tags' in this chapter.

Prioritizing and positioning labels

When you turn on dynamic labeling, ArcMap fits as many labels as possible—without overlap—within the available space on the map. As you pan and zoom, ArcMap automatically adjusts labels to fit the available space. For many maps, ArcMap default labeling will be adequate. If you need more control over which features are labeled and where labels are placed, you should work with label placement options, label priority, and label and feature weights. All of these properties work on a label class level.

Label placement options control the positioning of labels with respect to features. For example, you might specify that your city labels always be placed above and to the right of the cities. To learn more, see the ArcGIS Desktop Help. ▸

See Also

If you install the Maplex for ArcGIS extension, label placement, priority, and weights work differently. For more information, see Using Maplex for ArcGIS.

Setting label placement

1. Open the Label Manager by clicking the Label Manager button on the Labeling toolbar.
2. Click a label class in the Label Classes list.
3. Click the placement options you want from the Placement Properties box.
4. Optionally, click Properties to see the complete set of label placement properties.
5. Click OK.

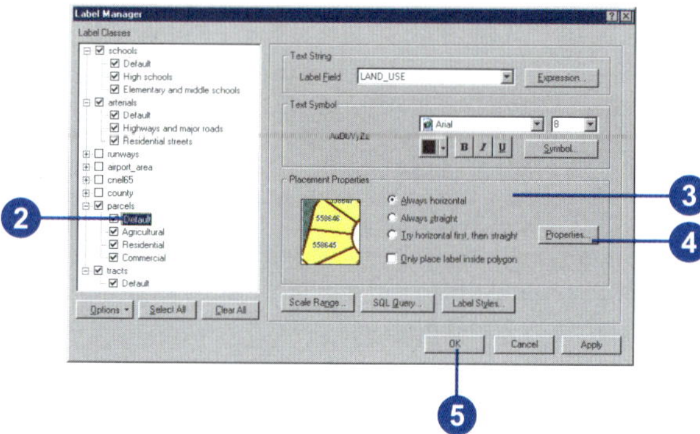

Setting label priority

1. Click the Label Priority button on the Labeling toolbar.
2. Click the label class or classes for which you want to change the label priority.

   Holding down the Shift or Ctrl keys allows you to select multiple label classes.
3. Click the arrow buttons to move a label class up or down in the list.

   Clicking the up arrows gives the class a higher priority, while clicking the down arrows gives the class a lower priority.
4. Click OK.

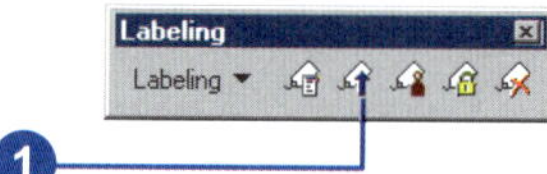

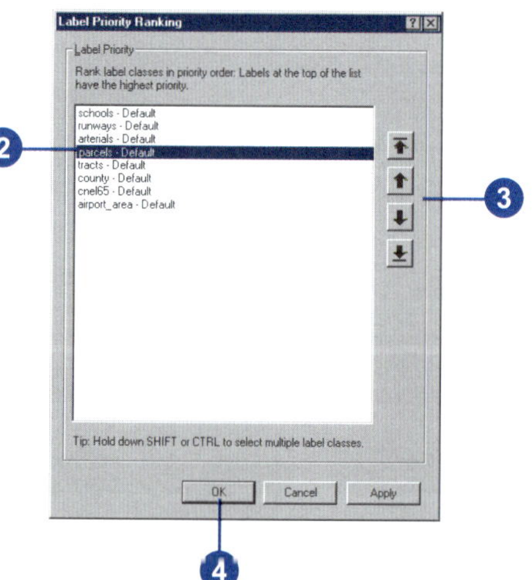

To increase the chance that more important features are labeled first, assign these features a higher label priority. For example, on a city street map, you'd probably assign a higher labeling priority to highways than residential streets.

Use label weights and feature weights to assign relative importances to labels and features to be used only when there is a conflict, that is, an overlap between a label and a feature. Ultimately, the final positioning of labels on your map is dependent on label and feature weights. In addition, when working with weights, ▶

Tip

Preventing labels from overlapping features

Setting a feature weight of High for point or line features ensures that no labels will be placed on top of these features. Setting a feature weight of High for polygon features ensures that no labels will be placed on the outline of these features.

Tip

Using feature weights

Feature weights other than None can dramatically slow labeling speed because ArcMap must evaluate the location of every feature before placing each label.

Setting label and feature weights

1. Click the Weights button on the Labeling toolbar.
2. Click the label or feature weight you want to change and click the desired weight from the dropdown menu.

   A label cannot overlap a feature with an equal or higher weight.
3. Click OK.

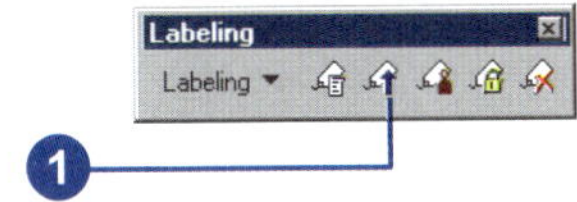

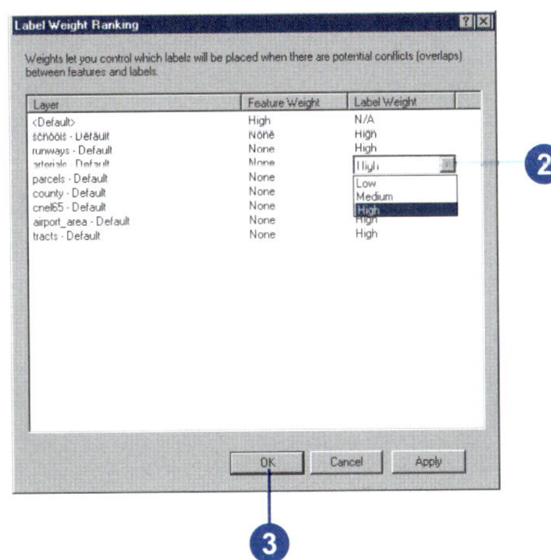

keep in mind that when you allow labels to overlap some features, generally more labels will be placed on your map because ArcMap has more room to place them.

Both labels and features can have a weight of None, Low, Medium, or High. The general rule is that a feature cannot be overlapped by a label with an equal or lesser weight. By default label weights are High and features have a default weight of None.

You can also work with duplicate label options. You can remove duplicates to avoid seeing many of the same label. You can opt to place duplicate labels for each feature or part of a multipart feature.

For more information and examples of how you might use label and feature weights, see the ArcGIS Desktop Help.

Tip

Avoiding label overlap with annotation

By default, annotation layers and map annotation groups are given a High feature weight to prevent labels from overlapping annotation text.

Working with duplicate labels

1. Open the Label Manager by clicking the Label Manager button on the Labeling toolbar.
2. Click a label class in the Label Classes list.
3. Click the Properties button in the Placement Properties box.
4. Click one of the Duplicate Labels options.
5. Click OK on all dialog boxes.

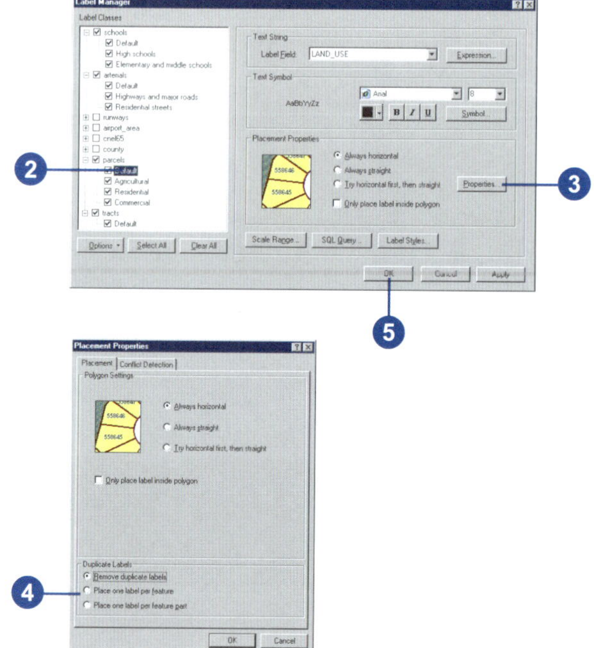

Converting labels to annotation

For additional control over label placement, you can convert dynamic labels to annotation so you can move the pieces of text around and place them exactly where you want. When you convert labels to annotation, you can choose to store the annotation in either a map document or a geodatabase. To learn more about the different types of annotation, see 'Working with annotation' in this chapter.

If you convert labels to map document annotation, the features that ArcMap couldn't automatically place are listed in the Overflow Annotation window. You can choose and place these individual annotation features on your map.

If you'd like to reuse some of your labeling work on other maps, you can save the labels in a geodatabase. For example, suppose you labeled cities and states with their names and stored the labels as an annotation feature class in a geodatabase. You can then load the data and the labels for display on another map.

Geodatabase annotation features that could not be placed are listed in the Unplaced Annotation window, which is opened ►

Converting labels to map document annotation

1. In the table of contents, right-click the layer you want to label and click Label Features. Work with your labels as described in this chapter.
2. Right-click the layer again, and click Convert Labels to Annotation.
3. Click In the Map to store your annotation in an annotation group in the map document.
4. Click Create Annotation For All features.

   If you don't want annotation for all features in the layer, click Features in current extent or Selected features to annotate just the currently selected features.
5. Optionally, type a new name for the annotation group.
6. Some labels may not currently display on the map because there is no room for them. To convert these labels, check the Convert unplaced labels to unplaced annotation box. This saves the unplaced labels in the map document, allowing you to position them later one at a time.
7. Click Convert.

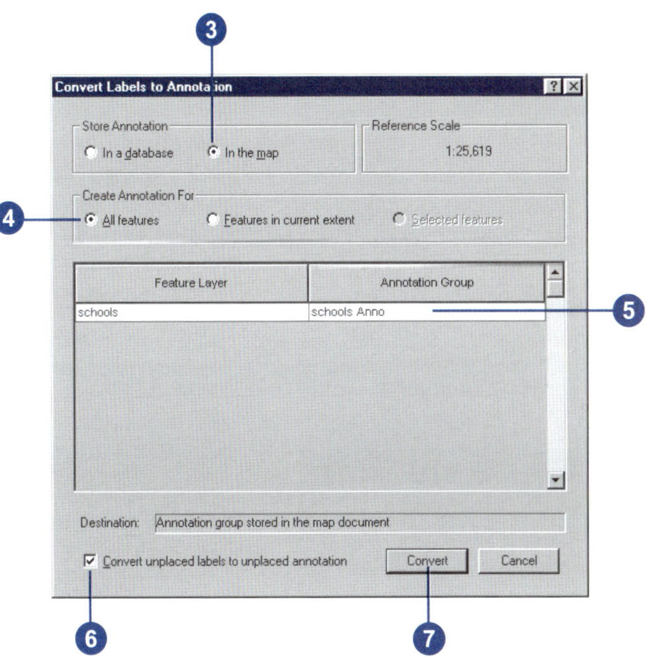

from the Annotation toolbar. To learn how to place and edit these features, see *Editing in ArcMap*.

You can also link annotation in a geodatabase directly to the feature it annotates. For more information on feature-linked annotation, see 'Working with annotation' in this chapter. To learn how to convert labels to feature-linked annotation, see *Building a Geodatabase*.

Tip

Seeing the Overflow Annotation window

You can also open the Overflow Annotation window from the View menu.

Tip

Converting labels to annotation for all layers in a data frame

You can also convert labels to annotation for all the labeled layers that are in a data frame. Click the data frame and click Convert Labels to Annotation to convert all the layers' labels at one time.

See Also

To add unplaced annotation to your map when using geodatabase annotation, use the Annotation toolbar. See Editing in ArcMap *for more information.*

Adding unplaced map document annotation to a map

1. If you have labels that could not be placed and you checked the box to convert them to unplaced annotation, the Overflow Annotation window appears.
2. In the Overflow Annotation window, right-click the annotation feature you want to place on your map and click Add Annotation.

   Once you place the annotation feature, you can use the Select Elements tool to reposition it on the map.
3. Repeat step 2 until you've placed all the annotation features you want on your map.

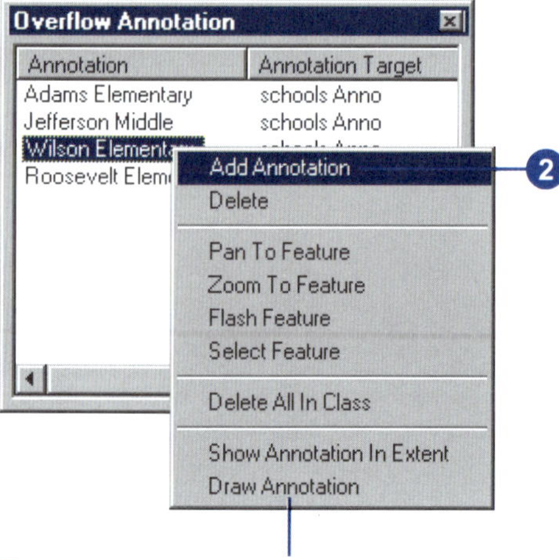

To see the unplaced annotation features, click Draw Annotation. The unplaced features are drawn in red by default.

Tip

Using geodatabases from previous ArcGIS versions

If you want to convert labels to geodatabase annotation with a geodatabase created in previous versions of ArcGIS, you'll need to upgrade your geodatabase to ArcGIS 9. To learn more, see Building a Geodatabase.

Converting labels to geodatabase annotation

1. In the table of contents, right-click the layer you want to label and click Label Features. Work with your labels as described in this chapter.
2. Right-click the layer again, and click Convert Labels to Annotation.
3. For Store Annotation, click In a database.
4. Specify the features you want to create annotation for.
5. To create feature-linked annotation, check the Feature Linked box. To create standard annotation, leave the box unchecked.
6. If you're creating standard annotation and want to add the annotation to an existing standard annotation feature class, check the Append box.
7. If you're creating feature-linked annotation, click the name of the new annotation feature class to change it. ▸

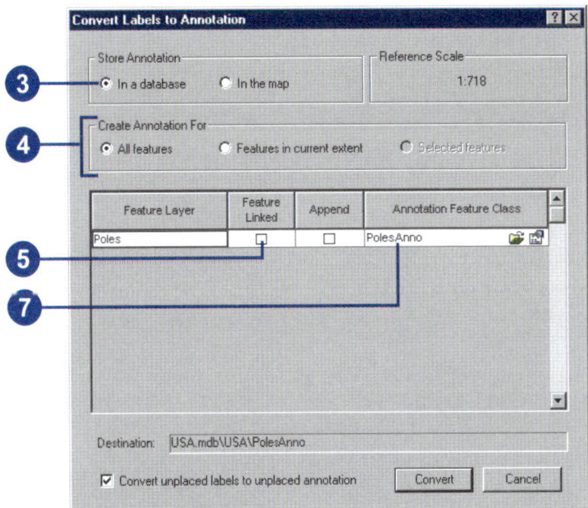

Tip

Converting label classes to annotation with ArcEditor, ArcInfo, and ArcView

Annotation classes are to geodatabase annotation feature classes as label classes are to a layer's labels. With an ArcEditor™ or ArcInfo license, the label classes will be converted into separate annotation classes within the annotation feature class. With ArcView®, you cannot create multiple annotation classes, so all of your label classes will be combined into a single annotation class.

See Also

After you've placed your geodatabase annotation features on the map, you may need to move them around or resize them. To learn how, see Editing in ArcMap.

8. If you're creating standard annotation, click the Open Folder button and specify the path and name of the new annotation feature class you will create or, if you're appending, the existing standard annotation feature class you're appending to.
9. If you're appending to an existing feature class, skip to step 15.
10. Click the Properties button.
11. Check the box to require edited annotation features to maintain reference to their associated text symbols stored in the feature class.
12. Specify additional editing behavior for the new annotation feature class.
13. If you are creating the new annotation feature class in an ArcSDE geodatabase and want to use a custom storage keyword, click Use configuration keyword, then choose the keyword you want to use (ArcInfo and ArcEditor only).
14. Click OK. ▸

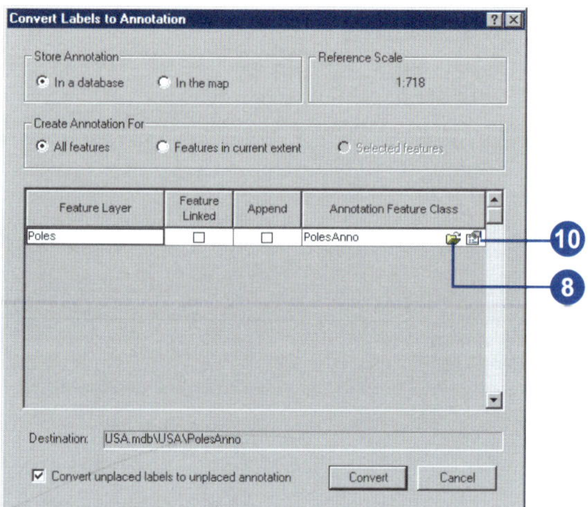

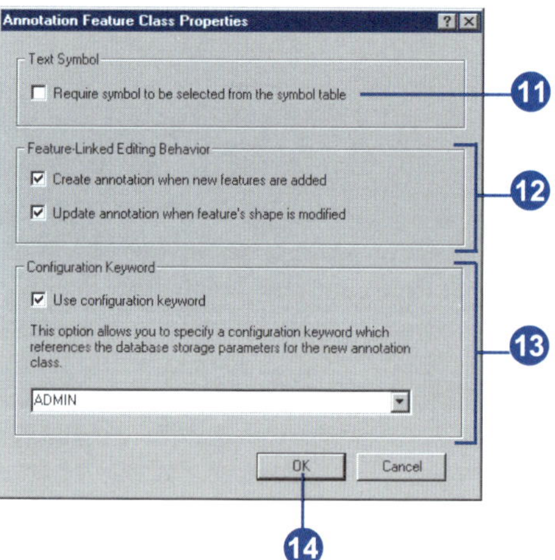

15. Some labels may not currently display on the map because there is no room for them.

    To convert these labels, check the Convert unplaced labels box. This saves the unplaced labels in the annotation feature class, allowing you to later position them one at a time in an ArcMap edit session.

16. Click Convert.

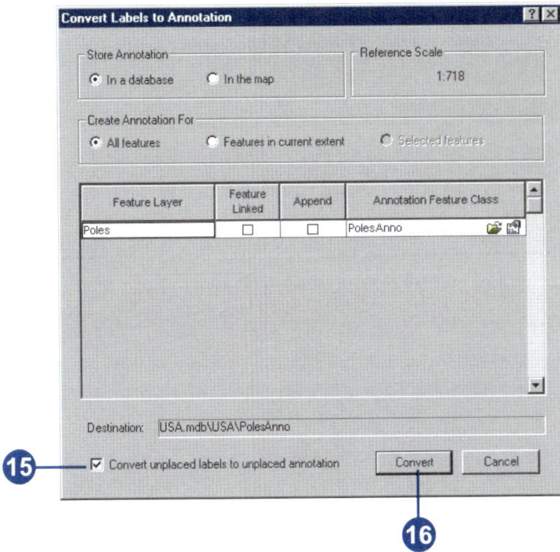

Working with annotation

What is annotation?

Annotation is one option in ArcGIS for storing text to place on your maps. With annotation, position, text string, and display properties are all stored together and are all individually editable. Adding dynamic labels is the other primary option for storing text. If the exact position of each piece of text is important, then you should store your text as annotation. ArcGIS fully supports two types of annotation: geodatabase annotation and map document annotation. ArcGIS also supports the display and conversion of other annotation types including ArcInfo coverage annotation and CAD annotation, although these types of annotation cannot be edited.

Labeling is the main alternative to annotation. A label's text and position are generated dynamically according to a set of placement rules. To learn more about labeling, see 'Working with labels' earlier in this chapter.

Although annotation is mainly used to persist pieces of text placed on or around a map, both geodatabase annotation and map document annotation also support the storage of graphics such as rectangles, circles, and lines as annotation. To learn more about graphics, see 'Working with graphics' earlier in this chapter.

Making a map with annotation

Follow these steps to use annotation in your maps:

1. Add your existing annotation to ArcMap.

   If you don't have annotation, you can label features in ArcMap and convert the labels to annotation. To learn more, see the sections on labeling in this chapter.

   If you have coverage, CAD, or other annotation formats and want your annotation to be editable or linked to features, convert them to geodatabase annotation using ArcToolbox. To learn more, see *Building a Geodatabase*.

2. Change the symbology of your geodatabase annotation using ArcMap editing tools. To learn more, see *Editing in ArcMap*.

   You can change symbology for most annotation formats using the Layer Properties dialog box. These changes are only in the current map unless you save them in a .lyr file.

   If you have map document annotation, use the Draw toolbar to change symbology. To learn more, see 'Working with graphics' earlier in this chapter.

3. Use the ArcMap editing tools to position geodatabase annotation. To learn more, see *Editing in ArcMap*.

   If you are creating annotation from labels, you can minimize manual positioning by working with the labeling options before converting the labels to annotation.

   If you have map document annotation, you can position it using the Draw toolbar. To learn more, see 'Moving, rotating, and ordering graphics' earlier in this chapter.

4. Manage your geodatabase annotation with ArcCatalog. To learn more, see *Building a Geodatabase*.

Geodatabase annotation

When creating new annotation or converting from existing annotation or labels, you can choose between geodatabase annotation and map document annotation. Geodatabase annotation is stored in geodatabase annotation feature classes. Conversely, map document annotation is stored in annotation groups in a particular map document. Geodatabase annotation is preferred if there are many pieces of annotation, if the annotation needs to be used outside a single map document, or if there will be several people concurrently editing the annotation.

Storing annotation in a geodatabase is similar to storing geographic features—lines, points, and polygons—in a geodatabase. You can add annotation stored in a geodatabase to any map, and it appears as an annotation layer in the ArcMap table of contents.

Like other feature classes in a geodatabase, all features in an annotation class have a geographic location, extent, and attributes. Annotation feature classes can be inside a feature dataset, or they can be standalone feature classes in a geodatabase. However, annotation is unique because, unlike simple features, each annotation feature has its own symbology.

Geodatabase annotation can be standard annotation or feature-linked annotation. *Standard annotation* elements are pieces of geographically placed text that are not formally associated with features in the geodatabase. For example, you might have a piece of standard annotation that represents a mountain range—the annotation simply marks the general area on the map. *Feature-linked annotation* is a special type of geodatabase annotation that is directly linked to the features that are being annotated by a geodatabase relationship class.

Feature-linked annotation

If you have an ArcEditor or ArcInfo license, you can create and edit geodatabase annotation that is linked directly to the features being annotated. If you have an ArcView license, you can view feature-linked annotation but not create or edit it.

Feature-linked annotation is similar to standard geodatabase annotation but also has some behavior that makes it similar to dynamic labeling.

- When a new feature is created, new annotation is automatically created. You can turn off this behavior when you create a new feature-linked annotation feature class.
- If you move a feature, the annotation for that feature moves with it. You can turn off this behavior when you create a new feature-linked annotation feature class.
- If you change an attribute of the feature that the annotation text is based on, the annotation text changes.
- If you delete the feature, the annotation is deleted.

An annotation class can be linked to only one feature class, but a feature class can have any number of linked annotation feature classes.

There are several ways to create feature-linked annotation. First, if you have defined a feature-linked annotation feature class, then as you create new features using the editing tools in ArcMap, annotation will be created for these features automatically. Second, you can also use the Annotate selected features command in ArcMap to add linked annotation to existing features. Finally, as you can with other types of annotation, you can convert labels to feature-linked annotation in ArcMap or use the ArcToolbox annotation conversion tools to create feature-linked annotation from coverage or CAD annotation.

To learn more about working with feature-linked annotation, see *Building a Geodatabase*.

Choosing between geodatabase annotation and map document annotation

Where should you store your annotation? The answer depends on how you plan to use it.

- If your text only applies to the current map, you might store your text as map document annotation in an annotation group. To learn more about map document annotation, see 'Storing graphics as annotation' earlier in this chapter.

- If you want to use your text in the current map and in others, you should store your text as geodatabase annotation in one or more annotation feature classes.
- If your data is stored in an enterprise geodatabase (that is, an ArcSDE geodatabase), you should store your annotation in that geodatabase to take advantage of versioning and the multiuser enterprise editing environment.
- If you want to use the specialized editing tools for creating and editing annotation in ArcMap, you should store your text as geodatabase annotation.
- If you have more than a few hundred pieces of text, as a general rule you should store your annotation in a geodatabase because ArcMap can access and display geodatabase annotation quicker than map document annotation. In addition, each piece of annotation stored as map document annotation will increase the size of your map document (.mxd).

Other types of annotation

ArcGIS also supports the display and conversion of several annotation formats including ArcInfo Workstation coverage, VPF, CAD, PC ARC/INFO, and SDE 3.x annotation. You can add these types of annotation directly to ArcMap and also change most annotation layer symbology properties. For these formats, however, you cannot change the symbology for individual pieces of annotation and you cannot edit the annotation positions or text strings. If you need these behaviors, convert your annotation to geodatabase annotation or map document annotation using the ArcToolbox annotation conversion tools. You can also use these tools to create coverage annotation from geodatabase annotation.

See 'Displaying annotation' in this chapter for more information on adding these annotation formats to your map. See *Building a Geodatabase* or the ArcGIS Desktop Help for more information on displaying these annotation formats and converting them to geodatabase annotation.

Displaying annotation

You can display all types of annotation in ArcMap for use in your maps—including geodatabase annotation, coverage, CAD (and other preexisting annotation formats), and map document annotation.

With the exception of the map document format, annotation is added to ArcMap as you would add other data. Annotation appears with other geographic data in the ArcMap table of contents, and you can change its display properties using the Layer Properties dialog box as you would for point, line, and polygon features.

Annotation is, however, different from simple point, line, and polygon features. Geodatabase annotation features store their own symbology, and to make permanent changes to their display, you use the ArcMap editing tools. To learn more, see *Editing in ArcMap*.

Likewise, you can change most display characteristics of coverage, CAD, SDE 3.x, and VPF annotation, but these formats internally store some display properties that cannot be changed in ArcMap. See ►

Displaying geodatabase annotation

1. In the table of contents, right-click the annotation layer name and click Properties.
2. Click the Symbology tab.
3. Choose one of these options:
    - Disable substitutions, to display with the original (geodatabase-stored) symbology
    - Substitute text color, to display with the original fonts, font sizes, and so on, with only color changed
    - Substitute individual symbols, to use symbols different from the original in the current layer
4. Optionally, click the Display tab to set a transparency level for your annotation and specify whether the annotation layer should draw based on its position in the table of contents.
5. Optionally, click the Annotation tab to see a summary of your annotation layer's properties.
6. Click OK.

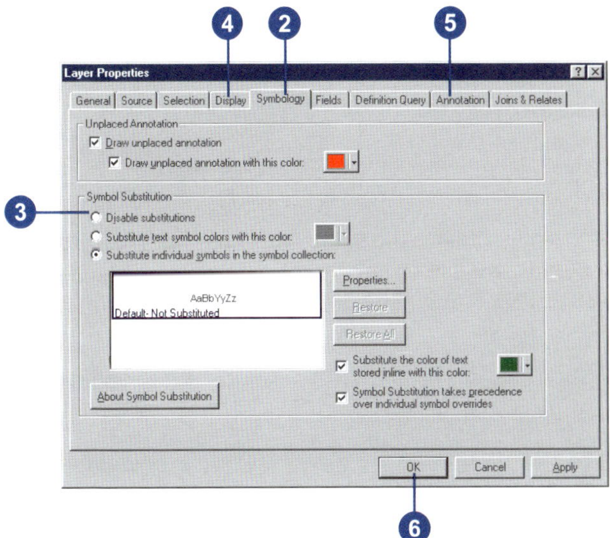

Building a Geodatabase to learn more about loading preexisting annotation formats to the geodatabase. By converting your annotation to the geodatabase, you will have more flexibility in ArcGIS.

Map document annotation is stored in map documents in annotation groups and not as separate data. To learn more about displaying annotation in this format, see 'Storing graphics as annotation' earlier in this chapter.

Tip

Working with coverage and CAD display properties

With coverage and SDE 3.x annotation, changing the size will only have an effect if $SIZE = 0. With CAD and VPF annotation, changing the size will have no effect. See the ArcGIS Desktop Help for more information on these annotation formats.

Displaying coverage and SDE 3.x annotation

1. In the table of contents, right-click the annotation layer name and click Properties.
2. Click the Symbols tab.
3. Click an entry in the list of symbol numbers ($SYMBOL) to see and modify its display properties.
4. To choose another text symbol or to change additional properties, click the Text symbol button.
5. Click OK when you're finished.

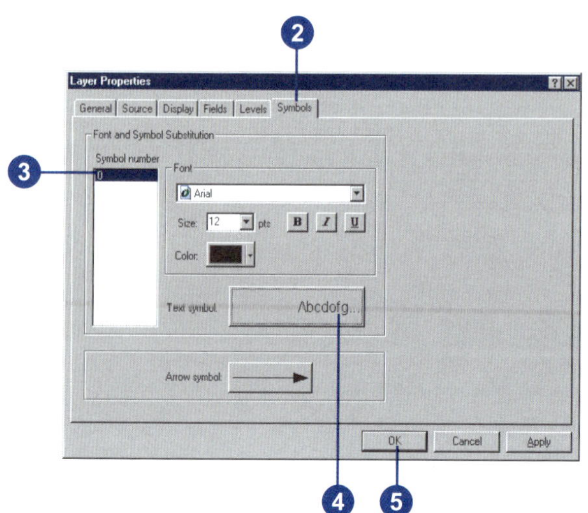

Displaying CAD and VPF annotation

1. In the table of contents, right-click the annotation layer name and click Properties.
2. Click the Fonts tab.
3. Click an entry in the list of symbol numbers to see and modify its display properties.
4. To choose another text symbol or to change additional properties, click the Text symbol button.
5. Click OK.

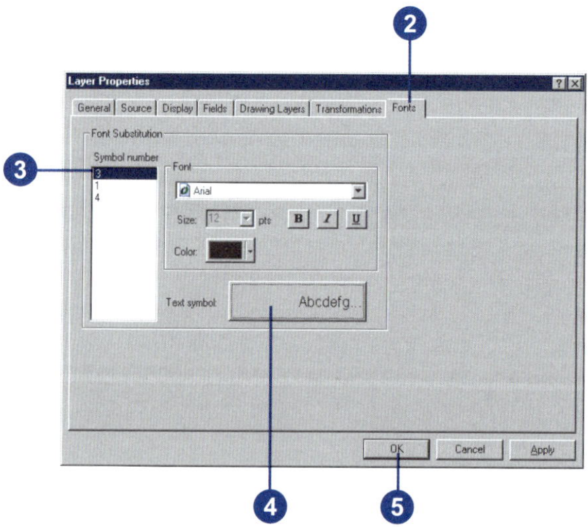

Using text formatting tags

Text formatting tags let you modify the formatting for a portion of text to create mixed-format text where, for example, one word in a sentence is underlined.

Text formatting tags can be used almost anywhere you see text—whenever you can specify both a text string and a text symbol. You can use tags in dynamic label expressions, annotation, legend text, map titles, and in the values of fields used to label features. Tags aren't recognized by the table of contents, Attribute table window, or Identify Results window, so tags added to field values appear unformatted as raw text in these windows.

For more information on text formatting tag syntax, see the ArcGIS Desktop Help. See 'Building label expressions' earlier in this chapter for using formatting tags in dynamic labels.

See Also

With geodatabase annotation, the Attributes dialog box has a formatted preview and lets you create mixed-format text without entering tag syntax. See Editing in ArcMap *for more information.*

Adding text with text formatting tags

1. Click the New Text button on the Draw toolbar.
2. Click the mouse pointer over the map display and type the text.

   As you type formatting tags into the string, you will see the tags as plain text.
3. Press the Enter key. The formatted text appears.

   If the string you entered has any syntax problems, formatting will be disabled and all the tags will appear as plain text.

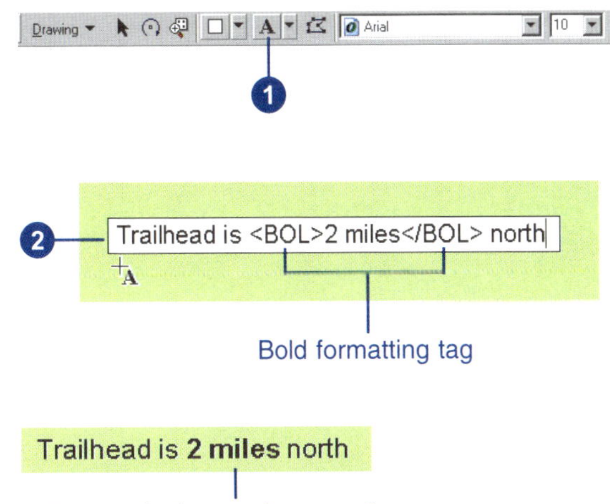

Editing text with formatting tags

1. Click the Select Elements button on the Draw toolbar and double-click the text element you want to edit.
2. Click the Text tab and modify the text shown in the Text box. Formatting tags will appear here as plain text.
3. Click OK to apply your changes and view any formatting changes on the display.

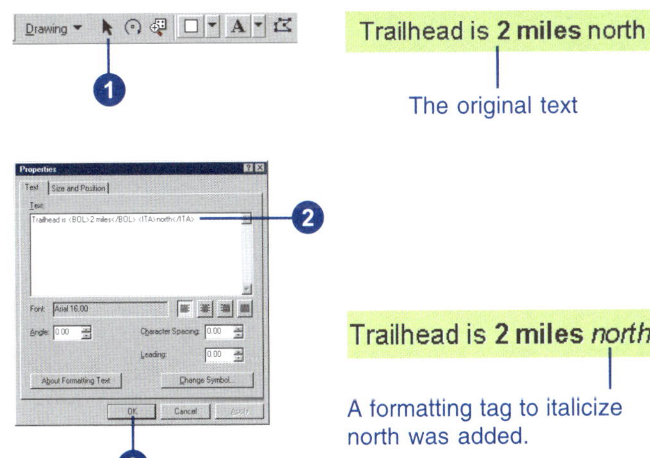

Working with styles and symbols 8

IN THIS CHAPTER

- **The Style Manager**
- **Controlling which styles are referenced in ArcMap**
- **Organizing style contents**
- **Saving the current styles**
- **Creating and modifying symbols and map elements**
- **Creating line symbols**
- **Creating fill symbols**
- **Creating marker symbols**
- **Creating text symbols**
- **Modifying and saving symbols and elements as you work**
- **Working with color**

Styles are a collection of predefined colors, symbols, properties of symbols, and map elements that allow you to follow a mapping standard and help promote consistency in your organization's mapping products.

Styles help define not only how data is drawn but also the appearance and placement of map elements and other cartographic additions on your map. Styles provide storage for your colors, map elements, symbols, and properties of symbols. Every time you choose and apply a particular map element or symbol, you are using the contents of a style.

Benefits of using styles include:

- Maintaining mapping standards for symbols, colors, patterns, methods of rendering distributions, relationships, and trends
- Letting your map communicate more effectively through familiar styles that enable people to easily explore, understand, and analyze a map
- Using a map template with referenced styles or groups of styles for ease in creating a map or map series
- Standardizing map symbolization so your maps will look the same when they are published or printed with different printers

In the previous chapters, you learned how to symbolize data and draw map elements and graphics. You are probably already familiar with most of the symbol and map element dialog boxes. In this chapter, you will learn how to select or create the symbols you want and save them in styles you can reuse to produce maps that meet your organization's needs.

The Style Manager

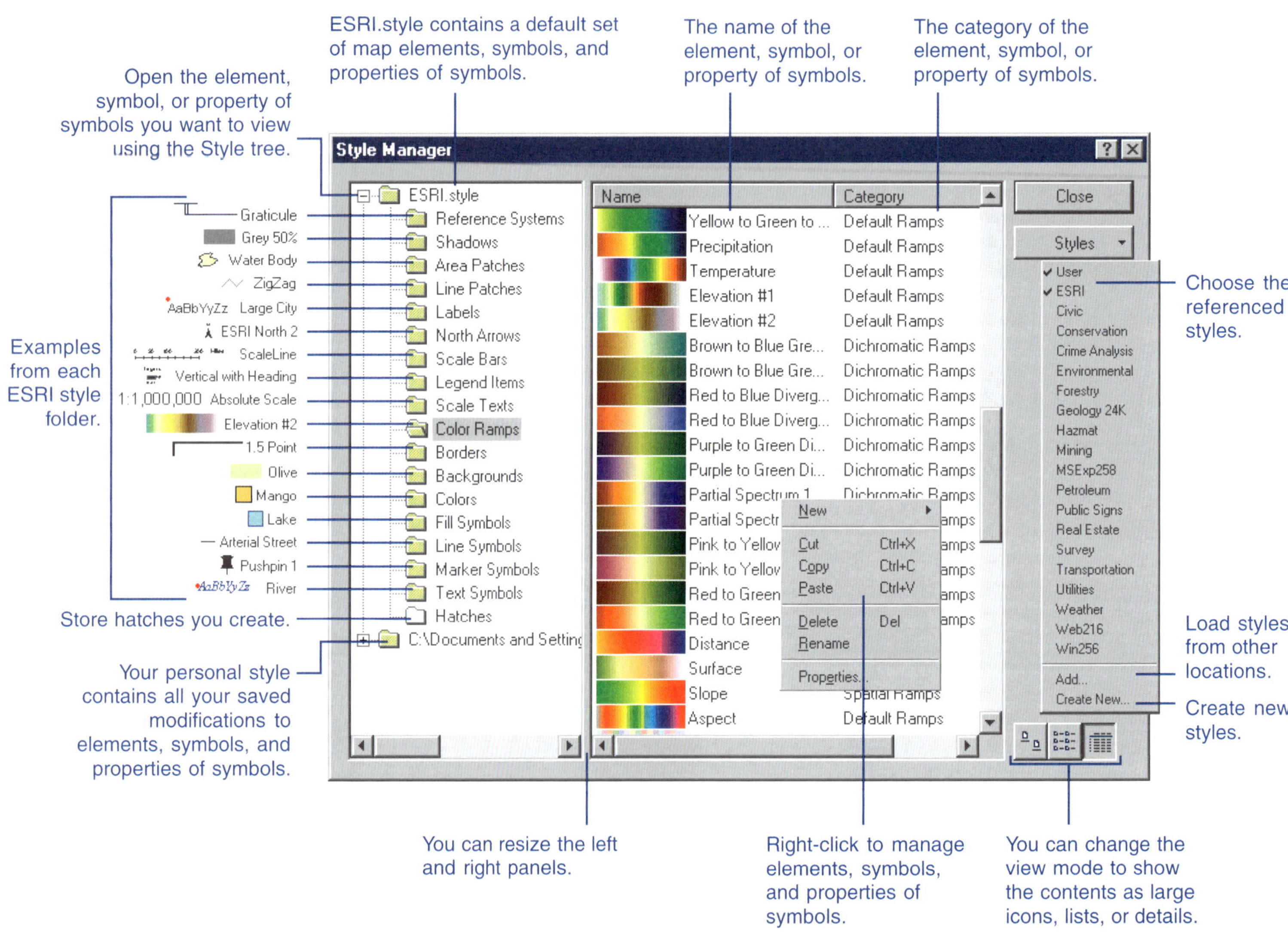

Controlling which styles are referenced in ArcMap

ArcMap by default displays a robust set of generic symbols and map elements in the ESRI style. To create maps for your application, you may need other styles. ArcMap also has a wide range of industry-specific styles to complement the ESRI style. When a style is referenced, its elements, symbols, and symbol properties are available for you to use in a map document.

You can add or remove styles at any time in an ArcMap session.

Tip

Are there other ways to reference styles?

You can reference styles in the Style Manager dialog box and in the Symbol Selector dialog boxes. However, the Symbol Selector dialog boxes only list the styles that contain the same symbols.

Tip

Can I install just the styles I need?

You can use the custom ArcMap installation to load only the styles you want to have accessible in ArcMap.

1. Click Tools, point to Styles, and click Style References.

   Your personal style and the ESRI style are referenced by default.

2. Check any additional styles you want to use.
3. Click the Add button to load more styles from other locations.

   Navigate to and click the styles you need.

4. Click OK.

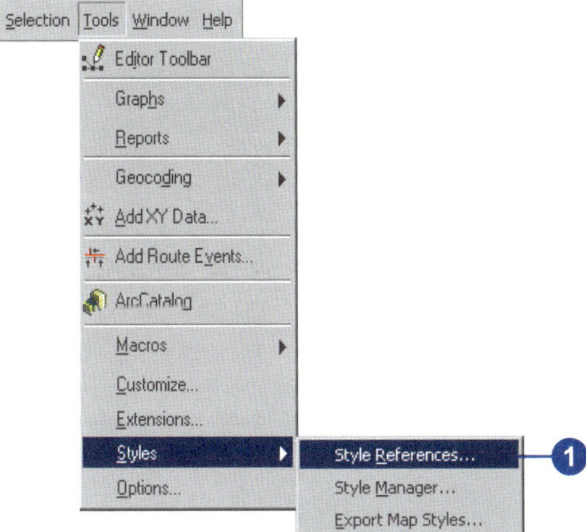

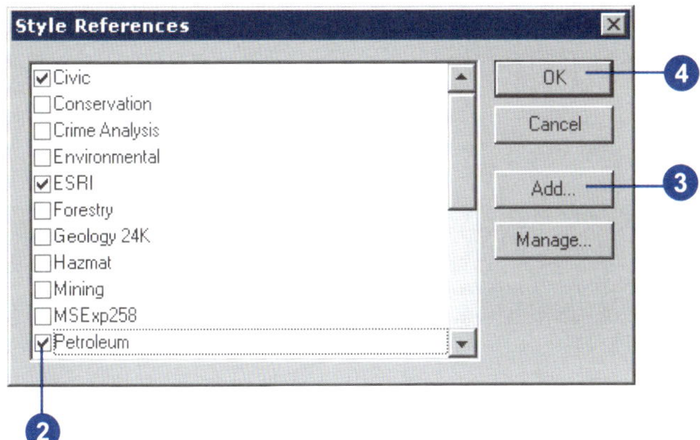

Organizing style contents

The *Style Manager* lets you organize styles and their contents, symbols, and map elements. You can copy, paste, rename, and modify any style content. You can also create new styles, symbols, or map elements.

New styles can be created by copying symbols or map elements from existing styles or your personal styles. Styles can be customized by deleting the symbols and map elements you don't need.

You can easily distinguish which folders contain elements and symbols, which can be modified, and which are empty.

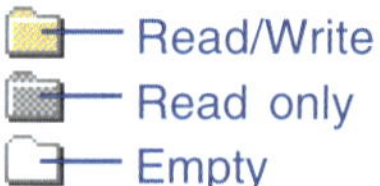

Tip

Finding the styles folders

By default, the ESRI styles folder is in the \Bin\Styles folder where ArcGIS is installed. Your personal style folder is in the Windows install location, for example, C:\Documents and Settings\<user>\Application Data\ESRI\ArcMap.

Creating a new style

1. Click Tools, point to Styles, and click Style Manager.
2. Click Styles and click Create New.
3. Navigate to the program's Styles folder.

   By default, the Styles folder is installed with ArcGIS in the ArcGIS\Bin\Styles folder.
4. Type the name for the new style you're creating.
5. Click Save.

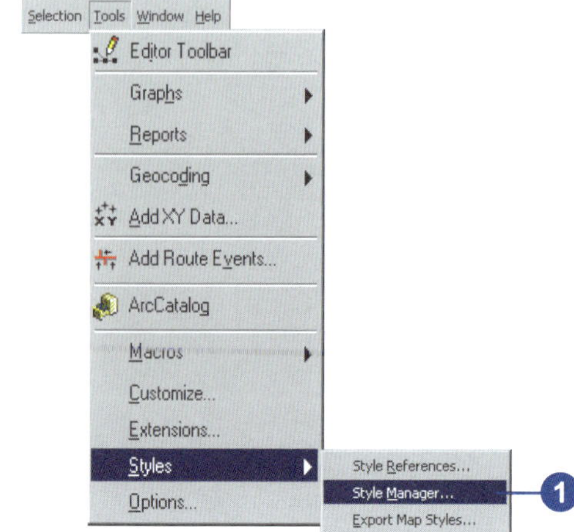

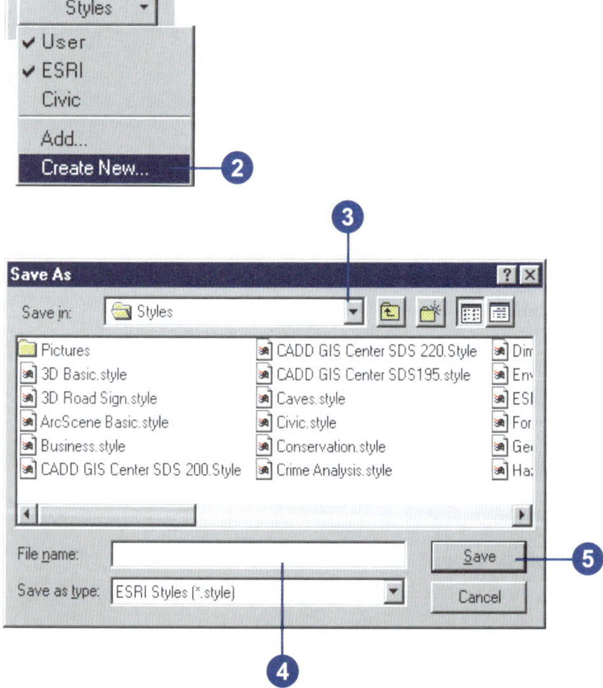

Tip

Using shortcut keys

You can use the standard keyboard shortcuts to cut, copy, paste, and delete styles.

Tip

Deleting style contents

Right-click any map element or symbol and click Delete from the context menu.

There is no undo for delete, so you may want to move the element to another folder as a backup rather than delete it.

Tip

What's a symbol category?

When you save a symbol, you can specify a name and a category for class distinction. The category can be used to differentiate drawing methods and other criteria. You can also create subcategories of styles. Categories can be viewed in the Style Manager dialog box.

See Also

For more information on route hatch styles, see the book Linear Referencing in ArcGIS.

Copying and pasting style contents

1. Click Tools, point to Styles, and click Style Manager.
2. Click the style folder whose contents you want to view.
3. Right-click an element and click Cut or Copy.
4. Click another style folder of the same type.
5. Right-click in the symbol contents window and click Paste.

   You can only paste into style folders that are the same type as the folder from which the element was copied.
6. Type a new name for the element.

   You can change the name later by using Rename on the context menu.

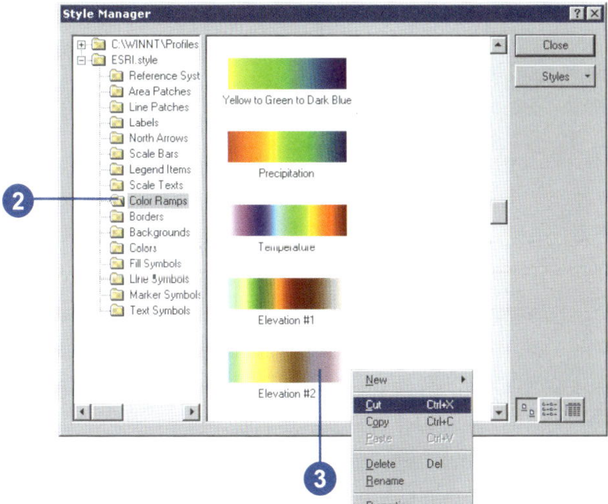

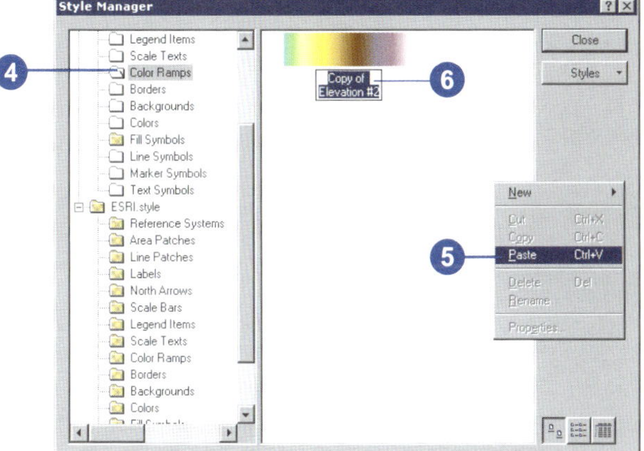

Saving the current styles

You can easily create a new style containing all the elements, symbols, and properties of symbols used in your map.

You can create and modify map elements and symbols as you design your map, then save everything into a style.

Exporting map styles allows you to save map elements and symbols from many styles into a single style.

Exporting the current map styles to a new style

1. Click Tools, point to Styles, and click Export Map Styles.
2. Navigate to where you want to save the new style.

   By default, the browser is set to the program's Styles folder.
3. Type a style name.
4. Click Save.

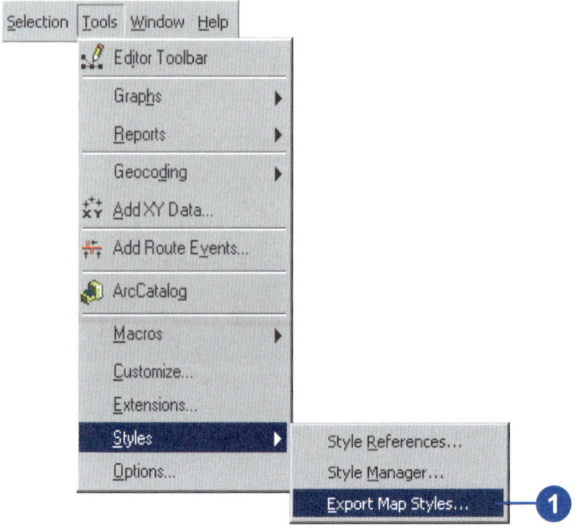

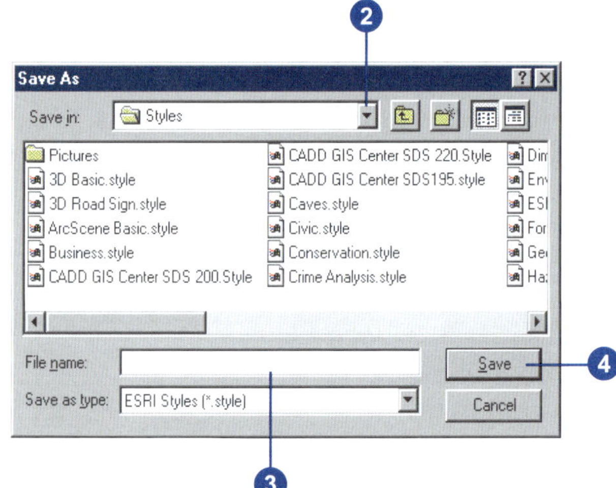

Tip

How are symbols stored in a map document?

The items you draw in ArcMap—symbols, map elements, and graphics—are copied into the map document. Therefore, you don't need the original referenced styles to open and draw the map again.

Tip

Can I create a style from a map without any referenced styles?

If you need to modify the symbols in a map and don't have the original styles, you can create a new style by exporting the existing map elements and symbols.

Creating and modifying symbols and map elements

You can use the Style Manager dialog box to create new symbols or to modify an existing symbol.

To set or modify properties of an element or symbol in a style, double-click it in the contents window of the Style Manager. This opens a property dialog box and allows you to define the appearance of the element or symbol and save changes within the style.

Since symbols and elements are organized by type, you create a new symbol or element by first identifying the type, then selecting the appropriate folder. When you choose to create a new symbol or element, the properties you can choose from are only associated with that symbol or element type. ▸

Tip

Using Microsoft Access to edit your styles

A style is a database. You can use Microsoft Access to edit names, check your spelling, control sorting, and complete other tasks.

Creating a new symbol in the Style Manager

1. In the style tree, click the symbol folder in which you want to create more symbols.
2. Right-click in the open space in the Symbol contents window, point to New, and click Line Symbol.

   Use the Symbol Property Editor to create the symbol you want and click OK.
3. Name the new symbol.

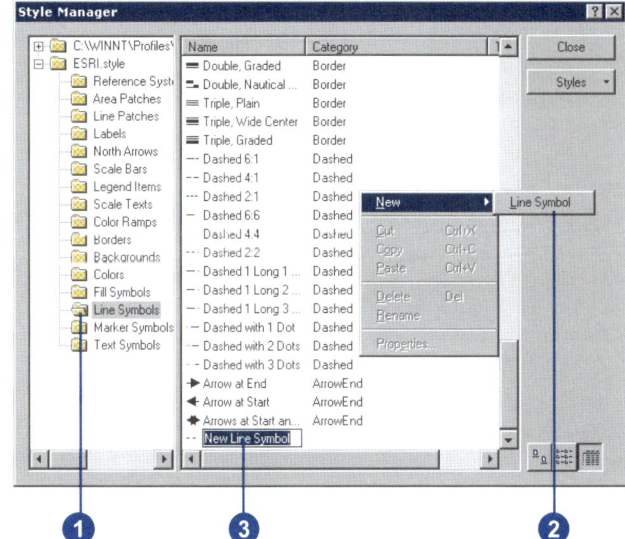

Layers appear in the list according to when they are drawn. Therefore, the bottom layer will be covered by the layer listed above it.

By adding or working with layers, you can enhance the symbols that already exist or combine them to make new symbols. The layers you use can be a mixture of different symbol types and properties, such as color, width, and offset. Turning on or off a layer affects which layers will be viewed in the preview. Whereas locking and unlocking a layer does not affect your capabilities in the symbol Properties Editor, it does affect the ability to change the color in the Symbol Selector you access through the Layer Properties. There, if you change the symbol color, only the unlocked layers will change to the new color.

Tip

Using a shortcut to the Symbol Property Editor dialog box

Instead of clicking the Properties button on the Symbol Selector dialog box, you can click the Preview window on the Symbol Selector dialog box.

Using the Symbol Property Editor to define symbols

1. Set the type of symbol.
2. Set the units of measure.
3. Click the tabs to set other symbol properties.

   The tabs vary depending on the symbol type.
4. You can preview the symbol's appearance.
5. You can zoom in and out on the preview using the preview mode options.
6. Click each layer to preview it.
7. Click to turn on or off drawing for a layer.

   This allows you to change a layer's appearance without deleting it.
8. Click to lock or unlock a layer's color.
9. Select layer options to add, delete, move up, move down, copy, and paste.
10. Click OK when you're finished.

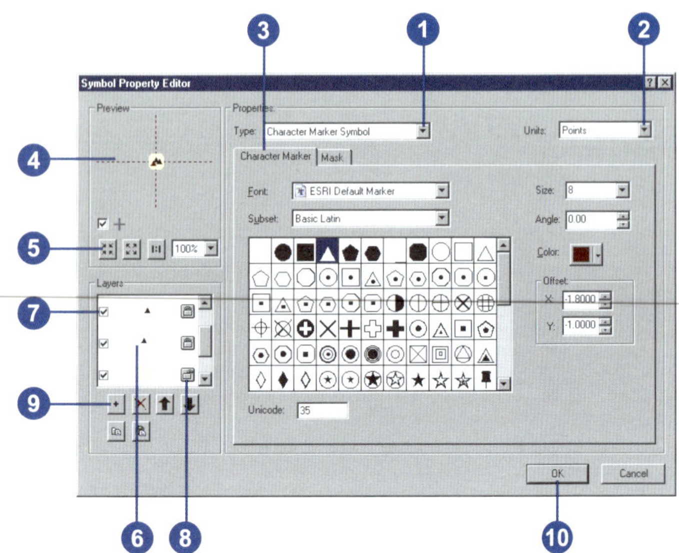

Creating line symbols

Line symbols are used to draw linear features such as transportation networks, water systems, boundaries, zonings, and other connective networks. Lines are also used to outline other features such as polygons, points, and labels. As graphics, lines can be used as borders, leaders for arrows and other annotation, and freehand drawing.

The examples here show you how to create some common line symbols: a cased road, a railroad, and arrowhead leaders.

The standard line types include:

- Simple—fast-drawing, one-pixel lines with predefined patterns or solid, wide lines
- Cartographic—straight-line template patterns and marker decorations
- Hash—hachures, template patterns, and marker decorations
- Marker—markers, template patterns, and other marker decorations
- Picture—formed by a .bmp (Windows bitmap) or .emf (Windows enhanced metafile) graphic

Any number of layers can be combined in a single line.

Creating an encased road symbol

1. In the line Symbol Property Editor dialog box, click the Type dropdown arrow and click Cartographic Line Symbol. Set the Units to Points.
2. Click the Color dropdown arrow and click a black shade.
3. Set the Width to 3.4 points.
4. Click Butt for Line Caps and Round for Line Joins.

   See the Preview for a change in the end of the line.
5. Click the Add a New Layer button.

   Cartographic Line Symbol should already be selected as the Type.
6. Click a shade of red from the color palette.
7. Set the Width to 2.6 points.
8. Click Butt for Line Caps and Round for Line Joins.
9. Click either of the options to see how the line symbol changes.
10. Click OK.

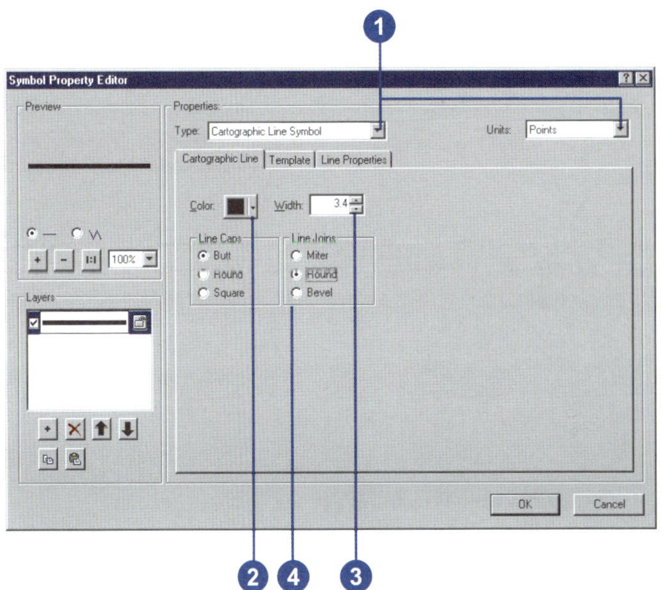

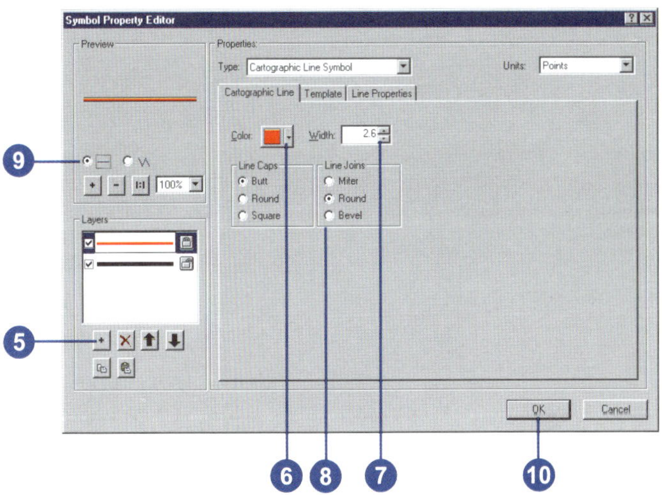

Tip

Why shouldn't parallel lines be cased road symbols?

Using a single wide layer produces clearer symbology where roads intersect and merge.

Tip

What properties of a hash line symbol can I modify?

You can set the color, width, caps, joins, pattern, and decoration of a hash line symbol.

Tip

Using the Template tab

The Template tab lets you design a common template for the symbol layers that will be layered. You can use the same template to stack and center line dashes with markers, or you can reverse the template pattern to center the marker in the gap. For hash lines, the template mark indicates how many dashes will occur in the pattern segment. You can also align multiple line layers using the template settings.

Creating a railroad symbol

1. In the line Symbol Property Editor dialog box, click the Type dropdown arrow and click Cartographic Line Symbol.
2. Click the Color dropdown arrow and click a gray.
3. Adjust the Width to .5 points.
4. Click the Add a New Layer button.
5. Click the Type dropdown arrow and click Hash Line Symbol.
6. Click the Hash Line tab and click Hash Symbol.
7. Click the Color dropdown arrow and click the gray shade.
8. Adjust the Width to .5 points.
9. Click OK.
10. Click the Template tab.
11. Slide the dark gray square to the 11th position to create a pattern length of 10 units.
12. Click the sixth position to add another hash.
13. Click the Cartographic Line tab.
14. Click the Color dropdown arrow and click the gray shade.
15. Adjust the Width to 4.
16. Click OK.

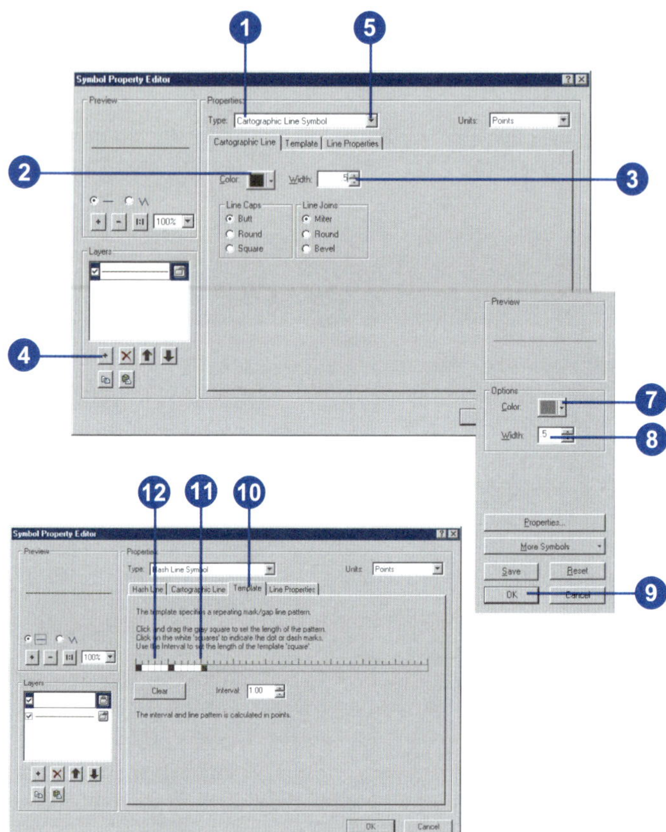

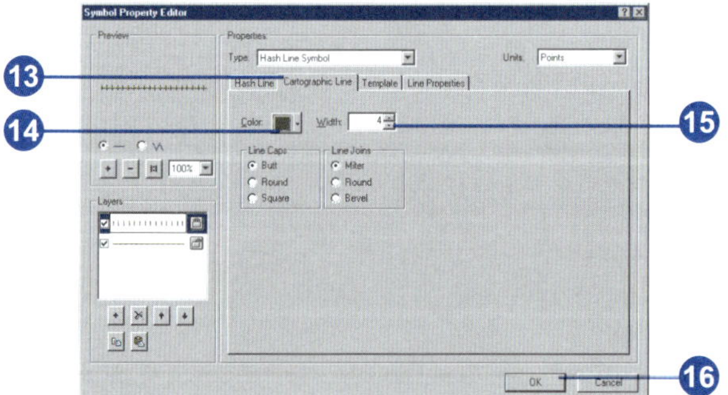

Tip

Creating leaders with markers other than arrows

You can use any marker to create start or end line decorations. You can adjust the length and height of the default arrow. The ESRI Arrowhead font contains additional arrow styles, but you can also use any marker for line decorations.

Tip

Why do my line arrows sometimes appear horizontal to the map?

You can toggle the arrowheads to follow the line orientation or stay at a fixed angle to the map. See step 5 in the task on this page.

Tip

Why should I lock or unlock a symbol layer?

The locking property controls whether or not the color option can be modified in the Symbol Selector dialog box. A locked layer can't be modified.

See Also

To learn how to automate map production by matching symbol names to data attributes, see Chapter 6, 'Symbolizing features'.

Creating arrowhead leaders

1. In the line Symbol Property Editor dialog box, click the Type dropdown arrow and click Cartographic Line Symbol.
2. Click the Line Properties tab.
3. Click the right-facing arrow.
4. Click Properties.
5. Click Rotate symbol to follow line angle.
6. Click Symbol and click Properties.
7. Click the Type dropdown arrow and click Character Marker Symbol.
8. Click the Font dropdown arrow and click ESRI Arrowhead.
9. Choose an arrowhead from the selection or type "37" in the Unicode box.
10. Click the Color dropdown arrow and click a green.
11. Set the Size to 18 points.
12. Click OK until you're back to the Symbol Property Editor dialog box.
13. Click the Cartographic Line tab.
14. Click the Color dropdown arrow and click a green shade.
15. Click Butt for Line Caps and Miter for Line Joins.
16. Click OK.

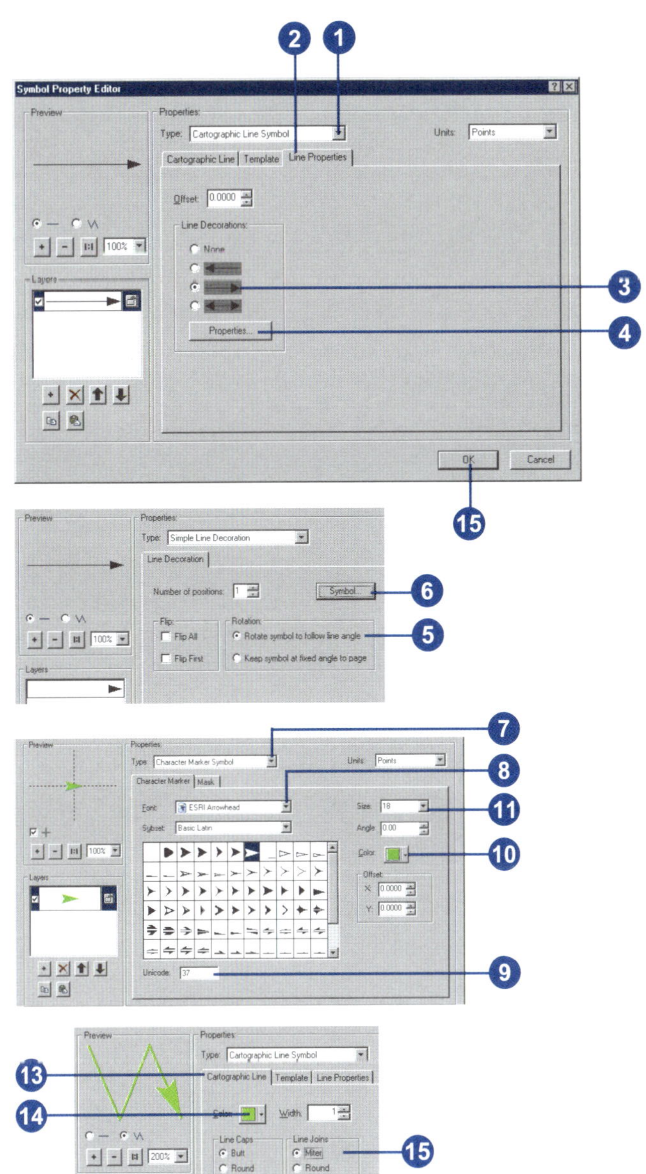

Creating fill symbols

Use fill symbols to shade polygonal features, such as countries, land uses, habitats, parcels, and footprints. Fills are also used in drawing graphic shapes and backgrounds, data frames, map elements, graphics, and text. In addition, a polygonal data layer can be drawn transparently.

The standard fill types are:

- Simple—fast-drawing solid fill with an optional outline
- Gradient—linear, rectangular, circular, or buffered color ramp fills
- Line—hatched lines at any angle, separation, or offset
- Marker—marker symbols drawn randomly or ordered
- Picture—continuous tiling of a .bmp (Windows bitmap) or .emf (Windows enhanced metafile) graphic

Any number of layers can be combined in a single fill.

Creating a solid fill

1. In the fill Symbol Property Editor dialog box, click the Type dropdown arrow and click Simple Fill Symbol.
2. Click the Color dropdown arrow and click the color you want.

   Or click More Colors and use the Color Selector dialog box to mix a new color.
3. Click OK.

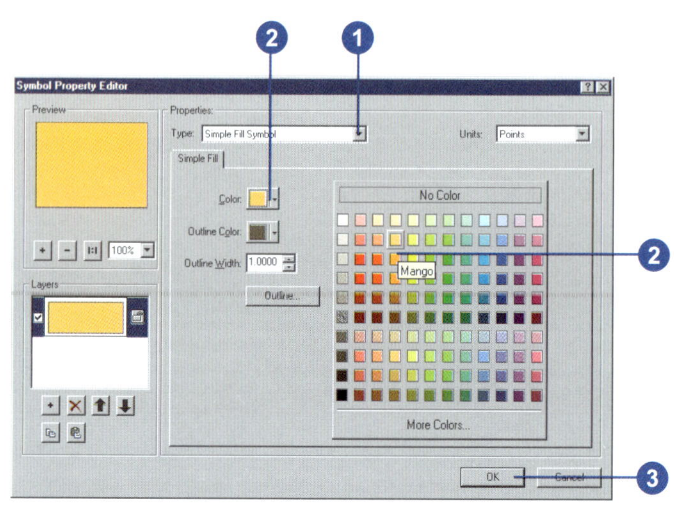

Adding a fill outline

1. Click Outline Color and click the color you want to use.
2. Set the Outline Width or click Outline to choose a predefined line symbol.

   Alternatively, use the line properties menu to create a new outline.
3. Click OK.

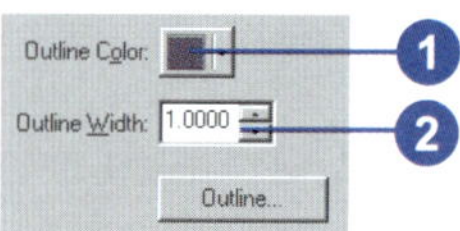

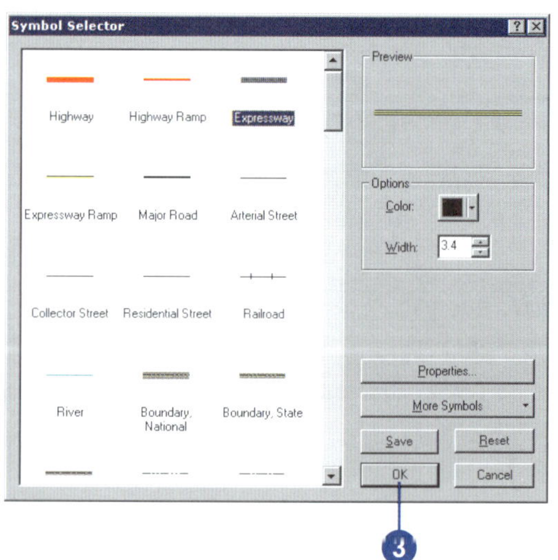

Tip

Do I have to use a predefined color ramp to create a gradient fill?

You can modify an existing color ramp or create a new one as you create the gradient symbol. See steps 5 through 9 in the task on this page.

See also

For more information on color, see the section 'Working with color' in this chapter.

Creating a gradient fill

1. In the fill Symbol Property Editor dialog box, click the Type dropdown arrow and click Gradient Fill Symbol.
2. Click the Style dropdown arrow and click Linear.
3. Adjust the number of color Intervals and the color stretch Percentage from start to end.
4. Click the Color Ramp Style dropdown arrow and choose another fill.
5. To modify the ramp, right-click the Color Ramp Style and click Properties.
6. If you want to modify your ramp, click Color 1, click the dropdown arrow, and choose a color for the first color of your gradient fill color ramp. Click Color 2 and choose the end color of your ramp.
7. Click OK.
8. Right-click Style and click Save to style.
9. Type a name for the new color ramp.

   The color ramp is stored in your personal style.
10. Click OK.
11. Click Outline and set the Width to 0 for no outline.
12. Click OK.

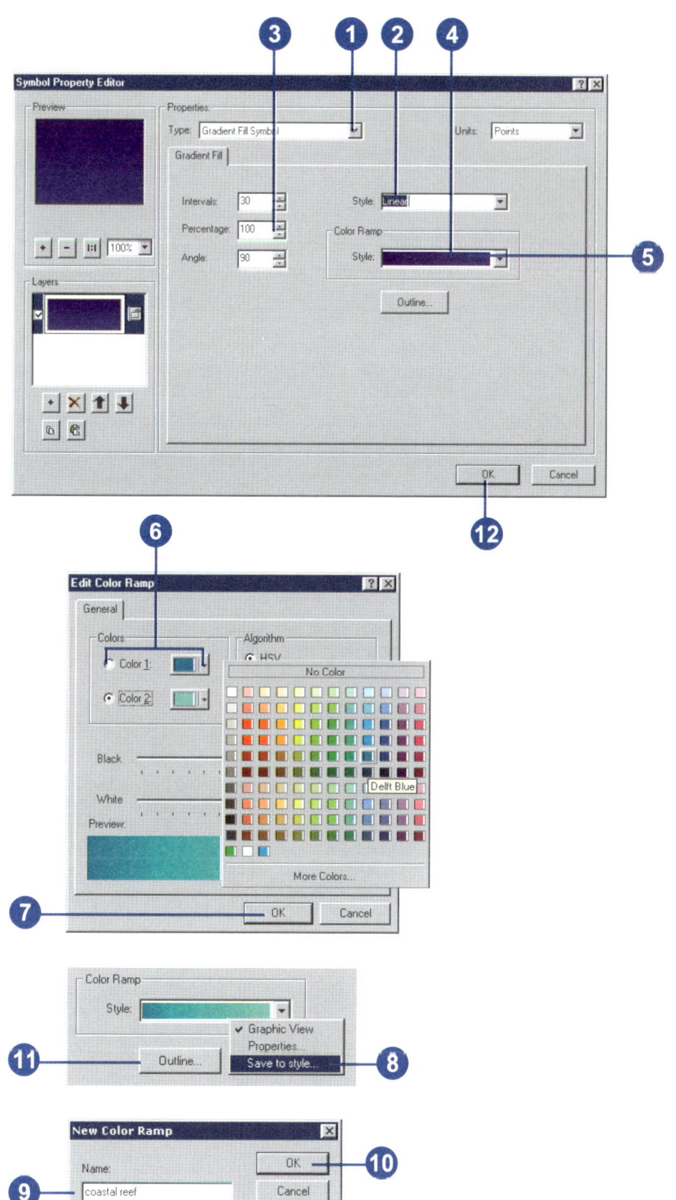

Tip

How do I create a transparent picture fill?

You can set the background or foreground color to "no color" to create a transparent picture.

Creating a random dot fill

1. In the fill Symbol Property Editor dialog box, click the Type dropdown arrow and click Marker Fill Symbol.
2. Click Random.
3. Click Marker.
4. Change the Color.
5. Change the Size to 3.
6. Click OK.
7. Click the Fill Properties tab.
8. Adjust the X and Y Separation to 5, 5 for a denser distribution.
9. Click OK.

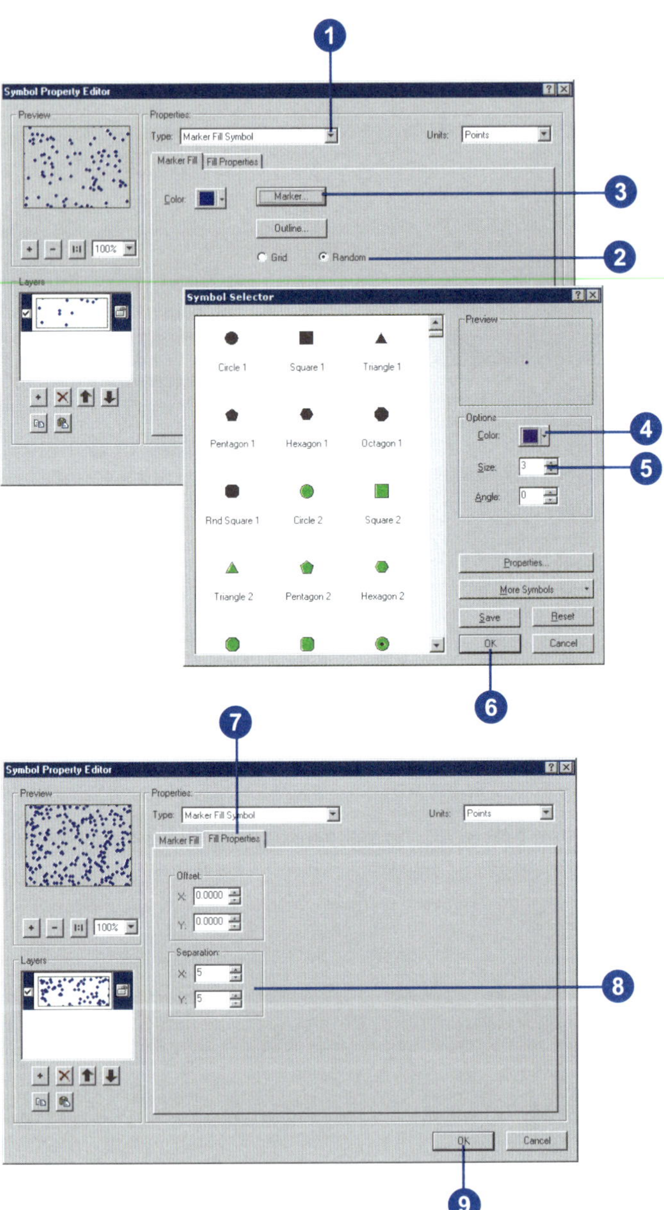

Tip

Is there more than one way to create a transparent overlay fill?

You can create a hatched line fill with alternating opaque and transparent hatches. You can also set the entire feature layer to a percentage of transparency. Combining these can achieve a variety of effects.

Tip

Can I mix symbols with different units in the same style?

You can use any units you prefer.

Creating an overlay fill

1. In the fill Symbol Property Editor dialog box, click the Type dropdown arrow and click Line Fill Symbol.
2. Click the Units dropdown arrow and click Inches.
3. Click Line and click Properties.
4. Click the Units dropdown arrow and click Inches.
5. Click the Color dropdown arrow and click an orange shade.
6. Set the Width to 0.05.
7. Click OK and click OK.
8. Adjust the Angle to 45.
9. Set the Separation to 0.1.
10. Click Outline and set the Width to 0. Click OK.
11. Click the Add a New Layer button.
12. Repeat steps 3 through 5, choose a darker orange, and set the Width to 0.01. Click OK twice.
13. Adjust the Angle to 45.
14. Set the Offset to 0.12.
15. Set the Separation to 0.1.
16. Click Outline and set it to 0. Click OK.
17. Click OK.

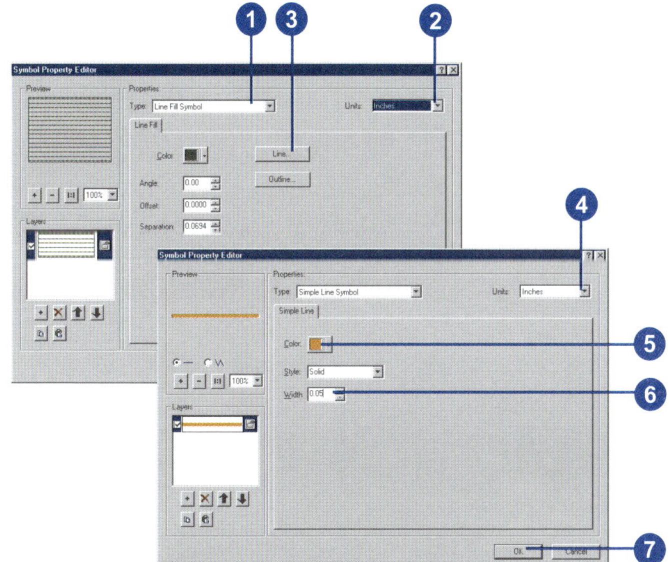

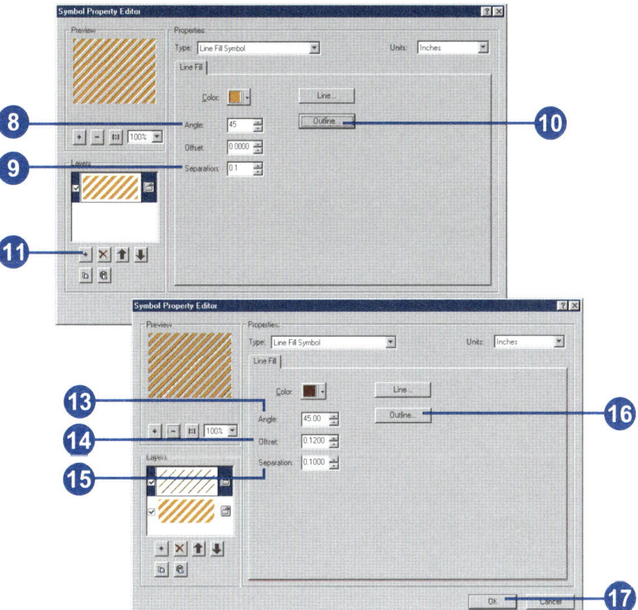

Creating marker symbols

Marker symbols are used to draw point features, labels, and other map annotations. They can be used in conjunction with other symbols to decorate line symbols and create fill patterns and text backgrounds. As graphics, they can add special cartographic elements.

The standard marker types are:

- Simple—fast-drawing set of basic glyph patterns with optional mask
- Character—a *glyph* from a TrueType® font
- Arrow—a glyph from a TrueType font
- Picture—a single .bmp (Windows bitmap) or .emf (Windows enhanced metafile) graphic

Any number of layers can be combined in a single marker. ▸

Tip

Can I use any TrueType font to create symbols?

You can use any text or display font in your system's font folder. You can also create your own TrueType fonts and copy them to your system's font folder.

Creating a character marker symbol from a TrueType font

1. In the marker Symbol Property Editor dialog box, click the Type dropdown arrow and click Character Marker Symbol.
2. Click the Font dropdown arrow and click ESRI Default Marker.
3. Set the Units to Points.
4. Select the marker symbol (99) from the window list.
5. Click the Mask tab.
6. Click Halo.
7. Click Symbol and create a white fill with a black outline width of 0.5. Click OK.
8. Adjust the Size of the halo to 2 points.
9. Click OK.

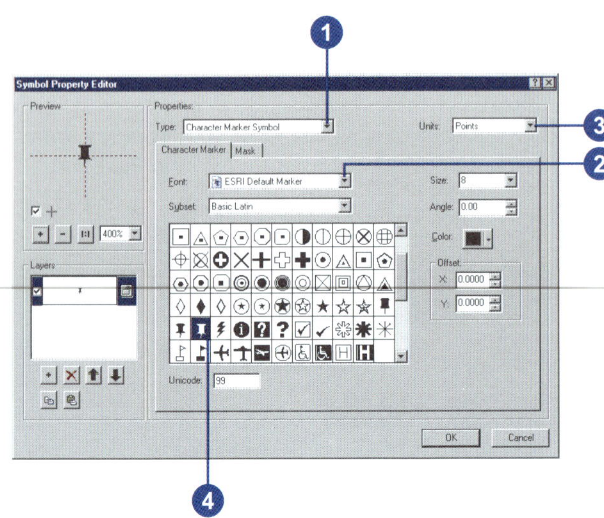

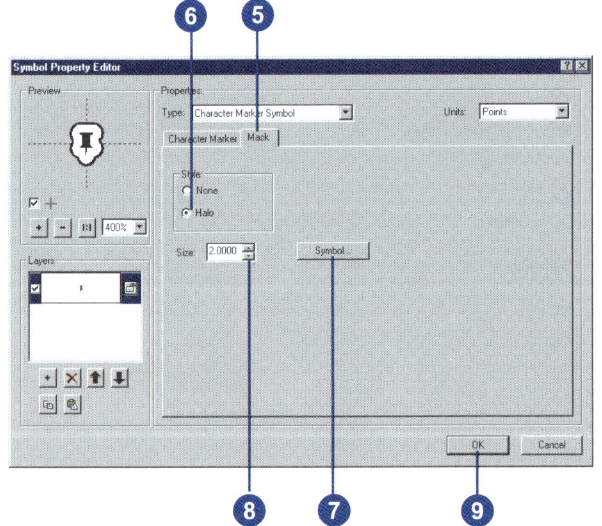

You can use outlines and halos with symbols. An outline uses a line symbol to surround the layer graphic, whereas a mask uses a fill symbol to draw a halo around all layers of the symbol. A halo can also have an outline as part of its definition. A mask is a halo created by designating a fill color with a specified width around the layer and an optional outline.

Tip

Isn't a North arrow just a character marker symbol?

Yes, a North arrow is created from a TrueType font. However, it also has unique properties that link it to its source data frame, and it can have a background and border.

Tip

Does ArcMap come with any existing arrowheads?

The ESRI Arrowhead font contains a variety of arrow shapes. For more information on how to access the font, see 'Creating a character marker symbol from a TrueType font' in this chapter.

Creating an arrow marker symbol

1. In the marker Symbol Property Editor dialog box, click the Type dropdown arrow and click Arrow Marker Symbol.
2. Click the Units dropdown arrow and click Points.
3. Click the Color dropdown arrow and click a red shade.
4. Set the Length to 21.6.
5. Set the Width to 9.
6. Click the Copy Layer button.
7. Click the Paste Layer button.
8. Click the Color dropdown arrow and click black.
9. Set the X Offset to -1.5.
10. Set the Y Offset to -2.
11. Click OK.

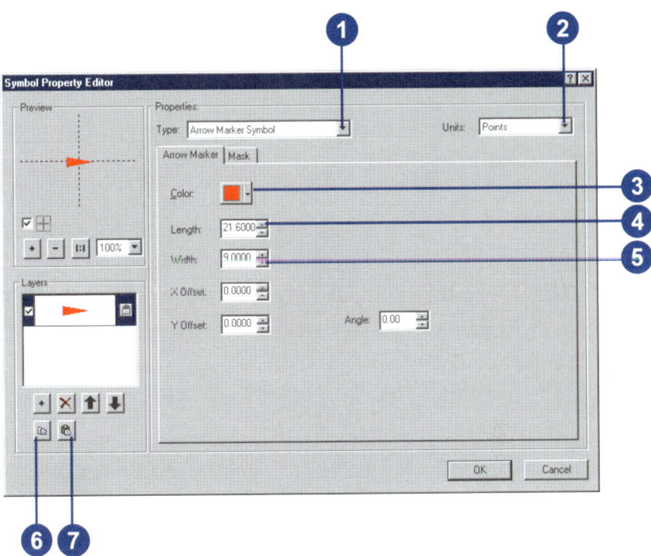

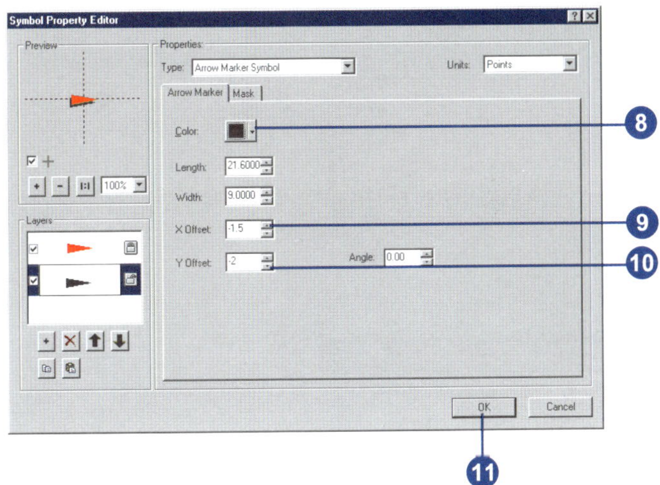

Creating a marker symbol from a picture graphic

1. In the marker Symbol Property Editor dialog box, click the Type dropdown arrow, and click Picture Marker Symbol.

   The Open dialog box automatically opens.

2. Click .bmp or .emf from the Files of type dropdown arrow.
3. Navigate to the location of the graphic and click the graphic file.
4. Click Open.
5. Set the Size.
6. Set the Angle.
7. Set the Background Color.
8. Set the Transparent Color.
9. Click OK.

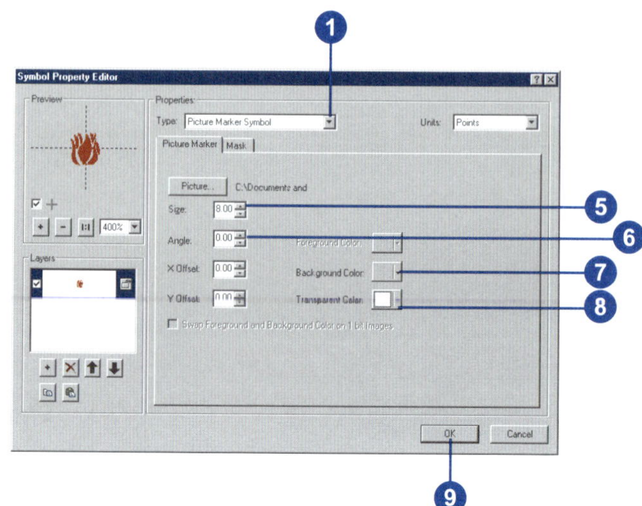

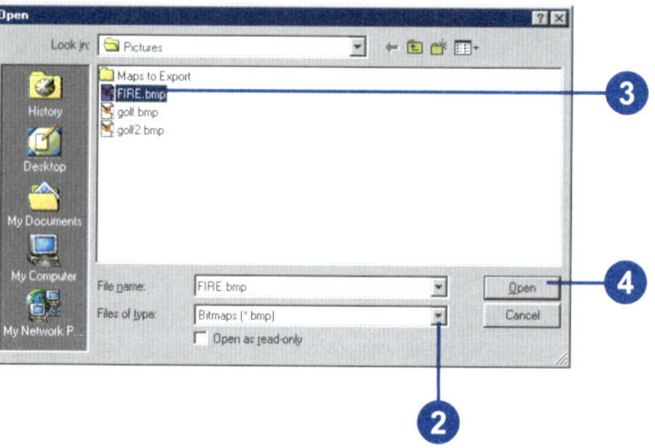

Tip

Does ArcMap come with any pictures?

The pictures used in the styles that come with ArcMap are stored in the \Bin\Styles\Pictures folder where ArcGIS is installed.

Tip

Creating your own pictures

You can create pictures with any graphics package that exports to .bmp or .emf format. You can also scan images, then use a graphics package to clean them up and save them as a .bmp or .emf file.

Tip

Is there a difference between a .bmp and a .emf?

A .bmp is a raster image, while .emf is a vector graphic with better clarity and scaling abilities. You can modify both foreground and background colors on one-bit .bmp pictures, but you can only modify the background color on multibyte .bmp and .emf pictures.

Tip

Swapping the foreground and background colors

Only the foreground color of a one-bit .bmp can be modified. Swapping colors toggles the color to be modified with the Symbol Selector.

Creating text symbols

Text symbols are used to draw labels and annotation that identify and add meaning to your data. Text is also used for titles, descriptions, callouts, legends, scalebars, grid and graticule labels, tables, and other textual and tabular information on your map.

You can create simple text symbols or add advanced formatting, backgrounds, fills, shadows, and halos to your text.

Text symbols consist of a single layer.

Creating a text symbol with a background

1. In the text Editor dialog box, click the General tab.
2. Click the Font and Size dropdown arrows to choose a font and font size. Click the Style buttons if you want to apply a font style. You can also change the color, angle, and offsets of your text.
3. Set the Vertical and Horizontal Alignment.
4. Click the Advanced Text tab.
5. Check Text Background and click Properties.
6. Click the Type dropdown arrow and click Line Callout.
7. Uncheck Leader and Accent Bar, and check Border.
8. Click the Symbol button to set the color and outline properties and click OK.
9. Set the Right and Left Margins.
10. Click OK in all dialog boxes.

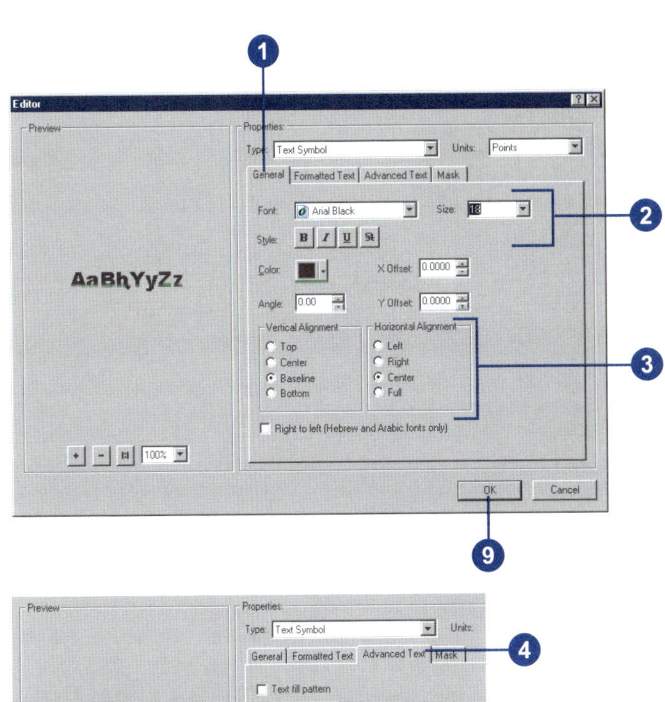

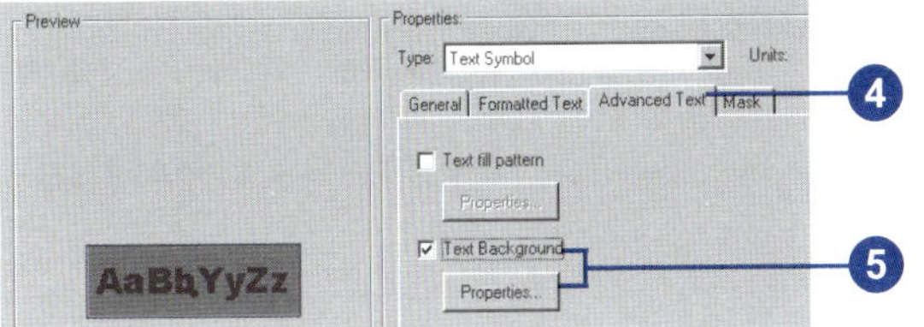

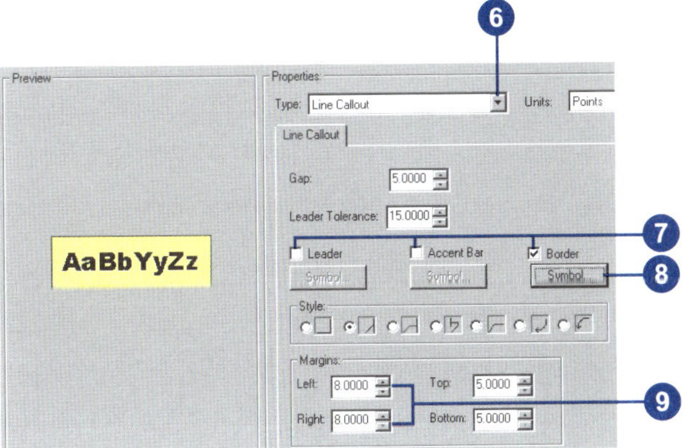

Tip

How are text and label properties different?

Text properties include options for changing the font, style, formatting, and effects. A label is drawn with text symbols but is derived from feature attributes and has additional properties for placement and conflict detection.

Tip

Drawing text without leader or accent lines

You can also use the Line Callout option to draw text with just a background. Simply check only the border option for the properties of your line callout.

Tip

Working with your text

A wide variety of effects can be achieved using different styles and toggling the leaders, accent bars, and borders. You also can enlarge the gap if the outline is too close to the accent bar or you can adjust the margins if your text is crowded in the background.

Creating a text callout symbol

1. In the text Editor dialog box, click the Advanced Text tab.
2. Check Text Background and click Properties.
3. Click the Type dropdown arrow and click Line Callout.
4. Click the third Style option.
5. Check Border, then click Symbol.
6. Click the Color dropdown arrow and click the color and outline you want. Click OK.
7. Click OK in all dialog boxes.

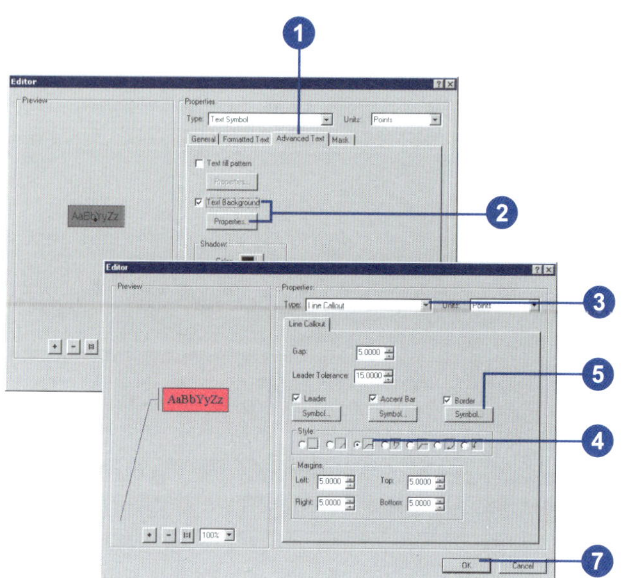

Creating a text symbol inside a marker

1. In the text Editor dialog box, click the Advanced Text tab.
2. Check Text Background and click Properties.
3. Click the Type dropdown arrow and click Marker Text Background.
4. Click Symbol.
5. Choose a marker and click OK.
6. Check Scale marker to fit text.
7. Click OK in all dialog boxes.

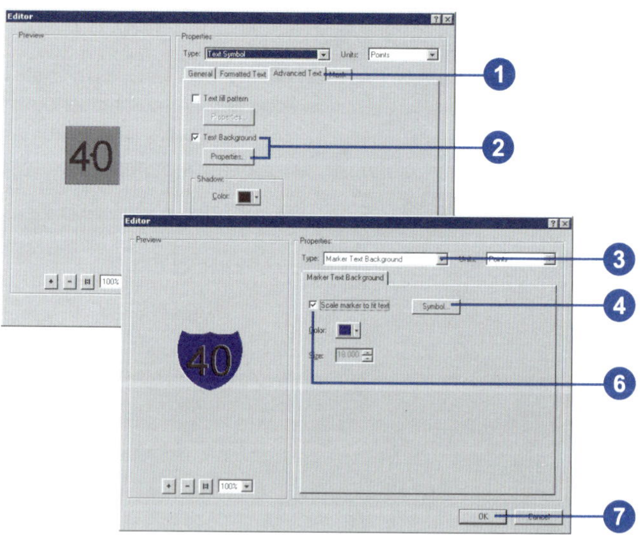

Tip

What units are used in the symbol dialog box menus?

The Symbol Selector dialog box menus and map element Properties dialog box menus use points. The Symbol Property Editor can be set to use points, inches, centimeters, or millimeters.

Creating a text symbol with a drop shadow

1. In the text Editor dialog box, click the General tab.
2. Click the Font and Size dropdown arrows to choose a font and font size. Click the Style buttons for font styles.
3. Click the Color dropdown arrow and choose a color.
4. Set the Vertical and Horizontal Alignment.
5. Click the Advanced Text tab.
6. Click Color and click a gray.
7. Set the X Offset to 2 and the Y Offset to -2.
8. Click OK in all dialog boxes.

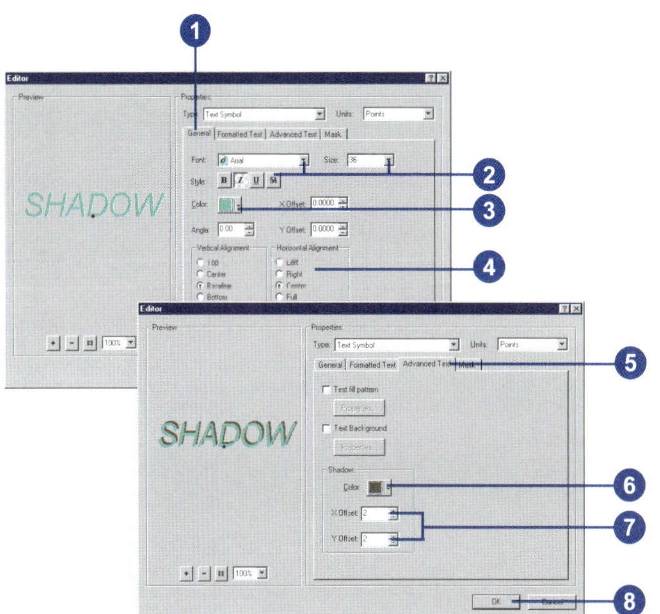

Creating a text symbol with a halo

1. In the text Editor dialog box, click the General tab.
2. Click the Font and Size dropdown arrows to choose a font and font size. Click the Style buttons for font styles.
3. Click the Color dropdown arrow and choose a color.
4. Click the Mask tab.
5. Click Halo.
6. Click Symbol and choose a fill and outline. Click OK.
7. Set a Size for the halo.
8. Click OK in all dialog boxes.

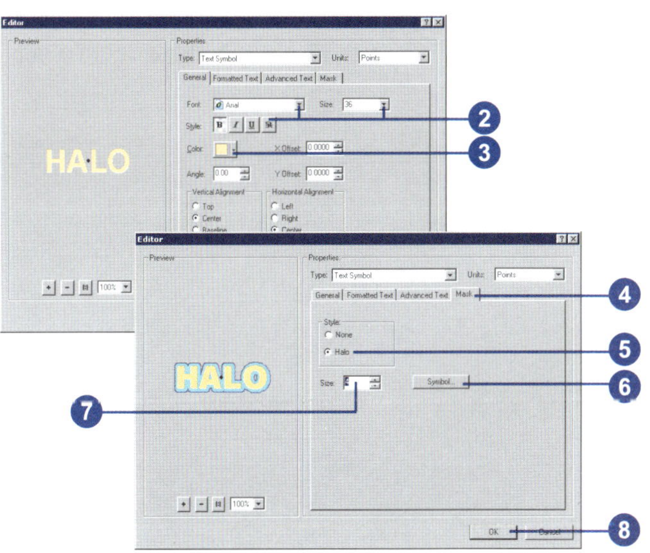

Creating a filled text symbol

1. In the text Editor dialog box, click the General tab.
2. Click the Font and Size dropdown arrows to choose a font and font size. Click the Style buttons if you want to apply a font style.
3. Set the Vertical and Horizontal Alignment.
4. Click the Formatted Text tab.
5. Set the Character Spacing.
6. Click the Advanced Text tab.
7. Check Text fill pattern and click Properties.
8. Choose a fill.
9. Set the Outline Width and Color and click OK.
10. Click OK in all dialog boxes.

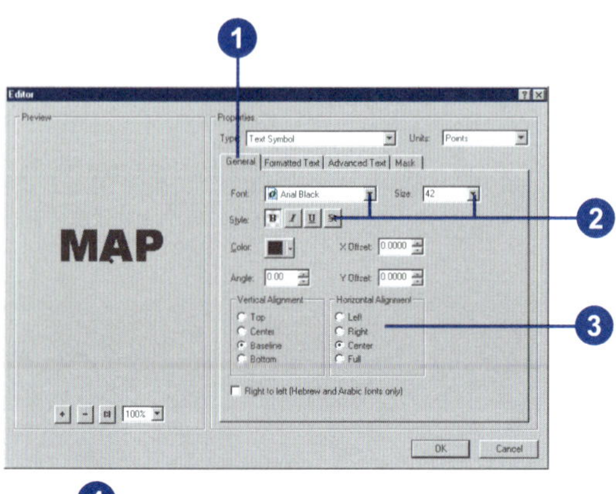

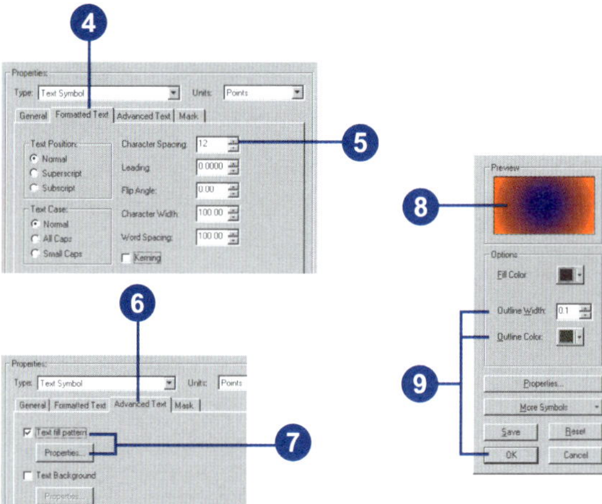

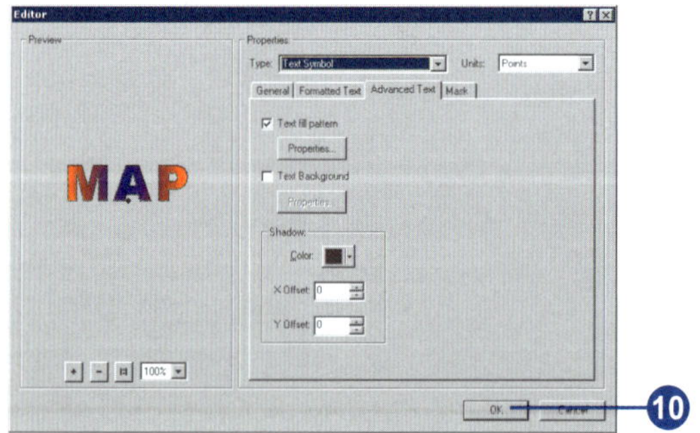

Modifying and saving symbols and elements as you work

As you compose your map, you may want to modify the symbols you've used to draw data and graphics.

When you save your changes from the Symbol Selector dialog box, the new symbols are stored in your personal style folder. Later, you can use the Style Manager to move them into another style. When you insert a map element on the map layout, you can modify its properties. Later, you may decide to make more changes and save them to use in another map.

You can also modify map elements as you work in ArcMap. Most map elements are composed of a mix of symbols and map elements. For example, the North arrow's graphic comes from a marker that is created from a font, and its frame comes from a border and background that are created from other symbols.

Saving symbols to a style using the Symbol Selector

1. In the table of contents, double-click the symbol you want to modify.

   The Symbol Selector dialog box is displayed.
2. Click a symbol.
3. If you want to make further simple modifications, use the Color and Width Options to set specific properties.
4. If you want to make further advanced modifications, click Properties to access the Symbol Editor dialog box and make the changes you want.
5. Click Save.
6. Type a Symbol Name.

   Your new symbol is saved in your personal style folder and appears in the Style contents window.
7. Click OK.
8. Click OK.

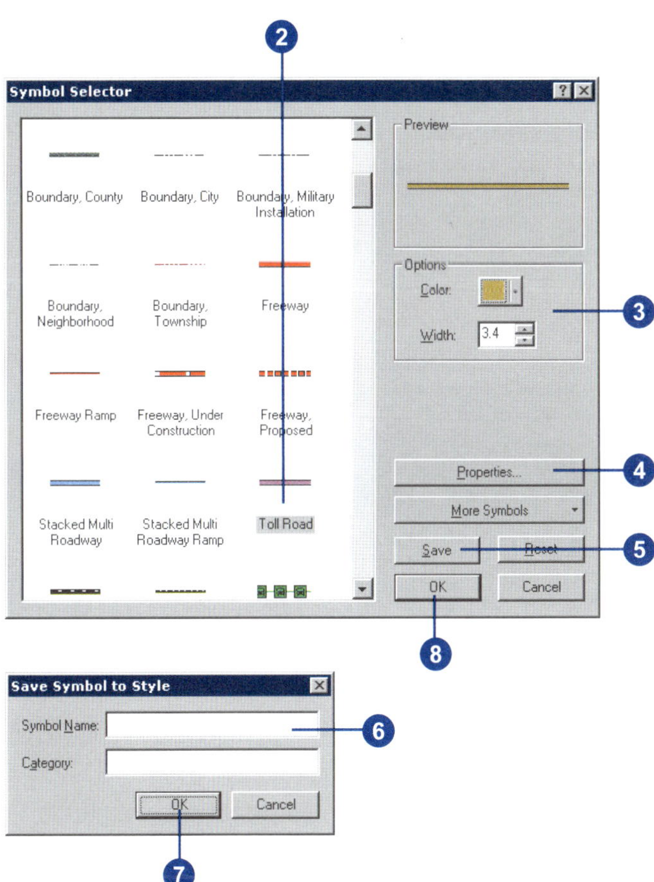

Tip

What's the difference between modifying symbols, graphics, and map elements?

Graphics and map elements have an additional property tab for size and position in relation to the page. Map elements also have a Frame tab for background and border properties.

Tip

Why can't I modify the color of a symbol?

Sometimes a symbol layer or layers are locked and cannot be modified. In this case, you can either click Properties and use the Symbol Editor dialog box to modify the colors you want. Or, if you find you will be modifying this color often, you may want to unlock the layer and save the symbol.

Modifying symbols used to draw map elements

1. In layout view, double-click the element you want to modify. This opens a Properties dialog box.
2. Click the Color dropdown arrow and click a new color.
3. Click the Style button.
4. Click Save.
5. Type a name and click OK.

   This element is stored in your personal style folder.
6. Click OK.
7. Click the Frame tab.
8. Click the Style Selector button to change the border style.
9. Click Properties.
10. In the Border dialog box, choose the desired properties and click OK.
11. Click Save.
12. Type a name and click OK.

    This style is stored in your personal style folder.
13. Click OK.
14. Repeat steps 8 through 13 to set the background style and properties.
15. Click OK.

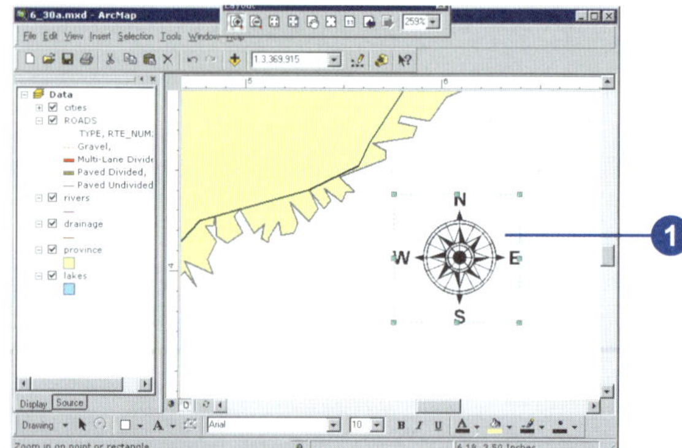

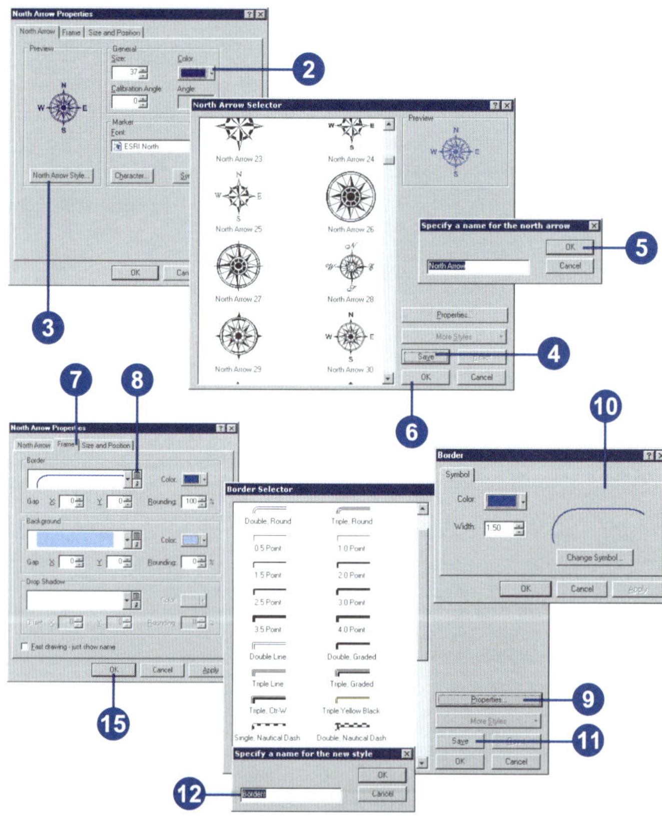

Working with color

Color is one of the fundamental properties of symbols and map elements. The color palette shows the colors from all the referenced styles. Your personal modifications are shown at the bottom of the palette. You can use a variety of dialog boxes to create colors. The Selector dialog boxes can be accessed from the color palette, while the Property dialog boxes are accessed from the Selector dialog boxes and the Style Manager.

ArcMap can define color in these five color models:

- RGB—red, green, blue
- CMYK—cyan, magenta, yellow, black
- HSV—hue, saturation, value
- Gray—gray shade ramp
- Names—ArcInfo color names

Tip

Using a null color

A null color lets you turn off outline drawing or create transparent areas in your symbols. However, a null color can't be used to knock out or block other colors.

Using the Style Manager to define colors

1. Click Tools, point to Styles, and click Style Manager.
2. Click the Colors folder to view its contents.
3. Right-click in the Contents window. Click New and choose a color model.
4. Click a color in the Color window or use the color model spinners to mix a color.
5. Click OK.
6. Type the name of the new color in the Contents window.

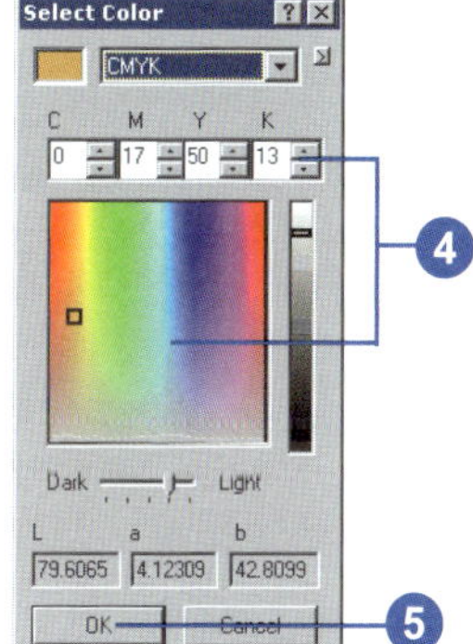

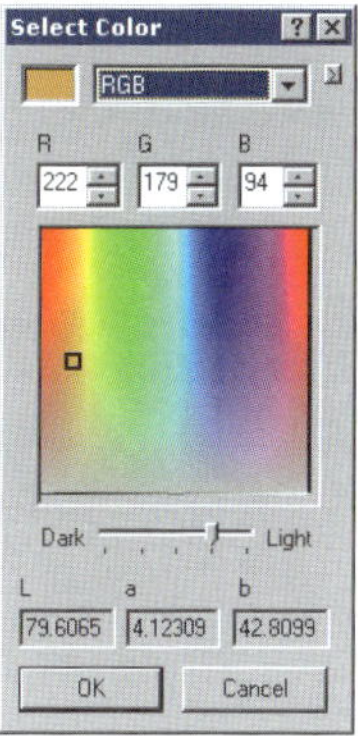

Using the Style Manager to define a null color

1. Right-click in the Contents window of the Colors folder. Click New and click Gray.
2. Click the arrow button and click Advanced Properties.
3. Check Color is Null.
4. Click OK.

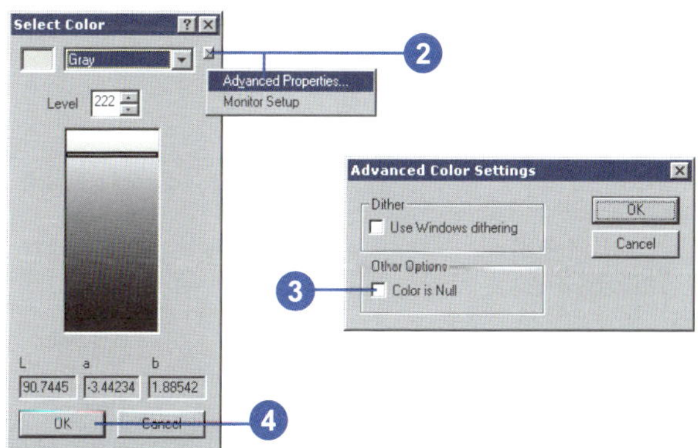

Tip

Identifying the colors on the palette

You can pause the mouse pointer over a color to see its name as a tip.

Tip

Does the color palette come from a style?

Yes, it is a combination of all the referenced styles. As you create custom colors, they are displayed on the palette.

Defining colors as you work

1. Click the Color dropdown arrow on a dialog box or right-click on a symbol in the table of contents.
2. Click a new color or click More Colors to view additional colors.
3. Use the Color Selector dialog box to mix a new color.

   You can toggle the color model with the arrow button menu choices or click the color preview to display the Property dialog boxes.
4. Click the arrow button and click Save Color.
5. Type a name for the new color.

   The color is saved in your personal style folder.
6. Click OK.

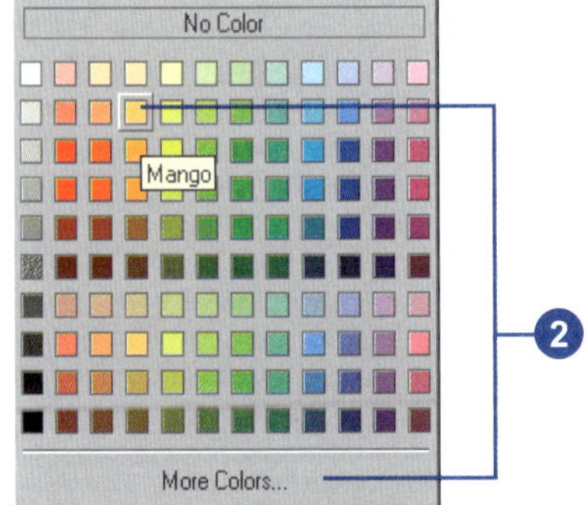

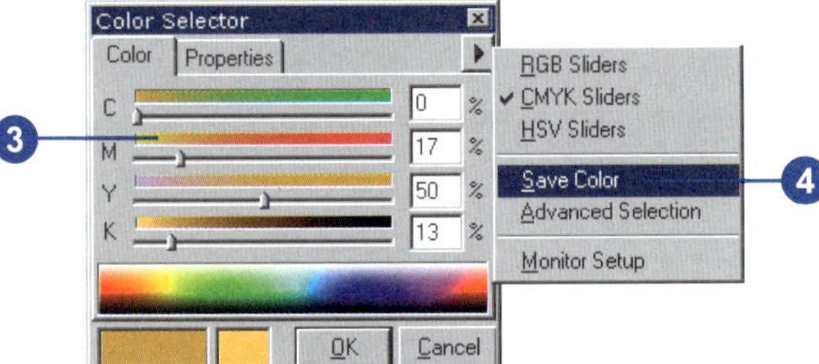

The new color is saved in your personal style folder and appears on the color palette.

Defining a null color as you work

1. Click the Color dropdown arrow on a dialog box.
2. Click More Colors.
3. Click the Properties tab.
4. Check Color is Null.
5. Click OK.

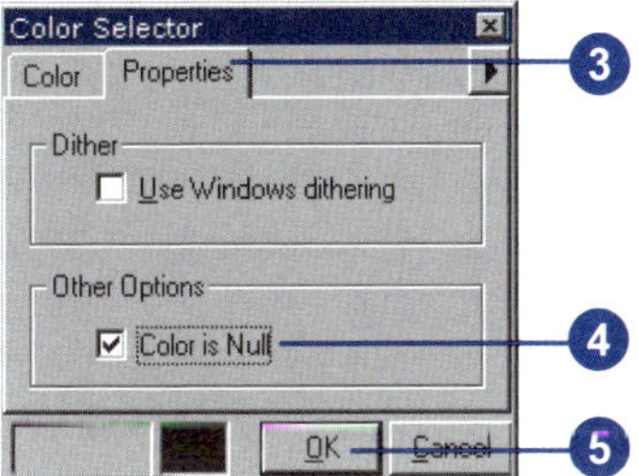

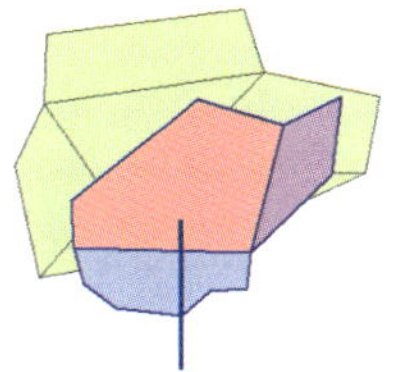

Polygon with an opaque fill color.

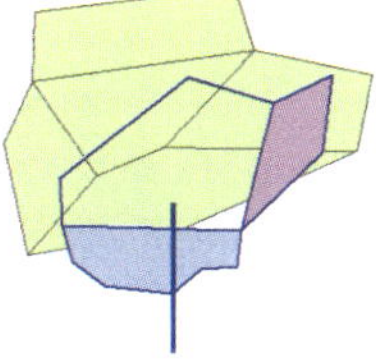

Polygon with a null fill color, making it display transparently.

Working with color ramps

ArcMap color ramps provide the means to apply a range of colors to a group of symbols. Color ramps are used, for example, in the Graduated colors layer symbology option. ArcMap has a range of color ramps already defined in the Color Ramp styles folder. Some are created for specific applications such as displaying elevation or precipitation. You can also create your own.

There are four types of color ramps. Algorithmic color ramps are a specific type of ramp that traverses the color spectrum between two colors. Preset color ramps provide the identical color ramp capability as provided in ArcView GIS 3.x. Random color ramps provide the user with the most distinct colors that traverse a color spectrum. Multipart color ramps are containers storing a sequence of any of the other three color ramps in any combination.

Using the Style Manager to define color ramps

1. Click Tools, point to Styles, and click Style Manager.
2. Click the Colors Ramps folder to view its contents.
3. Right-click in the Contents window. Click New and select Algorithmic Color Ramp.
4. In the Algorithmic Color Ramp Properties dialog box, click Color 1 and set the start color for the ramp.
5. Click Color 2 and set the end color for the ramp.
6. Adjust the value of black and white brightness throughout the ramp.
7. Click OK.
8. Type the name of the new color in the Contents window.

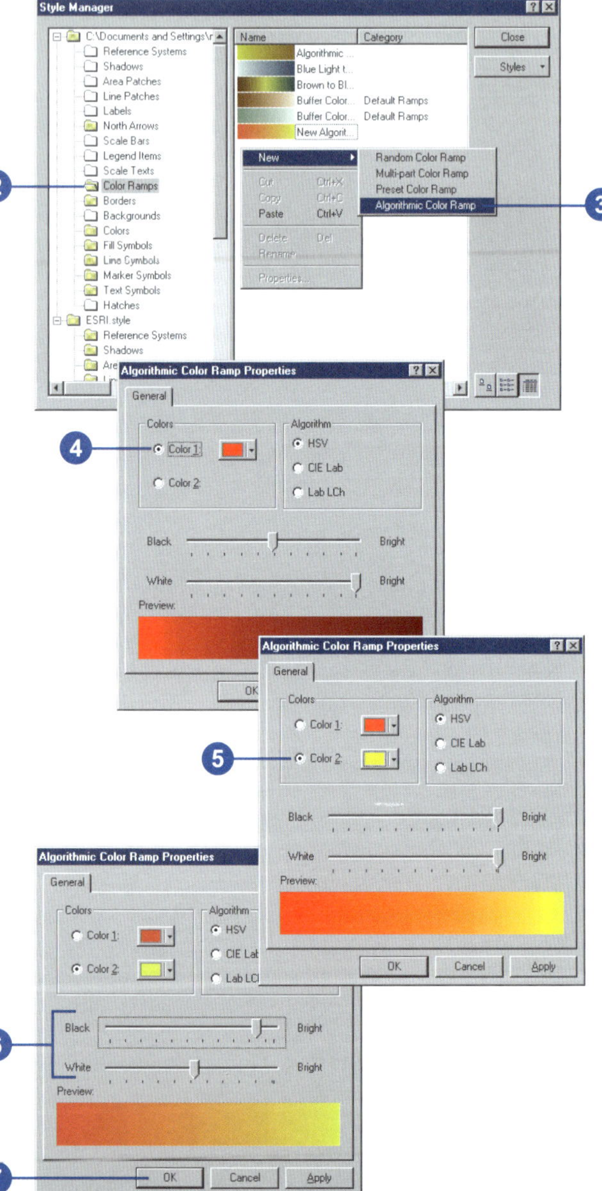

How do I adjust a multipart ramp to match my data?

When you're working with data, such as elevation, choose the Elevation color ramp in the styles folder. Copy and paste this to your personal styles folder. Double-click it to edit it, and remove the parts not in your data, such as water or mountains (at the top and bottom of the list).

Ramping colors in the Layer Properties dialog box Symbology tab

1. Double-click the layer in the table of contents.
2. Click the Symbology tab.
3. Click Quantities and click Graduated colors.
4. Choose a Value field and, if desired, a Normalization field.
5. Set the number of Classes or change the classification scheme by clicking Classify.
6. Double-click the top symbol in the Symbol column.
7. Set the color options. Click OK.
8. Double-click the middle symbol in the Symbol column and set the color options. Click OK.
9. Double-click the end symbol in the Symbol column and set the color options. Click OK.
10. Hold the Ctrl key and select each color-adjusted symbol.
11. Right-click one of the selected symbols and choose Ramp Colors.
12. Click OK in the dialog box.

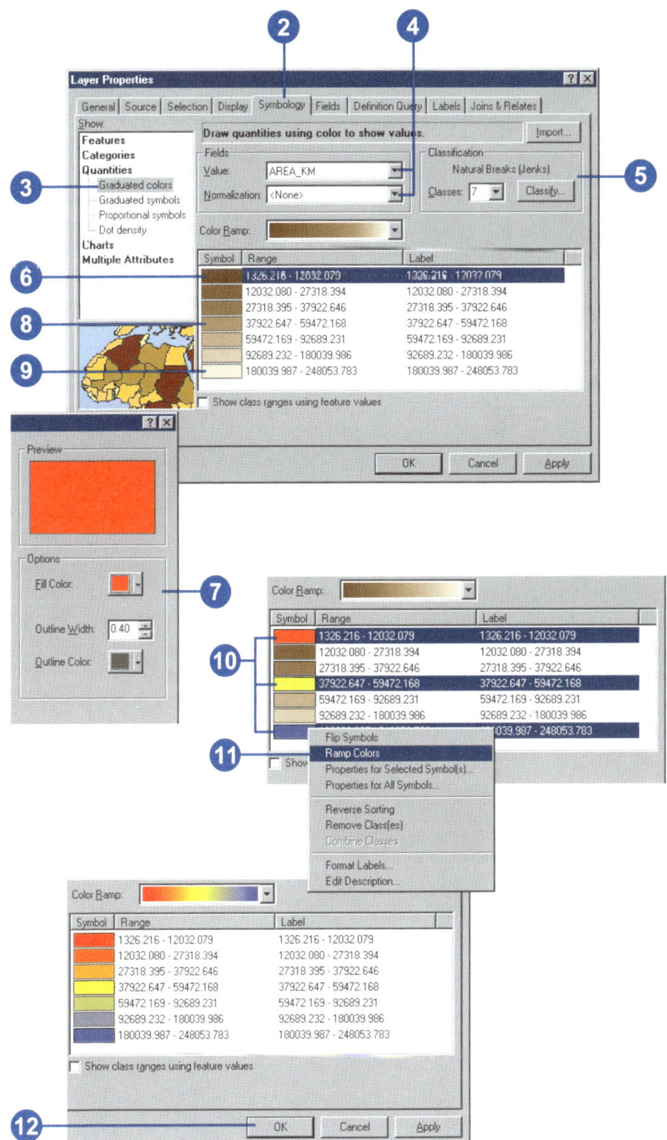

Working with rasters

9

IN THIS CHAPTER

Vector data—such as coverages and shapefiles—represent geographic features with lines, points, and polygons. Rasters—such as images and grids—represent geographic features by dividing the world into a regular pattern of discrete cells called *pixels*. Each pixel, short for picture element, represents an area, often has a geographic location, and has a value that represents the feature being observed. For example, the pixel values in an aerial photograph represent the amount of light reflecting off the earth's surface interpreted as trees, houses, streets, and so on, while the pixel values in a DEM represent elevations.

A raster can represent thematic data, such as land use or soils; *continuous* data including temperature, elevation, or spectral data such as satellite images and aerial photographs; or pictures, such as scanned maps, scanned drawings, or photographs of buildings. You'll generally display thematic and continuous rasters as data layers along with other geographic data on your map. Picture rasters, when displayed with your geographic data, can convey additional information about map features. You may also display a data layer representing a collection of raster datasets, called a raster catalog, which will be discussed further in this chapter.

Some rasters have a single *band* of data, while others have multiple bands. For example, a satellite image commonly has multiple bands representing different wavelengths, from the ultraviolet through the visible and infrared portions of the electromagnetic spectrum. This means that every pixel location has more than one value associated with it. On the other hand, a DEM has a single band of data, containing values of elevation, of which there is only one value.

Adding a raster dataset to your map

You can add all types of raster data to ArcMap: raster datasets, file-based raster catalogs, and geodatabase raster catalogs. Also, there are many different raster formats that can be added as a valid raster dataset. To see the complete list of valid raster formats, please refer to 'Supported raster formats' in the ArcGIS Desktop Help.

If you add raster data that covers the same geographic area as your map but in a different coordinate system, ArcMap uses the coordinate system of the first dataset added and performs a transformation on the fly.

If your raster dataset's coordinate system is not defined, you can use ArcCatalog to attach or define the coordinate system information. See 'Defining a raster's coordinate system' in the ArcGIS Desktop Help for more information.

If you need to transform a raster dataset, you must have georeferencing information or a world file and know its coordinate system. If your raster dataset does not have any georeferencing information ▸

Adding a single band raster dataset

1. Click the Add Data button on the Standard toolbar.
2. Click the Look in dropdown arrow and navigate to the folder that contains the raster dataset.

   If you want to add a single band of a multiband raster dataset, double-click the dataset to expose the individual bands.
3. Click the single band raster dataset.
4. Click Add.

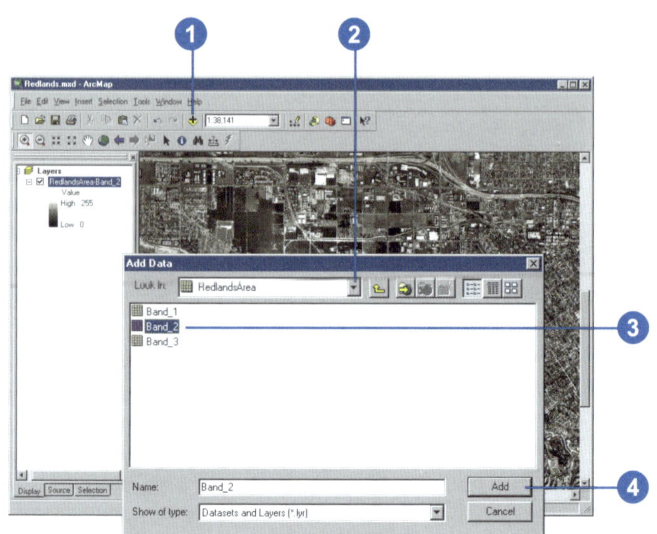

Adding a multiband raster dataset

1. Click the Add Data button on the Standard toolbar.
2. Click the Look in dropdown arrow and navigate to the folder that contains the raster dataset.
3. Click the multiband raster dataset.
4. Click Add.

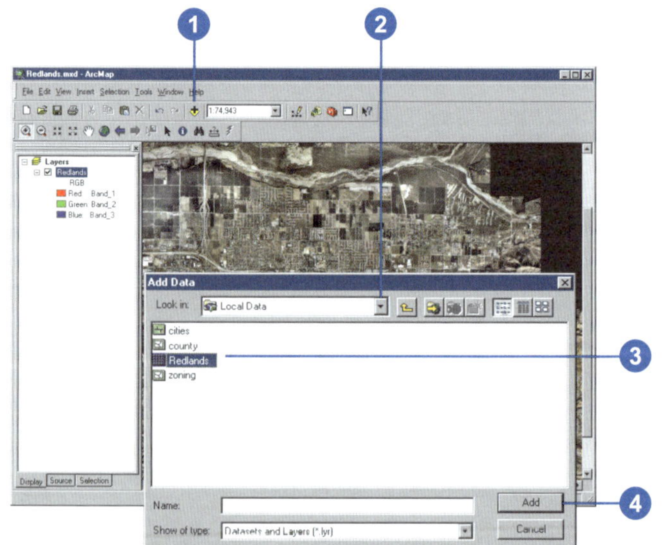

associated with it, you can georeference it in ArcMap. See 'Georeferencing a raster dataset' in this chapter.

Tip

Displaying raster dataset bands

When you add a raster layer with multiple bands, you can choose to display a single band of data or you can simply add the entire raster dataset.

Tip

Changing the defaults for drawing a multiband raster dataset layer

You can change the default red–green–blue (RGB) band combination that a multiband raster layer displays when it is loaded. Click the Tools menu, point to Options, and click the Raster tab.

Tip

Displaying pictures

If your raster is just a picture that doesn't align to any other geographic data, you can simply place it on a layout as a graphic element. Alternatively, you can create a hyperlink and associate it with a geographic feature on your map.

Adding a raster as a picture

1. Click the Insert menu on the Standard toolbar and click Picture.
2. Click the Look in dropdown arrow and navigate to the picture you want to add.
3. Click the picture.
4. Click Open.

   If you're in layout view, the picture is inserted on the layout. If you're in data view, the picture is inserted within the data frame.

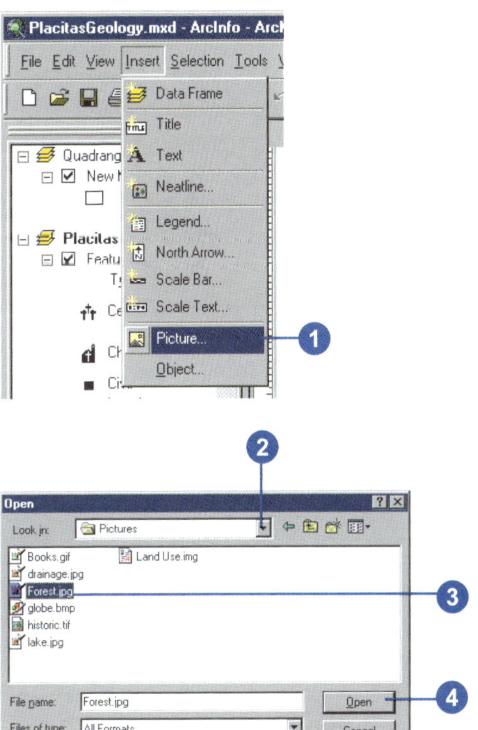

Using raster catalogs

Raster catalogs are used to display multiple adjacent or overlapping raster datasets without having to mosaic them into one larger file. Raster catalogs appear in the Add Data dialog box in ArcMap as ordinary tables or geodatabase raster catalogs. The raster datasets in the raster catalog are drawn in order—from the first to last record in the catalog's table.

A traditional file-based raster catalog can contain multiple raster dataset types, formats, resolutions, and file sizes. Any table format can be used to define a raster catalog, including text files and geodatabase tables.

Geodatabase raster catalogs can be used in the ArcMap selection environment. For more information on this, see 'Using the geodatabase raster catalog selection environment' later in this chapter.

Tip

Why is my raster catalog displayed as a wireframe?

A raster catalog displays as a wireframe if more than nine images are in the current display extent. You can change this default setting in the raster catalog Layer Properties dialog box under the Display tab.

Adding a raster catalog

1. Click the Add Data button on the Standard toolbar.
2. Click the Look in dropdown arrow and navigate to the folder that contains the raster catalog.

   NOTE: Traditional file-based raster catalogs have a table icon ( ), whereas geodatabase raster catalogs have their own icon ( ).
3. Click the raster catalog.
4. Click Add.

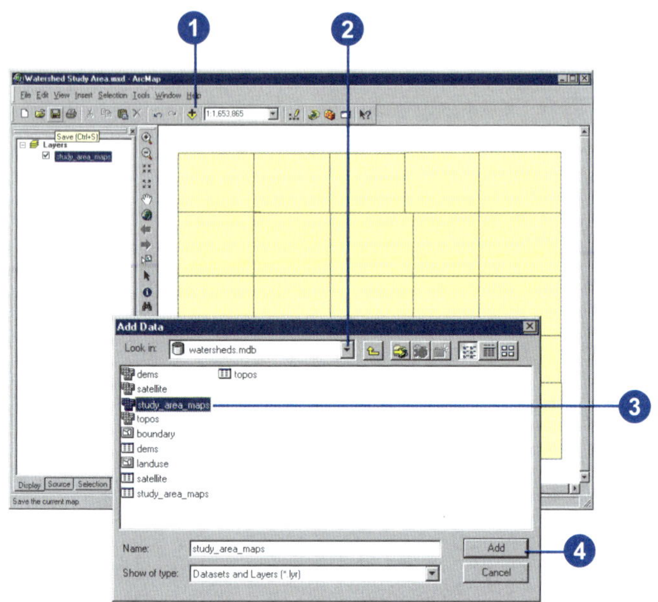

Rendering raster datasets and raster catalogs

Displaying rasters

How you display a raster depends on what type of data it contains and what you want to show. Some rasters have a predefined color scheme—a colormap—that ArcMap automatically uses to display them. For those that don't, ArcMap chooses an appropriate display method that you can adjust as needed.

If you want, you can change display colors, group data values into classes, or stretch values to increase the visual contrast. For multiband rasters, you can display any three bands together as an RGB composite. By trying different band combinations you may distinguish different features in the raster.

The raster resolution is the ratio of screen pixels to dataset pixels at the current map scale. Seeing the raster resolution allows you to determine if you are approaching the maximum resolution of the raster.

A ratio of 1:1 means you have reached the best display, or the maximum resolution of the raster, where every screen pixel is displaying exactly one raster cell. A ratio of 1:20 means that every screen pixel has to display 20 raster cells, so less detail will be seen in the raster layer. A ratio of 1:0.02 means that every screen pixel is displaying only a portion of a raster cell or that it takes many screen pixels to display a single raster cell.

Rendering raster datasets

Raster datasets can be displayed, or rendered, in your map in many different ways. Rendering involves the process of displaying your data visually.

When a raster dataset layer is added to ArcMap, it will be displayed with a default renderer, which is usually the most appropriate for the particular raster dataset layer. Generally, there are particular ways to display a raster dataset to take advantage of all its data. ArcMap allows you to choose from different drawing methods based on your display and analysis needs. Both individual raster datasets and raster catalogs provide similar display methods.

Methods of rendering raster data

There are several methods of displaying raster data. You can choose which renderer you want to use for your data on the Symbology tab of the Layer Properties dialog box.

- RGB Composite: Use RGB Composite for a multiband raster layer. You can draw a three-band composite of raster data, with the option of displaying fewer than three bands or changing the band combination.
- Unique Values: Use Unique Values when you want each value in the raster layer to be displayed individually. For instance, you may have discrete categories representing particular objects on the earth's surface, such as those in a thematic raster layer, which could display soil types or land use.
- Stretched: Use Stretched when you want to draw a single band of continuous data. The Stretched renderer displays continuous raster cell values across a gradual ramp of colors.
- Classified: Use Classified to display rasters by grouping cell values into classes. It is typical to use this type of thematic classification on continuous phenomena in a single-band raster layer.
- Colormap: Use Colormap for similar data as you would the Unique Values renderer, although the Colormap renderer uses a specified color scheme to display your data. This is only available when a raster dataset has an associated colormap.

Rendering raster catalogs

A raster catalog is a collection of raster datasets defined in table format, in which the records define the individual rasters that are included in the catalog. A raster catalog can be used to display a collection of adjacent rasters without having to mosaic them into a larger file. Raster catalogs can also be used to hold disparate, semioverlapping, or fully overlapping raster datasets.

By default, raster catalogs are displayed as a wireframe if more than nine images are in the current display extent. Otherwise, the actual raster data will be displayed. The use of the wireframe speeds up the display of raster catalogs. The default of nine images can be changed in the display properties of your raster catalog or on the Raster tab in the Options dialog box.

ArcMap has the ability to render each raster dataset member of a geodatabase raster catalog with its most appropriate renderer. The Symbology tab of a raster catalog's Layer Properties dialog box lists the renderers that are available for the catalog. This list of renderers can be edited to add or remove various renderers. Only the renderers in the list can be used in the rendering of the catalog. In the available renderers list, ArcMap places an asterisk next to each of the renderers that are currently active and are applied to one or more raster dataset members of the raster catalog. However, the active list can only be triggered when an image is displayed on the screen. This list will not be complete until the entire catalog has been viewed. The active renderers persist even after you change the display to another area, to the full extent, or back to a wireframe display.

The RGB Composite renderer

You use the RGB Composite renderer to display a multiband raster layer, such as satellite imagery or a color aerial photograph. This renderer gives you the options of displaying less than three bands or changing the band combination.

Tip

Turning bands on and off in the table of contents

In the table of contents, click the color square beside the band you want to turn off and uncheck Visible to turn it off.

Tip

Displaying the RGB value for a cell

To see the RGB value for a given cell, turn on MapTips for the layer. To do this, right-click the layer in the table of contents and click Properties. Check the Show MapTips box on the Display tab.

Drawing a multiband raster dataset as an RGB composite

1. In the table of contents, right-click the raster layer that you want to draw as an RGB composite and click Properties.
2. Click the Symbology tab.
3. Click RGB Composite.
4. Click the Bands dropdown arrow next to each color and click the band you want to display for that color.
5. If the raster dataset contains a background or border around the data that you want to hide, check Display Background Value and set the color to No Color. The cells will display transparently.
6. Optionally, click the Stretch Type dropdown arrow and click the stretch you want to perform.
7. Optionally, click Histograms to modify the stretch settings.
8. Click OK.

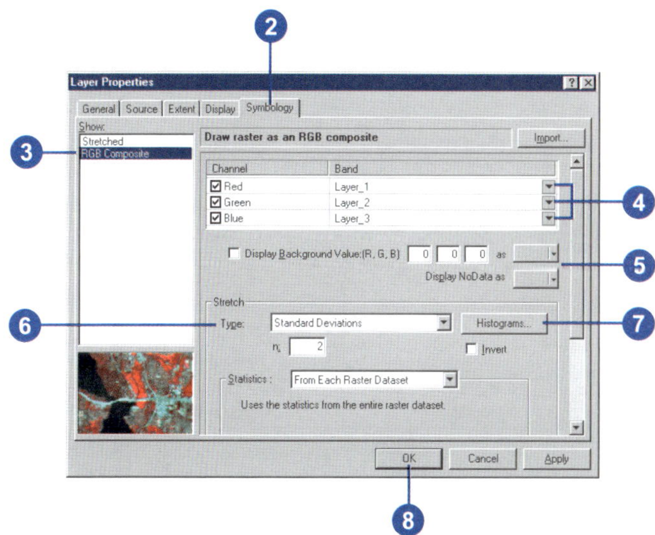

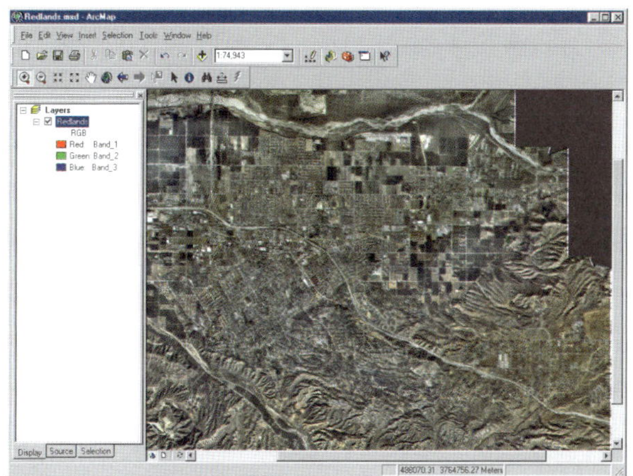

The Unique Values renderer

You use the Unique Values renderer when you want each value in the raster layer to be represented by an individual color. For example, a thematic raster layer, such as land use, contains specific values representing particular land use categories. You may want each of these categories to be displayed using an individual color.

The Unique Values renderer displays each value as a random color. If your data has a colormap, you can use the Colormap renderer to display your data with assigned colors.

Tip

The raster dataset draws too dark

You can alter the overall brightness or contrast of a raster dataset with the Effects toolbar. Alternatively, some raster renderers will allow you to stretch the data values to take advantage of the available colors.

Drawing raster datasets using the Unique Values renderer

1. In the table of contents, right-click the raster layer that you want to draw with a set of unique values and click Properties.
2. Click the Symbology tab.
3. Click Unique Values.
4. Click the Value Field dropdown arrow and click the field you want to map.
5. Click the Color Scheme dropdown arrow and click a color scheme.
6. Optionally, click a label and type in a more descriptive one.
7. Optionally, select a color or no color to display any NoData values.
8. Click OK.

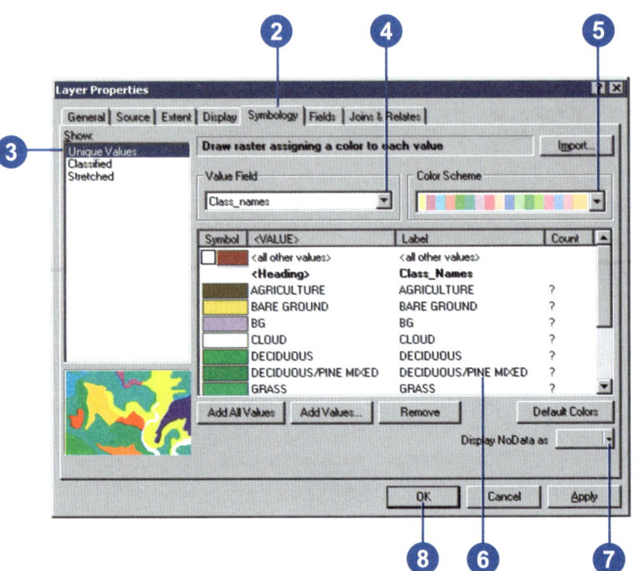

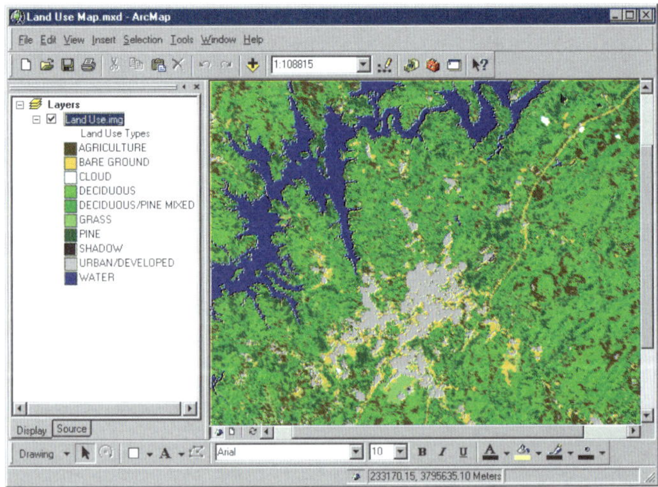

The Stretched renderer

The Stretched renderer displays continuous raster cell values across a gradual ramp of colors. Use the Stretched renderer when you want to draw a single band of continuous data. The Stretched renderer works well when you have a large range of values that you would like to display, such as in spectral imagery, aerial photographs, or elevation models.

You can choose from several different automatic stretches and a manual option when deciding exactly how to stretch the values.

Tip

Changing a raster dataset's color from the table of contents

You can quickly change the colors applied to a raster dataset by clicking the color ramp in the table of contents.

Tip

Using statistics

You can use the Statistics portion of the Symbology tab to examine and modify how your data displays.

Drawing raster datasets using the Stretched renderer

1. In the table of contents, right-click the raster layer that you want to display across a color ramp and click Properties.
2. Click the Symbology tab.
3. Click Stretched.
4. If your raster dataset has multiple bands, you may choose the band you want to stretch.
5. Optionally, click the Label boxes and type labels for the table of contents.
6. Click the Color Ramp dropdown arrow and click a color ramp.
7. Optionally, choose a color or no color to display any NoData values.
8. Optionally, click the Stretch Type dropdown arrow and click the type of Stretch, if any, you would like to perform.
9. Optionally, scroll down and click Statistics to change the statistics, if desired.
10. Click OK.

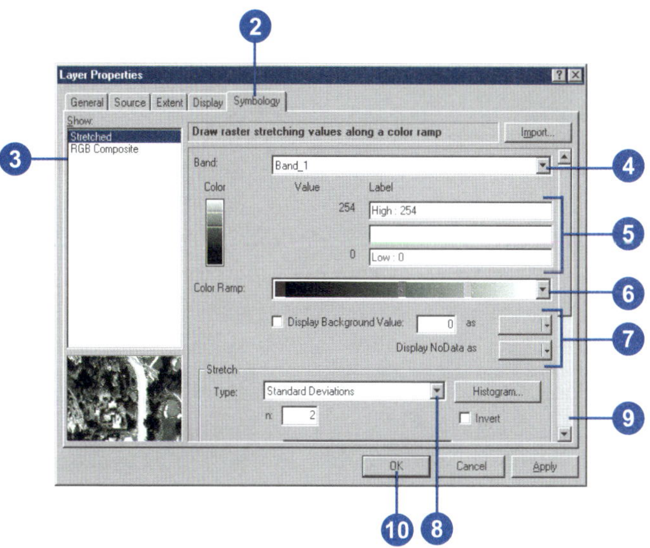

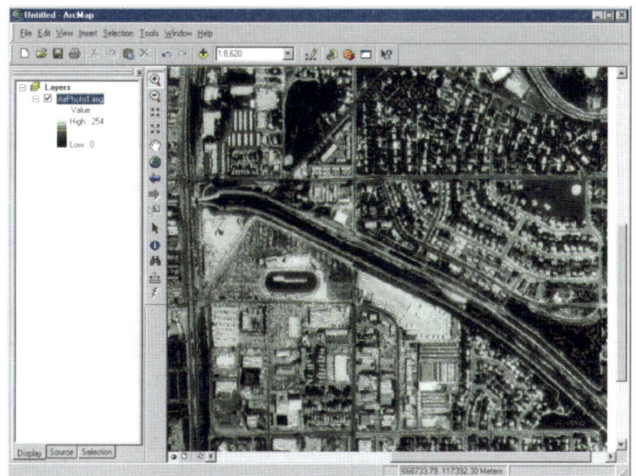

The Classified renderer

The Classified renderer is used with a single-band raster layer. The Classified method displays thematic rasters by grouping cell values into classes. It is typical to use this type of thematic classification on continuous phenomena, such as slope, distance, or suitability, where you want to classify the range into a small number of classes and assign colors to those classes.

You can choose a manual classification method or one of these standard methods:

- Equal Interval—The range of cell values is divided into equally sized classes, where you specify the number of classes.
- Defined Interval—You specify an interval to divide the range of cell values, and ArcMap determines the number of classes.
- Quantile—Each class contains an equal number of cells.
- Natural Breaks (Jenks)—The classes are based on natural groupings of values.
- Standard Deviation—Shows you the amount a cell's value varies from the mean.

Drawing raster datasets using the Classified renderer

1. In the table of contents, right-click the single band raster dataset layer that you want to display by grouping values into classes and click Properties.
2. Click the Symbology tab.
3. Click Classified.
4. Click the Value dropdown arrow and click the field you want to map. If your raster dataset does not have a table, the Value dropdown arrow is unavailable.
5. Optionally, click the Normalization dropdown arrow and click a field to normalize your data.
6. Click the Classes dropdown arrow and click the number of classes you want.
7. Click Classify and choose the classification method you want to use. You can choose a standard or a manual method of classification. Click OK.
8. Click the Color Ramp dropdown arrow and click a color ramp.
9. Optionally, choose a color or no color to display any NoData values.
10. Click OK.

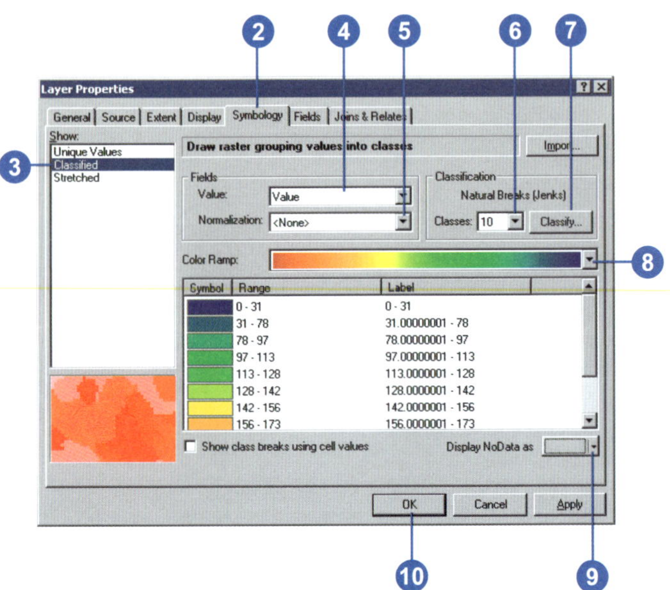

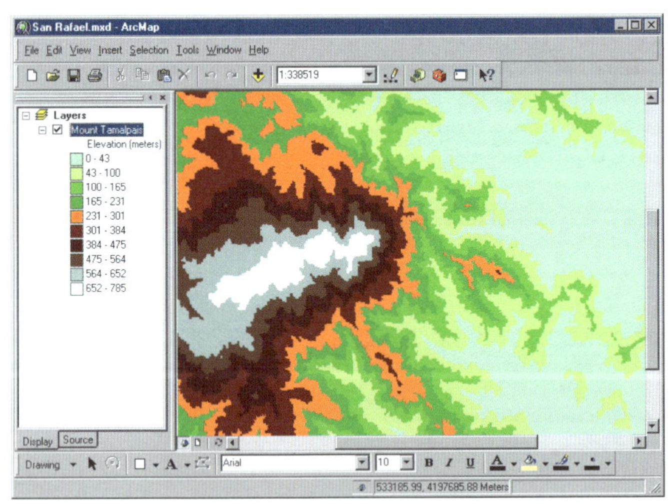

The Colormap renderer

You use the Colormap renderer much as you would the Unique Values renderer. However, you use the Colormap renderer when you choose to have the values in the raster layer represented by a prespecified color.

If your data has a colormap, you can use the Colormap renderer to display your data with assigned colors. The Colormap renderer displays automatically in the list of available renderers on the Symbology tab.

Tip

How do I know if my raster dataset has a colormap?

You can check to see if your raster dataset has a colormap in the Properties dialog box. Click the Source tab and see whether or not a colormap is present.

Tip

Where is the colormap information stored?

For GRID files, the colormap information is stored in a .clr file with the same name as the GRID. For most other formats, the colormap is embedded within the raster dataset.

Drawing thematic raster datasets that represent categories using the Colormap renderer

1. In the table of contents, right-click the raster layer that you want to draw with a colormap and click Properties.
2. Click the Symbology tab.
3. Click Colormap.
4. Click OK.

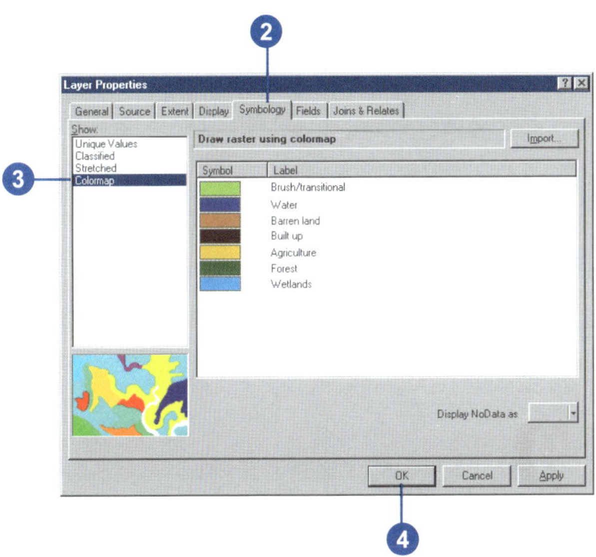

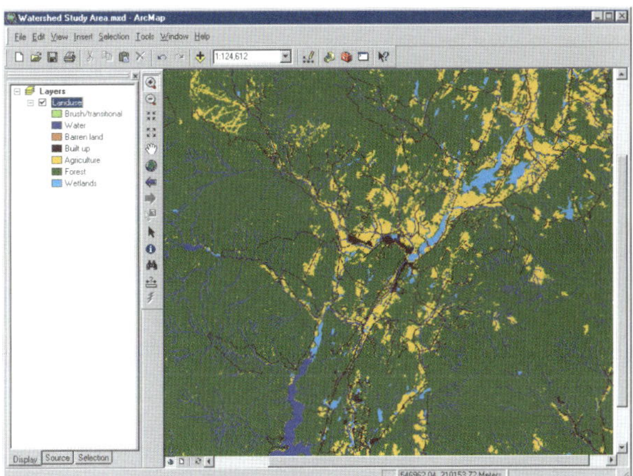

Raster resolution

The raster *resolution* refers to the size of the cells in a raster dataset and the ratio of screen pixels to image pixels at the current map scale. For example, one screen pixel may be the result of nine image pixels resampled into one—this is a raster resolution of 1:9. In this case, every screen pixel has to display nine raster cells, meaning the image is not as clear and detailed.

A resolution of 1:1, however, means that every screen pixel is displaying exactly 1 raster cell. If you zoom in closer than a raster resolution of 1:1, you won't see any more detail in that image.

Tip

Zooming to the raster resolution

Right-click the raster layer in the table of contents and click Zoom to Raster Resolution. You are now looking at your data at 1:1.

Tip

Drawing a raster transparently

Use the Effects toolbar to draw raster layers transparently over other layers on your map.

Displaying the raster resolution in the table of contents

1. In the table of contents, right-click the layer and click Properties.
2. Click the Display tab.
3. Check Display raster resolution in table of contents.
4. Click OK.

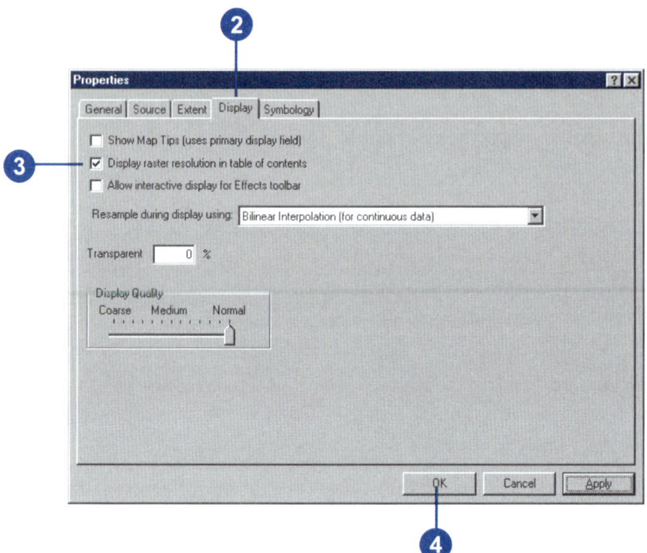

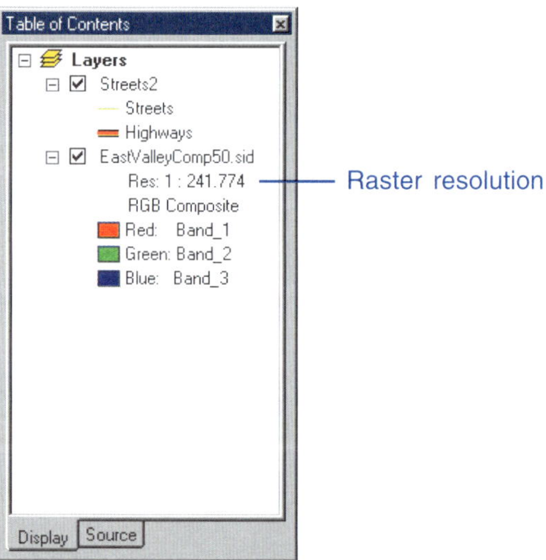

Ways to enhance raster display and efficiency

ArcMap provides alternatives that allow you to enhance the display of your raster layer.

Using faster drawing methods

You can reduce the time it takes to display a large raster dataset by creating *pyramids*. Pyramids record the original data in decreasing levels of resolution, created as a reduced resolution dataset (.rrd) file with the same filename as your raster dataset. ArcMap uses the appropriate level of resolution to quickly draw the entire dataset. You cannot build pyramids on a raster catalog; however, you can build them on each raster dataset within the raster catalog.

When you are working with raster catalogs, ArcMap can display your raster catalog as a wireframe (showing an outline of the dimensions for each raster dataset). To increase display efficiency, this occurs automatically if more than nine images are in the current extent. The default number of images can be manipulated within the Display tab of the raster catalog's Layer Properties dialog box.

Using the Effects toolbar

The Effects toolbar allows you to interactively adjust the brightness or contrast of a raster layer or make the raster layer display transparently. These enhancements are applied to the rendered screen display, not to the original raster dataset values. Brightness increases the overall lightness of the image—for example, making dark colors lighter and light colors whiter—while contrast adjusts the difference between the darkest and lightest colors. The transparency tool lets you see other data layers underneath the raster layer.

Applying contrast stretches

If your raster data represents continuous data, you can apply a contrast stretch to it based upon the statistics of the raster dataset. A stretch increases the visual contrast of the raster display. You might apply a stretch when your raster display appears dark or has little contrast. For example, images may not contain the entire range of values your computer can display; therefore, you could stretch the values of the image to utilize this range by using a contrast stretch. This may result in a crisper image, and some features may become easier to distinguish.

Image A

Image B

Example of a contrast stretch: Histogram A represents the pixel values in Image A. By stretching the values (shown in Histogram B) across the entire range, you can alter and visually enhance the appearance of the image (Image B).

Different stretches will produce different results in the raster display. You can experiment to find the best one for a particular raster dataset.

The types of stretches in ArcMap include a custom manual option or several standard methods. These standard stretches can be used with the RGB Composite or the Stretched renderers. The standard stretches are Standard Deviations, Minimum–

Maximum, Histogram Equalize, and Histogram Specification. You might use the Minimum–Maximum stretch to spread out tightly grouped values. The Histogram Equalize and Histogram Specification stretches obtain their values from your histogram manipulation. A two-standard deviation stretch is the default setting for raster datasets that have statistics, and it is used to brighten raster datasets, which normally appear dark.

With any of these stretch methods, you can examine and modify a histogram and see basic statistics (such as minimum, maximum, mean, and standard deviation) about your data. You could utilize these statistics if you are interested in emphasizing a particular value or examining distribution. When you adjust the histogram, you see multiple sets of vertical bars: the purple bars represent the current display values, and the gray bars represent your original values. When using multiband data in the RGB Composite renderer, the red, green, and blue bars represent the current display values.

Changing the display of the background

Backgrounds and outlines can often be the result of georeferencing your raster dataset. If your raster data has a background, border, or other *NoData* values, you can choose not to display them or choose to display them as a particular color.

These images are showing a NoData area with a black background and that same area using No Color.

Working with auxiliary files and raster proxy files

When a raster dataset is used in ArcGIS and auxiliary information (such as statistics, histograms, and pyramids) cannot be found inside the raster dataset or in an associated auxiliary file, ArcGIS will build and store it. ArcMap only has to build this information once and automatically reads it each time you display the raster dataset. Normally, auxiliary data is stored inside the raster dataset itself if the format allows it, or it can be written to auxiliary files next to the raster dataset.

If a raster dataset is marked with read-only permissions or if the folder it is located in is read-only, then auxiliary files are subsequently written to a different writable location, such as a temporary directory under a folder named "raster proxy files". These are known as raster proxy files.

Faster drawing with pyramids

You can reduce the time it takes to display a large raster dataset by creating pyramids. Pyramids are files that store the original data in decreasing levels of resolution. ArcMap uses the appropriate level of resolution to quickly draw the entire dataset.

As you zoom in, ArcMap displays layers with finer resolution. Performance is maintained because you're drawing successively smaller areas. Without pyramids, ArcMap has to query the entire raster dataset to determine the subset of cells that need to be displayed, thus taking longer to draw.

When you first add a raster dataset of more than 1,024 cells that does not have pyramids present, ArcMap automatically prompts you to create them. The pyramid file created is a reduced resolution dataset (.rrd) with the same filename as the dataset.

For uncompressed raster datasets, the minimum .rrd file size is approximately 8 percent of the size of the original raster dataset. However, depending on the compression technique used in the original raster file, the noncompressed .rrd file can be larger than the original file.

Changing the default setting for building pyramids

1. Click the Tools menu and click Options.
2. Click the Raster tab.
3. Click the General tab.
4. Click the choice that describes when you want to create pyramid layers.
5. Click OK.

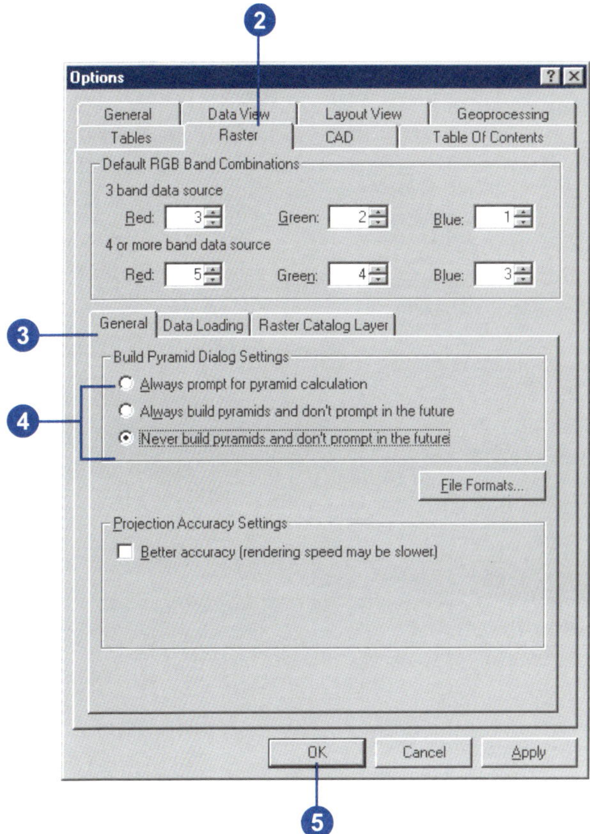

Using the Effects toolbar

The Effects toolbar in ArcMap allows a user to interactively enhance the display of their images. The effects that can be adjusted using slider bars include brightness, contrast, and transparency. Brightness increases the overall lightness of the image. The contrast adjusts the difference between the darkest and lightest colors. Transparency allows the user to see other data layers appearing underneath the raster layer.

Tip

Seeing the results of the Effects toolbar immediately

When you adjust the brightness, contrast, or transparency of your raster dataset, you can immediately see the results as you drag the slider bar. To enable this feature, right-click the raster layer in the table of contents and click Properties. Click the Display tab and check the option to Allow interactive display for Effects toolbar.

Improving the brightness or contrast of your raster layer

1. Click the View menu, point to Toolbars, and click Effects.
2. Click the Layer dropdown arrow and select the raster layer for which you want to change the brightness or contrast.
3. Click the brightness or contrast button.
4. Drag the slider bar to increase or decrease the brightness or contrast.

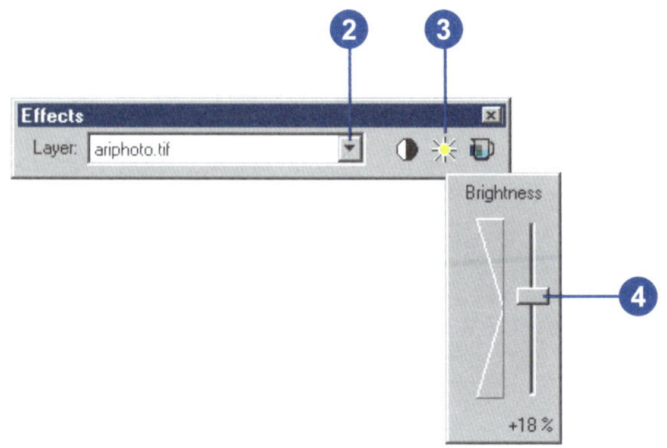

Contrast (left side): increase (top), decrease (bottom).

Brightness (right side): increase (top), decrease (bottom).

Drawing a raster layer transparently

1. Click the View menu, point to Toolbars, and click Effects.
2. Click the Layer dropdown arrow and click the raster layer you want to draw transparently.
3. Click the Adjust Transparency button.
4. Drag the slider bar to adjust the transparency.

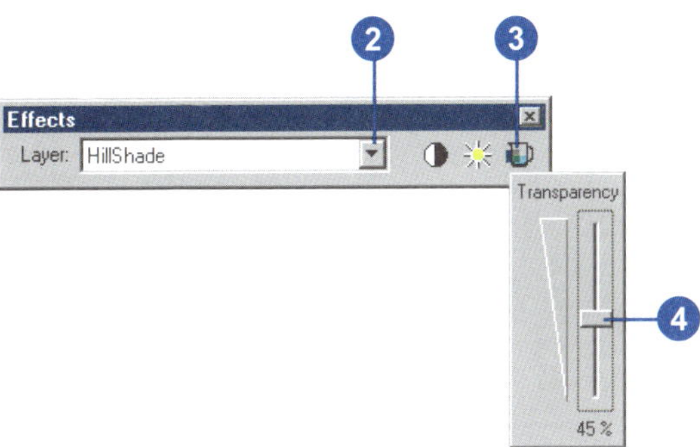

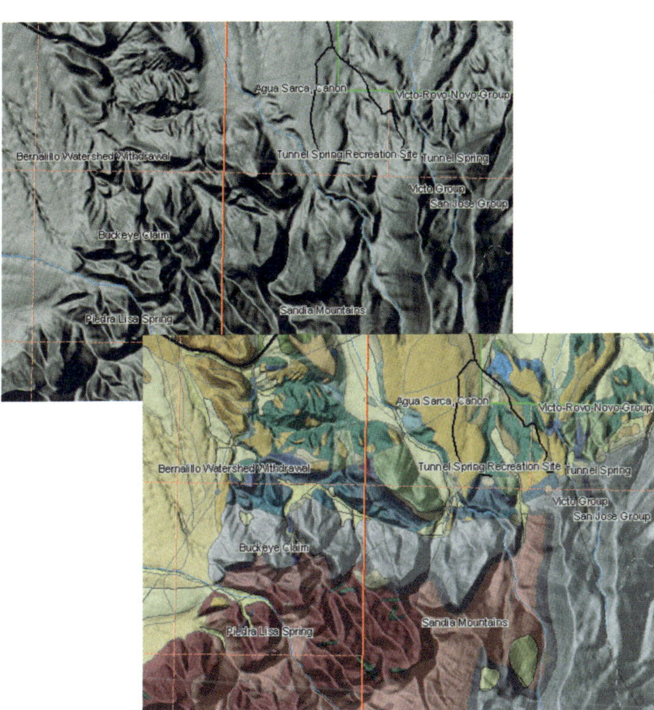

Without transparency (top), the hillshade obscures the underlying land use layer. With transparency (bottom), the underlying symbology appears through the hillshade, yielding a three-dimensional effect.

Applying contrast stretches

You can apply a contrast stretch to your continuous data. A stretch increases the visual contrast of the raster display. You might apply a stretch when your raster display appears dark or has little contrast.

Tip

Displaying the attributes of a cell

Use the Identify tool on the Tools toolbar and click the cell with attributes you're interested in seeing. The attribute information, including the cell value, will display.

See Also

For more information on stretches, see 'The Stretched renderer' in this chapter.

Stretching a raster dataset to improve the visual contrast

1. In the table of contents, right-click the raster layer that you want to display across a color ramp and click Properties.
2. Click the Symbology tab.
3. Click Stretched.
4. If your raster dataset has multiple bands, you can choose the band you want to stretch.
5. Optionally, click the Label boxes and type labels for the table of contents.
6. Click the Color Ramp dropdown arrow and click a color ramp.
7. Optionally, choose a color or no color to display any NoData values.
8. Optionally, click the Stretch Type dropdown arrow and click the type of Stretch, if any, you would like to perform.
9. Click OK.

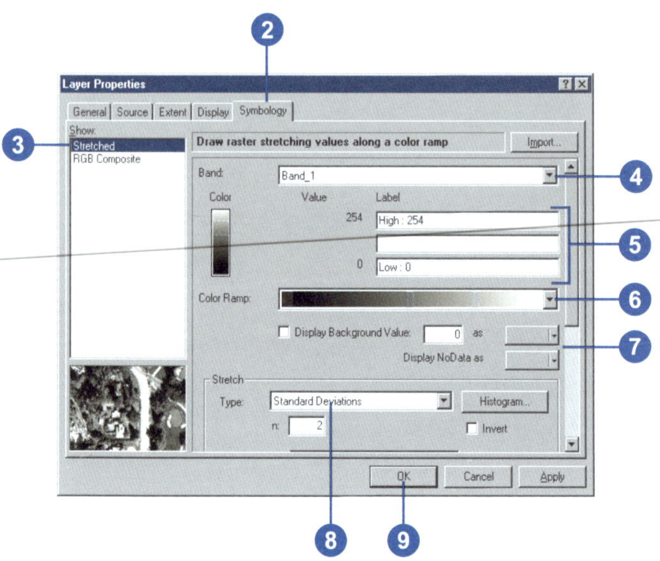

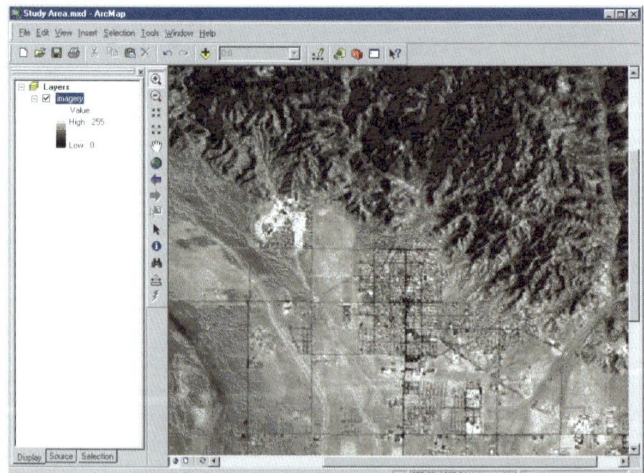

Changing the appearance of background values

Sometimes there are homogenous areas in a raster dataset that the user does not want to display. These can include borders, backgrounds, or other data considered to not have valid values. Sometimes these are expressed as NoData values, although other times they may have real values.

All renderers allow you to set the NoData value to a color or No Color, while the Stretched renderer allows you to identify a specific background value and display color or No Color.

Hiding background values

1. In the table of contents, right-click the raster dataset layer for which you want to increase the visual contrast and click Properties.
2. Click the Symbology tab.
3. Click the renderer you want to use.
4. Check Display Background Value and enter the correct background value in the box. You can also set NoData values as a color or no color.
5. Set the color to No Color.

   This will hide the background or border around the data in the raster layer.

   These cells will display transparently.
6. Click OK.

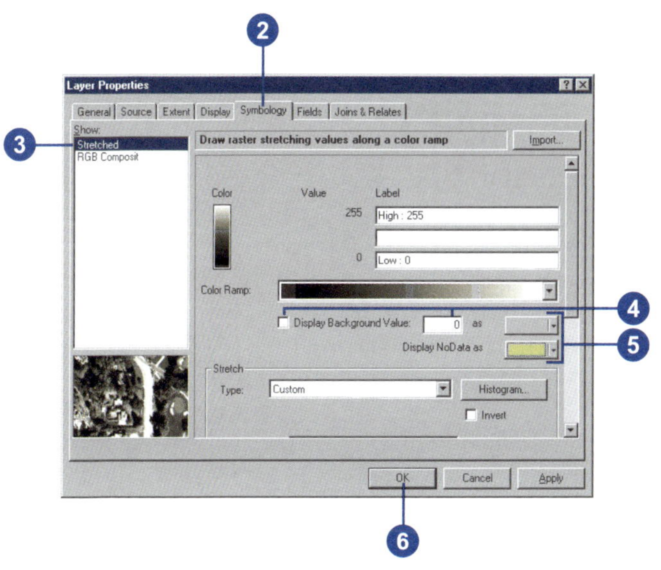

Using the geodatabase raster catalog selection environment

The selection environment allows the user to perform queries based on data within the raster catalog and its relationship to feature layers.

With the selection environment, you could:

- Select a set of raster datasets by cloud cover criteria.
- Select raster datasets by date.
- Select raster datasets based on location to features in a feature layer or graphic.

See Also

To learn more about using SQL to query data sets, see Chapter 13, 'Querying maps'.

See Also

To learn more about managing and working with raster catalogs, see Using ArcCatalog.

Selecting raster datasets by attributes

1. Click the Selection menu and click Select By Attributes.
2. Click the Layer dropdown arrow and click the layer containing the features you want to select.
3. Click the Method dropdown arrow and click a selection method.
4. Double-click a field to add the field name to the expression box.
5. Click an operator to add it to the expression.
6. Click the Get Unique Values button to see all the values.
7. Double-click a value to add it to the expression.
8. To check your syntax or criteria, click the Verify button.
9. Click Apply, then click Close.
10. To view the images, you may have to right-click the raster catalog and select Properties.
11. Click the Selection tab.
12. Check and draw rasters.
13. Click OK.

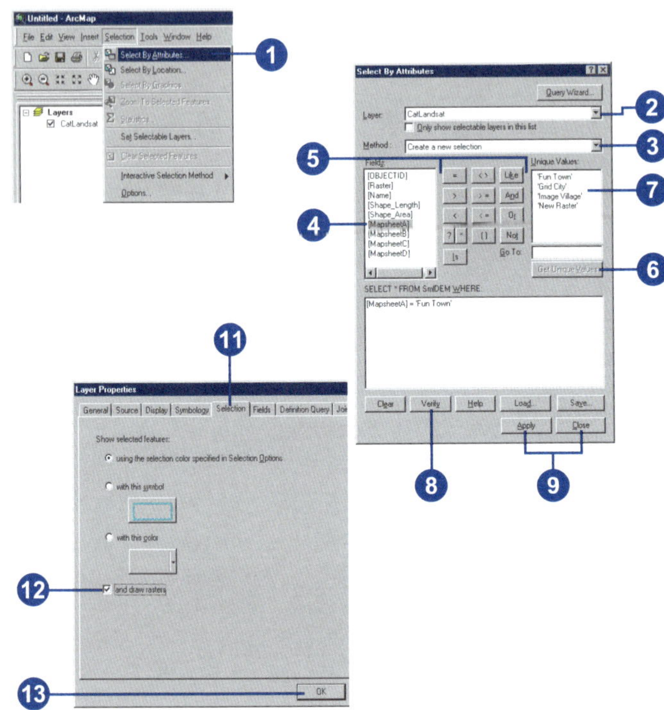

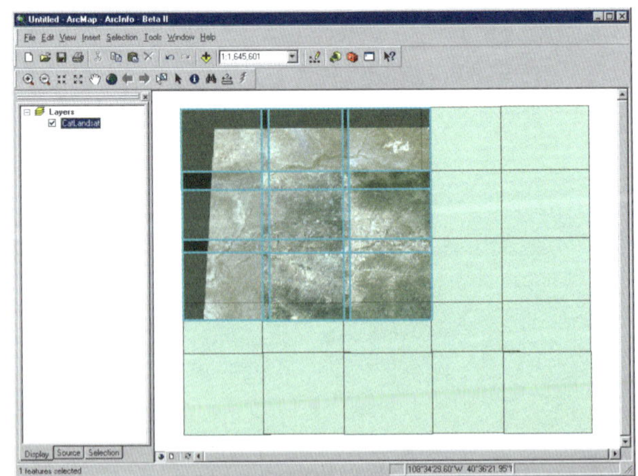

Tip

How do I view the selected raster datasets?

To view the selected raster datasets, you need to right-click the raster catalog in the table of contents and choose Properties. Next, click the Selection tab and click the and draw rasters check box. Click Apply to activate the change.

Selecting a raster catalog by location

1. Click the Selection menu and click Select By Location.
2. Click the first dropdown arrow and click a selection method.
3. Check the catalog layers you want the make the selection from.
4. Click the second dropdown arrow and click a selection method.
5. Click the third dropdown arrow and select the layer you want to use to search for the raster datasets.
6. To use only the selected features, check Use selected features.
7. Optionally, set a buffer by checking the Apply a buffer check box. Type the buffer distance and choose the appropriate units of measure from the dropdown arrow.
8. Click Apply.

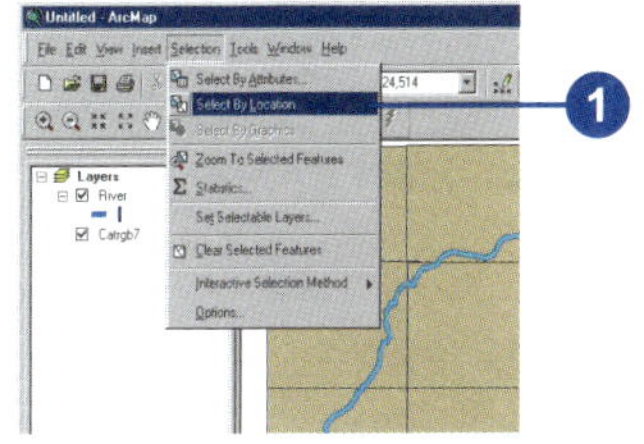

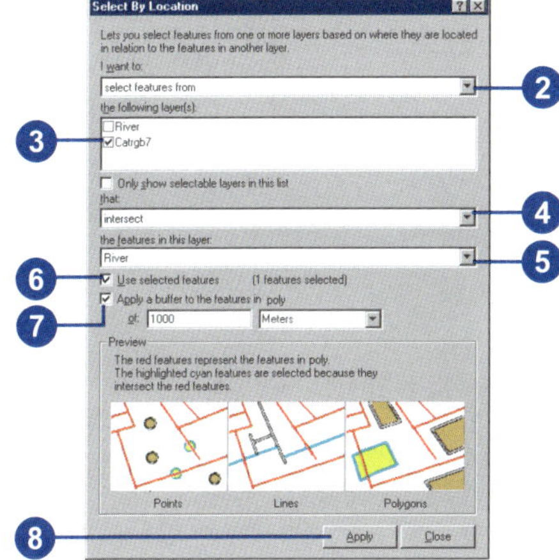

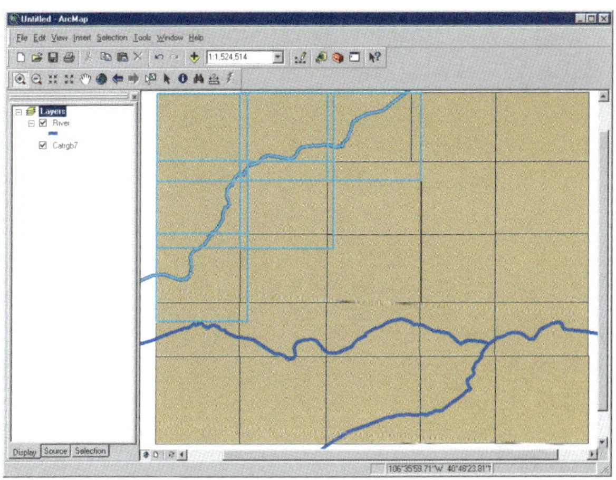

Projecting rasters on the fly

What is an on-the-fly projection?

ArcMap can perform what is commonly known as an on-the-fly projection. This means ArcMap can display data stored in one projection as if it were in another projection. The new projection is used for display and query purposes only. The actual data is not altered.

When is data projected on the fly?

Data is projected on the fly any time a data frame contains a layer whose coordinate system is defined as something different from the coordinate system definition of the data frame. A data frame's coordinate system can be defined manually or by adding data with a defined coordinate system.

ArcMap will not project data on the fly if the coordinate system for the dataset has not been defined. A dataset with an undefined coordinate system will simply be displayed in its native coordinate system. The coordinate system for any dataset can be defined using ArcCatalog.

The first layer added defines the data frame's coordinate system. This is true whether the data is projected or geographic. For example, if the first layer added contains a Lambert Conformal Conic projected coordinate system, all other layers will project on the fly to match this. Similarly, if the first layer added to the data frame contains data that uses a WGS84 geographic coordinate system, all other layers will adjust to match this. Even data that uses a projected coordinate system will unproject on the fly.

Issues with on-the-fly projection

On-the-fly projection uses a single polynomial transformation rather than a cell-to-cell projection. Inaccuracy with on-the-fly projection can be more of a concern for raster layers located at a latitude greater than 70 degrees north or south or for dataset blocks greater than one degree in extent.

Imprecision within projections on the fly can occur because of slight variations within a coordinate system due to its datum model. Other inaccuracies can result from topographic features. For example, a flat terrain area would have less distortion than an area with steep changes in elevation, such as a region with valleys and mountains. Orthorectified imagery corrects for this distortion.

The cells of a raster dataset will always be rectangular and of equal area with respect to the Cartesian coordinate system—map coordinate space—associated with the raster dataset. The shape and area a cell represents on the surface of the earth will never be constant across a raster dataset. Since the area on the face of the earth represented by the cells will vary across the raster dataset, the output cell size and the number of rows and columns may change when projected. Each projection treats the relationship between a three-dimensional world and a two-dimensional one differently. You should be aware of the properties and assumptions for each projection before selecting one.

About georeferencing

Raster data is commonly obtained by scanning maps or collecting aerial photographs and satellite images. Scanned map datasets don't normally contain spatial reference information (either embedded in the file or as a separate file). With aerial photography and satellite imagery, sometimes the locational information delivered with them is inadequate, and the data does not align properly with other data you may have. Thus, in order to use some raster datasets in conjunction with your other spatial data, you often need to align it, or *georeference* it, to a map coordinate system. A map coordinate system is defined using a map projection (a method by which the curved surface of the earth is portrayed on a flat surface).

When you georeference your raster dataset, you define its location using map coordinates and assign a coordinate system. Georeferencing raster data allows it to be viewed, queried, and analyzed with other geographic data.

How to align the raster dataset

Generally, you will georeference your raster dataset using existing spatial data (target data), such as a vector feature class, that resides in the desired map coordinate system. This assumes that there are features in your spatial data that are also visible in the raster—for example, street intersections or building corners.

The process involves identifying a series of *ground control points*—known x,y coordinates—that link locations on the raster dataset with locations in the spatially referenced data (target data). The control points are used to build a polynomial transformation that will convert the raster dataset from its existing location to the spatially correct location. The connection between one control point on the raster dataset (the "from point") and the corresponding control point on the aligned target data (the "to point") is called a link.

The number of links you need to create depends on the complexity of the polynomial transformation you plan to use to transform the raster dataset to map coordinates. However, adding more links will not necessarily yield a better registration. If possible, you should spread out the links over the entire raster dataset rather than concentrating them in one area. Typically, having at least one link near each corner of the raster dataset and a few throughout the interior produces the best results.

Generally, the greater the overlap between the raster dataset and target data, the better the alignment results, because you'll have more widely spaced points with which to georeference the raster dataset. For example, if your target data only occupies one quarter of the area of your raster dataset, the points you could use to align the raster dataset would be confined to that area of overlap. Thus, the areas outside the overlap area are likely not properly aligned.

Keep in mind that your georeferenced data is only as accurate as the data to which it was aligned. To minimize errors, you should georeference to data that is at the highest resolution and largest scale for your needs.

Transforming the raster dataset

When you've created enough links, you can transform—or *warp*—the raster dataset to permanently match the map coordinates of the target data. Warping uses a polynomial transformation to determine the correct map coordinate location for each cell in the raster.

Use a first-order—or affine—transformation to shift, scale, and rotate your raster dataset. This generally results in straight lines on the raster dataset mapped as straight lines in the warped raster dataset. Thus squares and rectangles on the raster dataset are commonly changed into parallelograms of arbitrary scaling and angle orientation.

A first-order transformation will probably handle most of your georeferencing requirements. With a minimum of three links, the mathematical equation used with a first-order transformation can exactly map each raster point to the target location. Any more than three links introduces errors, or residuals, that are distributed throughout all the links. However, you should add more than three links because if one link is positionally wrong, it has a much greater impact on the transformation. Thus, even though the mathematical transformation error may increase as you create more links, the overall accuracy of the transformation will increase as well.

The higher the transformation order, the more complex the distortion that can be corrected. However, transformations higher than third order are rarely needed. Higher-order transformations require more links and thus will involve progressively more processing time. In general, if your raster dataset needs to be stretched, scaled, and rotated, use a first-order transformation. If, however, the raster dataset must be bent or curved, use a second- or third-order transformation.

Interpreting the root mean square error

The degree to which the transformation can accurately map all control points can be measured mathematically by comparing the actual location of the map coordinate to the transformed position in the raster. The distance between these two points is known as the residual error. The total error is computed by taking the root mean square (RMS) sum of all the residuals to compute the RMS error. This value describes how consistent the transformation is between the different control points (links). Links can be removed if the error is particularly large, and more points can be added.

While the RMS error is a good assessment of the accuracy of the transformation, don't confuse a low RMS error with an accurate registration. For example, the transformation may still contain significant errors due to a poorly entered control point.

Resampling the raster dataset

While you might think each cell in a raster dataset is transformed to its new map coordinate location, the process actually works in reverse. During georeferencing, a matrix of "empty" cells is computed using the map coordinates. Then, each empty cell is given a value based on a process called *resampling*.

The three most common resampling techniques are nearest neighbor assignment, bilinear interpolation, and cubic convolution. These techniques assign a value to each empty cell by examining the cells in the ungeoreferenced raster dataset. Nearest neighbor assignment takes the value from the cell closest to the transformed cell as the new value. It's the fastest resampling technique and is appropriate for categorical or thematic data. Bilinear interpolation and cubic convolution techniques combine a greater number of nearby cells (4 and 16, respectively) to compute the value for the transformed cell. These two techniques use a weighted averaging method to compute the output transformed cell value and thus are only appropriate for continuous data such as elevation, slope, aerial photography, and other continuous surfaces.

Resampling is part of the georeferencing process; however, it can also be used to adjust the cell size of a raster dataset or can be part of a geoprocessing operation to alter the values of cells.

Should you rectify your raster dataset?

You can permanently transform your raster dataset after georeferencing it by using the Rectify command on the Georeferencing toolbar. Rectify creates a new raster dataset that is georeferenced to map coordinates. You can save this in ESRI GRID, TIFF, or ERDAS® IMAGINE® format.

ArcMap doesn't require you to rectify your raster dataset to display it with other spatial data. You might choose to rectify your raster dataset if you plan to perform analysis with it or want to use it with another software package that doesn't recognize the external georeferencing information created by ArcMap.

The Georeferencing toolbar

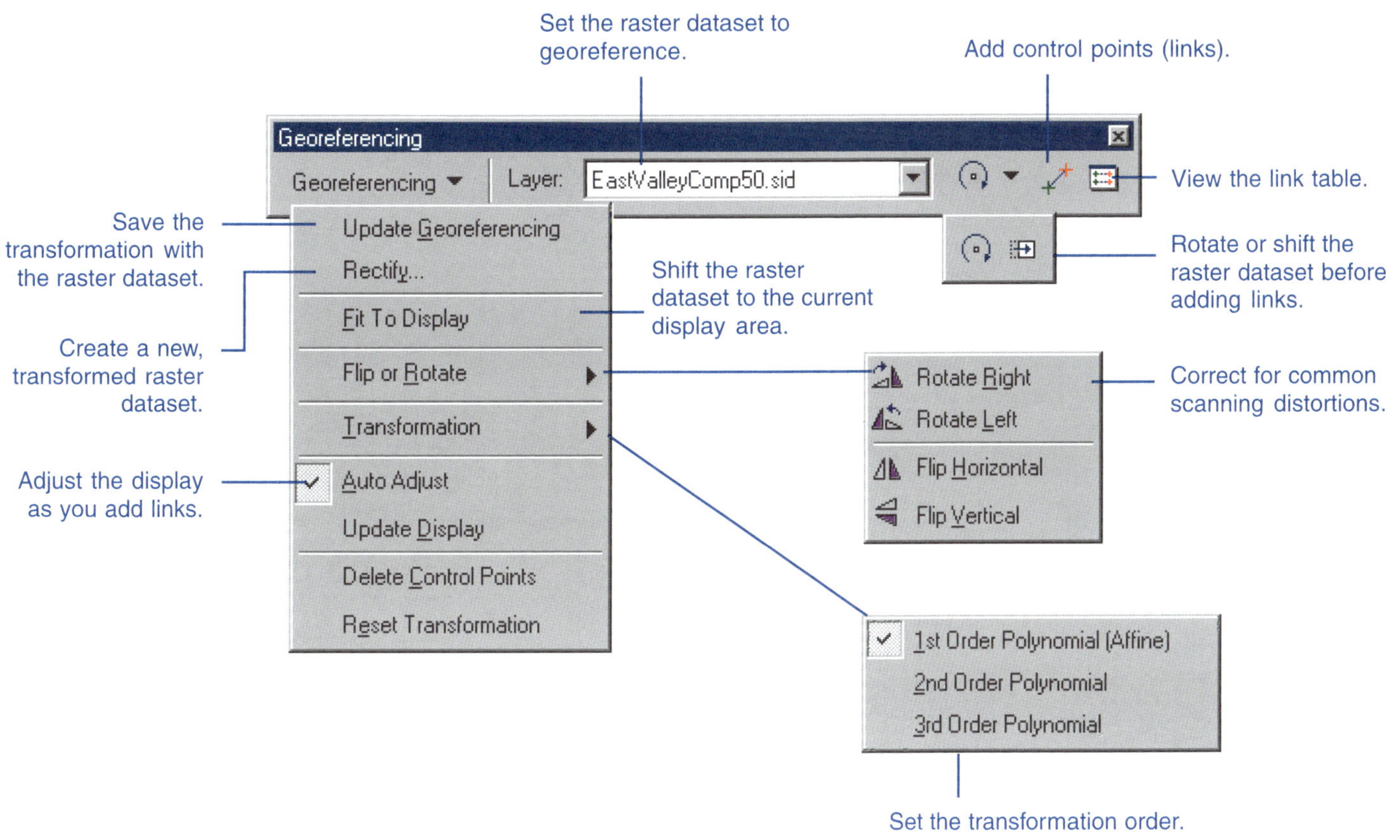

Georeferencing a raster dataset

The general steps for georeferencing a raster dataset are:

1. Add the raster dataset that you want to align with your projected data.
2. Add control points that link known raster dataset positions to known positions in map coordinates.
3. Save the georeferencing information when you're satisfied with the registration.

Tip

Displaying the Georeferencing toolbar

Right-click the Tools menu and point to Georeferencing.

Georeferencing a raster dataset

1. Add the layers residing in map coordinates and the raster dataset you want to georeference.
2. In the table of contents, right-click a target layer (the referenced dataset) and click Zoom to Layer.
3. From the Georeferencing toolbar, click the Layer dropdown arrow and click the raster layer you want to georeference.
4. Click Georeferencing and click Fit To Display.

   This will display the raster dataset in the same area as the target layers. You can also use the Shift and Rotate tools to move the raster dataset as needed. To see all the datasets, you may have to adjust their order in the table of contents.
5. Click the Control Points tool to add control points.
6. To add a link, click the mouse pointer on a known location on the raster dataset, then on a known location on the data in map coordinates (the referenced data). ▸

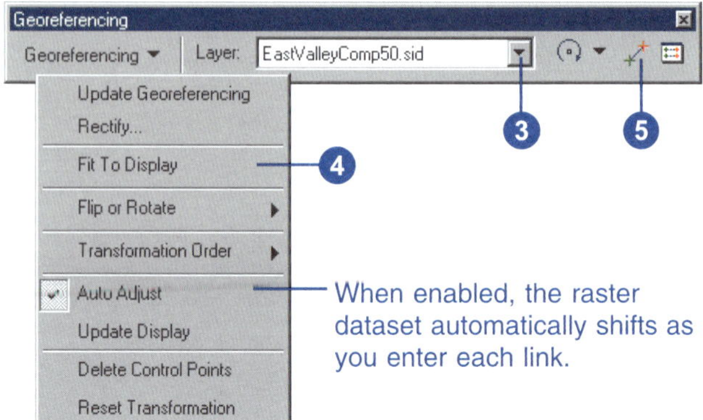

When enabled, the raster dataset automatically shifts as you enter each link.

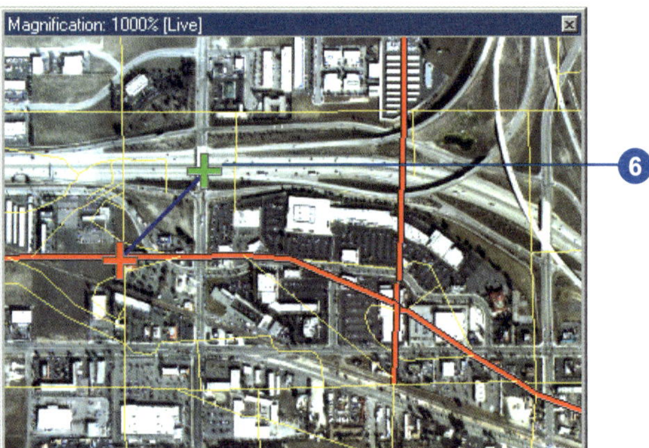

To create a link, click a control point on the raster dataset, then click the corresponding control point on the target data.

Tip

What features could I use as control points?

You could look for road intersections, land features, building corners, or other objects that you can identify and match in your raster dataset and aligned datasets.

Tip

Deleting a link

You can delete an unwanted link from the Link Table dialog box. Press the Esc key to remove a link while you're in the middle of creating it.

Tip

Using the georeferencing functions

The Rotate and Shift tools are no longer available after you place the first link.

Tip

Transforming the raster dataset permanently

You can permanently transform your raster dataset after georeferencing by using the Rectify command. Click Georeferencing and click Rectify.

You may find it useful to use a Magnification window to add in your links. When working with two raster datasets, you may want to open the Effects toolbar and adjust the transparency or turn layers on and off in the table of contents to view each image as you add your links.

7. Add enough links for the transformation order. You need a minimum of three links for a first-order transformation, six links for a second order, and 10 links for a third order.

8. Click View Link Table to evaluate the transformation.

   You can examine the residual error for each link and the RMS error. If you're satisfied with the registration, you can stop entering links.

9. Click Georeferencing and click Update Georeferencing to save the transformation information with the raster dataset.

   This creates a new file with the same name as the raster dataset, but with an .aux file extension. It also creates a world file for .tif and .img files.

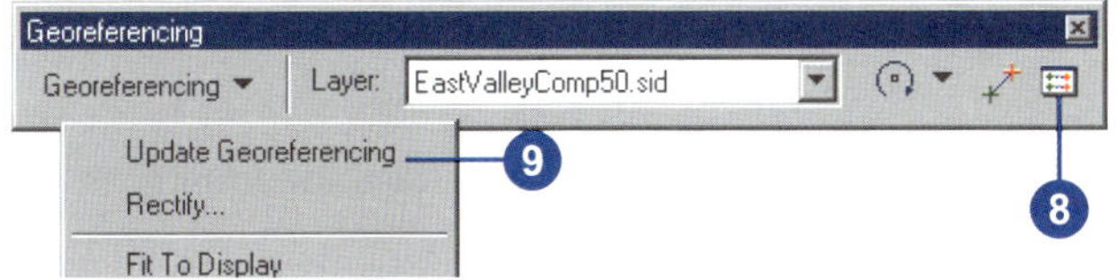

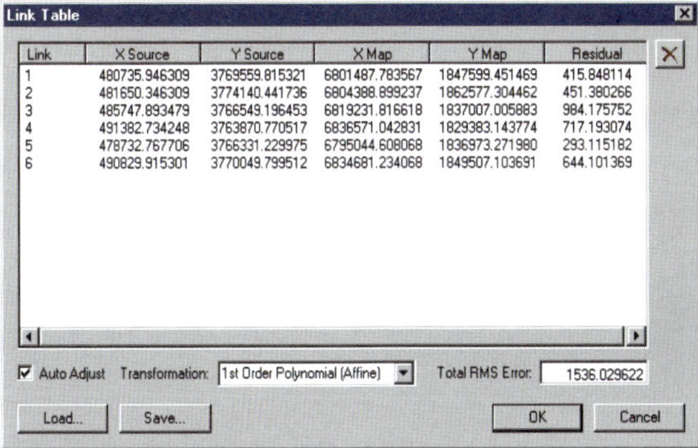

Link Table

Link	X Source	Y Source	X Map	Y Map	Residual
1	480735.946309	3769559.815321	6801487.783567	1847599.451469	415.848114
2	481650.346309	3774140.441736	6804388.899237	1862577.304462	451.380266
3	485747.893479	3766549.196453	6819231.816618	1837007.005883	984.175752
4	491382.734248	3763870.770517	6836571.042831	1829383.143774	717.193074
5	478732.767706	3766331.229975	6795044.608068	1836973.271980	293.115182
6	490829.915301	3770049.799512	6834681.234068	1849507.103691	644.101369

Auto Adjust Transformation: 1st Order Polynomial (Affine) Total RMS Error: 1536.029622

Load... Save... OK Cancel

Evaluate the links.

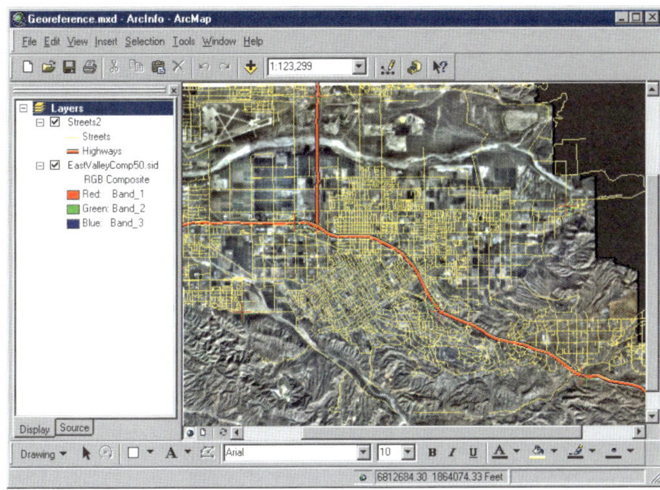

After updating georeferencing information, the raster dataset will align to other spatial data when added to a map.

Tip

What is my georeferenced raster dataset's coordinate system?

The coordinate system assigned to the raster dataset is the same as the coordinate system defined for the data frame.

Tip

Adding explicit values while georeferencing

You can add explicit values after you have clicked a control point on your raster dataset. Just right-click on your map and choose Input X and Y.

Entering explicit x,y map coordinates

1. Click the View Link Table button on the Georeferencing toolbar.
2. Click the Control Points tool.
3. Click the mouse on a known location in the unreferenced image to add the first coordinate in the link.
4. Right-click the image and choose Input X and Y.
5. Enter the reference coordinates in the Enter Coordinates dialog box.
6. Click OK.

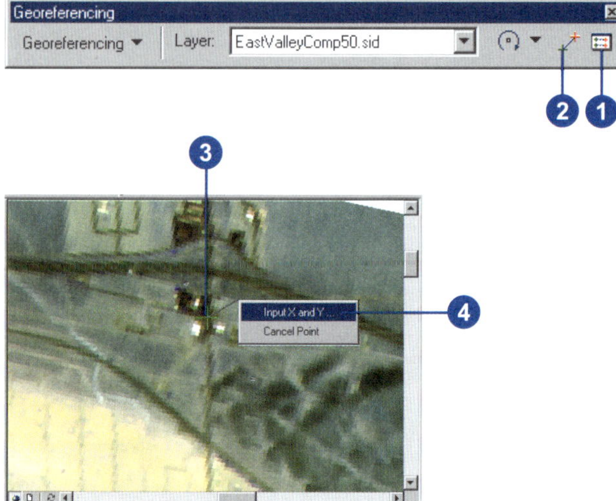

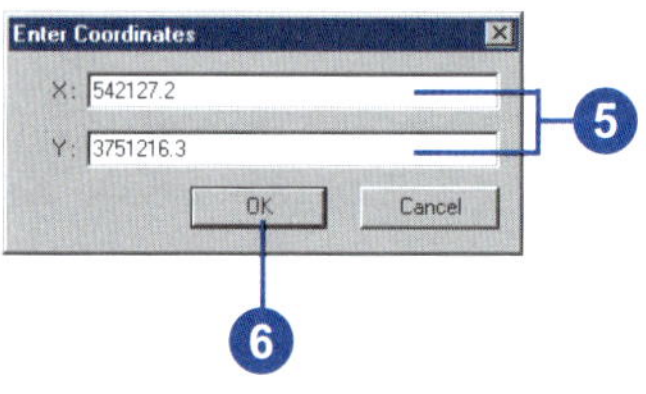

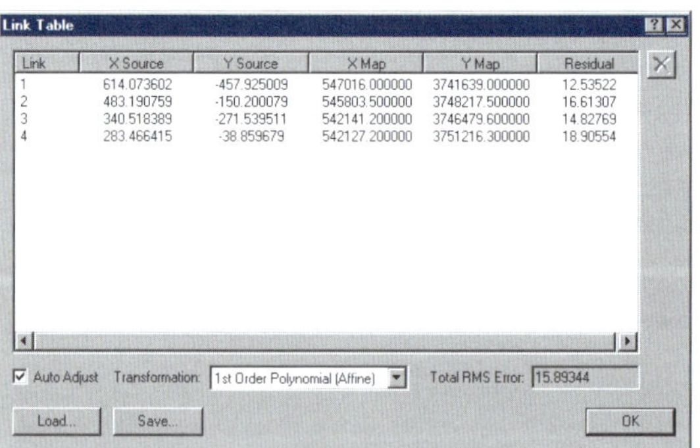

Querying data

Section 3

Working with tables

10

IN THIS CHAPTER

A *table* is a database component that contains a series of rows and columns, where each row, or record, represents a geographic feature—such as a parcel, power pole, highway, or lake—and each column, or *field*, describes a particular attribute of the feature—such as its length, depth, cost, and so on. Tables are stored in a database—for example, INFO™, Microsoft Access, dBASE®, FoxPro®, Oracle®, and SQL Server™.

You'll typically use tables in ArcMap to inspect the attributes of geographic features. From a table, you can identify features with particular attributes and select them on the map. Over time, you might also update the attributes to reflect changes to geographic features—for example, a new subdivision extends your parcel database, or the construction of a dam alters a river network.

Tables can also store information related to features such as warehouse inventories, monthly sales figures, and maintenance records. By joining this information to your spatial data, you can uncover new patterns and relationships that were not apparent before. For example, you might see which stores have the top monthly sales figures, what roads require maintenance in the near future, or where the largest number of endangered species is located.

Elements of a table

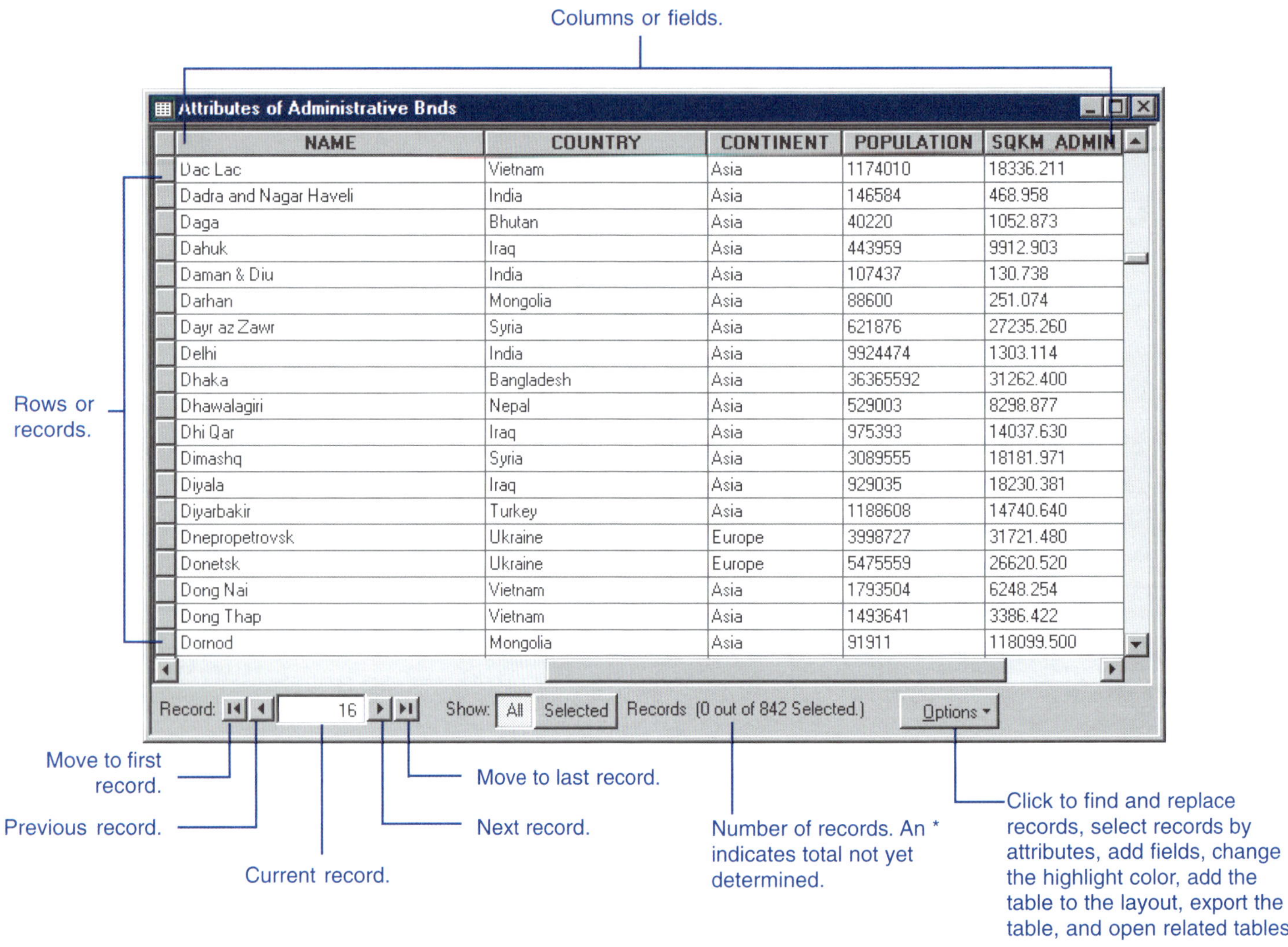

NAME	COUNTRY	CONTINENT	POPULATION	SQKM_ADMIN
Dac Lac	Vietnam	Asia	1174010	18336.211
Dadra and Nagar Haveli	India	Asia	146584	468.958
Daga	Bhutan	Asia	40220	1052.873
Dahuk	Iraq	Asia	443959	9912.903
Daman & Diu	India	Asia	107437	130.738
Darhan	Mongolia	Asia	88600	251.074
Dayr az Zawr	Syria	Asia	621876	27235.260
Delhi	India	Asia	9924474	1303.114
Dhaka	Bangladesh	Asia	36365592	31262.400
Dhawalagiri	Nepal	Asia	529003	8298.877
Dhi Qar	Iraq	Asia	975393	14037.630
Dimashq	Syria	Asia	3089555	18181.971
Diyala	Iraq	Asia	929035	18230.381
Diyarbakir	Turkey	Asia	1188608	14740.640
Dnepropetrovsk	Ukraine	Europe	3998727	31721.480
Donetsk	Ukraine	Europe	5475559	26620.520
Dong Nai	Vietnam	Asia	1793504	6248.254
Dong Thap	Vietnam	Asia	1493641	3386.422
Dornod	Mongolia	Asia	91911	118099.500

Opening a layer's attribute table

To explore the attributes of a layer on a map, open its *attribute table*. Once open, you can select features and find features with particular attributes.

You can open more than one table at a time. For example, you can view an attribute table of administrative boundaries and, at the same time, view the attribute table for cities.

Tip

Closing a table

You can close a table by clicking the Close button in the upper-right corner of the table window.

1. In the table of contents, right-click the layer for which you want to display a table.
2. Click Open Attribute Table.

   The layer's attribute table opens.

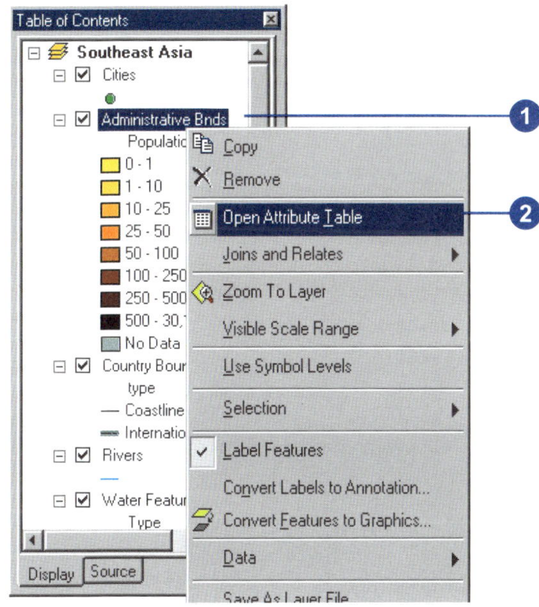

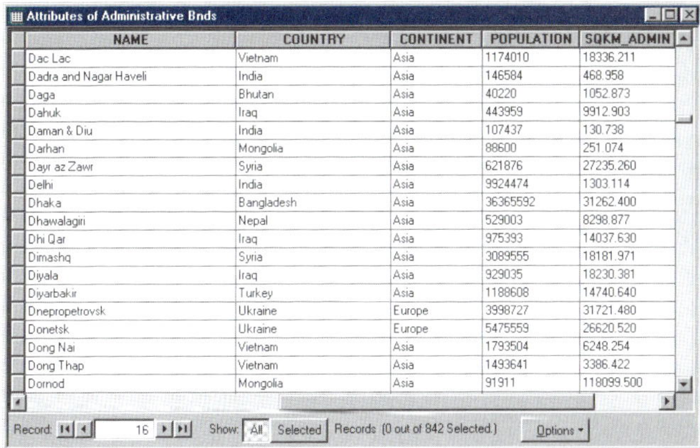

Attributes of Administrative Bnds

NAME	COUNTRY	CONTINENT	POPULATION	SQKM_ADMIN
Dac Lac	Vietnam	Asia	1174010	18336.211
Dadra and Nagar Haveli	India	Asia	146584	468.958
Daga	Bhutan	Asia	40220	1052.873
Dahuk	Iraq	Asia	443959	9912.903
Daman & Diu	India	Asia	107437	130.738
Darhan	Mongolia	Asia	88600	251.074
Dayr az Zawr	Syria	Asia	621876	27235.260
Delhi	India	Asia	9924474	1303.114
Dhaka	Bangladesh	Asia	36365592	31262.400
Dhawalagiri	Nepal	Asia	529003	8298.877
Dhi Qar	Iraq	Asia	975393	14037.630
Dimashq	Syria	Asia	3089555	18181.971
Diyala	Iraq	Asia	929035	18230.381
Diyarbakir	Turkey	Asia	1188608	14740.640
Dnepropetrovsk	Ukraine	Europe	3998727	31721.480
Donetsk	Ukraine	Europe	5475559	26620.520
Dong Nai	Vietnam	Asia	1793504	6248.254
Dong Thap	Vietnam	Asia	1493641	3386.422
Dornod	Mongolia	Asia	91911	118099.500

Record: 16 Show: All Selected Records (0 out of 842 Selected.) Options

Loading existing tabular data onto a map

Not all the *tabular data* associated with a layer has to be stored in its attribute table. Some data you may choose to store in separate tables—for instance, data that changes frequently such as monthly sales figures. You can add this tabular data directly to your map as a table and use it in conjunction with the layers on your map. These tables don't display on your map, but they are listed in the table of contents on the Source tab. You work with these tables as you would any table based on geographic features. For example, you can view the table, add new fields, create graphs, and join the table to other tables.

See Also

Joining a table to a layer allows you to visualize the information contained in a table. For more information, see 'Joining attribute tables' in this chapter.

1. Click the Add Data button.
2. Navigate to the table you want to add and click it.
3. Click Add.
4. Click the Source tab at the bottom of the table of contents.
5. Right-click the table and click Open.

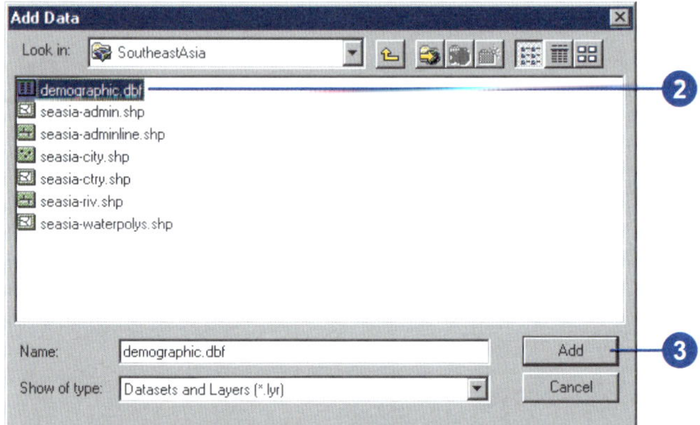

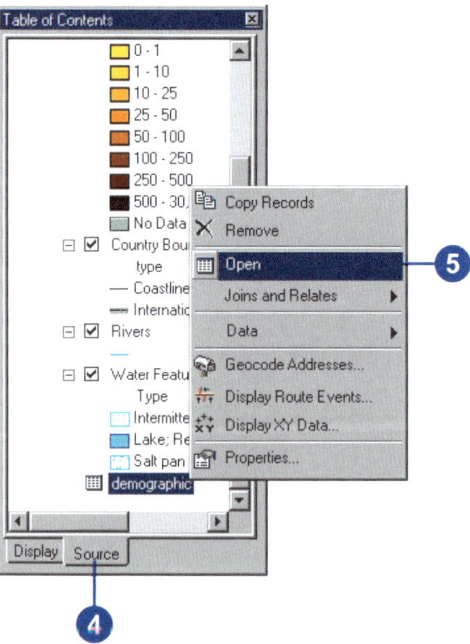

Arranging columns

When you open a table, you can rearrange its appearance. For example, you may want to widen or reduce the width of the visible columns, move a column, hide a column from being displayed at all, sort the table based on a selected field or fields, or freeze a field so you always see it as you scroll across the table.

Freezing a column is helpful when a table has many columns. It is often useful to see how the values in a certain column relate to the data in the rest of a table.

Freezing a column locks a column as the leftmost column in the table view. You can then use the horizontal scroll bar to see the other columns in the table. When you scroll, the frozen column remains in place while all other columns move. A frozen column is easily identified because it has a thick black line separating it from the other columns in the table.

Changing a column's width

1. Position the mouse pointer at the edge of the column you want to resize.

   The pointer's icon changes.

2. Click and drag the column's edge to the desired width.

   A black line indicates where the edge of the column will be located.

3. Drop the edge of the column.

   The column is resized.

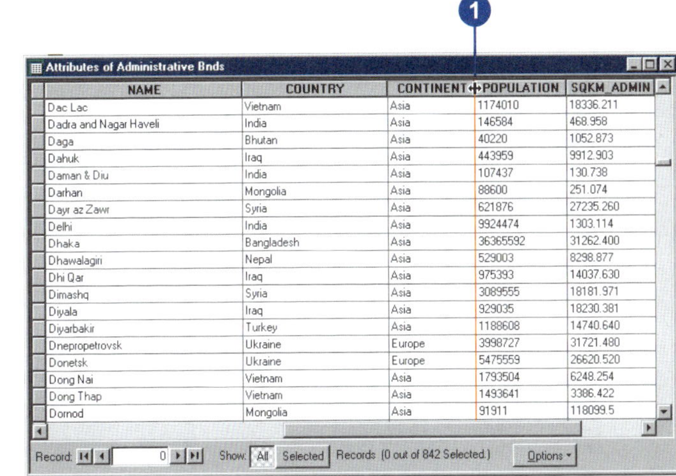

NAME	COUNTRY	CONTINENT	POPULATION	SQKM_ADMIN
Dac Lac	Vietnam	Asia	1174010	18336.211
Dadra and Nagar Haveli	India	Asia	146584	468.958
Daga	Bhutan	Asia	40220	1052.873
Dahuk	Iraq	Asia	443959	9912.903
Daman & Diu	India	Asia	107437	130.738
Darhan	Mongolia	Asia	88600	251.074
Dayr az Zawr	Syria	Asia	621876	27235.260
Delhi	India	Asia	9924474	1303.114
Dhaka	Bangladesh	Asia	36365592	31262.400
Dhawalagiri	Nepal	Asia	529003	8298.877
Dhi Qar	Iraq	Asia	975393	14037.630
Dimashq	Syria	Asia	3089555	18181.971
Diyala	Iraq	Asia	929035	18230.381
Diyarbakir	Turkey	Asia	1188608	14740.640
Dnepropetrovsk	Ukraine	Europe	3998727	31721.480
Donetsk	Ukraine	Europe	5475559	26620.520
Dong Nai	Vietnam	Asia	1793504	6248.254
Dong Thap	Vietnam	Asia	1493641	3386.422
Dornod	Mongolia	Asia	91911	118099.5

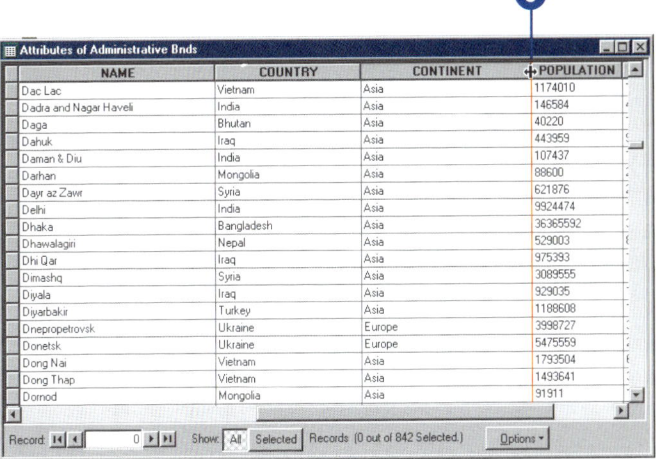

NAME	COUNTRY	CONTINENT	POPULATION
Dac Lac	Vietnam	Asia	1174010
Dadra and Nagar Haveli	India	Asia	146584
Daga	Bhutan	Asia	40220
Dahuk	Iraq	Asia	443959
Daman & Diu	India	Asia	107437
Darhan	Mongolia	Asia	88600
Dayr az Zawr	Syria	Asia	621876
Delhi	India	Asia	9924474
Dhaka	Bangladesh	Asia	36365592
Dhawalagiri	Nepal	Asia	529003
Dhi Qar	Iraq	Asia	975393
Dimashq	Syria	Asia	3089555
Diyala	Iraq	Asia	929035
Diyarbakir	Turkey	Asia	1188608
Dnepropetrovsk	Ukraine	Europe	3998727
Donetsk	Ukraine	Europe	5475559
Dong Nai	Vietnam	Asia	1793504
Dong Thap	Vietnam	Asia	1493641
Dornod	Mongolia	Asia	91911

Tip

Deselecting a column

To deselect all columns in the table, click Options and click Clear Selection or simply click a cell in the table.

Tip

Changing the selection color

By default, selected columns are highlighted in cyan. To change the selection color, click the Options button on the table, click Appearance, then click the color you prefer.

Rearranging a table's columns

1. Click the heading of the column you want to move.
2. Click and drag the column's heading.

   A red line indicates where the column will be positioned.
3. Drop the column.

   After you drop it, the column appears in the new position.

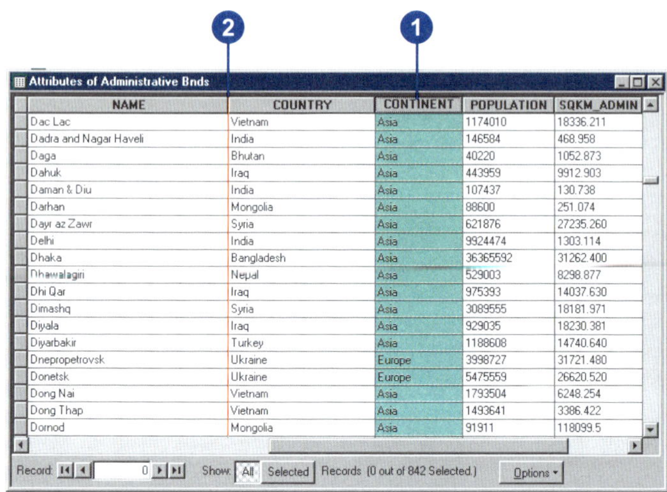

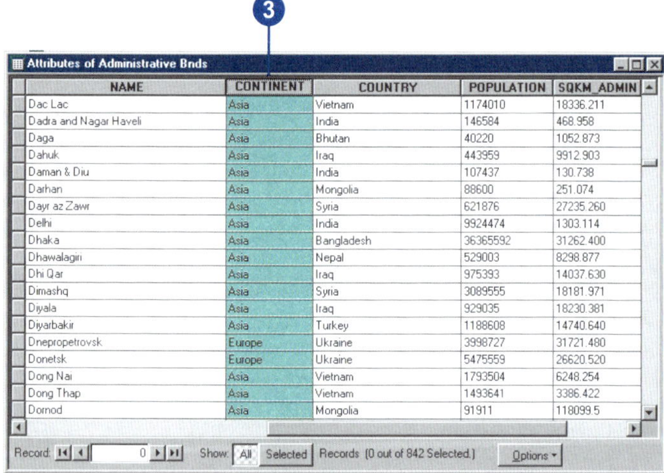

Tip

Unfreezing a column

Right-click the column heading and click Freeze/Unfreeze Column to unfreeze the column.

Tip

Hiding a column

Right-click the layer or table in the table of contents and click Properties. Click the Fields tab. Here you can set whether a field is visible or not.

Freezing a column

1. Click the heading of the column(s) you want to freeze.
2. Right-click the selected column's heading and click Freeze/Unfreeze Column to freeze the column.

   The column is now frozen.

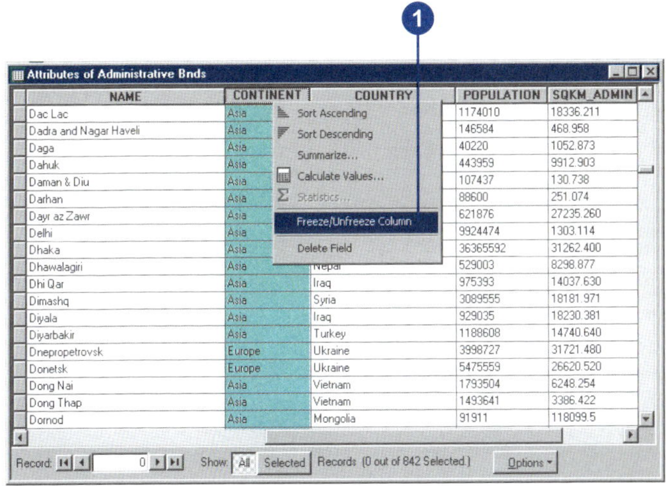

NAME	CONTINENT	COUNTRY	POPULATION	SQKM_ADMIN
Dac Lac	Asia		1174010	18336.211
Dadra and Nagar Haveli	Asia		146584	468.958
Daga	Asia		40220	1052.873
Dahuk	Asia		443959	9912.903
Daman & Diu	Asia		107437	130.738
Darhan	Asia		88600	251.074
Dayr az Zawr	Asia		621876	27235.260
Delhi	Asia		9924474	1303.114
Dhaka	Asia		36365592	31262.400
Dhawalagiri	Asia	Nepal	529003	8298.877
Dhi Qar	Asia	Iraq	975393	14037.630
Dimashq	Asia	Syria	3089555	18181.971
Diyala	Asia	Iraq	929035	18230.381
Diyarbakir	Asia	Turkey	1188608	14740.640
Dnepropetrovsk	Europe	Ukraine	3998727	31721.480
Donetsk	Europe	Ukraine	5475559	26620.520
Dong Nai	Asia	Vietnam	1793504	6248.254
Dong Thap	Asia	Vietnam	1493641	3386.422
Dornod	Asia	Mongolia	91911	118099.5

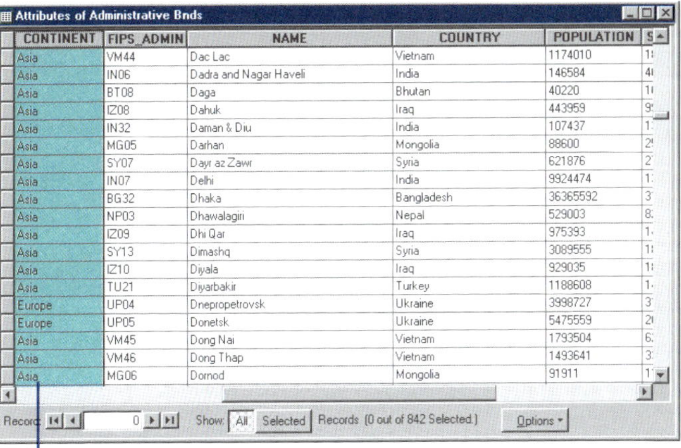

CONTINENT	FIPS_ADMIN	NAME	COUNTRY	POPULATION
Asia	VM44	Dac Lac	Vietnam	1174010
Asia	IN06	Dadra and Nagar Haveli	India	146584
Asia	BT08	Daga	Bhutan	40220
Asia	IZ08	Dahuk	Iraq	443959
Asia	IN32	Daman & Diu	India	107437
Asia	MG05	Darhan	Mongolia	88600
Asia	SY07	Dayr az Zawr	Syria	621876
Asia	IN07	Delhi	India	9924474
Asia	BG32	Dhaka	Bangladesh	36365592
Asia	NP03	Dhawalagiri	Nepal	529003
Asia	IZ09	Dhi Qar	Iraq	975393
Asia	SY13	Dimashq	Syria	3089555
Asia	IZ10	Diyala	Iraq	929035
Asia	TU21	Diyarbakir	Turkey	1188608
Europe	UP04	Dnepropetrovsk	Ukraine	3998727
Europe	UP05	Donetsk	Ukraine	5475559
Asia	VM45	Dong Nai	Vietnam	1793504
Asia	VM46	Dong Thap	Vietnam	1493641
Asia	MG06	Dornod	Mongolia	91911

The column has been frozen.

Controlling a table's appearance

You can tailor the look of the table window to suit your needs. For example, if you don't like the default table font, you can change it to another one and set the font size as well. Make changes for all tables or just one—each table can have its own individual settings.

You can also set the selection and highlight color for records in a table. Selected records are displayed with the selection color; the highlight color identifies a record when you're only viewing the selected records in the table.

Formatting a field can also enhance the look of a table. For example, you might set the number digits to display to the right of a decimal point or use scientific notation.

Setting the text font and size for a table

1. On the table window, click Options and click Appearance.
2. Click the Table Font dropdown arrow and click the font you want to use.
3. Click the Table Font Size dropdown arrow and click a point size.
4. Click OK.

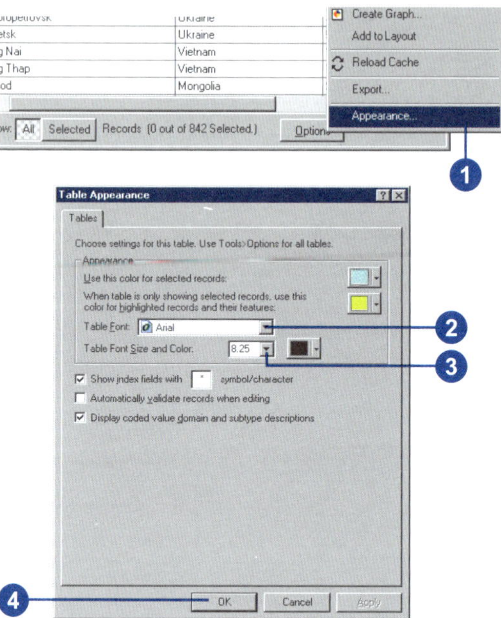

Setting the default text font and size for all tables

1. Click the Tools menu and click Options.
2. Click the Tables tab.
3. Click the Table Font dropdown arrow and click the font you want to use.
4. Click the Table Font Size dropdown arrow and click a point size.
5. Click OK.

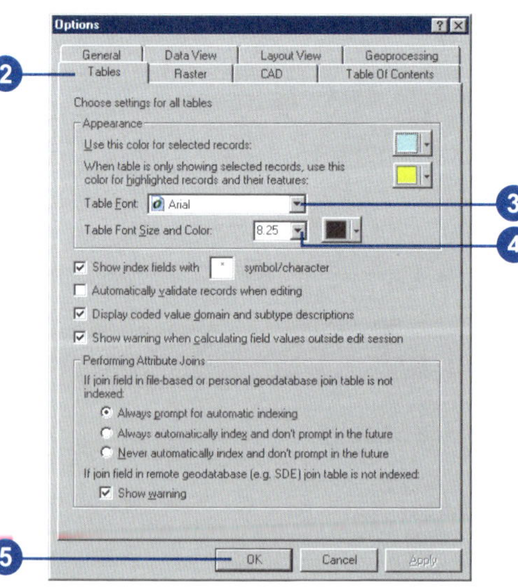

Tip

Selection vs. highlight color

The selection color shows the selected records in a table. When you're only viewing the selected records, clicking one highlights the feature with the highlight color.

Setting the selection and highlight color for a table

1. On the table window, click Options and click Appearance.
2. Click the selected records dropdown arrow and click the color you want to use.
3. Click the highlighted records dropdown arrow and click the color you want to use.
4. Click OK.

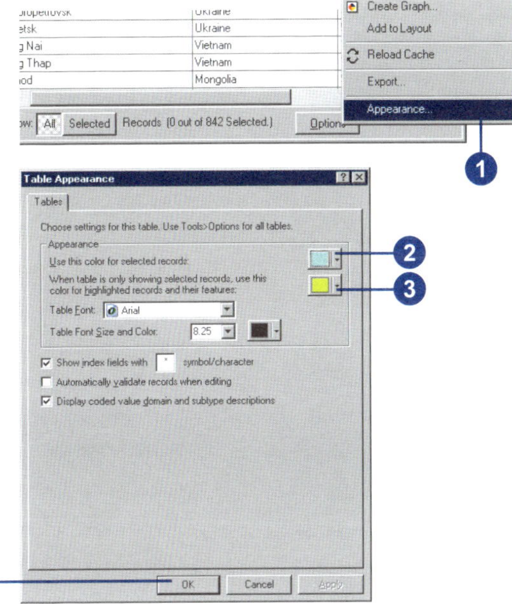

Setting the default selection and highlight color for all tables

1. Click the Tools menu and click Options.
2. Click the Tables tab.
3. Click the selected records dropdown arrow and click the color you want to use.
4. Click the highlighted records dropdown arrow and click the color you want to use.
5. Click OK.

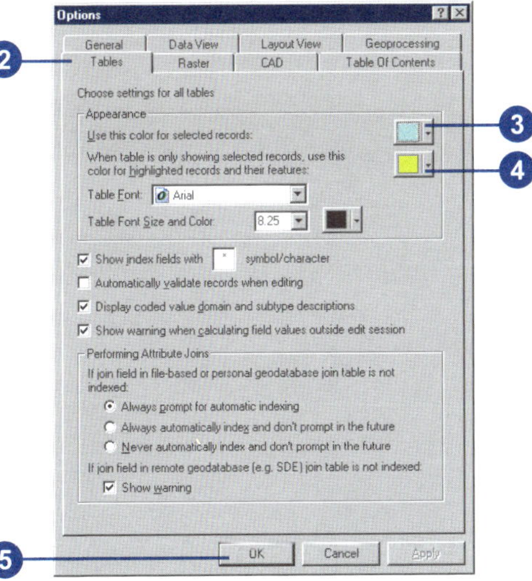

Formatting numeric fields

1. Right-click the layer or table in the table of contents and click Properties.
2. Click the Fields tab.
3. Click a numeric field in the list.
4. Click the button in the Number Format column.

   Only numeric fields have this button.
5. Set the number of decimal places, alignment, and so on.
6. Click OK on the Number Format dialog box.
7. Click OK when finished.

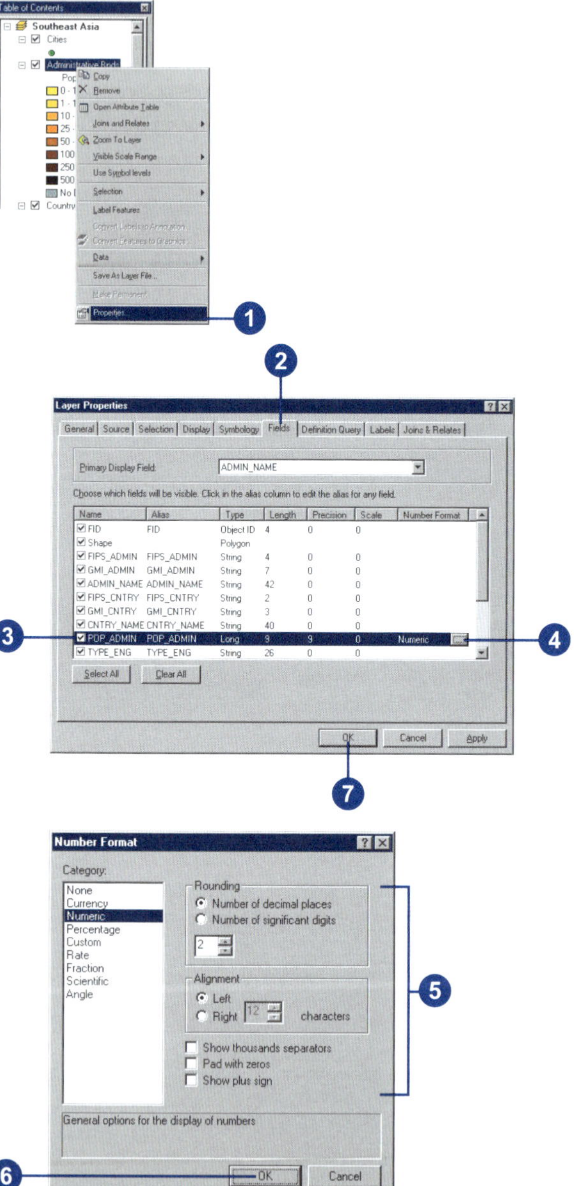

Locating and viewing records

You can use the navigation buttons at the bottom of the table window to quickly move to the next, previous, first, or last record in the table. If you know the specific record number, you can type that in as well.

When you want to find a record in a table that matches some numeric value or text string, you can search the table for that value in the selected fields or the entire table. Depending on the type of field—text or numeric—you have three different types of searches:

- Any part
- Whole field
- Start of field

Numeric fields are always searched using the whole field. If you're searching a text field, you can search for any part of the text string or the start of the field that matches the text you enter into the Find dialog box. The Find dialog box also gives you the option of searching up, down, or in all directions from the current position in the table.

Moving to a specific record number

1. Open the table.
2. Type the number of the record you want to move to and press Enter.

   The table scrolls to the record.

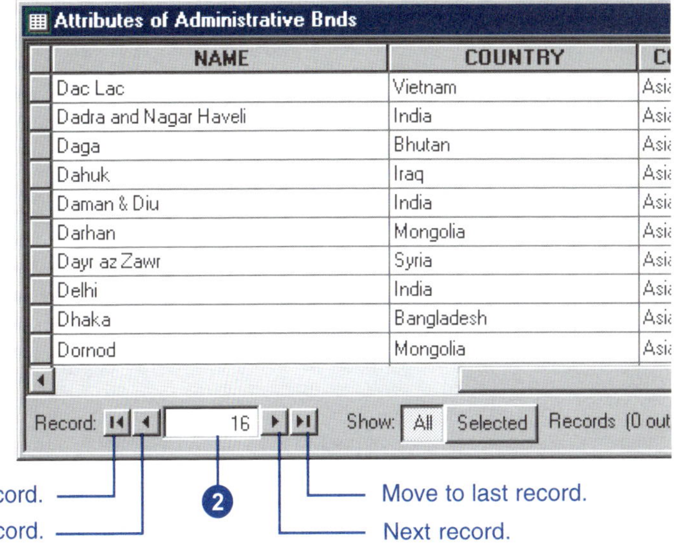

Viewing all or only the selected records

1. Open the table.
2. Click Show All to view all records or click Show Selected to view only the selected ones.

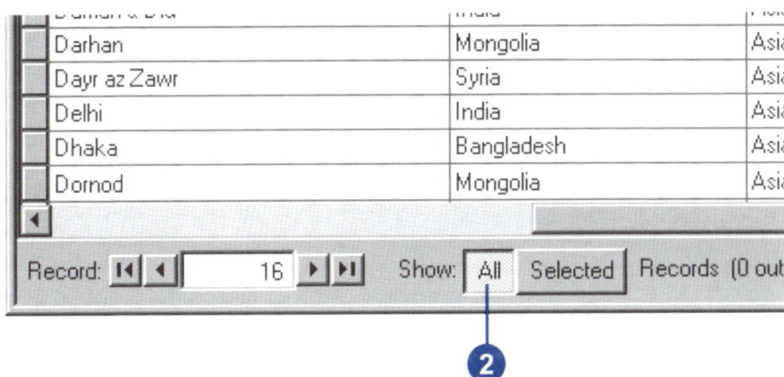

Tip

Matching the case

To match the capitalization of the text you type, check Match Case in the Find dialog box.

Tip

Search the whole table or just one field

To constrain the search to specific fields, check the box next to Search Only Selected Field(s).

Tip

Replacing text you find

In order to replace the text you find, you must be editing the table. For more information, see 'Editing attributes' in this chapter.

Finding records with particular attribute values

1. Click the heading of the column that contains the text for which you want to search.
2. Click Options and click Find & Replace.
3. Type the text you want to find in the Find what text box.
4. Click the Text Match dropdown arrow and click the type of search you want.
5. Click Find Next.

   The first record found containing your text is selected.
6. If you want to find another record containing the same text, click Find Next again.
7. Click Cancel to close the dialog box.

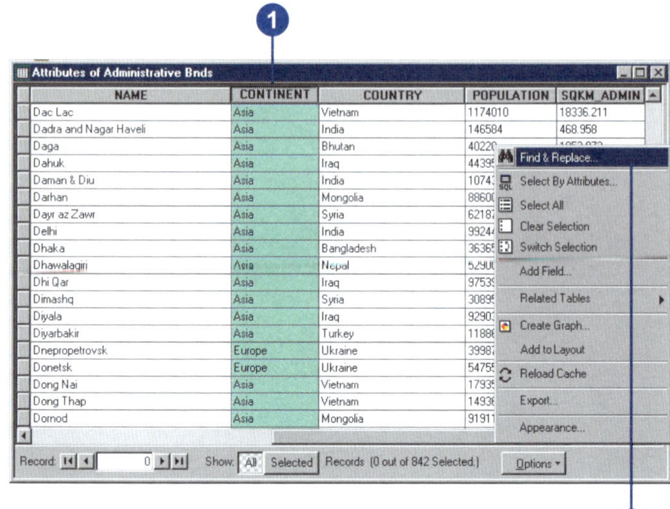

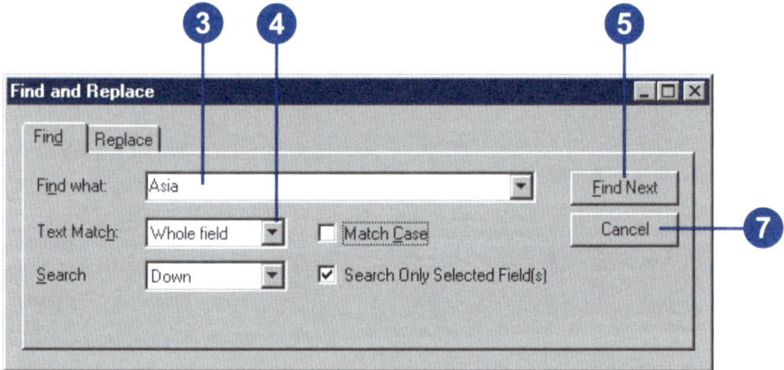

Sorting records

Sorting the records or rows in a table lets you derive information about its contents. For example, you could find the city with the largest population in Southeast Asia. After you sort a column's values in ascending order, the values appear ordered from A to Z or from 1 to n, where n is the largest number. Descending order arranges a column's values from Z to A or from n to 1.

Sometimes it's helpful to sort a table's rows by more than one column. For example, it might be more useful to sort the cities by country and by population—the effect is similar to producing a report. To sort by more than one column, you must first arrange the columns that you'll use for sorting. The sorting columns must be arranged in order from left to right, where the values in the column farthest to the left will be sorted first and the values in the column farthest to the right will be sorted last. The sorting columns are not required to be adjacent to each other; however, if they are, the order of the records is more obvious.

Sorting records by one column

1. Click the heading of the column whose values you want to use to sort the records.
2. Right-click the selected column's heading and click Sort Ascending or Sort Descending.

   The table's records are sorted.

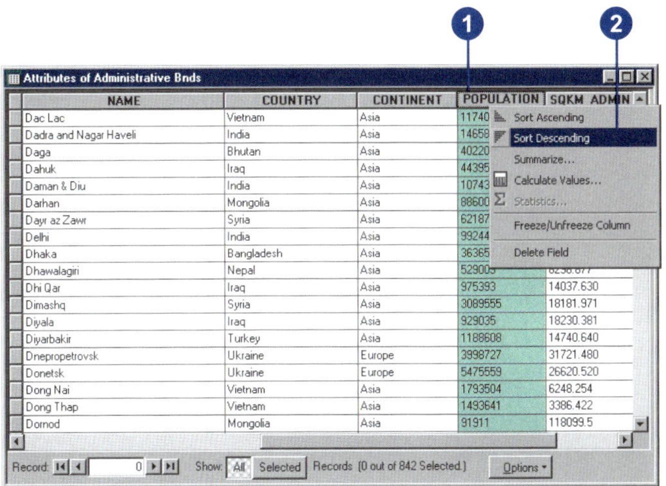

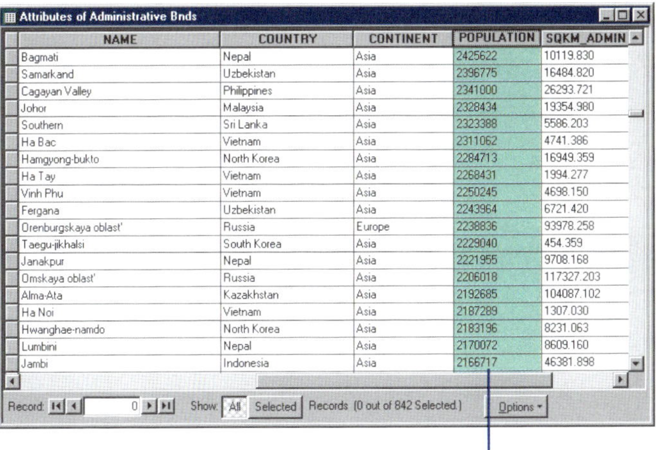

The records are sorted according to the selected column's values.

Tip

Selecting adjacent columns

Hold down the Ctrl key while selecting columns to select more than one.

Sorting records by more than one column

1. Rearrange the table's columns so the column whose values will be sorted first appears to the left of the column whose values will be sorted second.
2. Click the heading of the first column you want to use to sort the records.
3. Press the Ctrl key on the keyboard and click the second column's heading.
4. Repeat step 3 until you've selected all columns that will be used to sort the table's records.
5. Right-click a selected column heading and click Sort Ascending or Sort Descending.

   The table's records are sorted.

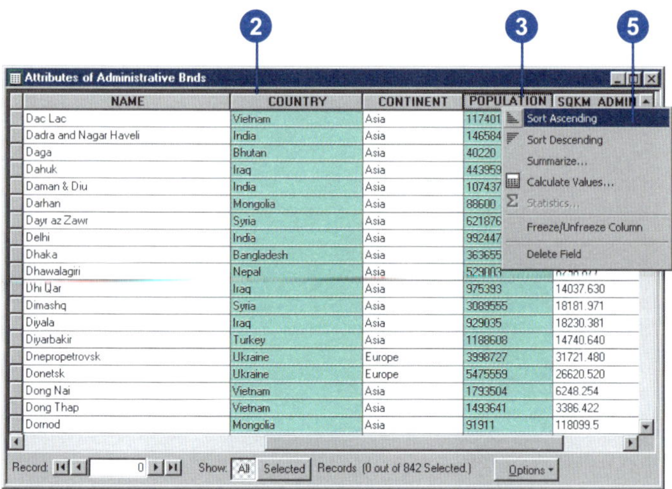

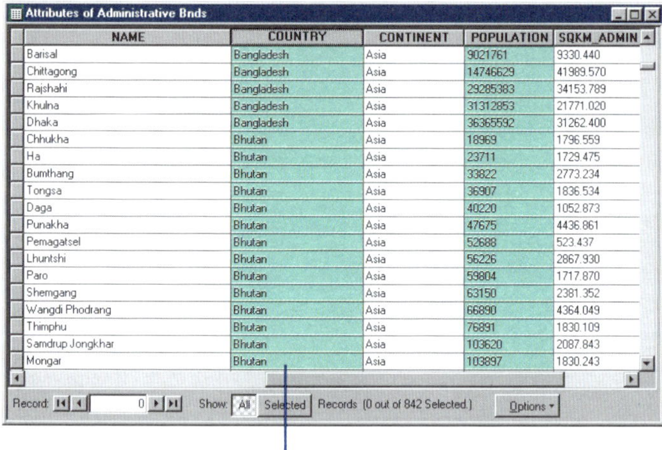

The records are sorted first by the left column's values, then by the right column's values.

Selecting records

There are various ways to select features in ArcMap. One way is to select features from an attribute table. From a table, you can interactively select records by pointing at them or by selecting those records that meet some criteria—for example, find all cities with a population greater than one million.

Once you've defined a selection, you'll see those features highlighted on your map. For example, suppose you wanted to find the locations of the five cities with the largest population. You would sort the records in the table in descending order based on population, then select the top five records in the table to see them highlighted on the map.

You can add to your selected set using any other ArcMap selection methods.

See Also

For more information on selecting map features, see Chapter 13, 'Querying maps'.

Interactively selecting records

1. Open the attribute table for a layer on your map.
2. Click the leftmost column in the table adjacent to the record you want to select.

   To select consecutive records, you can click and drag the mouse.
3. Press and hold the Ctrl key while clicking additional records.

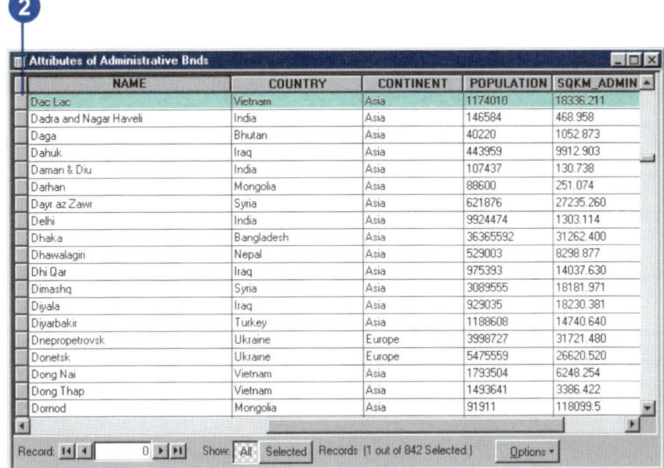

NAME	COUNTRY	CONTINENT	POPULATION	SQKM_ADMIN
Dac Lac	Vietnam	Asia	1174010	18336.211
Dadra and Nagar Haveli	India	Asia	146584	468.958
Daga	Bhutan	Asia	40220	1052.873
Dahuk	Iraq	Asia	443959	9912.903
Daman & Diu	India	Asia	107437	130.738
Darhan	Mongolia	Asia	88600	251.074
Dayr az Zawr	Syria	Asia	621876	27235.260
Delhi	India	Asia	9924474	1303.114
Dhaka	Bangladesh	Asia	36365592	31262.400
Dhawalagiri	Nepal	Asia	529003	8298.877
Dhi Qar	Iraq	Asia	975393	14037.630
Dimashq	Syria	Asia	3089555	18181.971
Diyala	Iraq	Asia	929035	18230.381
Diyarbakir	Turkey	Asia	1188608	14740.640
Dnepropetrovsk	Ukraine	Europe	3998727	31721.480
Donetsk	Ukraine	Europe	5475559	26620.520
Dong Nai	Vietnam	Asia	1793504	6248.254
Dong Thap	Vietnam	Asia	1493641	3386.422
Dornod	Mongolia	Asia	91911	118099.5

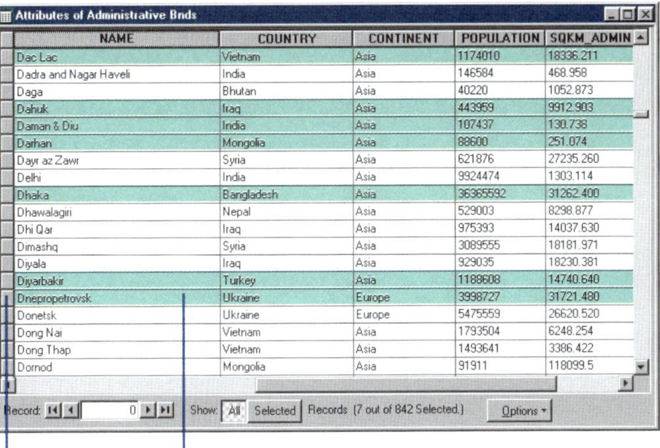

NAME	COUNTRY	CONTINENT	POPULATION	SQKM_ADMIN
Dac Lac	Vietnam	Asia	1174010	18336.211
Dadra and Nagar Haveli	India	Asia	146584	468.958
Daga	Bhutan	Asia	40220	1052.873
Dahuk	Iraq	Asia	443959	9912.903
Daman & Diu	India	Asia	107437	130.738
Darhan	Mongolia	Asia	88600	251.074
Dayr az Zawr	Syria	Asia	621876	27235.260
Delhi	India	Asia	9924474	1303.114
Dhaka	Bangladesh	Asia	36365592	31262.400
Dhawalagiri	Nepal	Asia	529003	8298.877
Dhi Qar	Iraq	Asia	975393	14037.630
Dimashq	Syria	Asia	3089555	18181.971
Diyala	Iraq	Asia	929035	18230.381
Diyarbakir	Turkey	Asia	1188608	14740.640
Dnepropetrovsk	Ukraine	Europe	3998727	31721.480
Donetsk	Ukraine	Europe	5475559	26620.520
Dong Nai	Vietnam	Asia	1793504	6248.254
Dong Thap	Vietnam	Asia	1493641	3386.422
Dornod	Mongolia	Asia	91911	118099.5

Selected records are highlighted in the table and on the map.

Tip

Saving and reusing selection expressions

You can save and reload selection expressions using the Save and Load buttons at the bottom of the Select By Attributes dialog box. This lets you quickly re-create a selected set of records by loading a saved expression.

Tip

How to build a query

Click Help on the Select By Attributes dialog box to learn about building a query.

Selecting records by attributes

1. Click Options in the table you want to query and click Select By Attributes.
2. Click the Method dropdown arrow and click the selection procedure you want to use.
3. Double-click the field from which you want to select.
4. Click the logical operator you want to use.
5. Click the Get Unique Values button, then scroll to and double-click the value in the Unique Values list you want to select. Alternatively, you can type a value directly into the text box.
6. Click Verify to verify your selection.
7. Click Close.

   Your selection is highlighted in the table.

   Use Apply if you intend to run more than one query or if you want to check your results before closing the Select By Attributes dialog box.

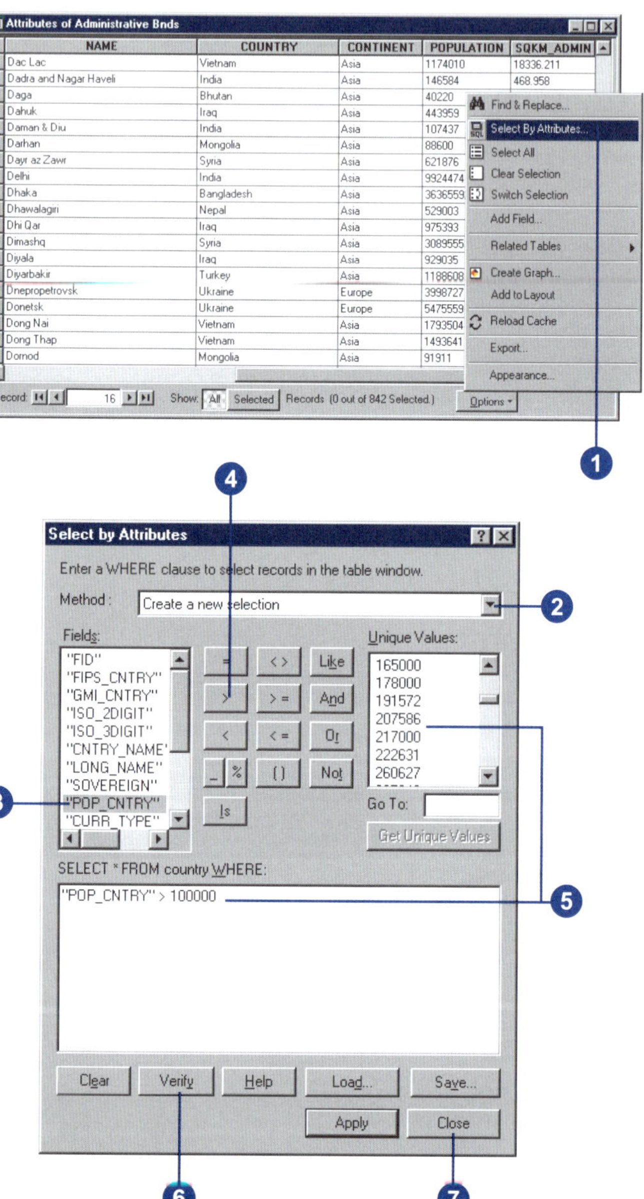

Tip

Selecting features

The Selection menu on the Standard toolbar contains additional tools for selecting features.

Selecting all records

1. Click Options in the table and click Select All.

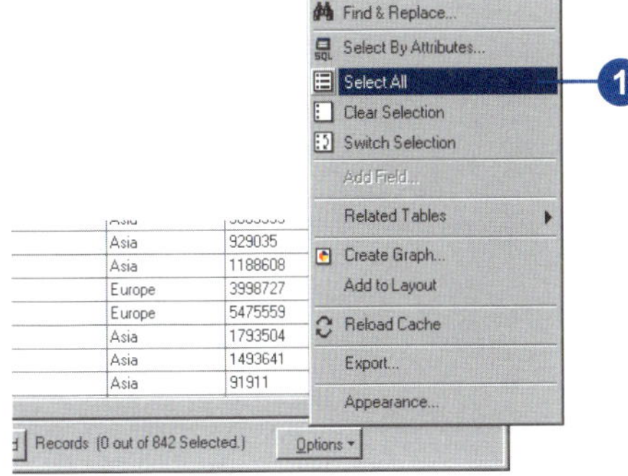

Clearing the selected set

1. Click Options in the table and click Clear Selection.

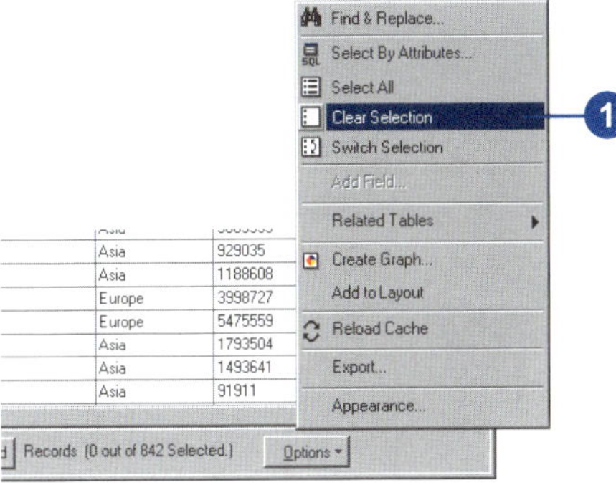

Switching the selected set

1. Click Options in the table and click Switch Selection.

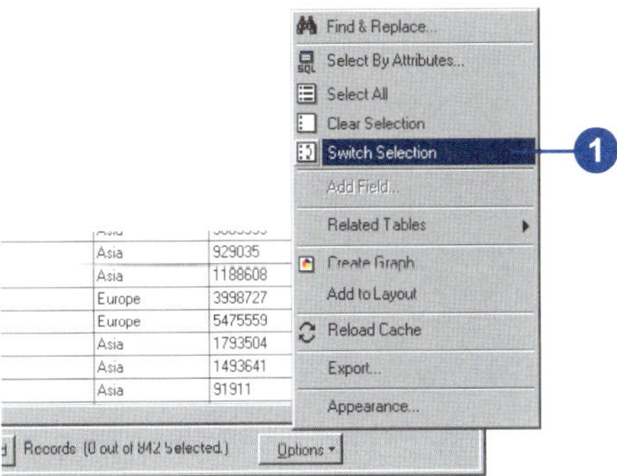

Exporting records

You can use ArcMap to export the records in a table to create a new table. For example, you might want to modify a table without altering the original records, pass a table along to another user, or create a new table with a particular set of records.

From ArcMap, you can export the selected records or all records in a table to create a new table. You can choose among several formats to export to, including dBASE, INFO, or geodatabase tables.

See Also

You can also export tables from ArcCatalog. For more information, see Using ArcCatalog.

1. Click Options in the table you want to export.
2. Click Export.
3. In the Export Data dialog box, click the Export dropdown arrow to choose to export Selected records or All records.

   This option is only available if records are selected in the table you want to export.
4. Click the Browse button and navigate to the folder or geodatabase in which you want to place the exported data.
5. Click the Save as type dropdown arrow and click the format to which you want to export the data. For example, click Personal Geodatabase tables.
6. Type a name for the exported table.
7. Click Save.
8. Click OK.

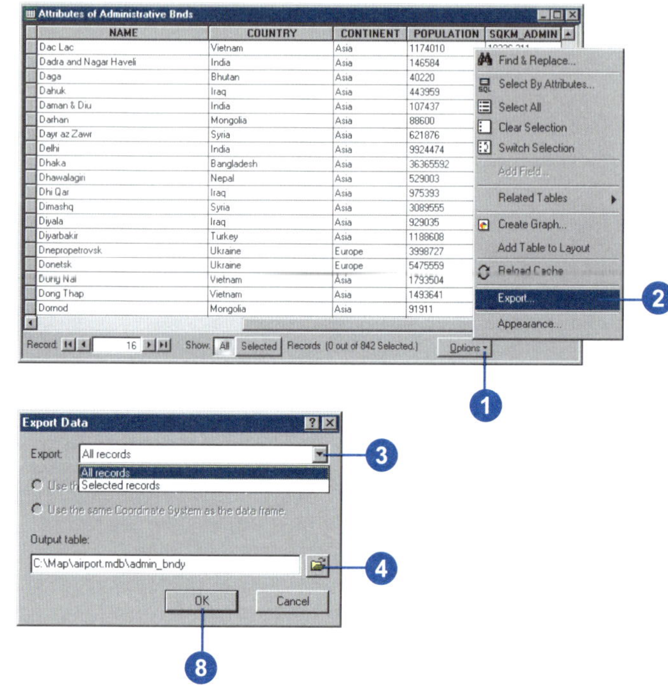

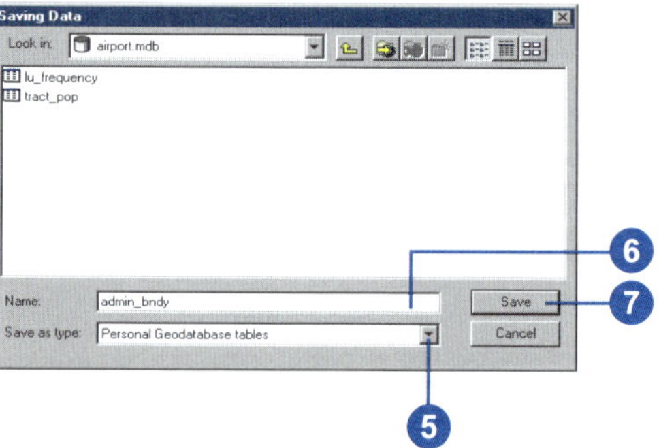

Summarizing data

Sometimes the attribute information you have about map features is not organized in the way you want—for instance, you have population data by county when you want it by state. By summarizing the data in a table, you can derive various summary statistics—including the count, average, minimum, and maximum value—and get exactly the information you want. ArcMap creates a new table containing the summary statistics. You can then join this table to the attribute table of a layer. Doing so lets you symbolize, label, or query the layer's features based on values for the summary statistics.

See Also

For more information on joining tables, see 'Joining attribute tables' in this chapter.

Summarizing data in a field

1. Right-click the field heading of the field you want to summarize and click Summarize.
2. Check the box next to the summary statistics you want to include in the output table.
3. Type the name and location of the output table you want to create or click the Browse button and navigate to a workspace.
4. Click OK.
5. Click Yes when prompted to add the new table to your map.

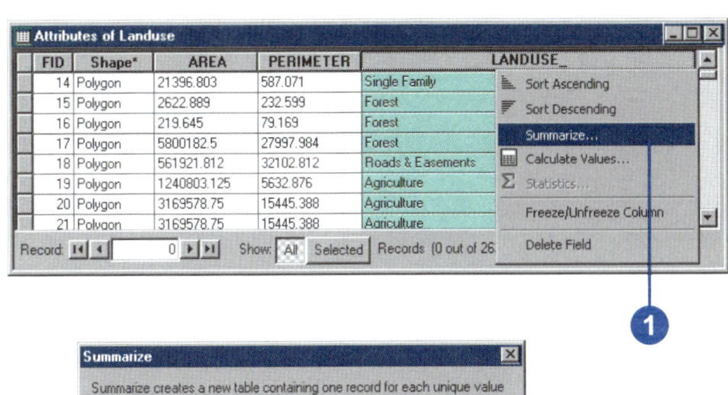

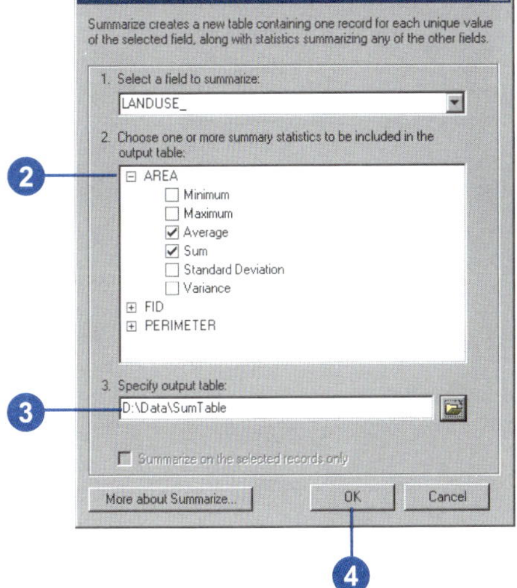

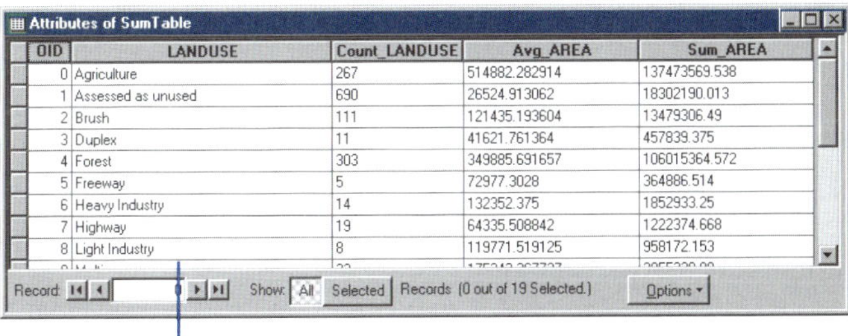

OID	LANDUSE	Count_LANDUSE	Avg_AREA	Sum_AREA
0	Agriculture	267	514882.282914	137473569.538
1	Assessed as unused	690	26524.913062	18302190.013
2	Brush	111	121435.193604	13479306.49
3	Duplex	11	41621.761364	457839.375
4	Forest	303	349885.691657	106015364.572
5	Freeway	5	72977.3028	364886.514
6	Heavy Industry	14	132352.375	1852933.25
7	Highway	19	64335.508842	1222374.668
8	Light Industry	8	119771.519125	958172.153

The new output table contains one record for each unique land use value and a field for each summary statistic you selected.

Adding and deleting fields

You can easily add or remove fields from a table in ArcMap as necessary. Most likely, you'll add or remove fields from data that you personally manage. Large organizations typically have large databases with well-defined database schemas that outline the contents—including fields—of the database. Unless you manage the database, it is unlikely that you'll be able to add or remove fields.

You can add or remove fields from a table as long as the following conditions are met:

- You have write access to the data.
- You're not currently editing the data in ArcMap.
- No other users or applications are accessing the data including other ArcMap or ArcCatalog sessions.

Tip

Why is Add Field unavailable?

The Add Field and Delete Field options are unavailable when you're editing the table.

Adding a field to a table

1. Click Options in the table you want to add a field to.
2. Click Add Field.
3. Type a name for the field.
4. Click the Type dropdown arrow and click the field type.
5. Set any other field properties, such as a field alias, as necessary.
6. Click OK.

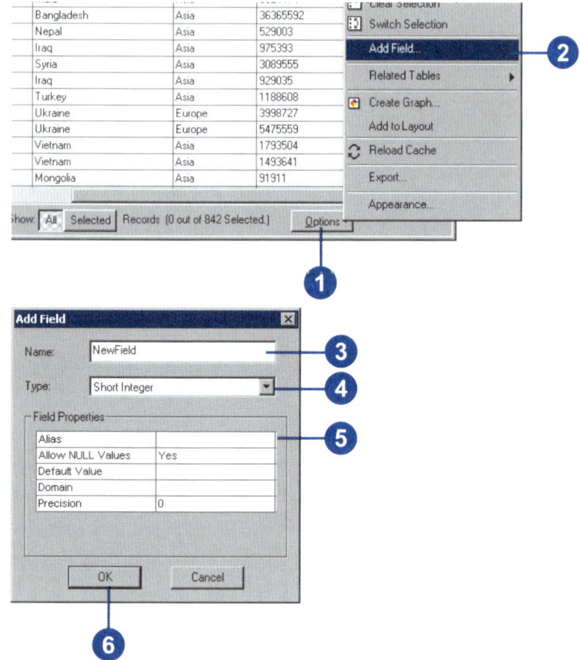

Deleting a field from a table

1. In the table window, right-click over the field header of the field you want to delete.
2. Click Delete Field.
3. Click Yes to confirm the deletion.

   Deleting a field cannot be undone.

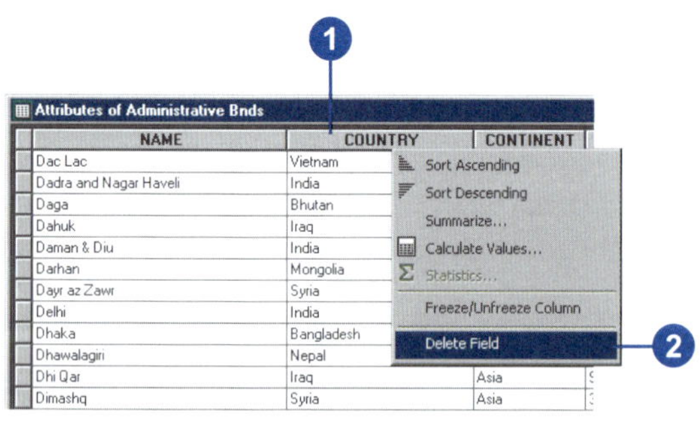

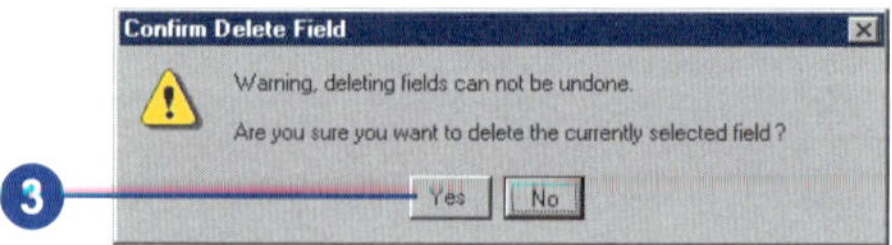

Editing attributes

Your database is only as good as the information it contains. Over time, you'll need to edit the information in your database to keep it accurate and up to date. ArcMap lets you edit the attributes of features displayed on your map and also the attributes contained in other database tables (for example, a table of monthly sales figures) that are not represented geographically on the map. You can edit any of the attribute values that appear in a table as well as add new records and delete records. You can also use the field calculator to change the attribute value of a field for several records at once.

As with editing map features in ArcMap, editing the attributes of features takes place within an *edit session*. You start an edit session by clicking Start Editing from the Editor menu on the Editor toolbar. Once you begin an edit session, you'll notice this icon next to the Options button on the table window, indicating that the table can be edited. In addition, those fields that you can edit will have a white background color to the field heading. ▸

Editing text in records

1. If you haven't started an edit session, click the Editor menu on the Editor toolbar and click Start Editing.
2. Open the table you want to edit.
3. Click the cell containing the attribute value you want to change.
4. Type the values and press Enter.

   The table is updated.

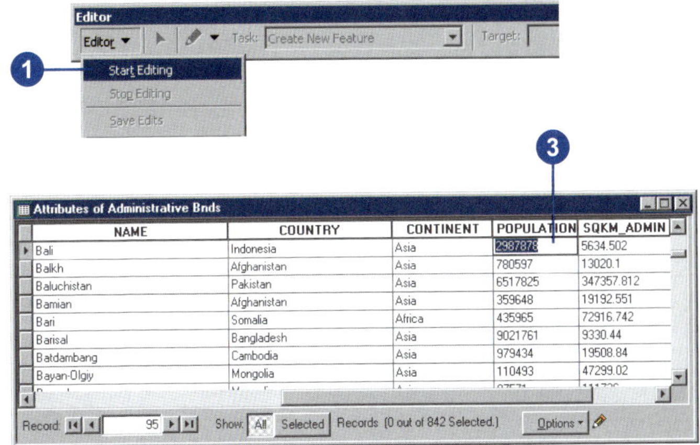

Adding new records

1. If you haven't started an edit session, click the Editor menu on the Editor toolbar and click Start Editing.
2. Open the table you want to edit.
3. Click Move to end of table.
4. Click a cell in the last, empty record and type in a new value.

   NOTE: Use these steps to add new records to tables that don't have associated geographic features. If you want to add features to your shapefile or geodatabase, use the Create New Feature task on the Editor toolbar.

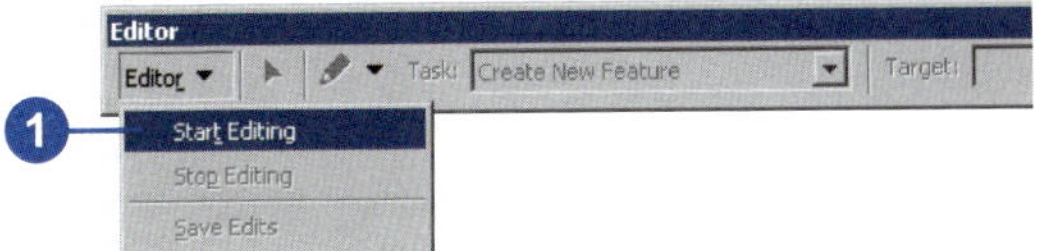

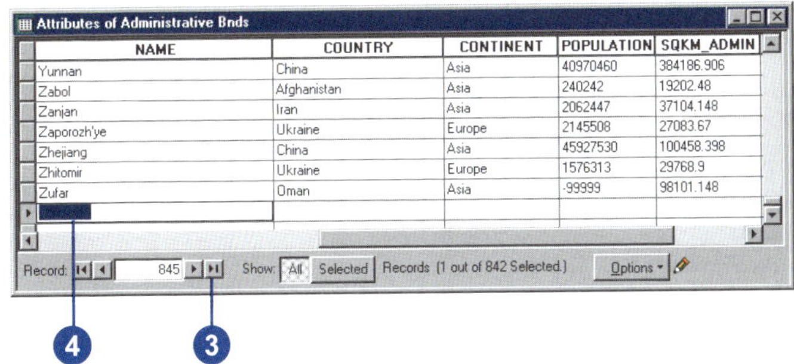

Now, you can make any of the attribute changes you need by clicking on a cell and typing a new attribute value. If you make a mistake, you can easily undo the edit by clicking Undo from the Edit menu.

Editing attributes through the table allows you to quickly make changes to several features (records) at once. When you're editing the attributes of specific map features, you may find it more convenient to do so using the Attributes dialog box, accessed from the Editor toolbar. This dialog box is tailored to updating the attributes of specific map features that you first point at with the mouse.

When you've completed your edits, you can save them and end the edit session.

Tip

Adding the Editor toolbar

To display the Editor toolbar, click Tools, then click Editor.

Tip

Navigating the cells in a table

You can navigate the cells in a table by pressing the Tab or arrow keys on your keyboard.

Deleting records

1. If you haven't started an edit session, click the Editor menu on the Editor toolbar and click Start Editing.
2. Open the table you want to edit.
3. Select the records you want to delete. Press and hold the Ctrl key while clicking to select more than one record.
4. Press the Delete key on the keyboard.

   Any geographic features associated with the records are also deleted.

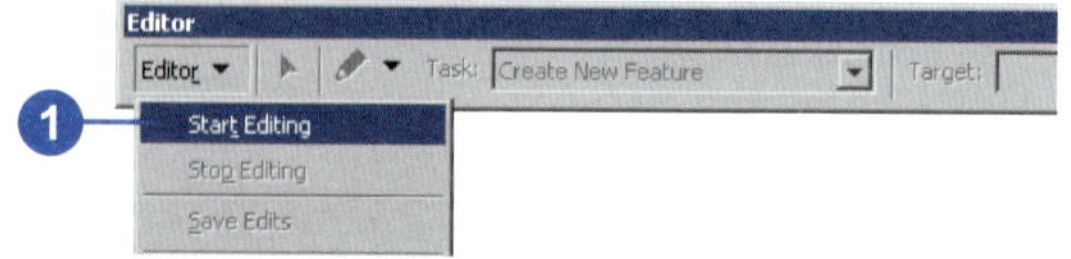

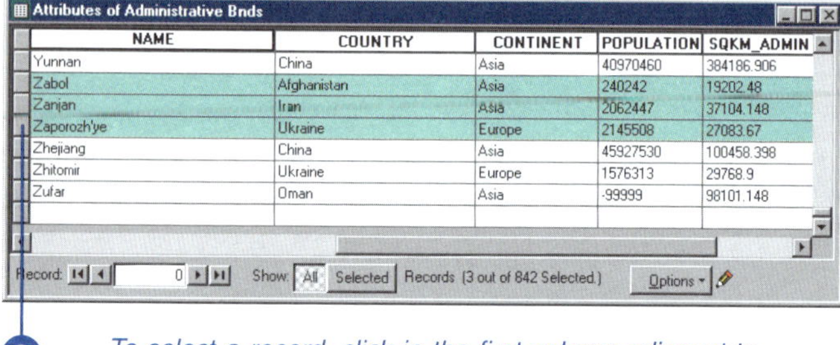

NAME	COUNTRY	CONTINENT	POPULATION	SQKM_ADMIN
Yunnan	China	Asia	40970460	384186.906
Zabol	Afghanistan	Asia	240242	19202.48
Zanjan	Iran	Asia	2062447	37104.148
Zaporozh'ye	Ukraine	Europe	2145508	27083.67
Zhejiang	China	Asia	45927530	100458.398
Zhitomir	Ukraine	Europe	1576313	29768.9
Zufar	Oman	Asia	-99999	98101.148

To select a record, click in the first column adjacent to the record you want to select.

Copying and pasting records

1. If you haven't started an edit session, click the Editor menu on the Editor toolbar and click Start Editing.
2. Open the table you want to edit.
3. Select the records you want to copy. Press and hold the Ctrl key while clicking to select more than one record.
4. Click Copy on the Standard toolbar.
5. Click Paste on the Standard toolbar. The new records are added at the end of the table.

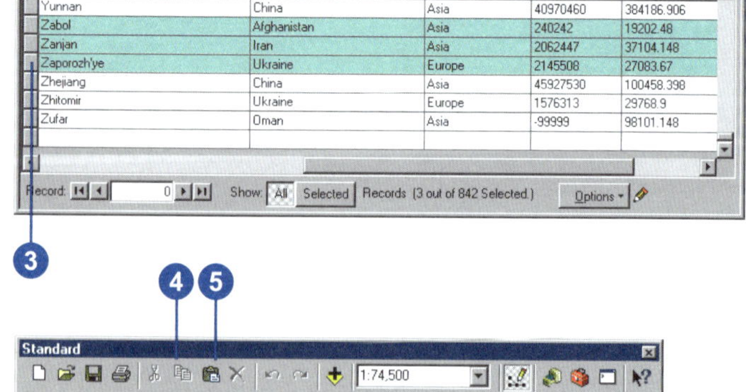

NAME	COUNTRY	CONTINENT	POPULATION	SQKM_ADMIN
Yunnan	China	Asia	40970460	384186.906
Zabol	Afghanistan	Asia	240242	19202.48
Zanjan	Iran	Asia	2062447	37104.148
Zaporozh'ye	Ukraine	Europe	2145508	27083.67
Zhejiang	China	Asia	45927530	100458.398
Zhitomir	Ukraine	Europe	1576313	29768.9
Zufar	Oman	Asia	-99999	98101.148

Making field calculations

Entering values with the keyboard is not the only way you can edit tables. In some cases, you might want to perform a mathematical calculation to set a field value for a single record or even all records. The ArcMap field calculator lets you perform simple as well as advanced calculations on any selected record.

The Field Calculator also lets you perform advanced calculations using VBA statements that process the data before calculations are made on the selected field. For example, using demographic data, you might want to find out the largest age group by percentage of the population for each county in the United States. You can create a script that preprocesses your data using logical constructs such as If...Then statements and Select Case blocks. This lets you perform sophisticated calculations quickly and easily.

Tip

Calculating fields outside an edit session

You can't undo a field calculation when performed outside of an edit session.

Making simple field calculations

1. If you haven't started an edit session, click the Editor menu on the Editor toolbar and click Start Editing.

   You can make calculations without being in an edit session; however, in that case, there is no way to undo the results.

2. Open the table you want to edit.

3. Select the records you want to update. If you don't select any, the calculation will be applied to all records.

4. Right-click the field heading for which you want to make a calculation and click Calculate Values.

5. Use the Fields list and Functions to build a calculation expression. You can also edit the expression in the text area below. Alternatively, you can just type in a value to set the field to. In this example, the string "Single Family" is used.

   NOTE: Use double quotes when calculating strings.

6. Click OK.

7. Don't forget to end your edit session. Click the Editor menu and click Stop Editing.

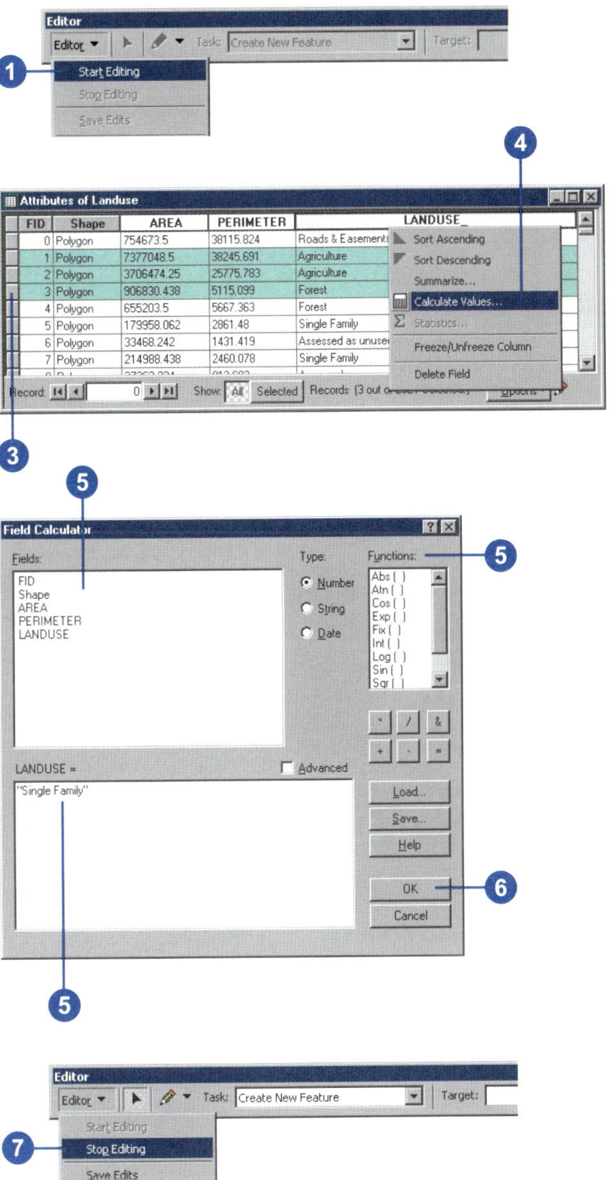

Tip

Reusing calculation expressions

After entering VBA statements, click Save to write them out to a file. The Load button will prompt you to find and select an existing calculation file.

See Also

For more information on VBA, consult any Visual Basic (VB) reference. The Visual Basic Editor (VBE)—accessed by clicking the Tools menu, pointing to Macros, and clicking Visual Basic Editor—also contains an online reference to VB commands.

Making advanced field calculations

1. If you haven't started an edit session, click the Editor menu on the Editor toolbar and click Start Editing.

   You can make calculations without being in an edit session; however, in that case, there is no way to undo the results.

2. Open the table you want to edit.
3. Select the records you want to update. If you don't select any, the calculation will be applied to all records.
4. Right-click the field heading for which you want to make a calculation and click Calculate Values.
5. Check Advanced.
6. Type VBA statements in the first text box.

   The VBA statements can include ArcMap methods. The VBA code shown in the figure gets the x coordinate of the centroid of each polygon in the layer and writes it out to a field called X_FIELD.

7. Type the variable or value that is to be written to the selected records.
8. Click OK.

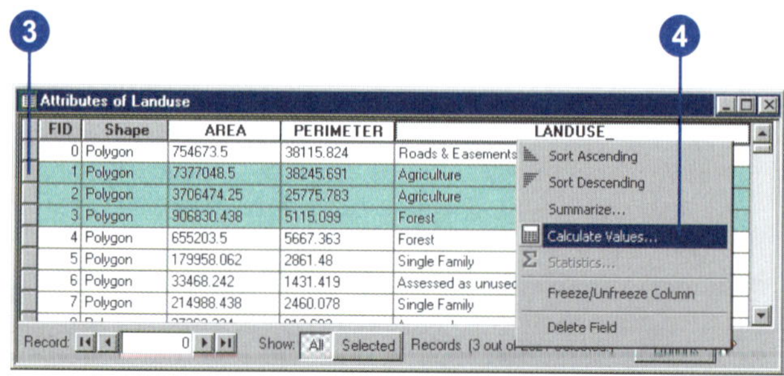

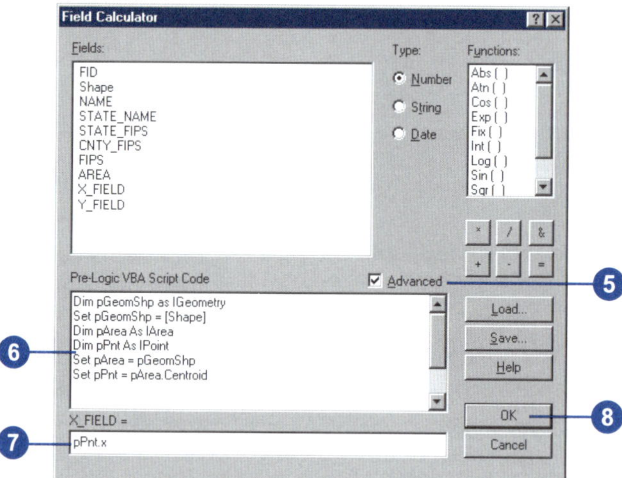

About joining attribute tables

Most database design guidelines advise organizing your database into multiple tables—each focused on a specific topic—instead of using one large table containing all the necessary fields. Having multiple tables prevents duplicating information in the database because you store the information only once in one table. When you need information that isn't in the current table, you can link the two tables together. In ArcMap, you can establish this kind of link by either joining or relating two tables together.

Joining the attributes from a table

Typically, you'll *join* a table of data to a layer's attribute table, extending the information you have about your geographic features. Joins are based on the value of a field that can be found in both tables. The name of the field does not have to be the same, but the data type has to be the same; you join numbers to numbers, strings to strings, and so on.

Suppose you obtain daily weather forecasts by county and generate weather maps based on this information. As long as the weather data is stored in a table in your database and shares a common field with your layer, you can join it to your geographic features and use any of the additional fields to symbolize, label, query, or analyze the layer's features.

When you join tables in ArcMap, you establish a one-to-one or many-to-one relationship between the layer's attribute table and the table containing the information you wish to join. The example above illustrates a one-to-one relationship between each county and the weather data. In other words, there's one record of weather data for each county.

Here's an example of a many-to-one relationship. Suppose you have a layer where each polygon is classified according to its land use type. The layer's attribute table only stores a land use code; a separate table stores the full description of each land use type. Joining these two tables together establishes a many-to-one relationship because many records in the layer's attribute table join to the same record in the table of land use descriptions. You might then use the more descriptive text when generating the legend for your map.

Shape	FID	County
Polygon	1	Atoka
Polygon	2	Kiowa
Polygon	3	Nowata

County	Rain	Total
Atoka	1.80	10.16
Kiowa	2.34	13.67
Nowata	1.62	11.90

Low Rainfall

High Rainfall

Symbolizing features based on joined rainfall data.

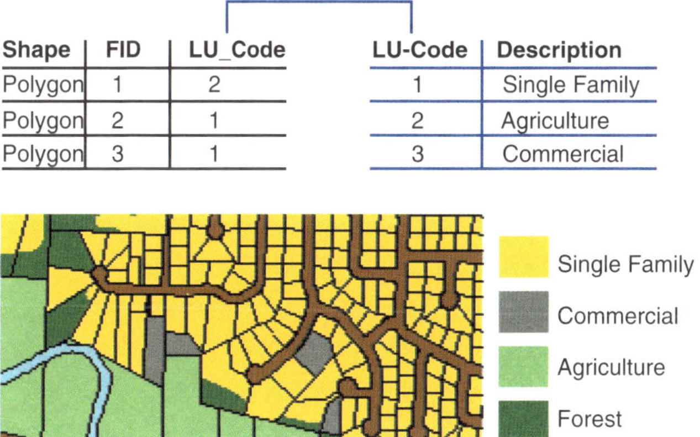

Shape	FID	LU_Code
Polygon	1	2
Polygon	2	1
Polygon	3	1

LU-Code	Description
1	Single Family
2	Agriculture
3	Commercial

Many polygons share the same land use description.

Summarizing your data before joining it

Depending on how your data is organized, you may have to start by summarizing the data in your table before you join it to a layer. When you summarize a table, ArcMap creates a new table containing summary statistics derived from your table. You can create various summary statistics including count, average, sum, minimum, and maximum.

For example, suppose you want to create weather maps by state instead of by county, but the weather information you have is organized by county. You could summarize the county data by state—for instance, finding the average rainfall for all counties within a state—then join the newly created output table to a state layer to create a weather map of rainfall by state.

State	County	Rain	Total
Oklahoma	Atoka	1.80	10.16
Oklahoma	Kiowa	2.34	13.67
Oklahoma	Nowata	1.62	11.90

State	Count	Avg_Rain	Max_Rain
Ohio	88	3.21	4.50
Oklahoma	77	2.56	3.86
Oregon	36	5.66	7.92

Sometimes it may be necessary to summarize your tabular data so it can be joined to your geographic data.

When to relate tables instead of joining them

You've seen how joining tables establishes a one-to-one or many-to-one relationship between a layer and a table. However, in some situations, you may want to establish a one-to-many or a many-to-many relationship between a layer and a table.

An example of a one-to-many relationship is building occupancy. One building, such as a shopping center, may be occupied by many tenants. You may want to join a table of tenants to the attribute table of a layer representing buildings. However, if you perform a join, ArcMap will only find the first tenant belonging to each building, ignoring additional tenants. In this case, you should *relate* the tables instead of joining them.

Unlike joining tables, relating tables simply defines a relationship between two tables. The associated data isn't appended to the layer's attribute table like it is with a join. Instead, you can access the related data when you work with the layer's attributes. For example, if you select a building, you can find all the tenants that occupy that building. Similarly, if you select a tenant, you can find what building it resides in (or several buildings, in the case of a chain of stores in multiple shopping centers—a many-to-many relationship).

Relates defined in ArcMap are essentially the same as simple relationship classes defined in a geodatabase, except that they are saved with the map instead of in a geodatabase. For more information on creating relationship classes, see 'Defining relationship classes', in *Building a Geodatabase*.

Joining data spatially

When the layers on your map don't share a common attribute field, you can instead join them using a *spatial join*. A spatial join joins the attributes of two layers based on the location of the features in the layers. With a spatial join, you can:

- Find the closest feature to another feature.
- Find what's inside a feature.
- Find what intersects a feature.

For example, you might want to tell customers where they can find the nearest retail store and how far away it is from them. Or a biologist might summarize information about endangered species sightings based on what region in a national park the observations were made.

For information on how to perform a spatial join, see Chapter 13, 'Querying maps'.

How are joins and relates saved in your map?

When you save a map containing joins and relates, ArcMap saves the definition of how the two attribute tables are linked rather than saving the linked data itself. The next time you open your map, ArcMap reestablishes the relationship (whether a join or relate) between the tables by reading the tables from the database. In this way, any changes to the source tables that have taken place since they were last viewed on the map are automatically included and reflected on the map.

If you want, you can make a permanent disk copy of a layer with joined data; simply export the layer. To export the layer, right-click it in the table of contents, point to Data, and click Export Data. This creates a new feature class with all of the attributes, including the joined fields, written out.

Joining attribute tables

Data comes from a variety of sources. Often, the data you want to display on your map is not directly stored with your geographic data. For instance, you might obtain data from other departments in your organization, purchase commercially available data, or download data from the Internet. If this information is stored in a table, such as a dBASE, INFO, or geodatabase table, you can associate it with your geographic features and display the data on your map.

ArcMap provides two methods to associate data stored in tables with geographic features: joins and relates. When you join two tables, you append the attributes from one onto the other, based on a field common to both tables. Relating tables defines a relationship between two tables—also based on a ▸

See Also

For information on spatial joins, see Chapter 13, 'Querying maps'.

Tip

Joining by relationship class

You can also join two tables using a predefined relationship class.

Joining the attributes in one table to another

1. Right-click the layer or table you want to join, point to Joins and Relates, and click Join.
2. Click the first dropdown arrow and click Join attributes from a table.
3. Click the second dropdown arrow and click the field name in the layer that the join will be based on.
4. Click the third dropdown arrow to choose the table to join to the layer. If the table is not currently part of the map, click the Browse button to search for it on disk.
5. Click the fourth dropdown arrow and click the field in the table to base the join on.
6. Click OK.

   The attributes of the table are appended to the layer's attribute table.

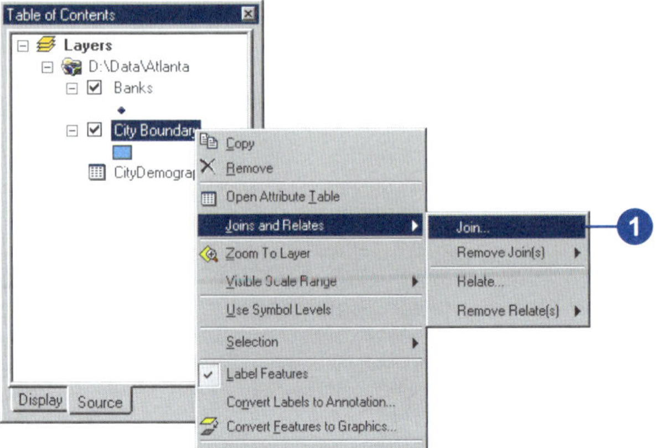

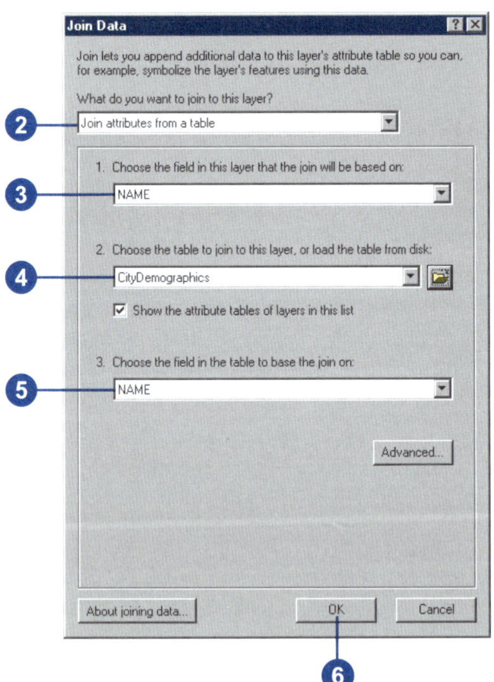

common field—but doesn't append the attributes of one to the other. Instead, you can access the related data when necessary.

You'll want to join two tables when the data in the tables has a one-to-one or a many-to-one relationship—for example, you have a layer showing store locations and you want to join a table of the latest monthly sales figures to it.

You'll want to relate two tables when the data in the tables has a one-to-many or many-to-many relationship—for example, your map displays a parcel database, and you have a table of owners. A parcel may have more than one owner and an owner may own more than one parcel.

Joins and relates are reconnected whenever you open the map. This way, if the underlying data in your tables changes, it will be reflected in the join or relate.

When you're finished using a join or relate, you can remove it.

Tip

Creating a new dataset from joined data

If you want to permanently save joined data with your geographic features, export the data to a new dataset. Right-click the layer in the table of contents, point to Data, and click Export data.

Removing a joined table

1. Right-click the layer containing a join you want to remove and point to Joins and Relates.
2. Point to Remove Join(s) and click the join you want to remove.

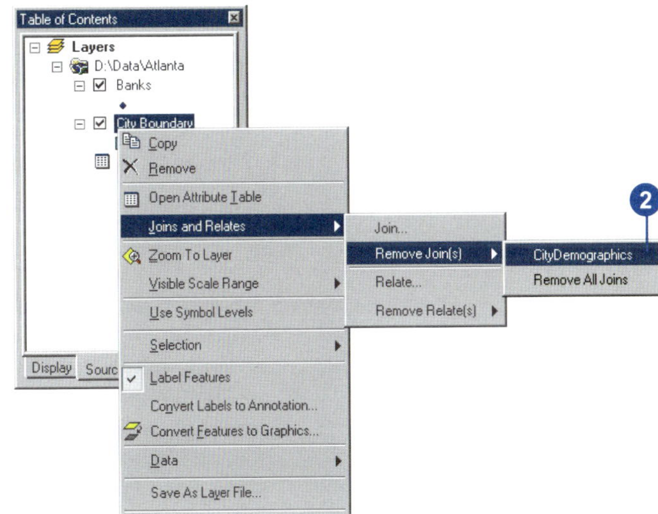

Managing joined tables

1. Right-click a layer or table in the table of contents and click Properties.
2. Click the Joins & Relates tab.

   All the joins for the layer or table are listed on the left side of the dialog box. You can add new joins or remove existing ones.

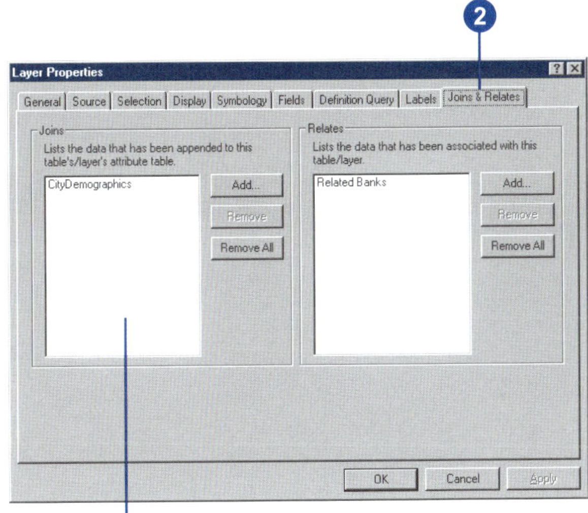

All joins for the layer or table are listed here.

Tip

You may not need to relate feature classes in geodatabases

If a feature class in a geodatabase participates in a relationship class, that relationship class will be immediately available for use. You don't need to relate the tables in ArcMap.

Relating the attributes in one table to another

1. Right-click the layer you want to relate, point to Joins and Relates, and click Relate.
2. Click the first dropdown arrow and click the field in the layer the relate will be based on.
3. Click the second dropdown arrow and click the table or layer to relate to, or load the table from disk.
4. Click the third dropdown arrow and click the field in the related table to base the relate on.
5. Type a name for the relate. You'll use this name to access the related data.
6. Click OK.

   The relate is now established between the two tables. The next topic discusses how to access records using the relate.

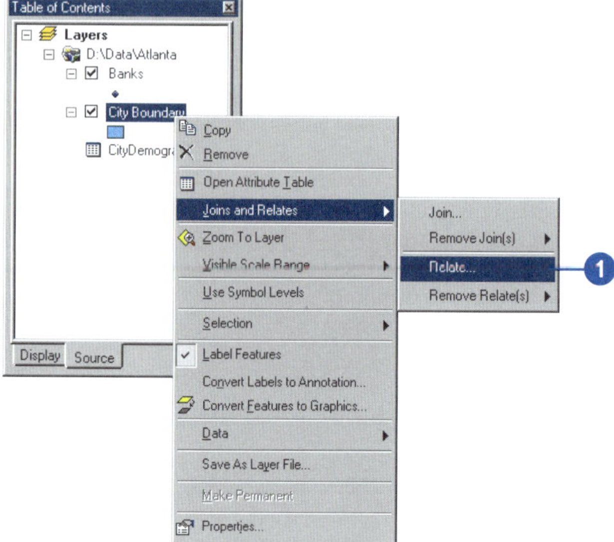

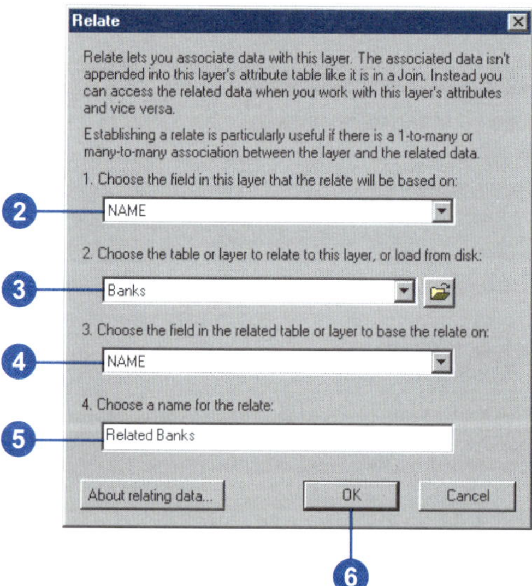

Tip

Relates work both ways

Once you define a relate, you can access the related records from either table participating in the relationship.

Tip

Accessing relationship classes

If your map contains layers from a geodatabase that participate in relationship classes, those relationship classes will be listed automatically along with any relates you define.

See Also

You must set up a relationship before you can access related records. For information on relating tables, see 'Relating the attributes in one table to another' on the previous page.

Accessing related records

1. Open the attribute table for which you've set up a relate.
2. Select the records in the table for which you want to display related records.
3. Click Options, point to Related Tables, and click the name of the relate you want to access.

   The related table displays with the related records selected.

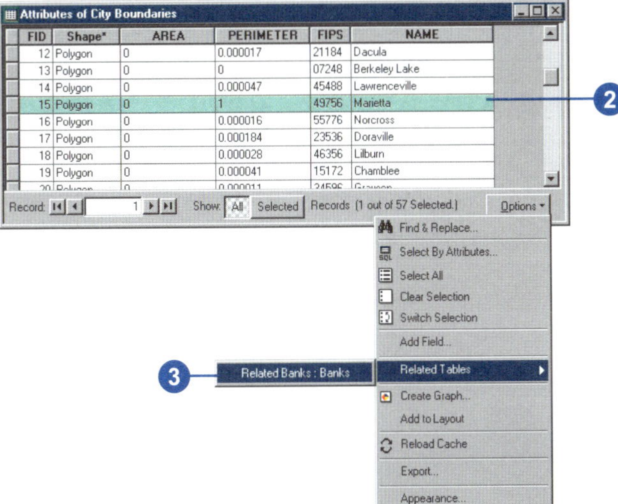

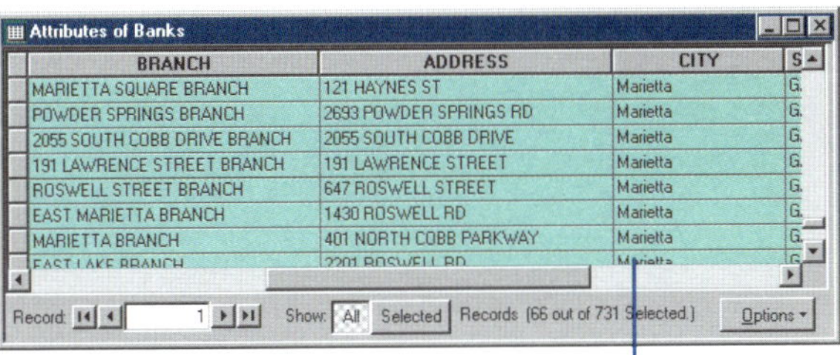

All banks in the city of Marietta are selected.

Removing a related table

1. Right-click the layer containing a relate you want to remove and point to Joins and Relates.
2. Point to Remove Relate(s) and click the relate you want to remove.

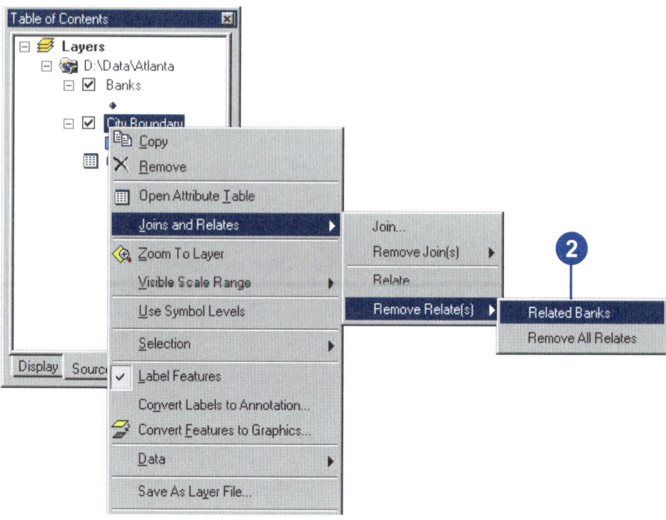

Managing related tables

1. Right-click a layer or table in the table of contents and click Properties.
2. Click the Joins & Relates tab.

   All the relates for the layer or table are listed on the right side of the dialog box. You can add new relates or remove existing ones.

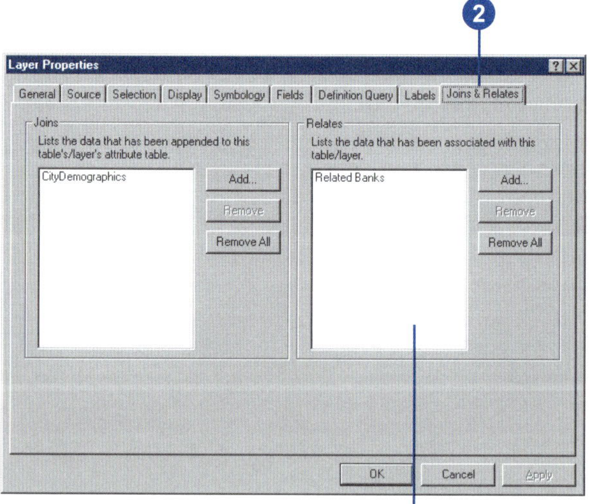

All relates for the layer or table are listed here.

Looking at data with graphs

11

IN THIS CHAPTER

- **Choosing which type of graph to make**
- **Creating a graph**
- **Displaying a graph**
- **Modifying a graph**
- **Creating a static copy of a graph**
- **Managing graphs**
- **Saving and loading a graph**
- **Exporting a graph**

Graphs present information about map features—and the relationship between them—in an attractive, easy-to-understand manner. They can show additional information about the features on the map or show the same information in a different way. Graphs complement a map because they convey information that would otherwise take some time to summarize and understand—for instance, you can quickly compare features to see which have more or less of a particular attribute.

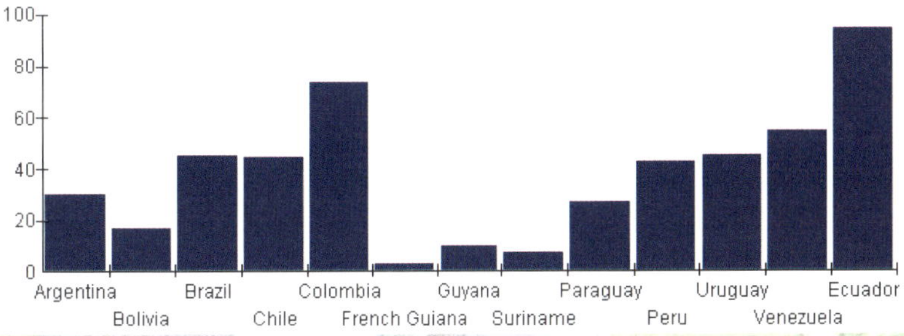

You can quickly see which countries have a high population density in this column graph.

The information displayed on a graph comes directly from the attribute information stored with your geographic data. Once created, you can easily add a graph to your map and print it out.

Choosing which type of graph to make

You can choose from several different types of graphs—both two- and three-dimensional. Some graphs are better than others at presenting certain kinds of information. Each graph has display properties that you can adjust to suit your needs. You can experiment with the various graph types and display properties to see what works best for you.

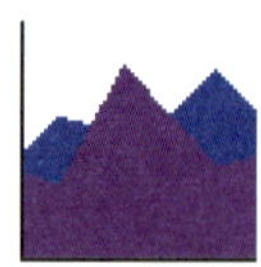

Area—An area graph consists of one or more lines drawn on an x,y grid with the area between the lines and the x-axis shaded. Like line graphs, area graphs show trends in values, but shading gives greater emphasis to differences in quantities. An area graph can be two- or three-dimensional.

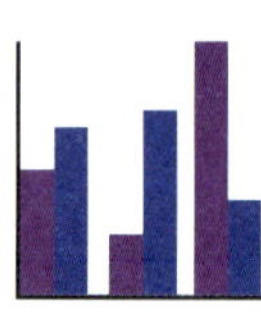

Bar and Column—Bar and column graphs consist of two or more parallel bars, each representing a particular attribute value. Use either of these graphs to compare amounts or to show trends—for example, monthly sales figures. Bar and column graphs can be two- or three-dimensional.

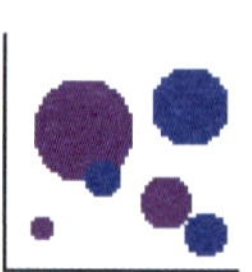

Bubble—The bubble graph lets you chart three variables in two dimensions. It's a variation of the scatter graph, where the size of the bubble represents a particular data value. For example, the size of the bubble might represent total population; the position along the y-axis, birth rate; and the position along the x-axis, death rate.

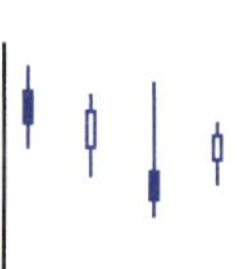

High–Low–Close—The high–low–close graph lets you chart a range of values on an x,y grid. The range is shown as a vertical bar, with a horizontal crossbar for the high, the low, and a normative value usually called the close. An alternate version, the open–high–low–close, adds a fourth crossbar for another value called the open.

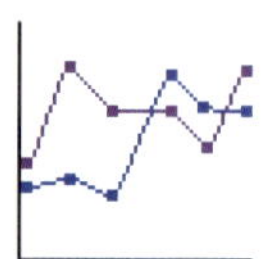

Line—A line graph consists of one or more lines, or sequences of symbols, drawn on an x,y grid. Line graphs show trends in value along a continuous scale. The three-dimensional version of the line graph is also available.

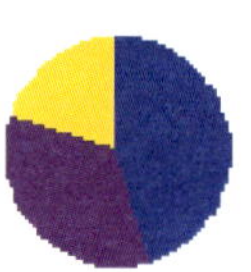

Pie—The pie chart consists of a circle, or pie, divided into two or more sections, or slices. Pie charts show relationships between parts and the whole and are particularly useful for showing proportions and ratios. You can highlight a pie slice by "exploding" it—moving it slightly away from the center. A pie chart can be two- or three-dimensional.

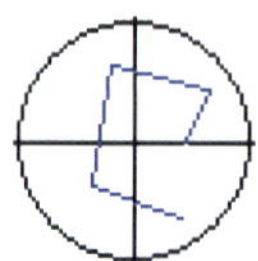

Polar—A polar graph is essentially a line graph drawn on a circular grid. The line relates values to angles. Polar graphs are useful primarily in mathematical and statistical applications.

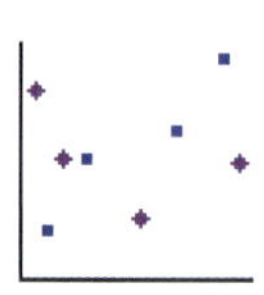

Scatter—A scatter graph plots x,y coordinates based on attribute values. The pattern may reveal a relationship between the values plotted on the grid. A scatter graph may also be three-dimensional, with data plotted along a z-axis.

Creating a graph

Before you create a graph, you should determine what you want to graph. You can graph all features or just the selected ones. Some graphs can effectively display only a limited amount of data, so choose your graph type appropriately. Alternatively, you might consider making more than one graph.

Most graphs are drawn on a grid whose scales are shown by two or three axes: x, y, and z. A data point displayed on the graph is defined by the intersection of two or more field values—for instance, birth rate plotted along the x-axis and death rate plotted along the y-axis on a scatter graph. Keep in mind that a data point doesn't necessarily appear as a point (or dot) on a graph. Depending on the type of graph, a single data point may be represented by a dot, a bar, a pie slice, or some other graphic.

For most graphs, you can choose more than one field value to plot along the axes. You can order the fields as necessary. For example, you can choose which bar in a bar graph represents which field. The order also defines how fields pair up for plotting data points on the graph. For example, suppose you want to graph two attribute fields for the x-axis and two for the y-axis on a scatter ▸

Creating a graph and adding it to a layout

1. Click the Tools menu, point to Graphs, and click Create.
2. Click the Graph type and subtype you want.
3. Click Next.
4. Click the dropdown arrow and click the layer or table you want to graph.
5. Check to graph only the selected features or records.
6. Check the fields you want to graph. You can use the arrow keys to order your columns.
7. Click an option to graph data series using Records or Fields.
8. Click Next. ▸

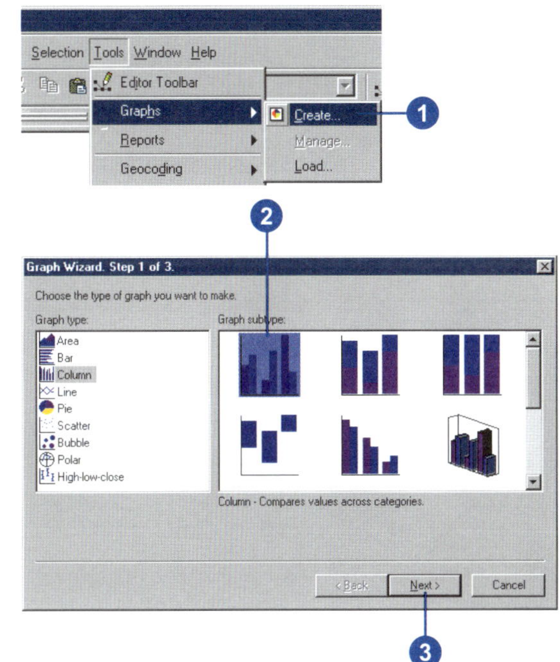

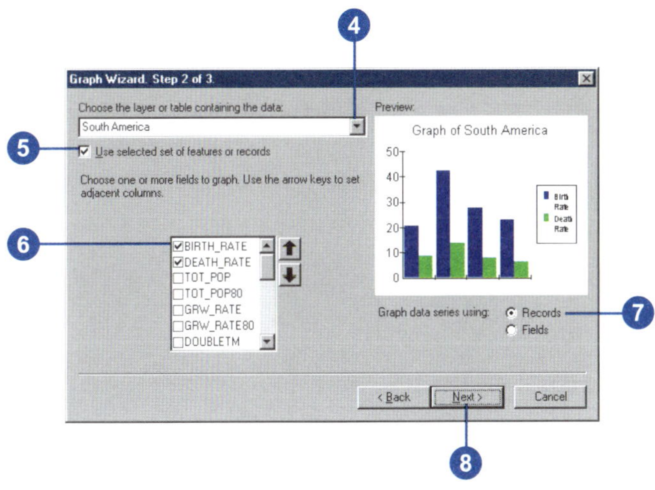

graph. The first field in the list for the x-axis pairs up with the first field in the list for the y-axis, and so on, to determine the location of the data points on the graph.

With some graph types, you can graph data using either records or fields. For instance, suppose you have data on birth and death rates by country. Graphing by record allows you to easily compare the birth and death rates for individual countries.

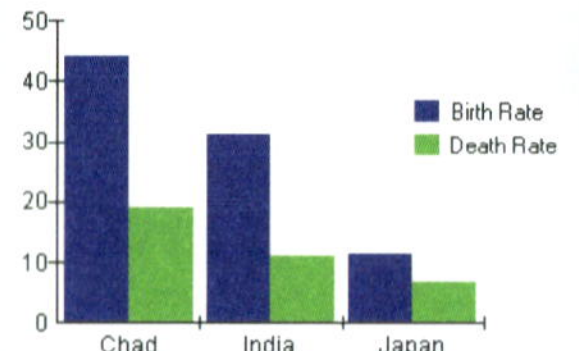

Graphing by field plots all birth and death rates together for all countries.

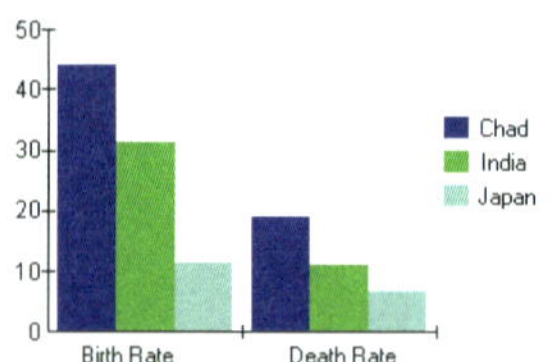

In some cases, you can also plot a secondary graph, called an overlay graph, over the primary one. An overlay graph is a line graph that uses the same x-axis as the primary graph.

9. Type a title for the graph.
10. Check Label X Axis With, then click the dropdown arrow and click a field.
11. Check Show Legend.
12. Check Show Graph on Layout.

    You can add the graph to the layout later if necessary.
13. Click Finish.

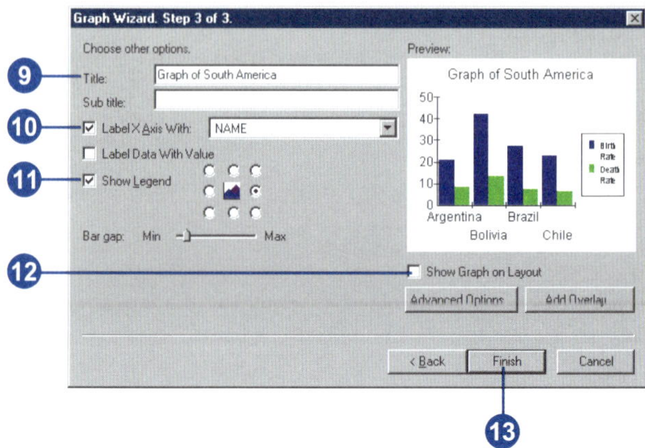

Adding an overlay graph

1. Right-click the title bar of the graph window and click Properties.
2. Click the Appearance tab.
3. Click Add Overlay. ▸

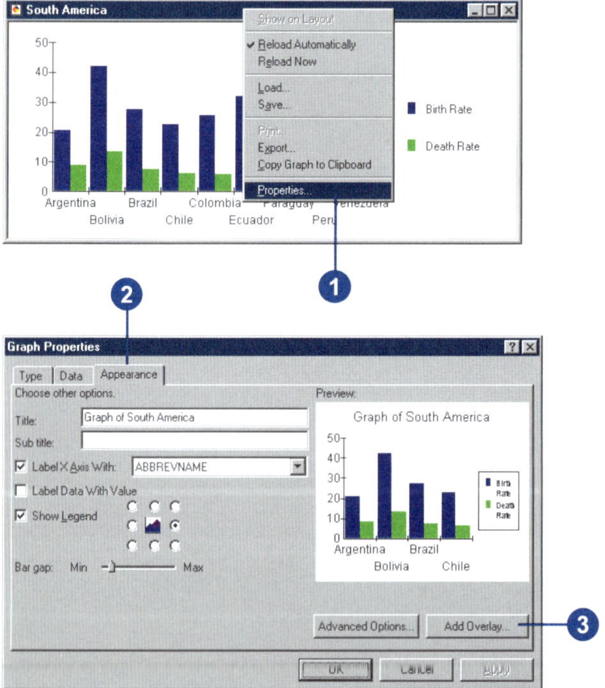

Tip

Why is the Add Overlay button unavailable?

You can only add an overlay graph to column, area, high–low–close, and scatter graphs.

4. Check Display overlay graph.
5. Click the type of overlay graph you want to add.
6. Click Next.
7. Click the Y axis field dropdown arrow and click the field you want to graph.
8. Click the option to add a second y-axis if the data values in the field you chose are not in the same range as the primary graph.
9. Click Next.
10. Click the Color dropdown arrow and set a line color for the graph.
11. Optionally, you can add statistical lines to the overlay. Check those you want to add.
12. Click Finish.

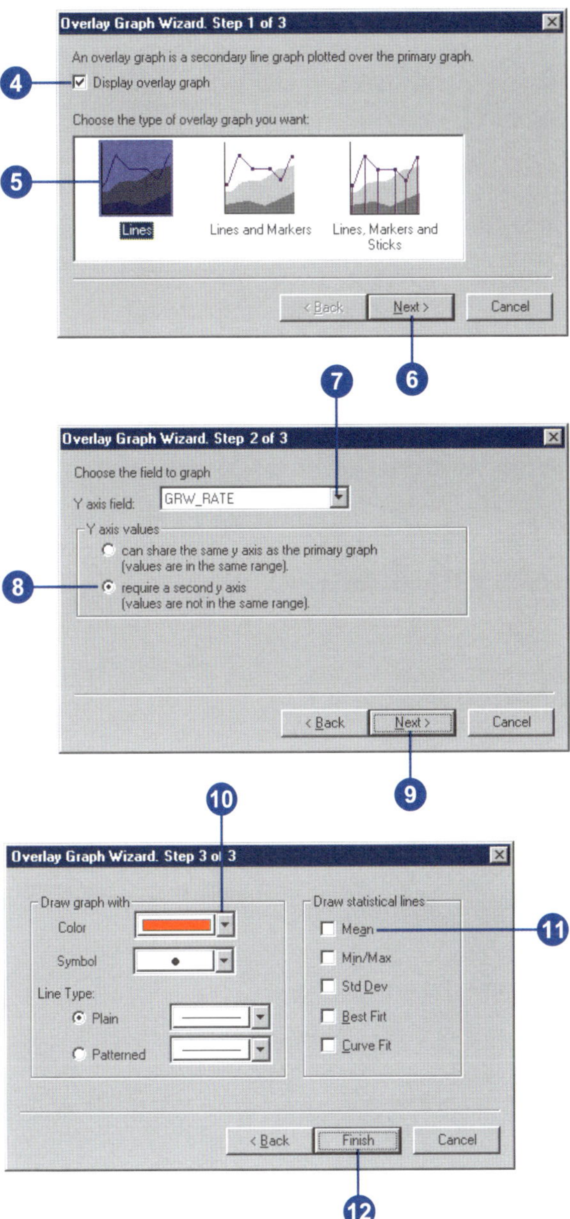

Displaying a graph

While working with the graphs on your map, you can choose to view them in separate windows alongside the ArcMap window, as map elements on the layout ready to be printed, or both.

Graphs are dynamic; they can update automatically as you change the selected set of features in a layer. Thus, as you browse a map selecting new features, the graph will update to reflect the new selected features.

Tip

Updating a graph on the layout

Graphs shown on the layout will automatically update as you change the selected set of features the graph is based on. If you want to create a static copy on the layout, copy and paste the graph on the layout.

Tip

You can only display a graph once on a layout

If you want to show different sets of data graphed the same way, you need to create multiple graphs. You can copy and paste a graph on a layout to create a static copy.

Adding a graph to the layout

1. Right-click the title bar of the graph window and click Show on Layout.

   If the graph is already on the layout, this option will be unavailable.

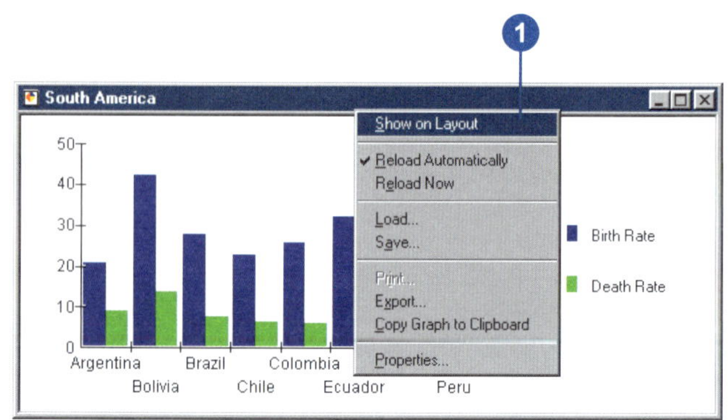

Updating a graph when the selection changes

1. Right-click the title bar of the graph window and click Reload Automatically.

   When Reload Automatically is checked, the graph will update whenever the selected set of features changes. Disabling this feature creates a static graph. This option works only when you create a graph using the selected features.

Modifying a graph

You can control most visual aspects of a graph to create an effective display of your data. For example, you can choose what type of graph you want to use, add titles, label axes, and change the color of graph markers—such as the bars in a bar graph.

Tip

Identifying features on a graph

Clicking on a data point on a graph—whether a bar on a bar graph or a point on a line graph—identifies the associated feature on the layer.

Changing the graph type

1. Right-click the title bar of the graph window and click Properties.
2. Click the Type tab.
3. Click the graph type you want to use.
4. Click the Graph subtype you want to use.
5. Click OK.

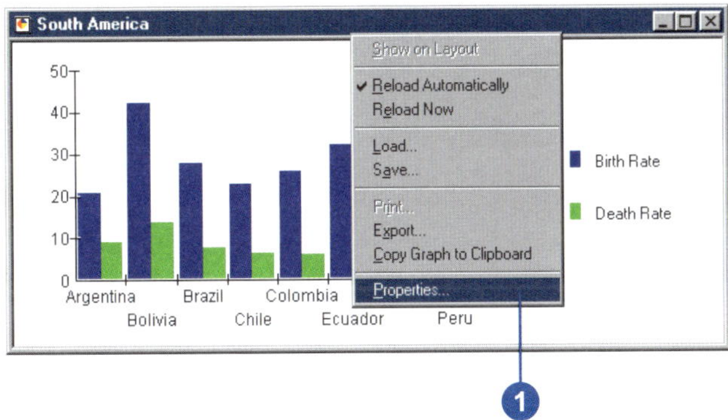

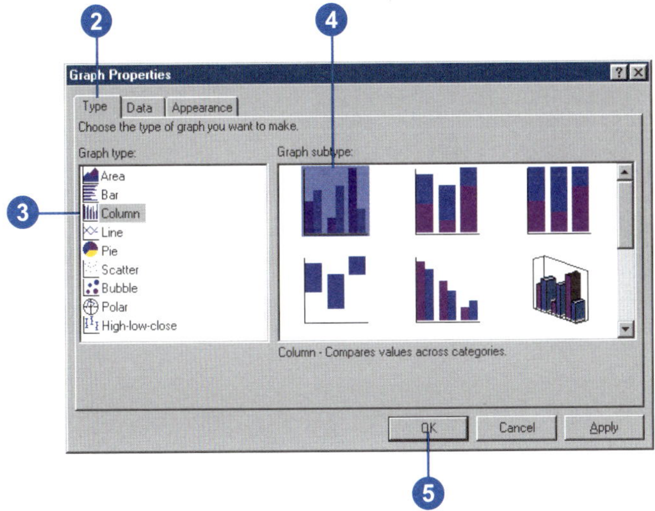

Tip

Changing the font and size of the title

To change the text font or text size of the graph title, click the Advanced Options button on the Appearance tab. Then, click the Fonts tab on the dialog box that appears.

Tip

Changing the text color of the title

To set the text color of the title, click the Advanced Options button on the Appearance tab. On the dialog box that appears, click the Background tab and change the text color.

Adding a title to a graph

1. Right-click the title bar of the graph window and click Properties.
2. Click the Appearance tab.
3. Type a title.
4. Click OK.

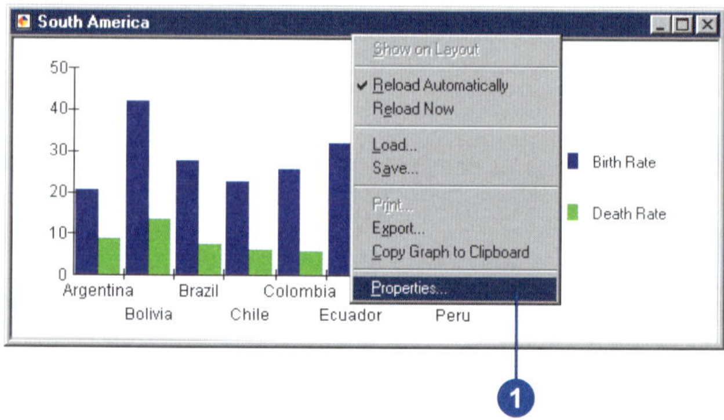

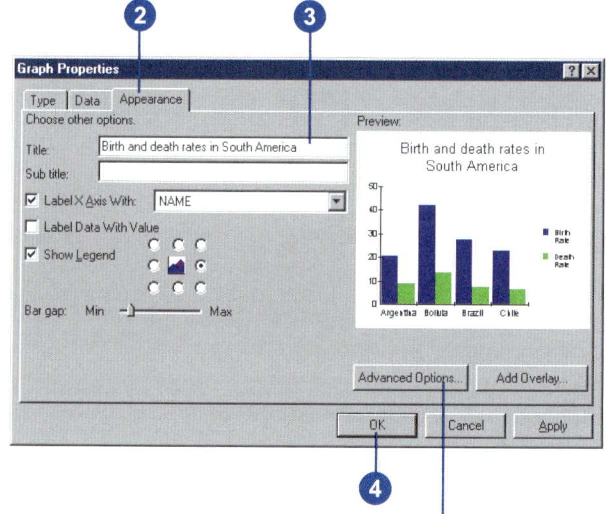

Click Advanced Options to set the text font, size, and color. Set the text font and size from the Font tab. Set the text color from the Background tab.

Tip

Why don't I see the color I want?

Click the Background tab on the Advanced Options dialog box and change the palette.

Tip

Setting point symbols

For graphs that plot points—for example, the scatter graph—you can also choose the symbol to plot points with. Click the Markers tab on the Advanced Options dialog box.

Changing graph marker colors

1. Right-click the title bar of the graph window and click Properties.
2. Click the Appearance tab.
3. Click Advanced Options.
4. Click the Markers tab.
5. Click the marker you want to change.
6. Click the Color dropdown arrow and click a color.
7. Click OK.

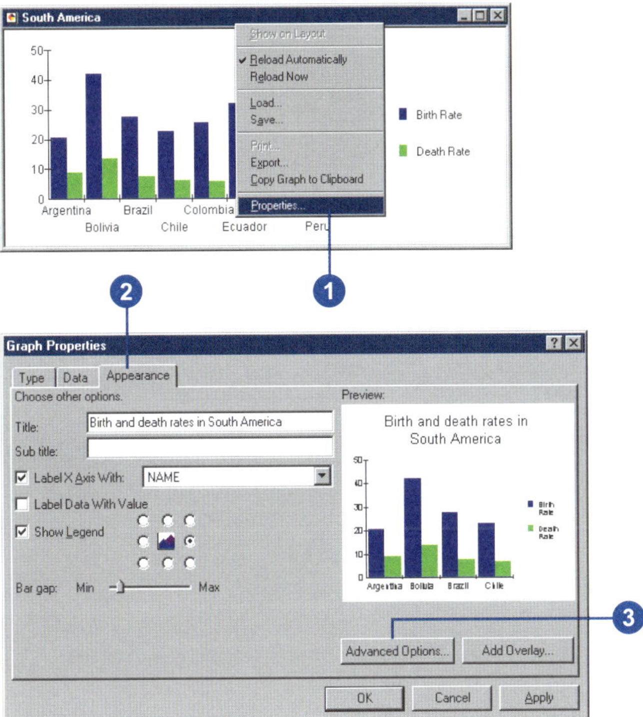

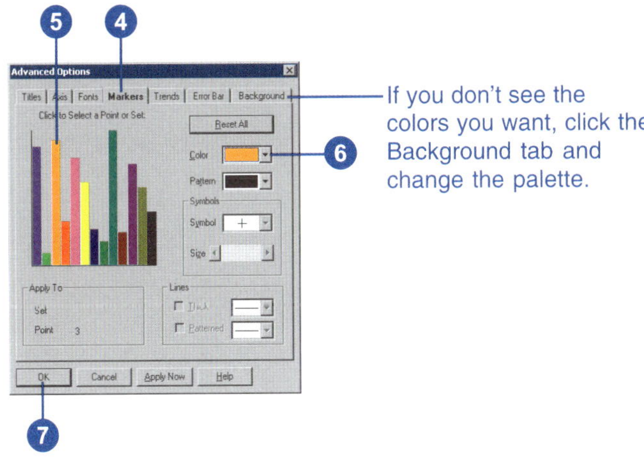

If you don't see the colors you want, click the Background tab and change the palette.

Tip

Changing the font of the legend

To change the text font or text size of the text in the legend, click the Fonts tab on the Advanced Options dialog box.

Tip

Changing the text color of the legend

The option for setting the text color of the legend is located on the Background tab on the Advanced Options dialog box.

Tip

Placing a border around a legend

You'll find the option to add or remove the border around a legend on the Background tab on the Advanced Options dialog box.

Adding a legend to a graph

1. Right-click the title bar of the graph window and click Properties.
2. Click the Appearance tab.
3. Check Show Legend and click a legend position.
4. Click OK.

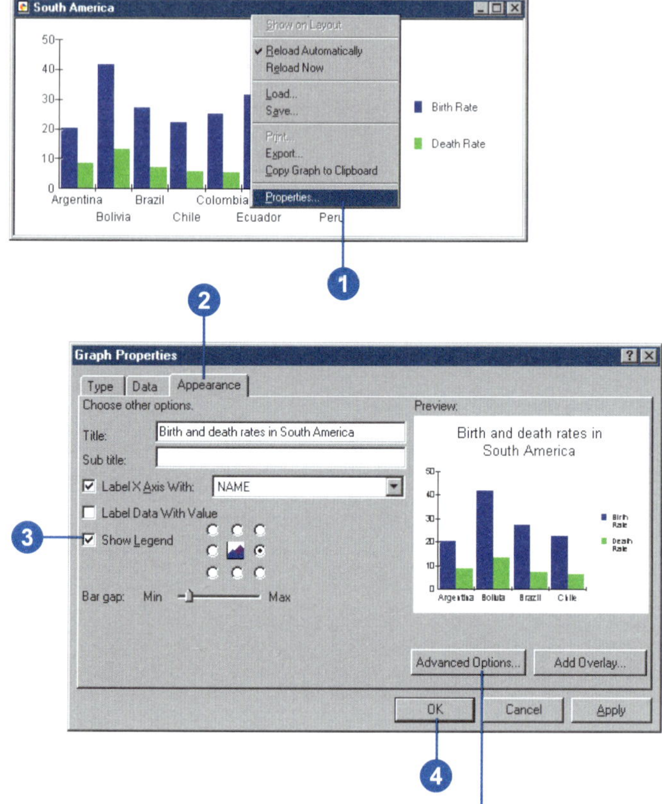

Click Advanced Options to set the text size, font, and color for the legend as well as add and remove borders. The text font and size controls are found on the Fonts tab. The text color and border controls are found on the Background tab.

Tip

Setting the text font for labeling axes

Click the Fonts tab on the Advanced Options dialog box. In Apply to, click Labels, then in Typeface, choose the font you want your labels displayed with.

Tip

Why don't I see the color I want?

Click the Background tab on the Advanced Options dialog box and change the palette.

Controlling the x-, y-, and z-axes of the graph

1. Right-click the title bar of the graph window and click Properties.
2. Click the Appearance tab.
3. Click Advanced Options.
4. Click the Axis tab.
5. Click the axis you want to modify.
6. Set the position of the axis. For example, click Variable, Left, or Right for the y-axis.
7. Set the scale, or numeric range, for the axis.

   Zero Origin: axis coordinates range from zero to maximum data value.

   Variable Origin: axis coordinates range is set to the actual data value range.

   User-Defined: you specify the coordinate range for the axis.
8. Optionally, check the boxes to show Tick Marks and Grids on your graph.
9. Click OK.

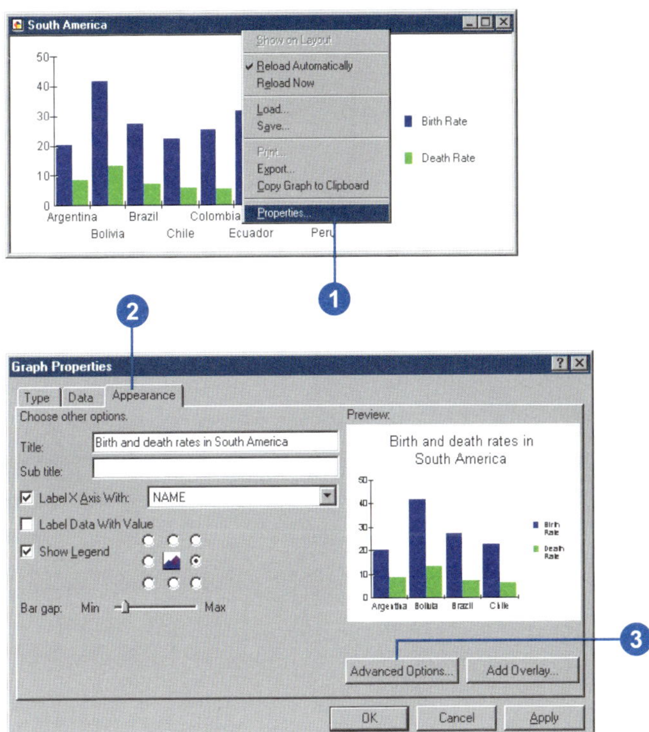

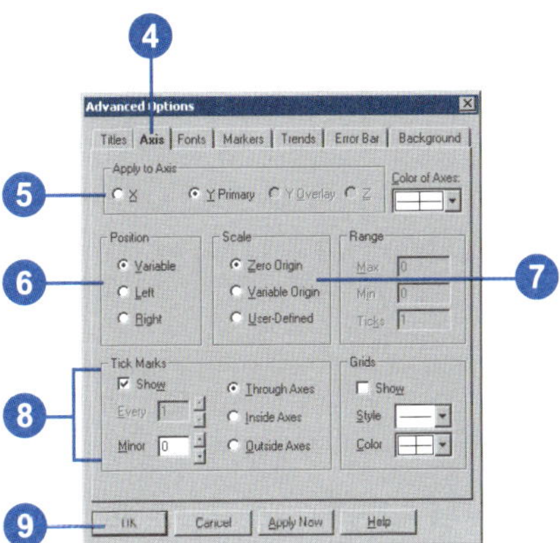

Tip

What are trend lines?

Trend lines are additional lines you can draw on top of your graph. A trend line can represent a statistical value, such as the mean or standard deviation, or it might be a limit line you define to highlight data values that fall outside a prescribed limit.

Drawing trend lines on a graph

1. Right-click the title bar of the graph window and click Properties.
2. Click the Appearance tab.
3. Click Advanced Options.
4. Click the Trends tab.

   Not all graph types support trend lines. If the tab is not available, the graph type does not support it.
5. Check All Sets to draw trend lines for each attribute value you're graphing.

   For example, if you're graphing birth and death rates, you can draw a mean line for both values or just one of them.
6. Check the line types you want to add to the graph.
7. Type a value to add your own Limit Lines (drawn along a specified y-axis value).
8. Click OK.

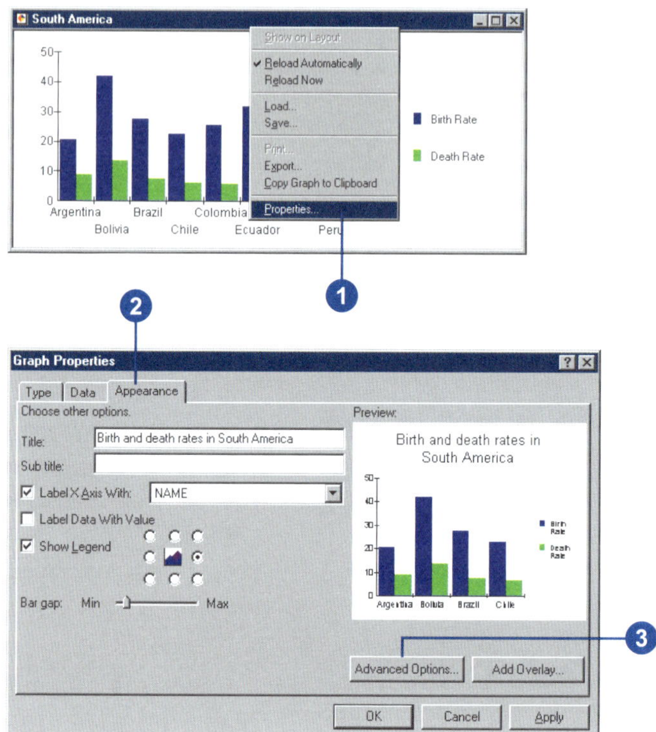

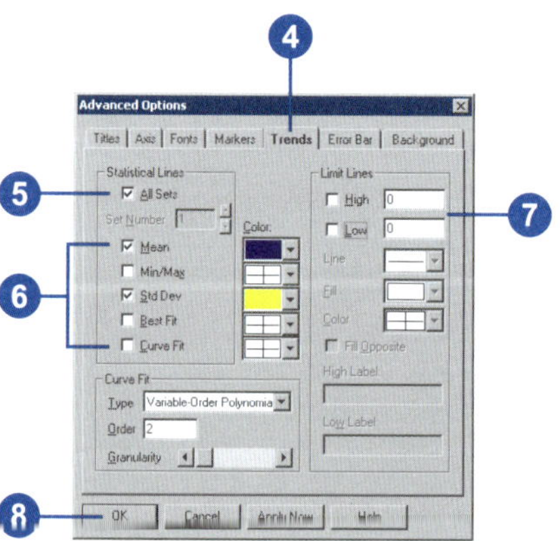

Creating a static copy of a graph

Graphs shown on a layout update automatically as you change the selected features in the layer the graph is based on. Sometimes you may want to create a static version of a graph to show how different features compare. To do this, you can copy and paste the graph on the layout.

Copying a graph on the layout creates a static version for display only. Although it looks just like the original graph, it is not listed in the Graph Manager, and you can only change certain appearance properties.

Tip

Copying a graph to the clipboard

From the graph window, you can copy a graph to the clipboard. This creates a bitmap image of the graph that you can then paste into other applications.

Copying and pasting a graph

1. Click the Select Elements tool on the Tools toolbar.
2. Click the graph on the layout you want to copy.
3. Click the Edit menu and click Copy.
4. Click the Edit menu and click Paste.

   The static copy of the graph appears on the layout.

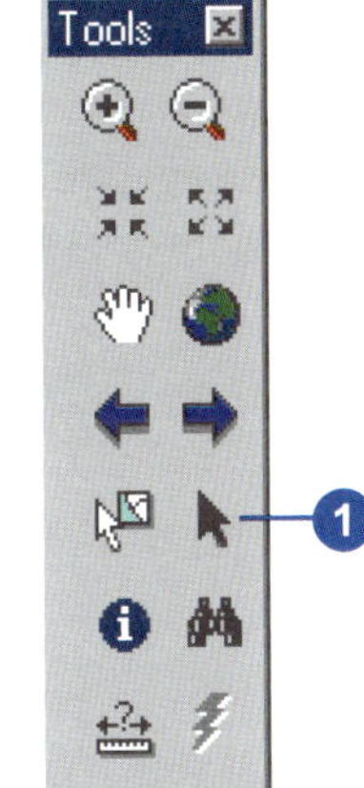

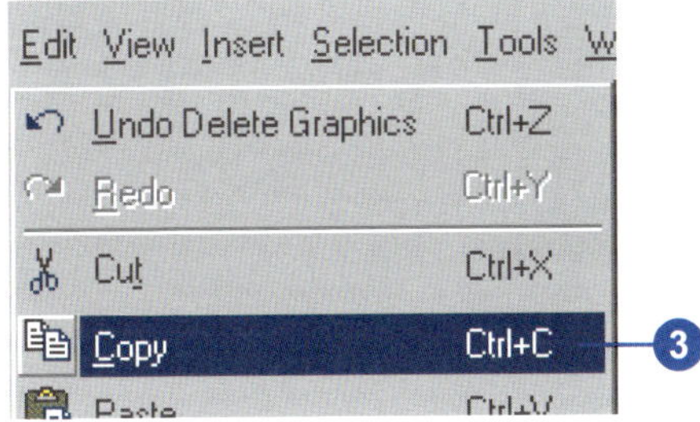

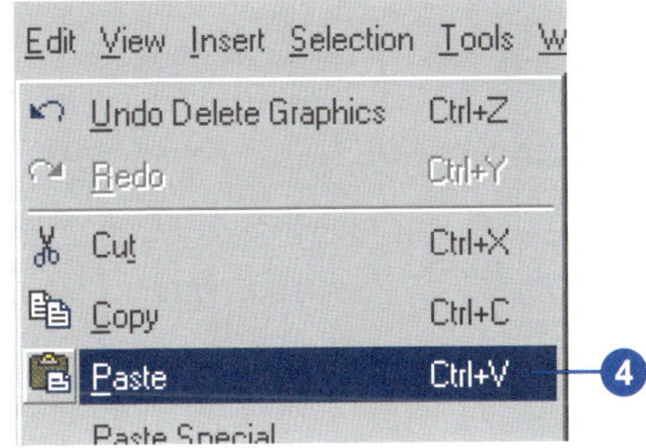

Managing graphs

Your maps may contain several graphs. To help you manage them, you'll use the Graph Manager. From this dialog box, you can open a graph, add it to the layout, rename it, and delete it.

If you remove the layer a graph is based on, the graph will still remain on the map. You need to explicitly remove the graph if you no longer want it. You can associate the graph with a new layer by displaying the Graph Properties and assigning a new layer.

Tip

How can I tell if a graph is on the layout?

If you see this icon next to the graph name in the Graph Manager, the graph is on the layout.

Opening a graph

1. Click the Tools menu, point to Graphs, and click Manage.
2. Click the graph you want to open.
3. Click Open.

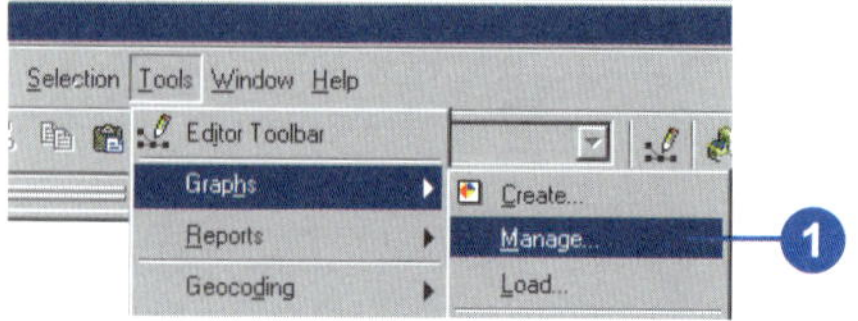

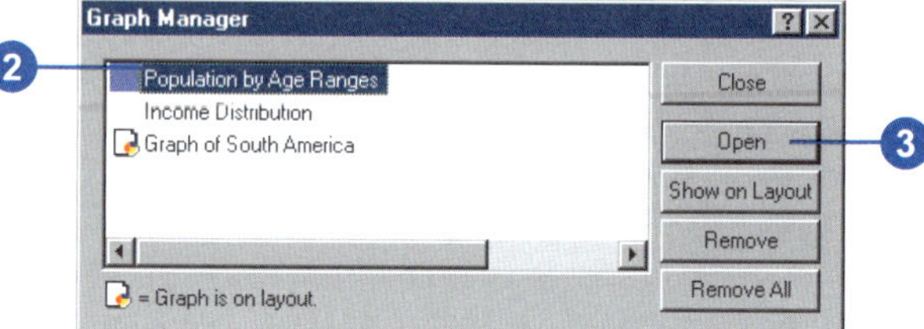

Removing a graph

1. Click the Tools menu, point to Graphs, and click Manage.
2. Click the graph you want to remove.
3. Click Remove.

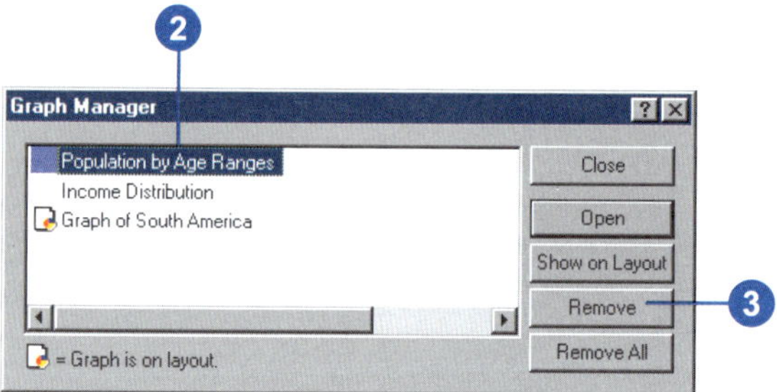

Renaming a graph

1. Click the Tools menu, point to Graphs, and click Manage.
2. Click the graph you want to rename.
3. Click again over the name.

   This will allow you to type a new name.
4. Type the new name.

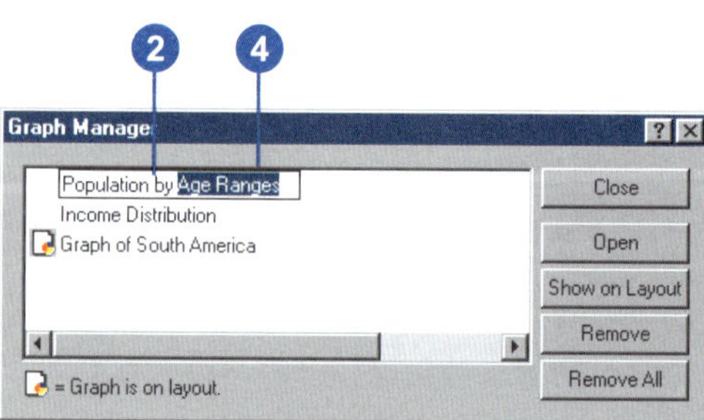

Saving and loading a graph

If you want to copy a graph that you've made on one map and put it on another, save it to a file on disk. That way, you can load the graph onto another map and place it as appropriate.

All the properties you've set on a graph are maintained when you save it to disk, including the type of graph, data the graph is based on, whether there's a selected set, what fields are being graphed, and so on. While you can preview a saved graph in ArcCatalog, you can only change the properties in ArcMap.

When you load a graph on a map that doesn't contain the layer the graph is based on, ArcMap will prompt you to add that layer to the map. If you choose not to, the graph will still display, but you won't be able to change the features shown on the graph.

Tip

Using a graph in another application

If you want to include a graph in another application, you can export the graph to these formats: Windows bitmap, JPG, Windows metafile, and PNG.

Saving a graph

1. Right-click the title bar of the graph window and click Save.
2. Click the Save in dropdown arrow and navigate to the location you want to save the graph to.
3. Type a name for the graph.
4. Click Save.

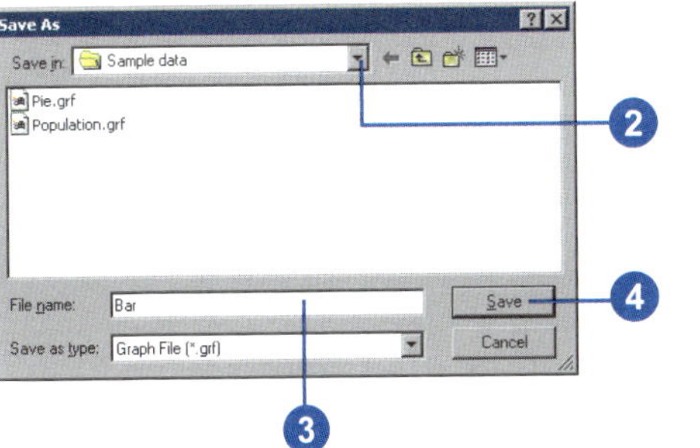

Loading a graph

1. Right-click the title bar of the graph window and click Load.
2. Click the Look in dropdown arrow and navigate to the location where the graph is stored.
3. Click the graph.
4. Click Open.

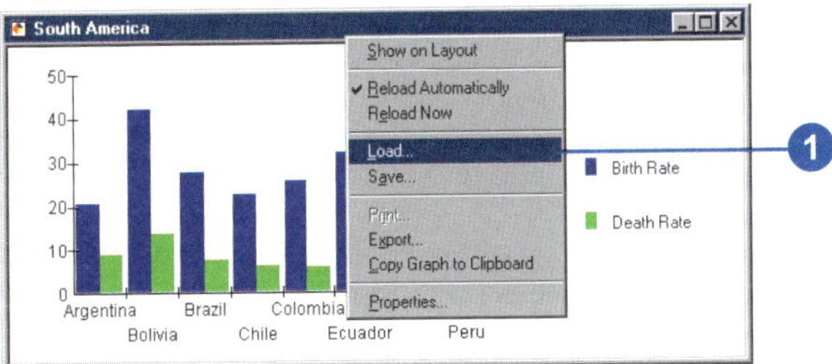

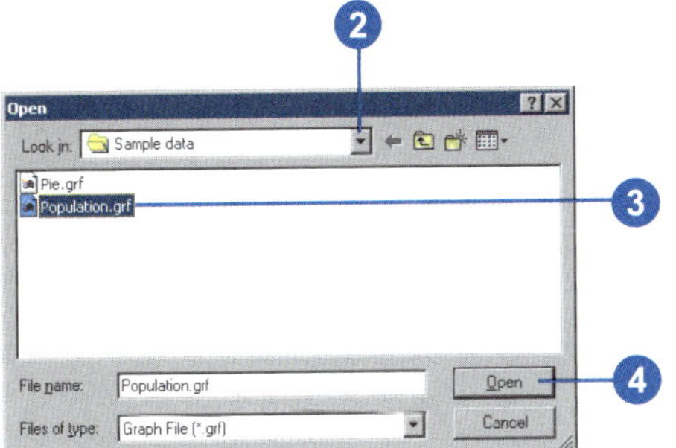

Exporting a graph

When you want to use a graph in another application, you can export it to one of these formats: bitmap (.bmp), JPEG (.jpg), PNG (.png), and Windows metafile (.wmf). For example, you might want to include a graph in a document you're distributing with your map.

Tip

Copying a graph to the clipboard

You can also copy a graph to the clipboard and paste it into another application. The graph is copied as a bitmap file (.bmp).

1. Right-click the title bar of the graph window and click Export.
2. Click the Save in dropdown arrow and navigate to the location where you want to save the exported graph.
3. Type a name for the graph.
4. Click the Save as type dropdown arrow and click the type of file you want to export.
5. Click Save.

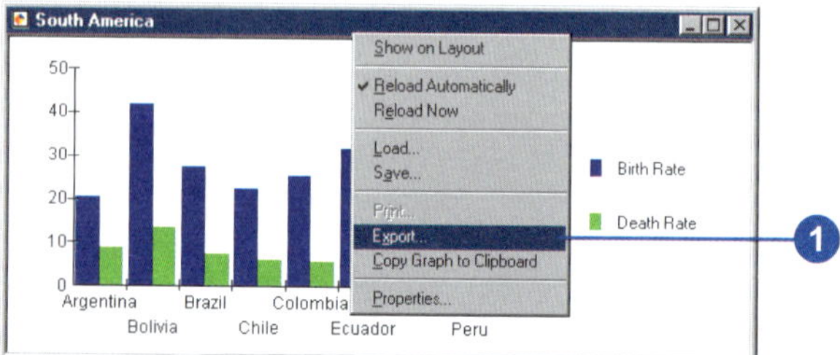

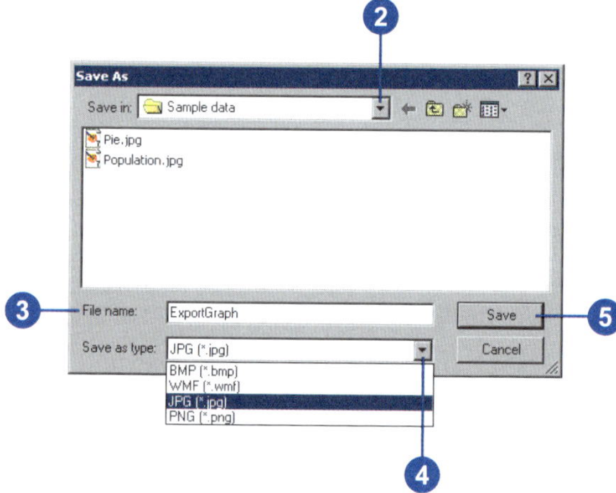

Creating reports

12

IN THIS CHAPTER

- **About reports**
- **Creating a simple report**
- **Setting the report type and size**
- **Working with fields**
- **Organizing report data**
- **Adding report elements**
- **Controlling the presentation**
- **Saving and loading a report**
- **Using Crystal Reports**

Reports present the facts and figures behind your analysis and are invaluable companions to the maps you're creating. Reports let you effectively display attribute information about map features in a tabular format that you control. The information displayed in a report comes directly from the attribute information stored with your geographic data.

South America - Population Growth Rate

Country	Population	Growth Rate	Birth Rate	Death Rate
Argentina	31883010	1.19	20.6	8.7
Bolivia	7110000	2.75	42	13.5
Brazil	147294000	2.01	27.4	7.6
Chile	12980000	1.71	22.7	6.3
Colombia	32335010	1.96	25.5	5.9
Ecuador	10329000	2.51	31.9	7.1
French Guiana	99000	-99	-99	-99
Guyana	800000	0.13	25.6	-99
Paraguay	4161000	2.9	34.3	6.3
Peru	21142000	2.2	30.5	8.6
Suriname	435800	2.11	31	7
Uruguay	3077000	0.55	17.1	9.7
Venezuela	19244000	2.6	29.6	5.3

A report created in ArcMap

ArcMap provides two methods for creating reports. Using the built-in ArcMap reporting tool, you can create reports that are stored directly with your map. Once created, you can add the report to your map layout and print it out. ArcMap also integrates with Crystal Decisions' Crystal Reports™ 9. Crystal Reports lets you quickly create presentation-quality reports to include with your map or distribute to others. The Report Designer provides a graphical interface for controlling the look of your report. (NOTE: Crystal Reports 9 Standard Edition is distributed with ArcMap. In order to access the reporting tools, you must have installed Crystal Reports.) Which one should you use? For a simple report, try the built-in ArcMap reporting tool. When you need a full-functioned graphical report building tool, try Crystal Reports.

About reports

What is a report?

A report presents tabular information about features on the map formatted in an attractive manner. Reports are derived from an attribute table on your map. You can choose which fields from your table you want to display and how you want to display them. Once you've created a report, you can place it on your map layout next to your geographic data or save it as a file—for example, a PDF file—for distribution. The figures below show a few examples of the kinds of reports you can create. This report displays records in a tabular form, where each record is represented by a row in the display. You can include the title, page numbers, the current date, summary statistics, and images.

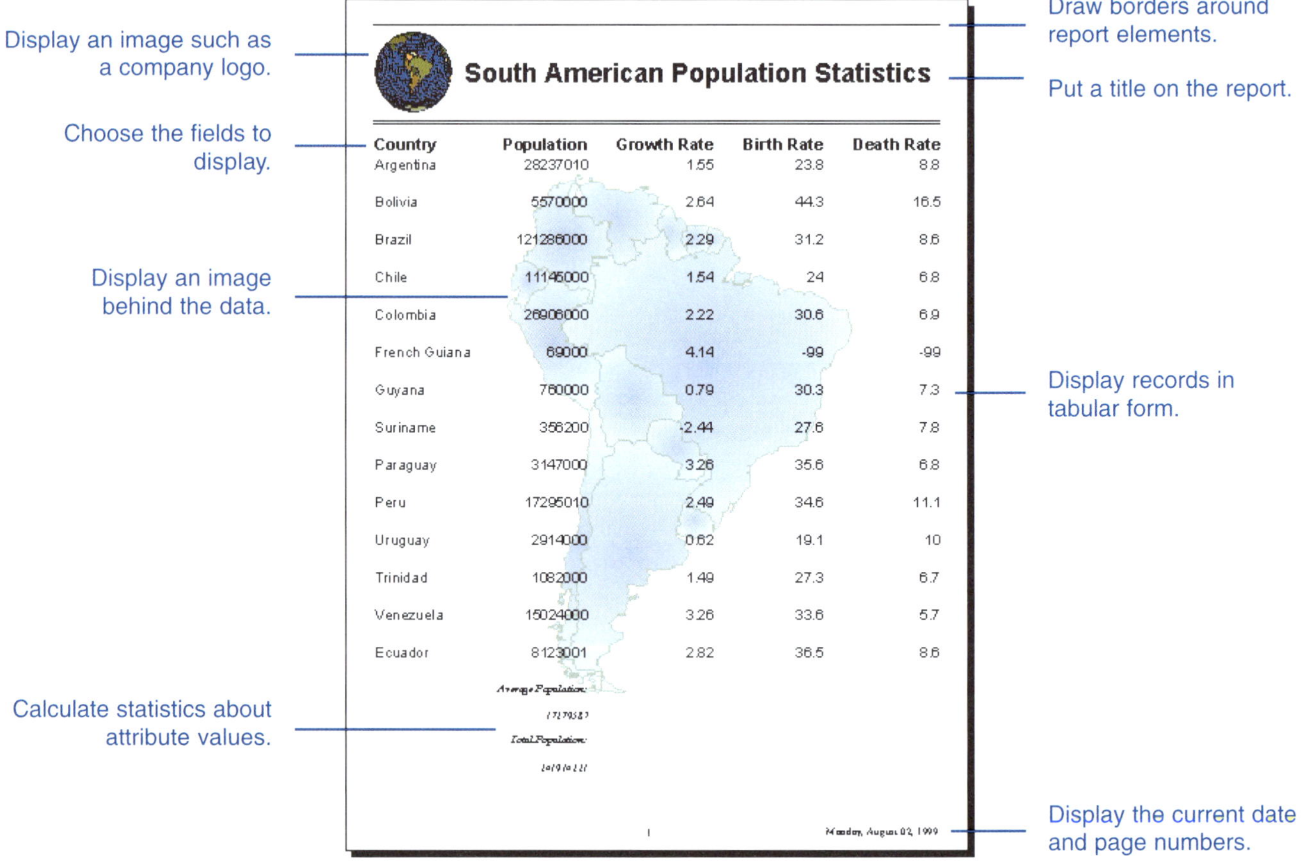

South American Population Statistics

Country	Population	Growth Rate	Birth Rate	Death Rate
Argentina	28237010	1.55	23.8	8.8
Bolivia	5570000	2.64	44.3	16.5
Brazil	121286000	2.29	31.2	8.6
Chile	11145000	1.54	24	6.8
Colombia	26906000	2.22	30.6	6.9
French Guiana	69000	4.14	-99	-99
Guyana	760000	0.79	30.3	7.3
Suriname	356200	-2.44	27.6	7.8
Paraguay	3147000	3.26	35.6	6.8
Peru	17295010	2.49	34.6	11.1
Uruguay	2914000	0.62	19.1	10
Trinidad	1082000	1.49	27.3	6.7
Venezuela	15024000	3.26	33.6	5.7
Ecuador	8123001	2.82	36.5	8.6

Average Population:

Total Population:

1 Monday, August 02, 1999

This report organizes the data in a single column, displaying field names and values vertically. The report groups cities by which country they're in and displays them alphabetically by name. By controlling the text font, size, and color and shading certain elements, you can create a report that highlights the information you want to convey.

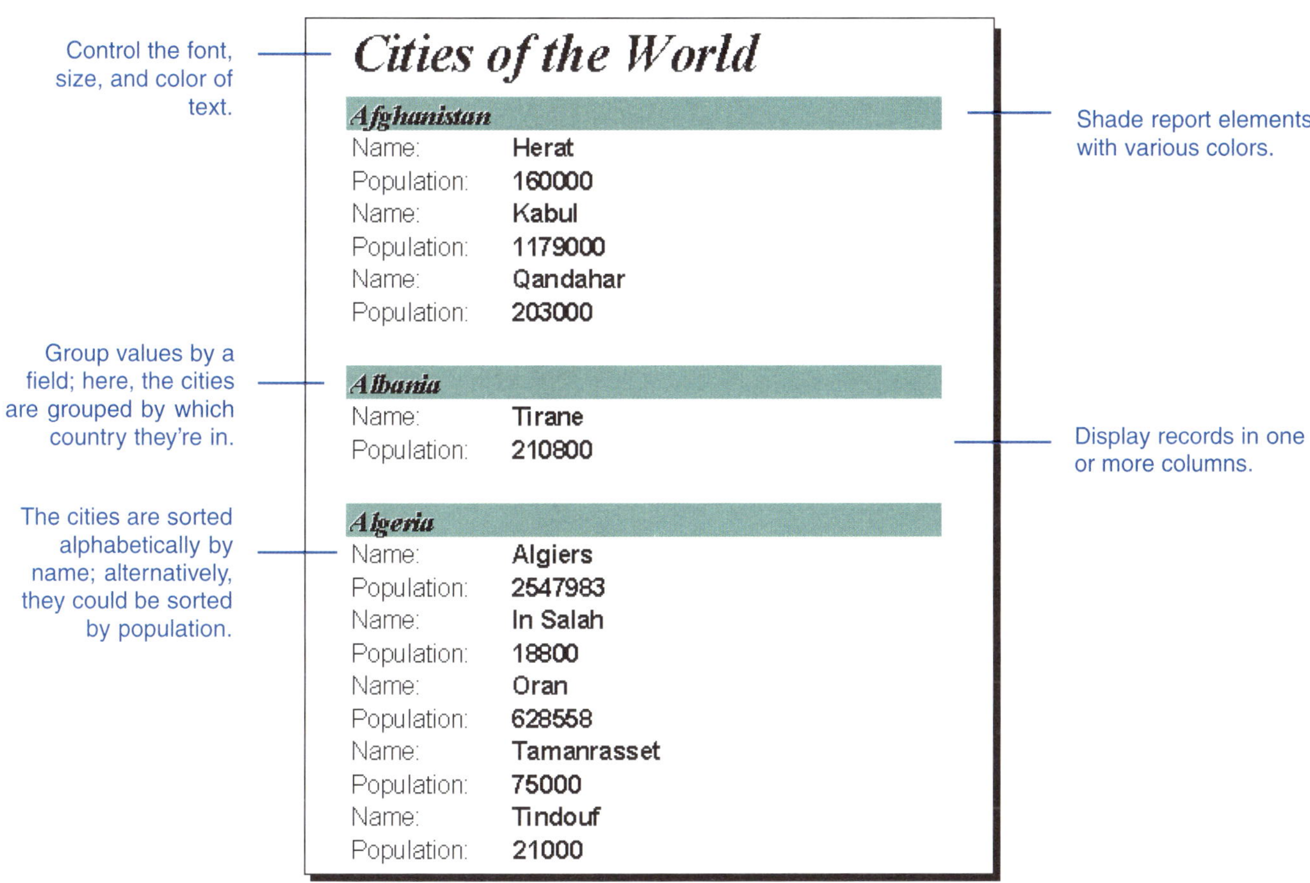

Report sections

Reports are divided into a series of sections; each one identifies a particular area on the report. You control how a report looks by manipulating the contents of a section and setting properties, such as its size and color. For example, the section at the top of the report typically contains the title and subtitle of the report; however, you don't have to include either of these report elements if you don't need them. The following figure highlights the various sections in a report with different colors.

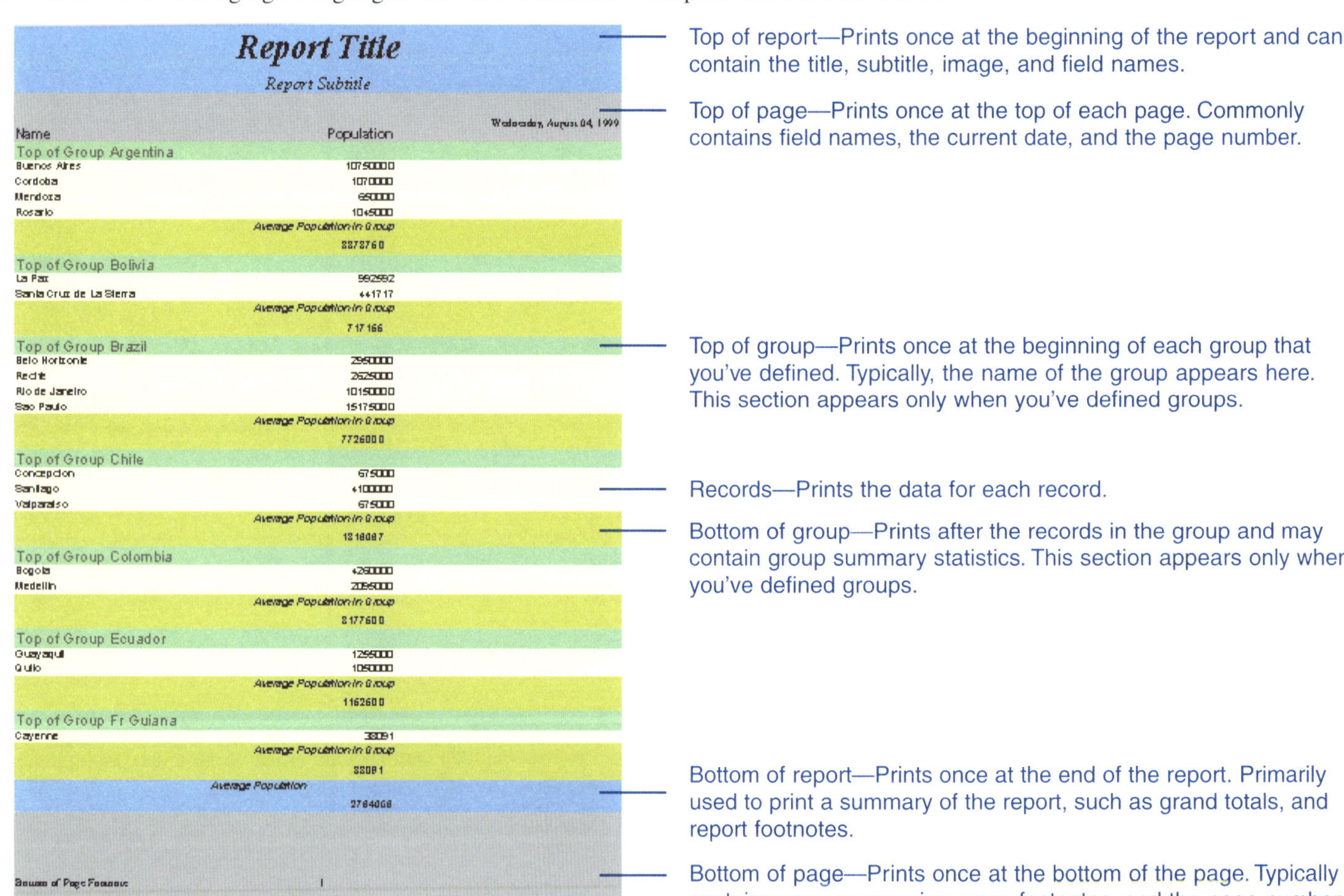

Working with sections

ArcMap automatically calculates the height of each section based on the height of the elements in it. An element can be a title, subtitle, column header, page number, and so on. For example, if you use a large point size for the title, ArcMap enlarges the section at the top of the report to accommodate the text.

Using a larger point size increases the height of the title and subsequently increases the height of the top of the report section.

If you want to manually control the height of a section, you can turn off the automatic sizing of the section and set the height explicitly. You will also need to make sure that the height of elements, such as the title, within a section are sized accordingly. Otherwise, they may be truncated.

By turning off the Autosize property, you can explicitly set the height of the section. Elements in that section, however, can become truncated.

The width is the same for all sections and is determined by the width you specify for the report.

Working with elements in a section

Within each section, you can control the size and position of an element within it. As with sections, you can let ArcMap automatically size elements or size them explicitly yourself. In the figure below, the title has an explicit height, width, and position set. It also has a border around it and a background color that's different from the background color of the section.

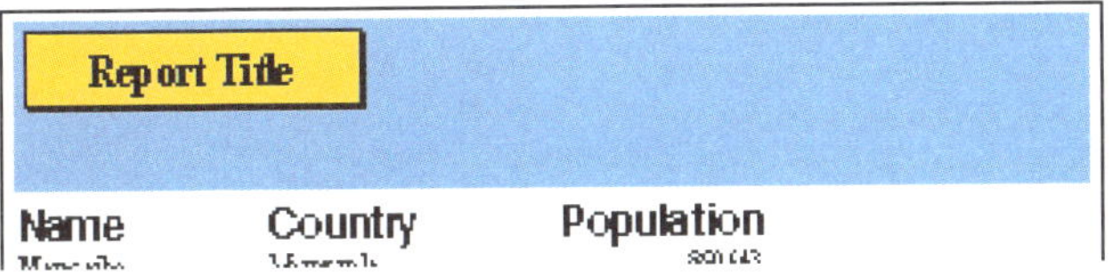

You can set the size, position, and color of an element.

Once the title is the right size, you can position it relative to other elements such as an image representing a company logo.

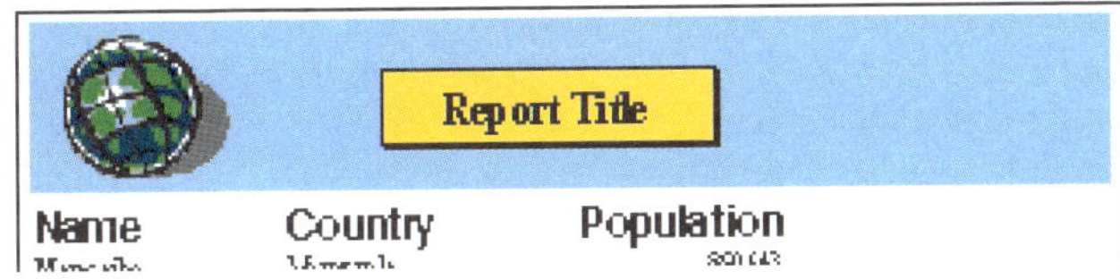

Creating a simple report

A report lets you organize and display the tabular data that's associated with your geographic features. Sometimes you'll want to print out a report to distribute with your map or, alternatively, you might put the report right on the map.

When you create your report, you choose what fields to display and whether you want to generate a report listing all features in a layer or only the selected ones. Once you've created your report, you can put it on the layout next to your map display or print it out.

A report has many properties that you set when you create it. For example, you can set a property to define what type of report you want—tabular or columnar. You can also set a particular page size and use a particular text font and color for the text on your report.

The set of steps to the right show you how to make a simple tabular report. Subsequent sections in this chapter focus on how to set report properties to achieve a particular result.

Creating a simple tabular report

1. Click the Tools menu, point to Reports, and click Create Report.
2. On the Fields tab, click the Layer/Table dropdown arrow and click the layer or table on which you want to base the report.
3. In the Available Fields list, double-click the fields you want to include in the report.
4. Check Use Selected Set if you want to create a report with only the selected features.
5. Click the arrow buttons to order the report fields.
6. Click the Sorting tab.
7. Click a field to sort in the Sort column.
8. Click the Display tab. ►

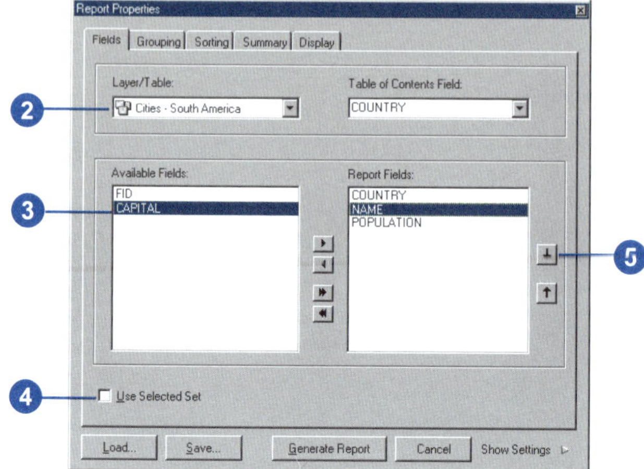

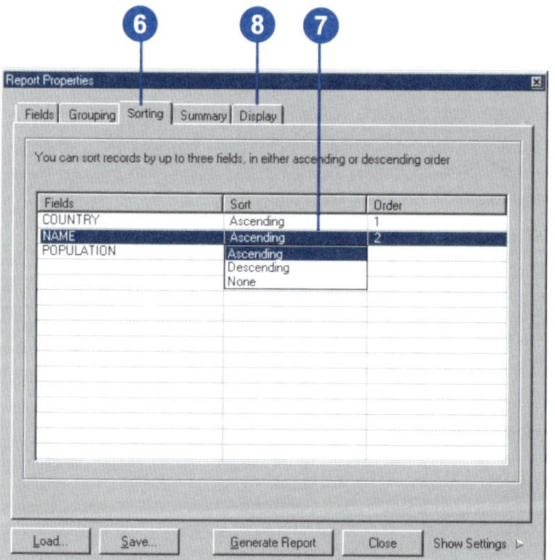

Tip

Adding a report to the map layout

Once you've created a report, you can display it on your map layout. By creating a report with a particular page size, you can ensure that it fits exactly where you want it on the layout.

Tip

Adding titles, subtitles, and page numbers to your report

You can add additional elements to your report—such as a title, page numbers, an image, and footnotes—to enhance the display. The Elements setting on the Display tab lists all the additional things you can add.

Tip

Shading records

To make the data in your reports easier to read, you can shade every other record with a color. On the Report Properties dialog box, click the Display tab. Click Report and click Records. Then set the Shade Records and Shade Color properties.

9. Under Settings, click Elements.
10. Check Title to add a title to the report.
11. Locate the Text property and type a title for the report.
12. Click the Font property and set the font and size of the title.
13. Click Show Settings to preview the report.
14. Click Generate Report.
15. At the top of the Report Viewer, click Add to add the report to the map layout.
16. Click OK.

    The report is added to the layout as a graphic element. Each page of the report is added as a separate graphic element on the layout.

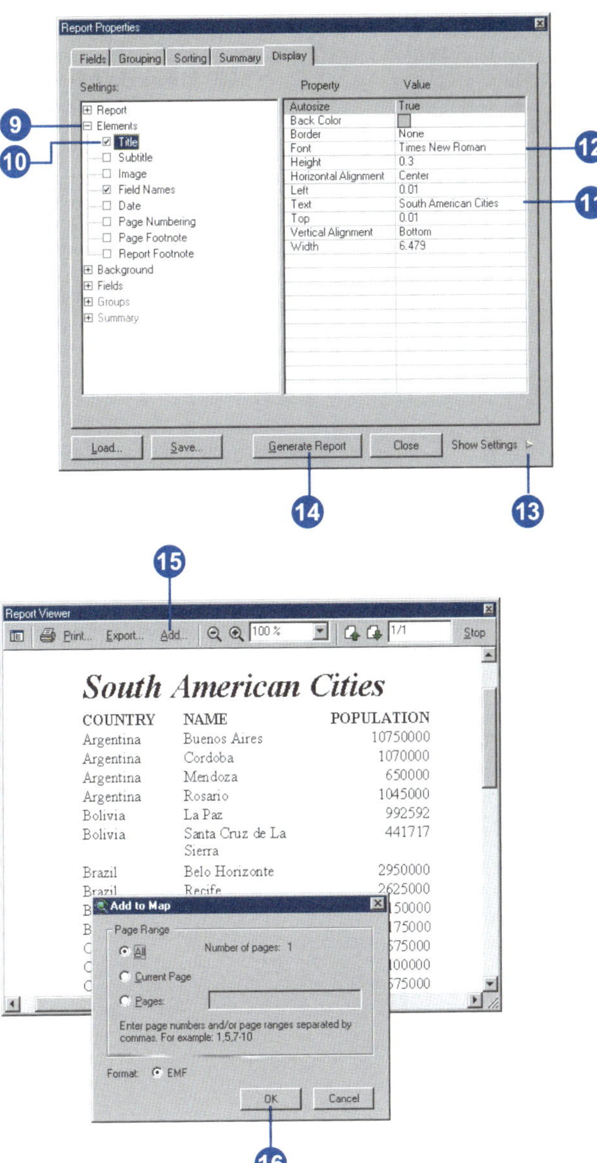

Setting the report type and size

You can make two different types of reports—tabular or columnar. A tabular report organizes data in rows and columns, much like a spreadsheet. Each row represents one record of data, and each column represents one field. A columnar report displays fields and their values vertically in columns, much like a newspaper column layout. With a columnar report, you can specify the number of columns you want to display.

When creating a report, you can specify the size you want. The size you choose will depend on how you plan to use the report. If you plan to print it out on paper, you'll typically use common paper sizes. If you plan to incorporate it onto your map layout, you'll want to set a page size that's as close to the actual size of the space available on the layout. That way, the text sizes you specify in the report will be the same on the map layout.

The page size you set initially determines the height and width of the report. You can, however, increase the width to accommodate the data you want to display. If the width exceeds the printer page size, the report will print on additional pages.

Setting the report type

1. Click the Display tab on the Report Properties dialog box.
2. Click Report.
3. Click the Style property, then click the dropdown arrow, and click Tabular or Columnar.
4. Optionally, for a columnar report, set the Column Count and Column Style properties.

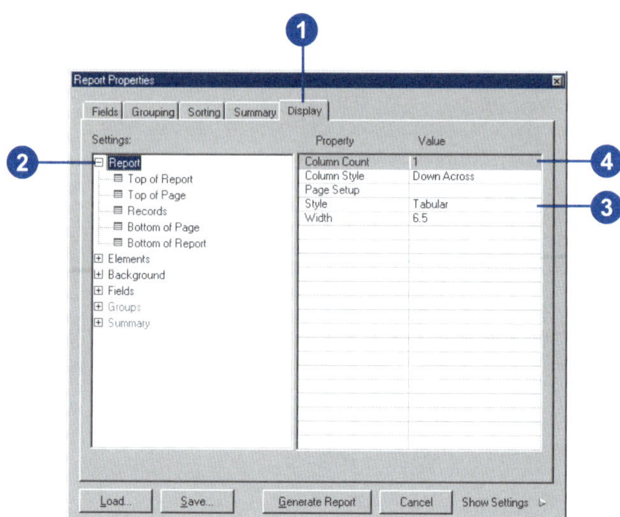

Setting the report width

1. Click the Display tab on the Report Properties dialog box.
2. Click Report.
3. Double-click the Width property and type a width in inches.

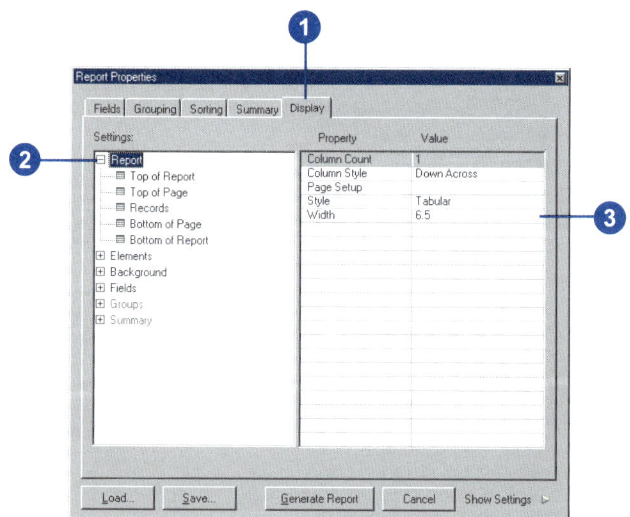

Setting the page size, orientation, and margin

1. Click the Display tab on the Report Properties dialog box.
2. Click Report.
3. Click the Page Setup property and click the button to display the Page Setup dialog box.
4. Click the Paper Size dropdown arrow and click the size you want.

   If you're incorporating the report on a layout, choose a size that's close to the available space on the layout.
5. Click Portrait or Landscape.
6. Type the Left, Right, Top, and Bottom margins.
7. Click OK.

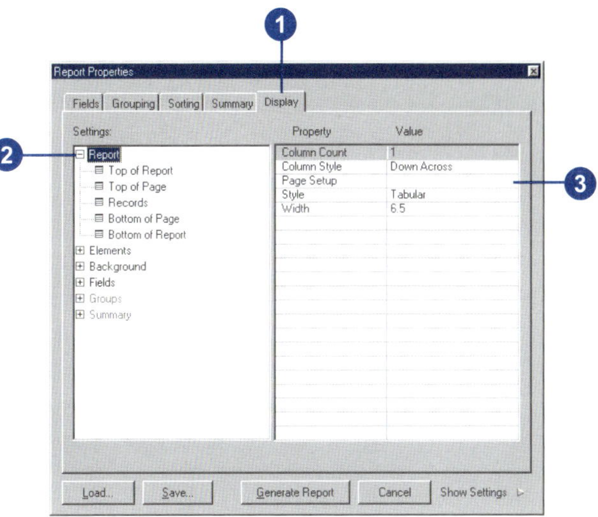

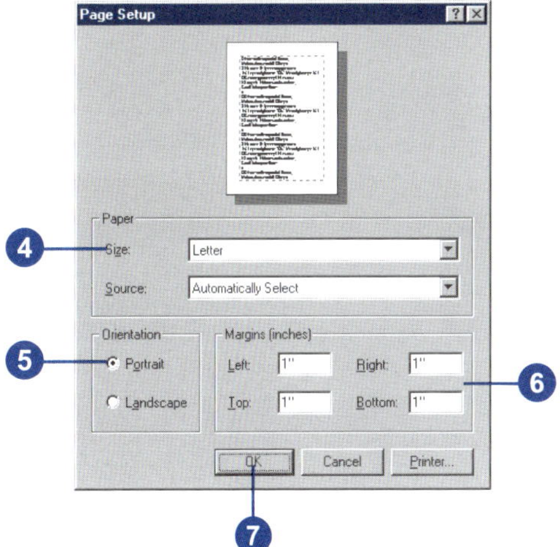

Tip

Setting the report height

While you can set the report width independently, the report height is determined by the paper size. However, by adjusting the page margins, you can precisely control the report size.

Tip

Composing the report for a layout

Compose your report at the size you want it to be on your layout. Choose the paper size so the report will fit in the available space. Then, you can adjust the margins of the page to get the exact size you need. This will ensure that the text on the report is sized correctly. You can still reduce or enlarge the report as it appears on the layout.

Working with fields

The information displayed on your report is based on the fields you choose to include. When you choose the fields, you can also set the order in which they will appear in the report. That way, you can have certain fields appear before others.

The field name displayed on the report is the same as its name in the database. However, as field names in a database are often abbreviations or cryptic descriptions of the attribute stored in the field, you can replace a name with an *alias*—your own descriptive text—to help clarify its meaning.

ArcMap automatically sets the display width of a field to accommodate the width of the data. However, you can set a width explicitly. You may also want to increase the width of the report to prevent the fields from wrapping on the page.

Ordering fields

1. Click the Fields tab on the Report Properties dialog box.
2. In the Available Fields list, double-click the fields you want to include in the report if you haven't done so already.
3. In the Report Fields list, click the field you want to move.
4. Click the up or down arrow buttons to move the field.

   The report displays the fields in the order you've specified.

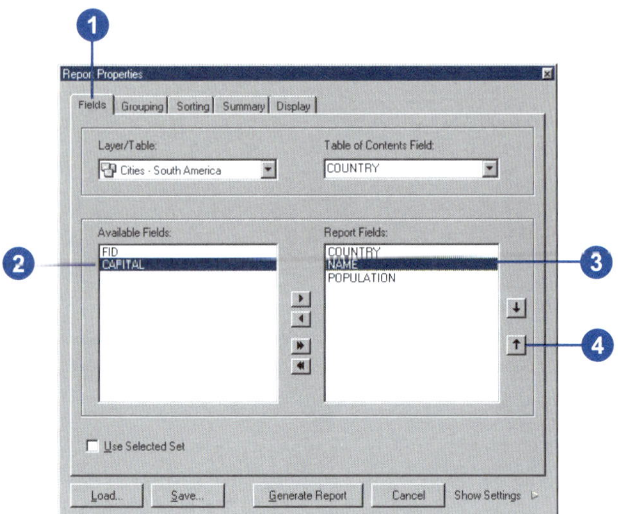

Displaying field names

1. Click the Display tab on the Report Properties dialog box.
2. Click Elements.
3. Check Field Names to display field names on the report.

   Field names appear as column headers in a tabular report and to the left of values in a columnar report.
4. Click the Section property and click the dropdown arrow. Click Top of Report or Top of Each Page. Top of Report displays the names once on the first page. This property is for tabular reports only.

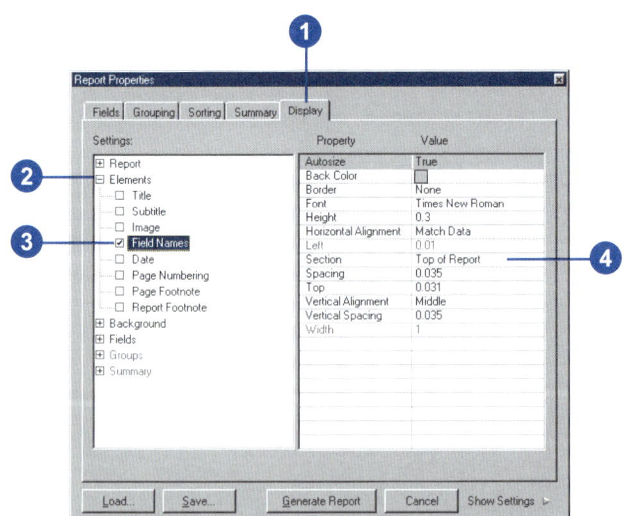

Tip

Setting field aliases

The field aliases you set while creating a report will only apply to the report.

Setting a field alias

1. Click the Display tab on the Report Properties dialog box.
2. Click Fields and click the field for which you want to set an alias.
3. Double-click the Text property and type the text you want to display.

   This field alias only displays on the report.

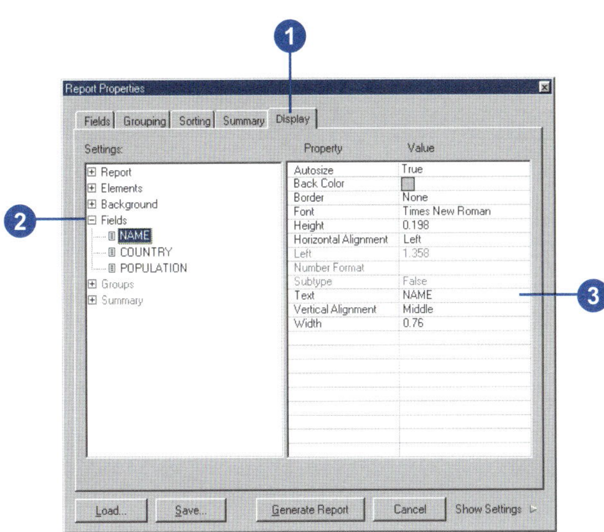

Tip

Setting your own field widths

ArcMap automatically determines how wide to display a field to accommodate the data. However, you can increase or decrease the width. If the width of all fields exceeds the width of the report, the fields will wrap.

Setting the display width of a field

1. Click the Display tab on the Report Properties dialog box.
2. Click Fields and click the field for which you want to set the width.
3. Double-click the Width property and type the width you want.

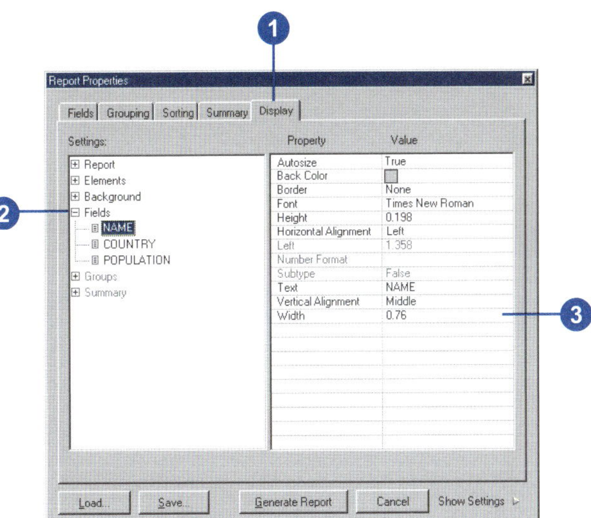

Tip

When to adjust the column spacing

Adjust the column spacing when you want to increase or decrease the horizontal distance between fields in a tabular report or between the field name and its value in a columnar report.

Increasing the space between columns in the report

1. Click the Display tab on the Report Properties dialog box.
2. Click Elements.
3. Check Field Names if it isn't already checked.
4. Double-click Spacing and type a horizontal distance in inches.

   The value you type sets the distance between the fields in a tabular report or the field name and its value in a columnar report.

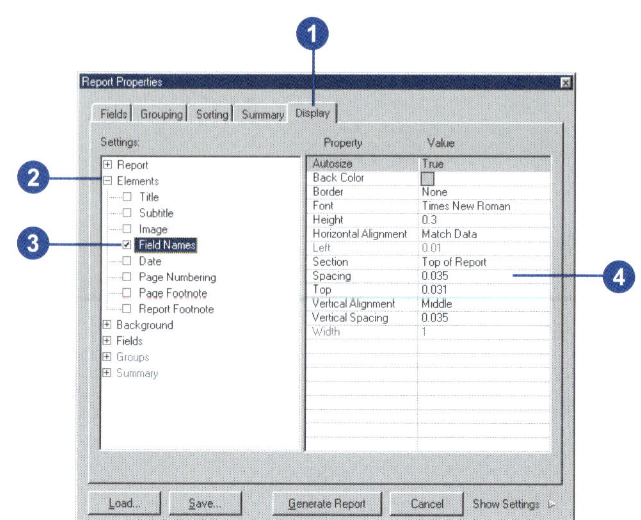

Tip

When to adjust the row spacing

Adjust the row spacing when you want to increase or decrease the vertical distance between records in the report.

Increasing the space between rows in the report

1. Click the Display tab on the Report Properties dialog box.
2. Click Report.
3. Click Records.
4. Click Autosize, click the dropdown arrow, then click False.
5. Double-click Height and type a height in inches for the row.

   Increasing the height increases the space between rows in the report.

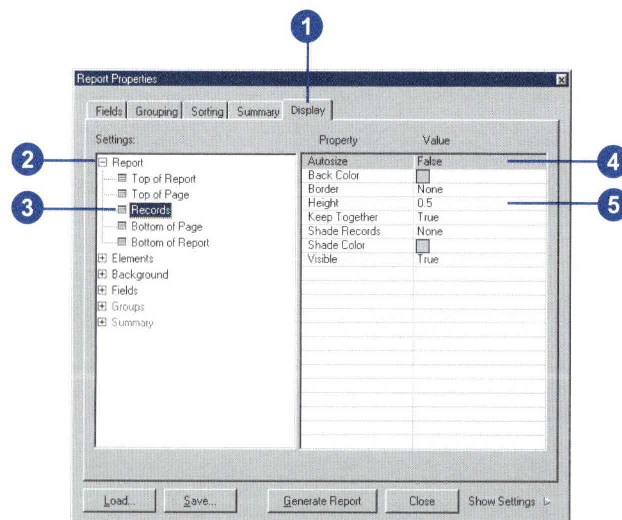

Tip

When to adjust the vertical spacing between fields in a columnar report

Field names and values stack on top of each other in a columnar report. Thus, a given record of data may occupy several rows in the report. When you want to increase or decrease the space between these fields, adjust the Vertical Spacing property.

Changing the vertical spacing between fields in a columnar report

1. Click the Display tab on the Report Properties dialog box.
2. Click Elements.
3. Check Field Names if it isn't already checked.
4. Double-click Vertical Spacing and type a distance in inches.

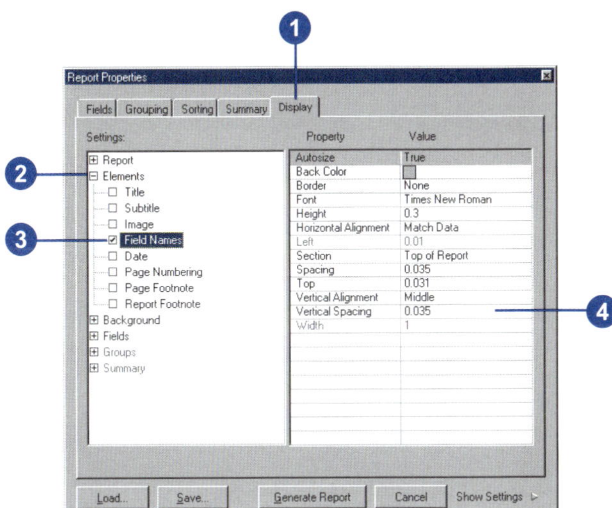

Organizing report data

One advantage of displaying your data in a report is that a report allows you to organize your data. For example, you can sort records based on the values in one or more fields—given a list of cities, you could sort them by total population. You can also group records together and calculate summary statistics. For example, you could group cities by what country they're in. This lets you easily see which city has the largest population in a given country. You can further calculate summary statistics—for example, compute the sum, average, count, standard deviation, and minimum and maximum values.

Tip

Sorting records using up to three fields

You can sort records using up to three fields in ascending or descending order.

Sorting records

1. Click the Sorting tab on the Report Properties dialog box.
2. Click a field to sort in the Sort column.
3. Click Ascending, Descending, or None.
4. If you want to sort other fields, click them, then click the sorting method.

   ArcMap sorts the fields based on the sort order. The figure to the right sorts cities alphabetically by country, then by their name.

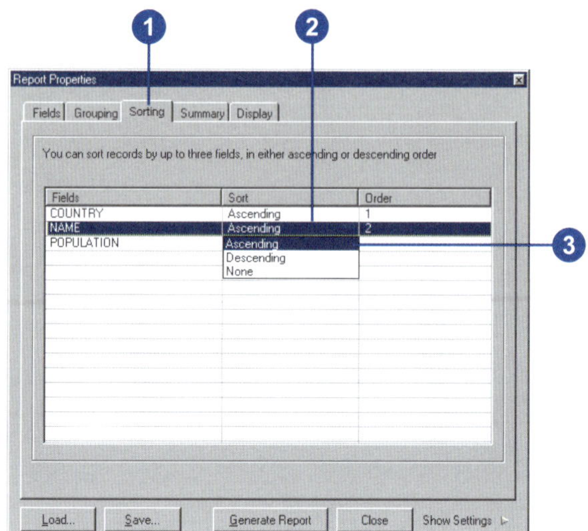

Grouping records

1. Click the Grouping tab on the Report Properties dialog box.
2. Double-click the field you want to use to group data.
3. Click Grouping Intervals and click the method for grouping data.
4. Click Ascending or Descending for the sort method.
5. Check Include Group Fields to repeat the group value in the report display.

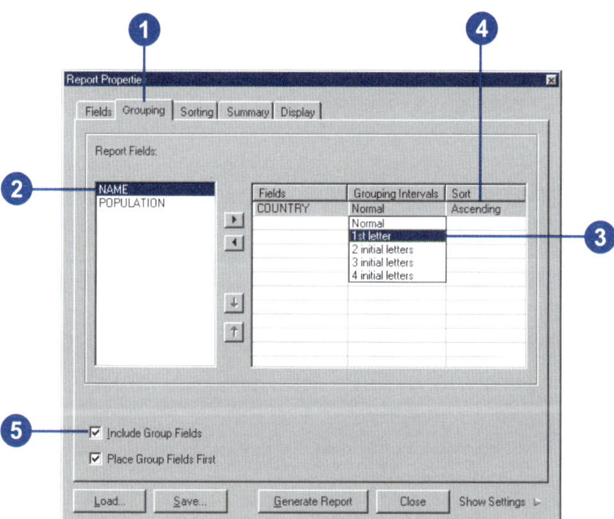

Tip

What summary statistics are available?

You can compute the average, count, minimum, maximum, standard deviation, and sum for any numeric fields on your report.

Tip

Where can you display summary statistics?

You can display summary statistics for a field at the end of the report, at the end of each page, and at the end of each group you've defined.

Tip

Shading records

To make the data in your reports easier to read, you can shade every other record with a color. On the Display tab, click Report and click Records. Then set the Shade Records and Shade Color properties.

Calculating summary statistics

1. Click the Summary tab on the Report Properties dialog box.
2. Click the Available Sections dropdown arrow and click the section you want the statistics to appear in.
3. For each numeric field, check the box that corresponds to the statistic you want to display.
4. To display statistics in each available section, repeat steps 2 and 3 for each section.

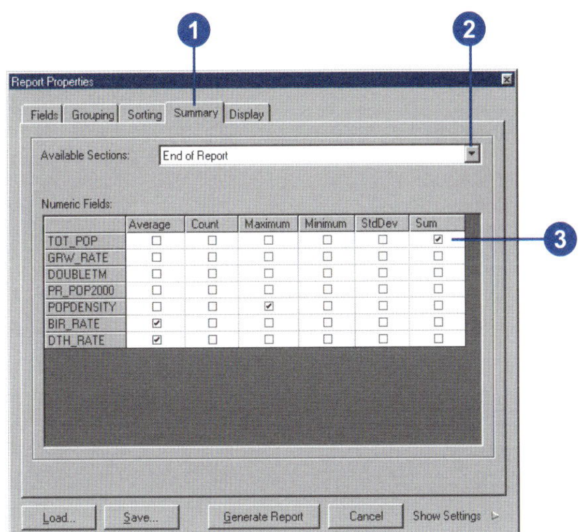

Adding report elements

To help you create attractive-looking reports, you can add these additional elements to your reports:

- A title
- A subtitle
- Page numbers
- The current date
- Images (e.g., a company logo)
- Footnotes

Once you've added an element, you can control the way it looks. For example, you might change the text font and size of the title and center it on the page.

This section describes how to add the particular element you want to your report while you're creating it.

Tip

Borders and shading

Accent report elements with borders and background colors.

Adding a title

1. Click the Display tab on the Report Properties dialog box.
2. Click Elements.
3. Check Title.
4. Double-click Text and type the text for the title.
5. Click Font and click the button to display the Font dialog box.
6. Set the font, size, font style, and color as desired and click OK.
7. Click Back Color and click the button to display the Color dialog box.
8. Click the color you want.
9. Click Border and click the button to open the Border Properties dialog box.
10. Click the border style you want and click OK.

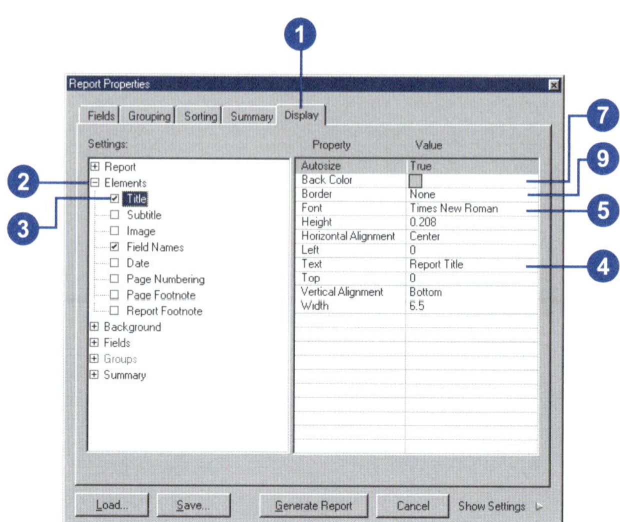

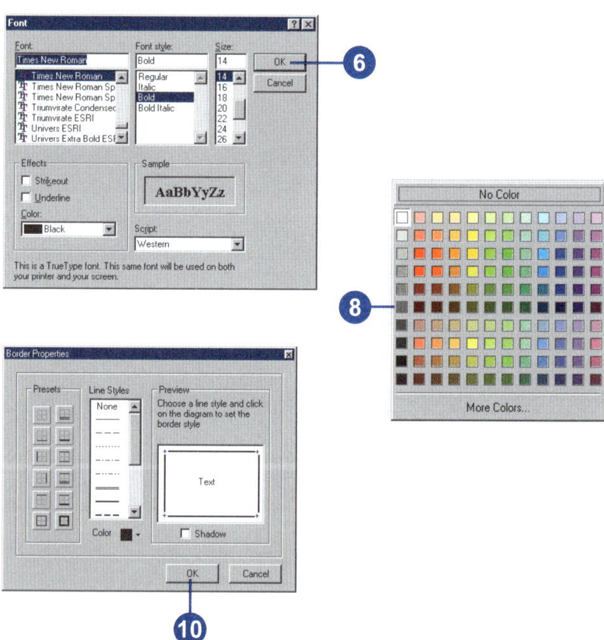

Tip

Setting the height of an element

In general, ArcMap automatically calculates the height of an element. For example, the height of the title is based on the font size you choose. However, you can also set the height explicitly. This is useful when you want to add additional space around the element. To set the height, set the Autosize property to False, then set the Height property to a height in inches.

Tip

Why isn't the element centered on the page properly?

Probably because you need to adjust the width of the element. The horizontal alignment of an element is based on its width, not the width of the report.

Adding a subtitle

1. Click the Display tab on the Report Properties dialog box.
2. Click Elements.
3. Check Subtitle.
4. Double-click Text and type the text for the subtitle.

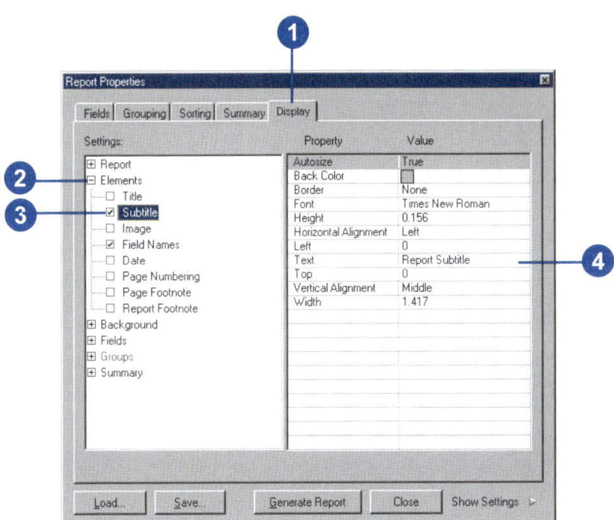

Adding page numbers

1. Click the Display tab on the Report Properties dialog box.
2. Click Elements.
3. Check Page Numbering.
4. Click Section, click the dropdown arrow, and click Top of Page or Bottom of Page.
5. Click Font and click the button to open the Font dialog box.
6. In the Font dialog box, set the font, size, style, and color as desired and click OK.

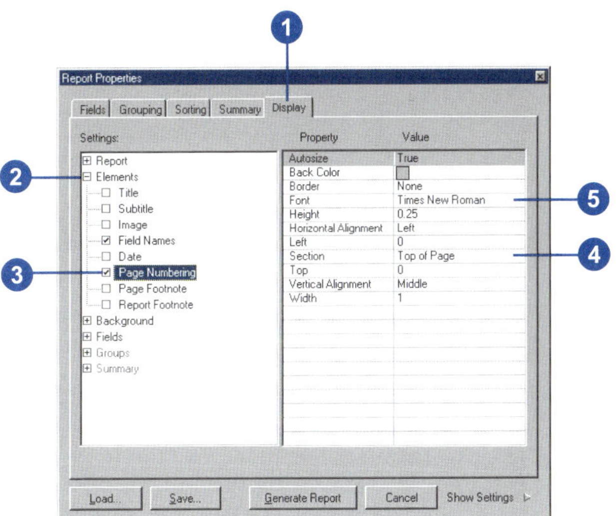

Tip

Positioning an element

Each element has its own height and width. An element, such as a title, is located in the section at the top of the report. You can position the upper-left corner of an element by adjusting the Top and Left properties of the element.

Adding the date

1. Click the Display tab on the Report Properties dialog box.
2. Click Elements.
3. Check Date.
4. Click Section, click the dropdown arrow, and click Top of Page or Bottom of Page.
5. Click Number Format, click the dropdown arrow, and click the date format you want.

   Dates can be represented as:
   - mm/dd/yy 1:00:00 AM
   - Monday, July 26, 1999
   - mm/dd/yy

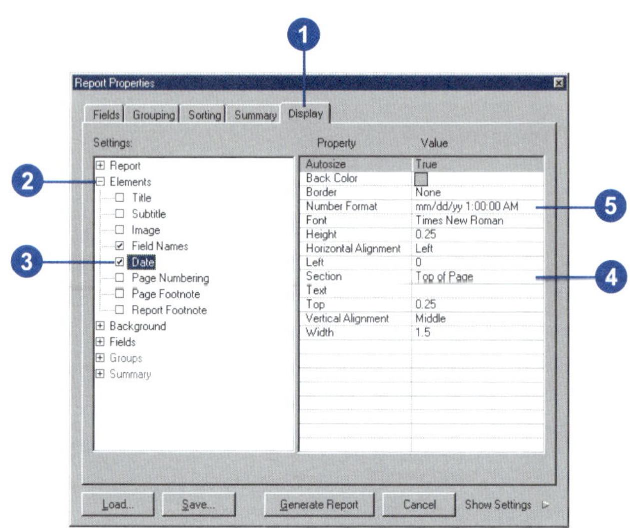

Tip

Adding footnotes

You can add footnotes to the bottom of each page or once at the end of the report.

Adding footnotes

1. Click the Display tab on the Report Properties dialog box.
2. Click Elements.
3. Check Page Footnote or Report Footnote.
4. Double-click Text and type the text for the footnote.
5. Click Font and click the button to open the Font dialog box.
6. In the Font dialog box, set the font, size, style, and color as desired and click OK.

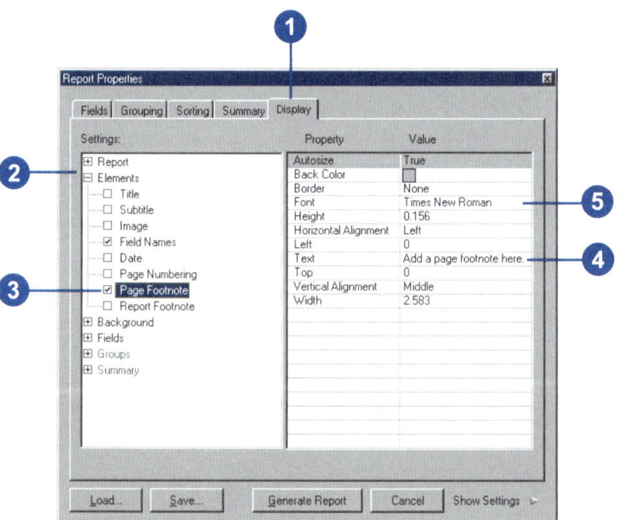

Tip

Getting an image to fit

Images aren't always sized exactly as you'd like them to be. Fortunately, when you add an image to your report, you can adjust its appearance to fit the available space. If it's too large, you can shrink it or crop it. If it's too small, you might stretch it instead.

Adding an image at the top of the report

1. Click the Display tab on the Report Properties dialog box.
2. Click Elements.
3. Check Image.
4. Click Picture, click the button to display the Open Image dialog box, and click the image you want to display.
5. Double-click Height and type a height in inches.
6. Double-click Width and type a width in inches.
7. Click Picture Display; click the dropdown arrow; and click Fit, Clip, or Stretch.

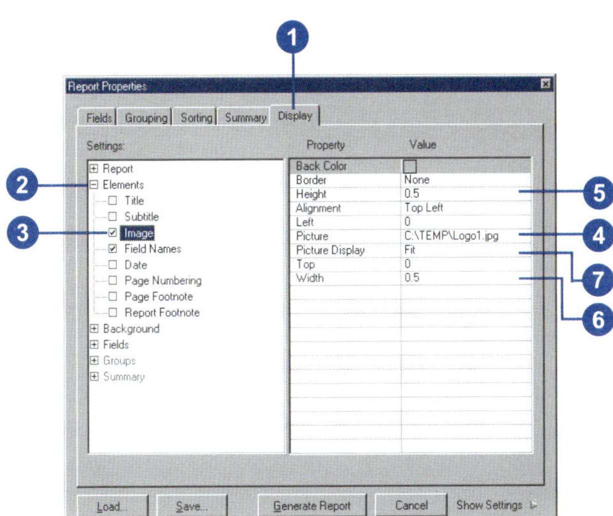

Adding an image in the background

1. Click the Display tab on the Report Properties dialog box.
2. Click Background.
3. Check Image.
4. Click Picture, click the button to display the Open Image dialog box, and click the image you want to display.

   The image will now appear behind the report data on all pages.

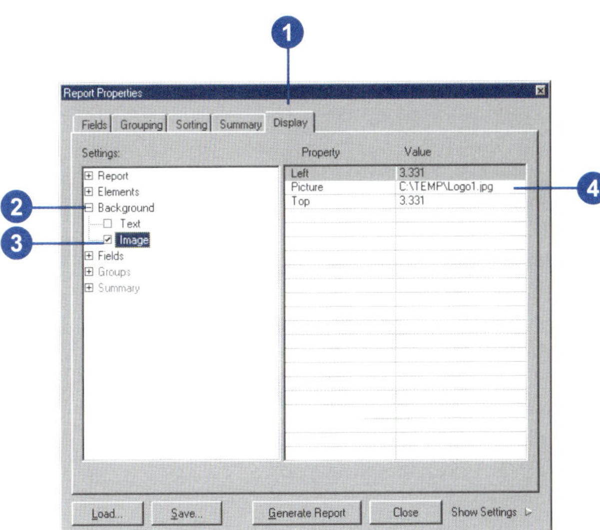

Tip

Positioning the background text and image

Use the Top and Left properties to position the upper-left corner of the background text or image on the report page.

Adding text in the background

1. Click the Display tab on the Report Properties dialog box.
2. Click Background.
3. Check Text.
4. Double-click the Text property and type the text string you want to appear in the background.

   The text displays behind the report data on all pages.
5. Click Font and click the button to open the Font dialog box.
6. In the Font dialog box, set the font, size, font style, and color as desired and click OK.

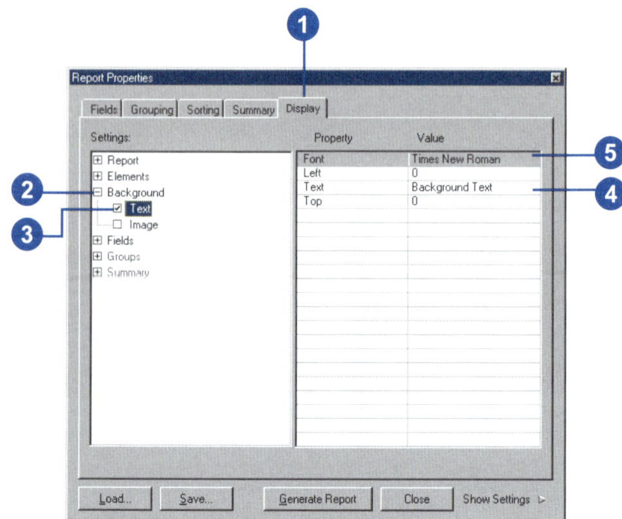

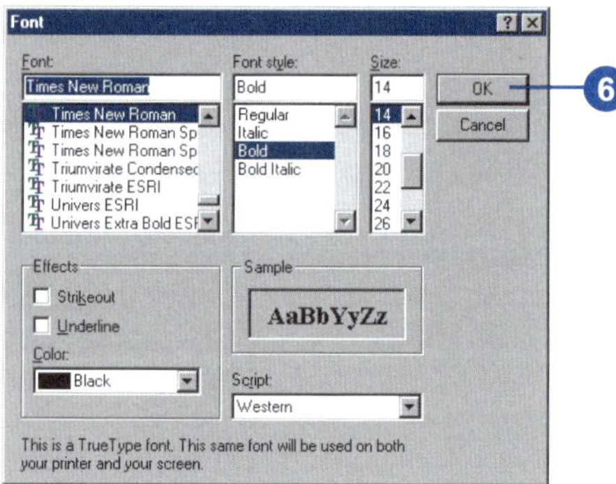

Controlling the presentation

As described earlier in this chapter, a report is made up of several sections, each of which defines a particular area on the report. For example, the section at the top of the report contains the title of the report; another section displays the records in the report.

Each section has properties that you can modify such as its color and whether or not it is displayed. A section also contains certain elements with their own properties that you can modify to change the visual appearance of the report. For example, you might choose to shade the records in your report.

By setting the properties of a section and the elements within it, you can create your own individual look. As you create your report, you can preview your settings to see what your report will look like.

Previewing a report

1. Click the Show Settings on the Report Properties dialog box.

   The dialog box expands to show you a preview of the report.

2. As you modify the report properties, click Update Settings to see the changes reflected on the preview.

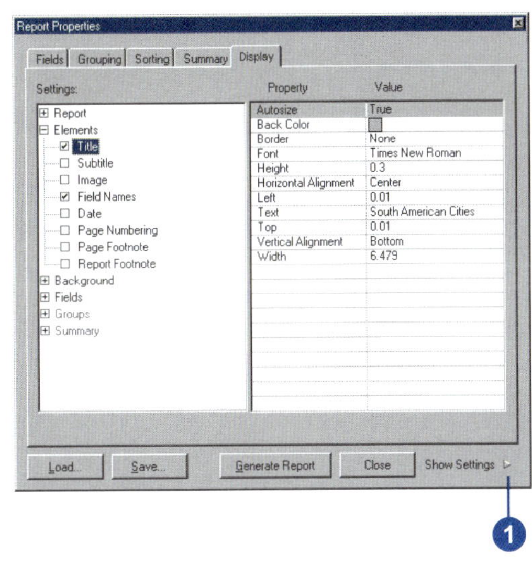

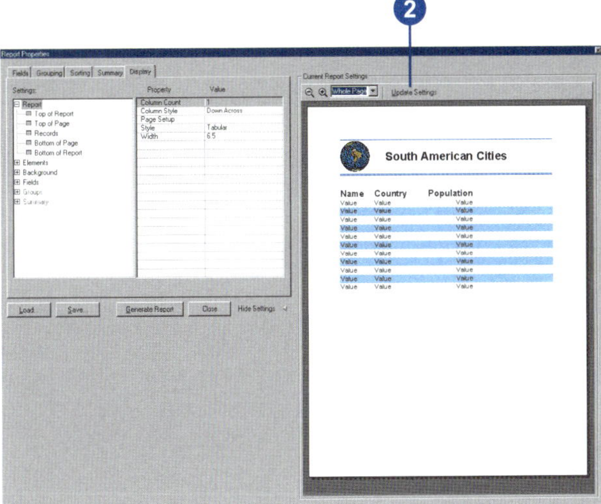

Tip

How to display more data on your report page

If you're not using a report section, such as the top or bottom of the page, you can hide it, thereby increasing the space available to display data on the report page.

Hiding a section

1. Click the Display tab on the Report Properties dialog box.
2. Click Report.
3. Click the report section you want to hide, for example, Top of Page.
4. Click Visible, click the dropdown arrow, and click False.

   The report section will not appear in the report.

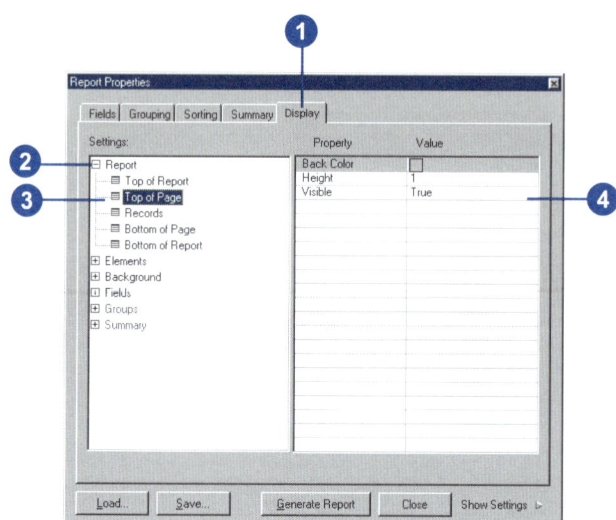

Tip

Using background colors

Each report element, such as the title, has its own background color. By using background colors, you can enhance the display of your report. In addition, you can apply a background color to report sections.

Setting the background color of a section

1. Click the Display tab on the Report Properties dialog box.
2. Click Report.
3. Click the report section you want to set a background color of, for example, Top of Page.
4. Click Back Color and click the button to open the Color dialog box.
5. Click the background color you want.

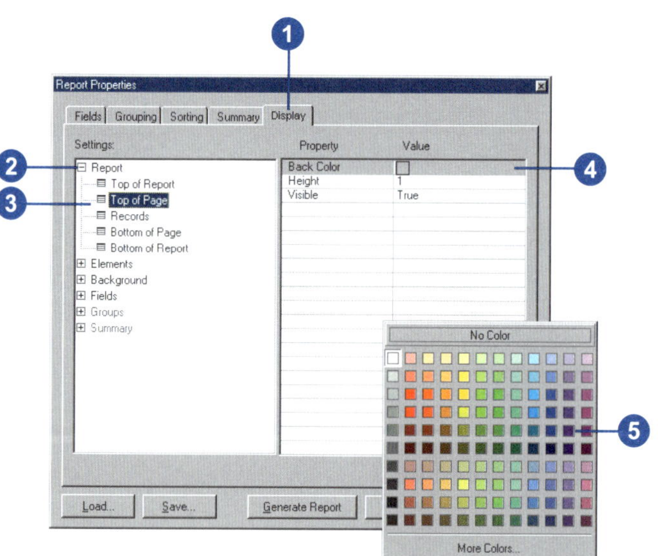

Tip

Making your data easier to read

You can enhance the readability of your report by shading records. This helps the reader to visually separate data in adjacent records. You can choose to alternately shade every other, every two, or every three records.

Shading records in the report

1. Click the Display tab on the Report Properties dialog box.
2. Click Report.
3. Click Records.
4. Click Shade Records, click the dropdown arrow, and click the option you want.
5. Click Shade Color and click the button to open the Color dialog box.
6. Click the color you want to shade records with.

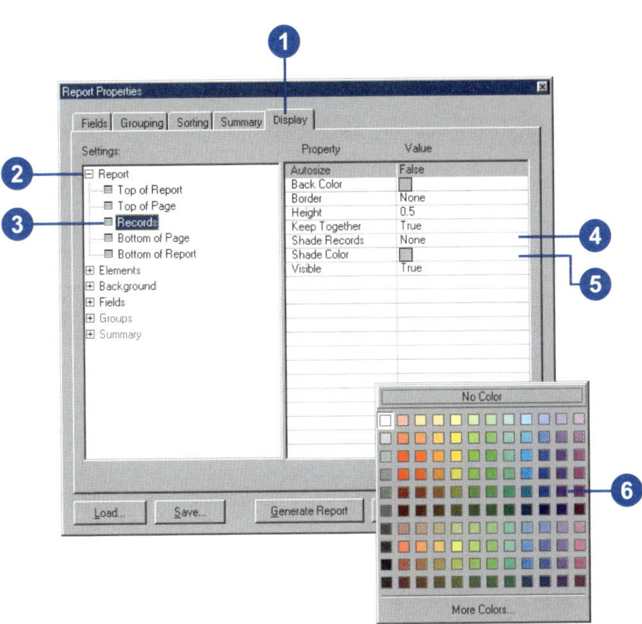

Saving and loading a report

If you want to copy a report that you've made on one map and put it on another, save it to a file on disk. Then you can load the report in another map and place it as appropriate. When you save a report to a file, you're creating a static copy that is not linked to the actual data from which you created the report. Thus, you won't be able to modify the report in any way.

You can also export a report to a different file type, including Adobe® Portable Document Format (PDF), Rich Text Format (RTF), and plain text (TXT).

Saving a report

1. Click Save on the Report Properties dialog box.
2. Click the Save in dropdown arrow and navigate to the location where you want to save the report.
3. Type a name for the report.
4. Click Save.

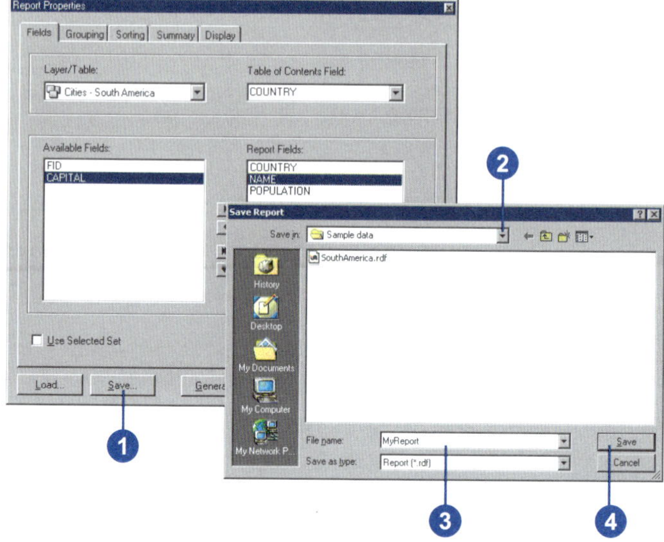

Loading a report

1. Click Load in the Report Properties dialog box.
2. Click the Look in dropdown arrow and navigate to the location where the report is stored.
3. Click the report.
4. Click Open.

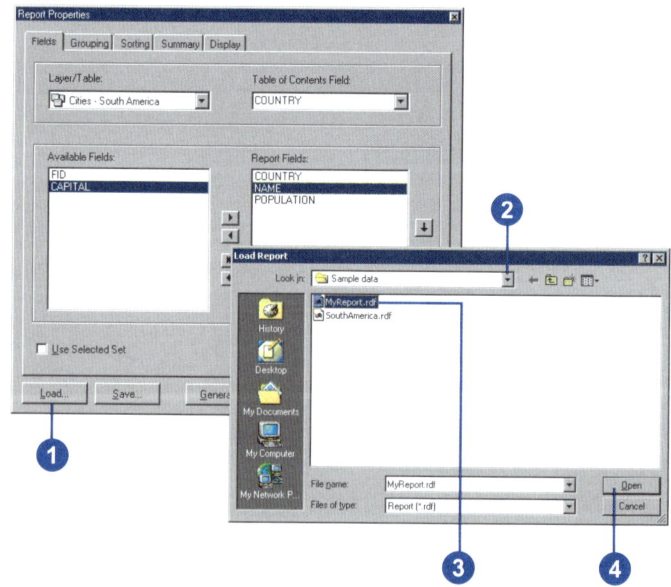

Tip

Export reports to PDF, RTF, or plain text

Once you create your report, you can include it on your map and also export it to several different file formats. Export to Adobe PDF, RTF, or plain text.

Exporting a report

1. Click Generate Report on the Report Properties dialog box.
2. Click Export on the Report Viewer window.
3. Click the Save in dropdown arrow and navigate to the location where you want to save the exported report.
4. Type a name for the report.
5. Click the Save as type dropdown arrow and click the type of file you want to export.
6. Click Save.

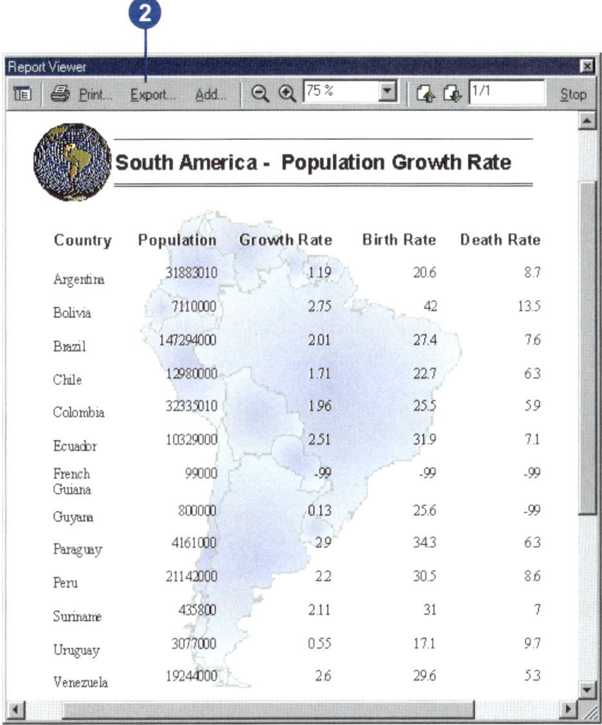

South America - Population Growth Rate

Country	Population	Growth Rate	Birth Rate	Death Rate
Argentina	31883010	1.19	20.6	8.7
Bolivia	7110000	2.75	42	13.5
Brazil	147294000	2.01	27.4	7.6
Chile	12980000	1.71	22.7	6.3
Colombia	32335010	1.96	25.5	5.9
Ecuador	10329000	2.51	31.9	7.1
French Guiana	99000	-99	-99	-99
Guyana	800000	0.13	25.6	-99
Paraguay	4161000	2.9	34.3	6.3
Peru	21142000	2.2	30.5	8.6
Suriname	435800	2.11	31	7
Uruguay	3077000	0.55	17.1	9.7
Venezuela	19244000	2.6	29.6	5.3

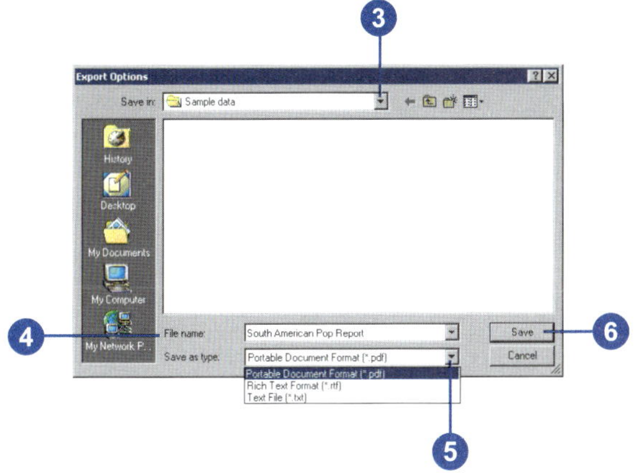

Using Crystal Reports

To help you create the reports you need, ArcMap has integrated with industry-leading Crystal Decisions' Crystal Reports. When you install ArcMap, you can optionally install Crystal Reports. Once installed, you'll have access to the ESRI Crystal Report Wizard that lets you create and view reports.

The reports you create with Crystal Reports are managed as files outside ArcMap. ArcMap simply passes the tabular information from the layers on your map to Crystal Reports. While you can access the reports you create from within ArcMap and display them in a separate window, you can't add them to your map layout. If you need to do this, use the built-in ArcMap reporting tool instead.

Creating a report

1. Click the Tools menu, point to Reports, and click Crystal Report Wizard.
2. Click Create a new report using these layers or tables.
3. Check the layers and tables you want to include in the report.
4. Type the name of the output geodatabase.
5. Click Next.

   This launches the Crystal Report 9 Wizard.
6. Double-click the fields you want to include in the report.
7. Click Finish to create the report.

   Optionally, click Next to set other report parameters.
8. Type a name for the report.
9. Click Save.

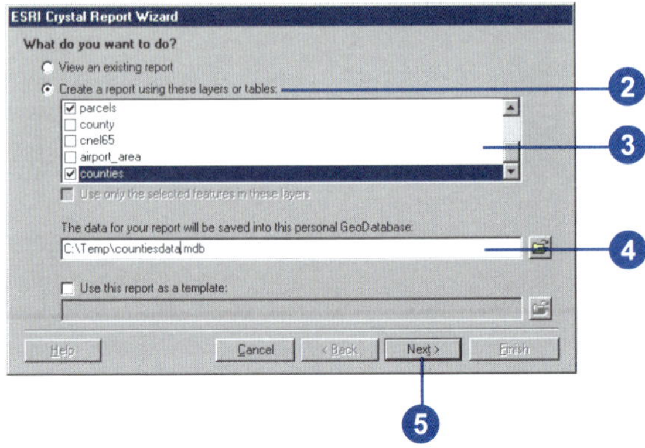

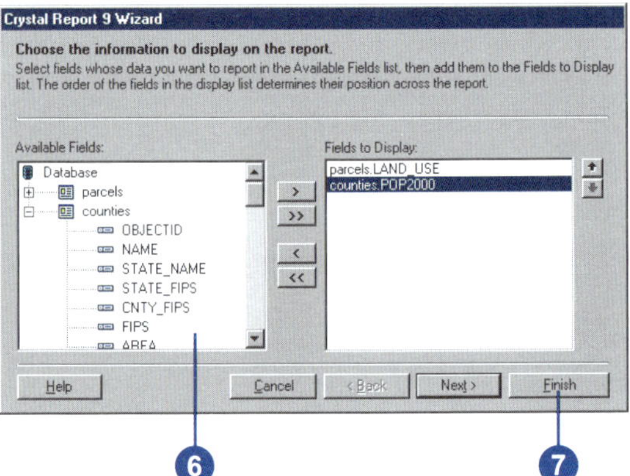

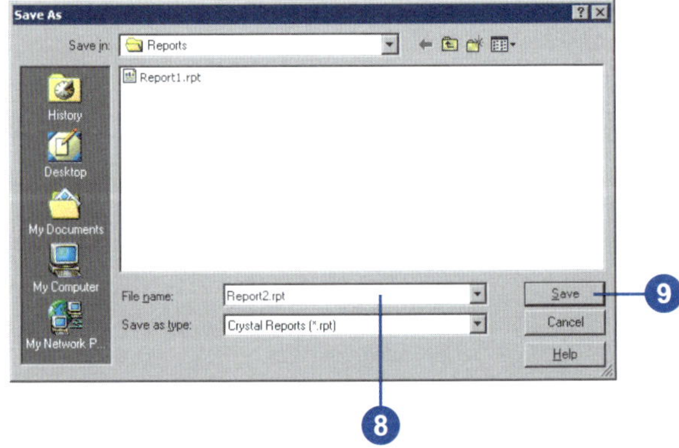

Tip

Can I add Crystal Reports to my map layout?

No. Instead, use the built-in ArcMap reporting tool to create your report and add it to the layout.

Viewing an existing report

1. Click the Tools menu, point to Reports, and click Crystal Report Wizard.
2. Click View an existing report.
3. Click Next.
4. Click the Look in dropdown arrow and navigate to the location of the report.
5. Click the report.
6. Click Open.

   The report appears in its own window.

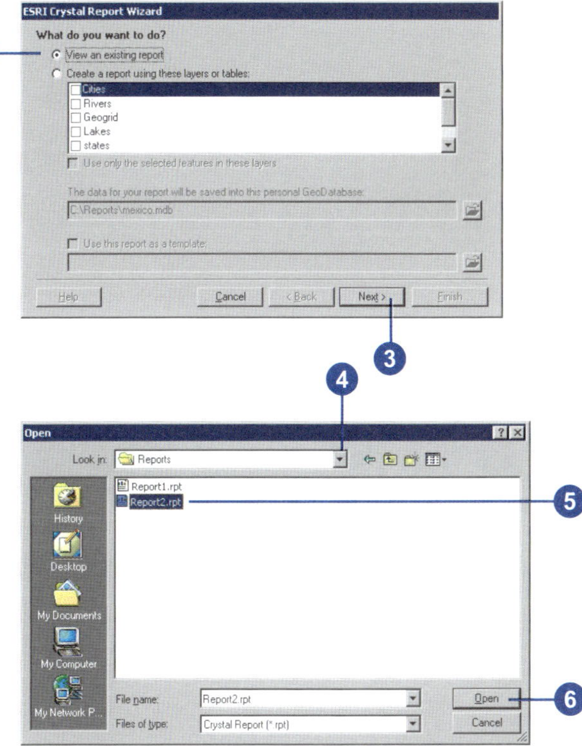

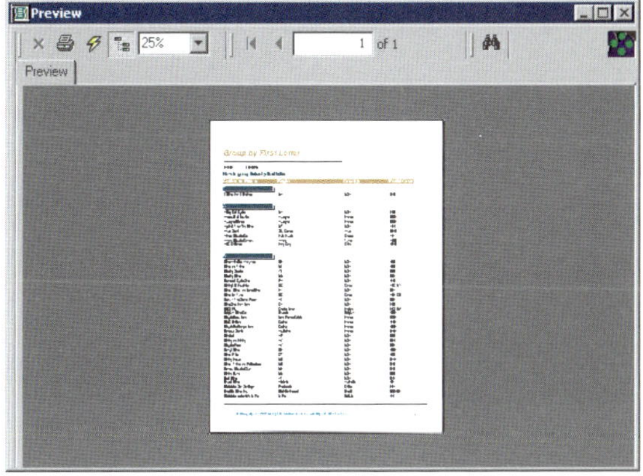

13 Querying maps

IN THIS CHAPTER

- **Identifying features**
- **Displaying a Web page or document about a feature**
- **Selecting features interactively**
- **Selecting features by searching with a SQL expression**
- **Building a SQL expression**
- **Ways to find features by their locations**
- **Selecting features by their locations**
- **Specifying how selected features highlight**
- **Displaying information about selected features**
- **Exporting selected features**
- **Joining the attributes of features by their locations**
- **Taking geoprocessing further**

Maps convey a great deal of information. You can learn a lot about an area just by looking at a map. Yet, sometimes those things that are most interesting and revealing are not immediately apparent by looking at a map. You can begin to discover new spatial relationships when you start asking questions such as: Where is...? Where's the closest? What's inside? and What intersects?

ArcMap provides a number of tools to help you find answers to these types of questions. With ArcMap, you can:

- Find out what a feature is by pointing at it. This may display additional information such as a picture or Web page.
- Find features with particular attributes such as cities with a population greater than one million.
- Find features with a particular spatial relationship. For instance, you can find the wildlife habitats within 50 kilometers of an oil spill or find all traffic accidents that occurred along a particular stretch of road.

Once you've found features, you can display their attributes and statistics, create reports and graphs, or export them to a new feature class.

Identifying features

When you view a map online, there's a lot of information you can get directly from it. If you want to know what a feature is, just pause your mouse over it to display a MapTip. A MapTip pops up on the screen providing a quick description such as a city name. If you want to know more about the feature, use the Identify tool to display all the attributes of the feature.

Tip

Using the Identify tool

You can hold down the Shift key and click the map to keep the existing results in the Identify Results dialog box.

Tip

Why can't I see the MapTips?

If you can't see MapTips even after you've enabled them, make sure that the layer is turned on and the features in the layer are not being hidden by features in overlapping layers.

Identifying features by pointing at them

1. Click the Identify button on the Tools toolbar. The Identify Results dialog box opens.
2. Click the mouse pointer over the map feature you want to identify.

   The features in the topmost layer under the pointer will be identified.

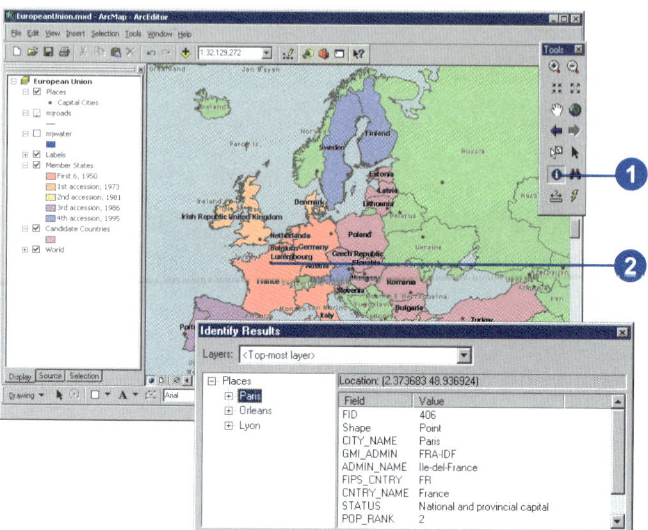

Displaying MapTips

1. In the table of contents, right-click the layer for which you want to display MapTips and click Properties.
2. Click the Display tab and check Show MapTips.
3. Click the Fields tab.
4. Click the Primary Display Field dropdown arrow and click the attribute field you want to display as the MapTip.
5. Click OK.
6. Move the mouse pointer over a feature to see the MapTip.

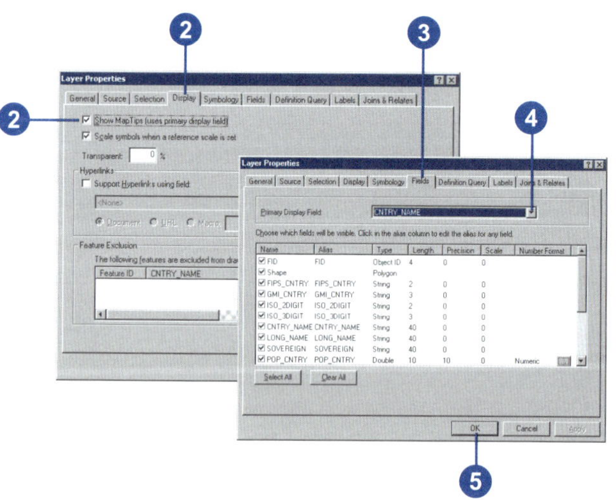

Displaying a Web page or document about a feature

Using *hyperlinks*, you can display documents—such as a text file or image—or Web pages accessed over the Internet by clicking a feature in your map. A hyperlink is a document path and name or a Web page address stored with a feature.

You can dynamically create hyperlinks as you browse your map, or alternatively, you can store hyperlinks with your data in an attribute field.

You can create hyperlinks easily and save them to the current ArcMap document or write them to a layer file by clicking features in the map and typing the hyperlinks.

If you would like to store the hyperlinks in the layer's attribute table, create a field in the feature attribute table and enter the hyperlinks for each feature. You must then instruct ArcMap to access the hyperlinks from this field.

Creating and accessing hyperlinks stored in a layer file or ArcMap document

1. Click the Identify button on the Tools toolbar.
2. Click a feature.
3. In the Identify Results window, right-click the feature you want to set a hyperlink for and click Add Hyperlink.
4. To add a hyperlink to a Web page, click Link to a URL and type a URL. To link to a document, click Link to a Document and type a pathname to the document.
5. Click the Hyperlink tool on the Tools toolbar and click a feature.

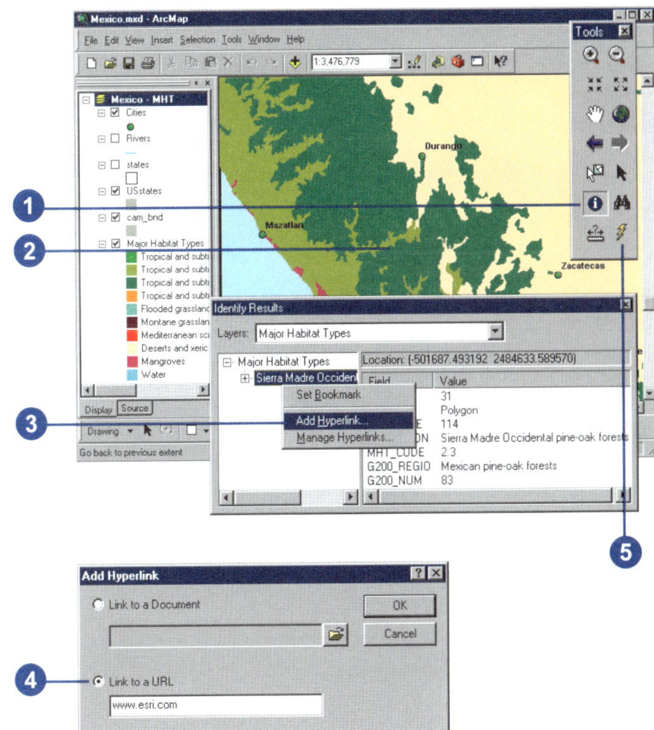

Displaying hyperlinked documents

If you specify a Web address as a hyperlink, ArcMap launches your default Web browser and displays the Web page. If you specify a document as a hyperlink, ArcMap opens that document in its native program. If the linked document is on a server, the document is only available to those with access to that server.

Using an attribute field as a hyperlink

1. In the table of contents, right-click the layer containing a field with hyperlinks and click Properties.
2. Click the Display tab.
3. Check Support Hyperlinks using field. Click the drop-down arrow and click a field.
4. Click Document or URL.
5. Click OK.
6. On the Tools toolbar, click the Hyperlink button.
7. Move the mouse pointer over a feature and click to display the hyperlink.

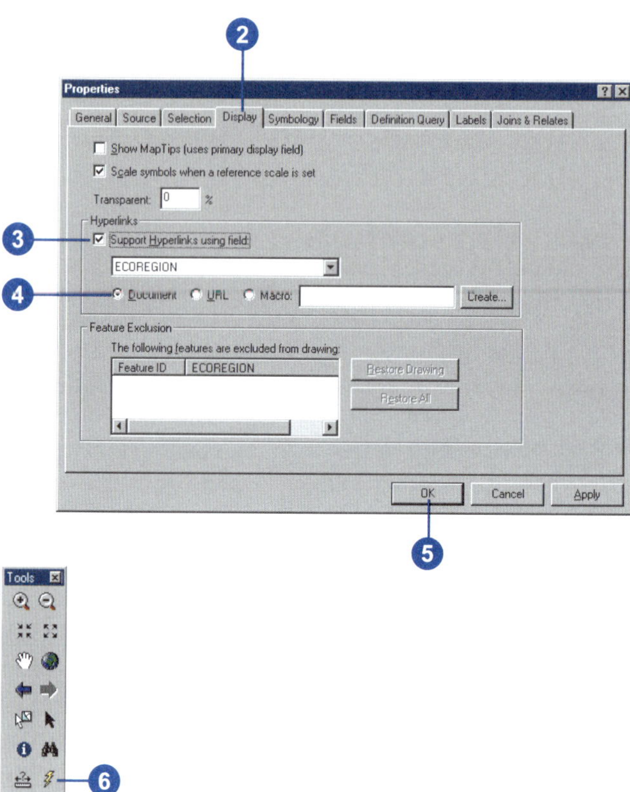

Selecting features interactively

You can select features with your mouse by clicking them one at a time or by dragging a box around them.

Before you select features with one of these methods, you can specify the layers you want to select from. Do this when the features you want to select overlap or are close to features from other layers. For example, if you have a layer of cities where many of the cities are on rivers, you can avoid selecting rivers by specifying that you want to select from the cities layer only.

You can also select features in the map by selecting their records in the attribute table. When you select feature records in a table, the feature highlights on the map.

Tip

Setting selectable layers

When you're setting selectable layers, you can check or uncheck all layers by holding down the Ctrl key while clicking one of the check boxes.

Making the Selection tab visible

1. Click Tools on the Main Menu and click Options.
2. Click the Table Of Contents tab.
3. In the Table Of Contents tab options, check the Selection box.
4. Click OK.

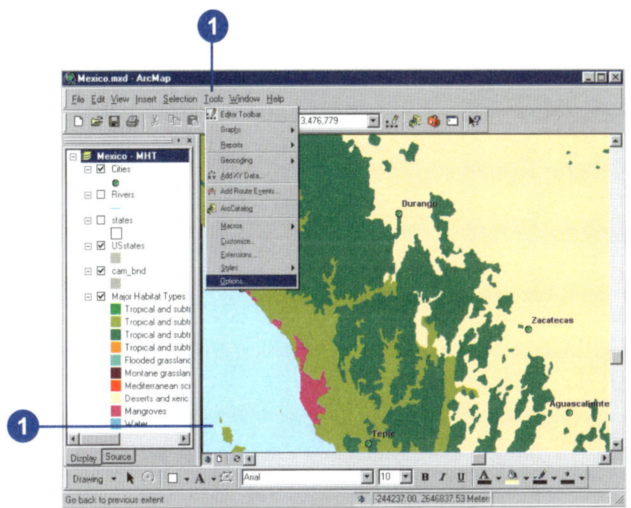

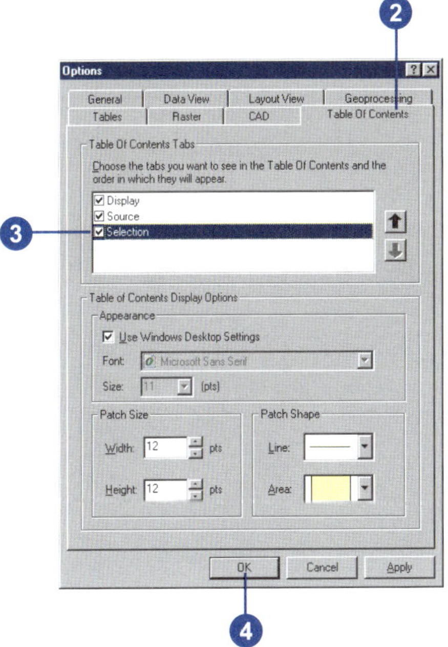

Selecting a feature by clicking it in the map

1. Click the Selection tab in the table of contents.
2. Click the layers you want to select from.
3. Click Selection, point to Interactive Selection Method, then click Create New Selection.
4. Click the Select Features tool.
5. Click the feature you want to select.
6. To select additional features, hold down the Shift key while clicking the features.

   To remove a feature from the selected set, click the Selection menu, point to Interactive Selection Method, and click Remove From Current Selection. Click a selected feature and it is deselected.

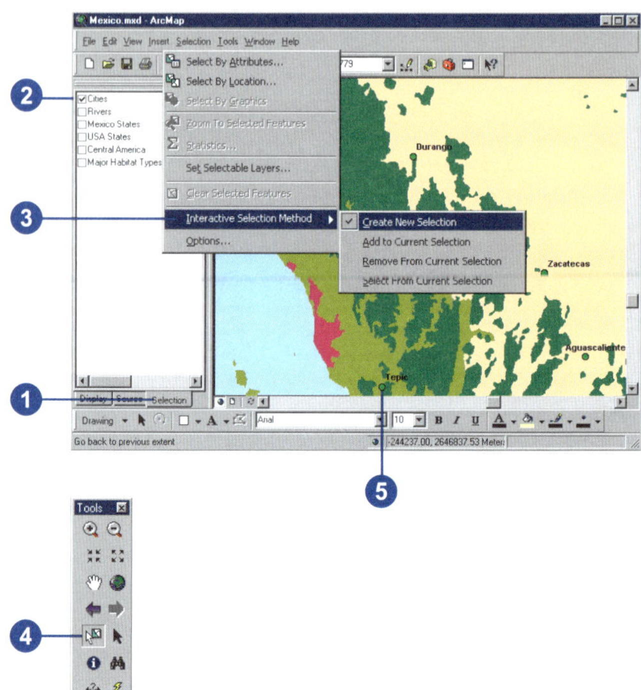

Tip

Deselecting all features

To deselect all selected features at once, click the map where there are no features or click a feature of a layer that is not selectable. If you are unable to do this, click the Selection menu and click Clear Selected Features.

Select features by dragging a box

1. Click the Selection tab in the table of contents.
2. Click the layers you want to select from.
3. Click Selection in the Main Menu, point to Interactive Selection Method, then click Create New Selection.
4. Click Selection in the Main Menu and click Options.
5. Specify how you'd like to select features with the box and click OK.
6. Click the Select Features button.
7. Click and drag a box around the features you want to select.
8. To select additional features, hold down the Shift key as you drag the box.

   To remove a feature from the selected set, click the Selection menu, point to Interactive Selection Method, and click Remove From Current Selection. Drag a box around the features you want to deselect.

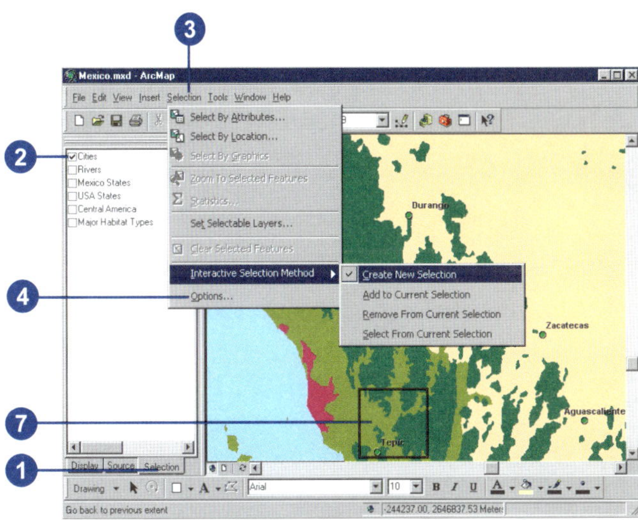

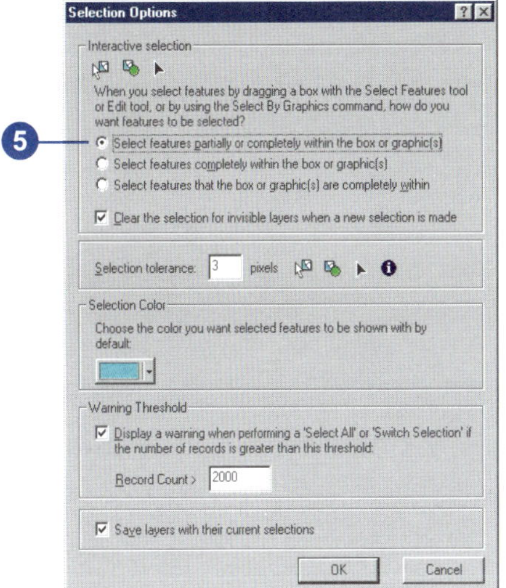

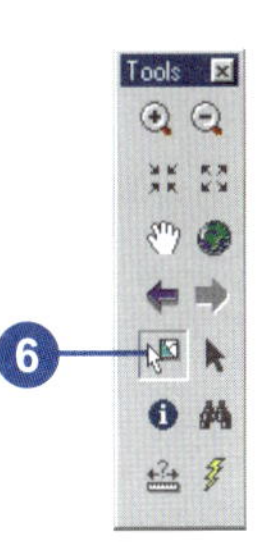

Tip

Selecting consecutive records in a table

To select consecutive records in a table, click and drag the mouse up or down.

Selecting a feature by clicking it in the table

1. Right-click a layer in the table of contents and click Open Attribute Table.
2. Select a feature in the table by clicking to the left of a record.
3. To select additional features, hold down the Ctrl key and click the features. To deselect a feature, hold down the Ctrl key and click the feature.

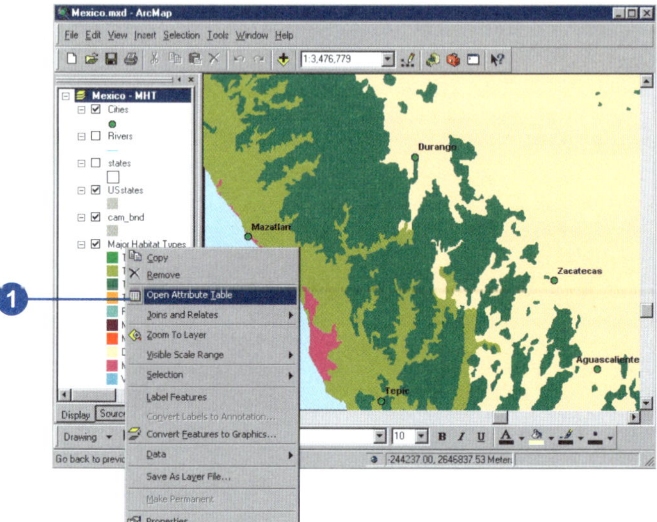

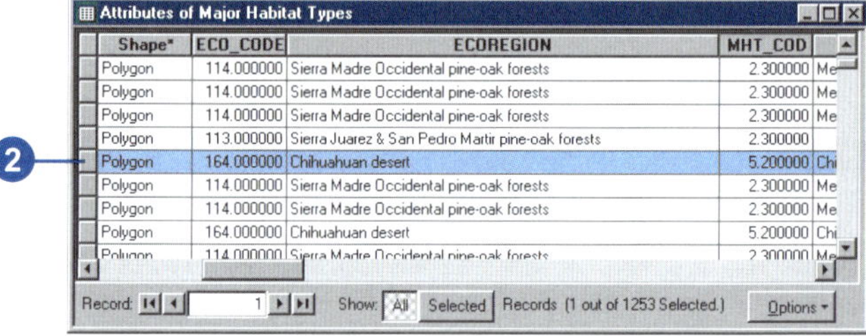

Selecting features by searching with a SQL expression

Structured Query Language is a powerful language you use to define one or more criteria that can consist of attributes, operators, and calculations. For example, imagine you have a map of customers and want to find those who spent more than $50,000 with you last year and whose business type is "Restaurant". You would select the customers with this expression: `Sales > 50000 AND Business_type = 'Restaurant'`.

When you search with SQL expressions, you can select features or table records in any data format supported by ArcMap. However, you format expressions differently depending on the format of the data you're querying. The following pages contain guidelines on how to build SQL expressions for different data formats.

Tip

Trying the Query Wizard

If you've never selected features by their attributes, try clicking the Query Wizard button to generate the selection expression.

1. Click Selection and click Select By Attributes.
2. Click the Layer dropdown arrow and click the layer containing the features you want to select.
3. Click the Method dropdown arrow and click a selection method.
4. Double-click a field to add the field name to the expression box.
5. Click an operator to add it to the expression.
6. Click Get Unique Values to see the values for the selected field.

   Double-click a value to add it to the expression.
7. To see if you're using proper syntax or if the criteria you've entered will select any features, click the Verify button.
8. Click Apply.

   The status bar at the bottom of the ArcMap window tells you how many features are selected.
9. If you're finished selecting features, click Close.

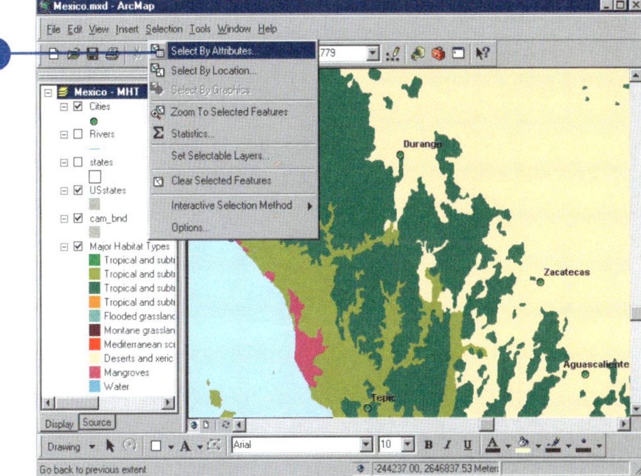

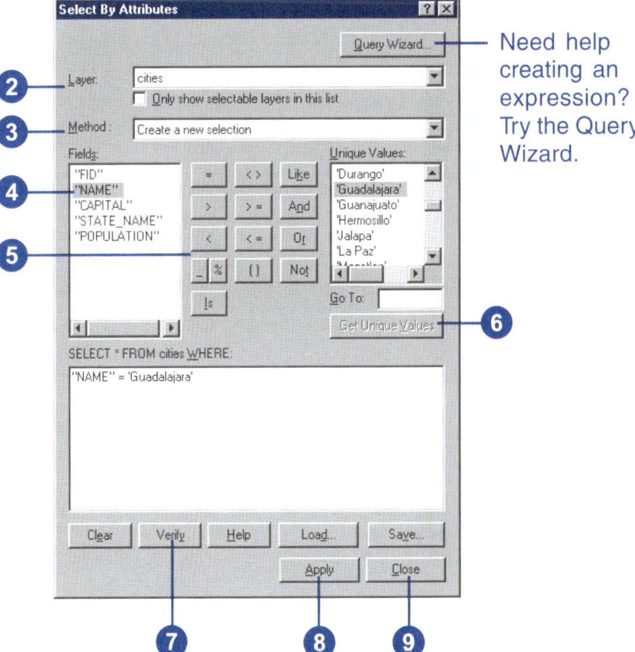

Need help creating an expression? Try the Query Wizard.

Building a SQL expression

Using SQL in ArcMap

SQL is a standard computer language for accessing and managing databases. It is used in ArcMap to select features with the Select By Attributes function or with the Query Builder dialog box used when setting a layer definition query.

SQL expressions are used in many parts of ArcMap and its extensions to define a subset of data on which to perform some operation. The interface to build these queries is the same throughout the program.

SQL expressions may also be used as parameters when selecting features programmatically.

SQL is only used in ArcMap to query the database. You cannot use SQL to insert, update, or delete data. The Select By Attributes dialog box only allows you to define a SELECT expression with a WHERE clause.

The following sections show you how to build basic SQL queries. For information on how to build more complex expressions, see SQL reference in the ArcGIS Desktop Help.

SQL versions

The SQL expression you need to create depends on the format of the data you're querying. This is because although SQL is a standard, database software may implement different versions of SQL.

ANSI SQL is used to query ArcSDE geodatabase data, whereas Jet SQL is used for personal geodatabase data. To query coverages, shapefiles, INFO tables, and dBASE tables, you use a limited version of SQL that doesn't support the many features and functions of ANSI and Jet SQL.

For instance, if you're querying ArcInfo coverages, shapefiles, INFO tables, or dBASE tables, field names must be enclosed in double quotes in the SQL expression:

```
"AREA"
```

If you're querying personal geodatabase data, enclose fields in square brackets:

```
[AREA]
```

If you're querying ArcSDE geodatabase data or data in an ArcIMS feature class or ArcIMS Image Service sublayer, don't enclose fields:

```
AREA
```

Some operators and keywords may also vary.

The dialog boxes in which you create SQL expressions will take care of listing the field names and values with the appropriate delimiters as well as selecting the relevant keywords or operators for you.

NOTE: The syntax shown in all the examples in this article is the Jet SQL for personal geodatabase data (Microsoft Access data).

A simple SQL expression

A basic SQL expression would look like:

```
SELECT * FROM states WHERE [STATE_NAME] =
'Alabama'
```

Simple expressions are similar to plain English and thus can be self-explanatory. This query expression would select the features containing Alabama, in a STATE_NAME field, in the states layer.

Because SQL can only be used in ArcMap to build queries, you only need to provide the part of the expression that follows the WHERE keyword.

Searching strings

Strings must always be enclosed within single quotes in queries.

For example:

```
[STATE_NAME] = 'California'
```

Strings in queries are case sensitive, except for personal geodatabase feature classes and tables. To make a case-insensitive search, you can use a SQL function to convert all values to the same case. For file-based data sources, such as shapefiles or coverages, use either the UPPER or LOWER function.

For example, the following query will select customers whose last name is stored as either Jones or JONES:

```
UPPER("LAST_NAME") = 'JONES'
```

Other data sources have similar functions. Personal geodatabases, for example, have functions named UCASE and LCASE that perform the same function.

Use the LIKE operator (instead of the = operator) with wildcards to build a partial string search.

For example, this query expression would select Mississippi and Missouri among the U.S. state names:

```
[STATE_NAME] LIKE 'Miss*'
```

* represents any group of characters.

For example, this query expression would find Catherine Smith and Katherine Smith:

```
[OWNER_NAME] LIKE '?atherine smith'
```

? represents any single character.

The wildcards to use are different depending on the database. The ones listed here apply only to data in a personal geodatabase. To learn about other wildcards, see the SQL reference in the ArcGIS Desktop Help.

These wildcard characters appear as buttons on the dialog box that you can click to enter the wildcard at the current cursor location in your expression. Only the wildcard characters that are appropriate to the data source of the layer or table you are querying are displayed.

If you use a wildcard character in a string with the = operator, the character is treated as part of the string, not as a wildcard.

You can use greater than (>), less than (<), greater than or equal to (>=), and less than or equal to (<=) operators to select string values based on sorting order.

For example, this query will select all the cities in a coverage with names starting with the letters M to Z:

```
"CITY_NAME" >= 'M'
```

Queries based on multiple character wildcards and sorting order may be impractical on large ArcSDE datasets.

The not equal (<>) operator can also be used when querying strings.

The NULL keyword

You can use the NULL keyword to select features and records that have null values for the specified field.

For example, to find cities whose 1996 population has not been entered, you can use:

```
[POPULATION96] IS NULL
```

or

```
[POPULATION] IS NOT NULL
```

The NULL keyword is always preceded by IS or IS NOT.

Searching numbers

You can query numbers using the equal (=), not equal (<>), greater than (>), less than (<), greater than or equal to (>=), and less than or equal to (<=) operators.

For example:

```
[POPULATION96] >= 5000
```

Numeric values are always listed using the point as the decimal delimiter regardless of your regional settings. The comma cannot be used as a decimal or thousands delimiter in a query.

Calculations

Calculations can be included in queries using these mathematical operators: `+ - * /`

Calculations can be between fields and numbers.

For example:

```
"AREA" >= "PERIMETER" * 100
```

Calculations can also be performed between fields.

For example, to find the countries with a population density of less than or equal to 25 people per square mile, you could use this query:

```
[POP1990] / [AREA] <= 25
```

Calculations between fields in a coverage or shapefile (or an INFO table or dBASE table) are not supported. You can include calculations between a field and a number.

Precedence parentheses

Queries are normally evaluated from left to right, but expressions enclosed in parentheses are evaluated first. You can either click to add parentheses into the query then enter the expression you want to be enclosed, or highlight the existing expression that you want to be enclosed and press the Parentheses button to enclose it.

For example:

```
[HOUSEHOLDS] > [MALES] * [POP90_SQMI] + [AREA]
```

will be evaluated differently from:

```
[HOUSEHOLDS] > [MALES] * ([POP90_SQMI] + [AREA])
```

Combining expressions

Complex queries can be built by combining expressions together with the AND and OR operators.

For example, the following query would select all the houses that have more than 1,500 square feet and a garage for three or more cars:

```
[AREA] > 1500 AND [GARAGE] > 3
```

When you use the OR operator, at least one expression of the two expressions separated by the OR operator must be true for the record to be selected.

For example:

```
[RAINFALL] < 20 OR [SLOPE] > 35
```

Use the NOT operator at the beginning of an expression to find features or records that don't match the specified expression.

For example:

```
NOT "STATE_NAME" = 'Colorado'
```

NOT expressions can be combined with AND and OR.

For example, this query would select all the New England states except Maine:

```
[SUB_REGION] = 'New England' AND NOT [STATE_NAME]
= 'Maine'
```

Querying dates

Querying date fields using the Query Builder or Select By Attributes dialog box

Some changes have been introduced in ArcGIS 9 for querying fields containing dates.

To query date fields, a format that is appropriate for the underlying data source is automatically shown in the unique values list or generated by the Query Wizard.

Most of the time you will only need to click the field, the operator, then the value to generate the proper syntax.

Ways to find features by their locations

With the Select By Location dialog box, you can select features based on their location relative to other features. You can use a variety of methods to select the point, line, or polygon features in one layer that are near to or overlap the features in the same or another layer.

Are crossed by the outline of

This method selects the features that are overlapped by the features of another layer. For example, selecting wilderness areas crossed by the outline of roads will select any wilderness area with a road crossing its boundaries.

Intersect

This method is similar to the Are crossed by the outline of method but also selects any features bordered by the reference features. For example, selecting wilderness areas intersected by roads will select any wilderness area with a road running within its boundaries or alongside it.

Are within a distance of

This method selects features near or adjacent to features in the same layer or in a different layer. For example, if you have a layer of clean and polluted wells, you can find all the clean wells within 500 meters of the polluted ones. Or, you could find the reservoirs and farms in other layers that are within this distance of the polluted wells. You can also use this option to find features adjacent to other features. For example, you may have already selected land parcels that your company might purchase, and now you want to get information about adjacent parcels. In this case you would select the parcels within zero distance of the ones you've already selected.

Have their center in

This method selects the features in one layer that have their center in the features of another layer.

Are completely within

This method selects features in one layer that fall completely inside the polygons of another. For example, you can select lakes completely within a forested area. To select features that are a distance from the edges of the polygon they fall inside, specify a buffer distance. For example, you can select lakes that are at least 500 meters within a forested area.

Completely contain

You can select polygons in one layer that completely contain the features in another layer. For example, you can select forested areas that have lakes completely within them. To select polygons that completely contain features a certain distance within them, specify a buffer distance. For example, you can select forested areas with lakes at least 500 meters within them by buffering the lakes.

Share a line segment with

This method selects features that share line segments, vertices, or nodes with other features.

Are identical to

This method selects any feature having the same geometry as a feature of another layer. The feature types must be the same—for example, you use polygons to select polygons, lines to select lines, and points to select points.

Contain

This method selects features in one layer that contains the features of another. This method differs from the Completely contain method in that the boundaries of the features can touch. For example, with the Contain method, a forest will contain a lake—and thus be selected—even if the border of the lake touches the border of the forest. The forest would not be selected using Completely contain because the borders touch.

Are contained by

This method selects features in one layer that are contained by the features in another. For example, you can select those cities that are contained by a county. This method differs from the Are completely within method in that the edges of the features can touch.

Touch the boundary of

If you are selecting features using a layer containing lines, this method selects lines and polygons that share line segments, vertices, or endpoints (nodes) with the lines in the layer. The lines and polygons will not be selected if they cross the lines in the layer.

If you are selecting features using a layer containing polygons, this method selects lines and polygons that share line segments or vertices with the polygon boundaries. The lines and polygons will not be selected if they cross the polygon boundaries.

You can't use this method to select point features.

Selecting features by their locations

Suppose you want to know how many homes were affected by a recent flood. Answering this question—and others like it—involves forming a spatial query. You want to find features based on where they are in relation to other features. For instance, if you mapped the flood boundary, you could then select all the homes that are within this area.

By combining queries, you can perform more complex searches. For example, suppose you want to find all the customers who live within a 20-mile radius of your store and who made a recent purchase so you can send them a promotional mailing. You would first select the customers within this radius (select by location), then refine the selection by finding those customers who have made a purchase within the last six months according to a date-of-last-purchase attribute (select by attribute).

1. Click Selection and click Select By Location.
2. Click the dropdown arrow and click a selection method.
3. Check the layers whose features you would like to select.
4. Click the dropdown arrow and click a selection method.
5. Click the dropdown arrow and click the layer you want to use to search for the features.
6. Check to use only the selected features.
7. Check Apply a buffer to the features in <layer> and set the distance within which to search for features.
8. Click Apply.

   ArcMap selects the features.
9. If you're finished selecting features click Close.

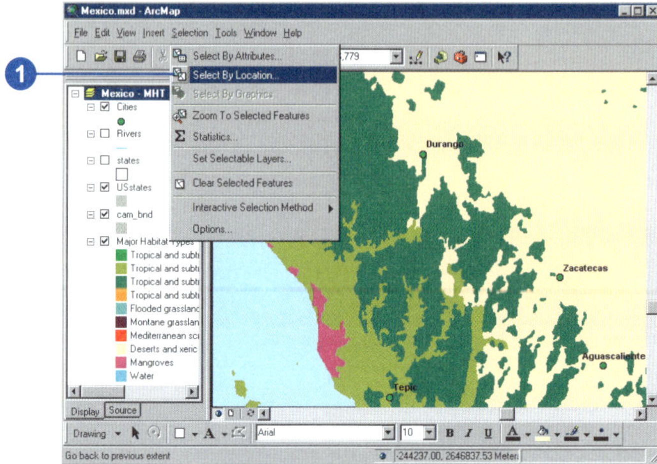

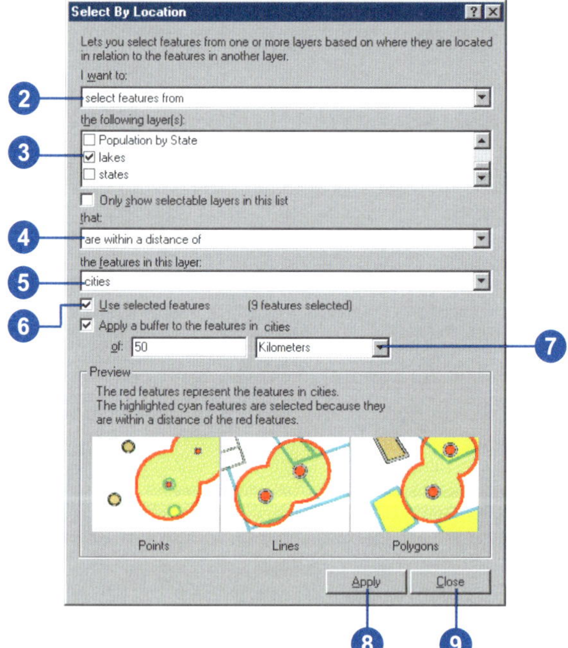

Specifying how selected features highlight

You can display selected features in any color or symbol. You can control this for all layers at once or for any specific layer.

Specifying how all layers highlight

1. Click the Selection menu and click Options.
2. Click the color box and click the color you want to use.

   For polygons, this is the color the edges highlight in. For points and lines, this is the color the entire feature highlights in.
3. Click OK.

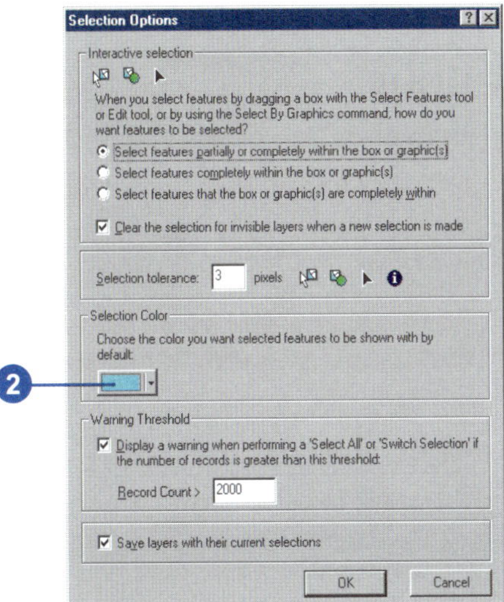

Specifying how a specific layer highlights

1. In the table of contents, right-click a layer and click Properties.
2. Click the Selection tab.
3. Click the third option and click the color you want.
4. Click OK.

   This setting overrides any setting you make with the Selection Options dialog box.

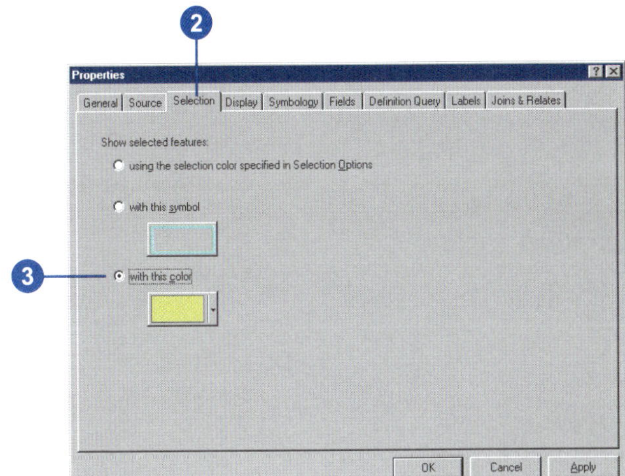

Displaying information about selected features

Once you've selected features, you can zoom to them or display their attributes or statistics. You can also create a report or graph of them. For information on how to create reports and graphs, see Chapter 11, 'Looking at data with graphs', and Chapter 12, 'Creating reports'.

Zooming to selected features

1. Right-click the layer in the table of contents that contains selected features.
2. Click Selection and click Zoom To Selected Features.

   ArcMap zooms to the selected features.

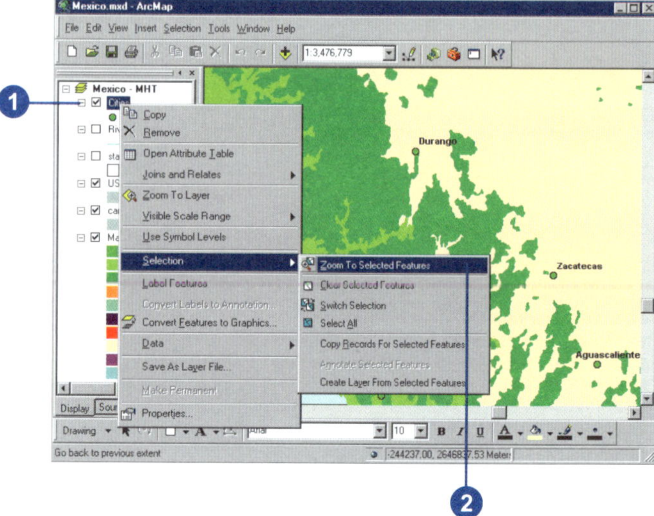

Displaying the attributes of selected features

1. In the table of contents, right-click the layer containing selected features and click Open Attribute Table.
2. Click Show Selected records.

   Records of the selected features display.

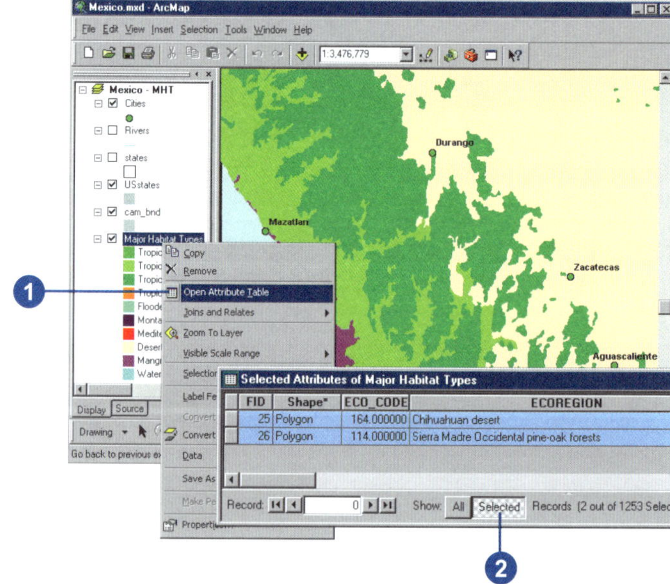

Displaying statistics

1. Select features using any selection method discussed in this chapter.
2. Click Selection and click Statistics.
3. Click the Layer dropdown arrow and click the map layer you want to see statistics about.
4. Click the Field dropdown arrow and click the field you want to see statistics about.

   The statistics are displayed in the Statistics window.

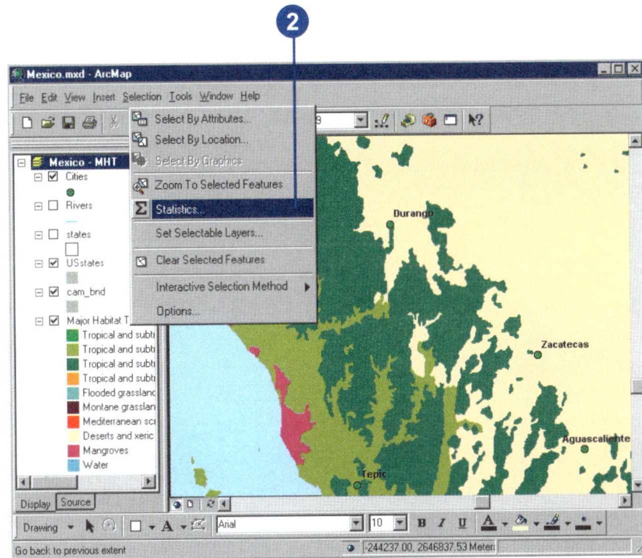

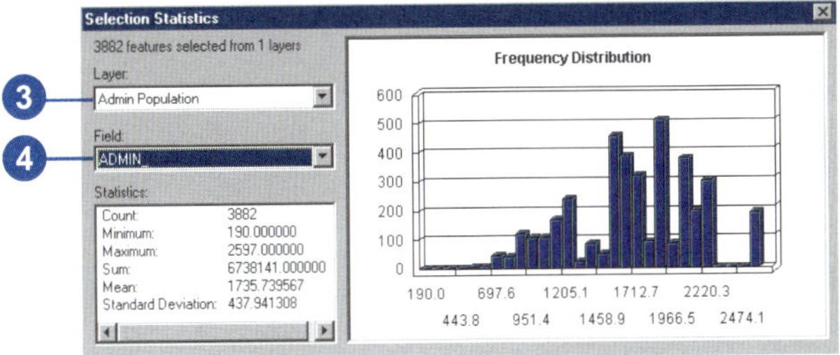

Exporting selected features

Suppose you've selected a set of features that meet some criteria and you want to work with these features outside of the current layer. You can export the selected features to a new shapefile or geodatabase feature class. Alternatively, instead of creating a new data source, you can simply create a new layer that only contains the selected features.

Exporting selected features to a new data source

1. Select features using any selection method discussed in this chapter.
2. Right-click the layer that contains the selected features, point to Data, and click Export Data.
3. Click the Export dropdown arrow and click Selected features.
4. Click Use the same Coordinate System as this layer's source data.
5. Click the Browse button and navigate to a location to save the exported data.
6. Type the name for the output source data.
7. Click the Save as type dropdown arrow and choose the output type.
8. Click Save.
9. Click OK.

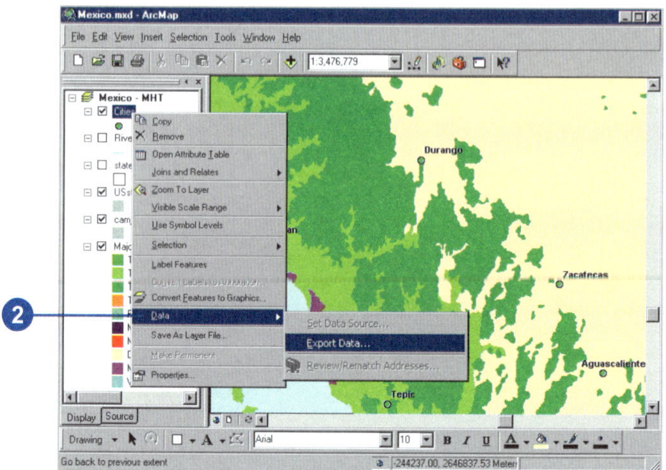

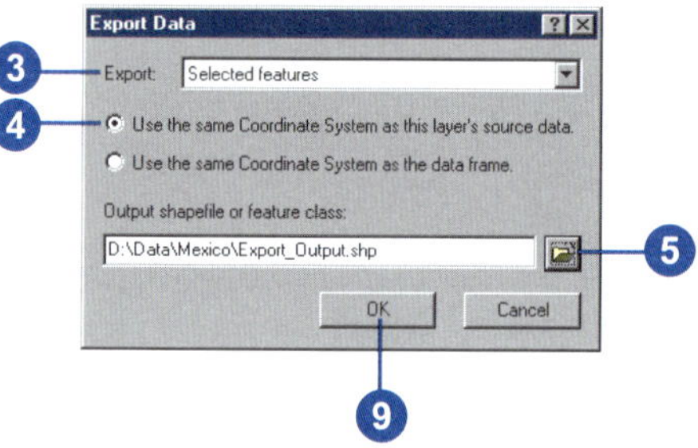

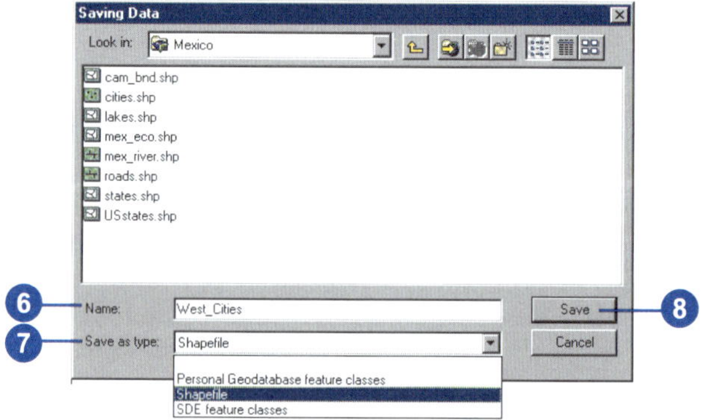

Creating a new layer from the selected features

1. Select features in a layer using any selection method discussed in this chapter.
2. Right-click the layer in the table of contents, point to Selection, and click Create Layer From Selected Features.

   ArcMap automatically names the new layer and adds it to the map.

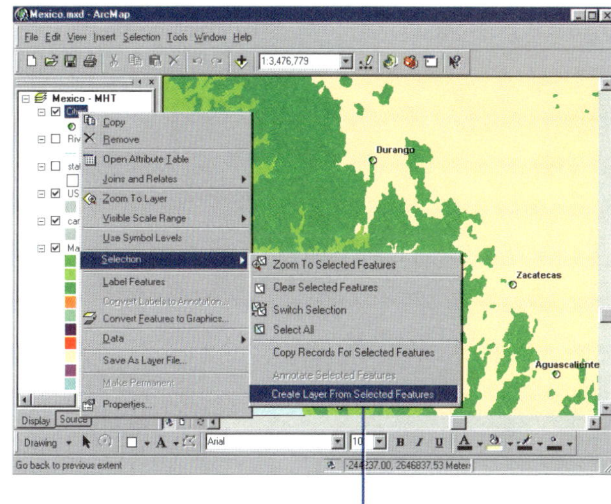

Joining the attributes of features by their locations

Often, what's most interesting about a map is not the individual layers on it, but the relationships between the features in those layers. For example, suppose you wanted to tell a customer where they can find the nearest branch office of your business, or you want to compare different wildlife species with information about the habitats they live in. These types of queries can be answered with a spatial join.

A spatial join joins the attributes of two layers based on the location of the features in the layers. With a spatial join, you can:

- Find the closest features to another feature.
- Find what's inside a feature.
- Find what intersects a feature. ▸

Tip

What is the nearest feature?

The nearest feature is defined as the feature that is geographically closest to another one. Proximity is based on a straight-line distance between features.

Finding the nearest feature

1. Right-click the layer to which you want to join attributes, point to Joins and Relates, and click Join.
2. Click the first dropdown arrow and click Join data from another layer based on spatial location.
3. Click the layer dropdown arrow and click the name of the layer whose attributes you want to join. If the layer is not currently part of the map, click the Browse button to search for it on disk.
4. Click the option to join the attributes of the feature closest to it.
5. Type the name of the output shapefile or feature class.
6. Click OK.

   A new layer is added to the map.

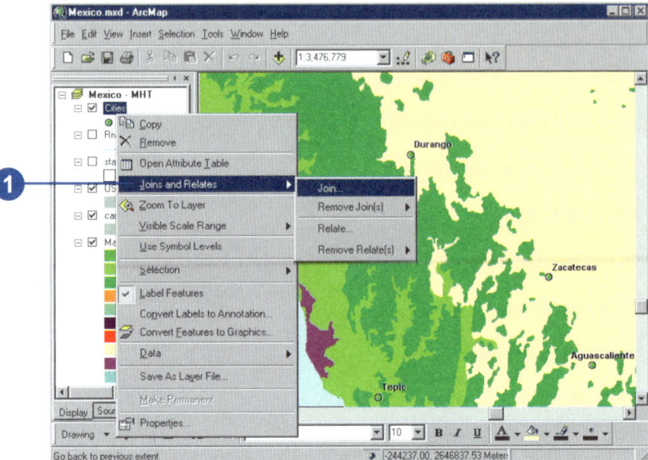

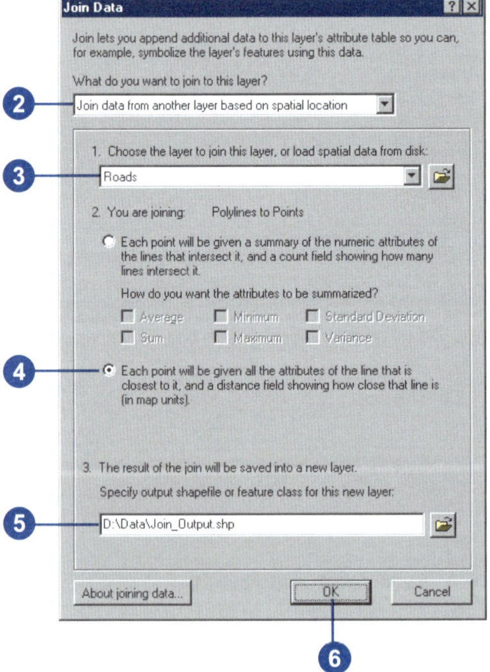

Like joining two tables by matching attribute values in a field, a spatial join appends the attributes of one layer to another. You can then use the additional information to query your data in new ways.

While you can also select features in one layer based on their location relative to another layer, a spatial join provides a more permanent association between the two layers because it creates a new layer containing both sets of attributes.

See Also

For more information on joining attribute tables, see Chapter 10, 'Working with tables'.

See Also

For more information on selecting features by location, see 'Selecting features by their locations' in this chapter.

Finding what's inside a polygon

1. Right-click the layer to which you want to join attributes, point to Joins and Relates, and click Join.
2. Click the first dropdown arrow and click Join data from another layer based on spatial location.
3. Click the layer dropdown arrow and click the name of the layer whose attributes you want to join. If the layer is not currently part of the map, click the Browse button to search for it on disk.
4. Click the option to join the attributes of the features that fall inside the polygon.
5. Check how you want to summarize attributes.

   In this example, the attributes of a city layer are being joined to a states layer. As more than one city will fall in a state, the attributes of the cities will be summarized. For instance, the output will contain the sum and average population for all cities that fall in a given state.
6. Type the name of the output shapefile or feature class.
7. Click OK.

   A new layer is added to the map.

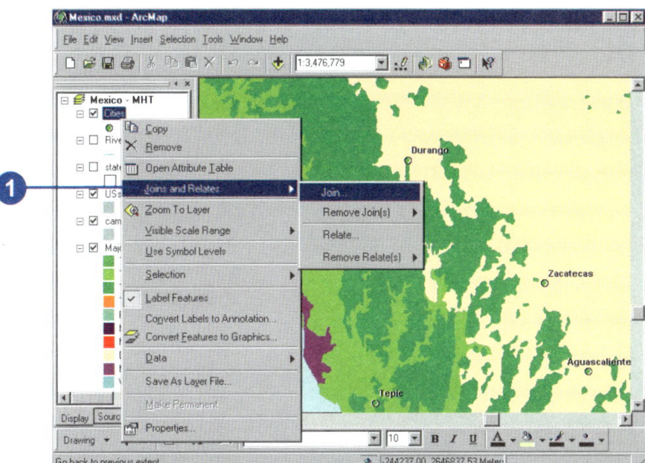

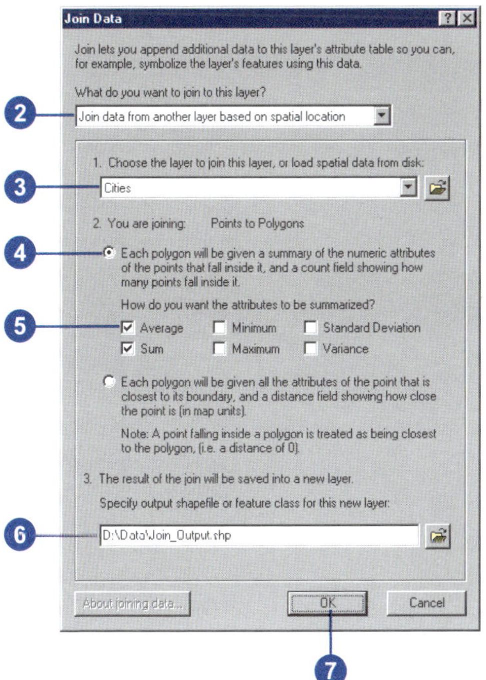

Finding what intersects a feature

1. Right-click the layer to which you want to join attributes, point to Joins and Relates, and click Join.
2. Click the first dropdown arrow and click Join data from another layer based on spatial location.
3. Click the layer dropdown arrow and click the name of the layer whose attributes you want to join. If the layer is not currently part of the map, click the Browse button to search for it on disk.
4. Click the option to join the attributes of the features that intersect it.
5. Check how you want to summarize attributes.

   In this example, the attributes of a roads layer are being joined to a rivers layer. You'll be able to find out how many roads cross each river and get a summary of the attributes of those roads.
6. Type the name of the output shapefile or feature class.
7. Click OK.

   A new layer is added to the map.

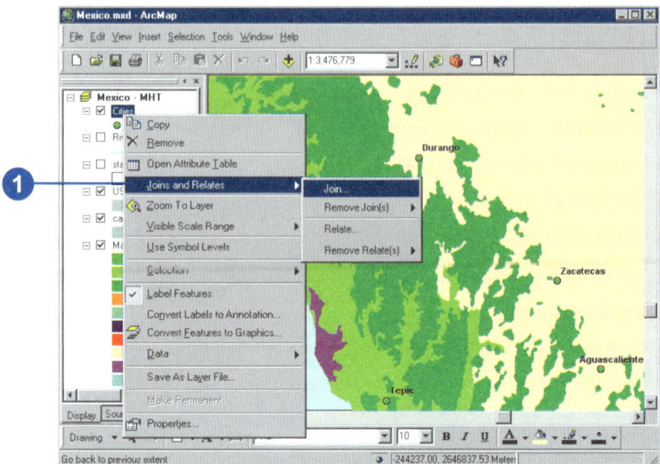

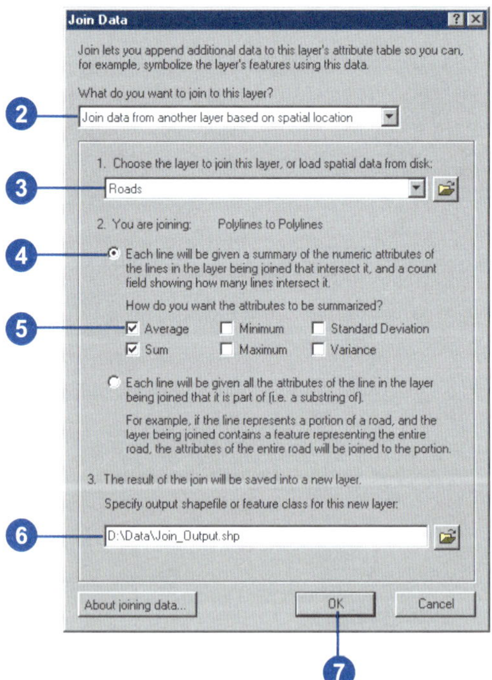

Taking geoprocessing further

Geoprocessing is a basic GIS function—it is the processing of geographic information. Any alteration or information extraction you wish to perform on your data involves a geoprocessing task. It can be a simple process, such as converting geographic data to a different format, or it can involve multiple processes linked together, such as those that project, clip, buffer, intersect, and export datasets. The geoprocessing environment within ArcGIS provides you with easy access to multiple tools or scripts that you can run via dialog boxes or a command line. It also allows you to build your own tools by creating scripts or building models.

Tools can be run from toolboxes contained within the ArcCatalog tree or the ArcToolbox window. The ArcToolbox window provides a shortcut to the tools you have stored in toolboxes that are themselves stored in folders or geodatabases on your computer's disk. Toolboxes contain *toolsets*, which are a group of geoprocessing tools. For example, the Analysis Tools toolbox contains several toolsets including the Overlay toolset. Within the Overlay toolset, you have access to the Intersect tool. Double-clicking this tool opens the tool's dialog box, where you can input values for parameters and run the tool.

The types of tools available are dependent on the ArcGIS Desktop license you are running (ArcView, ArcEditor, or ArcInfo) and which extensions you have.

To learn more about geoprocessing in ArcGIS, see *Using ArcGIS Geoprocessing Tools.*

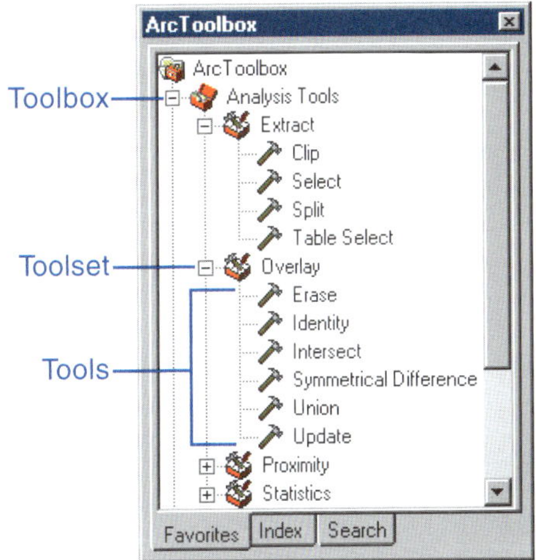

The ArcToolbox window

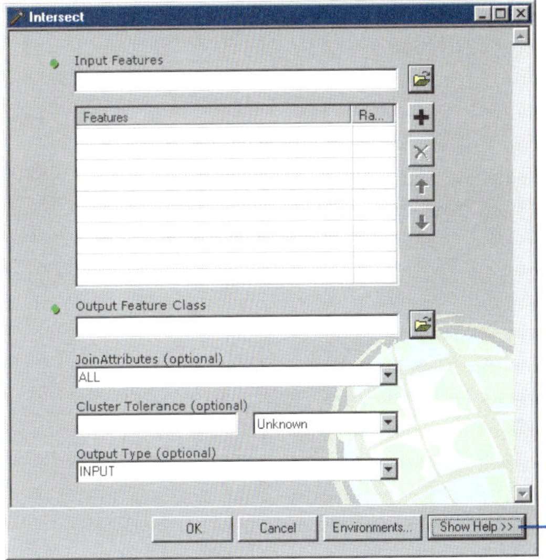

The Intersect tool dialog box

Analyzing geometric networks

14

IN THIS CHAPTER

- Geometric networks
- Opening a geometric network
- Symbolizing network features
- Adding network features
- Enabling and disabling features
- Adding the Utility Network Analyst toolbar
- Exploring the Utility Network Analyst toolbar
- Flow direction
- Displaying flow direction
- Setting flow direction
- Tracing on networks
- Tracing operations

The economic foundation of the modern world is its infrastructure—the collection of cables, pipelines, and wires that enables the movement of energy, commodities, and information. This infrastructure can be modeled as networks. ArcGIS provides a complete model for capturing, storing, and analyzing networks.

What can you do with networks in ArcMap?

ArcMap provides a rich set of tools that perform many common network analysis tasks on geometric networks. Some of the common types of analyses that you can perform on your network using ArcMap are:

- Trouble call analysis: determines the likely cause of a problem based on the location of customers with service problems.
- Isolation traces: determines which switches have to be opened in order to cut power to a part of the network.
- Contaminant tracing: determines whether a particular site is a possible cause of contamination.

Before you can work with geometric networks in ArcMap, you need to build a geometric network. While network features can be both created and edited in ArcInfo and ArcEditor, they are read-only in ArcView. To learn how to build a geometric network using ArcCatalog, see *Building a Geodatabase*.

Geometric networks

Networks consist of two fundamental components: edges and junctions. The edges and junctions in a network are topologically connected to each other. An *edge* is a type of network element that has a length and through which some commodity flows. Electrical transmission lines, pipes, and stream reaches are examples of edges. A *junction* occurs at the intersection of two or more edges and allows the transfer of flow between edges. Fuses, switches, service taps, and the confluence of stream reaches are examples of junctions. Edges connect to each other at junctions; the flow from edges in the network is transferred to other edges through junctions.

In ArcGIS, feature classes can participate together in a network. Feature classes representing transmission lines, switches, fuses, and transformers can all be part of the same network. Because the features have geometry and can be mapped, the network of features is called a *geometric network*. A geometric network contains the connectivity information between edges and junctions and defines rules of behavior such as which classes of edges can connect to a particular class of junction or to which class of junction two classes of edges must connect.

To create a geometric network, you use a wizard to specify which feature classes will participate in the network, or you create an empty network and add feature classes to it later. Once the network is created, it is maintained throughout the life cycle of the database. ArcGIS maintains the connectivity information whenever you edit the participating feature classes of a geometric network. The connectivity rules and relationships that you define in the geodatabase can be validated at any time to discover features that violate the established rules.

ArcGIS includes a variety of tools for analyzing networks and a rich set of objects for building custom networks with complex behavior. For more information on modeling networks in ArcGIS, see *Modeling Our World*. To learn how to create geometric networks and define connectivity rules, and for information on defining relationships between feature classes, see *Building a Geodatabase*. To learn about editing geometric networks, see *Editing in ArcMap*.

Opening a geometric network

Geometric networks are objects within the geodatabase. Geometric networks are automatically maintained by ArcGIS when their feature classes are edited.

In order to work with a geometric network in ArcMap, you must load a minimum of one feature class that participates in the network. If you want to work with only the feature classes that participate in the network—for example, if you are performing some analysis on the network—then you can load only these feature classes by loading the network object.

If you want to load all of the feature classes in the feature dataset that contain the network—for example, if you want to produce a printed map of the network—you can open the network by loading the feature dataset that contains the network.

See Also

For information on creating geometric networks, see Building a Geodatabase.

1. Open the document to which you want to add the network data or create a new document.
2. Click the Add Data button.
3. Navigate to the feature dataset in the geodatabase that contains the network you want to open.
4. Double-click the feature dataset to view the feature classes and geometric networks that it contains.
5. Click the geometric network and click Add.

   Your geometric network is added to ArcMap.

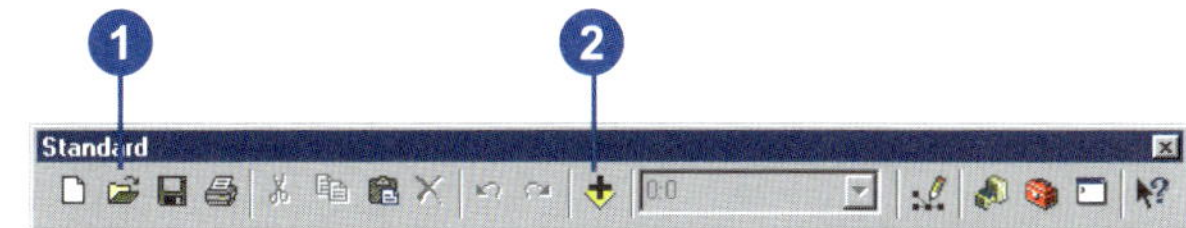

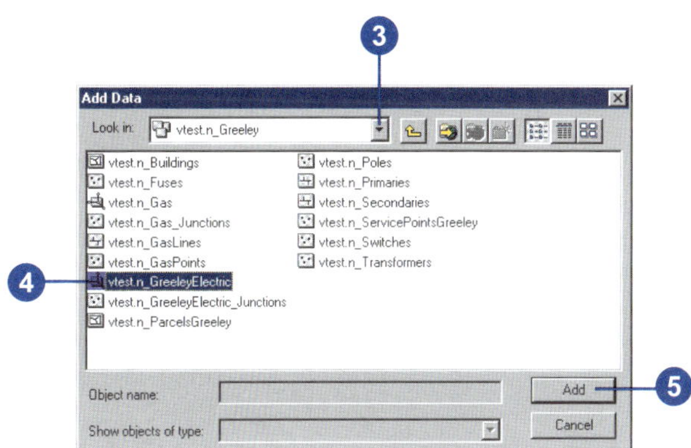

Opening a feature dataset containing a geometric network

1. Open the document to which you want to add the network data or create a new document.
2. Click the Add Data button.
3. Navigate to the feature dataset in the geodatabase that contains the network you want to open.
4. Click the feature dataset and click Add.

   The feature dataset containing your geometric network is added to ArcMap.

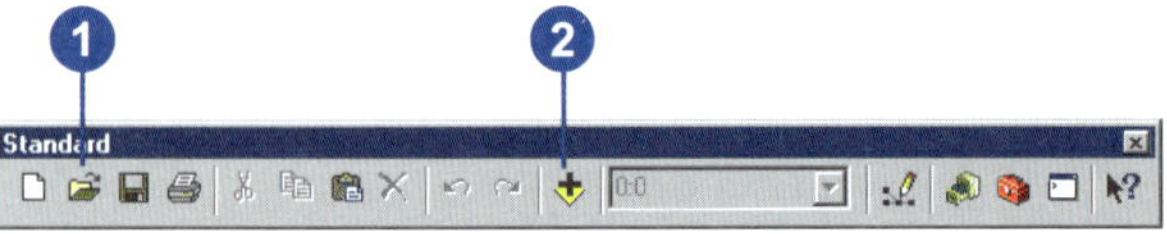

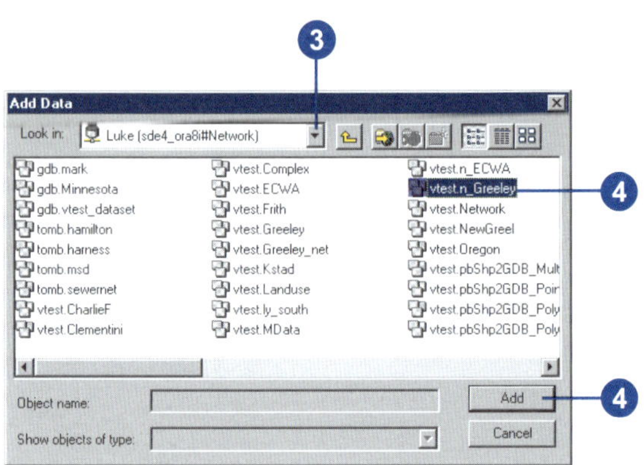

Symbolizing network features

You can use symbology in ArcMap to easily identify enabled or disabled features, sources, or sinks in your network.

All *network features* can be either enabled or disabled. Enabled features allow flow to pass through them, while disabled features do not. The status of each feature is stored in the Enabled field of the feature class's attribute table. The values in this field are defined by a coded value attribute domain and can only be 0 or 1. Features with a value of 1 are enabled, and those with a value of 0 are disabled. By symbolizing your features using this attribute, you can quickly tell which features are enabled and which are disabled.

A junction feature can act as a source or a sink (or neither). When you build a geometric network, you specify which feature classes contain sources or sinks. Those feature classes have an attribute named AncillaryRole that contains this information. The values in this field are defined by a coded value attribute domain. A value of 1 represents a source, and a value of 2 represents a sink. ▸

Displaying enabled and disabled features in a layer

1. In the table of contents, right-click the feature layer for which you want to display enabled and disabled features.
2. Click Properties.
3. Click the Symbology tab.
4. Click the Categories item in the list and click Unique values in the expanded list of items.
5. Click the Value Field drop-down arrow and click Enabled to use this attribute for the symbolization.
6. Click Add All Values.
7. To change the symbol for a particular value, double-click the symbol.
8. Click OK when you are finished formatting the symbols.

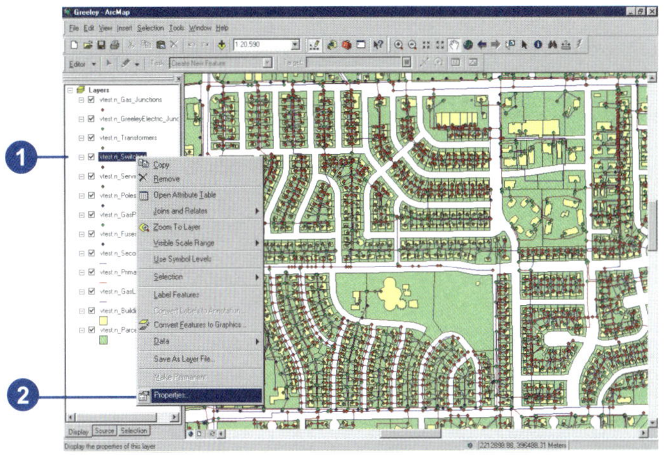

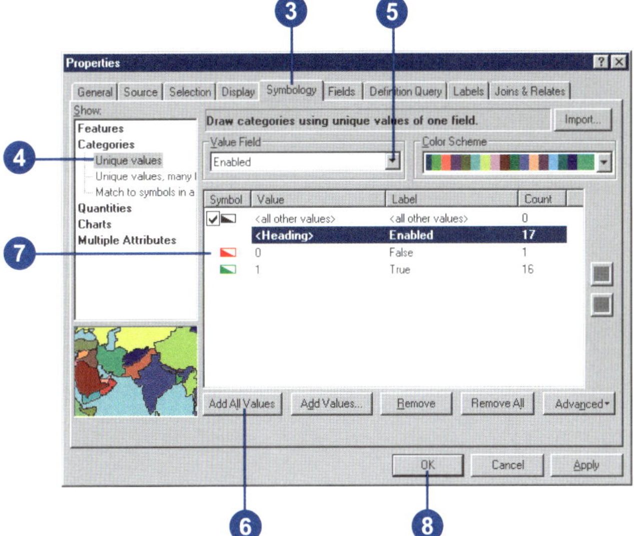

A value of 0 means that the feature is neither a source nor a sink. By symbolizing features using this attribute, you can quickly tell which junctions are sources and sinks.

For more information on attribute domains, see *Building a Geodatabase*.

Tip

Stopping map redraws

Each change made to the data view or layout view that affects the display of the data, such as docking a toolbar, maximizing the window, or changing symbology, causes the map to be redrawn. When working with large datasets, redrawing the map can take a considerable amount of time. If you are making multiple changes that will affect the view, you can stop the map from being redrawn by pressing the Esc key.

Displaying source and sink features in a layer

1. In the table of contents, right-click the feature layer for which you want to display source and sink features.
2. Click Properties.
3. Click the Symbology tab.
4. Click the Categories item in the list and click Unique values in the expanded list of items.
5. Click the Value Field drop-down arrow and click AncillaryRole to use this attribute for the symbolization.
6. Click Add All Values.
7. To change the symbol for a particular value, double-click the symbol.
8. Click OK when you are finished formatting the symbols.

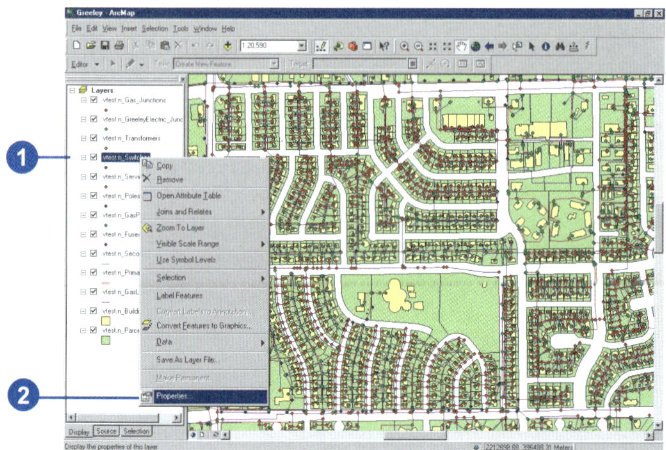

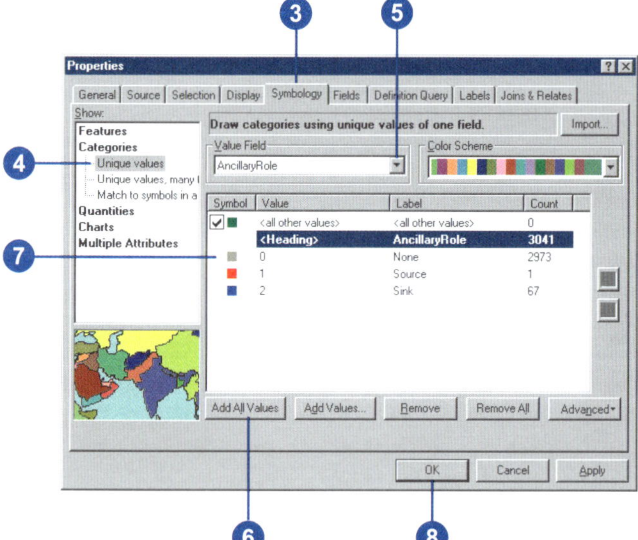

Adding network features

Adding features to a network is the same as adding features to any dataset. However, when features are added to a network, they connect topologically to other features in the network. These connections are automatically maintained inside the geodatabase and are based on geometric coincidence. For example, a junction must be coincident with an edge endpoint in order to be connected. To ensure geometric coincidence, use *snapping* when editing network features.

In this example, a new service is added to the network to provide water to a building. The new service connects to a water main on one end and snaps to a building on the other. In order to ensure that the new feature in the services feature class connects to the water main on one end and touches the building on the other, you must use snapping.

See Also

For information on snapping, see Editing in ArcMap.

1. Click Editor and click Start Editing.
2. Click Editor and click Snapping.
3. In the Snapping Environment window, set the appropriate snapping properties.
4. Close the Snapping Environment window. ▸

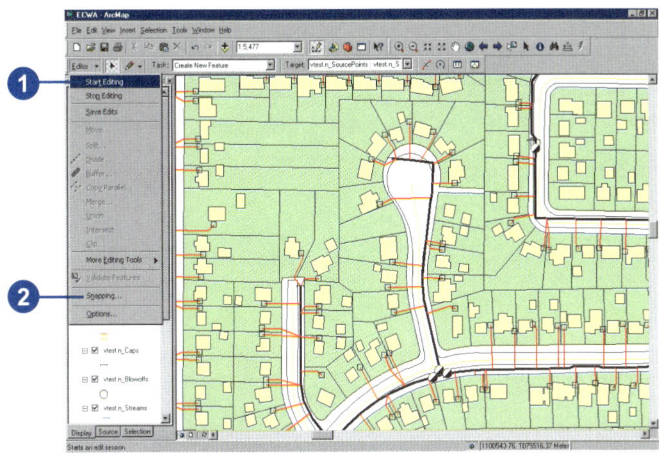

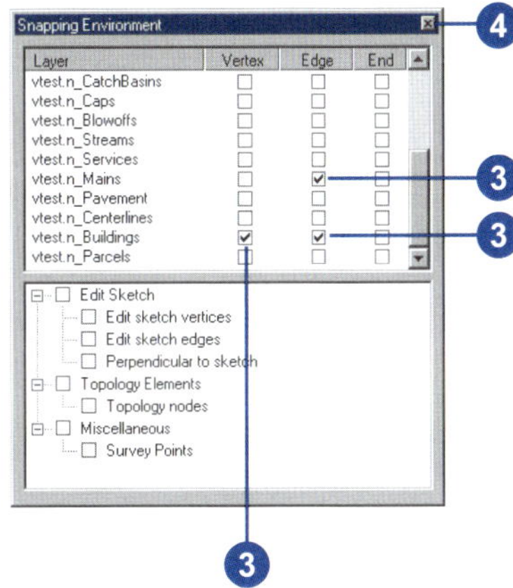

5. Click the Task dropdown arrow and click Create New Feature.
6. Click the Target dropdown arrow and click the layer to which you want to add a feature.
7. Click the Sketch tool.
8. Point to a position on the feature where the new feature is to connect. A target appears to show that snapping is on. Click to create the first vertex of the new feature.
9. Create the remaining vertices of the sketch and double-click to finish it.
10. Click Editor and click Stop Editing.
11. Click Yes to save your changes.

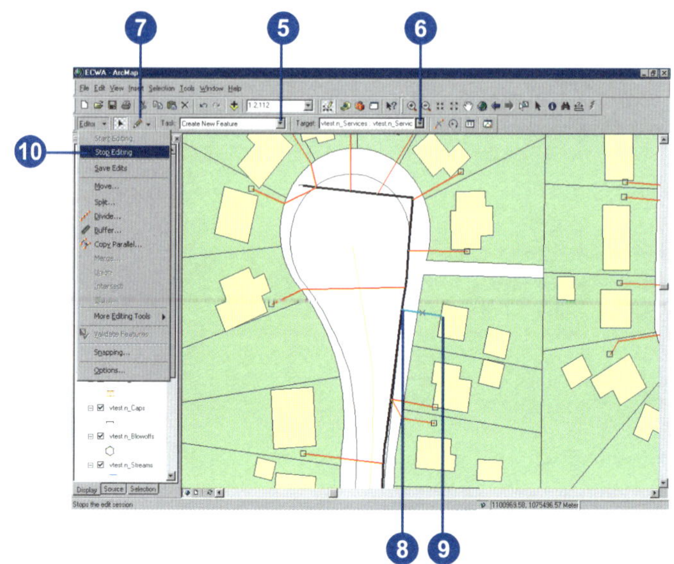

Enabling and disabling features

Any feature in a geometric network can be enabled or disabled. An *enabled feature* allows flow to pass through it, while a *disabled feature* does not. Disabling features allows you to treat features as if they were disconnected from the network, without actually removing the topological connections that they have to other features in the network. By default, all features in a geometric network are enabled when you create the network.

To enable or disable a feature, its Enabled attribute must be edited in the Attributes dialog box.

Tip

Enabling and disabling features

You can also enable or disable a feature by changing the value of its Enabled attribute in the attribute table. Open the attribute table for the feature's feature class, find the feature in the table, and edit the value for this attribute. For more information on working with attribute tables, see Chapter 10, 'Working with tables'.

1. Click Editor and click Start Editing.
2. Click the Edit tool and click the feature that you want to enable or disable.
3. Click the Attributes button.
4. Click in the Value column next to the Enabled property.

   A list box opens showing all of the valid codes for this attribute, as defined by the coded value attribute domain for the Enabled attribute.
5. Click True to enable the feature. Click False to disable the feature.
6. Click Editor and click Stop Editing.
7. Click Yes to save your changes.

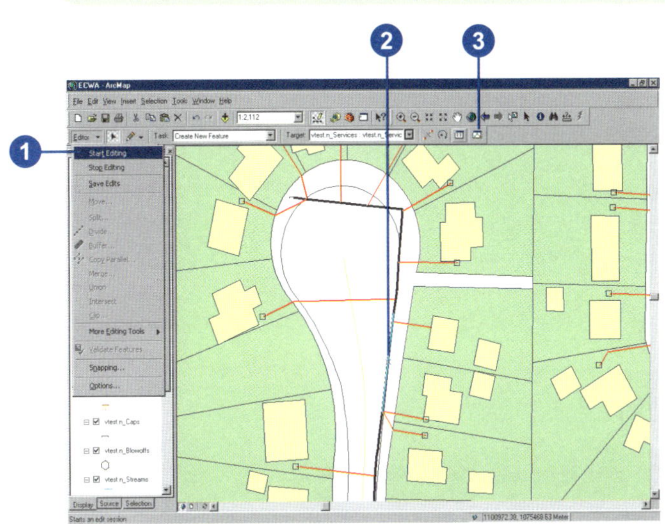

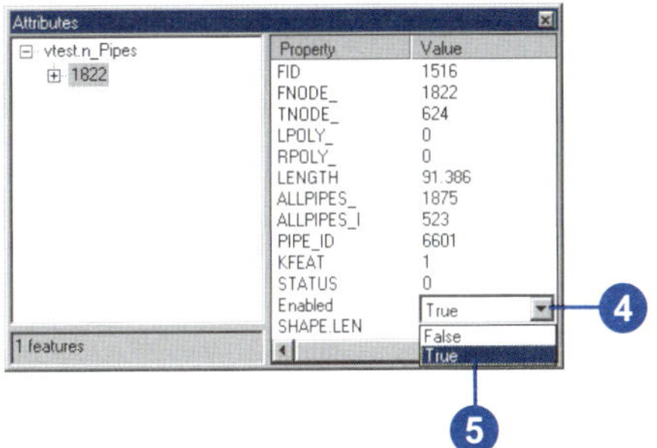

Adding the Utility Network Analyst toolbar

In order to use ArcMap to analyze your networks, you must load the Utility Network Analyst toolbar. This toolbar contains most of the tools needed to perform the analysis tasks presented later in this chapter.

Tip

Adding the toolbar

You can also add the toolbar by clicking the View menu, pointing to Toolbars, then clicking Utility Network Analyst.

1. Right-click the Main menu.
2. Click Utility Network Analyst.
3. Dock the toolbar to the ArcMap window.

   Now, each time you start ArcMap, the toolbar will be displayed.

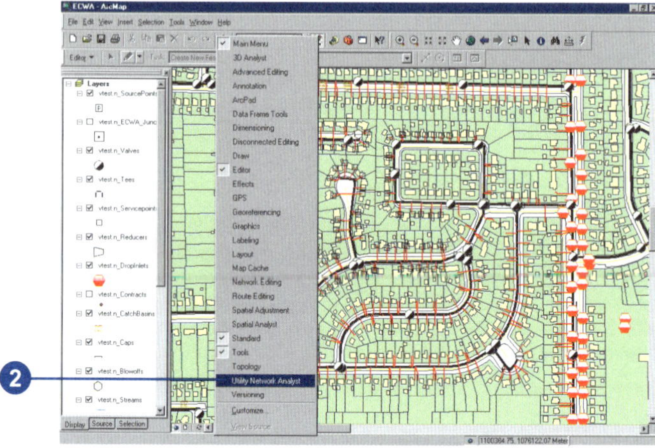

Exploring the Utility Network Analyst toolbar

The Utility Network Analyst toolbar is divided into two sections. The left side of the toolbar lets you choose a network with which to work and to set and display its flow direction. The right side of the toolbar lets you set up and perform trace operations on the current network (see 'Tracing on networks' in this chapter).

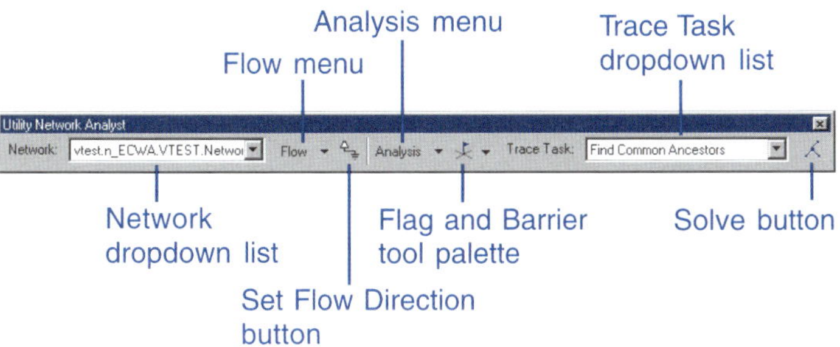

The Network dropdown list contains all of the geometric networks that are currently loaded in ArcMap. To work with a network in ArcMap—for example, set the flow direction or perform a trace operation—you must choose the network in this list.

The Flow menu contains items for displaying the flow direction of the features in the network. Clicking the Flow menu reveals three items: Display Arrows For, Display Arrows, and Properties.

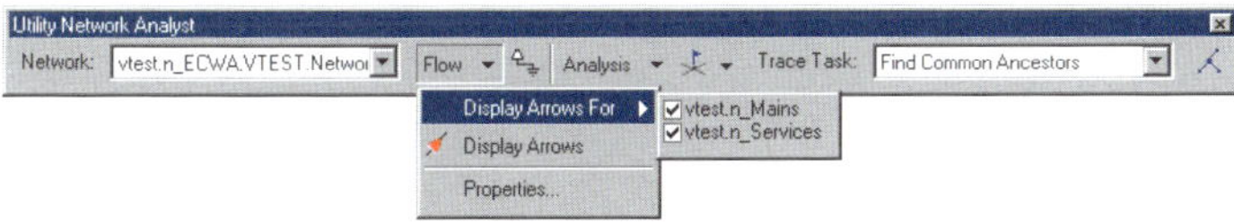

Clicking Display Arrows For produces a list of the edge feature classes in your network. By checking items in this list, you can specify for which layers to display flow direction. The Display Arrows command is a toggle button that turns on or off the display of flow direction arrows for your network. Clicking the Properties command opens the Flow Display Properties dialog box. The Arrow Symbol tab lets you specify the symbol, size, and color of the arrows used to indicate flow direction. The Scale tab lets you specify the scale range in which the arrows are displayed.

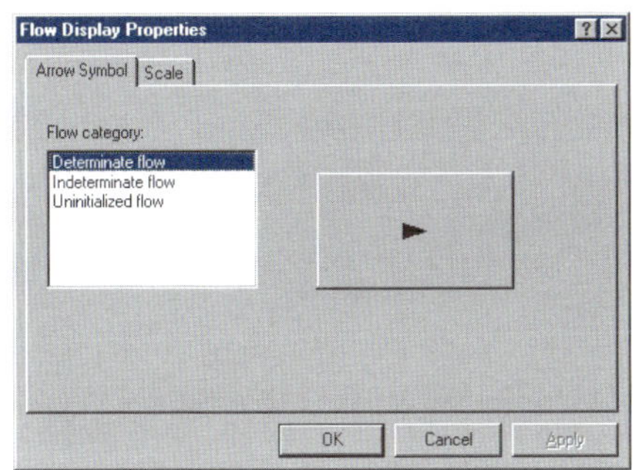

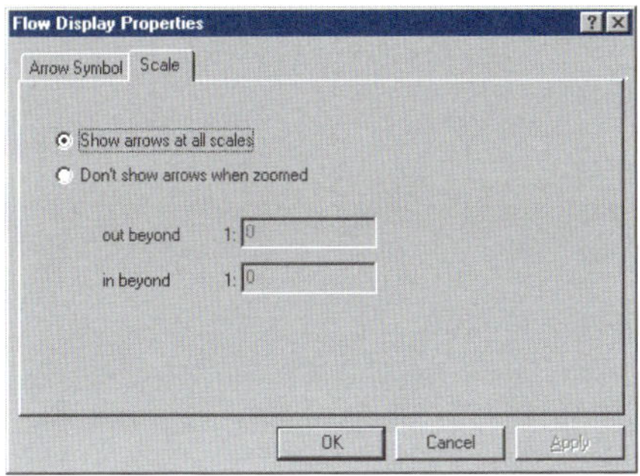

The Set Flow Direction button establishes flow direction in the network. This button is enabled when your network contains feature classes that you have designated as containing persistent sources and sinks.

The Analysis menu contains commands for setting up your network to perform trace operations. Clicking the Analysis menu reveals five commands: Disable Layers, Clear Flags, Clear Barriers, Clear Results, and Options.

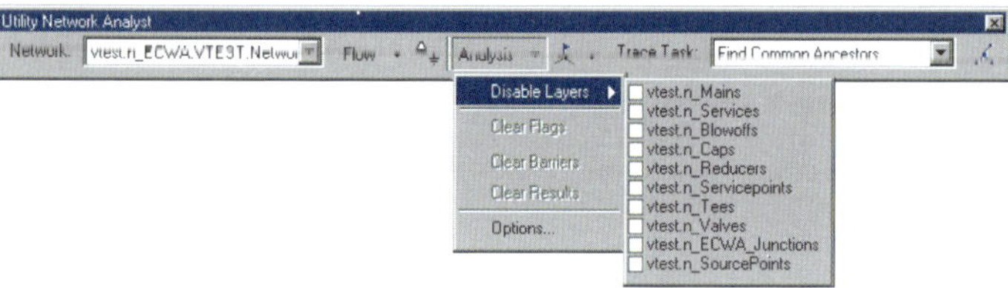

Clicking Disable Layers displays a list of feature classes contained in the geometric network. By checking feature classes in this list, you can disable feature classes in trace operations. This makes trace operations behave as if all of the features in those feature classes are disabled. The Clear Flags and Clear Barriers menu items remove flags and barriers, respectively, from the network. Clicking Clear Results clears the results of the previous trace operation.

Clicking Options opens the Analysis Options dialog box. This dialog box allows you to specify options for subsequent trace operations.

The General tab of the Analysis Options dialog box lets you specify on which features trace tasks are performed. You can perform trace tasks on all of the features in the network, only the selected features, or only the unselected features. You can specify whether or not trace tasks that consider flow direction include edges with indeterminate or unspecified flow direction. You can also use this tab to specify the snapping tolerance for placing flags and barriers on the map.

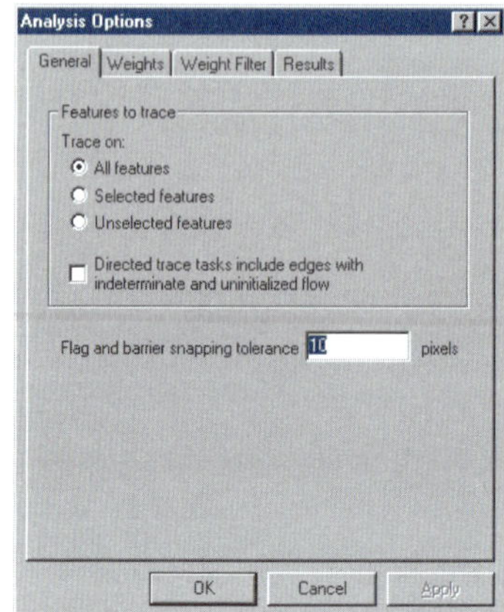

The Weights tab lets you specify which network weights are used when tracing. For example, the Find Path trace task uses weights to determine the cost of including a network feature in the result of the trace task.

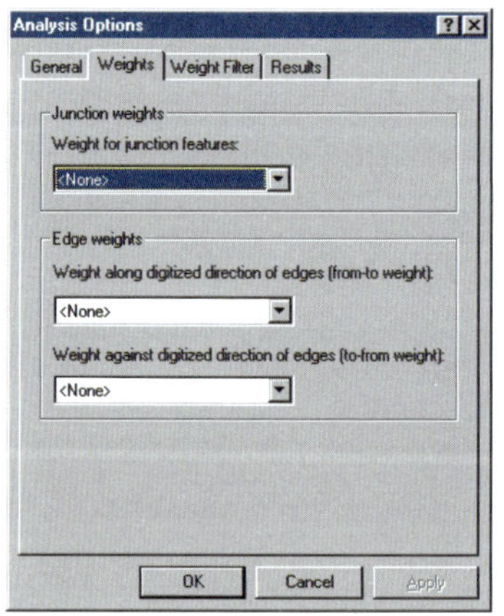

The Weight Filter tab lets you specify which network features can be traced based on weights assigned to the network features. For both edges and junctions, you can specify valid ranges of weights for features that may be traced.

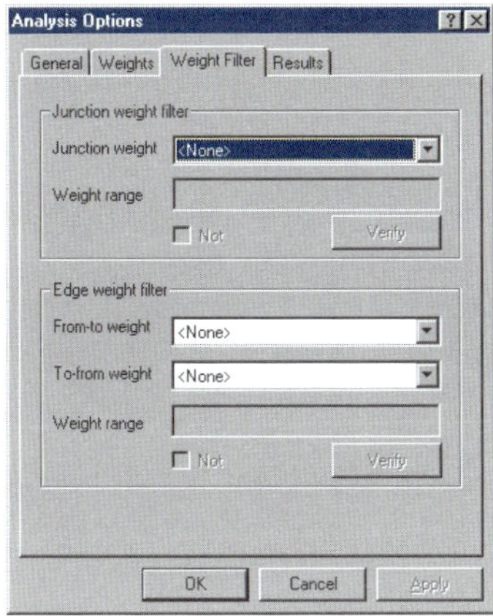

With the Results tab, you can determine in which format you want to receive the results of trace operations. Results can be given as drawings overlaid on the map or as a set of selected features. If you choose to draw the results, you can choose to render only the parts of complex edges that are actually traced, rather than the entire complex feature. You can also specify whether the results will include features that are traced during the operation or features stopping the trace. Finally, you can specify whether the results include both edge and junction features.

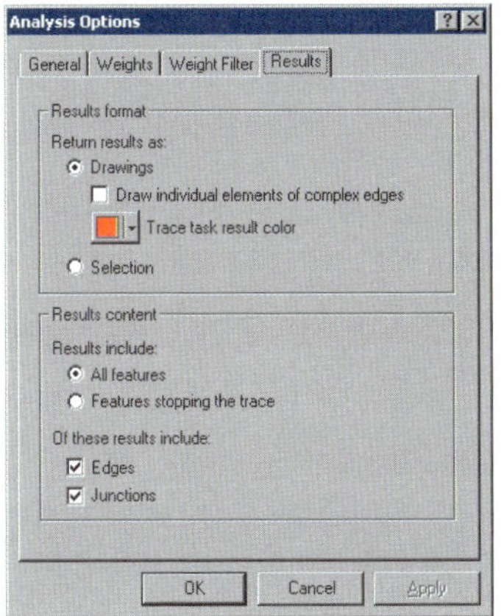

The Trace Task dropdown list contains a list of all the trace operations that you can perform using the Utility Network Analyst toolbar. ArcGIS comes with nine built-in trace operations.

While the Trace Task dropdown list is used to select the trace task, the Solve button is used to perform the trace operation once you have finished configuring your trace operation using the toolbar. The Solve button performs the trace operation that you selected in the Trace Task dropdown list; it does this using the parameters that you specified using the Analysis Options dialog box and the placement of flags and barriers on the network.

Flow direction

In network applications, knowing the flow direction along network edges is essential. Establishing the *flow direction* in a geometric network determines the direction in which commodities flow along each edge. The flow direction in a network is determined by the connectivity of the network, the locations of sources and sinks in the network, and the enabled or disabled state of features.

Sources and sinks drive flow through a geometric network. *Sources* are junction features that push flow away from themselves through the edges of the network. For example, in a water distribution network, pump stations can be modeled as sources, since they drive the water through the pipes away from the pump stations. *Sinks* are junction features that pull flow toward themselves from the edges in the network. For example, in a river network, the mouth of the river can be modeled as a sink, since gravity drives all water toward it. Flow moves away from sources and toward sinks. Because flow direction can be established with either sources or sinks, it usually suffices to specify only sources or only sinks in a network.

It is important to remember that disabled features are accounted for when setting flow direction. Disabling a feature makes it act as if flow cannot pass through it. Thus, disabling a feature means that the flow direction cannot be set for the disabled features or for those features that are connected to the sources or sinks exclusively through the disabled feature.

After you set the flow direction for your network, an edge has one of three categories of flow direction: determinate, indeterminate, or uninitialized.

Determinate flow direction

If the flow direction of an edge can be uniquely determined from the connectivity of the network, the locations of sources and sinks, and the enabled or disabled states of features, the feature is said to have *determinate flow*. Determinate flow for an edge is specified as either with or against the direction in which the feature was digitized.

Indeterminate flow direction

Indeterminate flow direction in a network occurs when the flow direction cannot be uniquely determined from the connectivity of the network, the locations of sources and sinks, and the enabled or disabled states of the features. Indeterminate flow commonly occurs for edges that form part of a loop, or closed circuit. It can also occur for an edge whose flow is determined by multiple sources and sinks, where one source or sink is driving the flow in one direction through the edge, but another source or sink is driving it in the opposite direction. For example, an edge that has a source at both of its ends will have indeterminate flow.

Uninitialized flow direction

Uninitialized flow direction in a network occurs in edges that are isolated from the sources and sinks in the network. This can happen if the edge is not topologically connected through the network to the sources and sinks or if the edge is only connected to sources and sinks through disabled features.

Specifying flow direction

All geometric networks that have flow have sources and sinks. In some cases, you may not know the locations of the sources and sinks, but you may know the flow direction. If this is the case, you must choose the junctions in your network to act as sources and sinks that produce the correct flow direction.

After setting the flow direction for your network, indeterminate flow may occur, even when you know the direction of flow. This is because the flow direction is determined by properties of the network or the features making up the network in addition to the connectivity or locations of sources and sinks. For example, in a water network, the flow direction in a pipe is determined by the difference in water pressure between the ends of the pipe. The pressure at each end of the pipe is affected by such things as the material out of which the pipe is made, the pipe diameter, the flow rate through the pipe, the physical configuration of the pipe (including any bottlenecks, valves, or sharp bends), the temperature of the water, the elevation of the ends of the pipe, and the connectivity of the network. Since ArcGIS deals with general networks (and not with domain-specific types of networks), this information is not used to set the flow direction. Thus, the flow direction may be set to indeterminate for some edges in these networks.

A set of similar variables exists in every domain. Developers can write custom flow-direction solvers that use these variables to find determinate flow direction in domain-specific networks.

Displaying flow direction

Network flow direction specifies the direction in which commodity flows through the network. ArcGIS stores this information for edge features in a network.

You can display the flow direction for edges using the Utility Network Analyst toolbar. You can display which edges have determinate flow direction, indeterminate flow direction, or uninitialized flow.

See Also

For more information on flow direction, see 'Flow direction' earlier in this chapter.

1. Click the Flow menu on the Utility Network Analyst toolbar.
2. Point to Display Arrows For.
3. Check the layers for which you want to display flow direction.
4. Click Properties.
5. Click the Arrow Symbol tab. Click a Flow category in the list and click the button to specify the size and color of the flow direction arrows. ▸

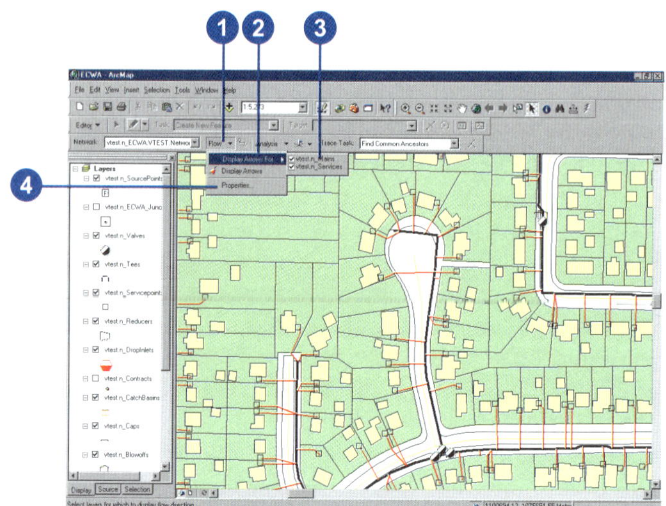

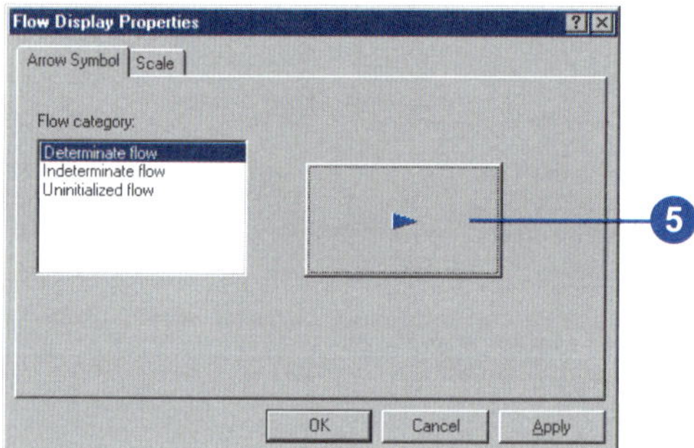

Tip

Clearing flow direction arrows

To remove the flow direction arrows, click Flow and click Display Arrows.

6. Click the Scale tab and specify the scales at which you want to display the flow direction arrows. To show the arrows at all scales, click Show arrows at all scales. To only show the arrows within a scale range, click Don't show arrows when zoomed and type the scale range limits in the text boxes.
7. Click OK.
8. Click Flow and click Display Arrows.

   The arrows symbolizing the flow direction are displayed.

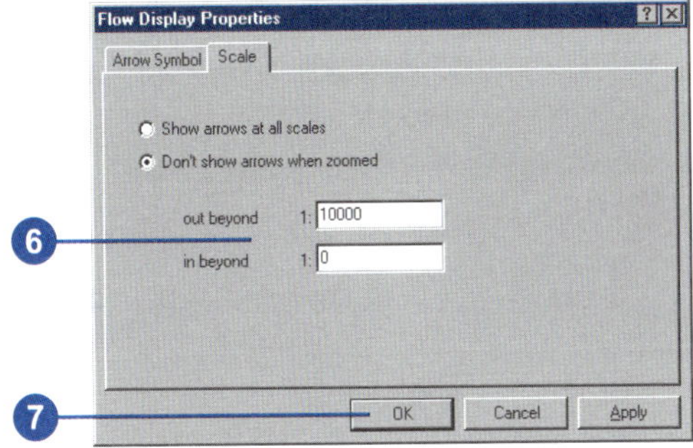

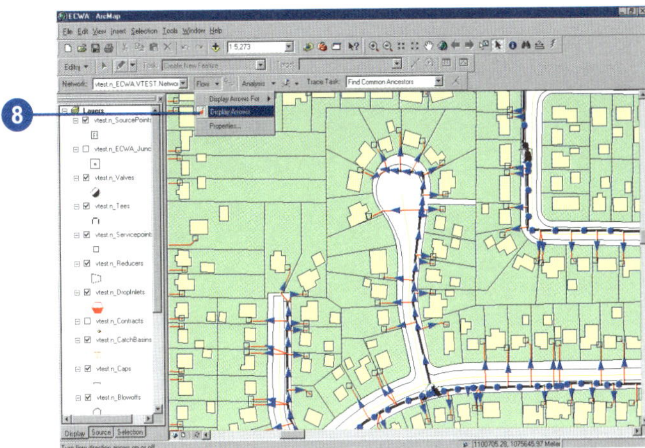

Setting flow direction

You can use ArcMap to maintain flow direction in your geometric network. ArcMap uses the network connectivity, the enabled or disabled state of network features, and the locations of sources and sinks to establish flow direction.

You must set flow direction whenever you:

- Create a new geometric network.
- Add or remove features from the network.
- Reshape features so as to change the network connectivity.
- Connect or disconnect features.
- Add or remove sources or sinks.
- Enable or disable features.
- Reconcile a version.

Setting the flow direction establishes the correct flow direction for the new network connectivity.

In order to set flow direction, your network must contain at least one junction feature class that you specified as containing sources and/or sinks.

Creating sources and sinks

1. Click the Editor menu and click Start Editing.
2. Click the Attributes button.
3. Click the Edit tool and click the feature that you want to set as a source or sink. This feature must belong to one of the feature classes that you specified as containing sources and sinks when you built your network.
4. In the Attributes window, click the Value column next to the AncillaryRole property.
5. Click Source or Sink to designate this feature as a Source or Sink. You can undo this later by clicking None in this list.
6. Click Editor and click Stop Editing.
7. Click Yes to save the edits to your network.

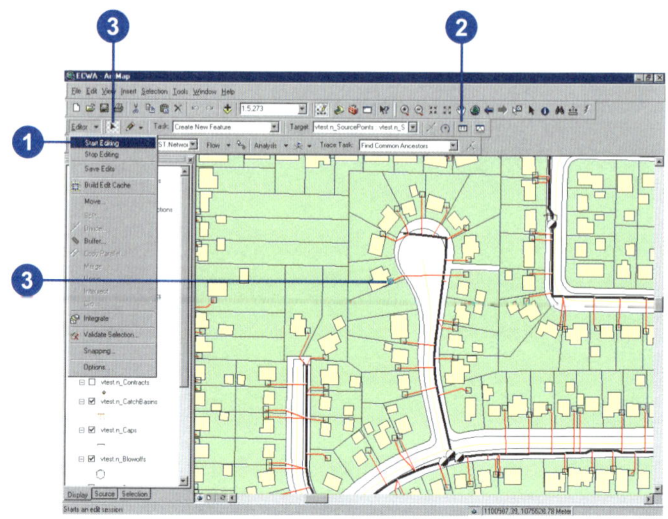

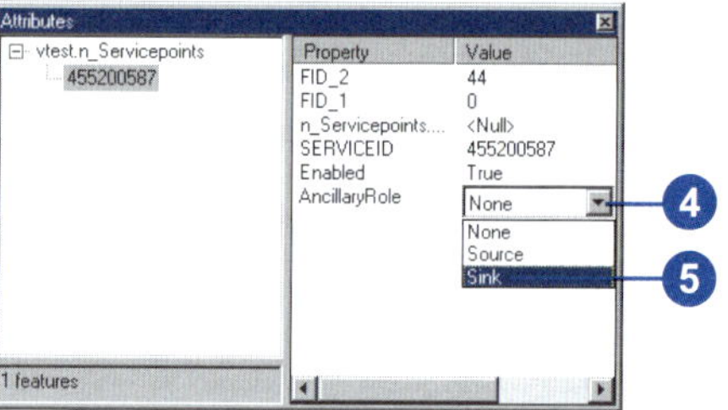

Tip

Versioning networks

You can use the versioning features of ArcGIS to create different versions of your network. Each version of the database that you create maintains its own set of information for your geometric network including the network connectivity, the enabled or disabled state of features, sources and sinks, and flow direction. With versions, for example, you could maintain one version of the network for use in trouble call analysis and maintain another version for planning maintenance and upgrades to the network.

For more information on versioning, see Building a Geodatabase.

Setting flow direction

1. Click the Editor menu and click Start Editing.
2. Click the Set Flow Direction button on the Utility Network Analyst toolbar. This sets the correct flow direction for your network.
3. Click Editor and click Stop Editing.
4. Click Yes to save the edits to your network.

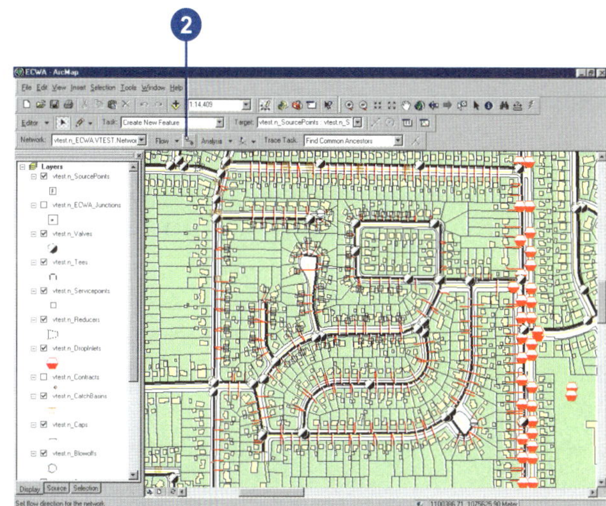

Tracing on networks

Network analysis involves tracing. The term *tracing* is used here to describe building a set of network elements according to some procedure. You can think of tracing as placing a transparency on top of a map of your network and tracing all the network elements that you want to include in your result onto the transparency.

When working with networks, tracing involves connectivity. A network element can only be included in a trace result if it is in some way connected to other elements in the trace result. The trace result is the set of network features that is found by the trace operation.

For example, suppose you want to find all of the features upstream of a particular point in a river network. Using a transparency placed over the map of the river network, you could trace over all of the branches of the river that were upstream of that point. What is drawn on the transparency after this would be your desired result.

Similarly, when you perform a trace operation in ArcMap, your result is a set of the network elements included in the trace. In ArcMap, your trace results can either be drawings on top of your map or a selection.

Flags and barriers

In ArcMap, *flags* define the starting points for traces. For example, if you are performing an upstream trace, you use a flag to specify where the upstream trace will begin. Flags can be placed anywhere along edges or on junctions. When performing the trace operation, ArcMap uses the underlying edge or junction feature as the starting point of the trace operation. Network elements connected to these edges or junctions are considered for inclusion in the trace result.

Barriers define places in the network past which traces cannot continue. If you are only interested in tracing on a particular part of your network, you can use barriers to isolate that part of the network. Like flags, barriers can be placed anywhere along edges or on junctions. When performing trace operations, ArcMap treats the underlying network features as if they are disabled, thus preventing the trace from continuing beyond these features.

Disabling features

Disabling features is a more permanent method of creating a barrier at a particular location. In a municipal water network, for example, if a water main has been opened and capped due to a road construction project, water cannot flow through that section of the water main. Disabling the network feature representing this water main would stop a trace at this feature.

Disabling feature layers

In some cases, disabling entire layers may be necessary. For example, by disabling the switches layer in an electrical distribution network and tracing from some point in the network, you can find the switches that need to be thrown to isolate this point from the network; these will be the features at which the trace operation is stopped.

Weights

Edges and junctions can have any number of weights associated with them. A *weight* is a property of a network feature typically used to represent a cost for traversing across an edge or through a junction. An example of an edge weight is the length of the edge. In a shortest path analysis, you would choose this weight if you wanted the resulting path to be of the shortest length. Another example is the resistance to traversing an edge in an electrical network. Using a resistance weight, the shortest path would be the path of least resistance.

When you build a network, you specify which attributes of edge and junction feature classes will become weights. You can use these weights to specify the cost of including a feature in the results of a trace operation. Of the trace tasks included with ArcGIS, only the Find Path, Find Path Upstream, and Find Upstream Accumulation trace tasks use weights to calculate the cost of the trace.

To find the cost using these trace tasks, you must specify which weights to use. For junction features, a single weight is used. For edge features, two weights can be used: one along the digitized direction of the edge feature (the from–to weight) and one against the digitized direction of the edge feature (the to–from weight). The digitized direction of an edge feature refers to the order in which the shape nodes of the feature are stored in the geodatabase. You can specify a different weight for each direction of an edge for cases where tracing an edge in one direction has a different cost from tracing it in the other direction.

Weight filters

You can use a weight filter to limit the set of network features that may be traced. A *weight filter* specifies which network features can be traced based on their weight values. A weight filter serves the same purpose as creating a selection of network elements based on a simple SQL query, except that the performance of the weight filter is much faster.

Using a weight filter, you specify valid or invalid ranges of weight values for network features that may be traced. As with using weights to represent the cost of including a feature in trace results, a single weight is used for junction features, and two weights may be used for edge features.

Traced features and features stopping the trace

When tracing using the Find Connected, Trace Downstream, or Trace Upstream trace tasks, you can return either the features that are traced or the features that stop the trace. Features that are traced are those that are actually traced over by the operation. Features that stop the trace are those features past which the trace cannot continue. Features that stop the trace include the following:

- Disabled features
- Features upon which barriers are placed
- Traced features that are only connected to one other feature (deadends)
- Features that have been filtered out with a weight filter

Using selections to modify trace tasks

When tracing, ArcMap lets you use selections to modify trace tasks in three main ways.

First, the Analysis Options dialog box lets you specify whether the trace operation is performed on all features in the network, on the selected features only, or on the unselected features only. Tracing on just the selected features means that unselected features act as barriers, while tracing on just the unselected features means that selected features act as barriers. By using selections in this manner, you could, for example, perform a trace operation to produce a set of barriers for a subsequent operation, or you could build a selection query to produce a set of network features upon which to perform a trace operation.

ArcMap also lets you specify which layers are selected when performing a trace operation. From the Selection menu in ArcMap, you can specify which layers can and cannot be selected. When ArcMap returns the results of a trace operation as a selection set, the settings you specify in the Selection menu are used to determine which features should be included in the selection set returned by the trace.

Finally, you can also use the interactive selection method—set through the Selection menu—to specify the behavior of the resulting selection set. You can create a new selection, add the results of your trace operation to the current selection, select the results of your trace operation from the current selection, or remove the results of your trace operation from the current selection.

By using the power of selections in ArcMap, you can use the simple trace tasks included with ArcMap to perform compound and complex trace operations.

Putting it all together

This section introduced some concepts you can use when constructing traces to perform on your network. You can return the trace results as a selection set, disable individual features or entire feature layers, place barriers on edges or junctions, include the traced features or the features stopping the trace, trace only on selected or unselected features, specify which layers to include in the results, and use different selection methods. All of these concepts can be used simultaneously when creating a trace result. Combining these concepts in trace operations allows you to execute powerful traces on your network.

Tracing operations

Using the Utility Network Analyst toolbar with your network, you can do the following:

- Trace downstream.
- Trace upstream.
- Find the upstream accumulation.
- Find an upstream path to the source.
- Find common ancestors.
- Find connected features.
- Find disconnected features.
- Find a path.
- Find loops.

You can use these simple tasks to perform many useful network analyses. You can also combine these with other features of ArcMap to perform complex network analysis operations.

To find all network elements that lie downstream of a given point in your network, use the Trace Downstream task.

To find all network elements that lie upstream of a given point in your network, use the Trace Upstream task. ▸

Adding flags and barriers

1. On the Utility Network Analyst toolbar, click the tool palette dropdown arrow.
2. Click the button representing the flag or barrier element that you want to add to the network.
3. Point to the edge or junction feature to which you want to add the flag or barrier.
4. Click to add the flag or barrier.

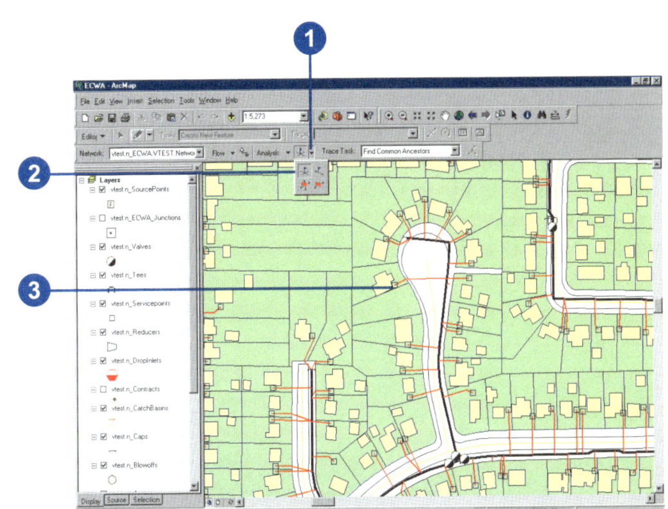

Tracing downstream

1. On the Utility Network Analyst toolbar, click the tool palette dropdown arrow and choose a flag tool.
2. Click to place flags at each point from which you want to trace downstream.
3. Click the Trace Task dropdown arrow and click Trace Downstream.
4. Click the Solve button.

   All of thc features downstream of your flags are displayed.

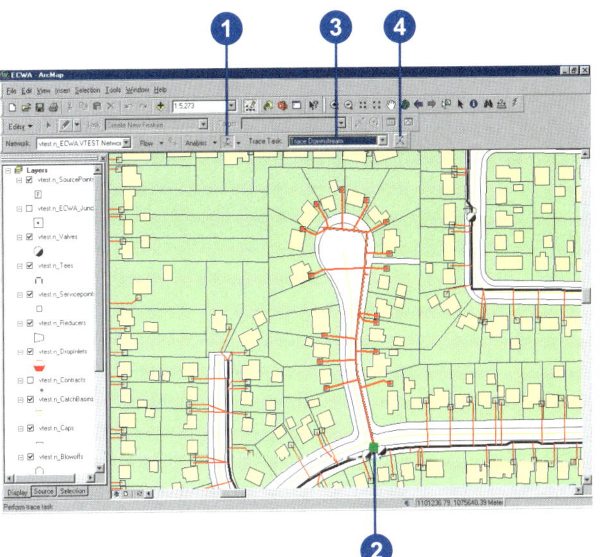

To find the total cost of all network elements that lie upstream of a given point in your network, use the Find Upstream Accumulation task.

To find an upstream path from a point in the network to the source, use the Find Path Upstream task.

To find the common features that are upstream of a set of points in your network, use the Find Common Ancestors task.

To find all of the features that are connected to a given point through your network, use the Find Connected task.

To find all of the features that are not connected to a given point through your network, use the Find Disconnected task.

To find a path between two points in the network, use the Find Path task. The path found can be just one of a number of paths between these two points—depending on whether or not your network contains loops.

To find loops in your network, use the Find Loops task. Loops can result in multiple paths between points in a network.

Tracing upstream

1. On the Utility Network Analyst toolbar, click the tool palette dropdown arrow and choose a flag tool.
2. Click to place flags at each point from which you want to trace upstream.
3. Click the Trace Task dropdown arrow and click Trace Upstream.
4. Click the Solve button.

   All of the features upstream of your flags are displayed.

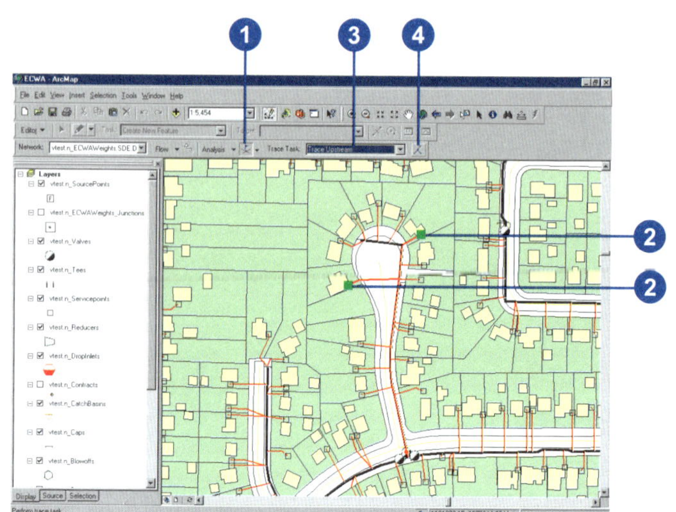

Finding the upstream accumulation

1. On the Utility Network Analyst toolbar, click the tool palette dropdown arrow and choose a flag tool.
2. Click to place flags at each point from which you want to find the upstream accumulation.
3. Click Analysis and click Options. ►

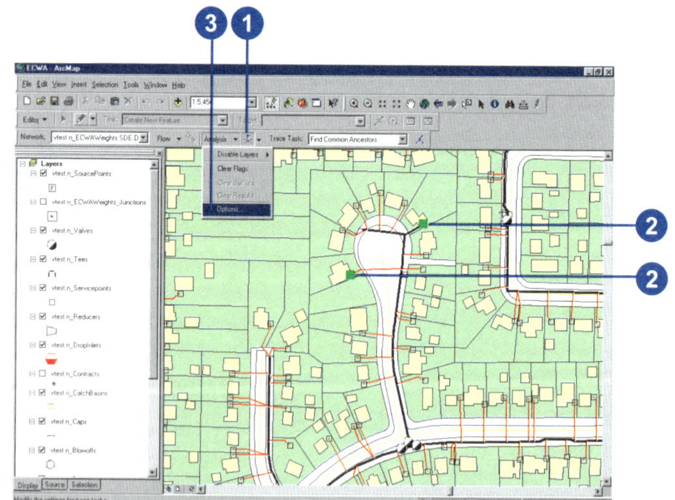

Tip

Finding the upstream accumulation without weights

By default, the Find Upstream Accumulation trace task does not use weights. If you do not use weights, the cost reported is the number of edge elements in the result.

4. Click the Weights tab.
5. Click the Junction weights dropdown arrow and click the name of the weight you want to use for junctions.
6. Click the from–to edge weight dropdown arrow and click the name of the weight you want to use for tracing edges along the digitized direction.
7. Click the to–from edge weight dropdown arrow and click the name of the weight you want to use for tracing edges against the digitized direction.
8. Click OK.
9. Click the Trace Task drop-down arrow and click Find Upstream Accumulation.
10. Click the Solve button.

    All of the features upstream of your flags are displayed, and the total cost of these features is reported in the status bar.

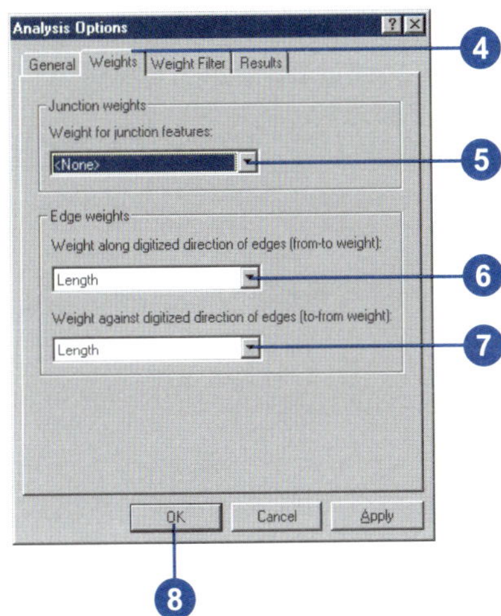

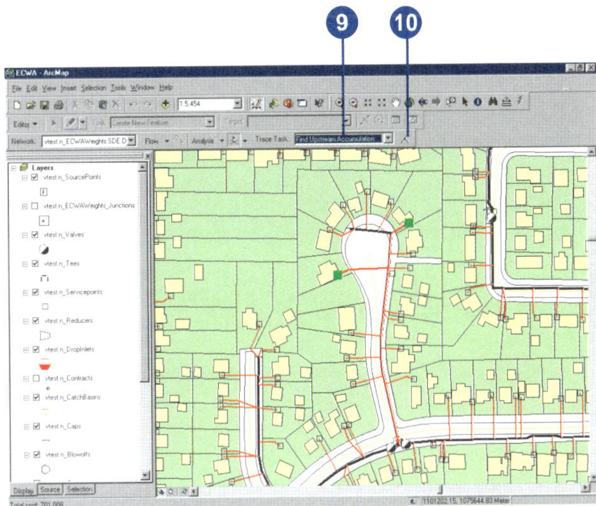

Tip

Finding an upstream path with weights

By default, the Find Path Upstream trace task does not use weights. If you use weights, the upstream path found to the source is the shortest path based on the weights you specify. To specify weights, follow steps 3 through 7 for 'Finding the upstream accumulation'.

Finding an upstream path to the source

1. On the Utility Network Analyst toolbar, click the tool palette dropdown arrow and choose a flag tool.
2. Click to place flags at each point for which you want to find an upstream path to the source.
3. Click the Trace Task drop-down arrow and click Find Path Upstream.
4. Click the Solve button.

   For each of your flags, an upstream path from the flag to the source is displayed.

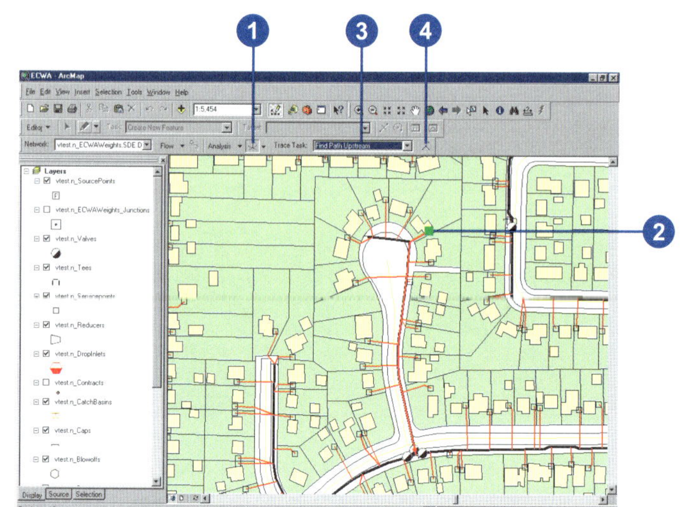

Finding common ancestors

1. On the Utility Network Analyst toolbar, click the tool palette dropdown arrow and choose a flag tool.
2. Click to place flags at each point for which you want to find the common ancestors.
3. Click the Trace Task drop-down arrow and click Find Common Ancestors.
4. Click the Solve button.

   The features that are upstream of all of your flags are displayed.

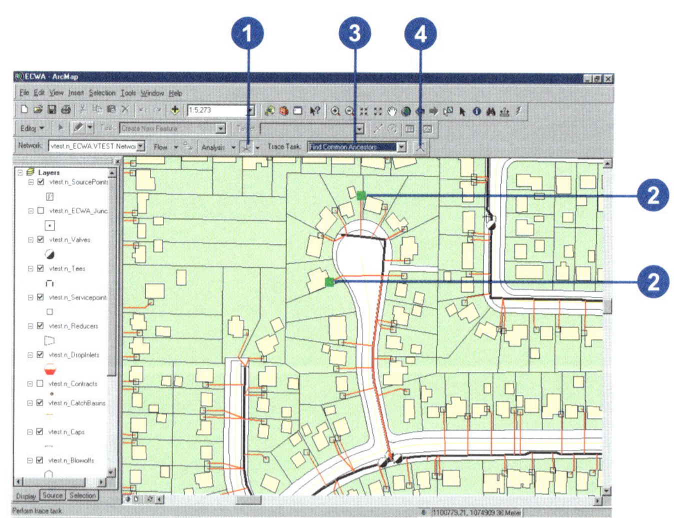

Tip

Using Find Connected or Find Disconnected

The Find Disconnected trace task always returns the features that the Find Connected trace task does not. The results of one of these trace tasks are often easier to view and analyze than the results of the other. For example, suppose you have a mostly connected network and you would like to check to make sure all of your network features are connected to each other. Performing a Find Disconnected trace task and checking to see if no features are returned is easier than performing a Find Connected trace task and making sure all of your features are returned.

Finding connected features

1. On the Utility Network Analyst toolbar, click the tool palette dropdown arrow and choose a flag tool.
2. Click to place flags at each point for which you want to find the connected features.
3. Click the Trace Task dropdown arrow and click Find Connected.
4. Click the Solve button.

   The features that are connected to the features on which you placed your flags are displayed.

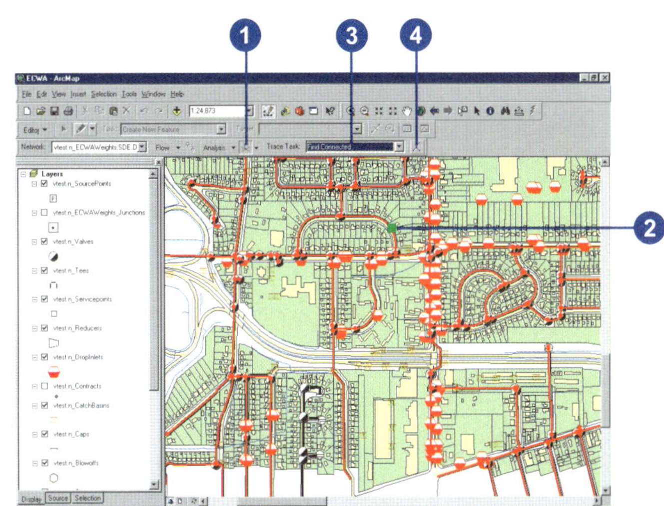

Finding disconnected features

1. On the Utility Network Analyst toolbar, click the tool palette dropdown arrow and choose a flag tool.
2. Click to place flags at each point for which you want to find the disconnected features.
3. Click the Trace Task dropdown arrow and click Find Disconnected.
4. Click the Solve button.

   The features that are not connected to the features on which you placed your flags are displayed.

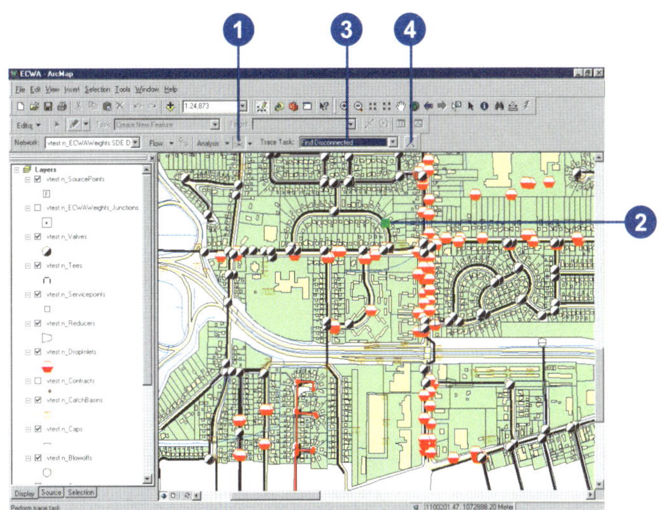

Isolating a point on the network

1. On the Utility Network Analyst toolbar, click the tool palette dropdown arrow and choose a flag tool.
2. Click on the map to place a flag at the point you want to isolate.
3. Click Analysis and click Disable Layers. Check the layer or layers containing the features that will be used to isolate this point.
4. Click Analysis and click Options.
5. Click the Results tab.
6. Click Selection in the Results format section.
7. Click Features stopping the trace in the Results content section.
8. Click OK. ▸

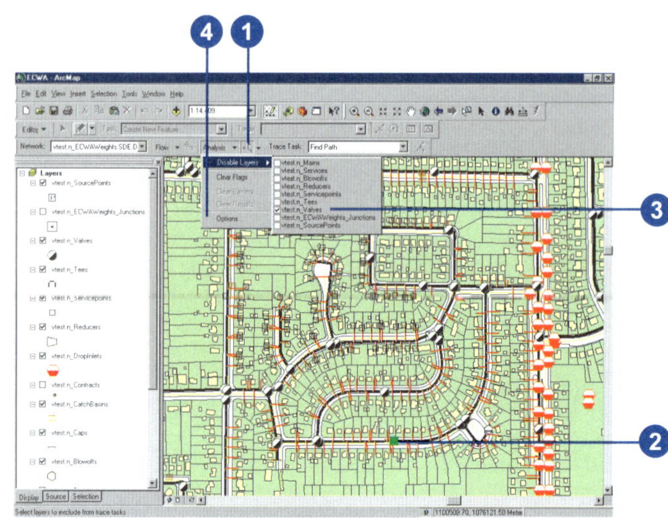

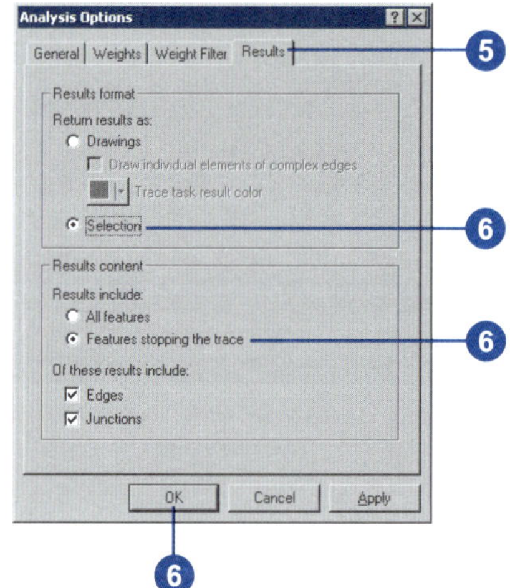

Tip

The Selection tab

You can choose the layers you want to select from using the Selection tab on the table of contents. If you don't see the Selection tab, click the Tools menu, click Options, and check Selection on the Table Of Contents tab.

9. Click Selection on the Main menu and click Set Selectable Layers.

   Or, click the Selection tab on the table of contents.

10. Click Clear All to uncheck all the layers.

11. Check the layers that contain the features that will be used to isolate your point in the network.

12. Click Close.

13. Click the Trace Task dropdown arrow and click Find Connected.

14. Click the Solve button.

    The features that are selected can be used to isolate your point in the network.

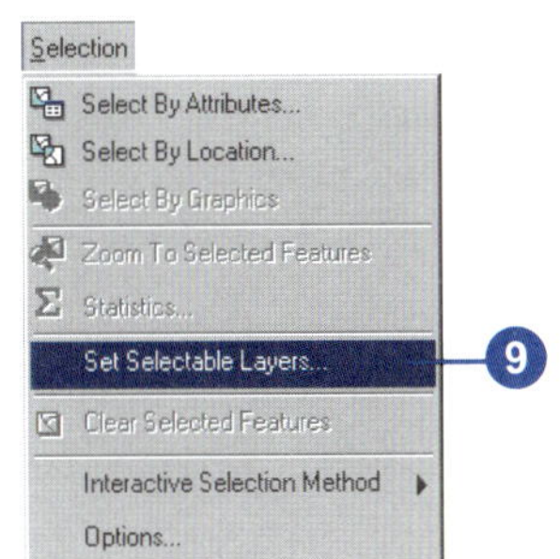

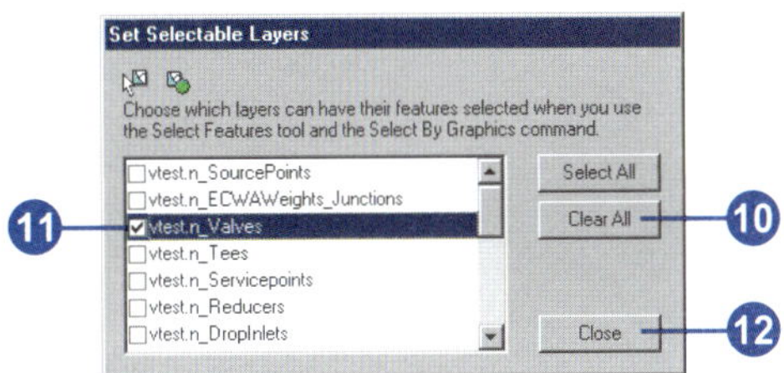

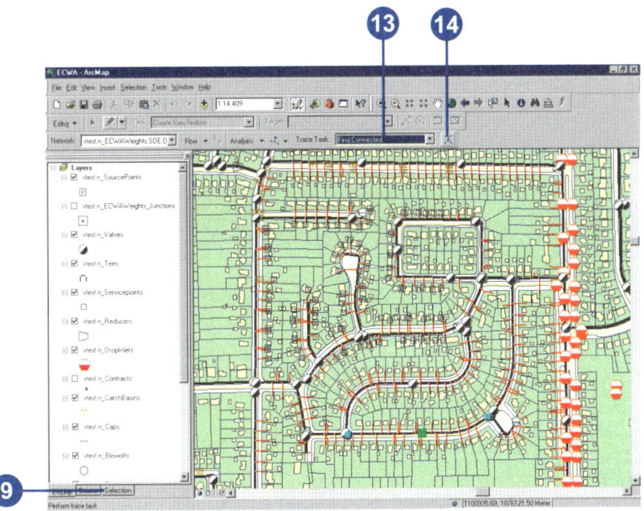

Tip

Weight filter syntax

When you create a range expression for a weight filter, you must use correct syntax. You can specify multiple valid or invalid ranges for each weight. You must delimit each range with commas. Each range can include a single value or a range of values. To specify a range of values, type a hyphen between the lower and upper bounds of the range—for example, "1-5, 10-22.2, 27".

Finding connected features using weight filters

1. On the Utility Network Analyst toolbar, click the tool palette dropdown arrow and choose a flag tool.
2. Click to place flags at each point for which you want to find the connected features.
3. Click Analysis and click Options.
4. Click the Weight Filter tab. Click the Junction weight dropdown arrow and click the name of the weight you want to use to filter junctions.
5. In the Weight range text box for junctions, type the expression you want to use to filter junctions.
6. Check the Not check box to exclude this range.
7. Click Verify to check the syntax of the junction weight filter. ►

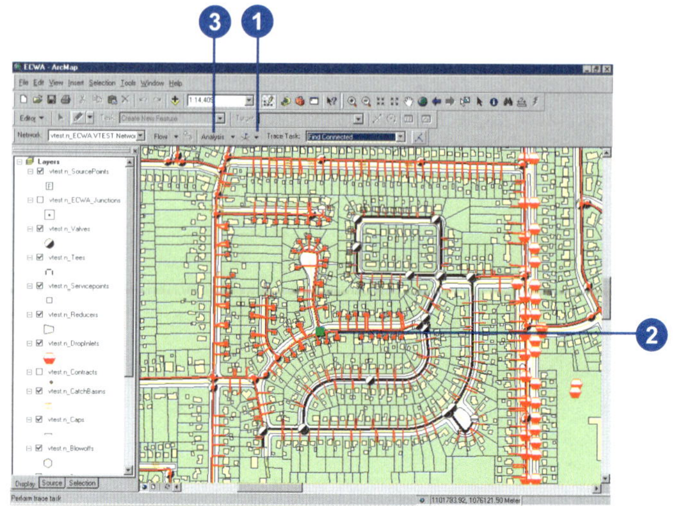

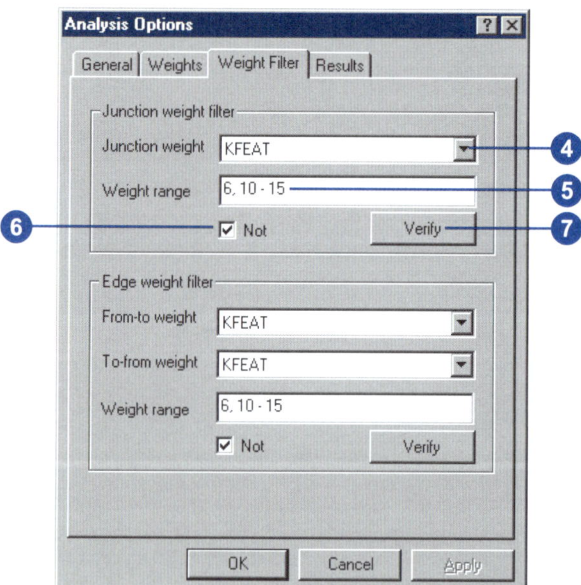

8. Click the From-to weight dropdown arrow and click the name of the weight you want to use to filter edges along their digitized direction.
9. Click the To-from weight dropdown arrow and click the name of the weight you want to use to filter edges against their digitized direction.
10. In the Weight range text box for edges, type the expression you want to use to filter edges.
11. Check the Not check box to exclude this range.
12. Click Verify to check the syntax of the edge weight filter.
13. Click OK.
14. Click the Trace Task dropdown arrow and click Find Connected.
15. Click the Solve button.

    The features that are connected to the features on which you placed your flags, using the weight filter you specified, are displayed.

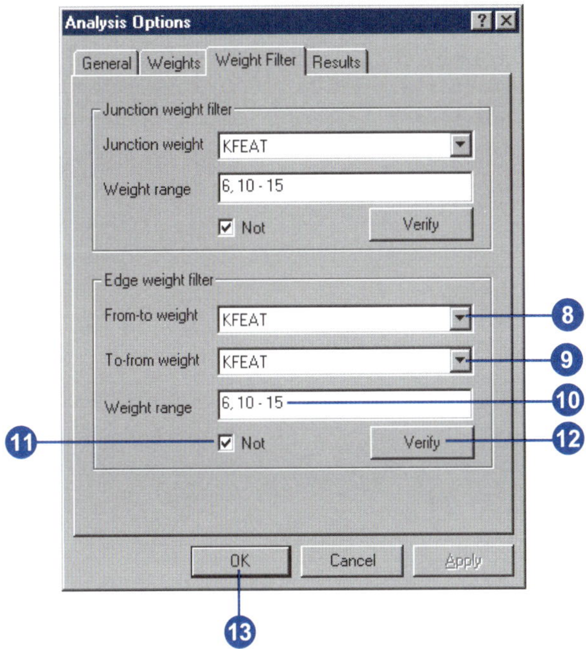

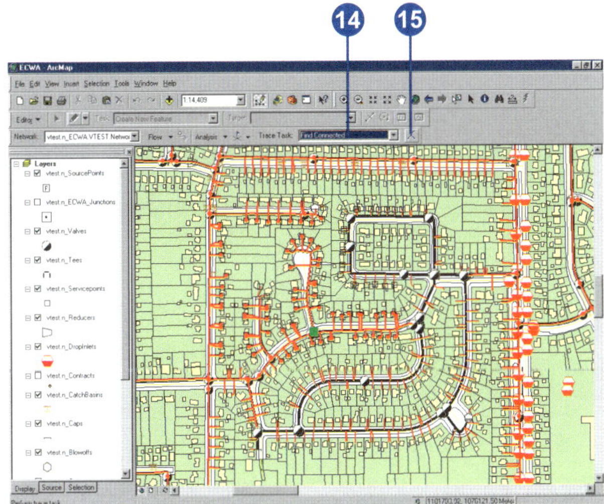

Tip

The Find Path trace task

When you use the Find Path trace task, the flags you place on the network must be either all edge flags or all junction flags. You cannot find a path among a mixture of edge and junction flags.

Tip

Finding paths without weights

By default, the Find Path trace task does not use weights. If you do not use weights, the shortest path based on the number of edge elements in the path is found.

Finding a path

1. On the Utility Network Analyst toolbar, click the tool palette dropdown arrow and choose a flag tool.
2. Click to place flags on the features among which you want to find a path.
3. Click the Trace Task dropdown arrow and click Find Path.
4. Click the Solve button.

   A path between the features on which you placed flags is displayed.

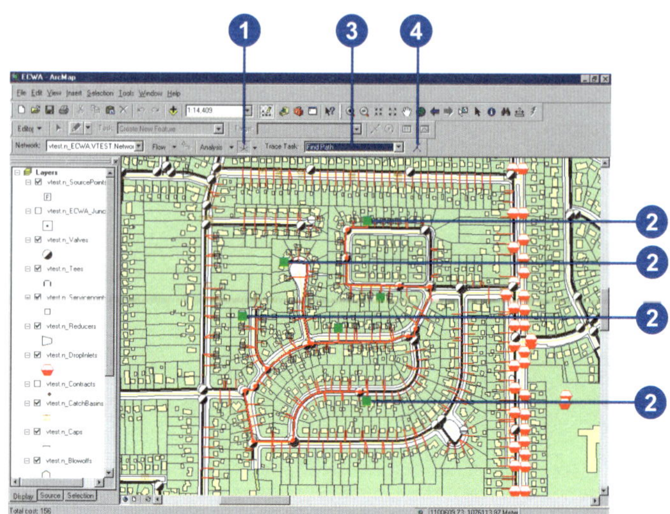

Finding the shortest path

1. On the Utility Network Analyst toolbar, click the tool palette dropdown arrow and choose a flag tool.
2. Click to place flags on the features among which you want to find a path.
3. Click Analysis and click Options. ►

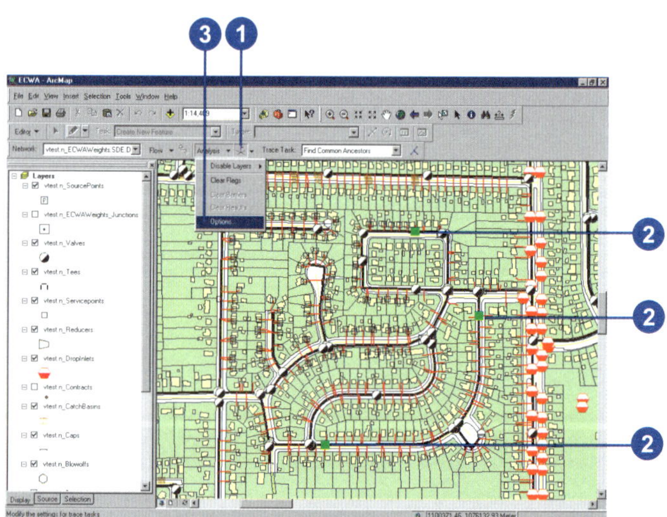

4. Click the Weights tab. Click the Junction weights dropdown arrow and click the name of the weight you want to use for junctions.
5. Click the from–to edge weight dropdown arrow and click the name of the weight you want to use for tracing edges along the digitized direction.
6. Click the to–from edge weight dropdown arrow and click the name of the weight you want to use for tracing edges against the digitized direction.
7. Click OK.
8. Click the Trace Task dropdown arrow and click Find Path.
9. Click the Solve button.

   The shortest path based on the weights you chose is displayed. The total cost of this path is reported in the status bar.

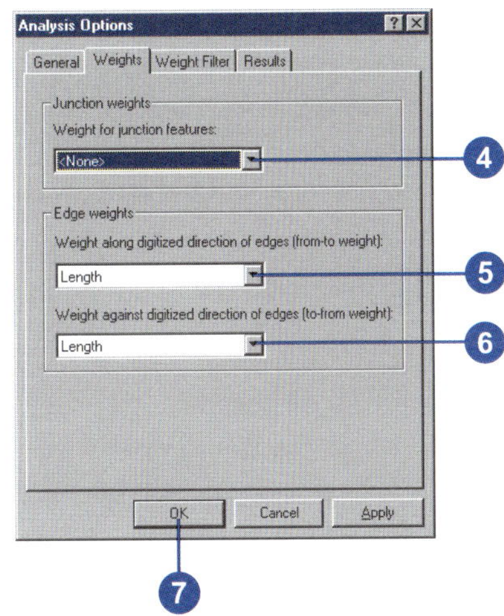

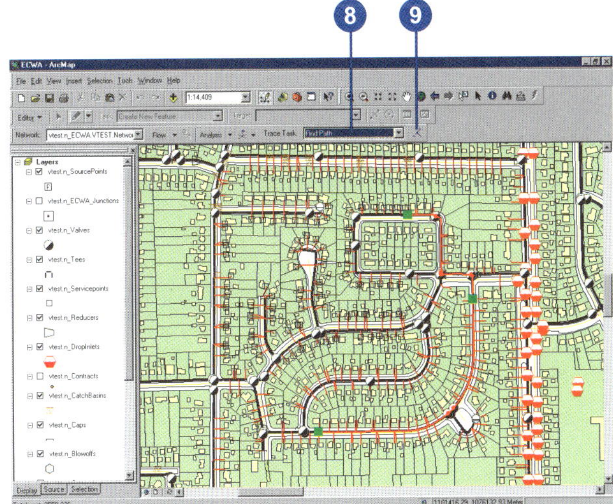

Tip

Finding a downstream path

Find a downstream path using a process similar to finding an upstream path.

Finding an upstream path

1. On the Utility Network Analyst toolbar, click the tool palette dropdown arrow and click the Add Junction Flag tool.
2. Click Analysis and click Options. Click the Results tab. Click Selection to return the results of trace tasks as selections.
3. Uncheck Junctions. This returns only edges in the results.
4. Click OK.
5. Click on the map to place a flag at the destination point.
6. Click the Trace Task dropdown arrow and click Trace Upstream.
7. Click the Solve button.
8. Click Analysis and click Clear Flags. ▸

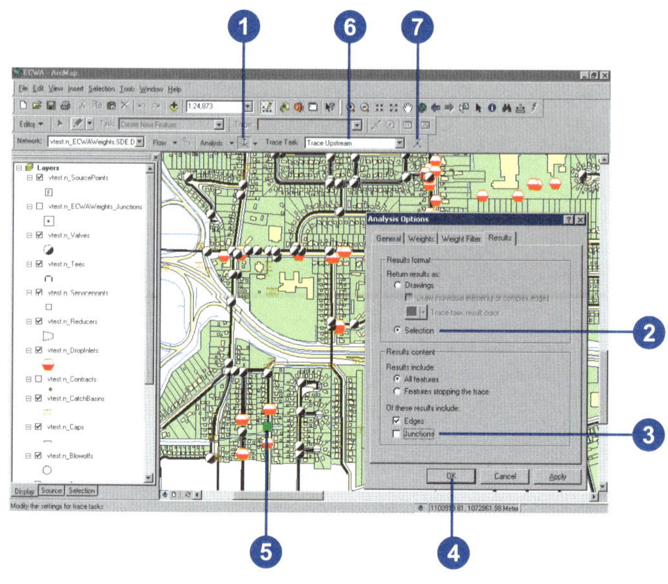

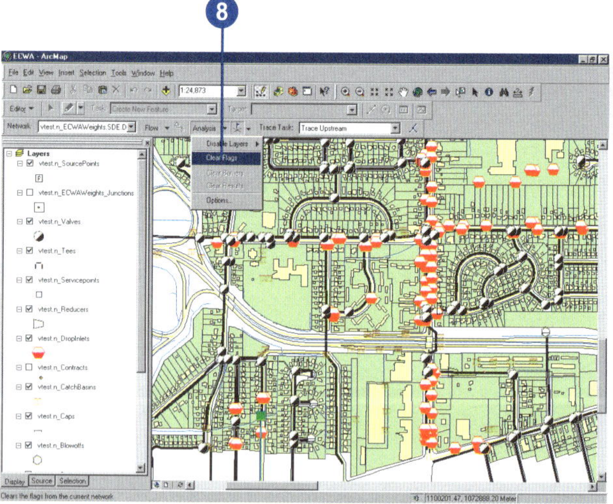

9. From the Main menu, click Selection. Point to Interactive Selection Method and click Add to Current Selection.
10. Click on the map to place a flag at the origin point.
11. Click the Trace Task dropdown arrow and click Trace Downstream.
12. Click the Solve button.
13. Click Analysis and click Options. Click the General tab and click Unselected features to treat the current selection as barriers.
14. Click OK. ▸

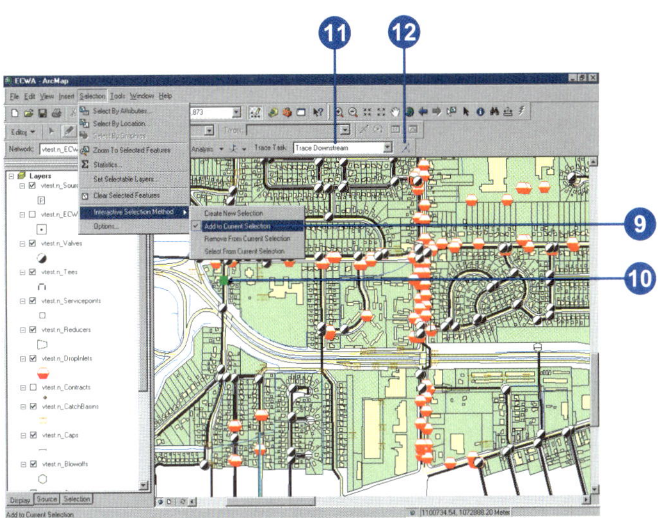

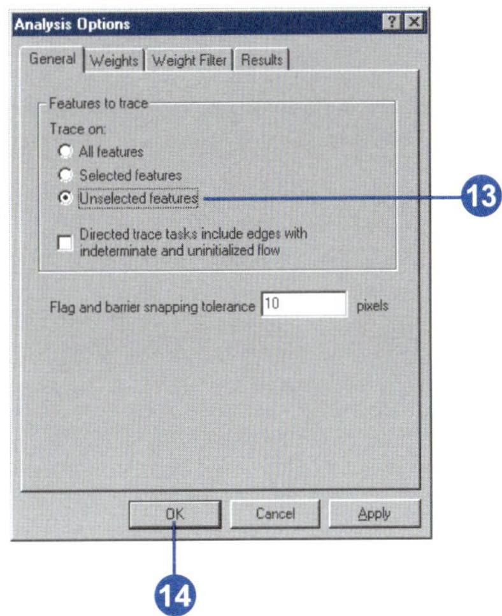

15. From the Main menu, click Selection. Point to Interactive Selection Method and click Create New Selection.
16. Click on the map to place a flag at the destination point.
17. Click the Trace Task dropdown arrow and click Find Path.
18. Click the Solve button.

    If it exists, the result will be an upstream path from the origin point to the destination point.

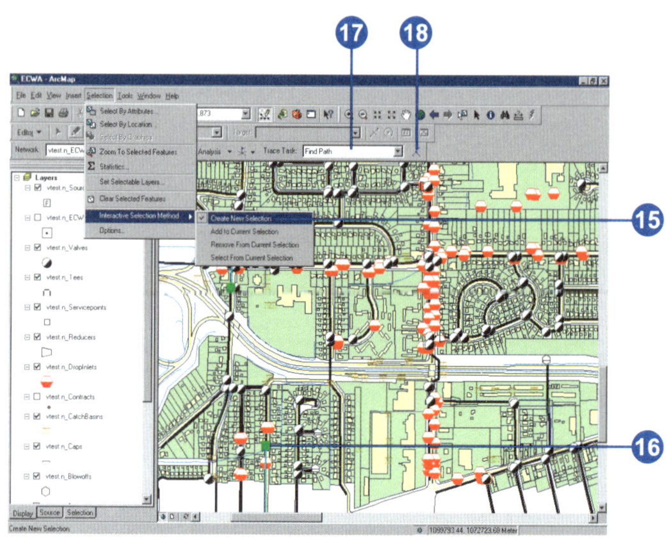

Finding loops

1. On the Utility Network Analyst toolbar, click the tool palette dropdown arrow and choose a flag tool.
2. Click to place at least one flag on each connected component in which you want to find the loops.
3. Click the Trace Task dropdown arrow and click Find Loops.
4. Click the Solve button.

    For each connected component on which you placed a flag, the features that loop back on themselves (i.e., can be reached from more than one direction) are displayed.

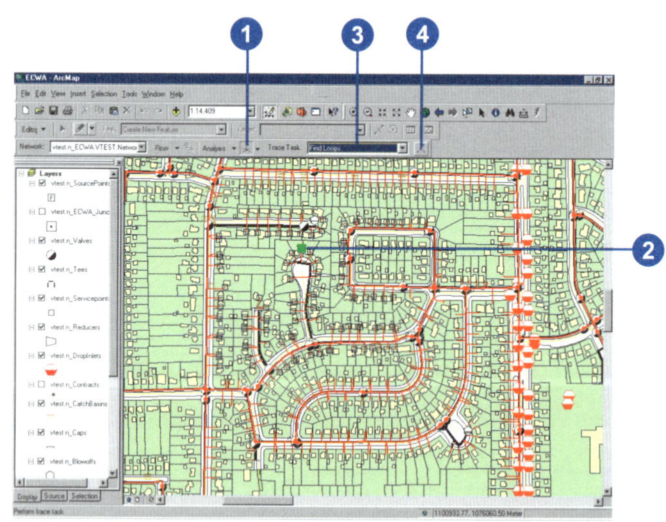

Map output

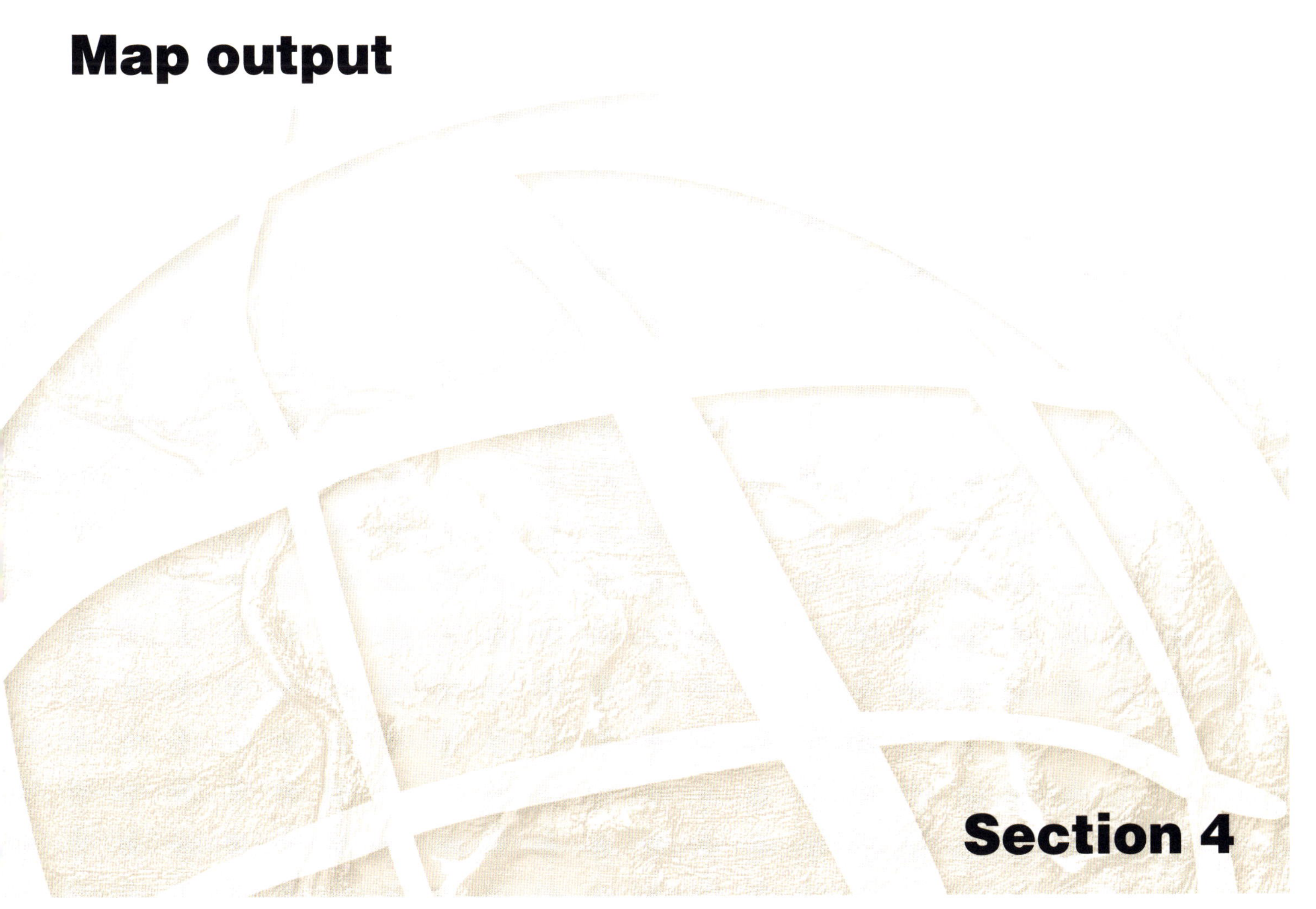

Section 4

Laying out and printing maps

15

IN THIS CHAPTER

Before you begin to symbolize data for a map, you'll need to think about how you want the map to look when it's printed or published.

You should consider questions such as:

- Will the map stand alone, or will it be part of a series of maps that share a similar design?
- What size will the printed version of the map be?
- How will the page be oriented?
- How many data frames will the map have?
- Will the map have other map elements such as a title, a North arrow, and a legend?
- Will the map contain graphs or reports to complement the geographic view of the data?
- How will scale be indicated on the map?
- How will the map elements be organized on the page?

If the map is part of a series, you may have a template to work from, or you may create a new template for the series. Map templates make it easy to produce maps that conform to a standard, and they save time by letting you do the layout work for all the maps in the series at once.

You can also use the map templates that come with ArcMap to quickly make a variety of styles of maps. You can use these templates to get ideas for your own maps, and you can modify them to suit your needs.

The Mexico and New Hampshire maps on this page and the next illustrate two different map *layouts* and show some of the ways you can use map elements to create a map.

Perhaps the most important part of a map is the geographic data. Geographic data is presented in the layout in a data frame. Simple maps usually have a single data frame, but maps can have multiple data frames.

The shape and orientation of the geographic features you're depicting may influence the size and shape of the data frame on the map as well as the orientation of the map on the page.

Aesthetic criteria, limitations of the media that you use to reproduce the map, and the number and size of other elements that you add to the map will bear upon your choice of page size and orientation.

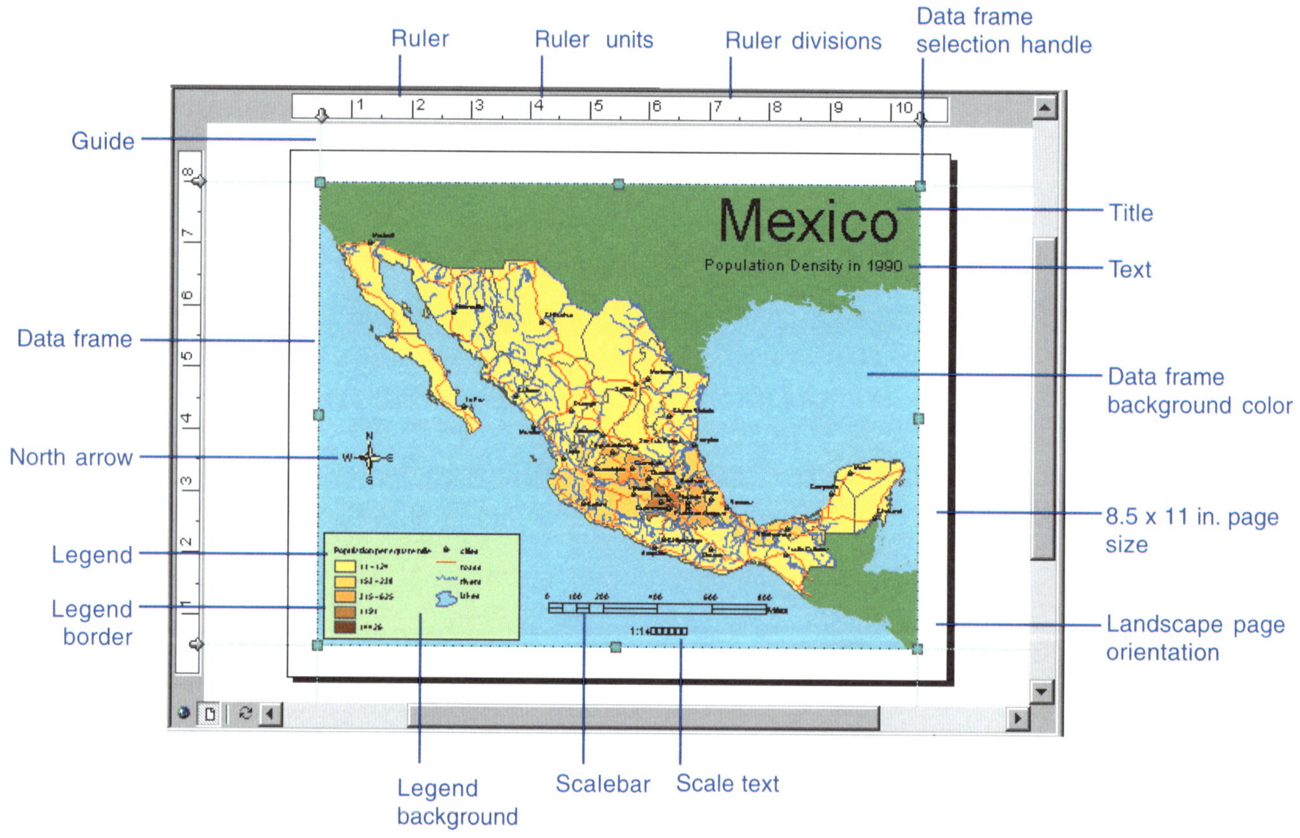

In addition to a data frame, most maps contain one or more other map elements. These include titles, North arrows, legends, scalebars, scale text, graphs, reports, text labels, and graphics.

One challenge of cartography is to arrange the elements of the map on the page to create a useful, visually pleasing map. ArcMap includes adjustable rulers, guides, and grids that can help you position elements precisely where you want them.

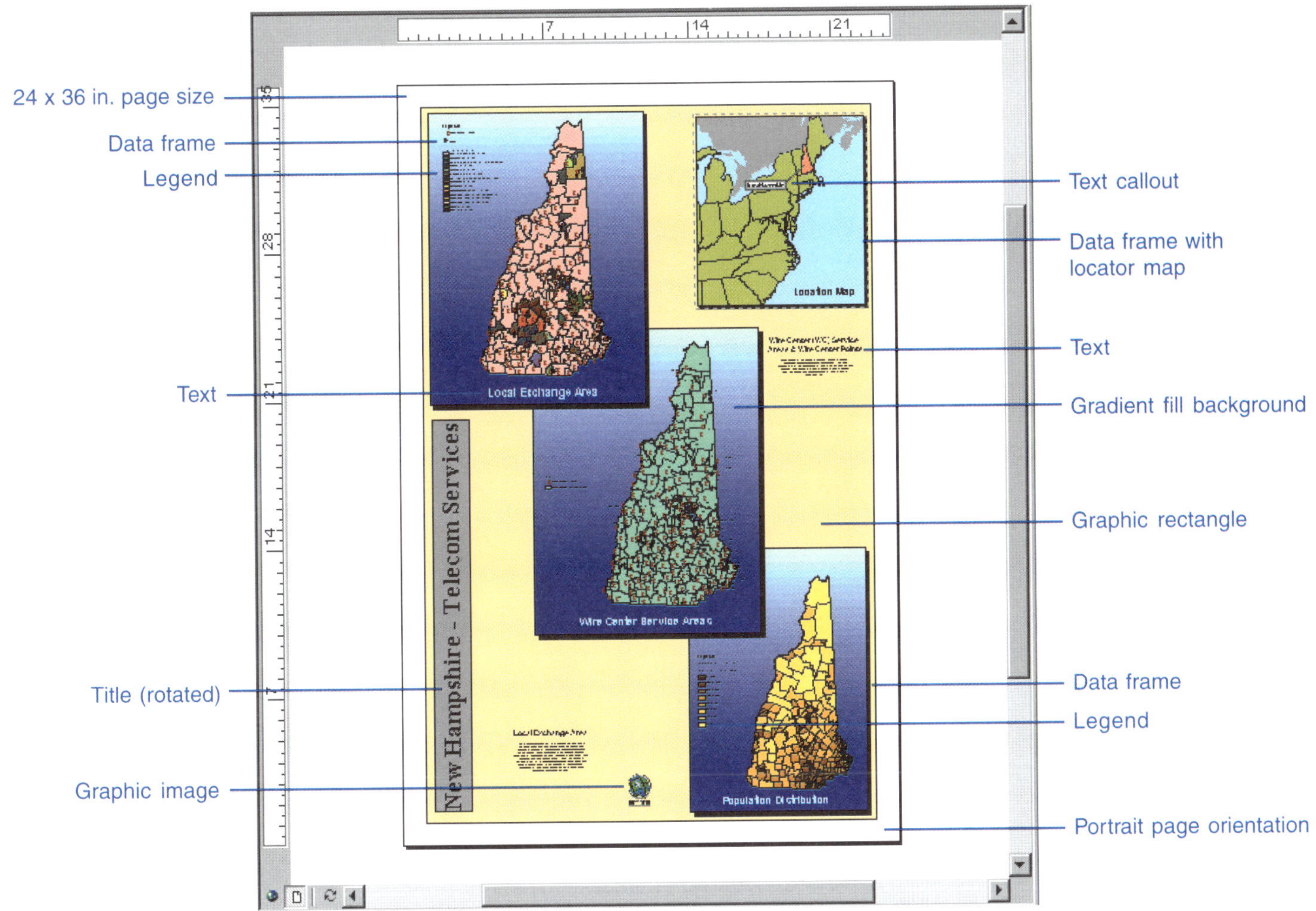

About map templates

If you are creating a series of maps and it is important that they have the same look and feel, you can use a map template to standardize the layout. If the series contains the same background data, you can include that data in the template. Using a template can save you time since you don't have to manually reproduce the common parts of the maps.

You can also use the map templates that come with ArcMap to quickly make a decent looking map, with a minimal amount of layout work on your part. Just choose a template that has the look you want, add your data, and make whatever changes you want, and your map is done.

Like maps and layers, templates can be shared within an organization to increase productivity and standardize the maps that the organization produces. You can use a template to store layout, data, and customization of the ArcMap interface that you want to be able to use over and over again.

You can modify existing maps or templates and save them as new templates, or you can create new maps of your own and save them as templates.

Map templates are ArcMap documents that ArcMap recognizes as templates. When you start a new map using a template, ArcMap reproduces the template on a new map document and keeps the original template document intact. Map templates have the file extension .mxt to differentiate them from map documents (.mxd).

Using map templates

If you want to make a map using a template, start a new map, choose the template that you want to use, then start adding layers to the map. Later, if you want to change the layout, you can apply a new template.

The Normal template

ArcMap uses a special template called the *Normal template* (Normal.mxt) to store information about the default user interface, for example, the state—visible or hidden, docked or free-floating—of each of the ArcMap toolbars. This information is recorded automatically in the Normal template when you change it, so when you start ArcMap (whether you saved the map you were working on or not), the toolbars look the same as they did when you quit.

When you add custom toolbars or tools to ArcMap, you can save the changes to the Normal template or to the current map. If you save changes to the interface in the Normal template, they will be reflected in all the maps that you open. If you save changes to another map or template, they will only appear when you open that map or template.

Starting a map from a template

Map templates make it easy to reuse the same layout or even the same data on a series of maps. You can use the templates that come with ArcMap to make maps quickly—all you need to do is add data, a title, and any other supporting information that you choose.

When you open a template, you get a new untitled ArcMap document plus any layout or data that's saved in the template.

Tip

Starting a new map using a shortcut

If you click the New Map File button on the Standard toolbar, you can start a new, empty map.

New Map File button

Tip

Storing customization in templates

In addition to layout and data, map templates (like maps) can store customization of the ArcMap user interface such as custom toolbars and tools.

Opening a template when you first start ArcMap

1. Start ArcMap.
2. Click Start using ArcMap with A template.
3. Click OK.
4. Click a template.
5. Click OK.

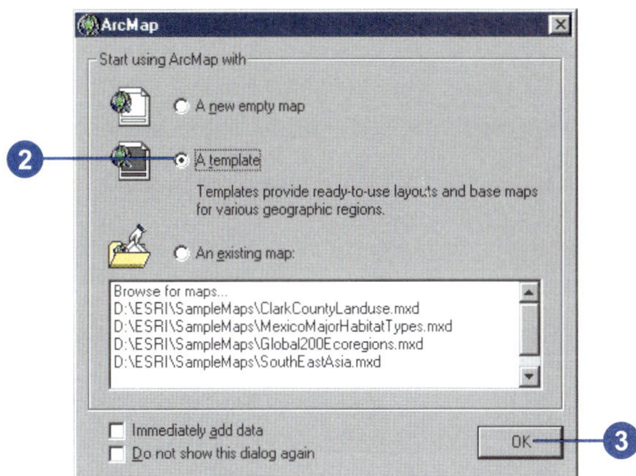

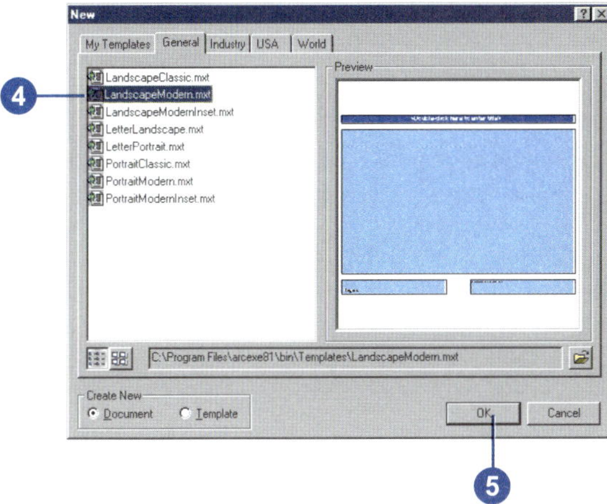

Opening a new template while in ArcMap

1. Click File and click New.
2. Click a template and click OK.

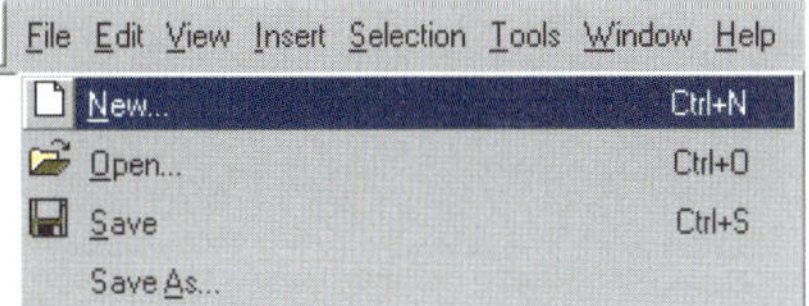

Saving a map as a template

If you create a map that you'd like to use as a template, or if you modify an existing template and want to use it again, you can save it as a template.

You can save a map template anywhere on your network. When you want to use the template, you can open it from ArcCatalog or ArcMap.

If you save a template in the ArcMap Templates folder (by default in the folder \Bin\Templates where you've installed ArcGIS), it will show up in the list of templates on the New map document dialog box. You can also create subfolders in this folder, and they'll show up as separate tabs on this dialog box—when you click each tab you'll see the templates in that folder. If you work with many different templates, this is a great way to organize them.

Tip

Changing a map template

To change an existing template, open the template (.mxt) file and make the necessary edits directly on it.

Saving a template

1. Click File and click Save As.
2. Click the Save as type dropdown arrow and click ArcMap Templates.
3. Navigate to the folder where you want the template saved.
4. Type a name for the new template.
5. Click Save.

   NOTE: You can only save a map as a template if your map was built using Normal.mxt as the base template. If it wasn't, click the Edit menu and click Select All Elements while in layout view. Then copy and paste the elements into a new empty map. Then you can save the new map as a template.

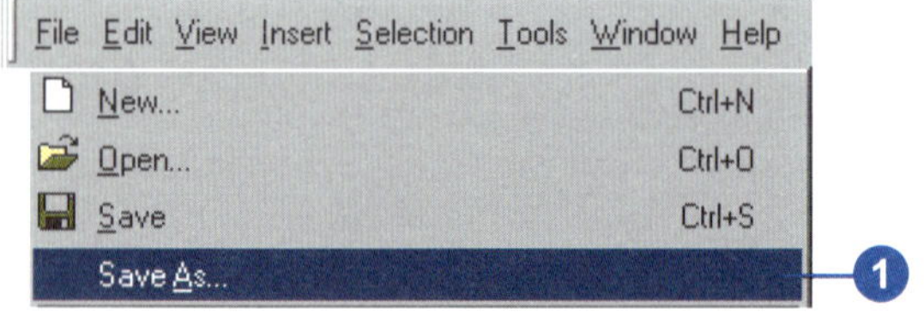

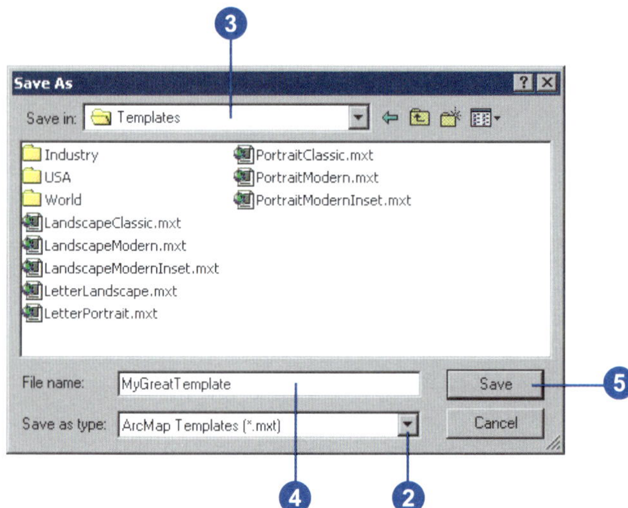

Saving a template so it will appear in a new tab

1. Click File and click Save As.
2. Click the dropdown arrow and click ArcMap Templates.
3. Navigate to the Templates folder.
4. Click the New Folder button.
5. Type the name of the new folder—this name will appear on the New map document dialog box as a tab.
6. Double-click the new folder.
7. Type the name of the new template.
8. Click Save.

   Next time you start a map from a template, you'll see a new tab with your template on the New map document dialog box.

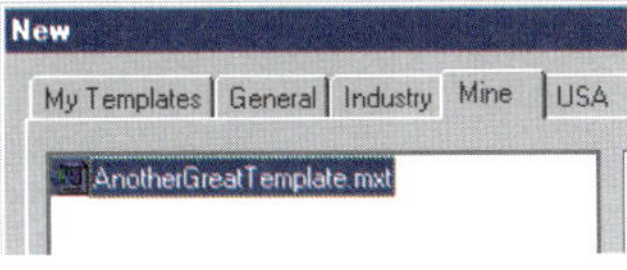

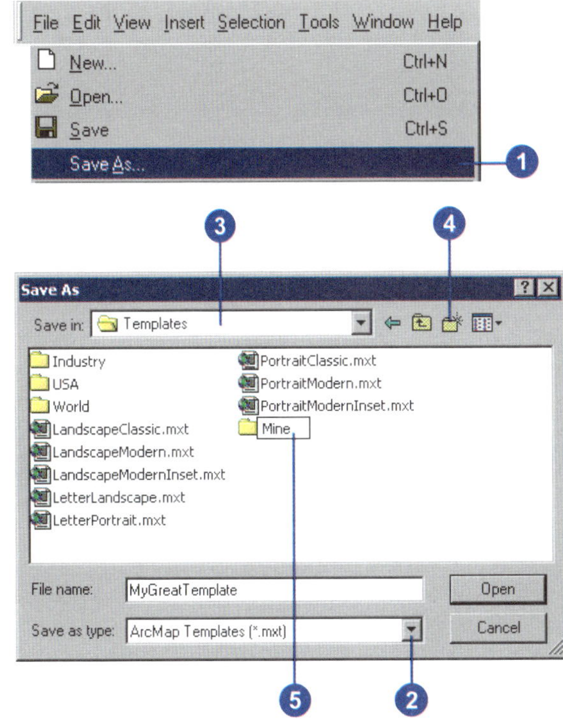

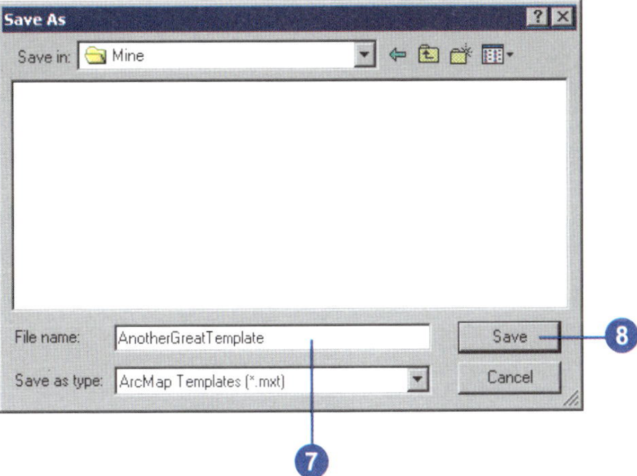

Setting up the page

When you create a map for printing or publication, you'll work on the virtual page in layout view.

If you intend to print or export a graphic of a map, you should plan the size of the map. Will the map be printed on a small page or a large one? What printer will you use and which print engine will be most efficient for your map content? Will it be viewed close up or at a distance?

ArcMap makes it easy for you to change the size of the page if needed, but it is wise to have the final product in mind when you begin designing the map.

If the virtual page doesn't match the page size and orientation you've planned for your map, you can change the page setup. By default, the virtual page size is the same as your system printer's default page size, but you can set the page to be one of many standard sizes, or you can define a custom page size for your map.

You can set the page size, page orientation, printer, printer engine, and the visibility of printer margins in the layout from the Page and Print Setup dialog box. ▸

Switching to layout view

1. Click View and click Layout View.

   OR

   Click the Layout View button on the bottom left corner of the map display area.

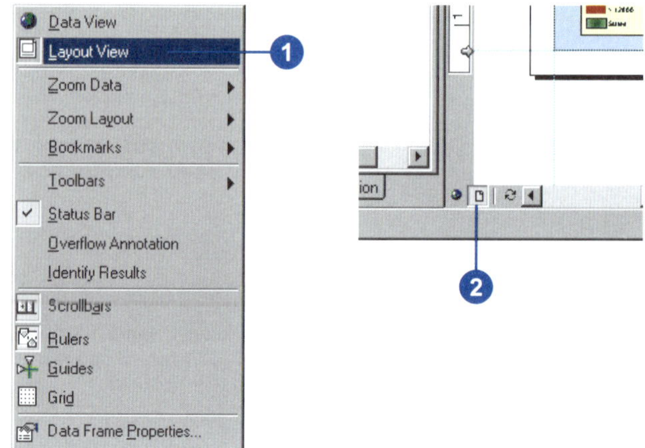

Setting up the page size and printer properties

1. Right-click the virtual page and click Page and Print Setup.

   You can also open the Page and Print Setup dialog box from the File menu or the Print dialog box. ▸

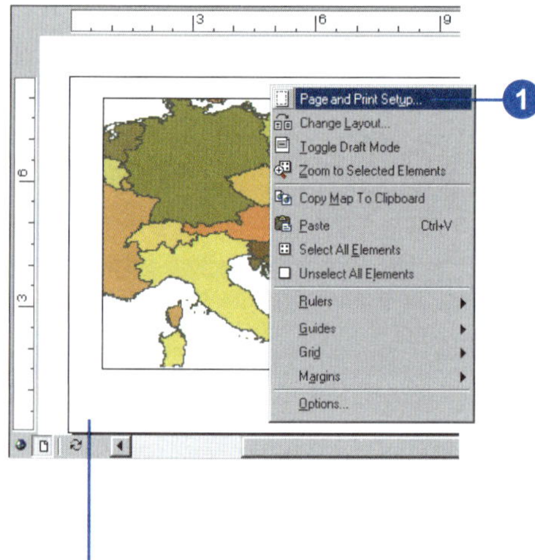

Right-click outside the selected data frames to get the Page context menu.

The page setup is important because it affects the size of the features, symbols, labels, and other text as well as other map elements.

Tip

Changing the page size later

While it's best to set the page size before you begin creating a map, you can make changes to the page size later if necessary. ArcMap will automatically rescale map elements to fit the new page size. You can turn off the rescaling function by unchecking the box on the Page and Print Setup dialog box if you'd prefer to adjust the size and shape of your map elements by hand.

2. Click the Name dropdown arrow and click the printer you want to use.
3. Click the Paper Size dropdown arrow and click the page size that's right for your map.
4. Click OK.

   Because the Use Printer Paper Settings box is checked, the map Width and Height text boxes are updated with the new page size, and the Page Orientation is set accordingly.

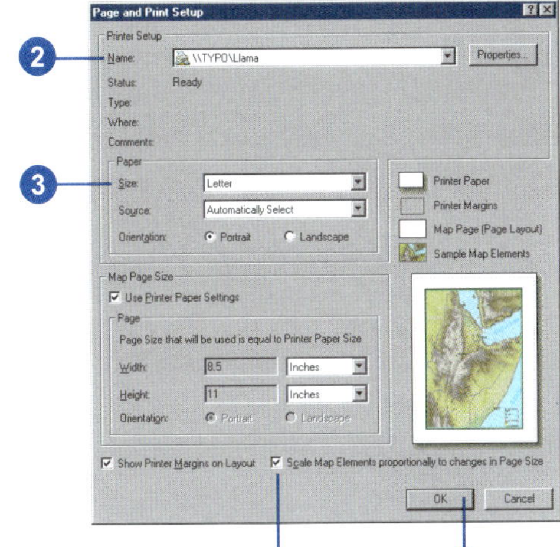

Check to automatically rescale map elements when you change the page size.

Making the map page size independent of the system printer

1. Uncheck Use Printer Paper Settings on the Page and Print Setup dialog box.
2. Click the Standard Sizes dropdown arrow and click the page size that's right for your map.
3. Optionally, to define a custom page size, type the page size for your map in the Width and Height text boxes.
4. Click OK.

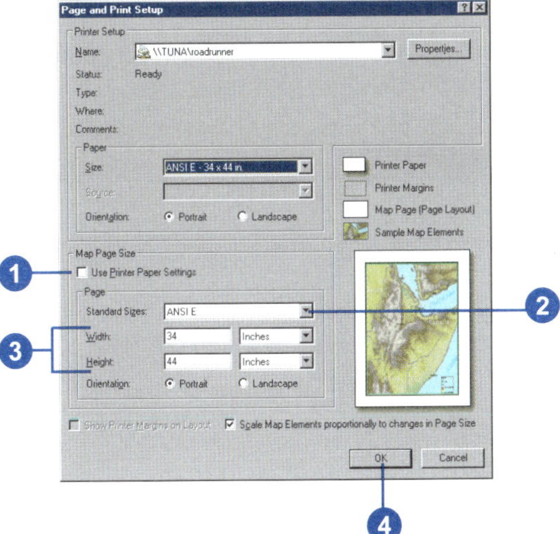

Tip

Changing the page orientation later

While it's a good idea to set up the page orientation before you begin laying out your map, you can change the page orientation at any time. If Scale Map Elements is turned on, your map elements will be adjusted to fit the new orientation.

Tip

Why show printer margins?

It's useful to show the printer's built-in margins on the layout so you won't place map elements on a nonprinting part of the page.

If you're working on a map where the virtual page size is not set to be the same as your printer's page size, you'll be unable to use the printer's margins. However, you can use guides to define the map's margins.

See Also

For more information on using guides, see 'Using rulers, guides, and grids' in this chapter.

Setting the page orientation

1. On the Page and Print Setup dialog box, click a Paper or Page Orientation (Landscape or Portrait) to set the page orientation.
2. Click OK.

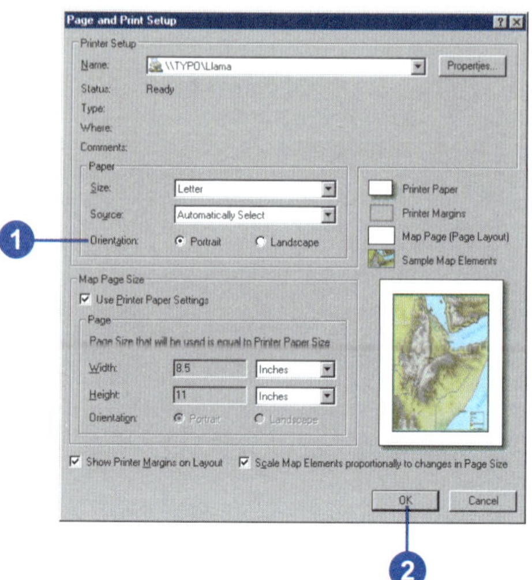

Showing or hiding printer margins

1. Check Show Printer Margins on Layout in the Page and Print Setup dialog box.
2. Click OK.

Printer margins shown by light gray dotted line

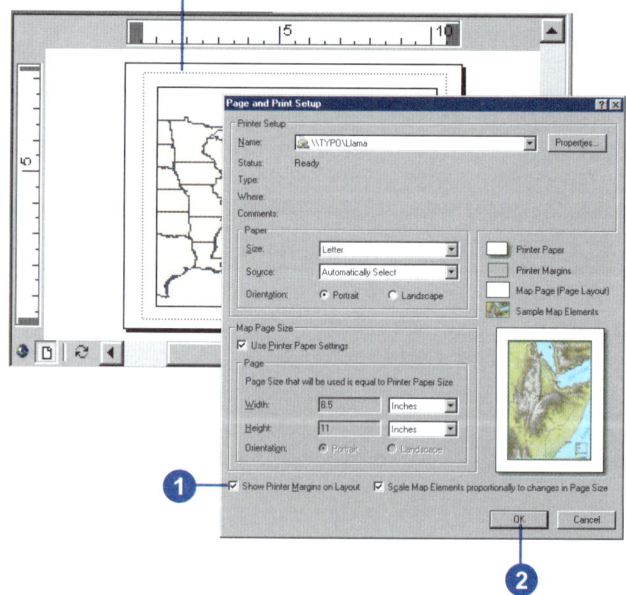

Customizing data frames

In layout view you see geographic data in a data frame on the virtual page. You can use the data frame to emphasize the geographic data on the map, for example, by adding a border, a background, or a drop shadow.

To help locate geographic features, you can add *grids* to your data frame. Grids subdivide the data frame by latitude and longitude, projected linear units, or a specified number of rows and columns.

Tip

Why rename a data frame?

When you just have one data frame, its name isn't that important. However, if you make a map with multiple data frames, it can be convenient to name them.

Renaming a data frame

1. Click the data frame in the table of contents.
2. Wait a moment, then click the data frame a second time.
3. Type a new name for the data frame.

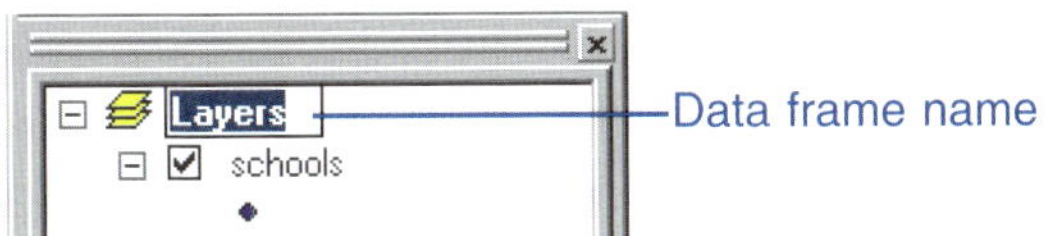

Adding a border to a data frame

1. In the table of contents, right-click the data frame and click Properties.
2. Click the Frame tab.
3. Click the Border dropdown arrow and click a symbol.
4. Click the Color dropdown arrow and click a color.
5. Type an X and Y gap to offset the border from the edge of the data frame.
6. Type a Rounding percentage to round the corners of the border.
7. Click OK.

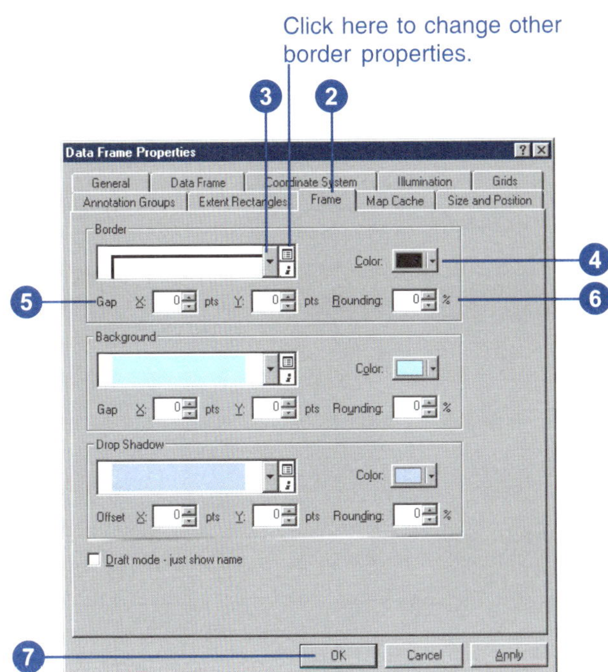

Adding a background to a data frame

1. In the table of contents, right-click the data frame and click Properties.
2. Click the Frame tab.
3. Click the Background dropdown arrow and click a background.
4. Click the Color dropdown arrow and click a color.
5. Type an X and Y gap to offset the background from the edge of the data frame.
6. Type a Rounding percentage to round the corners of the background.
7. Click OK.

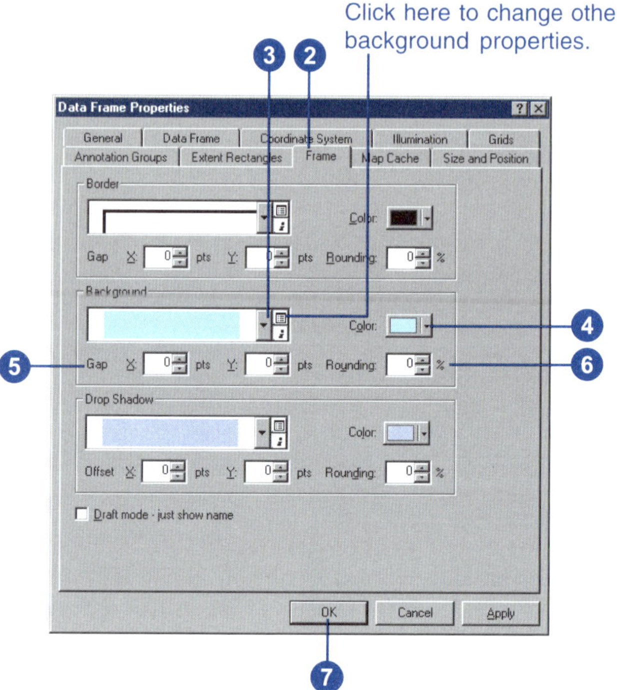

Adding a drop shadow to a data frame

1. In the table of contents, right-click the data frame and click Properties.
2. Click the Frame tab.
3. Click the Drop Shadow dropdown arrow and click a drop shadow.
4. Click the Color dropdown arrow and click a color.
5. Type an X and Y offset to shift the drop shadow away from the border of the data frame.
6. Type a Rounding percentage to round the corners of the drop shadow.
7. Click OK.

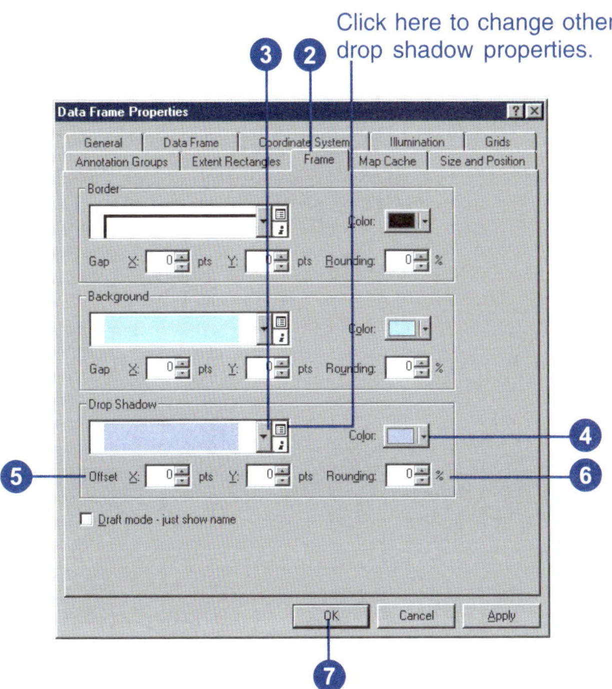

Using rulers, guides, and grids

You can use rulers, guides, and grids in layout view to align map elements on the page.

Rulers show the size of the page and map elements on the final printed map. Guides are straight lines that you can use to align map elements on the page. A grid is a grid of reference points on the layout that you can use to position map elements.

You can use each of these layout aids as visual indicators of element size and position. You can also turn on snapping to force map elements to snap to the rulers, guides, or grid. Snapping makes it easy to align map elements precisely.

You can use guides in layout view to align map elements on the page. ▸

Tip

How can I tell exactly where the pointer is?

You can use the readouts in the lower-right corner of the ArcMap window to find the geographic position and page position of the pointer.

Turning rulers on and off

1. Right-click the page.
2. Point to Rulers and click Rulers.

   The rulers are on by default.

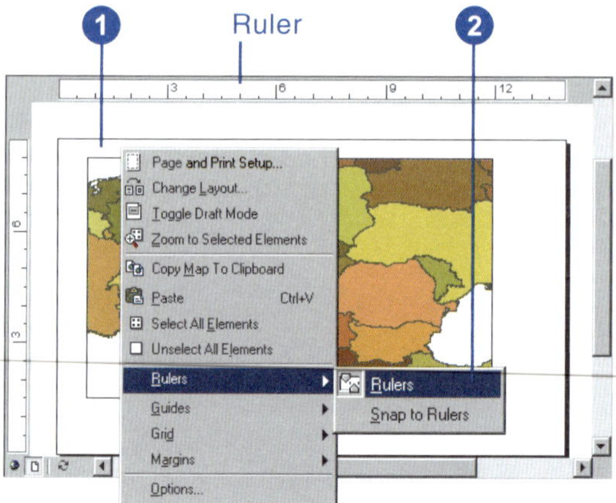

Snapping to rulers

1. Right-click the page.
2. Point to Rulers and click Snap to Rulers.

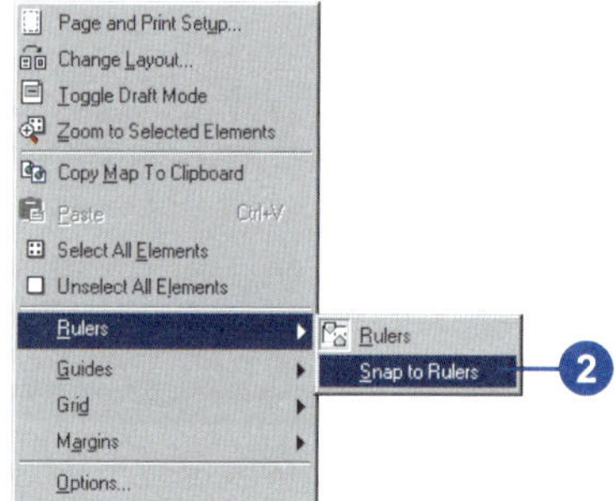

Although you can see guides on the virtual page in layout view, they will not show up when you print your map.

You can use a snapping grid in layout view to align map elements on the page.

Although you can see the grid on the virtual page in layout view, it will not show up when you print your map.

Tip

What units can I use to measure the page?

You can set the units of measure shown on the rulers to points, centimeters, or inches. You can also change the number of divisions per unit.

Setting the units and divisions on rulers

1. Right-click the ruler.
2. Click Options.

   If the ruler is not showing in your layout view, then click Tools on the Main menu, click Options, and click the Layout View tab.
3. Click the Units dropdown arrow and click a unit of measure.
4. Click the Smallest Division dropdown arrow and click the size of the smallest division.
5. Click OK.

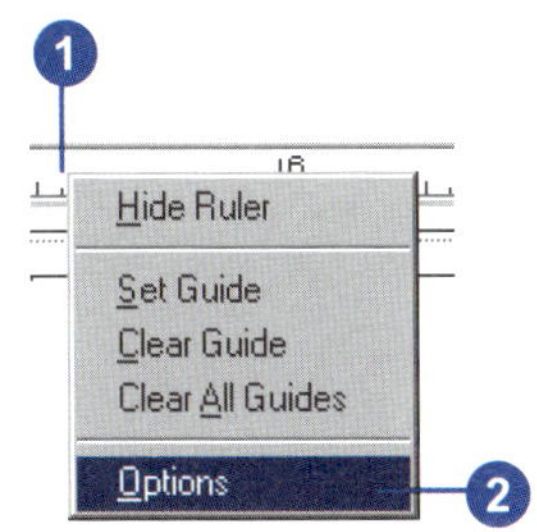

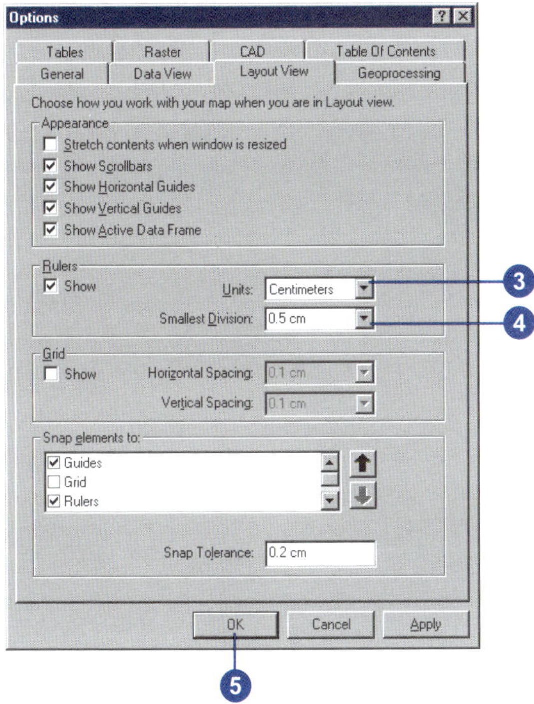

Turning guides on and off

1. Right-click the page.
2. Point to Guides and click Guides.

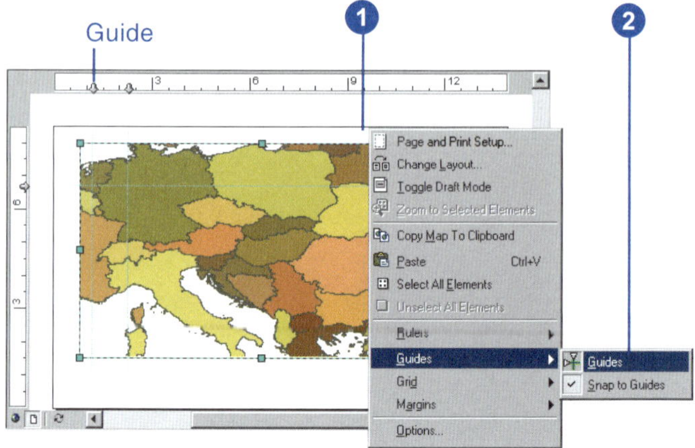

Snapping to guides

1. Right-click the page.
2. Point to Guides and click Snap to Guides.

   When you move map elements to the vicinity of a guide, the map element will snap to the guide.

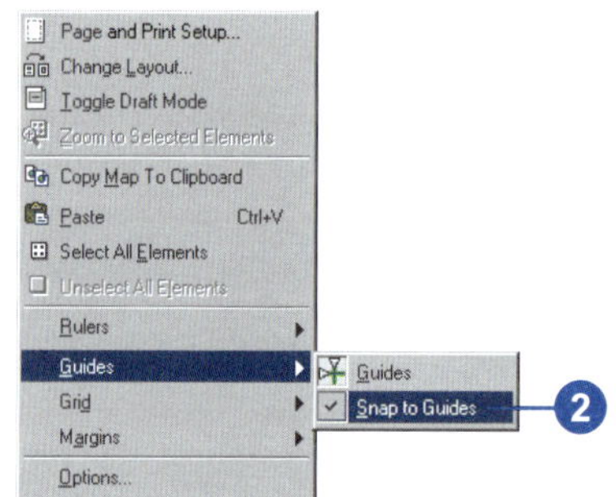

Adding a guide

1. Click the ruler at the location where you want to add a guide. It doesn't have to be exact, because you can move the guide.

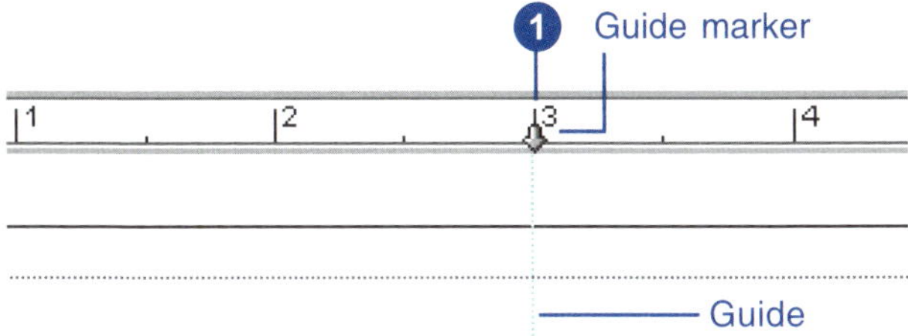

Moving a guide

1. Point to a guide marker on the ruler. Your pointer will change.
2. Click and drag the guide marker to a new location on the ruler.

   The guide appears with a dotted line until you release the mouse button.

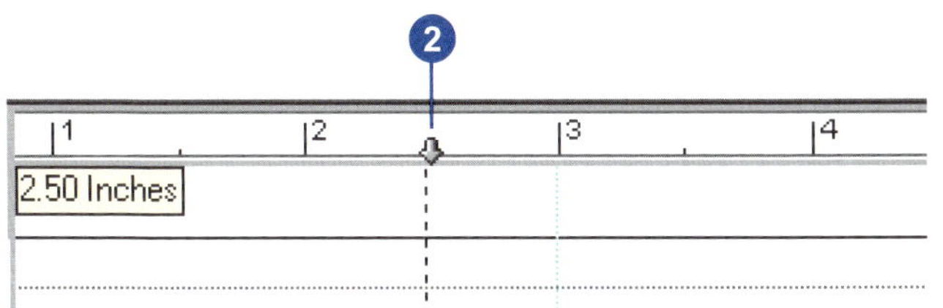

Removing a guide

1. Point to a guide marker on the ruler. Your pointer will change.
2. Right-click the guide marker and click Clear Guide.

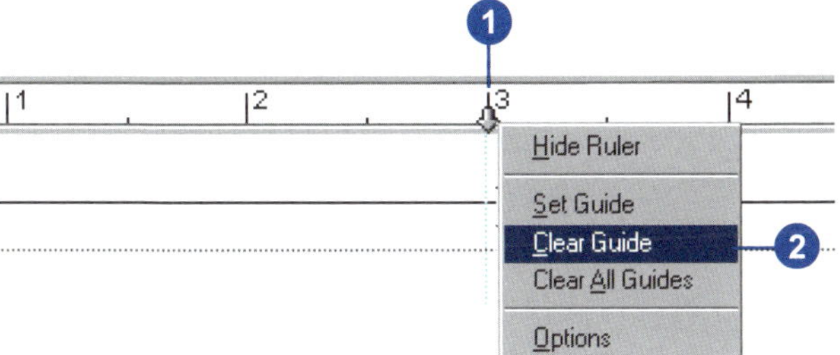

Tip

Using guides to set margins on the page

You can use guides to set margins for a map that is not the same size as your printer's page size.

The guides provide a visual indication of the map's margins to help you avoid positioning map elements in the margins.

Turning on snapping to guides will provide an additional cue that an element is near a margin.

See Also

For more information on margins, see 'Showing or hiding printer margins' in this chapter.

Removing all guides from a ruler

1. Right-click the ruler.
2. Click Clear All Guides.

   All of the guides are removed from the ruler.

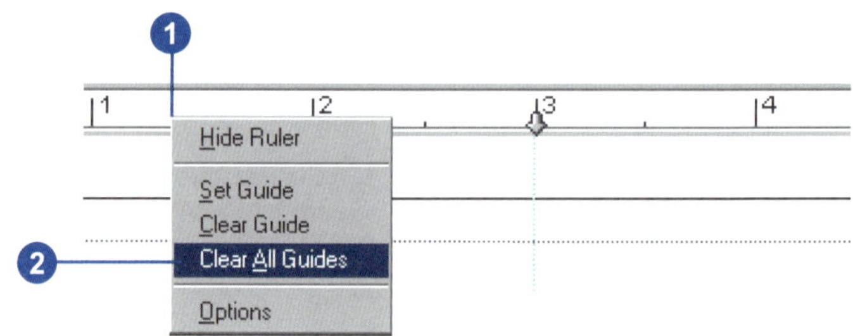

Using guides to define a map's margins

1. Click the rulers to add guides where you want the map's margins to be.
2. Optionally, move the guides to fine-tune their position.

Tip

Snapping to the grid

The snapping grid appears as a grid of dots on the virtual page. You can use the grid as a purely visual reference in placing map elements, or you can turn on snapping to grid. If you turn on snapping, the position of elements on the page will be constrained by the vertices of the snapping grid.

Tip

Why change the grid size?

You can change the size of the grid to allow more or less freedom in positioning map elements when snapping to grid is enabled.

Turning the grid on and off

1. Right-click the page.
2. Point to Grid and click Grid.

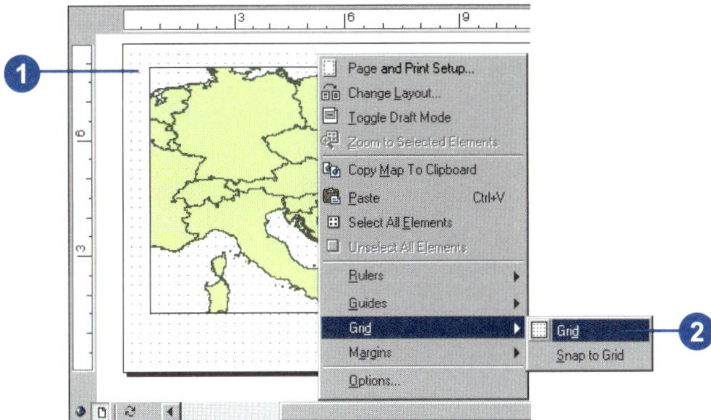

Snapping to the grid

1. Right-click the page.
2. Point to Grid and click Snap to Grid.

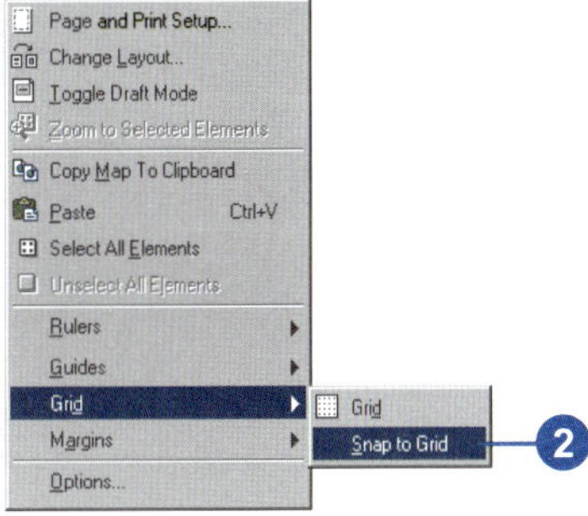

Changing the grid size

1. Click Tools and click Options.
2. Click the Layout View tab on the Options dialog box.

   The Options dialog box appears. ▸

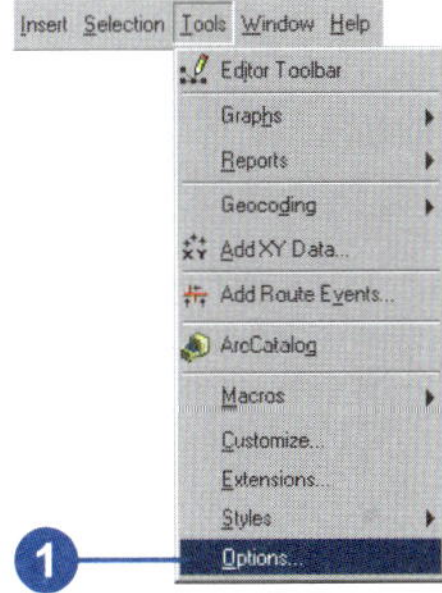

Tip

Changing the snapping order

The order in which guides, grids, and rulers appear in the Snap elements to list determines which one an element will snap to. For example, in the event that the element is within the snapping tolerance of both a guide and a grid point, it will snap to whichever is on top—the guide or the grid point.

You can use the up and down arrow keys to change the snapping order.

3. Click the Horizontal Spacing dropdown arrow and click a number of units to specify the horizontal spacing of the snapping grid.
4. Click the Vertical Spacing dropdown arrow and click a number of units to specify the vertical spacing of the snapping grid.
5. Click OK.

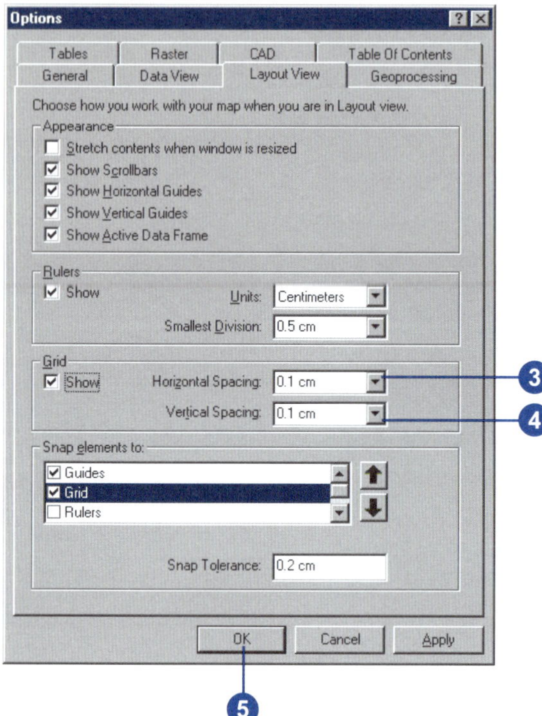

Changing the snapping tolerance

1. Right-click the page.
2. Click Options.
3. Type a number of units for the snapping tolerance.
4. Click OK.

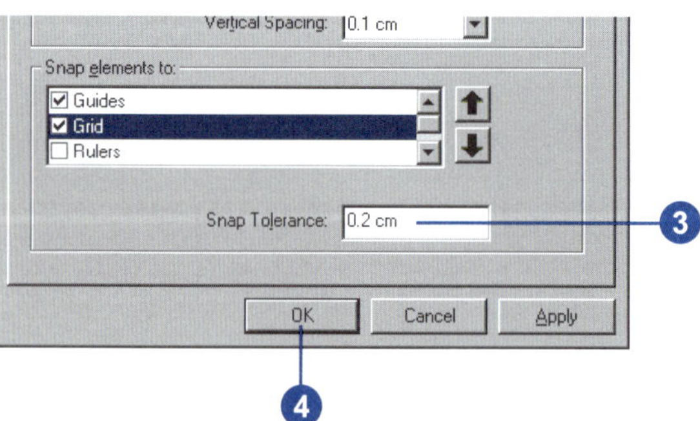

Adding data frames

A map is composed of one or more data frames (and data) arranged on the page, plus one or more other map elements.

Simple maps usually have only a single data frame. Sometimes you want to show more data than a single frame can conveniently hold. If that's the case, you may decide to add another data frame to the map.

You can use additional data frames in different ways, for example, to show insets and overviews or to allow map readers to compare different representations of the same area.

Adding a new data frame to a map

1. Click Insert on the Main menu and click Data Frame.

   You can add any data to the new data frame.

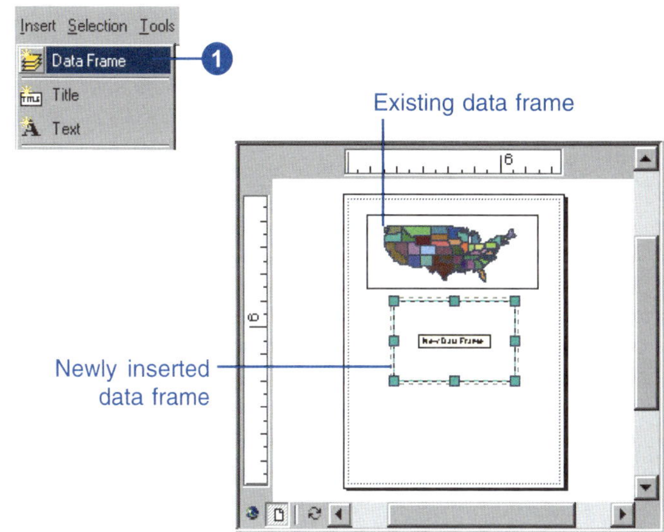

Tip

Choosing how to add a data frame

If you want to show two different layers of data in two data frames, it may be quicker to add a new data frame to the map than to copy an existing data frame.

If your data frames have layers in common, it may be more convenient to add the common layers to a single data frame than to duplicate the data frame.

Duplicating a data frame

1. In Layout View, click the data frame to select it.
2. Click Edit and click Copy. ▸

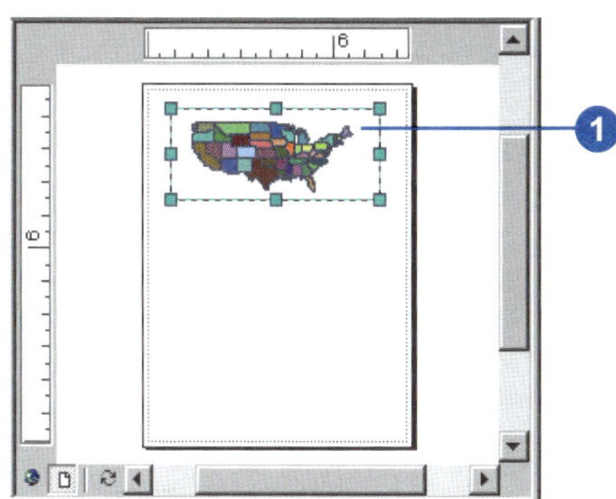

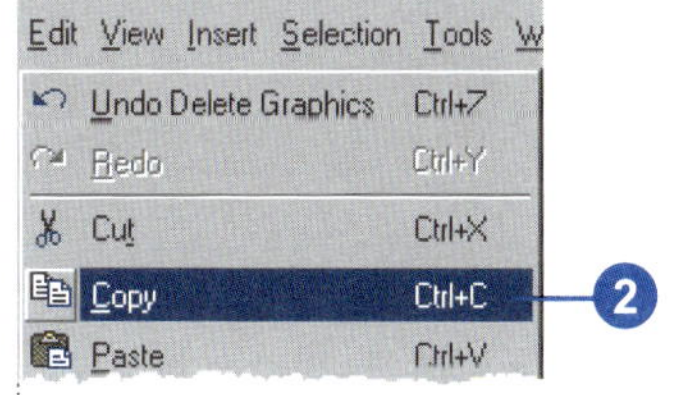

3. Click Edit and click Paste.
4. Click the copy, located on top of the original data frame, and drag it to a new place on the page.

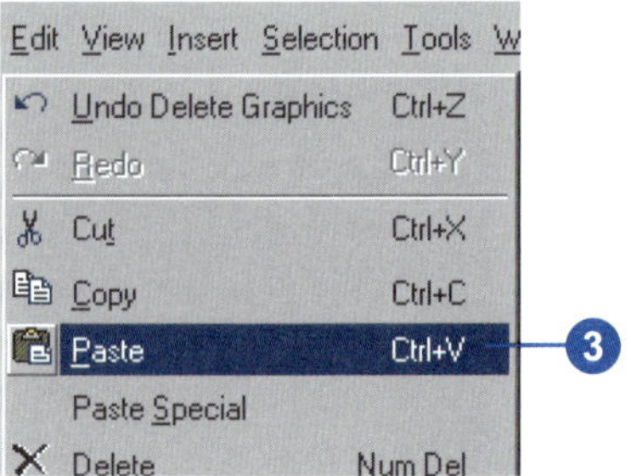

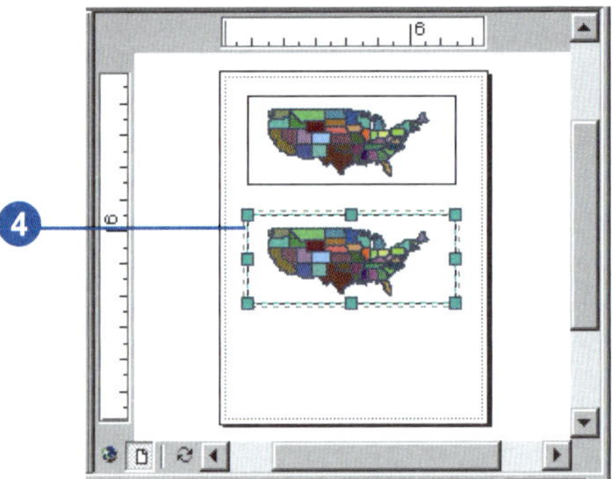

Resizing a data frame

1. Click the data frame to select it.
2. Click a selection handle and drag it to change the size of the data frame.

   Hold Shift while resizing to maintain a 1:1 size ratio.

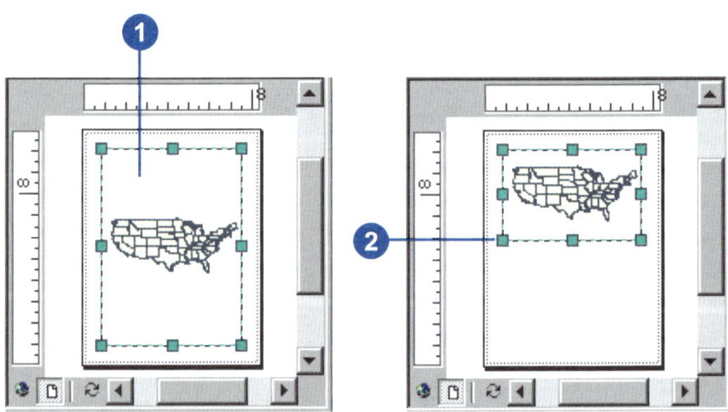

Tip

Showing a data frame's position with an extent rectangle

You can use a data frame with a large extent to provide context for another data frame—for example, showing the location of a state within a country. If the area on the map is familiar to your audience, it may not be necessary to add any more information.

Sometimes, the area that you show in the detail data frame does not have a commonly recognized outline. In this case, it may be useful to show its position with an extent rectangle.

Using one data frame to show the location of another

1. Click the overview data frame to select it.
2. Right-click the overview data frame and click Properties.
3. Click the Extent Rectangles tab.
4. Click the detail data frame (in this case called My Data Frame) in the Other data frames list and click the right arrow button to send it to the Show extent rectangle for these data frames list.
5. Click Frame to choose a border for the extent rectangle. ▸

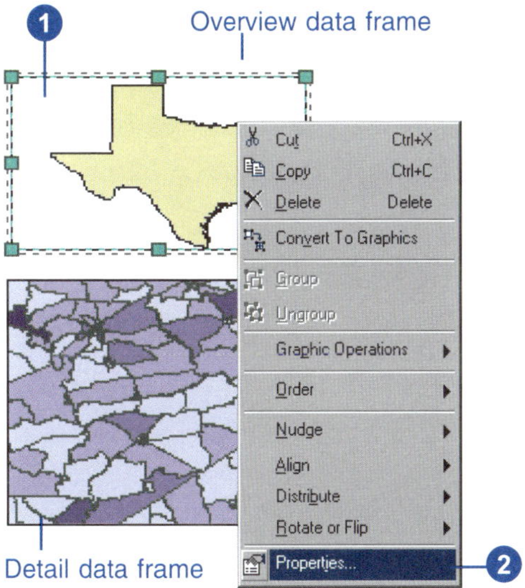

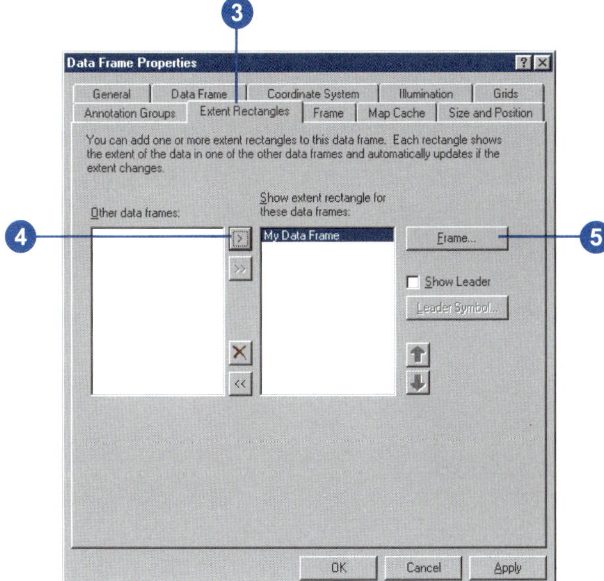

Tip

Showing multiple extents

You can use extent rectangles to show the positions of several different data frames on a single data frame.

Tip

Changing extents after you add an extent rectangle

You can change the extent of either data frame when you have an extent rectangle—the rectangle will automatically be updated to reflect the new relationship of the data frames.

6. Click the Border dropdown arrow and click a border.
7. Click OK on the Frame Properties dialog box.
8. Click OK on the Data Frame Properties dialog box.

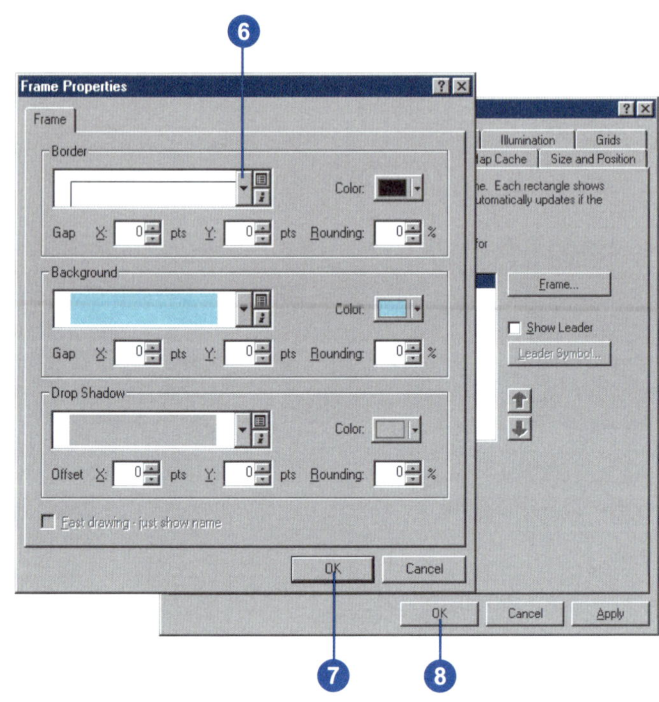

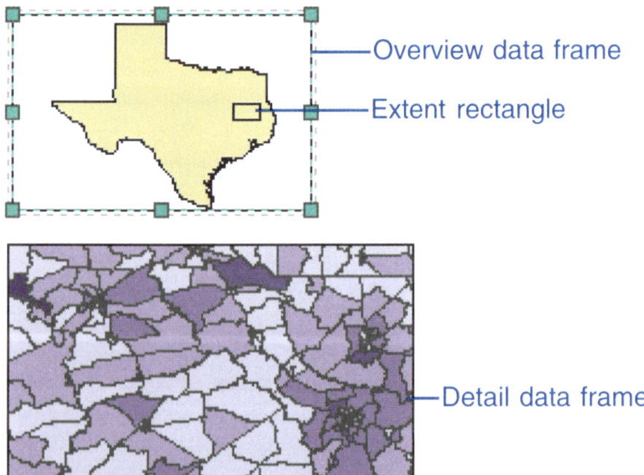

Adding map elements related to data frames

Some map elements—such as North arrows, scalebars, scale text, and legends—are related to the data in data frames. North arrows indicate the orientation of the map. *Scalebars* provide a visual indication of the sizes of features and distances between features shown on the map. Scale text indicates the scale of the map and features on the map. A legend tells a map reader what the symbols used to represent features on the map mean. ▸

Tip

Resizing a map element

Map elements aren't always the size you want when they're added to a map. You can change the size of map elements by selecting them and dragging the selection handles.

Dragging a handle away from an element enlarges it. Dragging a handle toward an element reduces it.

Adding a North arrow

1. Click Insert and click North Arrow.
2. Click a North arrow.
3. Click OK.
4. Click and drag the North arrow into place on your map.
5. Optionally, resize the North arrow by clicking and dragging a selection handle.

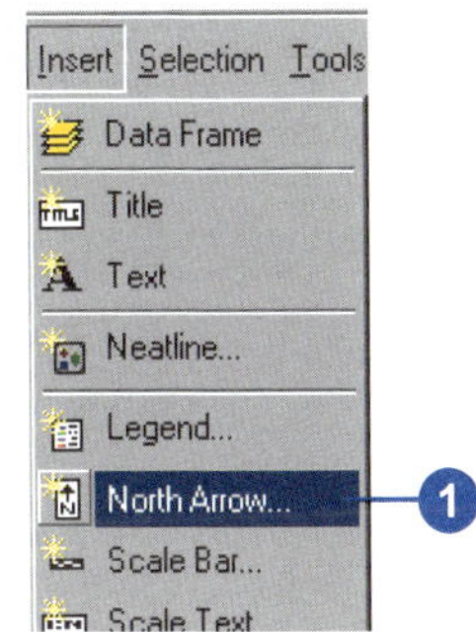

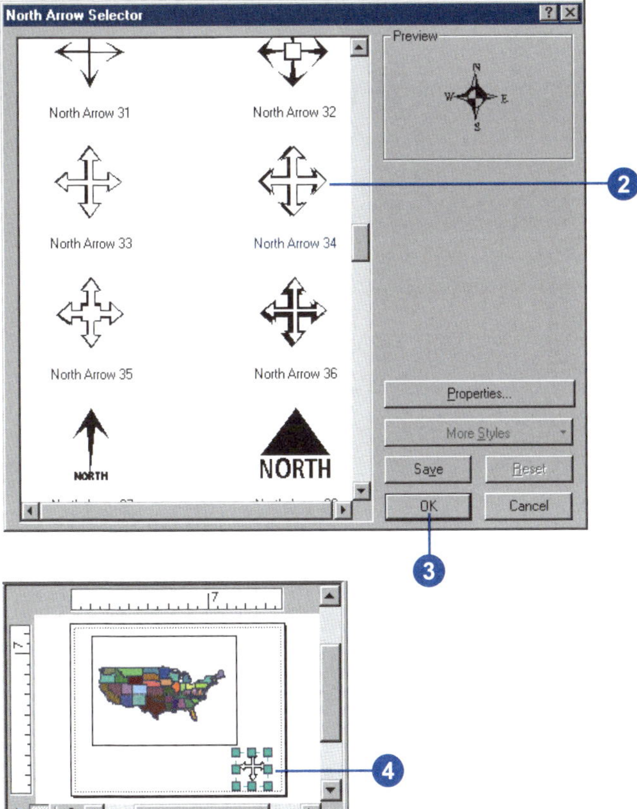

You can use a scalebar to represent the scale of your map.

A scalebar is a line or bar divided into parts and labeled with its ground length, usually in multiples of map units such as tens of kilometers or hundreds of miles.

If the map is enlarged or reduced, the scalebar remains correct. ▸

Adding a scalebar

1. Click Insert and click Scale Bar.
2. Click a scalebar.
3. Optionally, click Properties to modify the scalebar's properties.
4. Click OK.
5. Click and drag the scalebar into place on your map.
6. Optionally, resize the scalebar by clicking and dragging a selection handle.

   If you resize the width, the distance measures along the scalebar are recalculated. If you resize the height, the height of the bar is altered and the text size is altered accordingly.

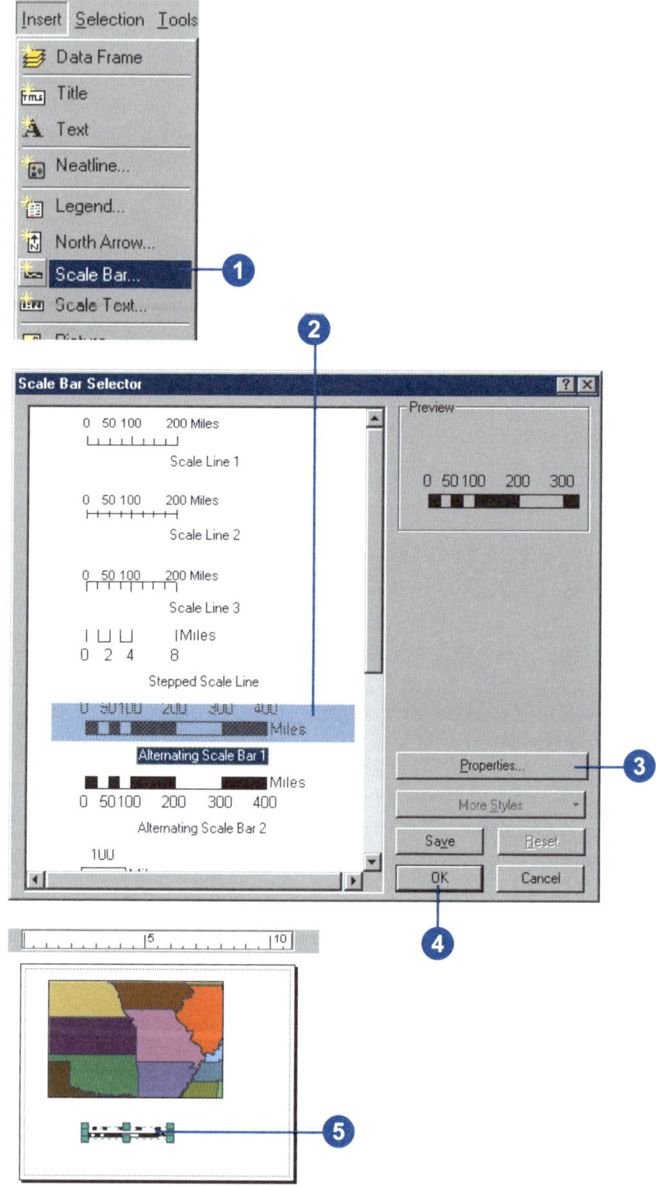

When you add a scalebar to a map, the number and size of the divisions might not be exactly as you would like them. For example, you might want to show four divisions rather than three or show 100 meters per division instead of 200.

You might also want to change the units that the scalebar shows or adjust how those units are represented.

You can adjust many characteristics of a scalebar from the Scale Bar properties dialog box. ▸

Tip

Changing the units label

By default, the units label on a scalebar is the same as the scalebar units. Sometimes you might want to change the label of the scalebar, for example, from Kilometers to km. Just type the new scalebar label in the Label text box.

Tip

Why can't I see the Size and Position tab or the Frame tab?

You can only change the size and position and frame of an element after it has been placed on the map. If you click Properties while inserting a map element, you won't see these tabs.

Customizing a scalebar's scale and units

1. Right-click the scalebar and click Properties.
2. Click the Scale and Units tab.
3. Click the arrow buttons to set the Number of divisions.
4. Click the arrow buttons to set the Number of subdivisions.
5. Click the When resizing dropdown arrow and click how you want the scalebar to respond when the map scale changes.

   Adjust division value—the division value will vary with the map scale. The number of divisions and the width of the scalebar remain constant.

   Adjust number of divisions—the number of divisions will vary with the map scale. The division value and the width of the scalebar remain constant.

   Adjust width—the width of the scalebar will vary with the map scale. The division value and number of divisions remain constant.
6. Choose the units for the scalebar.
7. Click Symbol and choose a text style for the scalebar labels.
8. Click OK.

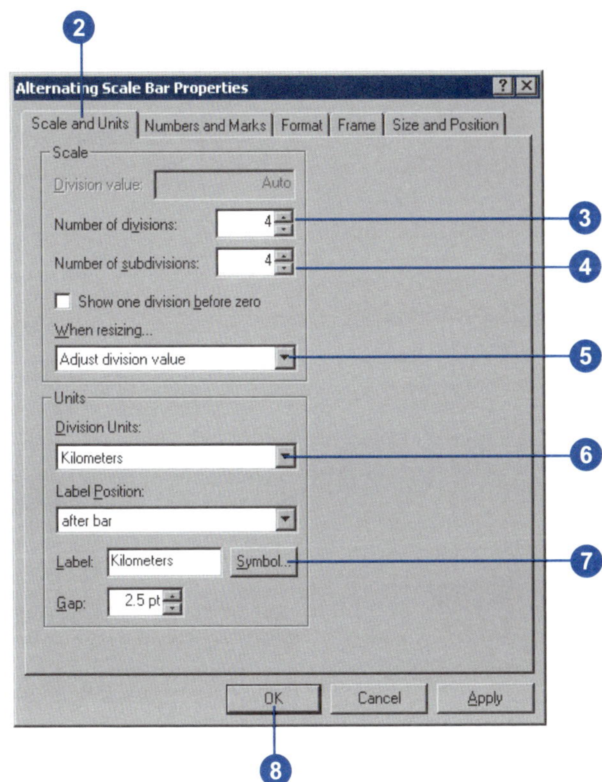

When you add a scalebar to a map, the number labels and tic marks might not be exactly as you would like them. For example, you might want to label the endpoints of the scalebar but not the divisions, or you might want larger tic marks at the major divisions of the bar than at the minor ones.

You might also want to change the size or color of the font that the numbers are drawn in. ▸

Customizing a scalebar's numbers and marks

1. Right-click the scalebar and click Properties.
2. Click the Numbers and Marks tab.
3. Click the Numbers Frequency dropdown arrow to choose where along the bar to place the numbers.
4. Click the Numbers Position dropdown arrow to choose where to place numbers relative to the bar.
5. Click the Marks Frequency dropdown arrow to choose where along the bar to place tic marks.
6. Click the Marks Position dropdown arrow to choose where to place tic marks relative to the bar.
7. Click the Division Height arrow buttons to increase or decrease the height of division tic marks.
8. Click the Subdivision Height arrow buttons to increase or decrease the height of subdivision tic marks.
9. Click OK.

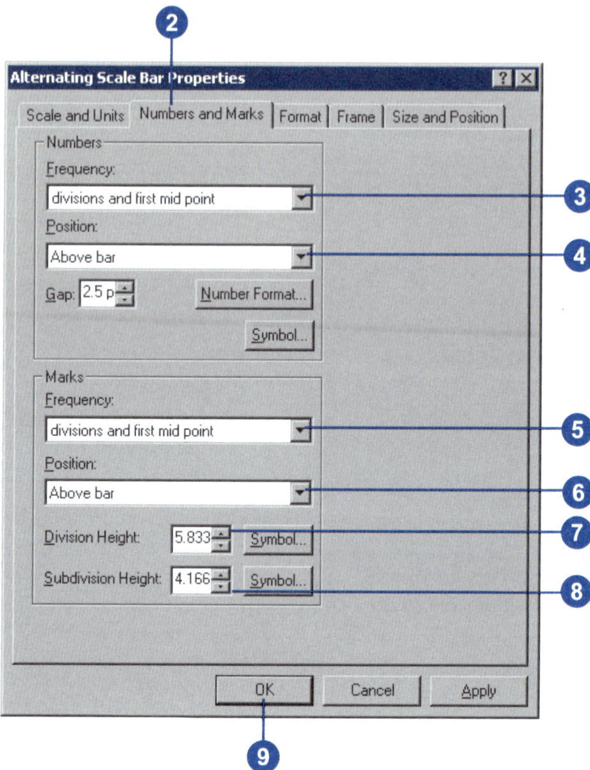

You can also represent the scale of your map with scale text.

Scale text tells a map reader how many ground units are represented by a map unit—for example, "one centimeter equals 100,000 meters".

One drawback of scale text is that if a printed copy of the map is duplicated at another scale (enlarged or reduced), the scale text will be in error. Scalebars do not suffer this limitation.

Many maps have both scale text and a scalebar to indicate the map scale. ►

Adding scale text

1. Click Insert and click Scale Text.
2. Click a sample of the style of scale text to add to the map.
3. Optionally, click Properties to customize the scale text.
4. Click OK.
5. Click and drag the scale text into position on your map.
6. Optionally, resize the scale text by clicking and dragging a selection handle.

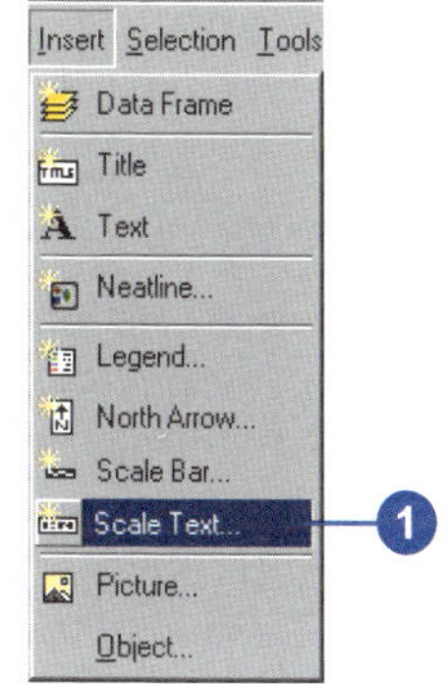

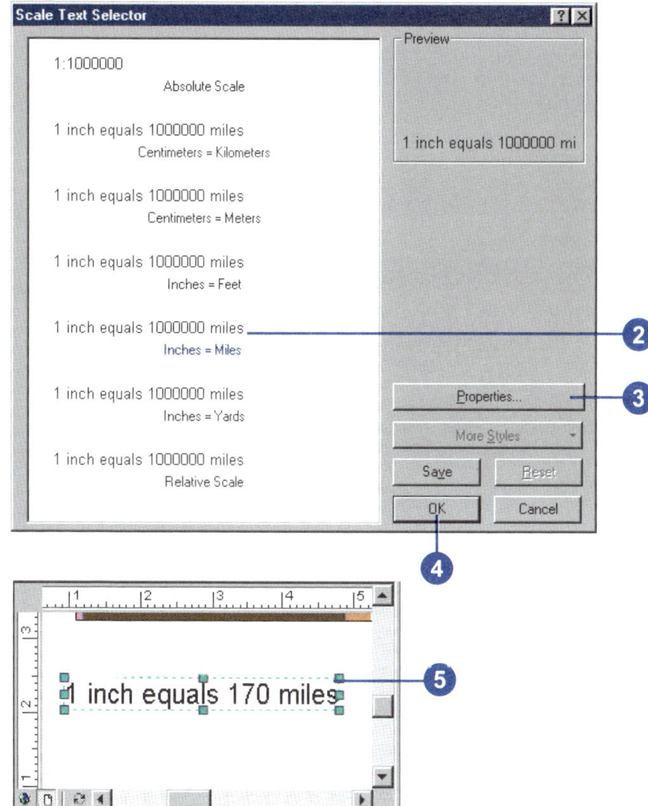

You can use a legend to tell map readers the meaning of the symbols you've used on the map.

Legends consist of examples of the symbols on the map with a label containing some explanatory text.

When you use a single symbol for the features in a layer, the layer is labeled with the layer's name in the legend.

When you use multiple symbols to represent features in a single layer, the field you used to classify the features becomes a heading in the legend, and each category is labeled with its value.

Tip

Enable the Legend Wizard

You can insert legends with the help of the Legend Wizard. To enable the wizard, click the Tools menu and click Options. Click the General tab and check Show Wizards when available.

Adding a legend

1. Click Insert and click Legend.

   The Legend Wizard appears. If you don't see it, enable the Legend Wizard as described in the tip to the left.

2. By default, all the layers on the map will appear as legend items in the legend. To remove a legend item, click it, then click the left arrow button.

3. Use the up and down arrow buttons to order the legend items.

4. Click Next. ►

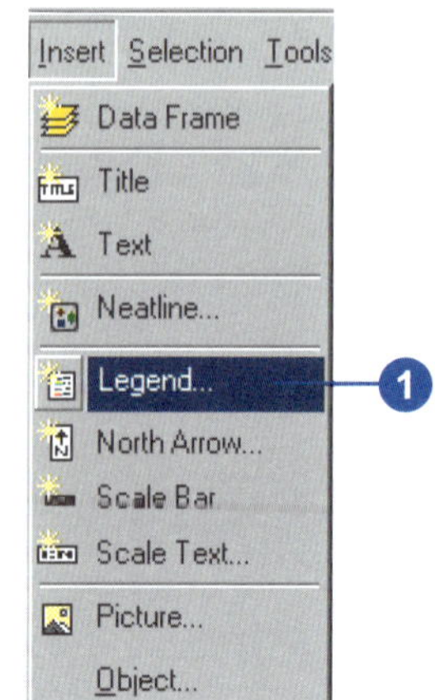

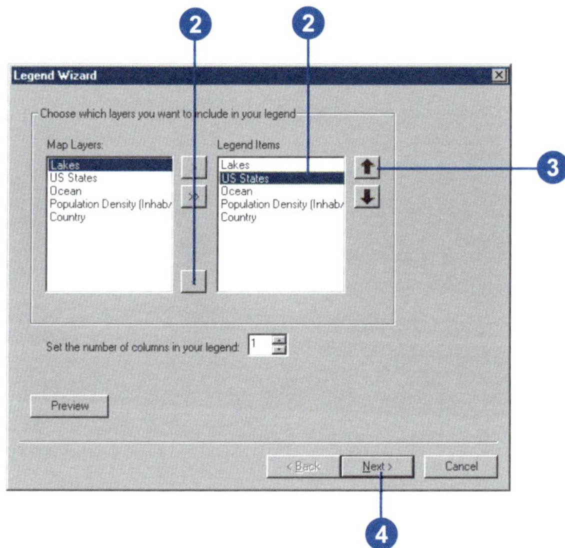

Tip

Changing legend labels

You can edit the text of the labels that appear in the legend by changing the text in the ArcMap table of contents or on the Symbology tab of the Layer Properties dialog box.

Tip

Legends on maps with multiple data frames

When you have more than one data frame, inserting a legend adds a legend for the selected data frame.

Each legend corresponds to a single data frame, although you can arrange multiple legends together as a single legend for a complex map.

5. Type a title for the legend.
6. Set the text color, size, and font as desired.
7. Click Next.
8. Click the Border dropdown arrow and click a border.
9. Click the Background dropdown arrow and click a background.
10. Click the Drop Shadow dropdown arrow and click a drop shadow.
11. Click Next. ▸

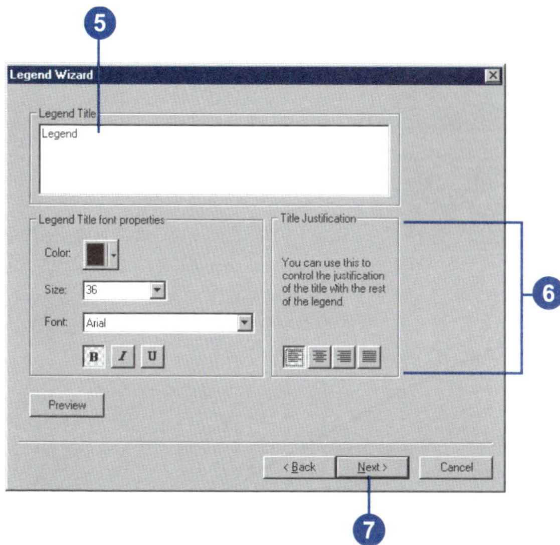

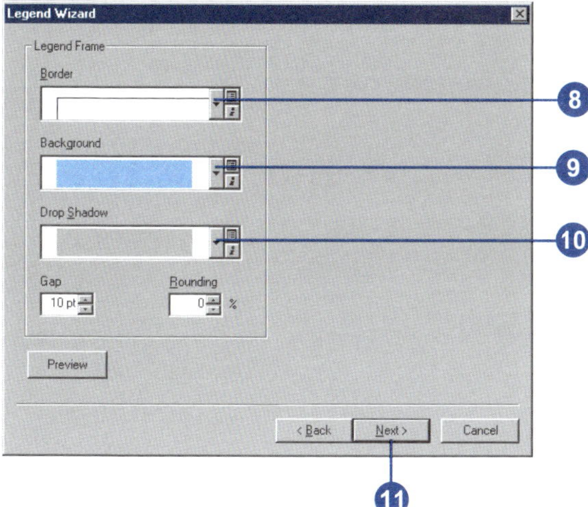

Tip

Editing the legend

You can edit your legend after it's been created. Select the legend in the Layout View and double-click to bring up the Legend Properties.

12. Click a Legend Item in the list to modify the symbol patch.
13. Set the Patch properties as desired.
14. Click Next.
15. Set the Spacing between legend elements by typing a value into the appropriate box.
16. Click Finish.

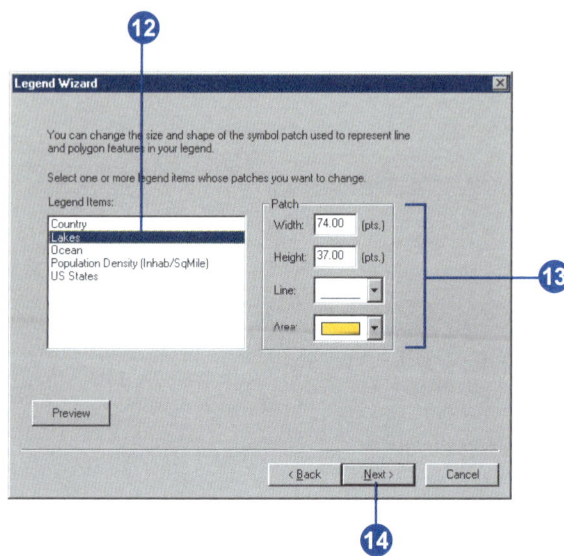

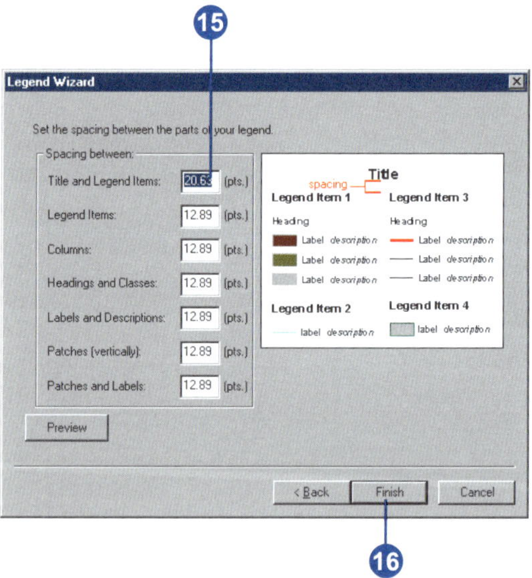

Legends have patches that show examples of the map symbols. By default, the legend patches are points, straight lines, or rectangles that match the map symbols. You can customize the legend patches so areas are represented with patches of another shape or so rivers are drawn with a sinuous rather than straight line.

Tip

Changing a single layer's legend patch

If you have two layers with the same geometry in a data frame—for example, a layer of roads and a layer of streams—you can set their legend patches independently, so roads are shown with a straight line and rivers with a sinuous line.

In the Legend Items list, click the Item for the layer you want to change. Right-click and click Properties. Click the General tab and check the Override default patch box. Click the Patch dropdown arrow and click a new patch.

Changing the patches in a legend

1. Double-click the legend on the map to open the Legend Properties dialog box.
2. Click the Legend tab.
3. Click the dropdown arrow to select a new patch shape.
4. Click OK.

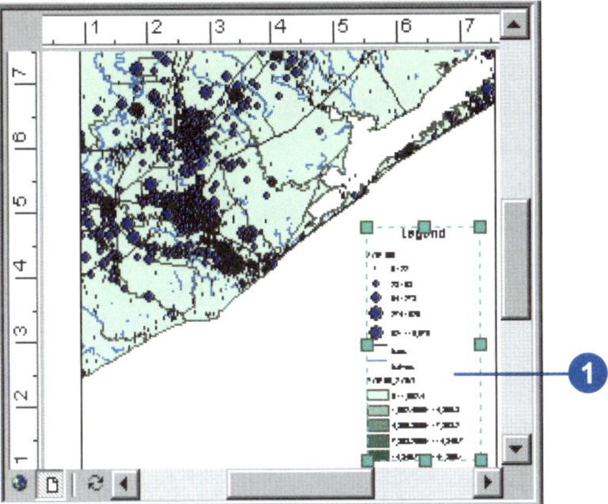

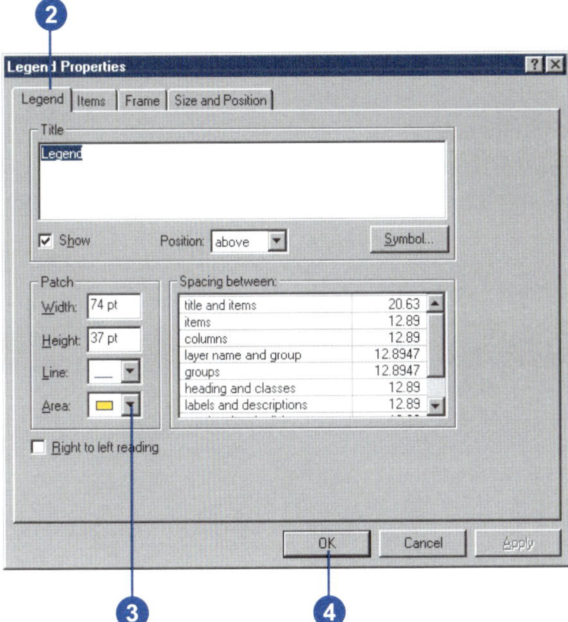

Tip

Using a Description in your legend

If you want an additional description of single symbol layers to appear in your legend, choose a legend item style that includes a Description. To add a description, right-click the layer you want to add descriptive text to in the table of contents, click Properties, and click the Symbology tab on the Layer Properties dialog box. If you are using the Single symbol method, click the Description button. If you are using a drawing method other than Charts, right-click a symbol after you have specified its symbology options and click Edit Description. The text you enter will appear next to that symbol in the legend—the text won't, however, appear in the table of contents.

Tip

Typing a legend Description that spans multiple lines

If you want your Description to be on more than one line, insert a line break by pressing the Ctrl and Enter keys together in the Description for Legend dialog box.

See Also

For more information on symbolizing layers and adding Descriptions, see Chapter 6, 'Symbolizing features'.

Changing the items in a legend

1. Right-click the legend on the map and click Properties.
2. Click the Items tab.
3. Click a legend item in the Legend Items list.
4. Click the up and down arrows to move the item up or down in the legend.
5. Optionally, click Style and change the Item's style in the legend.
6. Optionally, check Place in new column to place the highlighted item in a new column.
7. Optionally, change the number of columns in the legend for the highlighted legend item by clicking the up and down arrow keys.
8. Optionally, remove an item from the legend by clicking it and clicking the left arrow key.
9. Click OK.

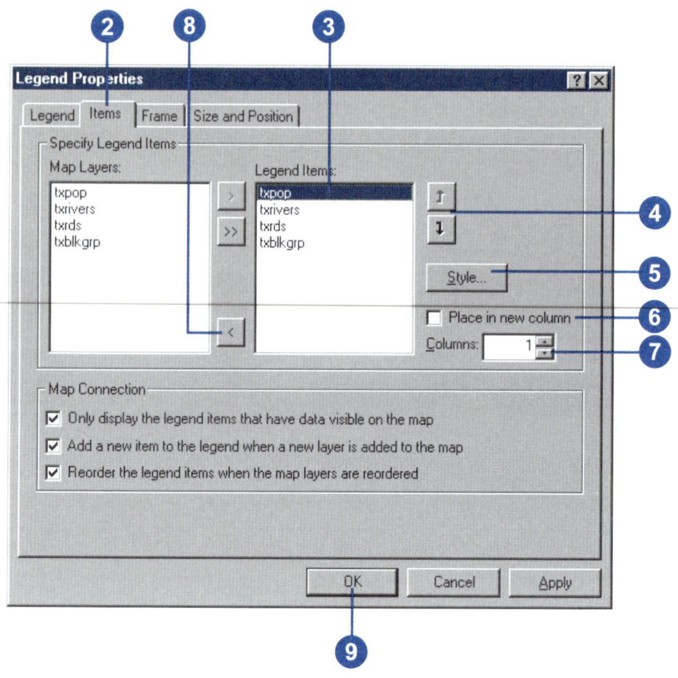

Certain map elements—including scalebars, scale text, North arrows, legends, and data frames—can have frames.

You can use frames to set map elements apart from other elements or from the background of the map.

You can also use frames to visually link map elements to other parts of the map by using similar frames for related elements.

Tip

Why can't I see the Frame tab?

You can only set the frame properties after an element has been placed on the map.

Tip

Framing grouped elements

If you group some elements together, you can right-click the group and set a frame for the group.

Framing a map element

1. Right-click the element on the map and click Properties.
2. Click the Frame tab.
3. Click the Border dropdown arrow and click a border.
4. Click the Background dropdown arrow and click a background.
5. Click the Drop Shadow dropdown arrow and click a drop shadow.
6. Click OK.

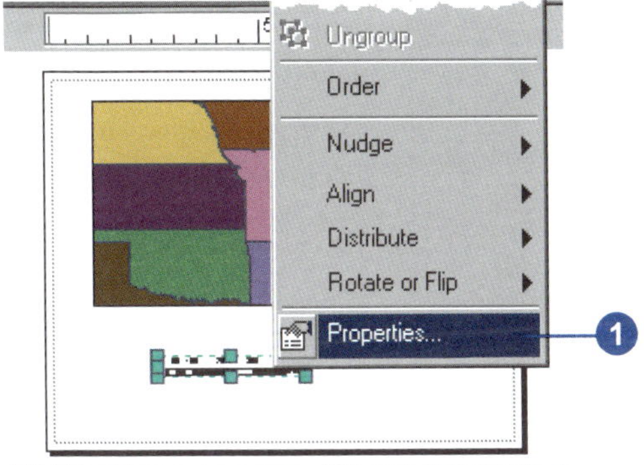

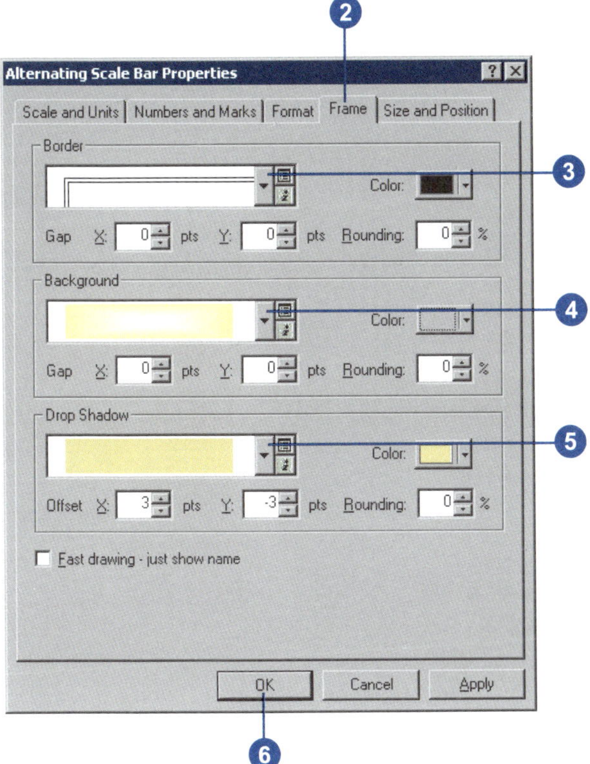

Tip

Why convert a map element to graphics?

You might want to convert a map element, such as a legend, to graphics if you want more precise control over each element that comprises the element. Once you convert a map element to graphics, you can't reconstruct the map element from the individual pieces.

Converting map elements to simple graphics

1. Right-click an element, such as a legend, and click Convert To Graphics.

   The element is now a grouped graphic.

2. Right-click the graphic and click Ungroup.

   The individual graphics that comprise the map element can now be edited separately.

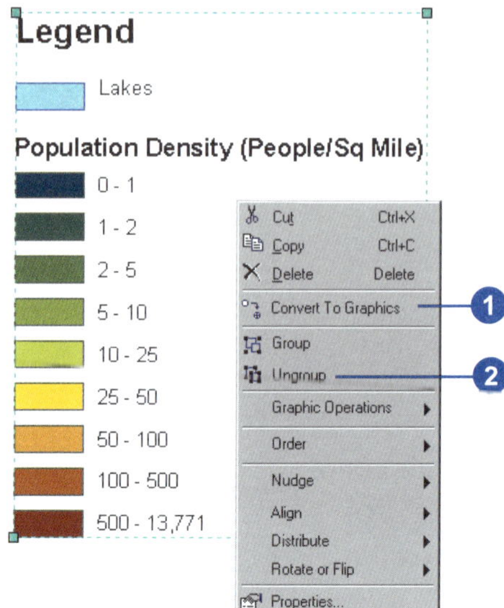

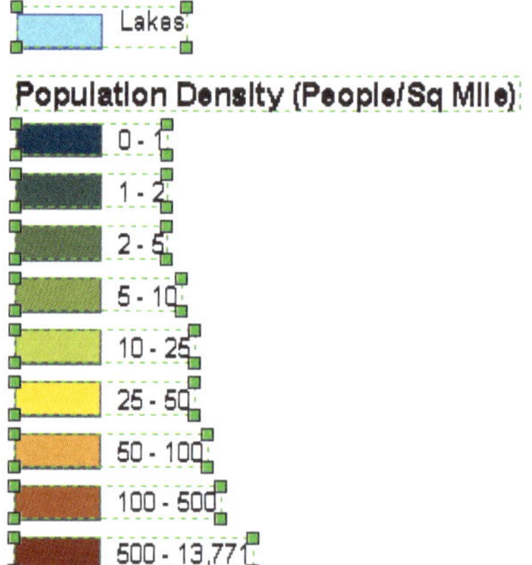

After ungrouping, you can edit the individual graphic elements.

Creating grids and graticules

ArcMap contains several types of grids and *graticules*—also called reference systems—that can be added to a map in layout view.

Tip

What type of grid should I display?

If the data you're mapping covers a large area of the earth's surface, you can show graticules that represent lines of latitude and longitude.

If you're mapping a region, such as a country, you can show a measured grid that references a particular projected coordinate system.

If you're mapping a local area, such as a study area, you can show a reference grid that divides the data frame into squares that you can reference by row and column.

Tip

Why don't I see the Grids and Graticules Wizard?

You need to enable wizards. From the Tools menu, click Options, click the General tab, then check Show Wizards when available.

Adding a graticule

1. In the table of contents, double-click the data frame for which you want to add a graticule.
2. Click the Grids tab on the Data Frame Properties dialog box.
3. Click New Grid.

   The Grids and Graticules Wizard should appear. If not, see the tip to the left.
4. Click Graticule.
5. Type a name for the new grid.
6. Click Next.
7. Click an Appearance option.
8. Type the Intervals you want.
9. Click Next. ▸

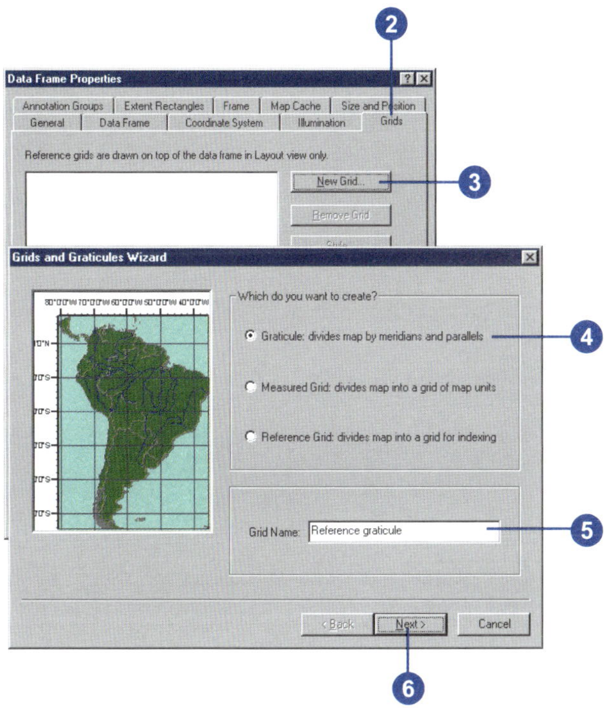

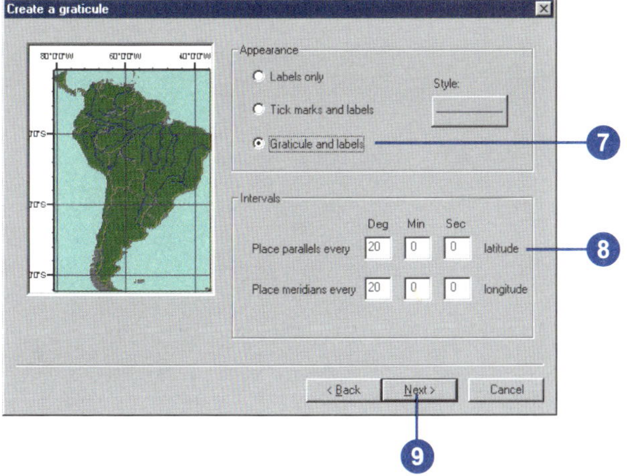

Tip

Working without the Grids and Graticules Wizard

You can build a grid or graticule without the wizard. In the Data Frame Properties dialog box, select the Grids tab, then click New Grid and click Properties. Here you can design your reference system.

10. Check the Axes you want and set how they should appear.
11. Click the Font button beside Text Style to set the text style.
12. Click Next.
13. Click the Graticule Border you want.
14. Check Place a border outside the grid.
15. Click to specify whether the graticule is static or updates with changes to the data frame.
16. Click Finish.

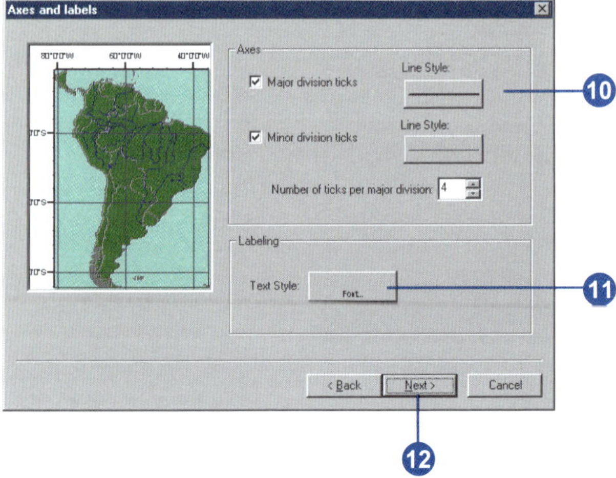

Tip

Why don't I see the Grids and Graticules Wizard?

You need to enable wizards. From the Tools menu, click Options, then check Show Wizards when available.

Adding a measured grid

1. In the table of contents, double-click the data frame you want to add a measured grid to.
2. Click the Grids tab on the Data Frame Properties dialog box.
3. Click New Grid.

   The Grids and Graticules Wizard should appear. If not, see the tip to the left.
4. Click Measured Grid.
5. Type a name for the new grid.
6. Click Next.
7. Click an Appearance option.
8. Click Properties to set a coordinate system for the grid that differs from that of the data frame.
9. Type the Intervals you want.
10. Click Next. ▸

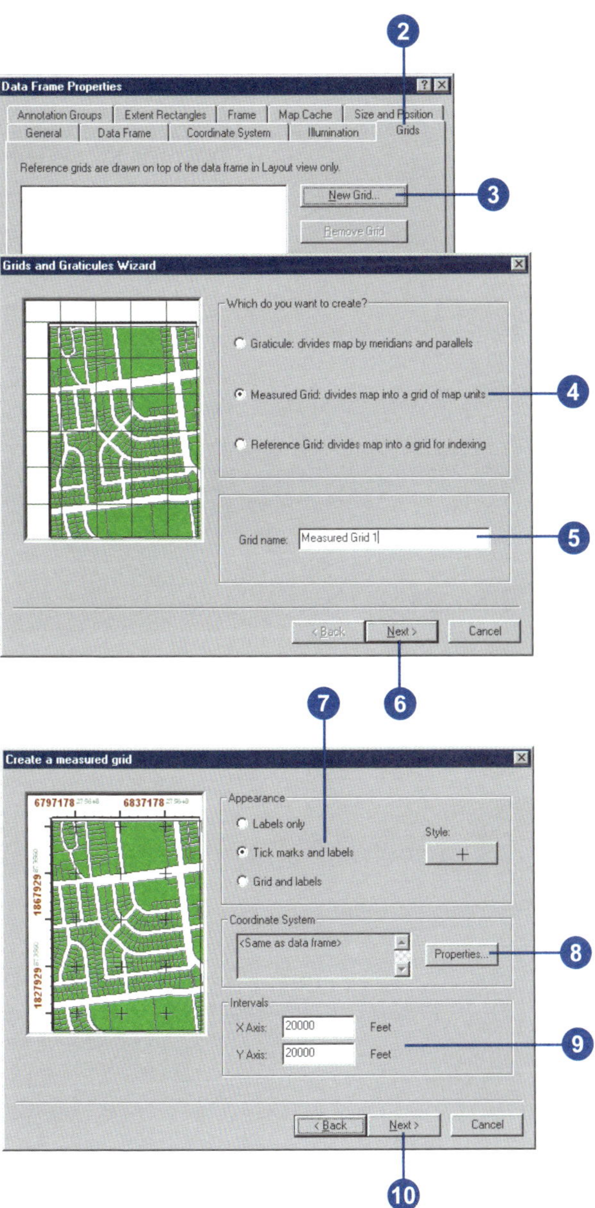

11. Check the Axes you want and set how they should appear.
12. Click the Font button beside Text Style to set the text style.
13. Click Next.
14. Click the Measured Grid Border you want.
15. Check Place a border outside the grid.
16. Click to specify whether the measured grid is static or updates with changes to the data frame.
17. Click Finish.

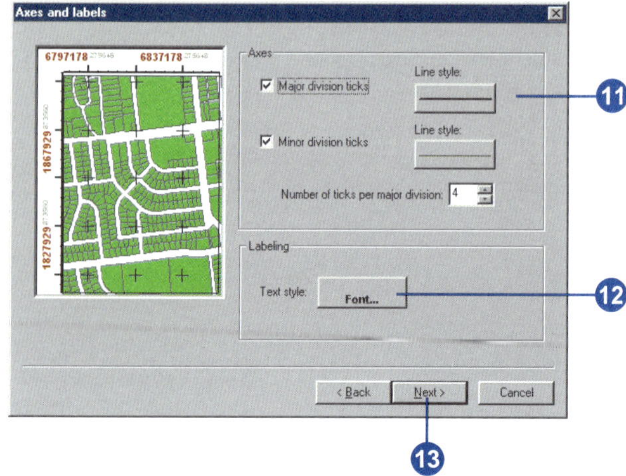

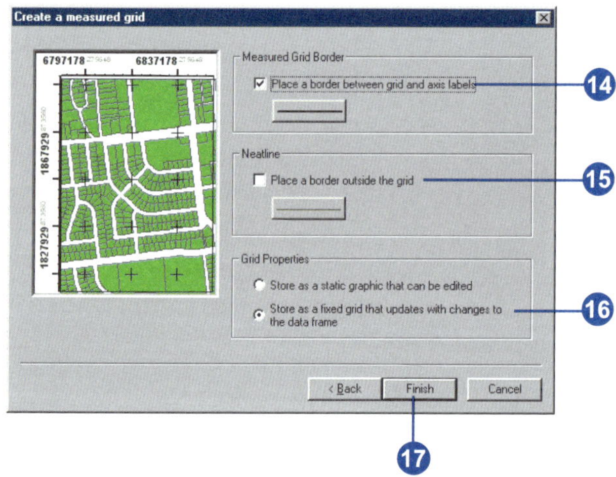

Tip

Why don't I see the Grids and Graticules Wizard?

You need to enable wizards. From the Tools menu, click Options, then check Show Wizards when available.

Adding a reference grid

1. In the table of contents, double-click the data frame you want to add a reference grid to.
2. Click the Grids tab on the Data Frame Properties dialog box.
3. Click New Grid.

   The Grids and Graticules Wizard should appear. If not, see the tip to the left.
4. Click Reference Grid.
5. Type a name for the new grid.
6. Click Next.
7. Click an Appearance option.
8. Type the Intervals you want.
9. Click Next. ►

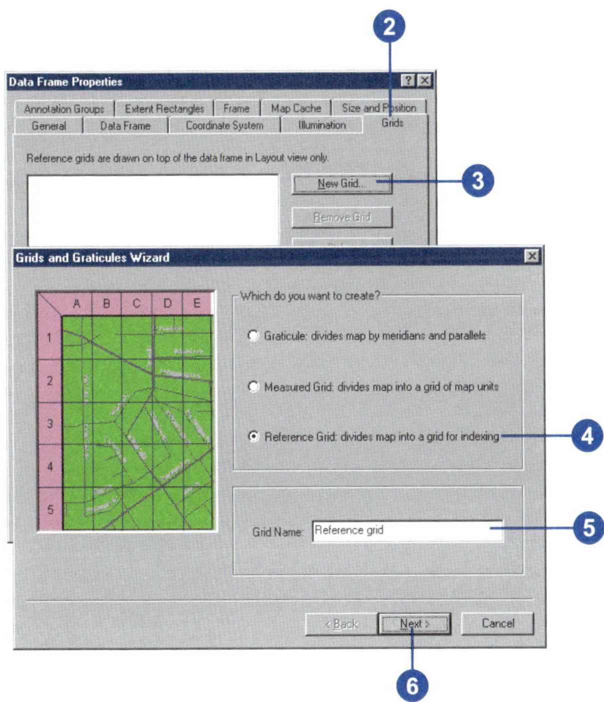

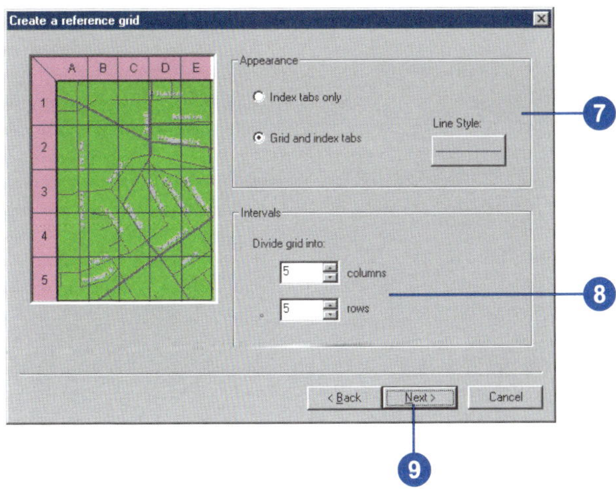

10. Set the Tab Style.
11. Set the Tab Configuration.
12. Click Next.
13. Check Place a border between grid and axis labels.
14. Check Place a border outside the grid.
15. Click to specify whether the graticule is static or updates with changes to the data frame.
16. Click Finish.

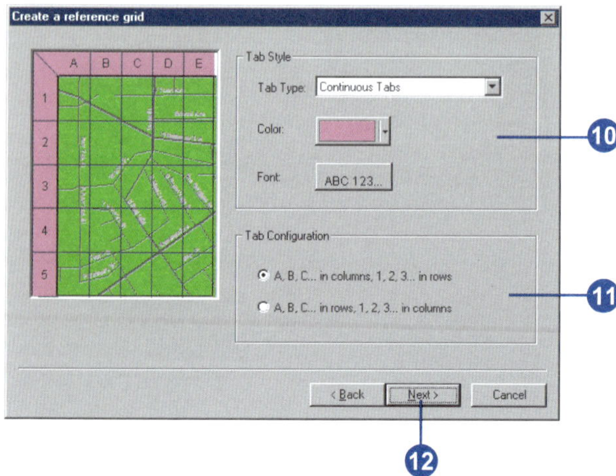

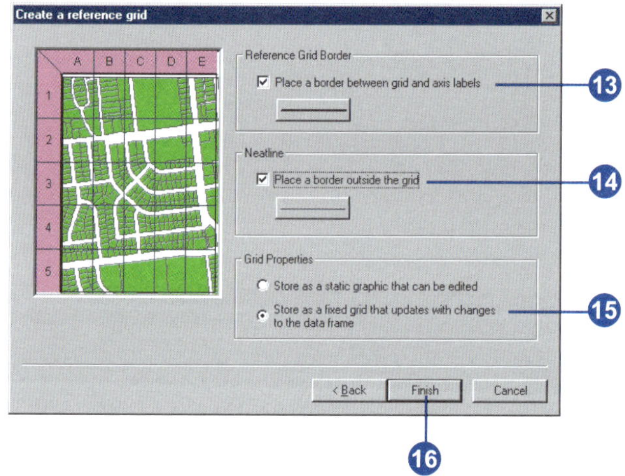

Adding other map elements

In addition to those elements that are related to data frames, there are elements that provide additional information, frame or group the elements of the map, or ornament the map.

Most maps have a title. A title communicates the topic of the map to a casual viewer and provides a way to refer to a map.

Many maps have graphic elements in addition to the geographic data on the map.

Graphics can be used to ornament a map, group related parts of a map together, identify a map with an organization, or emphasize a part of the map. You can use graphic rectangles to frame a group of other map elements.

If you want to frame an individual map element, right-click it, click Properties, and click the Frame tab—you can use this method to choose borders and backgrounds for legends, North arrows, data frames, scalebars, scale text, and data frames.

Maps can have pictures or graphic images in addition to the geographic data on the map. ▸

Adding a title

1. Click Insert and click Title.
2. Type a title for the map.
3. Click and drag the title into place on your map.
4. Optionally, modify the appearance of the title text.

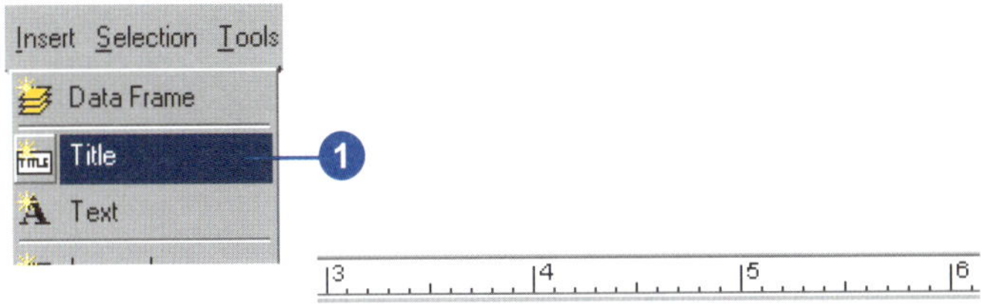

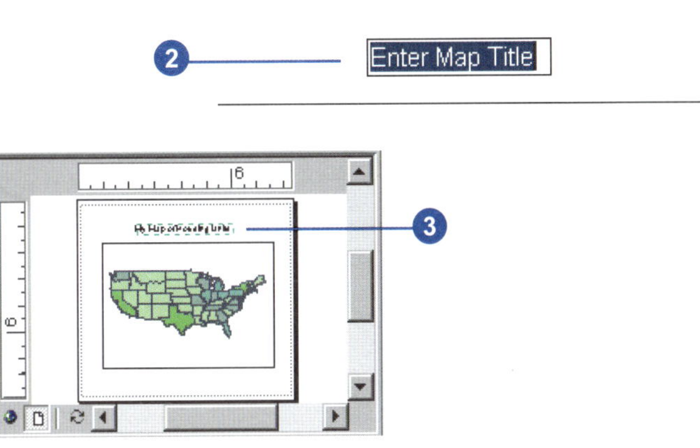

Modifying a title

1. With the title selected, click the Font dropdown arrow and click a font in the Draw toolbar.
2. Click the Font Size dropdown arrow and click a font size.
3. Click Bold, Italic, or Underline to change the style of the text.
4. Click the Text Color dropdown arrow and click a color.

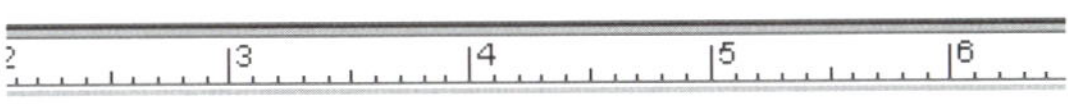

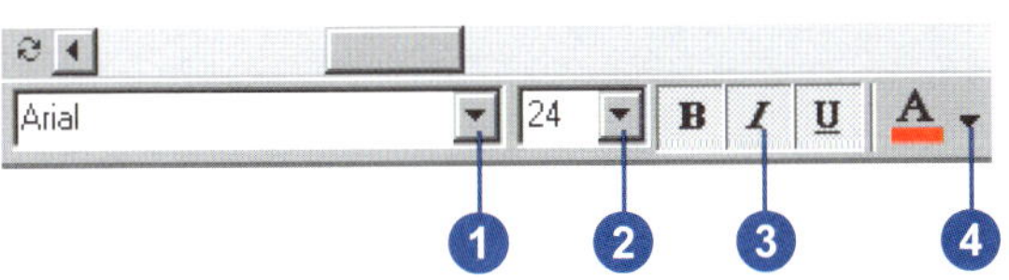

You might add a graphic image of your company's logo to indicate the source of a map or add a nation's flag to a map to indicate its subject.

You can also ornament a map by placing representative images of places, people, or objects found in an area on a map.

While most of the data on a map is usually geographic data presented in data frames, maps can also contain reports and graphs that support or complement the geographic data. Reports and graphs are two alternative ways of presenting complex tabular information; they can make your map more informative or persuasive.

Tip

Mixing your own colors

If you don't find the exact color you want in the array of colors on the Fill Color dropdown menu, you can mix your own. Click the dropdown arrow and click More Colors.

Adding a graphic element

1. Click the graphics dropdown arrow on the Draw toolbar.
2. Click the New Rectangle button.
3. Click the map and drag a box where you want the rectangle.

   The graphic element appears on the map.

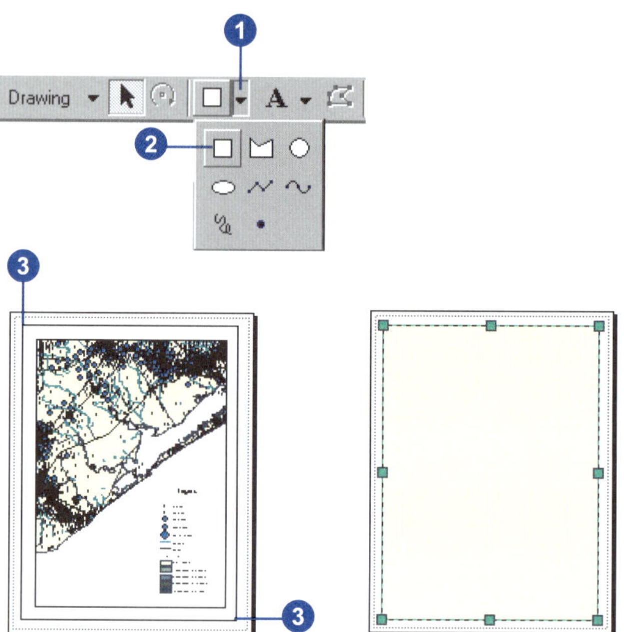

Applying color to a graphic element

1. Click a graphic element to select it.
2. Click the Fill Color dropdown arrow on the Draw toolbar.
3. Click a color.

   The fill color is applied to the graphic element.

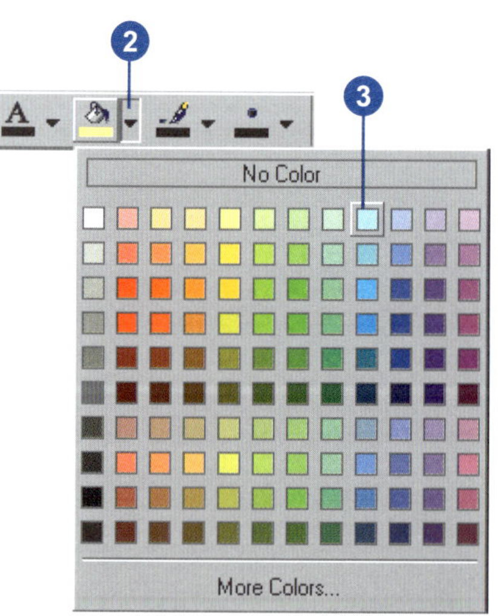

Applying a line color to a graphic element

1. Click a graphic element to select it.
2. Click the Line Color drop-down arrow on the Draw toolbar.
3. Click a color.

   The line color is applied to the graphic element.

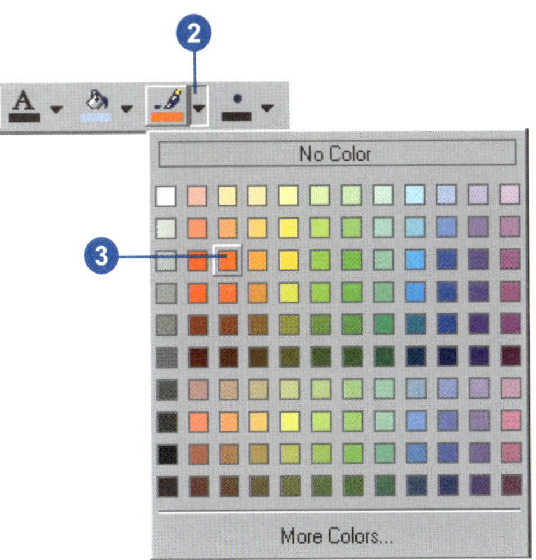

Placing a graphic element behind other elements

1. Click the graphic element to select it.
2. Right-click the graphic element, point to Order, and click Send to Back.

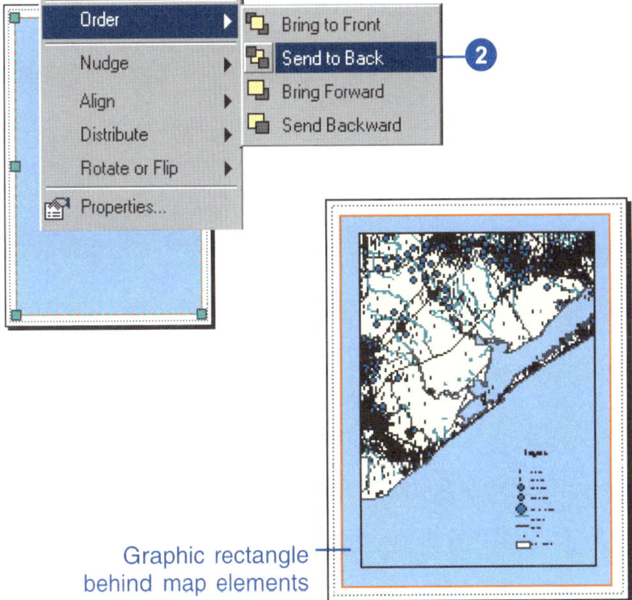

Adding a neatline

1. Click the Insert menu and click Neatline.
2. Click the Placement option you want.
3. Check Group neatline with element(s) if you want to group the elements with the neatline.
4. Click the Border dropdown arrow and click the type of neatline you want.
5. Click OK.

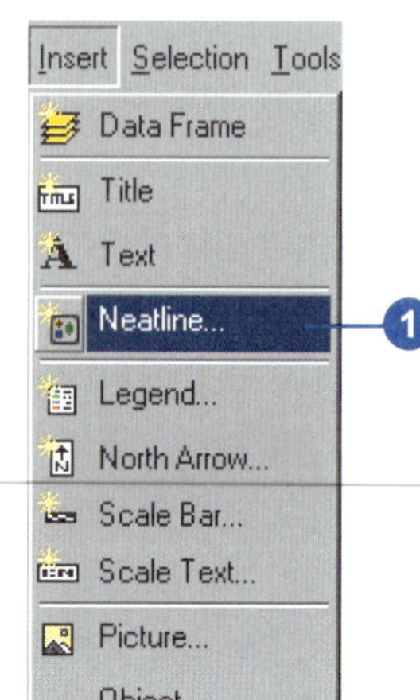

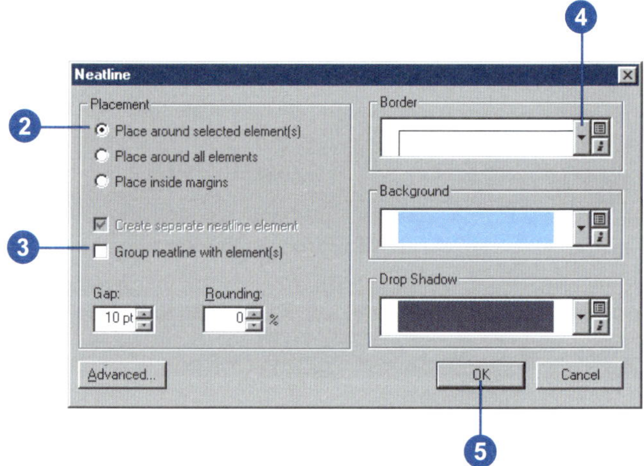

Tip

Stretching picture graphics

ArcMap keeps the same ratio of width to height (the aspect ratio) when you resize picture graphics, so they won't be distorted when you change their size. If you want to make a graphic wider or taller without changing its other dimension, right-click the graphic, click Properties, then uncheck Maintain aspect ratio on the Picture tab. You can then stretch the graphic.

Adding a picture

1. Click Insert and click Picture.
2. Navigate to the folder that contains the picture.
3. Optionally, select the type of picture that you want to add.
4. Click the picture that you want to add.
5. Click Open.
6. Click and drag the picture into position on your map.
7. Optionally, resize the picture by clicking a selection handle and dragging it.

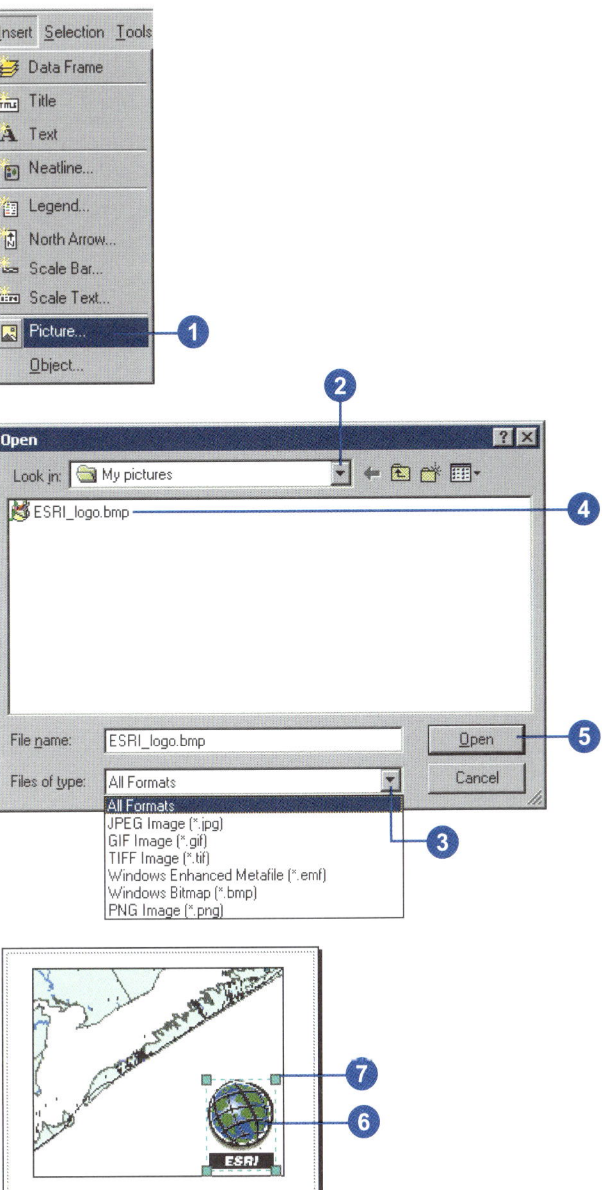

Aligning and grouping map elements

Sometimes you'll have multiple elements that you want aligned on the map. For example, you might want the left edges of two legends to be aligned. You can quickly align map elements by using the Align tools.

If you have multiple elements that you've placed relative to each other and you want them to maintain the same relationship when they're moved, you can group them together. The map elements will be treated as a single element until you ungroup them.

Tip

Snapping to the map's margins when you align elements

The Align tools can either align selected map elements with each other or with the margin of the map. If you click Align to Margins, the Align tools will snap selected elements to the margins of the map. Click Align to Margins again to make the Align tools align selected elements with each other.

Aligning map elements

1. Click one of the map elements to select it.
2. Hold the Shift key and click the other map element.

   Now both elements are selected.
3. Right-click one of the selected elements, point to Align, and click Align Center.
4. Click the aligned map elements and drag them into position on your map.

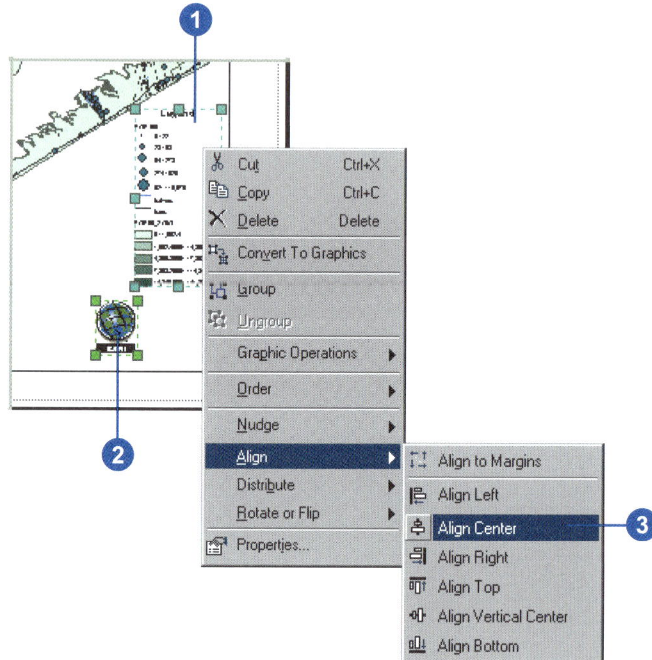

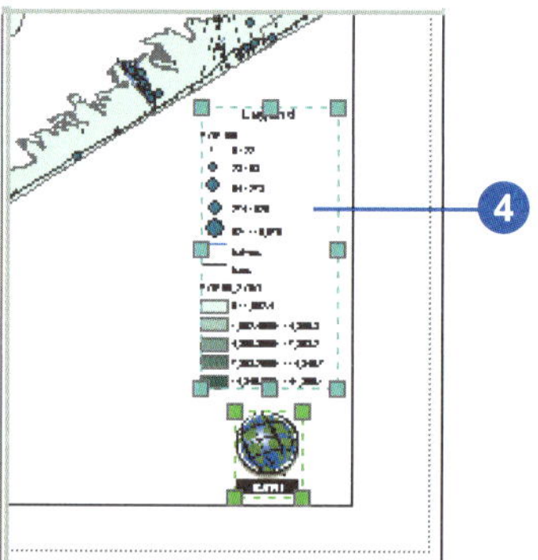

Tip

Ungrouping map elements

To ungroup map elements, select the group, right-click it, and click Ungroup.

Tip

Resizing groups of elements

If you group some elements, you can click the group and drag a selection handle to resize the group, just as you would a graphic.

The font size of text in a group will be scaled to match the new size of

See Also

For more information on aligning, distributing, positioning, and ordering map elements, see Chapter 7, 'Working with graphics and text'.

Grouping map elements

1. Click one of the map elements to select it.
2. Hold the Shift key and click the other map element.

   Now both elements are selected.
3. Right-click one of the selected elements and click Group.

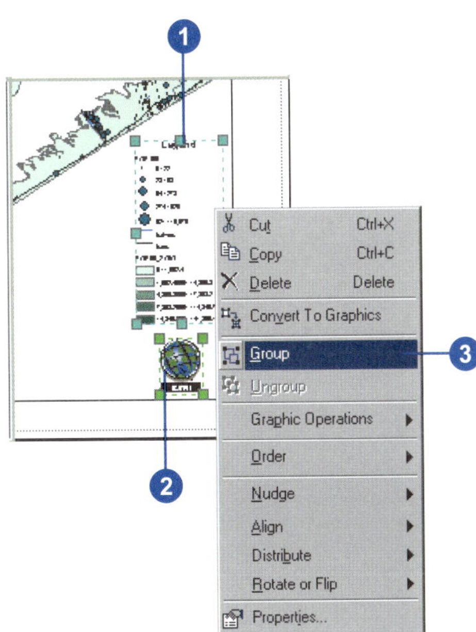

Printing a map

Once you've created a map, you'll probably want to print it.

If you want to print a map and you are not using the printer's settings, or you want to verify the map placement on a page, you may want to preview your map before you print it.

When printing a map that is set to be different from the printer page size, there are several scenarios.

If the map is smaller than your printer page size, go ahead and print or choose a smaller page size on your printer.

If the map is larger than the printer's default page size, you have several options. You can change the page size of the printer, change the printer that you're using, or change the page size of the map. ▸

Tip

Previews of maps with many features or raster data layers

Large maps with many features or raster layers may look excessively dark in the small print preview window. This is because feature sizes and rasters are not resampled for the preview. This will not affect the way your map appears when it is printed at full scale. Turn off most layers when previewing.

Previewing and printing a map

1. Click File and click Print Preview.
2. Examine the preview. If it looks right, click Print. ▸

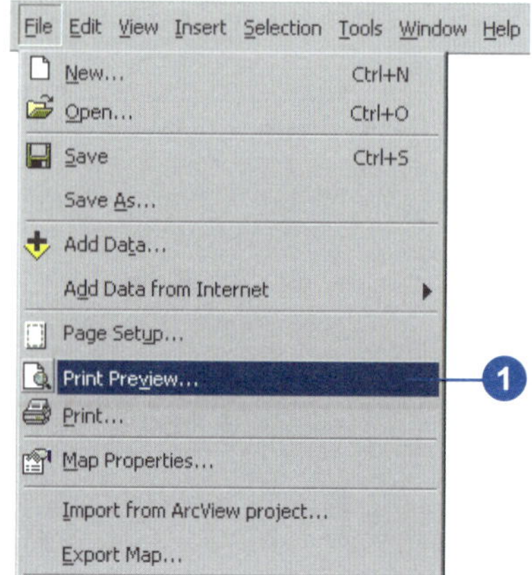

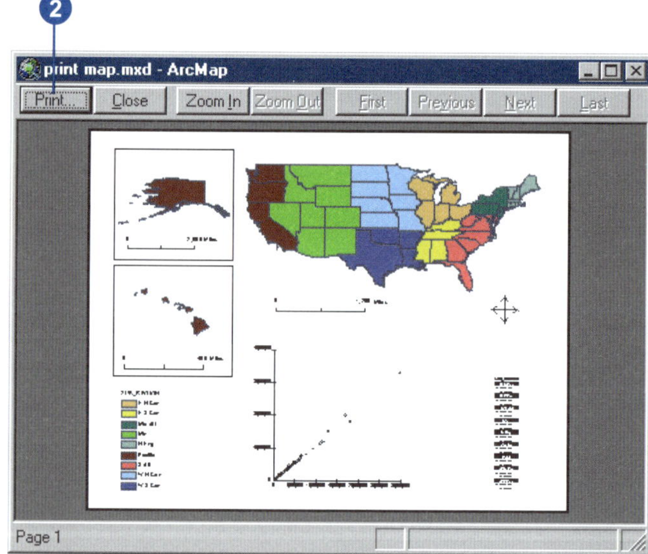

If your printer will accommodate a large page but its default page size is smaller than your map, change the Paper Size on the Page and Print Setup dialog box.

If your printer cannot accommodate a page the size of the map, you can print the map as tiles on separate pages, scale the map to the printer page size, or simply proceed with printing and clip the map to the printer's page size.

If you have another printer on the network that will print your large map on a single page, you can switch to that printer. Simply choose the new printer and the correct page size on the Page and Print Setup dialog box.

You can also change the page size of the map on the Page and Print Setup dialog box. See 'Setting up the page' earlier in this chapter.

Tip

Set page size the same as printer

If you have the page size set to be the same as the printer page, printing your map will be simple—just click File, click Print, then click OK.

3. Verify that you're printing to the correct printer with the printer engine you want.
4. Optionally, click the Setup button to display the Page and Print Setup dialog box and choose another printer engine. Click OK.
5. Click OK.

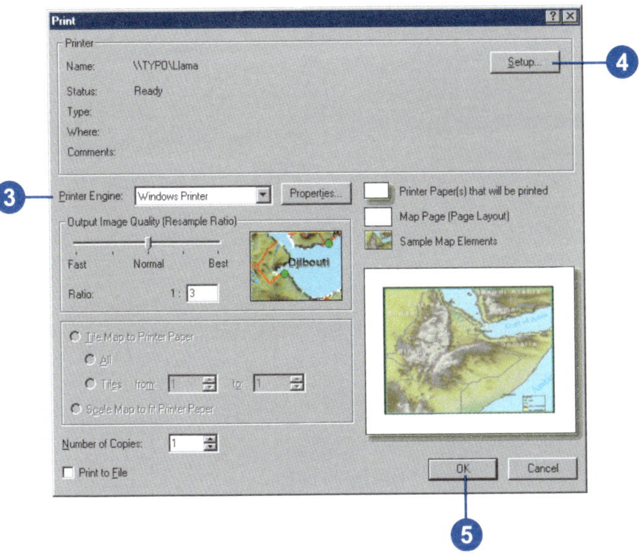

Tiling a map that's bigger than your printer page

1. Click File and click Print Preview.

   If the map page size is larger than the printer page size, you will see a portion of the map in the preview.

2. Click Print. ▸

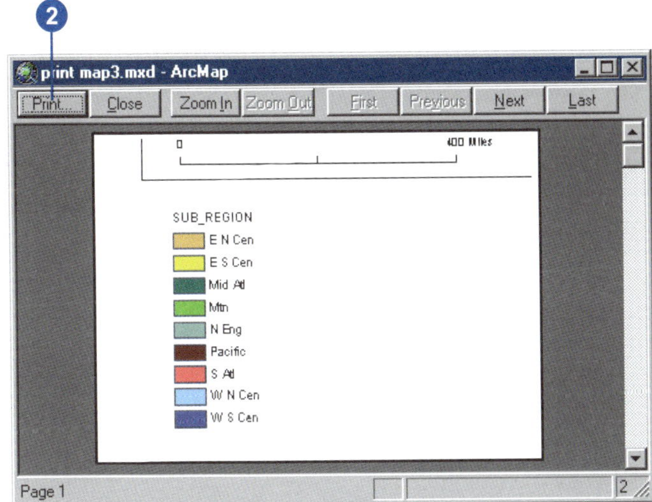

Tip

Using ArcPress

For large or complex maps or maps with rasters or transparent layers, the Windows and PostScript® printer engines may produce print files that are larger and more complex than your printer can handle.

If your printer is unable to print your map, you should purchase the ArcPress™ printer engine to use with ArcMap.

3. Click Tile Map to Printer Paper.
4. Click OK.

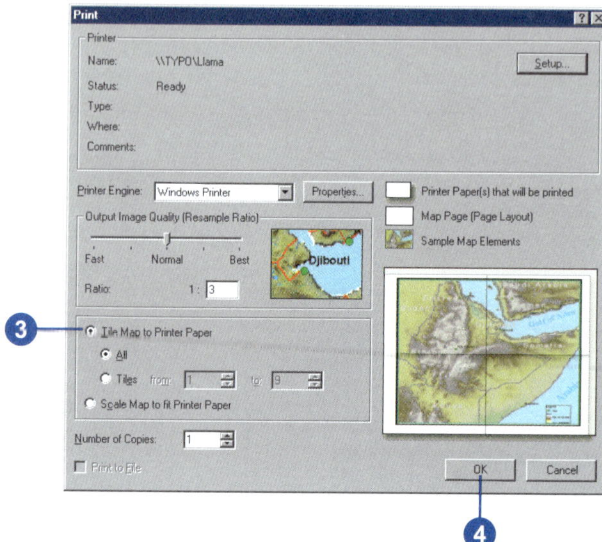

Printing multiple copies

1. Click File and click Print. ▸

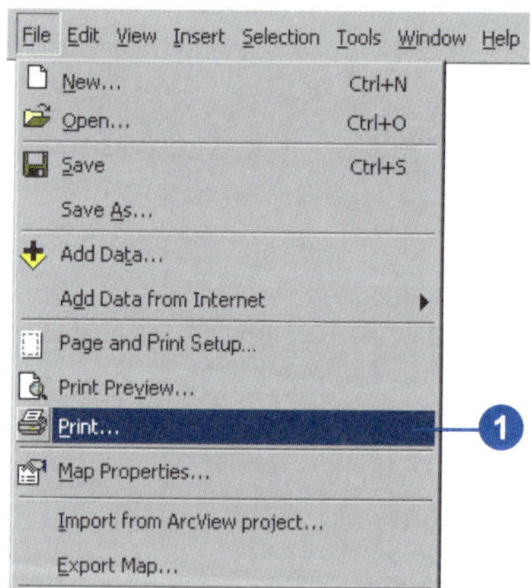

Tip

Setting the page orientation

If you have the page size set to be the same as the printer page, the page orientation will match the best fit for the selected printer and printer engine.

If you have a custom page size, the map's page orientation will be rotated according to the width and height of your map page. It's best to set the printer page orientation to match, or your map may be clipped.

2. Type the number of copies you want to print.
3. Click OK.

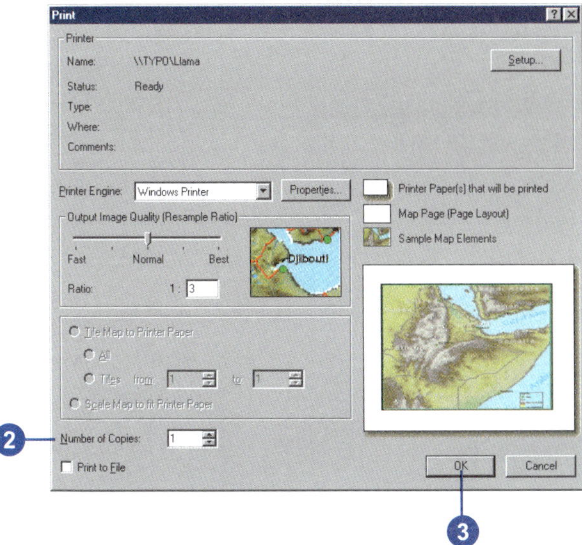

Printing to a file

1. Click File and click Print.
2. Check Print to File.
3. Optionally, click the Setup button to display the Page and Print Setup dialog box and choose another printer engine. Click OK.
4. Click OK. ▸

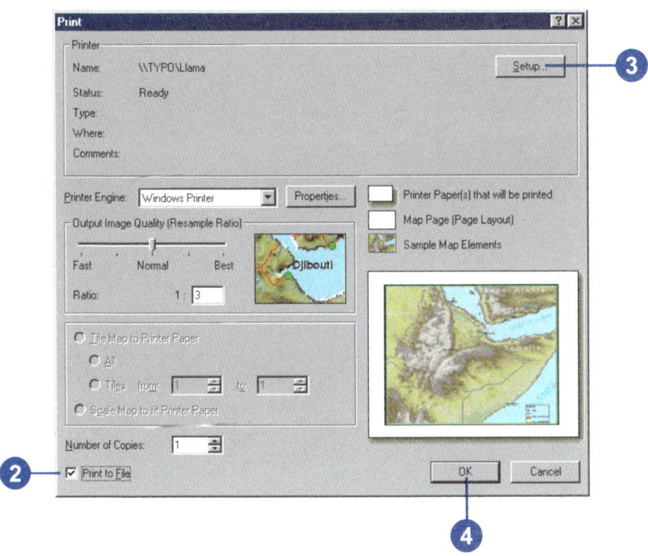

Tip

Print filename extensions

*If you choose the Windows printer engine, you'll create a Windows printer file, *.prn. If you choose the PostScript printer engine, you'll create a PostScript file, *.ps. If you choose the ArcPress printer engine, you'll create a raster file with the extension of the ArcPress driver you've selected.*

5. Navigate to where you want to save the output printer file.
6. Type a name for your print file.
7. Click Save.

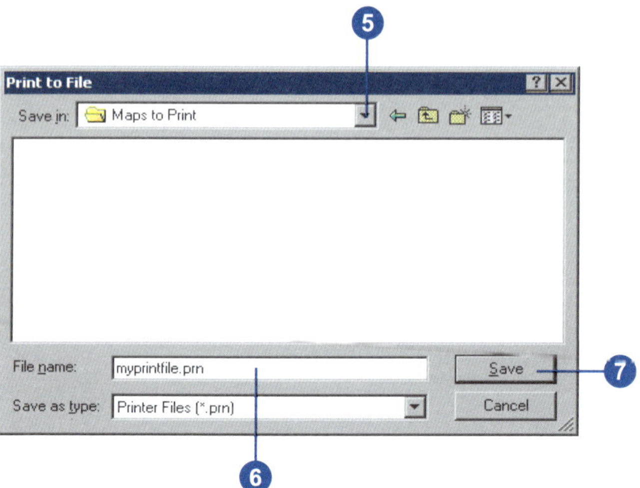

Changing the layout

A quick way to change the way your map looks is to change the layout so that it matches the layout of a template. To change the layout, use the Change Layout tool. This tool is ideally suited for maps that you haven't spent much time arranging and adding map elements to. For example, suppose you created an empty map, added a few layers to it, and symbolized the layers the way you wanted them to look. You could then use the Change Layout tool to arrange the map elements as defined in a layout.

When you change the layout, the resulting map will contain only those map elements that are defined in the template. For instance, if your original map has a legend but the template you want to use doesn't, the new layout will not contain a legend. In addition, any map elements will be formatted as they are defined on the template, not on the original map. Thus, if you've spent a long time setting the properties of the map elements, you probably won't want to use the Change Layout tool, as the settings will not carry over to the new layout.

Changing the layout

1. Click the Change Layout tool on the Layout toolbar.
2. Click the tab containing the template you want to use to change the layout.
3. Click the template you want to use.
4. Click Finish if the number of data frames on the map matches the number of data frames in the template.

   Click Next if the number of data frames on the map doesn't match the number of data frames in the template and proceed to the next step to arrange the data frames on the map.
5. Click the data frame you want to position on the map.
6. Click Move Up or Move Down to change its position in the list.

   Order the data frames to position them appropriately on the layout. If your map has more data frames than the template provides space for, the extra data frames will be positioned at the lower-left corner of the map.
7. Click Finish.

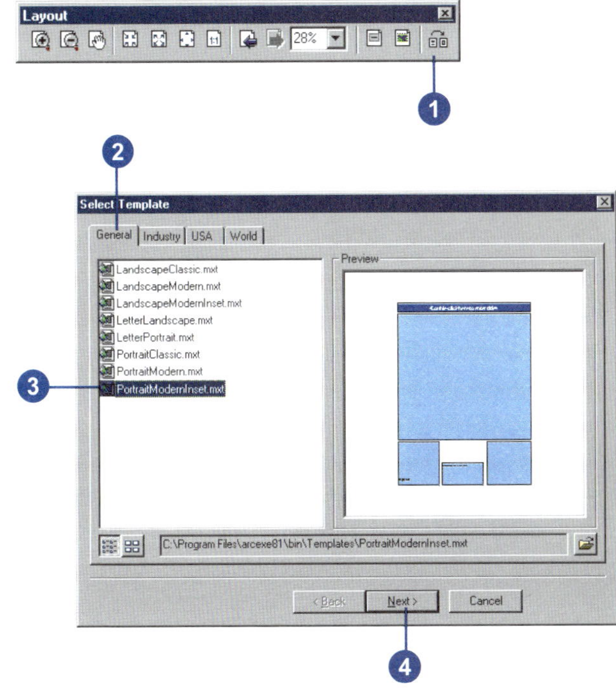

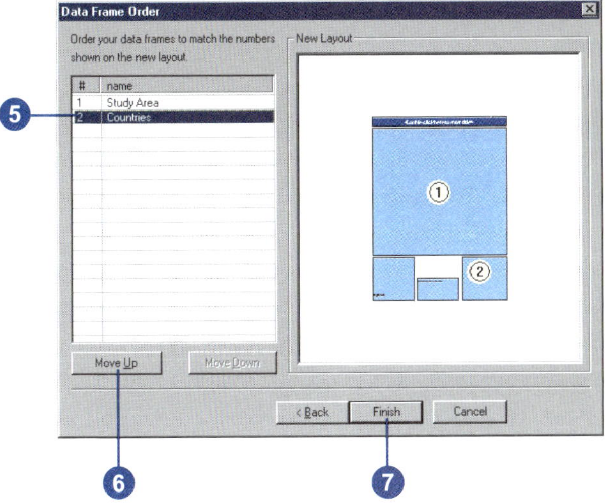

Exporting a map

Once you've created a map, you may want to export it from a map document to another file type.

You can export maps as several types of image formats. These include .emf, .eps, .ai, .pdf, .svg, .bmp, .jpg, .png, .tif, .gif.

EMF (Enhanced Windows Metafiles) files are Windows native vector, or vector and raster, graphics. They are useful for embedding in Windows documents because they can be resized without distortion.

EPS (Encapsulated PostScript) files are primarily used for vector graphics and printing.

AI (Adobe Illustrator®) files preserve most layers: annotation, labels, and data frame graphic text are exported to a layer called "Graphics"; map surround elements and graphic text elements are exported to an "Extras" layer; and raster data is exported to another "Extras" layer. All other data layers retain the same layer name as in ArcMap.

PDF files are designed to be consistently viewable across different platforms. They are commonly used for distributing documents on the Web.

SVG (Scalable Vector Graphics) files are XML-based files specifically designed for viewing on the Web. SVG can contain both vector and raster information. This is a good choice for displaying maps on a Web page because they are rescalable and more easily edited than raster files.

BMP files are simple, native Windows raster images. They do not scale as well as EMF files.

JPEG (Joint Photographic Experts Group) files are compressed image files. They are commonly used for images on the Web because they are more compact than many other file types.

PNG (Portable Network Graphics) files are a compressed format for the Web. They support 24-bit color and have the ability to define a transparent color.

TIFF (Tagged Image File Format) files can store pixel data at several bit depths and can be compressed with any of a selection of compression techniques. They are the best choice for importing into image editing applications across operating systems.

GIF (Graphics Interchange Format) files are a standard raster format for use on the Web. GIFs cannot contain more than 256 colors (8 bits per pixel), which makes them smaller than other file formats. They are a good choice for maps that contain a limited number of colors. GIF files also have the ability to define a transparent color.

Exporting a map

1. Click File and click Export Map.
2. Navigate to where you want to save the export file.
3. Click the Save as type dropdown arrow and click the type of file that you want to export.
4. Type a name for the export file.
5. Optionally, click the arrow to expand the Options and set the parameters for the file type that you chose.
6. Click Save.

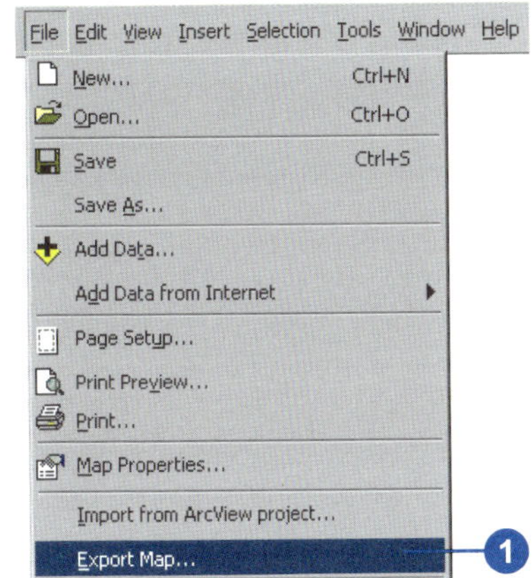

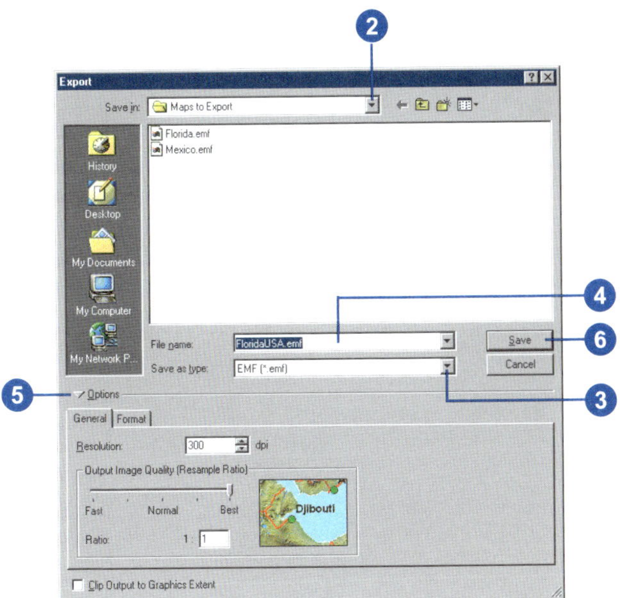

Customization

Section 5

Customizing ArcMap

16

IN THIS CHAPTER

- Basic user interface elements
- Hiding and showing toolbars
- Creating custom toolbars
- Changing a toolbar's contents
- Modifying context menus
- Changing a command's appearance
- Creating shortcut keys
- Saving customizations in a template
- Setting toolbar options
- Creating, editing, and running macros
- Creating custom commands with VBA
- Working with UIControls
- Locking documents and templates
- And more

Although ESRI end user applications are designed to be flexible and easy to use, you may want the ArcMap interface to reflect your own preferences and the way you work. If you work in a larger organization, others may want you to develop a customized work environment for them. As a developer, you'll be glad to learn that many of the customization tasks you may be asked to perform can be handled without writing a single line of code; in fact, you may be able to instruct others on how to use the customization environment themselves to create the look and feel they want on their own. You can change or create toolbars, menus, keystrokes, and so on, to help you get your work done in the most efficient way. Not only can you change the way the existing work environment is organized, but you'll also be able to provide additional functionality by linking code you or others have written to menu commands or tools. This chapter shows that the customization environment for ArcMap is rich with possibilities.

Basic user interface elements

ArcMap has a Main menu and a Standard toolbar, which appear by default. Both are referred to as *toolbars*, although the Main menu toolbar contains menus only. Toolbars can contain menus, buttons, tools, combo boxes, and edit boxes; these are different types of commands. Whether it's built into the application or it's something you've created yourself, code is associated with each *command*. All commands execute in generally the same manner, although you use each type differently when interacting with the application.

- *Menus* arrange other commands into a list. A context menu is a floating menu that pops up at the location of the pointer when you press the right mouse button.
- *Buttons* and *menu items* run a script when you click them.
- *Tools* require interaction with the display before an action is performed—that is, before their script is run. The Zoom In tool is a good example of a tool—you click or drag a rectangle over a map before seeing the contents of the rectangle in more detail.
- *Combo boxes* let you choose an option from a dropdown list. For example, in ArcMap you can choose the font used for the text you add to the map.
- *Text boxes* or *edit boxes* let you type in text. For example, in ArcMap you can type the scale at which you want to view the map.

The tasks in this book apply to all types of commands unless otherwise noted. Each task's description will note any exceptions that apply to specific types of commands.

Docking toolbars

Any toolbar can be docked at the top or bottom or to the left or right of the ArcMap window. Alternatively, toolbars can float on the desktop while functioning as part of the application. When you dock a toolbar, it is moved and resized with the application's window. To prevent a toolbar from docking, hold down the Ctrl key while dragging it.

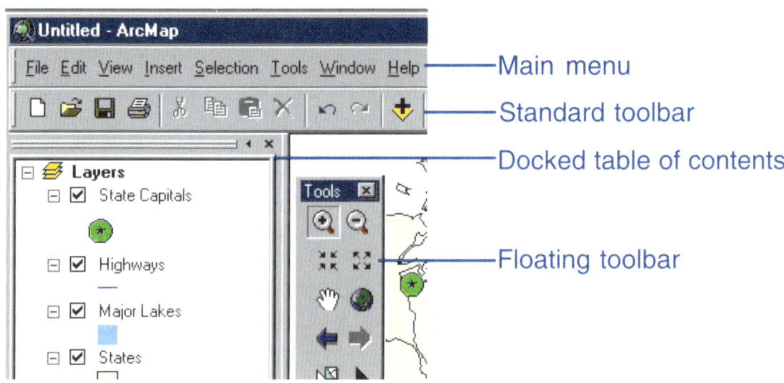

The table of contents in ArcMap is docked on the left by default, but you can dock it elsewhere in the window or have it float on the desktop if you prefer.

Changing the way the application looks

Whether you want to position toolbars in a specific area of the application, group commands in a way that works best for you, add new macros or load add-ins that you've gotten from another source, load styles, or always work with the same geographic data, you'll find that you can customize ArcMap in numerous ways. One of the principal ways to tailor the applications to suit your needs is to use the Customize dialog box to change menus and toolbars. You can carry out many of the tasks described in the rest of this chapter by starting with this dialog box. The Customize dialog box resembles and has many of the same properties as the equivalent dialog box in Microsoft Office applications. If you've used any of these applications, the environment will be familiar to you.

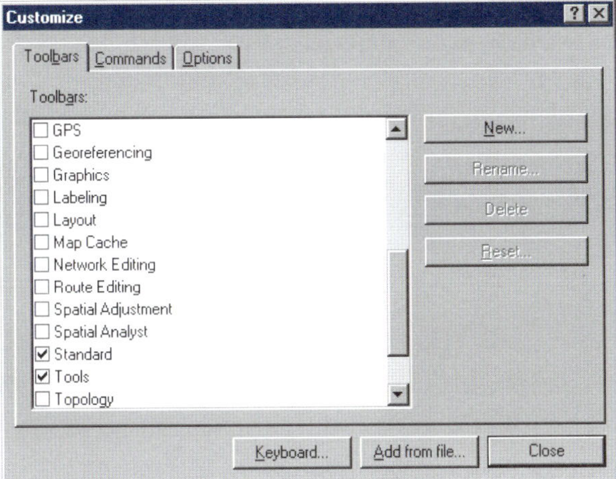

The Toolbars tab of the Customize dialog box

When you open the Customize dialog box, you can modify existing menus, toolbars, and context menus with simple drag-and-drop techniques. Afterward, if you prefer, you can return the menus and toolbars built into ArcMap to their default settings. You can also create your own menus, toolbars, and context menus.

Where to save your changes

When you make customization changes to the user interface in ArcMap, you can save your changes in one of the following three places:

- The current map document. There is always a map document open in ArcMap.
- A base template. A kind of map document that provides a quick way to create a new map. Templates can contain data, a custom interface, and a predefined layout that arranges map elements, such as North arrows, scalebars, and logos, on the virtual page. Map templates have a .mxt file extension. There may not be a base template loaded in ArcMap.
- The Normal template. A special template that is automatically loaded in ArcMap. This template stores any personal settings you have made to the user interface that you want loaded every time you use ArcMap.

When you first start ArcMap after installing the software, a Normal template is automatically created and put in your profiles location, which is one of the following folders depending on your operating system:

Windows NT

C:\WINNT\Profiles\<your username>\Application Data\ESRI\ArcMap\Templates\

Windows 2000 and XP

C:\Documents and Settings\<your username>\Application Data\ESRI\ArcMap\Templates\

This is the default out-of-the-box Normal template that contains all the standard toolbars and commands and places the toolbars and the table of contents in their default positions. Any customizations that you save in your Normal template get saved in this file. If you want to make changes that appear every time you open ArcMap, save them in the Normal template.

You might want to make changes that only appear when working with a particular map. For example, you might want your custom query and analysis toolbar to appear only in specific maps. In this case you would choose to save your customization in the current document. By default, all of your changes are saved in the Normal template. After you save a change in the current document, however, all subsequent changes will be saved in the current document by default.

Suppose you've created more than just a custom toolbar—you've created an entire environment with custom tools and macros that are used only when you edit a dataset's features. You can save this environment as a customized template. When you create a new map document, you can choose to base it on the Normal template or your custom template. For more information on saving customized templates, see 'Saving customizations in a template' later in this chapter.

Suppose there are custom toolbars or tools your administrator wants to make accessible to everyone in your organization. Your administrator could create a customized Normal template and allow everyone in your organization to use that Normal template instead of the default Normal template. To accomplish this, your administrator would customize the Normal template and copy that Normal.mxt file to the \ArcGIS\Bin\Templates folder. Everyone would then start with this Normal template instead of the default Normal template. The following is an explanation of how this works.

If there is no Normal.mxt file in your profiles location when you start ArcMap, the application will look in the \ArcGIS\Bin\Templates folder where ArcMap is installed. If a Normal.mxt file exists in the \ArcGIS\Bin\Templates folder, that file will be copied to your profiles location and is subsequently treated as your personal Normal template. Therefore, you start off with a copy of your organization's customized Normal template, but from that point on you are able to save your own customizations to it.

If a Normal.mxt file is not found in your profiles location or in the \ArcGIS\Bin\Templates folder, then a new default Normal.mxt file is created and placed in your profiles location.

Hiding and showing toolbars

In addition to the Main menu and the Standard toolbar, ArcMap has other toolbars that contain commands to help you perform a group of related tasks. ArcMap has buttons on the Standard toolbar for quickly displaying its most commonly used toolbars. You can hide or show toolbars using the Toolbars list on the View menu or the Customize dialog box. A check mark next to a toolbar name indicates it's visible. Note that the Main menu appears in the toolbars list but cannot be hidden. After checking a toolbar, the application displays it as a floating toolbar on the desktop. If the toolbar was previously turned on, it returns to its last position. Toolbar position and visibility are saved in the Normal template.

Tip

Shortcut to the Toolbars list

You can access the Toolbars list without using the View menu. Simply right-click any toolbar or the status bar.

Tip

Hiding floating toolbars

To quickly hide a floating toolbar, click its Close button.

Hiding and showing toolbars from the View menu

1. Click View and point to Toolbars.
2. Check a toolbar to show it.

   Uncheck a toolbar to hide it.

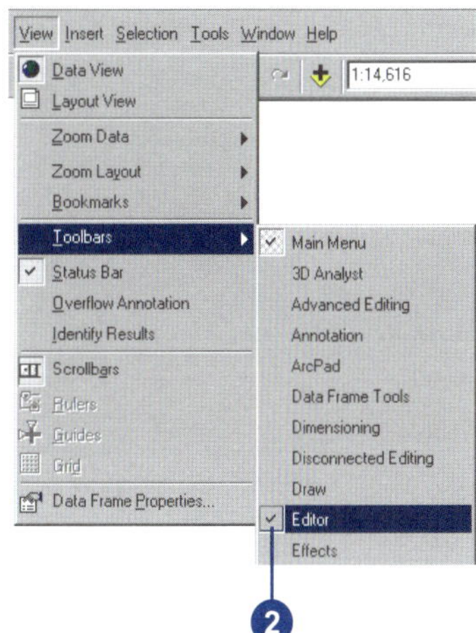

Toggling toolbars from the Customize dialog box

1. Click the Tools menu and click Customize.
2. Click the Toolbars tab.
3. Check a toolbar to show it.

   Uncheck a toolbar to hide it.
4. Click Close.

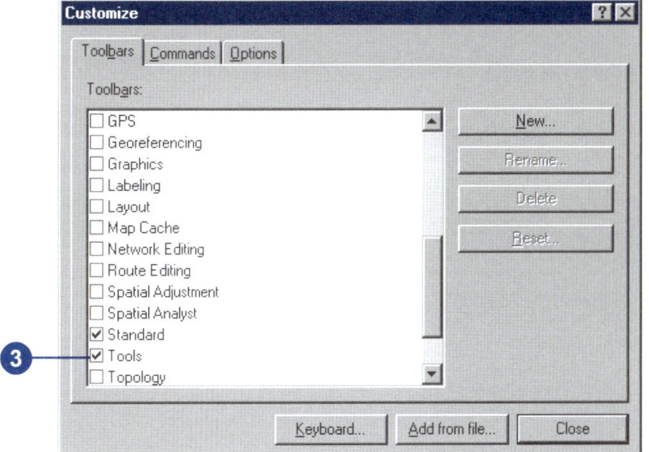

Creating custom toolbars

Several toolbars are provided with ArcMap, but you may want to create a new toolbar with buttons to run your custom scripts. You can rename or delete a toolbar created in ArcMap with the New button of the Customize dialog box; on the other hand, if the toolbar is built into the application or is part of an ActiveX® dynamic link library (DLL) that you added with the Add from file button, it cannot be renamed or deleted.

Creating a new toolbar

1. Click the Tools menu and click Customize.
2. Click the Toolbars tab.
3. Click New.
4. Type in the name of your new toolbar.
5. Click the dropdown arrow of the Save in combo box and choose the template in which this toolbar will be saved.
6. Click OK.

   The new, empty toolbar appears in the Toolbars list and is displayed in the application as a floating toolbar.
7. Click Close.

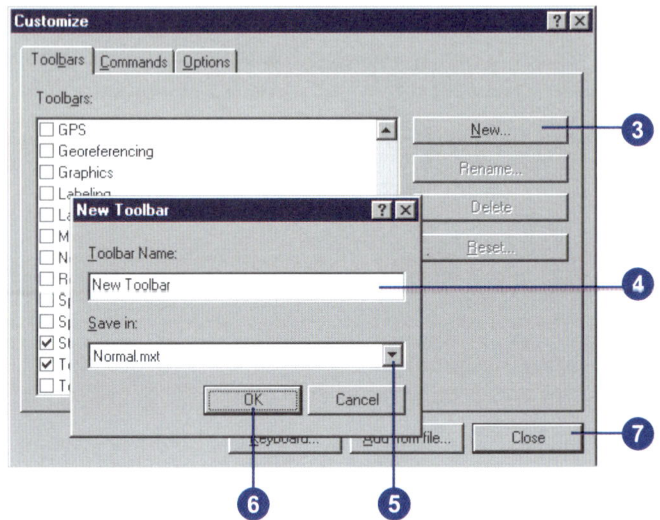

Renaming a toolbar

1. Click the Tools menu and click Customize.
2. Click the Toolbars tab.
3. Click the toolbar you want to rename.
4. Click Rename.
5. Type the name of your toolbar.
6. Click OK.
7. Click Close.

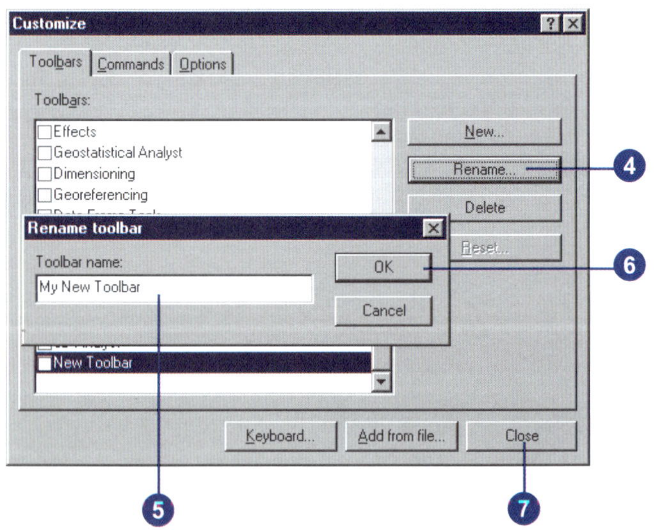

Deleting a toolbar

1. Click the Tools menu and click Customize.
2. Click the Toolbars tab.
3. Click the custom toolbar that you want to delete.
4. Click Delete.

   The toolbar is removed from the Toolbars list.
5. Click Close.

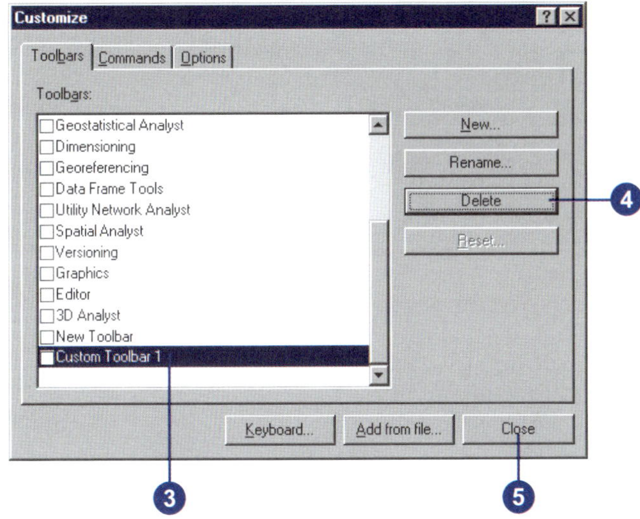

Changing a toolbar's contents

You can modify the contents of any toolbar by adding, moving, and removing commands. Grouping commands together on a toolbar can help to visually separate commands used for different tasks such as browsing and querying. After modifying a built-in toolbar, you can return it to its original contents; you might want to do this if you accidentally remove a command from the toolbar.

Tip

About the Save in combo box

The Save in combo box appears on the Commands tab, in the New toolbar dialog box, in the Reset toolbar dialog box, and in the Customize Keyboard dialog box. Use this setting to specify whether the change you are about to make will be saved in the Normal template, another template, or the current document.

Adding a command to a toolbar or menu

1. Click the Tools menu and click Customize.
2. Click the Toolbars tab.
3. Make sure the toolbar you want to change is checked.
4. Click the Commands tab.
5. In the Save in combo box, click the dropdown arrow and choose the template in which the changes to the toolbar will be saved.
6. Click the category that contains the command you want to add.
7. Click the command you want to add.
8. Drag the command you want to add to any location on the target toolbar.
9. Repeat steps 6 through 8 until all the commands you want are added.
10. Click Close.

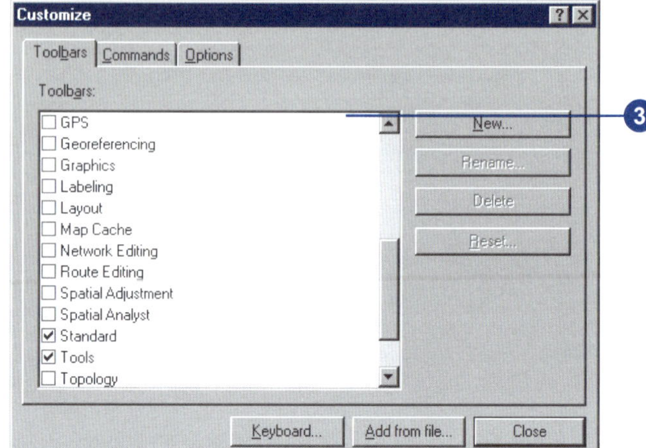

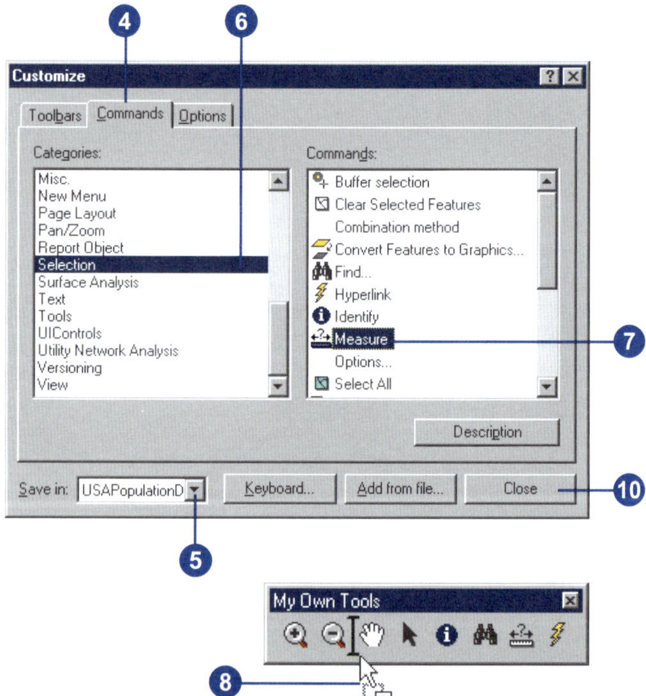

Tip

Creating access keys

All menus on the Main menu and their commands have an underlined character in their caption called an access key. It lets you access the menu from the keyboard by holding down Alt and pressing the underlined letter. To create an access key, place an ampersand (&) in front of a letter in the menu's (or the command's) caption.

Adding a new, empty menu to a toolbar

1. Show the toolbar you want to add a new, empty menu to.
2. Click the Tools menu and click Customize.
3. Click the Commands tab.
4. Click New Menu in the Categories list.
5. Click and drag the New Menu command from the Commands list and drop it on the toolbar.

   An empty menu called "New Menu" appears in the toolbar.
6. Right-click New Menu in the toolbar.
7. Type an appropriate caption for the menu in the text box.
8. Press Enter.
9. Click Close.

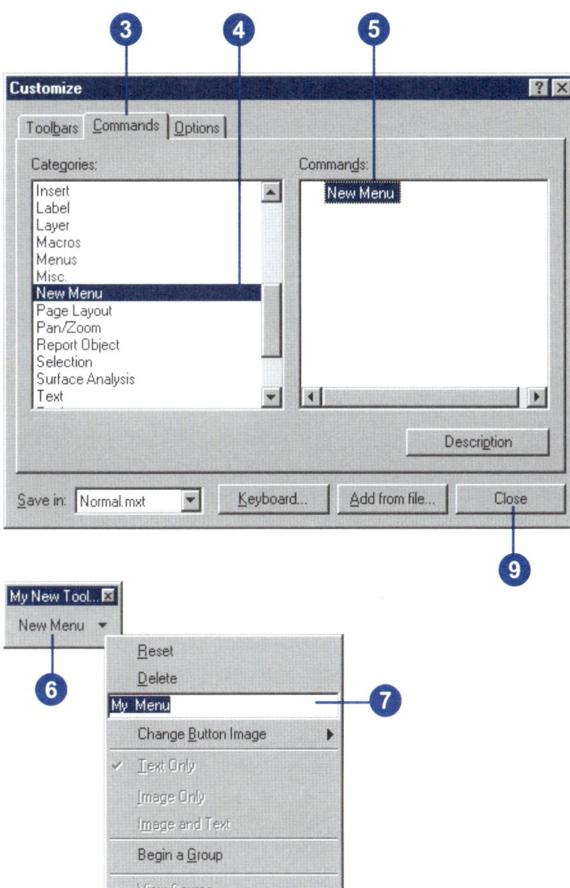

Modifying context menus

ArcMap contains several *context menus* to provide easy access to commands appropriate to the task at hand. By right-clicking, you'll see the built-in context menus. You can add a command to any of the listed context menus should your work require it.

Adding a command to a context menu

1. Click Tools, then click Customize.
2. Click the Toolbars tab.
3. Check the Context Menus toolbar.
4. Click Context Menus on the Context Menus toolbar.

   A list of all the context menus in the application appears.
5. Click the arrow for the context menu to which you want to add a command.

   The context menu's commands are listed.
6. Click the Commands tab in the Customize dialog box.
7. Click the category that contains the command you want to add to the menu.
8. Click and drag the command from the Commands list and drop it on the context menu.

   The command appears in the the context menu.
9. Click Close in the Customize dialog box.

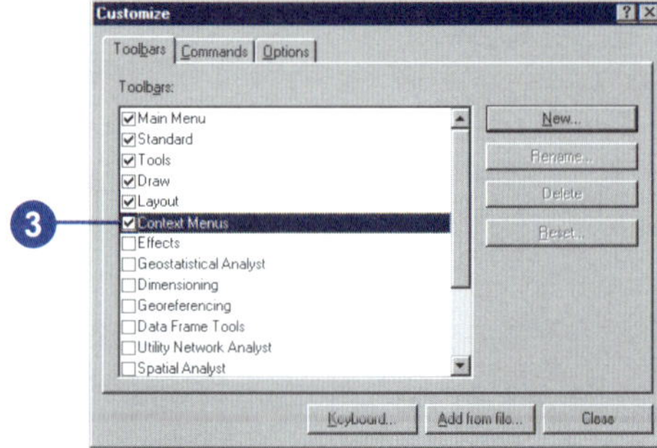

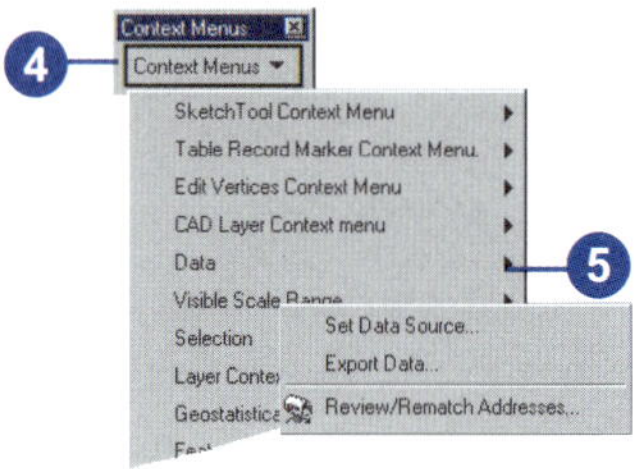

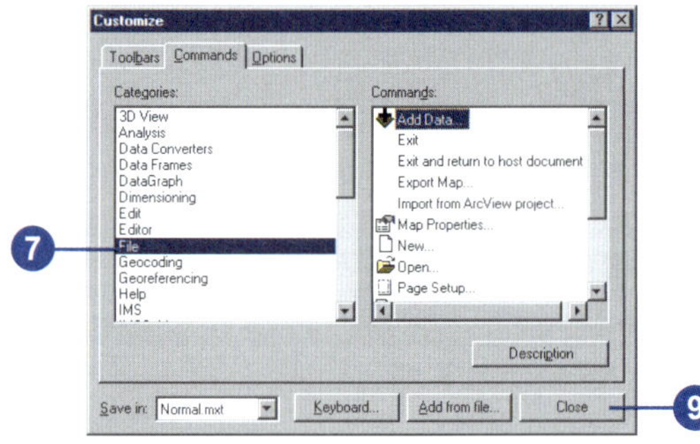

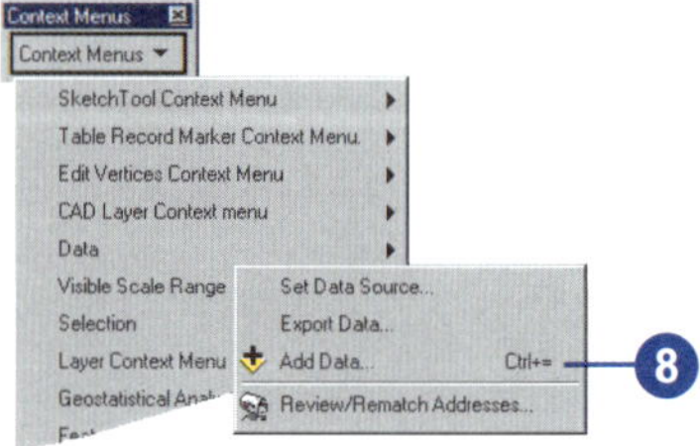

Moving a command

1. Show the toolbar with the command you want to move.
2. If you're moving the command to another toolbar, show the destination toolbar.
3. Click the Tools menu and click Customize.
4. Drag the command to its new position and drop it.

   The command appears in the new position.
5. Click Close in the Customize dialog box.

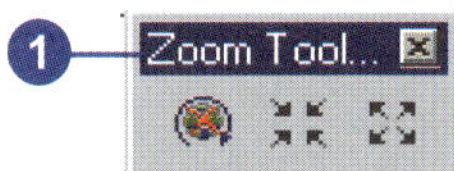

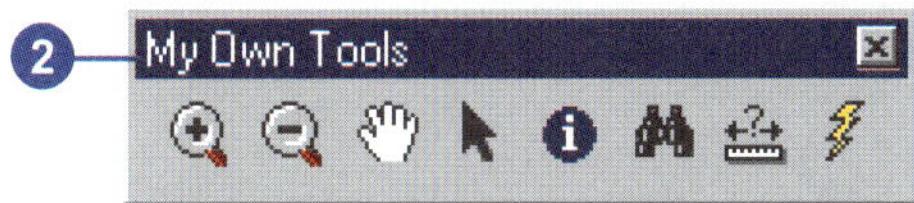

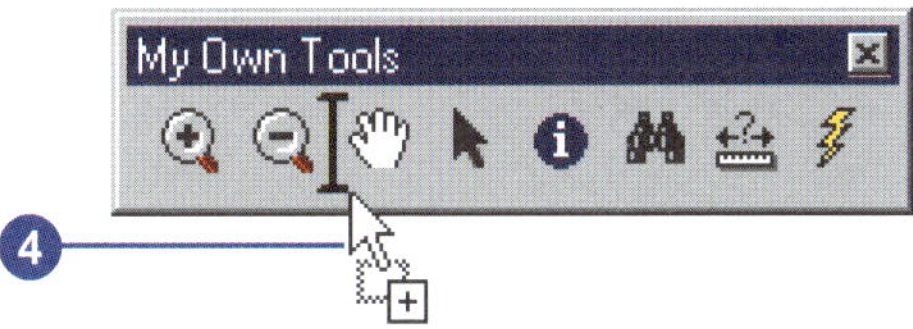

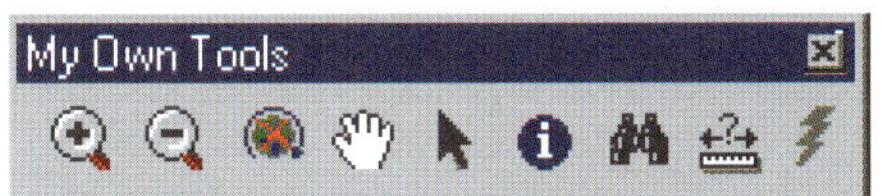

Tip

Removing commands

When you remove a command from a toolbar, you're not permanently deleting it—you're just making it unavailable on the toolbar. The command still appears in the commands list in the Customize dialog box. Later, you can always add the command to the same toolbar or to a different one.

Removing a command

1. Show the toolbar containing the command that you want to remove.
2. Click the Tools menu and click Customize.
3. Click and Drag the tool you want to remove from the toolbar.

   The mouse pointer changes to a linc through a circlc.
4. Drop the command.
5. Click Close in the Customize dialog box.

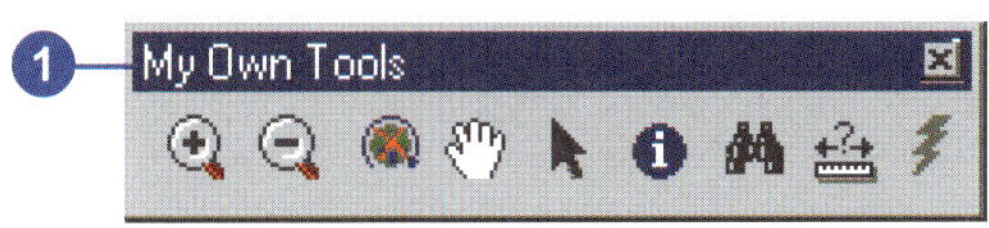

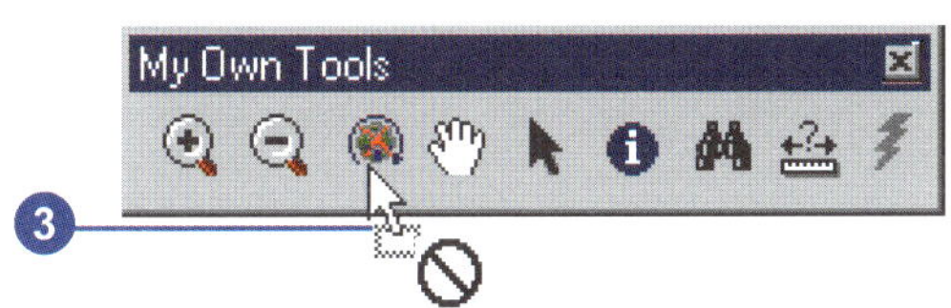

Tip

Why open the Customize dialog box?

Even though you don't make use of it in an operation such as grouping commands, you must display the Customize dialog box to place the application in a state in which you can make changes to its user interface.

Grouping commands

1. Show the toolbar containing the commands that you want to group together.
2. Click the Tools menu and click Customize.
3. On the toolbar, right-click the command located to the right of where the grouplng bar should be placed.
4. Click Begin a Group.

   A grouping bar appears in the toolbar.
5. Click Close in the Customize dialog box.

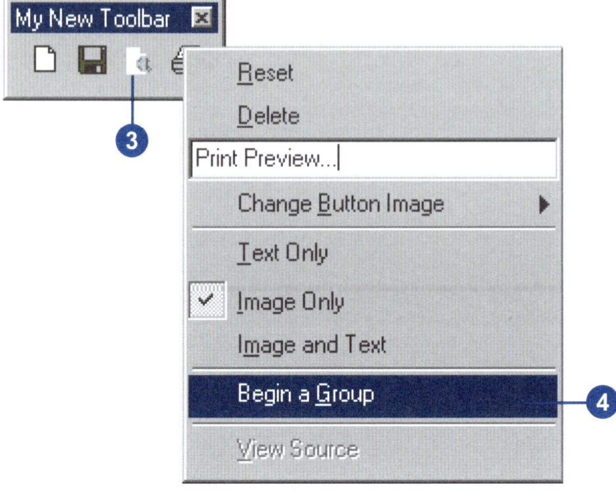

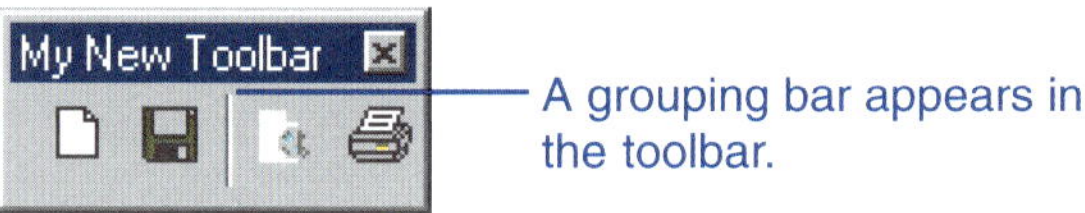

A grouping bar appears in the toolbar.

Resetting a built-in toolbar

1. Click the Tools menu and click Customize.
2. Click the Toolbars tab.
3. Click the built-in toolbar that you want to reset.
4. Click Reset.
5. Click the dropdown arrow and choose the template in which the changes to the toolbar settings were made.
6. Click OK.
7. Click Close.

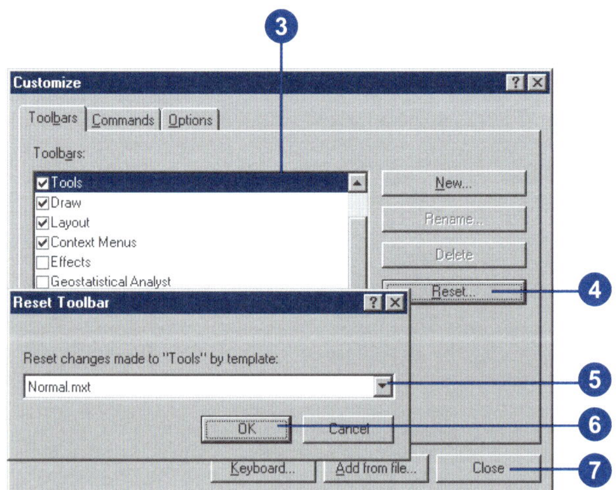

Changing a command's appearance

You can modify the display type, caption, and image of a menu, button, or tool without programming. By default, a button or tool dropped directly onto a toolbar has the display type Image Only, while it has the display type Image and Text when dropped onto a menu. Menus can only have the display type Text Only. The caption is the text that appears with the appropriate display types. Menus and their contents can be accessed from the keyboard by holding down Alt and pressing the underlined letter. Create one of these *access keys* by typing an ampersand (&) in front of a letter in the caption.

Other properties, such as ToolTip and Message, can only be modified with programming. When you hold the mouse pointer over a command, its ToolTip displays as a short message in a floating yellow box. A command's Message displays in the status bar.

Changing the display type

1. Show the toolbar containing the command whose display type you want to change.
2. Click the Tools menu and click Customize.
3. On the toolbar, right-click the command you want to change.
4. Check Image Only to display only the command's image.

   Check Text Only to display only the command's caption.

   Check Image and Text to display both.
5. Click Close in the Customize dialog box.

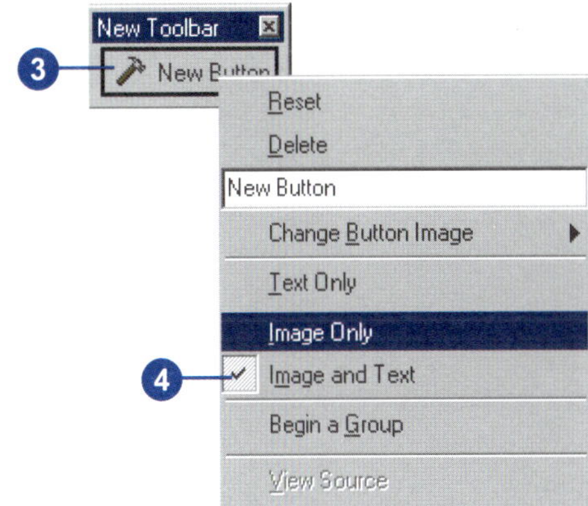

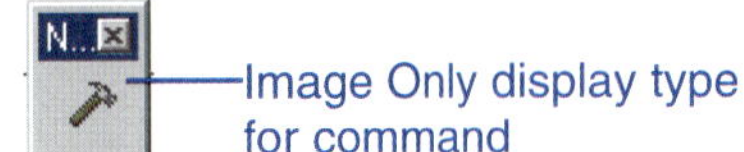

Image Only display type for command

Changing the caption

1. Show the toolbar containing the command whose caption you want to change.
2. Click the Tools menu and click Customize.
3. In the toolbar, right-click the command you want to change.
4. Type a new caption in the text box on the context menu.
5. Press Enter.

   The new caption is applied.
6. Click Close in the Customize dialog box.

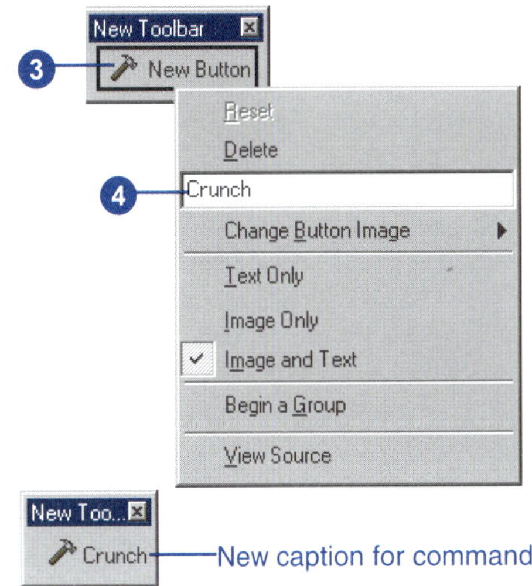

New caption for command

See Also

To learn how to set the properties of commands in code, see the ArcObjects™ Developer Help.

Changing the icon

1. Show the toolbar containing the command whose image you want to change.
2. Click the Tools menu and click Customize.
3. In the toolbar, right-click the command you want to change.
4. Point to Change Button Image.
5. Click one of the images displayed. Or click Browse, navigate to a custom image, and click Open.

   The new image is applied. It appears in the toolbar if the display type is Image Only or Image and Text.
6. Click Close in the Customize dialog box.

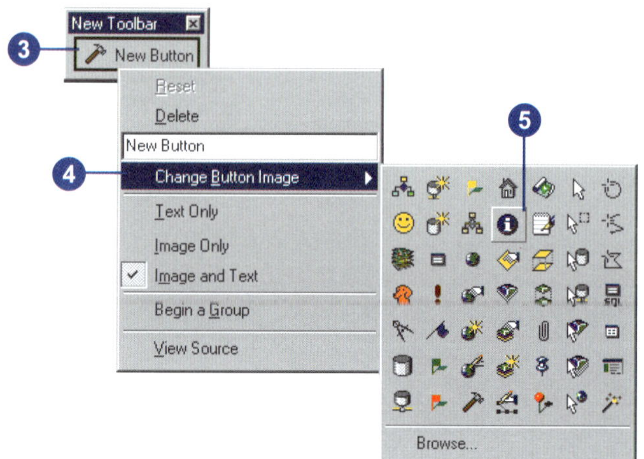

Resetting a built-in command

1. Show the toolbar with the command you want to reset.
2. Click the Tools menu and click Customize.
3. In the toolbar, right-click the command you want to change.
4. Click Reset.

   The command returns to its default settings.
5. Click Close in the Customize dialog box.

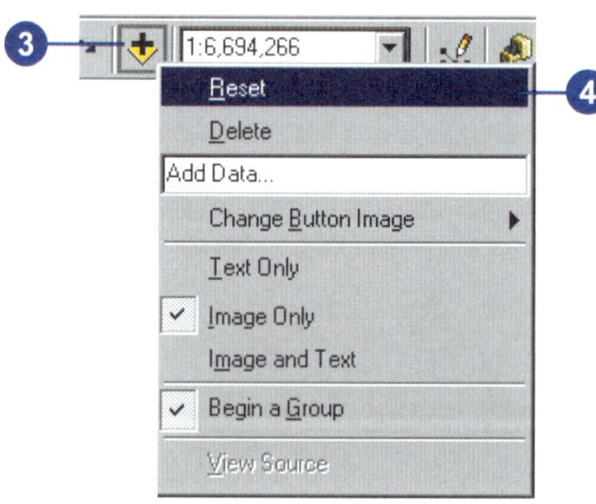

Creating shortcut keys

When you access a menu from the keyboard using its access key, the menu opens and you can see its contents. In contrast, a command's *shortcut key* executes the command directly without having to open and navigate the menu first. For example, Ctrl + C is a well-known shortcut for copying something in Windows. One command can have many shortcuts assigned to it, but each shortcut can only be assigned to one command. A command's first shortcut is displayed to its right when the command appears in a menu.

Assigning a shortcut key

1. Click the Tools menu and click Customize.
2. Click Keyboard.
3. Click the category containing the command you want to modify.
4. Click the command to which you want to add a keyboard shortcut.
5. Click in the Press new shortcut key text box, then press the keys on the keyboard that you want to use for a shortcut.

   If those keys have been assigned to another command, that command's name will appear below.
6. Click the dropdown arrow and choose the template in which the shortcut key will be saved.
7. Click Assign if the keys aren't currently assigned to another command.

   The new shortcut appears in the Current Key/s list.
8. Click Close in the Customize Keyboard dialog box.
9. Click Close in the Customize dialog box.

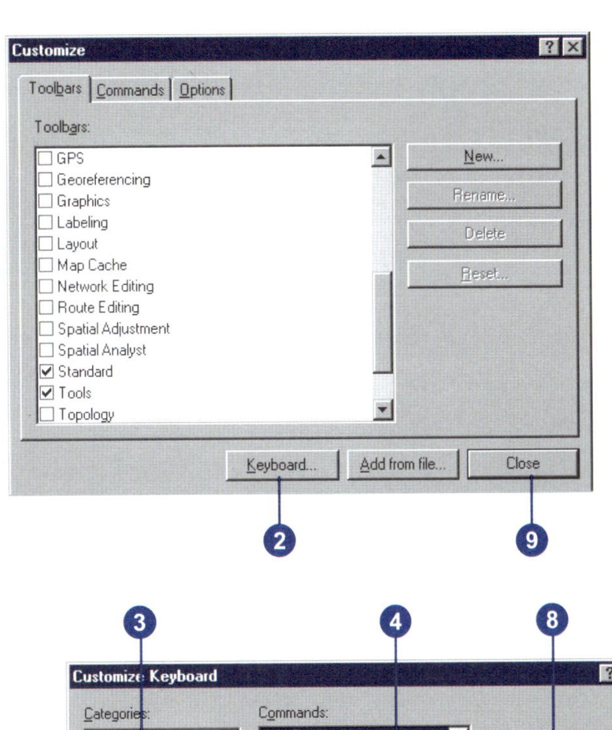

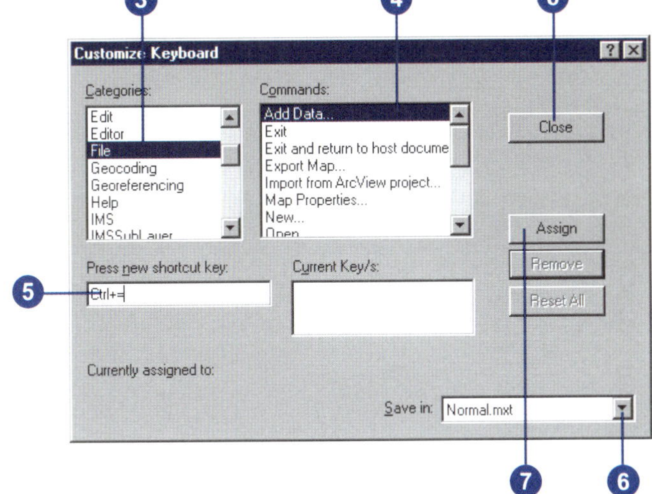

Removing a keyboard shortcut key

1. Click the Tools menu and click Customize.
2. Click Keyboard.
3. Click the category that contains the command you want to modify.
4. Click the command from which you want to remove a keyboard shortcut.
5. Click the dropdown arrow in the Save in combo box and choose the template from which to delete the shortcut key setting.
6. Click the shortcut in the Current Key/s list that you want to delete.
7. Click Remove.
8. Click Close in the Customize Keyboard dialog box.
9. Click Close in the Customize dialog box.

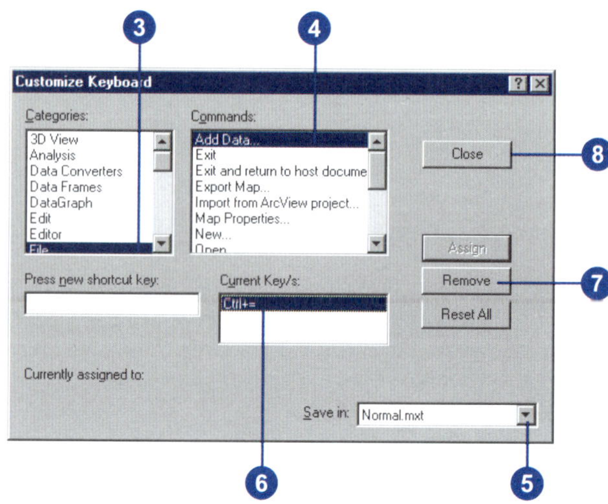

Resetting built-in shortcut keys

1. Click the Tools menu and click Customize.
2. Click Keyboard.
3. Click the dropdown arrow in the Save in combo box and choose the template whose shortcut keys will be reset.
4. Click Reset All.
5. Click Yes when asked if you want to reset your shortcuts.
6. Click Close in the Customize Keyboard dialog box.
7. Click Close in the Customize dialog box.

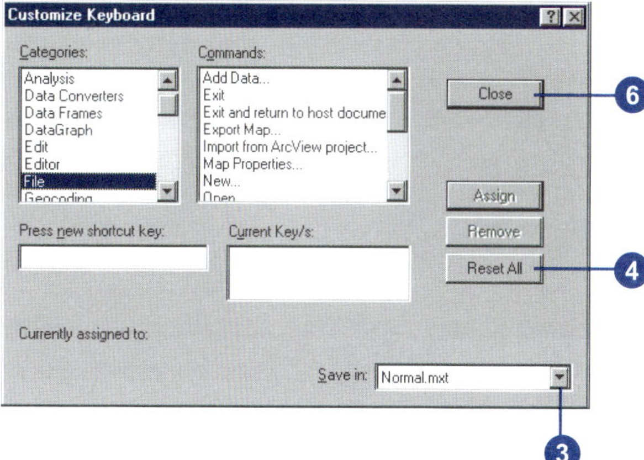

Saving customizations in a template

If you create a map that contains customizations or code you'd like to use as the basis for other maps, or if you modify an existing template and want to use it again, you can save it as a template. The template will contain all customizations that were made graphically as well as any modules created in the VBE.

You can save a map template anywhere on your network. When you want to use the template, you can open it from ArcMap.

If you save a template in the ArcMap Templates folder, the \ArcGIS\Bin\Templates folder where you have installed ArcMap, it will show up in the list of templates on the New map document dialog box. You can also create subfolders in this folder, and they'll show up as separate tabs on this dialog box—when you click each tab you'll see the templates in that folder.

Saving a template

1. Click File and click Save As.
2. Click the dropdown arrow and click ArcMap Templates.
3. Navigate to the folder where you want the template saved (for example, the default templates folder \ArcGIS\Bin\Templates).
4. Type a name for the new template.
5. Click Save.

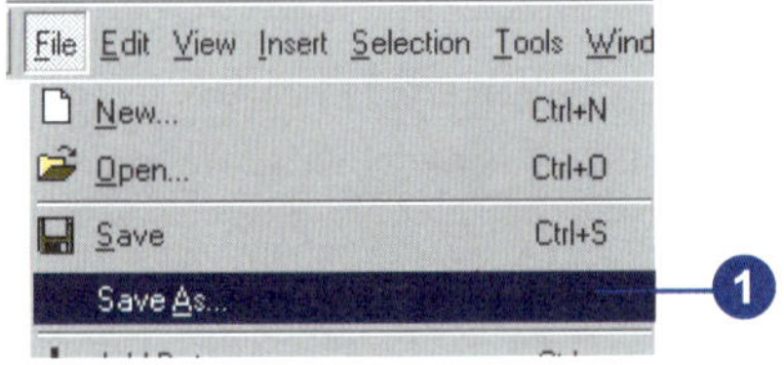

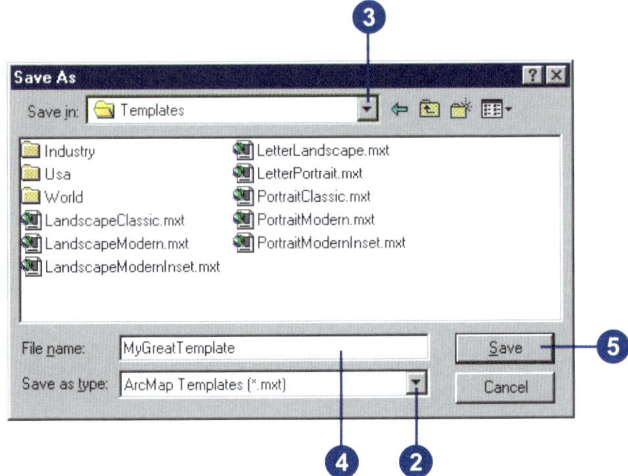

Saving a template so it will appear in a new tab

1. Click File and click Save As. ▸

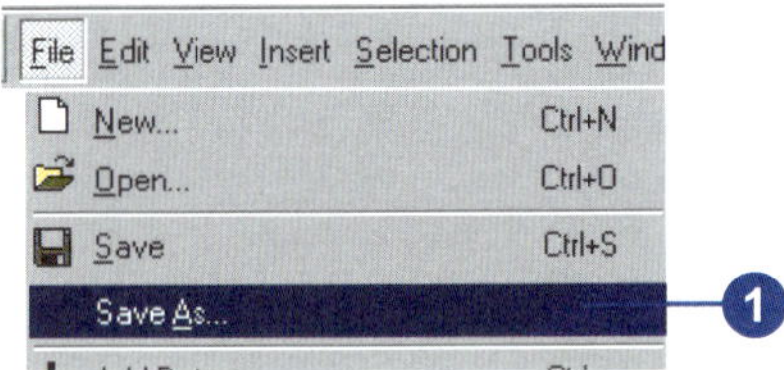

Tip

Regenerating Normal.mxt

If you save customizations to the ArcMap Normal.mxt template, and later decide that you'd like to reset the entire template to its original settings, delete the file. ArcMap will regenerate Normal on startup if it is missing.

Tip

Making changes to an existing template

If you want to open a template as a document in order to make changes to it, use Open in the File menu. If you open the template using the Open an existing map option on the Startup dialog box or by double-clicking on the template filename, you'll create a new document based on your template instead of just opening your template as a document. This new document has a reference to your template. If you were to try to save this document as your template again, you'll get a save error because essentially you are trying to create a template that references itself.

2. Click the dropdown arrow and click ArcMap Templates.
3. Navigate to the Templates folder.
4. Click the New Folder button.
5. Type the name of the new folder—this name will appear on the New map document dialog box as a tab.
6. Double-click the new folder.
7. Type the name of the new template.
8. Click Save.

   The next time you start a map from a template, you'll see a new tab with your template on the New map dialog box.

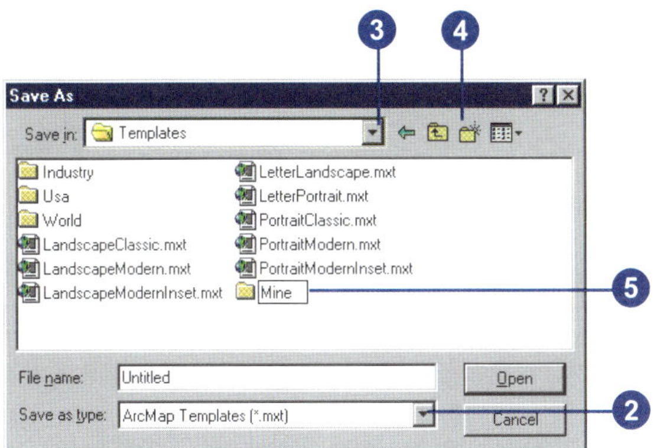

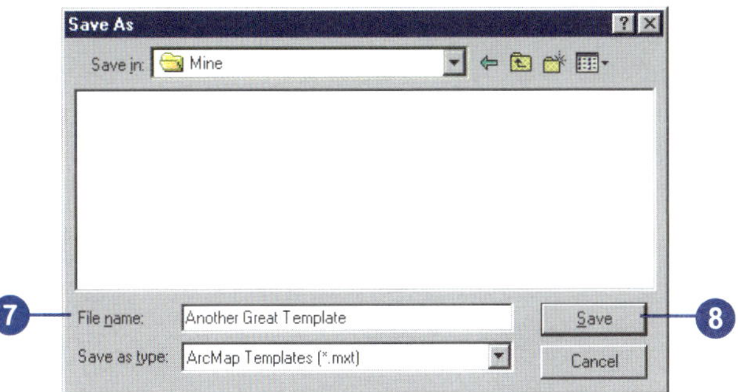

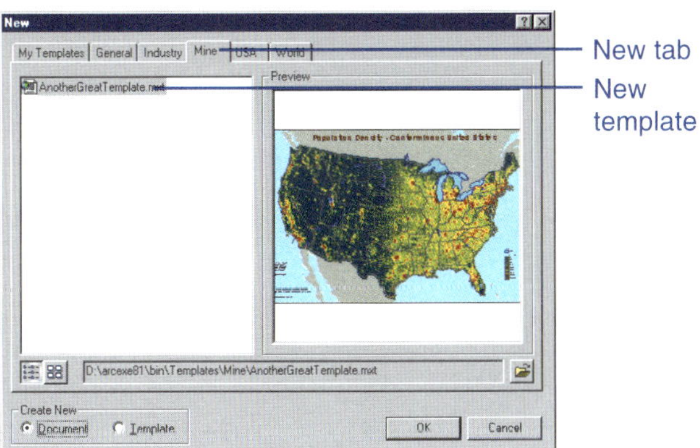

Changing where customization changes are saved by default

By default, all customization changes you make to the user interface get saved to the Normal template, unless you select either another template or the current document in the Save in combo box before making the change. You may prefer to save all of your customizations to the current document by default. This means that unless you specify differently in the Save in combo box, all of your changes will get saved in the current document.

Tip

Default document or template

Once you choose the document as the default for customization changes, this default setting will be in effect every time you start ArcMap—that is, until you check Save customizations to Normal template, making the Normal template the default template again.

Saving customizations to the document by default

1. Click the Tools menu and click Customize.
2. Click the Options tab.
3. Uncheck Save customizations to Normal template by default.

   Changes are now saved to the current document by default.
4. Click Close.

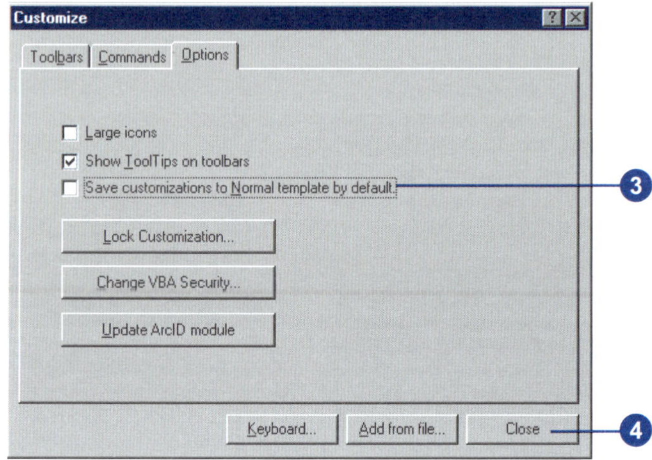

Setting toolbar options

The Options tab in the Customize dialog box lets you specify the size of icons on commands and whether ToolTips will appear on all the toolbars in ArcMap when you hold the mouse pointer over a command.

Later, this chapter shows how the Options tab also provides a means to lock or unlock the Customize dialog box, the Macros dialog box, and the Visual Basic Editor. In addition, you can use the Options tab to change VBA security and update the status of the Normal template's ArcID module.

Displaying toolbars with large icons

1. Click the Tools menu and click Customize.
2. Click the Options tab.
3. Check Large icons to display large icons for a toolbar's commands.
4. Click Close.

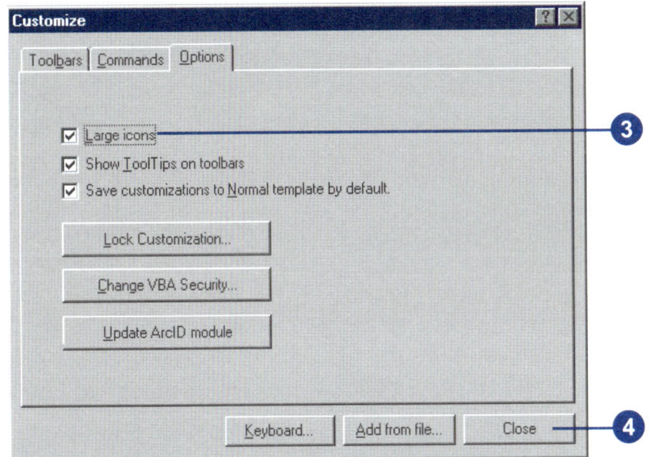

Showing ToolTips on toolbars

1. Click the Tools menu and click Customize.
2. Click the Options tab.
3. Check Show ToolTips on toolbars to display ToolTips for the commands on a toolbar.
4. Click Close.

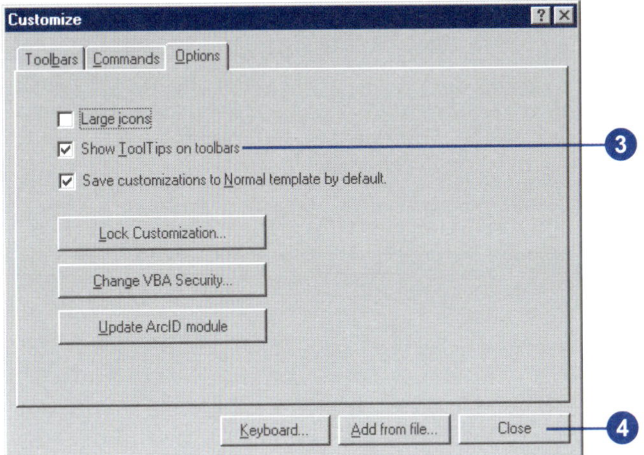

Creating, editing, and running macros

ArcMap comes with *VBA*. VBA is not a standalone program. It provides an integrated programming environment, the VBE, that lets you write a VB *macro* and debug it right away in ArcMap. A macro can integrate some or all of VB's functionality with the extensive object library available through ArcMap. The ESRI Object Library is always available to you in the VBA environment.

In the VBE in ArcMap, there can be up to three VBA projects. The document, the base template, and the Normal template all have a VBA project. The VBA project for the Normal template is called Normal (Normal.mxt). The VBA project for the current document is called Project (<NameofDoc>.mxd). The VBA project for the base template is called TemplateProject (<NameofTemplate>.mxt). Macros can be stored in any of these VBA projects depending on where you want the code to be available. ▸

Creating a macro in the Visual Basic Editor

1. Click the Tools menu, point to Macros, then click Macros.
2. Click the dropdown arrow in the Macros in combo box, then click the document or template in which you want to create this macro.
3. Type the name of the macro you want to create in the Macro name text box.
4. Press the Enter key or click Create.

   The stub for a Sub procedure for the macro appears in the Code window.

   If you don't specify a module name, the application stores the macro in a module named NewMacros.
5. Type the code for the macro.
6. Click the VBE File menu and click Save Project.

Preceding the name of a macro with a module's name and a dot stores it in the specified module.

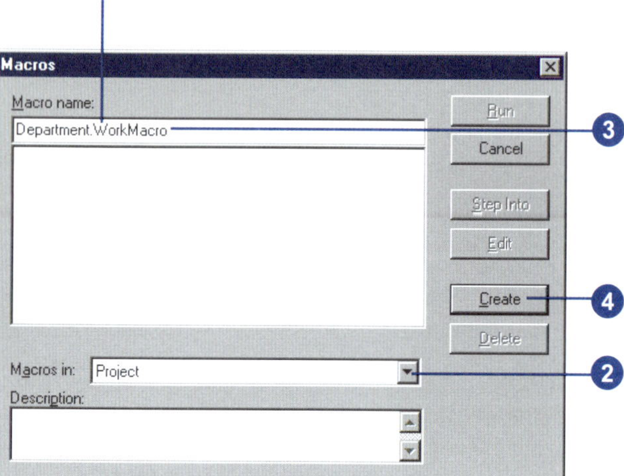

When you create a macro, you're creating a VB Sub Procedure. The procedure's name is the name you assign to the macro. You add code to the procedure in a Code window just as you would in VB. You can organize your macros in different modules; each module has its own Code window. To add your macro to a specific module, type the module name before the macro's name, for example, "Department.WorkMacro". If the module doesn't exist, a new module with that name is created for you and added to the VBA project. Similarly, if you provide a name for a new macro, but don't specify which module to store it in, a new module "NewMacros" is created. Using modules makes it easier to share your VB code with others. You can export a module to a .bas file from, and import a .bas file to, your VBA project.

Editing a macro in the Visual Basic Editor

1. Click the Tools menu, point to Macros, then click Macros.
2. In the list below the Macro name text box, click the name of the macro you want to edit.
3. Click Edit.

   The code that's been written for the macro appears in the Code window.
4. Edit the code.
5. Click the VBE File menu and click Save Project.
6. Click Close to close the VBE.

Running a macro in the Visual Basic Editor

1. Click the Tools menu, point to Macros, and click Visual Basic Editor.
2. In the VBE Project window, double-click ThisDocument or the module containing the macro that you want to run.

   The Code window for that module appears.
3. Position the cursor inside the appropriate Sub procedure.
4. Click the VBE Run menu and click Run Sub/UserForm.

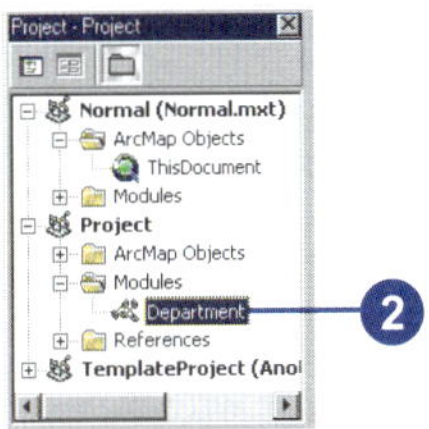

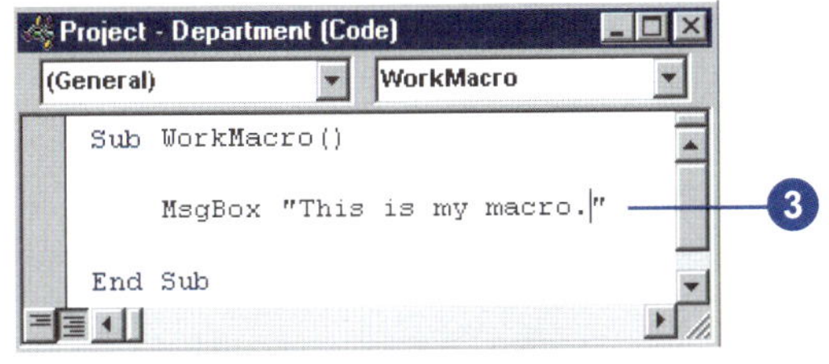

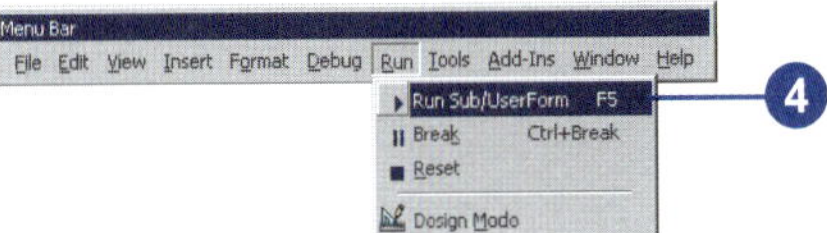

Running a macro in the Macros dialog box

1. Click the Tools menu, point to Macros, then click Macros.
2. Click the Macros in dropdown arrow and choose the document or template containing the Macro you want to run.
3. Type the name of the macro you want to run or select it from the list that appears.
4. Click Run.

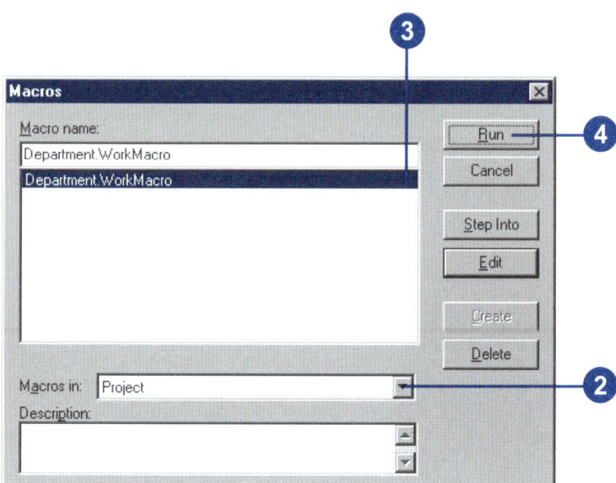

Adding a macro to a toolbar or menu

1. Show the toolbar to which you want to add a macro.
2. Click the Tools menu and click Customize.
3. Click the Commands tab.
4. Click Macros in the Categories list.
5. Click and drag the macro from the Commands list and drop it on the toolbar.
6. Click Close.

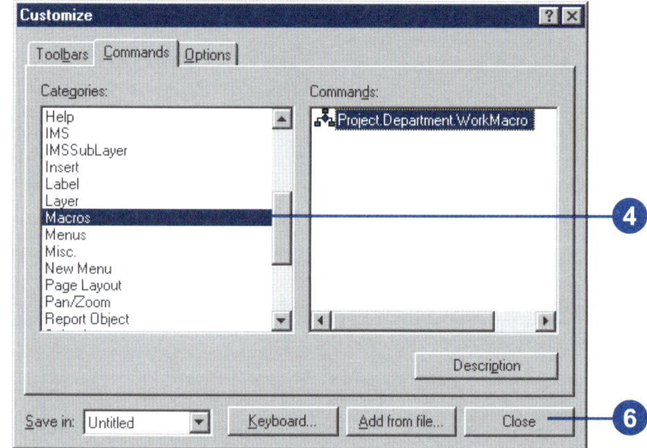

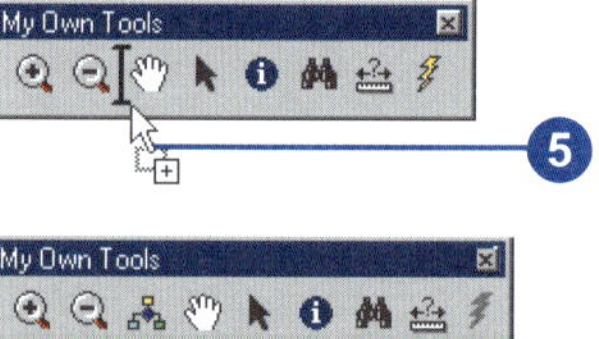

Creating custom commands with VBA

An extensive object library is available in VBA in ArcMap. For example, ArcMap exposes a Map, a PageLayout, a LineFillSymbol, and so on. The ArcObjects Developer Help describes the classes, interfaces, properties, methods, and enumerations that are available in the development environment that is built into ArcMap.

Toolbars and commands are *Component Object Model (COM) objects*, too. To be a command, the object must meet a basic set of requirements for all commands. To be a tool, the object must also satisfy tool requirements. The customization environment makes it easy for you to create custom commands with VBA. You create a new button, tool, combo box, or edit box (collectively called *UIControls*) in the Customize dialog box, then attach code to that object's control events. After you have created it, you can drag this new control onto a toolbar.

Creating a new command

1. Show the toolbar to which you want to add a new command.
2. Click the Tools menu and click Customize.
3. Click the Commands tab.
4. Click the dropdown arrow on the Save in combo box and click the document or template in which the new command will be saved.
5. Click UIControls in the Categories list.
6. Click New UIControl.
7. Click the type of UIControl you want to create.
8. Click Create to create the control without attaching code to it.

   The name of the control appears in the commands list. You can add code for the control at another time. If you want to start adding code to the control right away, click Create and Edit and skip to step 13.
9. Click the newly created UIControl, click it again to activate in-place editing, and type a new name for the UIControl. ▸

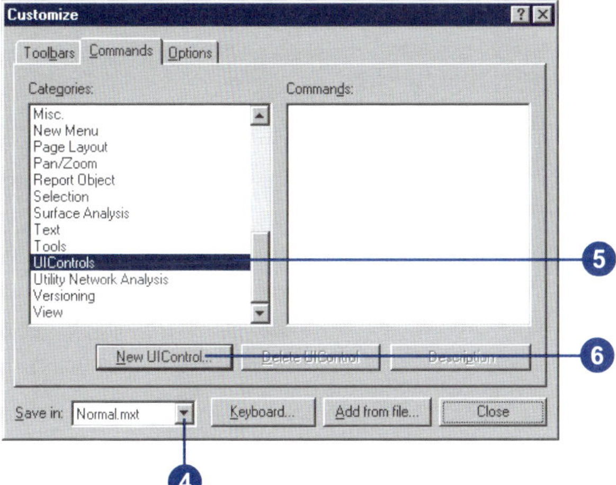

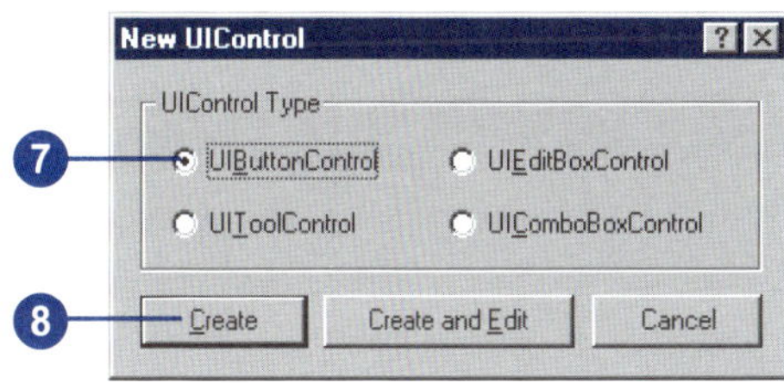

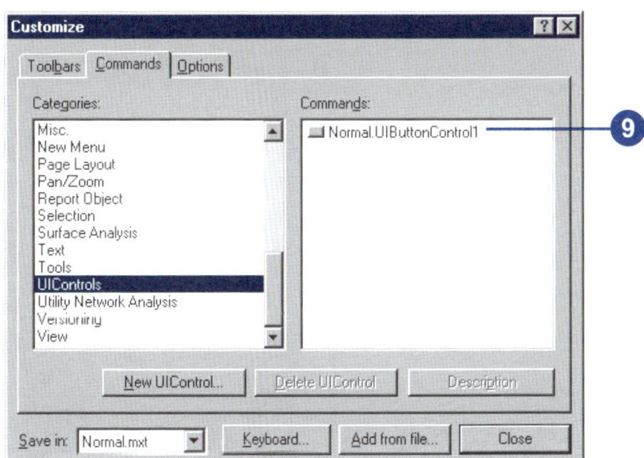

10. Click and drag the newly created UIControl and drop it on a toolbar or menu.
11. On the toolbar or menu, right-click the control to set its image, caption, and other properties.
12. Right-click the new control and click View Source.

    The Visual Basic Editor appears, displaying the control's code in the code window.
13. Click the Procedures\Events dropdown arrow and click one of the control's event procedures.
14. Type code for the event procedure.
15. Repeat steps 13 and 14 until all the appropriate event procedures have been coded.
16. Click Save in the Visual Basic Editor.
17. Click the Close button in the Visual Basic Editor.
18. If you clicked Create and Edit in step 8, open the Customize dialog box, click the Commands tab, and drag the newly created UIControl from the Commands list to a toolbar or menu.
19. Click Close in the Customize dialog box.

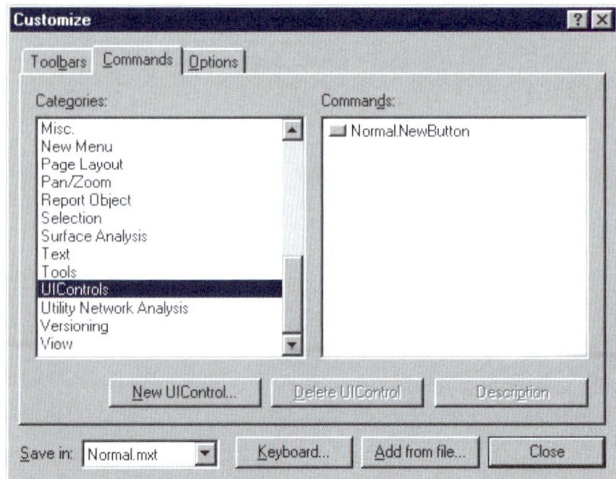

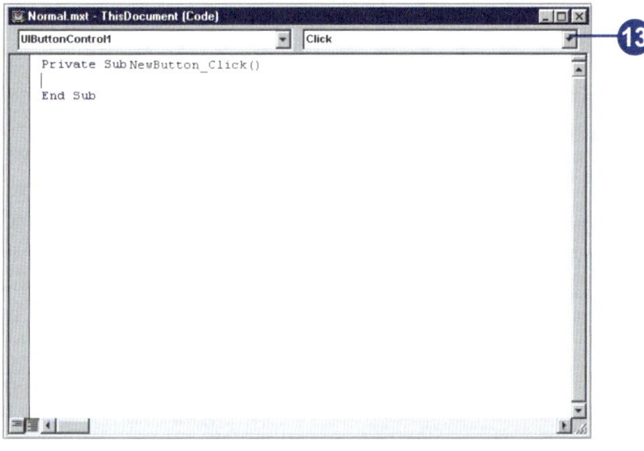

Working with UIControls

If you create a macro and add it to a toolbar, you've essentially customized what happens when you click the button. UIControls, however, provide a way to enrich an application in addition to button clicks and menu selections. The event procedures associated with these controls allow you to respond to user interaction and update controls based on the state of the application. If you create a combo box or an edit box, you might be able to avoid using a dialog box to get information.

The UIButtonControl works similarly to the built-in buttons that come with the application. Typically, you use a UIButtonControl to start, end, or interrupt an action or series of actions. You can write code to set whether it appears enabled or as if it is pressed in. You can also set its ToolTip, provide a message that will appear in the status bar, and respond to its Click event.

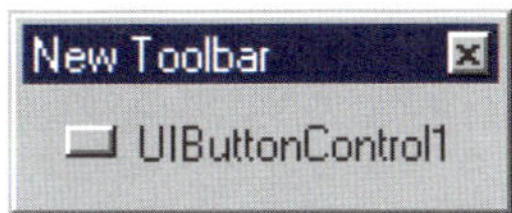

The UIToolControl works similarly to the built-in tools that come with the application. Typically, you use a UIToolControl to perform some type of interaction with the display. You can write code to toggle whether the tool appears as enabled or set its ToolTip. You can respond to mouse and key events. In addition, you can have it respond when the user selects the tool, double-clicks it, or right-clicks it. The UIToolControl can respond when the map refreshes or when the tool is deactivated.

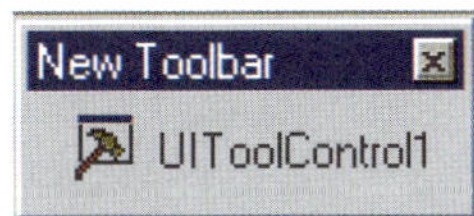

The UIComboBoxControl works similarly to the combo boxes that appear as part of the interface. It combines the features of a text box and a list box. Typically, you use a UIComboBoxControl to provide a set of choices from which a selection can be made. You can also type into the edit box portion of the control. The combo box has methods that allow you to populate its list or remove individual or all items. Several properties associated with the combo box let you work with items, return the index of the selected item, return the text at a specific index, return the text in the control's edit box, and determine how many items are in the control. In addition, you can respond to several events including when the user makes a change in the edit portion of the control or when a change to the selection occurs. As with the UIButtonControl, you can set the control's ToolTip and provide a status bar message.

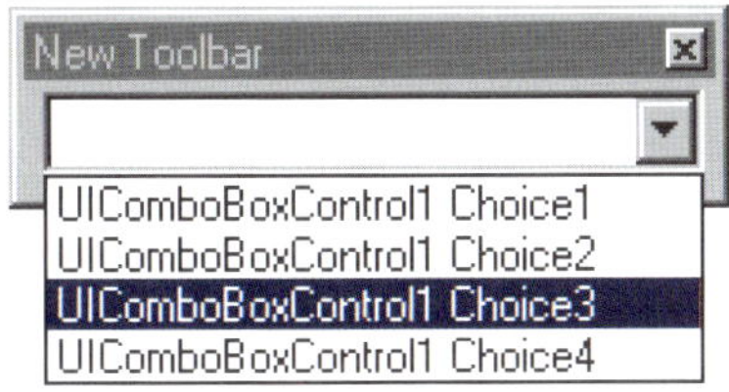

The UIEditBoxControl works similarly to the edit boxes that appear as part of the interface. Typically, you use a UIEditBoxControl to display information entered by the user. The control can also display data derived from an external source. You can use its Clear method to remove its contents, and its Text property contains the text that's displayed. You can specify whether the control appears as enabled. In addition, you can respond to when the user makes a change or presses a key. As with the UIButtonControl, you can set this control's ToolTip and provide a status bar message.

Adding custom commands

You don't have to use VBA to create custom commands. In fact, in some cases your custom commands may require you to use another development environment. You can create custom objects in any programming language that supports COM. Custom commands or toolbars created outside VBA are often distributed as ActiveX DLLs. After adding a custom object into ArcMap, you can use it as you would any built-in command.

1. Click the Tools menu and click Customize.
2. Click Add from file.
3. Navigate to the file containing the custom command.
4. Click the file and click Open.

   The Added Objects dialog box appears, reporting which new objects have been registered with ArcMap.
5. Click OK.
6. Click the Toolbars tab.
7. Check the toolbar to which you want to add the custom command.
8. Click the Commands tab.
9. Click the custom command's category in the Categories list.
10. Click and drag the command from the Commands list and drop it on the toolbar.
11. Click Close.

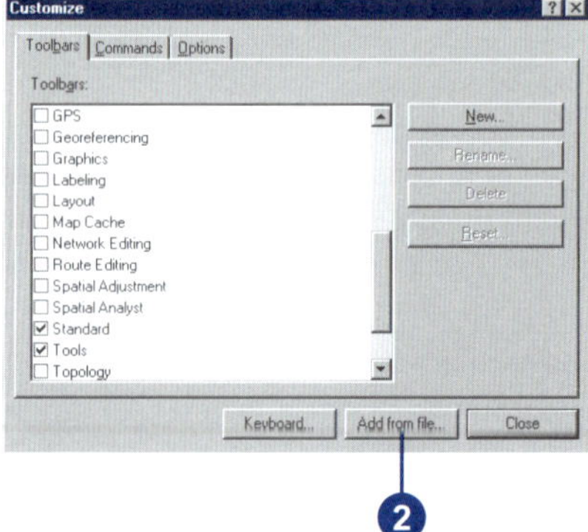

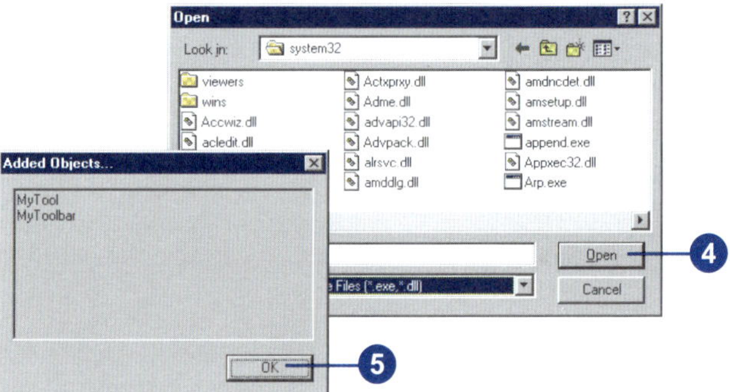

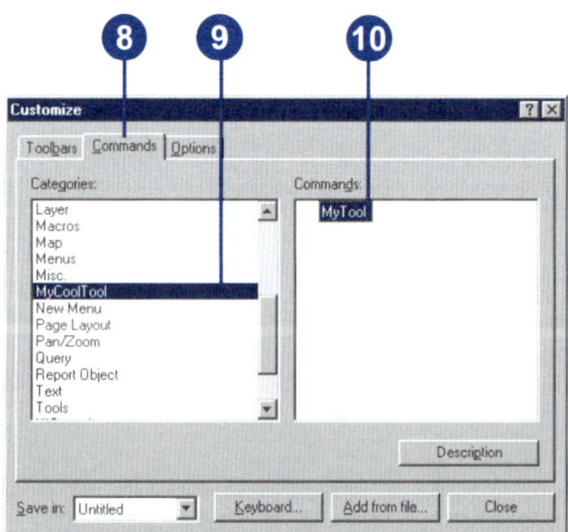

Updating the ArcID module

If you are developing applications or writing macros that make use of COM objects that you've added in, you can update the Normal.ArcID module to include newly added commands. In this way, you'll be able to refer to the COM objects you've added in by name when using a method such as CommandBars.Find. Updating also allows you to keep Normal.ArcID up to date should you remove a command.

1. Click the Tools menu and click Customize.
2. Click the Options tab.
3. Click Update ArcID module.
4. Click Close.

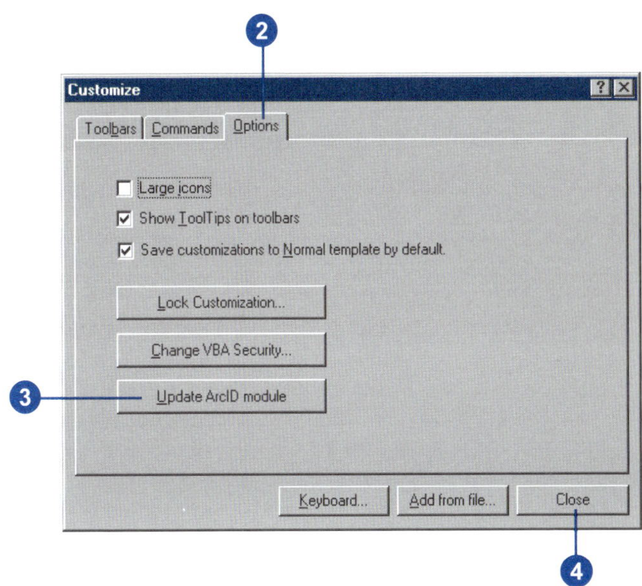

Using VB's code completion feature, you can list the commands that have been added to the ArcID module.

Locking documents and templates

To protect proprietary or sensitive information or work in progress or to prevent others from changing the way you've customized a document or template, you can use the Lock Customization facilities provided in the Options tab of the Customize dialog box. In addition, Lock Customization prevents access to the Macros dialog box and the Visual Basic Editor. To lock individual documents or templates while still allowing access to the Visual Basic Editor, you can use the Protection tab of the VBA Project Properties dialog box. This dialog box lets you password protect the ArcMap document or template you've saved.

See Also

You can create your own customization filter to control what aspects of the customization environment other users will have access to. For an example written in VB, see the ArcObjects Developer Help.

Locking customization

1. Click the Tools menu and click Customize.
2. Click the Options tab.
3. Click Lock Customization.
4. In the dialog box that appears, enter a password that has at least five alphanumeric characters and confirm it.

   To use the Customize dialog box, the Macros dialog box, or the Visual Basic Editor with the document or template subsequently, the correct password must be supplied.
5. Click OK in the Lock dialog box.
6. Click Close.

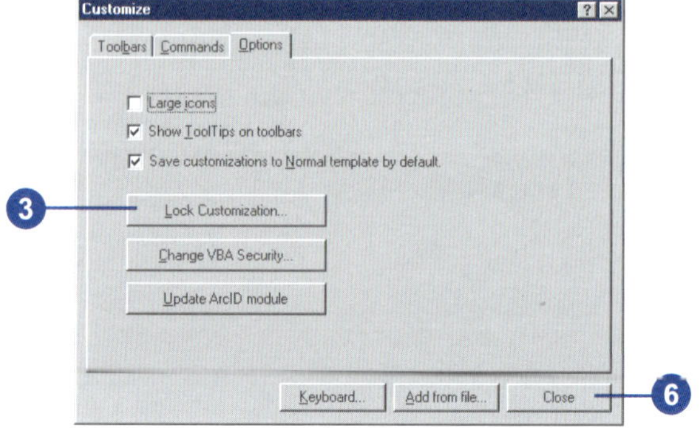

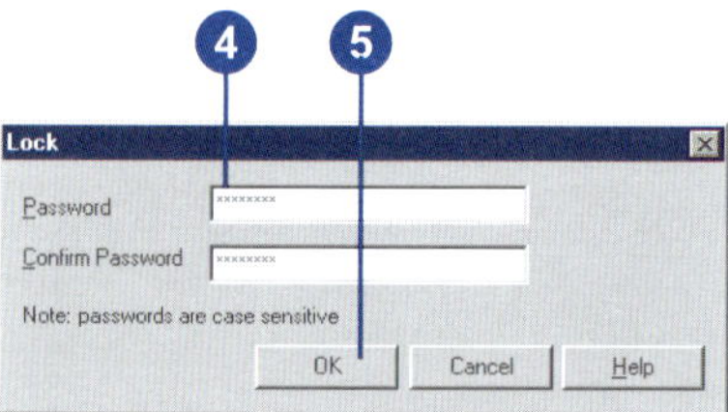

Unlocking customization

1. Click Tools, then click Customize.

   The Unlock dialog box appears.
2. Enter the password to unlock your selection.
3. Click OK.

   The Customize dialog box appears. If you specify an incorrect password, a message appears.

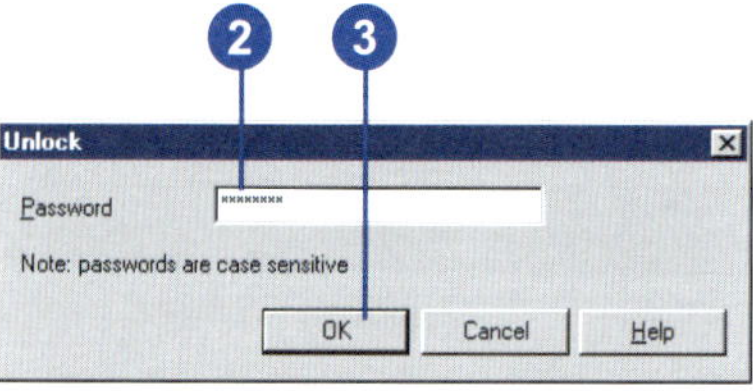

Locking documents and templates

1. Click the Tools menu, point to Macros, then click Visual Basic Editor.
2. In the Project Explorer, right-click the project or template you want to lock and click Project Properties.
3. Click the Protection tab.
4. Check Lock project for viewing to lock the project so it cannot be viewed or edited.
5. Type a password and confirm it.
6. Click OK in the Project Properties dialog box.
7. Click Save Project.

   The next time someone opens the Project and attempts to view the Project Properties, they'll be prompted for a password.

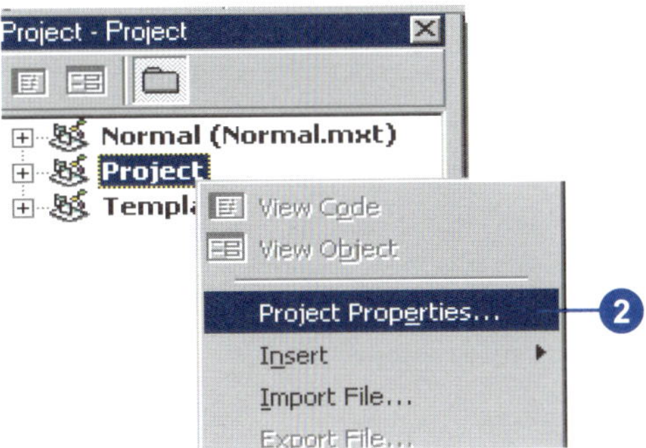

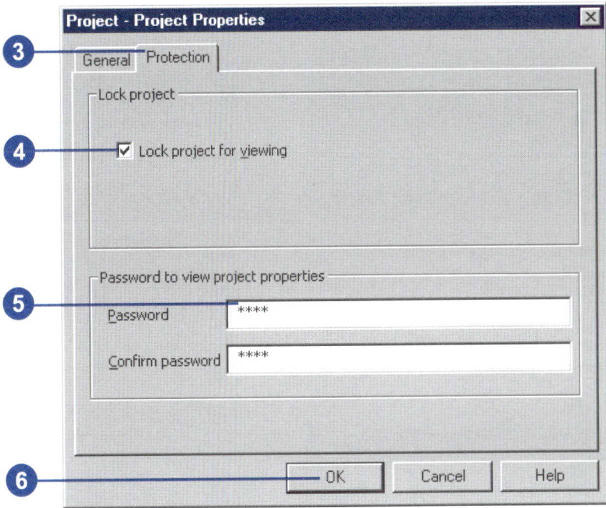

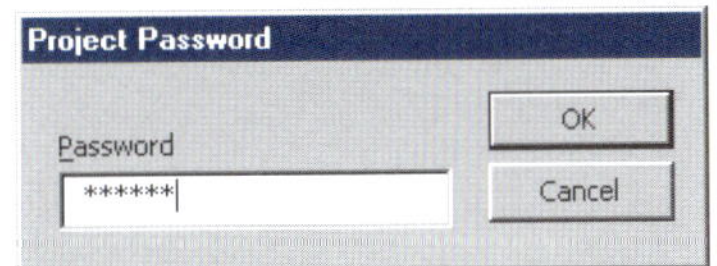

Changing VBA security

A macro virus is a type of computer virus that's stored in a macro within a document, template, or add-in. When you open such a document or perform an action that triggers a macro virus, the macro virus might be activated, transmitted to your computer, and stored in your Normal template. From that point on, every document you open could be automatically "infected" with the macro virus; if others open these infected documents, the macro virus is transmitted to their computers.

ESRI applications offer the levels of security described in the Security dialog box to reduce the chances of macro viruses infecting your documents, templates, or add-ins.

See Also

To learn more about using digital signatures and about how security levels and digital signatures work together, see the Microsoft Visual Basic Help in the VBE.

1. Click the Tools menu and click Customize.
2. Click the Options tab.
3. Click Change VBA Security.
4. Click the level of security you want.
5. Click the Trusted Sources tab to see a list of the names of organizations or individuals whose signed macros will be allowed to run.

   When you check the Always trust macros from this source check box in the Security Warning dialog box, which appears when you open a document or template with macros, this digital certificate is added to the Visual Basic for Applications Trusted Sources list.
6. Click OK.
7. Click Close.

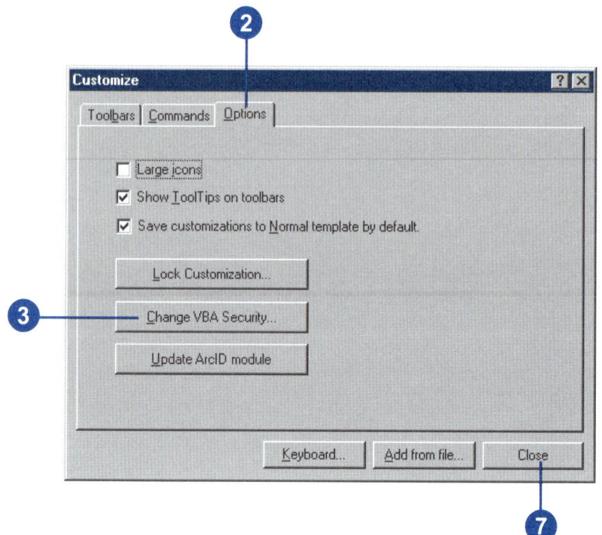

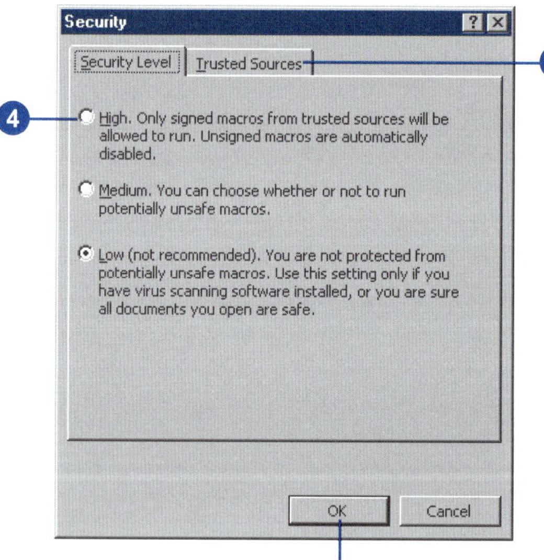

Glossary

access key

A keyboard shortcut that allows a user to access the contents of the Main menu by holding down the Alt key and pressing the underlined letter on the menu or menu command item. An access key is created by placing an ampersand (&) in front of the appropriate letter in the command's caption.

active data frame

The data frame currently being worked on—for example, the data frame to which layers are being added. The active data frame is highlighted on the map, and its name is shown in bold text in the table of contents.

alias

An alternative name specified for fields, tables, and feature classes that is more descriptive and user-friendly than the actual name of these items. On computer networks, a single e-mail alias may refer to a group of e-mail addresses. In database management systems, aliases can contain characters, such as spaces, that can't be included in the actual names.

analysis

The process of identifying a question or issue to be addressed, modeling the issue, investigating model results, interpreting the results, and possibly making a recommendation.

annotation

Descriptive text used to label features on or around a map. Information stored for annotation includes a text string, a position at which it can be displayed, and display characteristics.

annotation feature class

A geodatabase feature class that stores text or graphics that provide additional information about features or general areas of a map (annotation). An annotation feature class may be linked to another feature class, so edits to the features are reflected in the annotation (feature-linked annotation). Annotation in a geodatabase is edited during an edit session, using the tools on the Annotation toolbar. See also annotation group.

annotation group

A container within a map document for organizing and managing text or graphics that provide additional information about features or general areas of a map. Annotation groups allow control of the display of different sets of annotation. Annotation stored in a map document is edited with the tools on the Drawing toolbar. See also annotation feature class.

annotation target

In ArcMap, the annotation group or feature class in a map document where new annotation will be created when using the New Text tools on the Draw toolbar or when copying and pasting annotation. Annotation created with the Annotation Edit tools is created in the current Editing target, not in the annotation target.

ArcInfo workspace

A file-based collection of coverages, grids, triangulated irregular networks (TINs), or shapefiles stored as a directory of folders in the file system.

aspect

The compass direction in which a topographic slope faces, usually expressed in terms of degrees from the north. Aspect can be generated from continuous elevation surfaces. The aspect recorded for a TIN face is the steepest downslope direction of the face. The aspect of a cell in a raster is the steepest downslope direction of a plane defined by the cell and its eight surrounding neighbors.

attribute

1. Information about a geographic feature in a GIS, generally stored in a table and linked to the feature by a unique identifier. For example, attributes of a river might include its name, length, and average depth.

2. In raster datasets, information associated with each unique value of raster cells.

3. Cartographic information that specifies how features are displayed and labeled on a map; the cartographic attributes of a river might include line thickness, line length, color, and font.

attribute table

A database or tabular file containing information about a set of geographic features, usually arranged so each row represents a feature and each column represents one feature attribute. In raster datasets, each row of an attribute table corresponds to a certain region of cells having the same value. In a GIS, attribute tables are often joined or related to spatial data layers, and the attribute values they contain can be used to find, query, and symbolize features or raster cells.

Attributes dialog box

In ArcMap, a dialog box that displays attributes of selected features for editing.

background

See NoData.

band

A set of adjacent wavelengths or frequencies with a common characteristic, such as the visible band of the electromagnetic spectrum.

barrier

1. An object that is placed on a map to specify the point in a network past which a trace cannot continue. A location in a linear network through which nothing can flow.

2. A line feature used to keep certain points from being used in the calculation of new values when interpolating a grid or creating a triangulated irregular network (TIN). The line can represent a cliff, ridge, or some other interruption in the landscape. Only the sample points on the same side of the barrier as the current processing cell will be considered.

bookmark

See spatial bookmark.

buffer

A zone around a map feature measured in units of distance or time. A buffer is useful for proximity analysis.

button

An icon that runs a command, macro, or custom code when clicked. Buttons can be added to any menu or toolbar. When they appear in a menu, buttons are referred to as menu commands.

CAD

A computer-based system for the design, drafting, and display of graphical information. Also known as computer-aided drafting, such systems are most commonly used to support engineering, planning, and illustrating activities.

cartography

The art and science of expressing graphically, usually through maps, the natural and social features of the earth.

cell

1. The smallest unit of information in an image, raster, or grid. In a map, each cell represents a portion of the earth, such as a square meter or square mile, and usually has an attribute value associated with it, such as soil type or vegetation class. Cells are usually square or rectangular in shape, although hexagonal and circular areas have also been used.

2. A pixel.

chart

1. A map for air or water navigation.

2. A graphic representation of tabular data; a diagram showing the relation between two or more variable quantities, usually measured along two perpendicular axes. Also referred to as a graph.

3. In ArcView GIS 3, one of the five types of documents that can be contained within a project file. A chart visually presents the attribute data contained in a table.

choropleth map

A thematic map in which quantitative spatial data is symbolized with shading or graduated colors.

class

1. A group or category of attribute values; a set of entities possessing certain common attribute values.

2. Pixels in a raster file that represent the same condition.

classification

The process of sorting or arranging entities into groups or categories based on their attribute values; on a map, the process of representing members of a group by the same symbol, usually defined in a legend.

color ramp

A range of colors used to show ranking or order among classes on a map.

column

The vertical dimension of a table. Each column stores the values of one type of attribute for all of the records, or rows, in the table. All the values in a given column are of the same data type—for example, number, string, BLOB, date.

COM

See Component Object Model (COM).

combo box

A user interface tool that combines the features of a text box and a dropdown list. For example, the Location combo box in ArcCatalog allows the selection of an item in the Catalog tree by typing its path or choosing its path from a dropdown list.

command

1. An instruction to a computer program, usually one word or concatenated words or letters, issued by the user from a control device such as a keyboard, or read from a file by a command interpreter.

2. A menu, menu item, button, combo box, or text box on a toolbar.

Component Object Model (COM)

A binary standard that enables software components to interoperate in a networked environment regardless of the language in which they were developed. Developed by Microsoft, COM technology provides the underlying services of interface negotiation, life cycle management (determining when an object can be removed from a system), licensing, and event services (putting one object into service as the result of an event that has happened to another object).

computer-aided design

See CAD.

context menu

List menus that pop up when the right mouse button is clicked in Windows applications. Some keyboards also have an application key that opens context menus.

continuous raster

A raster that represents the world with a set of values that vary continuously to form a surface. For example, a raster digital elevation model and an interpolated chemical concentration surface are continuous rasters.

contour line

A line drawn on a map that connects points of equal elevation above a datum, usually a mean sea level.

control

In mapping, a system of points with established horizontal and vertical positions that are used as fixed references for known ground points or specific locations. The establishment of controls is one of the first steps involved in digitizing.

coordinate system

A fixed reference framework superimposed onto the surface of an area to designate the position of a point within it; a reference system consisting of a set of points, lines, and/or surfaces; and a set of rules, used to define the positions of points in space in either two or three dimensions. The Cartesian coordinate system and the geographic coordinate system used on the earth's surface are common examples of coordinate systems.

coverage

A data model for storing geographic features using ArcInfo software. A coverage stores a set of thematically associated data considered to be a unit. It usually represents a single layer, such as soils, streams, roads, or land use. In a coverage, features are stored as both primary features (points, arcs, polygons) and secondary features (tics, links, annotation). Feature attributes are described and stored independently in feature attribute tables. Coverages cannot be edited in ArcGIS.

current task

During editing in ArcMap, a setting in the Current Task dropdown list that determines the task with which the sketch construction tools (Sketch, Arc, Distance–Distance, and Intersection) will work. The current task is set by clicking a task in the Current Task dropdown list.

data

Any collection of related facts arranged in a particular format; often, the basic elements of information that are produced, stored, or processed by a computer.

data frame

A map element that defines a geographic extent, a page extent and a coordinate system, and other display properties for one or more layers in ArcMap. A dataset can be represented in one or more data frames. In data view, only one data frame is displayed at a time; in layout view, all a map's data frames are displayed at the same time. Many cartography texts use the term map body to refer to what ESRI calls a data frame.

data source

Any geographic data. Data sources may include coverages, shapefiles, rasters, or feature classes.

data view

An all-purpose view in ArcMap and ArcReader for exploring, displaying, and querying geographic data. This view hides all map elements, such as titles, North arrows, and scalebars.

database

One or more structured sets of persistent data, managed and stored as a unit and generally associated with software to update and query the data. A simple database might be a single file with many records, each of which references the same set of fields. A GIS database includes data about the spatial locations and shapes of geographic features recorded as points, lines, areas, pixels, grid cells, or TINs, as well as their attributes.

dataset

Any organized collection of data with a common theme.

decimal degrees

Values of latitude and longitude expressed in decimal format rather than in degrees, minutes, and seconds.

DEM

See digital elevation model (DEM).

determinate flow direction

A conclusively definitive line or course in which something is issuing or moving in a stream. For an edge feature, this occurs when the flow direction can be ascertained from the topology of the network, the locations of sources and sinks, and the enabled or disabled states of features.

digital elevation model (DEM)

The representation of continuous elevation values over a topographic surface by a regular array of z-values, referenced to a common datum. Typically used to represent terrain relief.

digital terrain model (DTM)

See digital elevation model (DEM).

digitizing

The process of converting the geographic features on an analog map into digital format using a digitizing tablet, or digitizer, which is connected to a computer. Features on a paper map are traced with a digitizer puck, a device similar to a mouse, and the x,y coordinates of these features are automatically recorded and stored as spatial data.

directory

An area of a computer disk that holds a set of data files and/or other directories. Operating systems use directories to organize data. Directories are arranged in a tree structure, in which each branch is a subdirectory of its parent branch. The location of a directory is specified with a pathname, for example, C:\gisprojects\shrinking_lemur_habitat\grids.

disabled feature

A network object or shape representing a geographic object through which flow is impossible.

discrete raster

A raster that typically represents phenomena that have clear boundaries and attributes that are descriptions or categories. Each cell in a discrete raster stores an integer value that represents a feature. In a grid of land cover, for example, the value 1 might represent forested land, the value 2 urban land, and so on.

disk

A storage medium consisting of a round, flat, spinning plate coated with a magnetic material for recording digital information.

display units

The mode in which measurements, dimensions of shapes, and distance tolerances and offsets are rendered on a computer screen or on a printed map. Although they are stored with consistent units in the dataset, users can choose the units in which coordinates and measurements are displayed—for example, feet, miles, meters, or kilometers.

docking

Moving a floating toolbar or window to a fixed location in the graphical user interface.

edge

1. A line between two nodes, points, or junctions that forms the boundary of one or more faces of a spatial entity. In an image, edges separate areas of different tones or colors. In topology, an edge defines lines or polygon boundaries; multiple features in one or more feature classes may share edges.

2. In a geometric network, a linear feature (for example, a pipeline in a sewer system). A network edge can be simple or complex. A simple edge is always connected to exactly two junction (point) features, one at each end. A complex edge is always connected to at least two junction features at its endpoints, but it can be connected to additional junction features along its length.

3. In a TIN, a line segment between nodes (sample data points). Edges store topologic information about the faces that they border.

edit box

See text box.

edit cache

See map cache.

edit session

In ArcMap, the environment in which spatial and attribute editing take place. After starting an edit session, a user can modify feature locations, geometry, or attributes. Modifications are not saved unless the user explicitly chooses to save them.

Editor toolbar

In ArcMap, a set of tools that allows the creation and modification of features and their attributes.

enabled feature

In geodatabases, a network feature that allows flow to pass through it.

ESRI Standard Label Engine

In ArcMap, the software used to place labels.

event

A geographic location stored in tabular rather than spatial form. Event types include address events, route events, xy events, and temporal events. Address events are features that can be located based on address matching with a street network or other address identifier, such as ZIP Codes or lot numbers. Route events are linear, continuous, or point features occurring along a base route system. Xy events are simple coordinate pairs that describe the location of a feature, such as a set of latitude and longitude degrees. Temporal events are used to describe observations through time of particular objects or groups of objects.

extent

The coordinate pairs defining the minimum bounding rectangle (xmin, ymin and xmax, ymax) of a data source. All coordinates for the data source fall within this boundary.

extent rectangle

A rectangle that is displayed in one data frame, showing the size and position of another data frame.

feature

1. A group of spatial elements which together represent a real-world entity. A complex feature is made up of more than one group of spatial elements: for example, a set of line elements with the common theme of roads representing a road network.

2. A representation of a real-world object on a map. Features can be represented in a GIS as vector data (points, lines, or polygons) or as cells in a raster data format. To be displayed in a GIS, features must have geometry and locational information.

feature attribute table

See attribute table.

feature class

A collection of geographic features with the same geometry type (such as point, line, or polygon), the same attributes, and the same spatial reference. Feature classes can stand alone within a geodatabase or they can be contained within shapefiles, coverages, or other feature datasets. Feature classes allow homogeneous features to be grouped into a single unit for data storage purposes. For example, highways, primary roads, and secondary roads can be grouped into a line feature class named "roads". In a geodatabase, feature classes can also store annotation and dimensions.

feature dataset

A collection of feature classes stored together that share the same spatial reference; that is, they have the same coordinate system, and their features fall within a common geographic area. Feature classes with different geometry types may be stored in a feature dataset.

feature weight

A ranking system that indicates whether features from a given feature class may be covered by a label in cases where the label cannot be placed in free space. Feature classes with lower weights will tend to have labels placed over their features before feature classes with higher weights. Polygon feature classes have two types of weights: boundary weights and interior weights.

feature-linked annotation

Annotation that is stored in the geodatabase with links to features through a geodatabase relationship class. Feature-linked annotation reflects the current state of features in the geodatabase: it is automatically updated when features are moved, edited, or deleted.

field

1. A column in a table that stores the values for a single attribute.

2. The place in a database record, or in a graphical user interface, where data can be entered.

3. A synonym for surface.

See also attribute, column.

flag

1. An indicator ,such as a signal, symbol, character, or digit, used as an identifier.

2. In ArcMap, an object that is placed on a network to specify the starting point for a tracc task.

flow direction

The route or course followed by commodities proceeding through edge elements in a network.

font point size

A unit of measure for fonts, nearly equal to 1/72 of an inch.

GDB

See geodatabase.

geocoding

The process of assigning x,y coordinate values to street addresses or ZIP Codes so that they can be displayed as point features on a map. In a GIS, address geocoding requires a reference dataset that contains address attributes for the area of interest.

geodatabase

An object-oriented data model introduced by ESRI that represents geographic features and attributes as objects and the relationships between objects, but is hosted inside a relational database management system. A geodatabase can store objects such as feature classes, feature datasets, nonspatial tables, and relationship classes.

geographic coordinates

A measurement of a location on the earth's surface expressed in degrees of latitude and longitude.

geographic information system (GIS)

An arrangement of computer hardware, software, and geographic data that people interact with to integrate, analyze, and visualize the data; identify relationships, patterns, and trends; and find solutions to problems. The system is designed to capture, store, update, manipulate, analyze, and display the geographic information. A GIS is typically used to represent maps as data layers that can be studied and used to perform analyses.

geometric network

Topologically connected edge and junction features that represent a linear network such as a road, utility, or hydrologic system.

geoprocessing

A GIS operation used to manipulate data stored in a GIS workspace. A typical geoprocessing operation takes an input dataset, performs an operation on that dataset, and returns the result of the operation as an output dataset. Common geoprocessing operations are geographic feature overlay, feature selection and analysis, topology processing, and data conversion. Geoprocessing allows for definition, management, and analysis of information used to form decisions.

georeferencing

Assigning coordinates from a known reference system, such as latitude/longitude, UTM, or State Plane, to the page coordinates of a raster (image) or a planar map. Georeferencing raster data allows it to be viewed, queried, and analyzed with other geographic data.

GIS

See geographic information system (GIS).

global positioning system (GPS)

A constellation of 24 radio-emitting satellites deployed by the U.S. Department of Defense and used to determine location on the earth's surface. The orbiting satellites transmit signals that allow a GPS receiver anywhere on earth to calculate its own location through triangulation. The system is used in navigation, mapping, surveying, and other applications in which precise positioning is necessary.

glyph

A type of symbol; a stylized figure or graphic used to convey information nonverbally, representing single variables or the relationship between many variables through attributes such as size, color, shape, and orientation. A glyph could be a pie chart symbol or an arrow.

GPS

See global positioning system (GPS).

graph

See chart.

graphic text

Text added in ArcMap layout view that exists in page space and is stored in a map document. Graphic text does not move if the extent or scale is changed.

graticule

1. A network of longitude and latitude lines on a map or chart that relates points on a map to their true locations on the earth.

2. A glass plate or cell with a grid or cross wires on it that rests in the focal plane of the eyepiece of a telescope, used to locate and measure celestial objects.

grid

In cartography, any network of parallel and perpendicular lines superimposed on a map and used for reference. These grids are usually named after the map's projection; for example, Lambert grid and Transverse Mercator grid.

grid cell

See cell.

ground control point

A point on the ground whose location has been determined by a horizontal coordinate system or a vertical datum. A ground control point links a location on an ungeoreferenced raster to a location in map coordinates.

group layer

A group of several layers that appear and act as a single layer. Group layers make it easier to organize a map, assign advanced drawing order options, and share layers for use in other maps. Data referenced by a layer must share the same coordinate system.

hatches

In linear referencing, a series of vertical line or marker symbols displayed on top of features at an interval specified in route measure units.

heads-up digitizing

Manual digitization of features by tracing a mouse over features displayed on a computer monitor, often as a method of vectorizing raster data.

hillshading

1. Shadows drawn on a map to simulate the effect of the sun's rays over the land.

2. On a raster, a simulated shadow effect achieved by assigning an illumination value from 0 to 255 to each cell according to a specified azimuth and altitude for the sun. Hillshading creates a three-dimensional effect.

hyperlink

A reference (link) from some point in one electronic document or resource to some point in another document or resource or another place in the same document. Activating the link, usually by clicking on it with the mouse, causes the browser to display the target of the link.

identify

In ArcGIS, a tool that, when applied to a feature (by clicking on it), opens a window showing that feature's attributes.

image

A raster-based representation or description of a scene, typically produced by an optical or electronic device such as a camera or a scanning radiometer. Common examples include remotely sensed data (for example, satellite data), scanned data, and photographs. An image is stored as a raster dataset of binary or integer values that represent the intensity of reflected light, heat, sound, or any other range of values on the electromagnetic spectrum. An image may contain one or more bands.

indeterminate flow direction

A flow direction that is unknown or undiscoverable. Indeterminate flow direction occurs when flow direction cannot be determined from the topology of the network, the locations of sources and sinks, and the enabled or disabled states of features.

join

1. Appending the fields of one table to those of another through an attribute or field common to both tables. A join is usually used to attach more attributes to the attribute table of a geographic layer.

2. The process of connecting two or more separate spatial entities. If two line segments are joined, they become one spatial object for further processing.

junction

A node joining two or more arcs.

label

1. Text placed next to a feature on a map to describe or identify it.

2. To run the labeling process in ArcMap; to begin dynamically placing attribute-driven text for map features.

label class

A category of labels that represents features with the same labeling properties. For example, in a roads layer, label classes could be created to define information and style for each type of road: interstate, state highway, county road, and so on.

label engine

See ESRI Standard Label Engine.

label expression

A statement that determines the label text. Label expressions typically concatenate or modify the contents of one or more fields, and may add additional text strings to create more informative labels. They can contain Visual Basic script or JScript to add logic, text processing, and formatting for the labels.

Label Manager

The tool used to display and set labeling properties for the currently active data frame. The Label Manager is accessible through the Labeling toolbar.

label priority

A ranking system that determines the order in which labels will be placed on a map. Labels with higher priority will be placed before labels with lower priority. Labels placed last will have a greater chance of being crowded out or placed in an alternate position.

label weight

An ESRI Standard Label Engine ranking system that indicates whether labels from a given label class may be covered by another label in cases where label placement conflicts occur. Labels with higher weight are less likely to be overlapped than labels with lower weight.

LAN

See local area network (LAN).

layer

1. A set of references to data sources such as a coverage, geodatabase feature class, raster, and so on that defines how the data should be displayed on a map. Layers can also define additional properties, such as which features from the data source are included. Layers can also be used as inputs to geoprocessing tools. Layers can be stored in map documents (.mxd) or saved individually as layer files (.lyr).

2. A standalone feature class in a geodatabase managed with SDE 3.

layout

The arrangement or overall design of elements on a digital map display or printed map, possibly including a title, legend, North arrow, scalebar, and geographic data.

layout view

In ArcMap and ArcReader, the view for laying out a map. Layout view shows the virtual page upon which geographic data and map elements, such as titles, legends, and scalebars, are placed and arranged for printing. See also data view.

legend

The reference area on a map that lists and explains the colors, symbols, line patterns, shadings, and annotation that have been used on the map to code the various elements and data values. The legend includes a sample of each symbol with text describing what it means. Legends often include the map's scale, origin, and projection.

legend patch shape

The geometric shape of either a line or a polygon that is used to represent a specific kind of feature in a legend and in the ArcMap table of contents.

local area network (LAN)

Communications hardware and software that connect computers in a small area, such as a room or a building. Computers in a LAN can share data and peripheral devices, such as printers and plotters, but have no necessary link to outside computers.

macro

A computer program, usually a text file, containing a sequence of commands that are executed as a single command. Macros are used to perform commonly used sequences of commands or complex operations.

magnifier window

A secondary window in ArcMap data view that shows a magnified view of a small area without changing the map extent. Moving the magnifier window around will not affect the current map display.

map

1. A graphic depiction on a flat surface of the physical features of the whole or a part of the earth or other body, or of the heavens, using shapes to represent objects and symbols to describe their nature; at a scale whose representative fraction is less than 1:1. Maps generally use a specified projection and indicate the direction of orientation.

2. Any graphical representation of geographic or spatial information.

3. The document used in ArcMap to display and work with geographic data. In ArcMap, a map contains one or more layers of geographic data, contained in data frames, and various supporting map elements, such as a scalebar.

map cache

A setting used in ArcMap that allows temporary storage of geodatabase features from a given map extent in the desktop computer's RAM, which may result in performance improvements in ArcMap for editing, feature rendering, and labeling.

map display

A graphic representation of a map on a computer screen.

map document

In ArcMap, the file that contains one map, its layout, and its associated layers, tables, charts, and reports. Map documents can be printed or embedded in other documents. Map document files have a .mxd extension.

map element

A graphic or piece of text stored in a map document in the layout, in an annotation group, or in a geodatabase annotation feature class. For example, a circular graphic, a city name, or a North arrow stored in the layout are all map elements.

map feature

See feature.

map projection

See projection.

map scale

See scale.

map surround

See map element.

map template

In ArcMap, a kind of map document that provides a quick way to create a new map. Templates can contain data, a custom interface, and a predefined layout that arranges map elements, such as North arrows, scalebars, and logos, on the virtual page. Map templates have a .mxt file extension.

map units

The ground units—for example, feet, miles, meters, or kilometers—in which the coordinates of spatial data are stored.

MapTip

In ArcGIS, a user-assistance component that displays an onscreen description of a map feature when the mouse is paused over that feature.

menu

A list displayed on the computer screen that presents a selection of available commands from which a user can choose an operation to be performed.

menu item

An item in a list of commands displayed on a menu.

metadata

Information about the content, quality, condition, and other characteristics of data. Metadata for geographical data may document its subject matter; how, when, where, and by whom the data was collected; accuracy of the data; availability and distribution information; its projection, scale, resolution, and accuracy; and its reliability with regard to some standard. Metadata consists of properties and documentation. Properties are derived from the data source (for example, the coordinate system and projection of the data), while documentation is entered by a person (for example, keywords used to describe the data).

multipart feature

A feature that is composed of more than one physical part but only references one set of attributes in the database. For example, in a layer of states, the state of Hawaii could be considered a multipart feature. Although composed of many islands, it would be recorded in the database as one feature.

multipoint feature

A feature that consists of more than one point but only references one set of attributes in the database. For example, a system of oil wells might be considered a multipoint feature, since there is a single set of attributes for multiple well holes.

neatline

The border delineating and defining the extent of geographic data on a map. It demarcates map units so that, depending on the map projection, the neatline does not always have 90-degree corners. In a properly made map, it is the most accurate element of the data; other map features may be moved slightly or exaggerated for generalization or readability, but the neatline is never adjusted.

network

1. A set of edge, junction, and turn elements and the connectivity between them; also known as a logical network. In other words, an interconnected set of lines representing possible paths from one location to another. A city streets layer is an example of a network.

2. In computing, a group of computers that share software, data, and peripheral devices, as in a local area network (LAN) or wide area network (WAN).

network feature

One of the topologically connected edge or junction features, representing a linear network, such as a road, utility, or hydrologic system, that composes a geometric network.

network trace

A function that follows connectivity in a geometric network. Specific kinds of network tracing include finding features that are connected, finding common ancestors, finding loops, tracing upstream, and tracing downstream. See also geometric network.

NoData

In a raster, the absence of a recorded value. While the measure of a particular attribute in a cell may be zero, a NoData value indicates that no measurements have been taken for that cell at all. See also null value.

Normal template

The template that is automatically loaded in ArcMap and contains all the standard toolbar and command default settings. User interface customization that is saved in the Normal template is loaded each time ArcMap is launched.

normalization

The process of dividing one numeric attribute value by another in order to minimize differences in values based on the size of areas or the number of features in each area. For example, normalizing (dividing) total population by total area yields population per unit area, or density.

North arrow

A map symbol that shows the direction of north on the map, thereby showing how the map is oriented.

null value

The absence of a recorded value for a geographic feature. A null value differs from a value of zero in that zero may represent the measure of an attribute, while a null value indicates that no measurement has been taken. See also NoData.

object

1. In GIS, a digital representation of a discrete spatial entity. An object may belong to an object class and will thus have attribute values and behavior in common with other defined elements.

2. In computing, a piece of software that performs a specific task and is controlled by another piece of software, called a client. For example, an object is often the interface by which an application program accesses an operating system and other services.

operator

The symbolic representation of a process or operation performed against one or more operands in an expression, such as "+" (plus, or addition) and ">" (greater than). When evaluated, operators return a value as their result. If multiple operators appear in an expression, they are evaluated in order of their operator precedence. For example, in the expression "1 + 2", the plus operator (+) adds the operands 1 and 2 together and returns the value 3. The operators implemented in ESRI software represent a small subset of the hundreds associated with programming, mathematics and other fields. The applications found in ESRI software products support different operators depending on their intended use.

overview window

A secondary window in ArcMap data view that shows the full extent of the data, without changing the map extent. A red box in the window represents the current map extent.

parametric curve

A curve that is defined mathematically rather than by a series of connected vertices. A parametric curve has only two vertices, one at each end.

path

In computing, the location of a computer file, given as the drive, directories, subdirectories, and filename, in that order.

pixel

1. The smallest element of a display device, such as a video monitor, that can be independently assigned attributes, such as color and intensity. The term is an abbreviation for "picture element".

2. In remote sensing, the fundamental unit of data collection. A pixel is represented in a remotely sensed image as a rectangular cell in an array of data values.

3. The smallest unit of information in an image or raster map. Usually square or rectangular, pixel is often used synonymously with cell.

projected coordinate system

A reference system used to locate x,y and z positions of point, line, and area features in two or three dimensions. A projected coordinate system is defined by a geographic coordinate system, a map projection, any parameters needed by the map projection, and a linear unit of measure.

projected coordinates

A measurement of locations on the earth's surface expressed in a two-dimensional system that locates features based on their distance from an origin (0,0) along two axes, a horizontal x-axis representing east–west and a vertical y-axis representing north–south. Projected coordinates are transformed from latitude and longitude to x,y coordinates using a map projection. See also geographic coordinates.

projection

A method by which the curved surface of the earth is portrayed on a flat surface. This generally requires a systematic mathematical transformation of the earth's graticule of lines of longitude and latitude onto a plane. It can be visualized as a transparent globe with a light bulb at its center casting lines of latitude and longitude onto a sheet of paper. Generally, the paper is either flat and placed tangent to the globe (a planar or azimuthal projection), or formed into a cone or cylinder and placed over the globe (cylindrical and conical projections). Every map projection distorts distance, area, shape, direction, or some combination thereof.

property

An attribute of an object defining one of its characteristics or an aspect of its behavior. For example, the Visible property affects whether a control can be seen at run time. You can set an item's properties using its Properties dialog box.

pyramid

In raster datasets, a reduced resolution layer that copies the original data in decreasing levels of resolution to enhance performance. The coarsest level of resolution is used to quickly draw the entire dataset. As the display zooms in, layers with finer resolutions are drawn; drawing speed is maintained because fewer pixels are needed to represent the successively smaller areas.

query

A request that selects features or records from a database. A query is often written as a statement or logical expression.

raster

A spatial data model that defines space as an array of equally sized cells arranged in rows and columns. Each cell contains an attribute value and location coordinates. Unlike a vector structure, which stores coordinates explicitly, raster coordinates are contained in the ordering of the matrix. Groups of cells that share the same value represent geographic features. See also vector.

raster band

See raster dataset band.

raster catalog

A collection of raster datasets defined in a table of any format, in which the records define the individual raster datasets that are included in the catalog. A raster catalog is used to display adjacent or overlapping raster datasets without having to mosaic them together into one large file.

raster cell

See cell.

raster dataset band

One layer in a raster dataset that represents data values for a specific range in the electromagnetic spectrum, such as ultraviolet, blue, green, red, infrared, or radar, or other values derived by manipulating the original image bands. A raster dataset can contain more than one band. For example, satellite imagery commonly has multiple bands representing different wavelengths of energy from along the electromagnetic spectrum. See also band.

raster snapping

See snapping.

reference map

A map designed to convey where things are in relation to each other.

reference scale

The scale at which symbols will appear on the page at their true size, specified in page units. As the extent is changed, the text and symbols will change scale along with the display. Without a reference scale, the symbols will look the same at all map scales. ArcGIS Desktop uses reference scales for annotation groups, geodatabase annotation feature classes, geodatabase dimension feature classes, and data frames.

relate

An operation that establishes a temporary connection between records in two tables using an item common to both.

relief shading

See hillshading.

resampling

The process of extrapolating new cell values when transforming rasters to a new coordinate space or cell size. The three most common resampling techniques are nearest neighbor assignment, bilinear interpolation, and cubic convolution.

resolution

1. The area represented by each cell or pixel in a raster.

2. The smallest spacing between two display elements, expressed as dots per inch, pixels per line, or lines per millimeter.

3. The detail with which a map depicts the location and shape of geographic features. The larger the map scale, the higher the possible resolution. As scale decreases, resolution diminishes and feature boundaries must be smoothed, simplified, or not shown at all; for example, small areas may have to be represented as points.

route event

See event.

scale

The ratio or relationship between a distance or area on a map and the corresponding distance or area on the ground, commonly expressed as a fraction or ratio. A map scale of 1/100,000 or 1:100,000 means that one unit of measure on the map equals 100,000 of the same unit on the earth. See also scalebar.

scalebar

A map element that shows the map scale graphically.

segment

A line that connects vertices. For example, in a sketch of a building, a segment might represent one wall.

select

To choose from a number or group of features or records; to create a separate set or subset.

selectable layers

Layers from which features can be selected in ArcMap with the interactive selection tools. Selectable layers can be chosen using the Set Selectable Layers command in the Selection menu or on the Selection tab in the table of contents.

selected set

A subset of features in a layer or records in a table that is chosen by the user.

selection anchor

In an ArcMap editing session, a small "x" located in the center of selected features. The selection anchor is used in the snapping environment, or when rotating, moving, and scaling features.

shapefile

A vector data storage format for storing the location, shape, and attributes of geographic features. A shapefile is stored in a set of related files and contains one feature class.

shared boundary

A segment or boundary common to two features. For example, in a parcel database, adjacent parcels will share a boundary. Another example might be a parcel that shares a boundary on one side with a river. The segment of the river that coincides with the parcel boundary would share the same coordinates as the parcel boundary.

shared vertex

A vertex common to multiple features. For example, in a parcel database, adjacent parcels will share a vertex at the common corner.

shortcut key

A command's shortcut key executes the command directly without first having to open and navigate a menu. For example, Ctrl + C is a well-known shortcut for copying a file in Windows.

sink

A junction feature that pulls flow toward itself from the edges in the network. For example, in a river network, the mouth of the river can be modeled as a sink, since gravity drives all water toward it. See also source.

sketch

In ArcMap, a shape that represents a feature's geometry. Every existing feature on a map has this alternate form, a sketch, that allows visualization of that feature's composition, with all vertices and segments of the feature visible. When features are edited in ArcMap, the sketch is modified, not the original feature. A sketch must be created in order to create a feature. Only line and polygon sketches can be created, since points have neither vertices nor segments.

sketch constraint

In ArcMap editing, an angle or length limitation that can be placed on segments created using the Sketch tool.

sketch operation

In ArcMap, an editing operation that is performed on an existing sketch. Examples are Insert Vertex, Delete Vertex, Flip, Trim, Delete Sketch, Finish Sketch, and Finish Part. All of these operations are available from the Sketch context menu.

slope

The incline, or steepness, of a surface. The slope of a TIN face is the steepest downhill slope of a plane defined by the face. The slope for a cell in a raster is the steepest downhill slope of a plane defined by the cell and its eight surrounding neighbors. Slope can be measured in degrees from horizontal (0–90) or percent slope, which is the rise divided by the run, multiplied by 100. A slope of 45 degrees equals 100 percent slope. As slope angle approaches vertical (90 degrees), the percent slope approaches infinity.

snapping

An automatic editing operation in which points or features within a specified distance or tolerance of other points or features are moved to match or coincide exactly with each other's coordinates.

snapping tolerance

In an ArcMap editing session, the distance within which the pointer or a feature will snap to another location. If the location being snapped to (vertex, boundary, midpoint, or connection) is within that distance, the pointer will automatically snap. Snapping tolerance can be measured using either map units or pixels.

source

A junction feature that pushes flow away from itself through the edges of the network. For example, in a water distribution network, pump stations can be modeled as sources, since they drive the water through the pipes away from the pump stations. See also sink.

spatial analysis

1. The study of the locations and shapes of geographic features and the relationships between them. Spatial analysis is useful when evaluating suitability, when making predictions, and for gaining a better understanding of how geographic features and phenomena are located and distributed.

2. The process of modeling spatial data and examining and interpreting the results. Spatial analysis is useful for evaluating suitability and capability, for estimating and predicting, and for interpreting and understanding. There are four traditional types of spatial analysis: topological overlay and contiguity analysis; surface analysis; linear analysis; and raster analysis.

spatial bookmark

In ArcMap, a shortcut created by the user that identifies a particular geographic location to be saved for later reference.

spatial data

1. Information about the locations and shapes of geographic features and the relationships between them, usually stored as coordinates and topology.

2. Any data that can be mapped.

spatial join

A type of table join operation in which fields from one layer's attribute table are appended to another layer's attribute table based on the relative locations of the features in the two layers.

spatial overlay

The process of superimposing layers of geographic data that occupy the same space in order to study the relationship between them.

SQL

See Structured Query Language (SQL).

standard annotation

Annotation that is stored in the geodatabase, consisting of geographically placed text strings that are not associated with features in the geodatabase.

stretch

A display technique that increases the visual contrast between cells of similar value when applied to raster datasets.

Structured Query Language (SQL)

A syntax for defining and manipulating data from a relational database. Developed by IBM in the 1970s, SQL has become an industry standard for query languages in most relational database management systems.

style

An organized collection of predefined colors, symbols, properties of symbols, and map elements. Styles promote standardization and consistency in mapping products.

Style Manager

The tool used to create new styles and edit existing ones. The Style Manager displays the contents of all the styles that are currently referenced by the map. It also contains personal and additional styles that may be used in ArcMap.

surface

A geographic phenomenon represented as a set of continuous data, such as elevation or air temperature over an area or the boundary between two distinct materials or processes. A clear or sharp break in values of the phenomenon (breaklines) indicates a significant change in the structure of the phenomenon, such as a cliff, not a change of geographic feature. Surfaces can be represented by models built from regularly or irregularly spaced sample points on the surface, or by contour lines, isolines, bathymetry, or the like.

symbol

A graphic representation of a geographic feature or class of features that helps identify it and distinguish it from other features on a map. For example, line symbols represent arc features; marker symbols, points; shade symbols, polygons; and text symbols, annotation. Many characteristics define symbols including color, size, angle, and pattern.

symbol level drawing

A setting that determines the drawing order of features based on their symbols. When symbols have more than one layer, symbol level drawing can be used to specify the order in which each layer of the symbol is drawn.

symbology

The set of conventions, or rules, that defines how geographic features are represented with symbols on a map. A characteristic of a feature may influence the size, color, and shape of the symbol used.

table

1. A set of data elements arranged in rows and columns. Each row represents an individual entity, record, or feature, and each column represents a single field or attribute value. A table has a specified number of columns but can have any number of rows.

2. In ArcView GIS 3, one of the five types of documents that can be contained within a project file. A table stores attribute data.

table of contents

A list of data frames and layers on a map that show how the data is symbolized.

tabular data

Descriptive information, usually alphanumeric, that is stored in rows and columns in a database and can be linked to map features. See also table.

target layer

In an ArcMap editing session, the layer to which edits will be applied. The target layer must be specified when creating new features and modifying existing features.

template

See map template.

text box

An entity that displays text entered by a user or derived from another source for editing purposes.

text formatting tag

Tags used with text in ArcGIS that allow formatting to be modified for a portion of a text string. This allows the creation of mixed-format text where, for example, one word in a sentence is underlined. Text formatting tags adhere to XML syntax rules, and can be used most places where both a text string and a text symbol can be specified. The tags are most commonly used with labels, annotation, and graphic text.

thematic map

A map designed to convey information about a single topic or theme, such as population density or geology.

TIN

Triangulated irregular network. A vector data structure used to store and display surface models. A TIN partitions geographic space using a set of irregularly spaced data points—each of which has an x-, y-, and z-value. These points are connected by edges into a set of contiguous, nonoverlapping triangles, creating a continuous surface that represents the terrain.

TOC

See table of contents.

tool

1. An entity in ArcGIS that performs such specific geoprocessing tasks as clip, split, erase, or buffer. A tool can belong to any number of toolsets and/or toolboxes.

2. A command that requires interaction with the user interface before an action is performed. For example, with the Zoom In tool, you must click or draw a box over the geographic data or map before it is redrawn at a larger scale. Tools can be added to any toolbar.

toolbar

A set of commands that allows users to carry out related tasks. The Main Menu toolbar has a set of menu commands; other toolbars typically have buttons. Toolbars can float on the desktop in their own window, or they may be docked at the top, bottom, or sides of the main window.

toolset

In geoprocessing, a group of tools that perform similar tasks.

ToolTip

In ArcGIS, the description of a tool that is displayed onscreen when the mouse is paused over that tool.

topological association

The spatial relationship between features that share geometry, such as boundaries and vertices. When a boundary or vertex shared by two or more features is edited using the topology tools in ArcMap, the shape of each of those features is updated.

topology

1. In geodatabases, a set of governing rules applied to feature classes that explicitly defines the spatial relationships that must exist between feature data.

2. In an ArcInfo coverage, the spatial relationships between connecting or adjacent features in a geographic data layer (for example, arcs, nodes, polygons, and points). Topological relationships are used for spatial modeling operations that do not require coordinate information.

3. The geometric relationships, determined mathematically, between connecting or adjacent features in a geographic dataset. Topology may include information about connectivity, direction, length, adjacency, and polygon definition. Topology makes most types of geographic analysis possible because it allows analysis of spatial relationships between features.

tracing

The process of building a set of network features based on some procedure.

true curve

See parametric curve.

UIControl

A custom button, tool, text box, or combo box created with Visual Basic for Applications (VBA).

uninitialized flow direction

For an edge feature in a network, a condition that occurs when the edge feature is not topologically connected through the network to sources and sinks or if the edge feature is only connected to sources and sinks through disabled features.

VBA

Visual Basic for Applications. The embedded programming environment for automating, customizing, and extending ESRI applications, such as ArcMap and ArcCatalog. It offers the same tools as Visual Basic (VB) in the context of an existing application. A VBA program operates on objects that represent the application, and can be used to create custom symbols, workspace extensions, commands, tools, dockable windows, and other objects that can be plugged into the ArcGIS framework.

vector

A coordinate-based data model that represents geographic features as points, lines, and polygons. Each point feature is represented as a single coordinate pair, while line and polygon features are represented as ordered lists of vertices. Attributes are associated with each feature, as opposed to a raster data model, which associates attributes with grid cells. See also raster.

vertex

One of a set of ordered x,y coordinate pairs that defines a line or polygon feature.

visible scale range

The extent between user-specified limits (minimum and maximum scale range) that a map scale must fall within for layers to be drawn. If a map scale falls outside the specified thresholds, the layer will not be drawn.

VPF feature class

See feature class.

warp

Transforming a raster to map coordinates.

weight

1. A number that tells how important a variable is for a particular calculation. The larger the weight assigned to the variable, the more that variable will influence the outcome of the operation.

2. A property of a network feature typically used to represent a cost for traversing across an edge or through a junction. For example, in a water network, a certain amount of pressure is lost when traveling the length of a pipe due to surface friction. Weights are calculated based on some attribute of each feature, such as length or volume.

weight filter

A specification of which network features can be traced based on their weight values.

wizard

An interactive user interface that helps a user complete a task one step at a time. It is often implemented as a sequence of dialog boxes that the user can proceed through, filling in required details. A wizard is usually used for long, difficult, or complex tasks.

Index

A

B

C

D

E

F

G

J

K

L

M

N

O

P

Q

R

S

T

U

V

W

X

Z

ArcGIS® 9

Using ArcMap™

ESRI® ArcMap™ software, part of the suite of integrated applications in ArcGIS® Desktop—ArcInfo™, ArcEditor™, and ArcView®—is used to display and query maps, create publication-quality hard-copy output, develop custom mapping applications, and perform many other map-based tasks. ArcMap also includes a fully integrated editor that can work with versioned multiuser geodatabases implemented within a commercial relational database management system (RDBMS), personal geodatabases, and shapefiles. ArcMap provides an easy and natural transition from viewing a map to editing its geometry.

Using ArcMap shows you how to put ArcMap to work immediately. Whether you're just beginning with mapping and geographic information system (GIS) technology or you're a power user, this book makes it easy to identify a task you need to perform—from basic to advanced—and shows you how to get it done.

You'll learn how to

- Put your geographic information on a map.
- Effectively display your geographic data.
- Create and update geographic data.
- Make great looking, publication-quality maps.
- Build interactive displays that link charts, tables, reports, photographs, and the Internet to your data.
- Understand relationships such as "Where is?" "How much?" and "What if?"
- Develop custom map-based applications tailored to your needs.

Begin by following the quick-start tutorial. The tutorial gives you an overview of what you can do with ArcMap and shows you how to make your first map. If you prefer, jump right in and experiment on your own. When you have questions, you'll find concise, step-by-step answers that are fully illustrated to help you complete a task.

ESRI • 380 New York Street • Redlands, CA 92373-8100
909-793-2853 • FAX 909-793-5953 • www.esri.com

9781589480988